plans 186–187   The thickness of aluminum sheets 274–275   The tensile strength of steel rods 275–277   The weight of drug dosages 278–279   The lengths of metal parts 267–268   The variance in tablet weights 315–317   The variance in radar-altimeter readouts 319–320   The variance in aircraft fuel-tank-lid diameters 320–321

PUBLIC ADMINISTRATION
The 1970 draft lottery 27   Deciphering secret codes 60–62   Honest weights and the normal curve of error 210–211   The decision to seed hurricanes 211–212, 516–517   The trial of the Pyx 292   The weight of raisin boxes 285–286   Antitrust pork barrel 289–290   Do nursing homes discriminate against the poor? 291–292   Housing-code violations 322–324   Revenue sharing and the flypaper effect 439–440   The role of time series in social experiments 472

ECONOMICS
Profit survey of fast-food outlets 236   Estimating mean family income 242   Estimating mean managerial salaries 243–244   Testing the association between education and income 369–371, 373–375, 378, 382–383, 385, 388, 389–391, 394–395, 396   The inflation rate vs. its predictability 400   Predicting changes in the GNP 401   Snowfall and unemployment 402   Regressing income on education and job experience 419–421, 424–428, 429, 430–431, 432   Housing prices and Proposition 13 437–438   The price of heroin and the incidence of crime 438–439   Forecasting the avionics sales of Butler Aviation 458–461, 463–464, 466–467, 470–471   The dangers of extrapolation 474

OTHER FIELDS
Biology   Were Mendel's data fudged? 328–329

Geology   Drilling for oil 153

History and Law   Deciding authorship 64–65   Probability in court 146–147

Mathematics and Statistics   Dubious sampling: the *Literary Digest* case 27–28   How to lie with statistics 65–67   The crooked die 152   The paradox of the Chevalier de Méré 147–148   Bernoulli process: coin tossing 173–175, 177   The Slutsky-Yule effect 473

Medicine   Smoking and heart disease 281–282, 402–403   The cavity-fighting ability of toothpastes 286–287, 352   The never-ending search for new and better drugs 291

Psychology   Standard scores 101–102   The ESP mystery 154

ESSENTIALS OF STATISTICS

Heinz Kohler
Amherst College

Scott, Foresman and Company
Glenview, Illinois

London

Boston

Library of Congress Cataloging-in-Publication Data

Kohler, Heinz, 1934-
    Essentials of statistics / Heinz Kohler.

p.   cm.
    Includes indexes.
    ISBN 0-673-18530-3
    1. Statistics.   I. Title.
    QA276.12.K64   1988
    519.5—dc19   87-17913
                                        CIP

Acknowledgments

Cover photo by MOZO Photo/Design.

Table 1.2 on p. 15 and Table 2.1 on 39: From "The 100 largest U.S. multinationals," *Forbes,* July 5, 1982, pp. 126–28. Copyright © 1982 by Forbes, Inc. Reprinted by permission.

Table 2.C on p. 68 from "Teachers' salaries/Average pay in public schools during the 1984–85 school year" from *The New York Times,* August 31, 1985. Copyright © 1985 by The New York Times Company. Reprinted by permission.

Photo on p. 28: *Biometrika.*

Table 1.A on p. 31 from "1985 Mutual Funds 'Honor Roll.'" Excerpted by permission of *Forbes* magazine, September 16, 1985. Copyright © Forbes Inc., 1985.

Table 3.A on p. 106 from "Three Measures of the Top 15 in R&D spending." Reprinted from the July 5, 1982 issue of *Business Week* by special permission, Copyright © 1982 by McGraw-Hill, Inc.

Table 3.F on p. 108 from "The 500 Largest Industrial Corporations Outside the U.S." in *Fortune,* August 23, 1982, p. 183. Copyright © 1982 by Time, Inc. Reprinted by permission.

Table 3.G on p. 108 from "Public Utility Finance and Economic Waste" by Glenn P. Jenkins, *Canadian Journal of Economics,* August 1985. Reprinted by permission.

Photo on p. 154: The Granger Collection.

Photo on p. 187: Offentliche Bibliothek der Universitat, Basel, Switzerland.

Table on p. 254 from *The Design and Analysis of Industrial Experiments,* edited by Owen L. Davies, 1960. Reprinted by permission of the publishers, The Longman Group Limited.

Photo on p. 255: The Granger Collection.

Table on p. 290: From "Antitrust Pork Barrel" by Roger L. Faith, Donald R. Leavens, and Robert D. Tollison, *Journal of Law and Economics,* October 1982, pp. 349–42. Copyright © 1982 by the University of Chicago. All rights reserved. Reprinted by permission of the University of Chicago Press.

Photo on p. 293: Courtesy of the Statistical Laboratory, Iowa State University, Ames, Iowa 50011.

Photo on p. 294: Courtesy, University of California, Berkeley, Department of Statistics.

Photo on p. 295: Keystone Press Agency, Ltd.

Photo on p. 329: *Biometrika.*

Photo on p. 403: National Portrait Gallery, London.

Application 3 from p. 439: "The Price of Heroin and the Incidence of Crime" by George F. Brown, Jr., and Lester P. Silverman, *Journal of the American Statistical Association,* September 1974. Copyright © 1974 by Journal of the American Statistical Association. Reprinted by permission.

Photo on p. 440: Courtesy of the Statistical Laboratory, Iowa State University, Ames, Iowa 50011.

Figure on p. 516: Reprinted by permission of the Harvard Business Review. An exhibit from "Decision Analysis Comes of Age" by Jacob W. Ulvila and Rex V. Brown (September/October 1982). Copyright © 1982 by the President and Fellows of Harvard College; all rights reserved.

Figure on p. 517 from "The Decision to Seed Hurricanes" by R. A. Howard, J. E. Matheson, D. W. North, *Science* 176, 4040 (June 16, 1972): 1196. Copyright © 1972 by The American Association for the Advancement of Science. Reprinted by permission.

Photo on p. 519: Courtesy, Columbia University, Office of Public Information.

Appendix Table A on p. T-2: From *A Million Random Digits with 100,000 Normal Deviates* by The Rand Corporation. Copyright 1955 by The Rand Corporation. Reprinted by permission.

Appendix Table G on p. T-16, Appendix Table I on pp. T-18 to T-20, and Appendix Table J on p. T-21 from *Biometrika Tables for Statisticians,* Vol. 1, edited by E. S. Pearson and H. O. Hartley, p. 146, pp. 171–73, and pp. 192–93. Reprinted by permission.

Appendix Table H on p. T-17 is taken from part of Table IV of Fisher & Yates: *Statistical Tables for Biological, Agricultural and Medical Research* published by Longman Group Ltd., London (previously published by Oliver and Boyd Ltd., Edinburgh) and by permission of the authors and publishers. The author is grateful to the Literary Executor of the late Sir Ronald A. Fisher, F.R.S., to Dr. Frank Yates, F.R.S., and to Longman Group Ltd., London, for permission to reprint a part of Table IV from their book (6th Edition, 1974).

Figures on pages 63, 399–400, 401, 403, and 472 from *Statistics: A Guide to the Unknown,* edited by Judith M. Tanur. Copyright © 1972 by Holden-Day, Inc. Reprinted by permission of Holden-Day, Inc.

PREFACE

TO THE INSTRUCTOR

This text is designed for an introductory course in statistics that is, above all else, *applications-oriented*. As the detailed listing in the front endpapers shows, the applications found in this book are drawn from such diverse fields as accounting, advertising, finance, general management, marketing, personnel, production and operations management, quality control, public administration, economics, and a variety of other fields. Many of these applications have been integrated into the basic structure of chapters as "Practice Problems" or "Applications." Others have been set off as self-contained "Analytical Examples" or somewhat less rigorous "Close-Ups." Altogether, 114 such examples illustrate the usefulness of statistics in numerous aspects of our lives. The end-of-chapter problems (which number 600) thoroughly reinforce the text treatment; the *Student Workbook,* and the *Instructor's Manual* abound with almost 1500 additional applications. Also, most chapters contain one or more biographical sketches of the scholars who developed the statistical theory being discussed. The 16 "Biographies", also found near the ends of chapters, provide a sense of history; they also lend human interest to the theoretical material and highlight some of the controversies that have divided statisticians in the past and, in some cases, still do so in the present.

Following the introductory chapter that discusses the nature of the discipline and the *collection* of data by means of complete and partial surveys, this book is divided into five parts. Part One, "Descriptive Statistics," deals with the *presentation* of data by means of tables, graphs, and numerical summary measures (Chapters 2 and 3). Part Two, "Analytical Statistics: The Foundations of Inference," introduces the theory of probability (Chapter 4) and then discusses the most important discrete and continuous probability distributions (the binomial distribution in Chapter 5, the normal distribution in Chapter 6). Part Three, "Analytical Statistics: Basic Inference," introduces sampling distributions and estimation (Chapter 7) and hypothesis testing using classical and chi-square techniques (Chapters 8–9). Part Four, "Analytical Statistics: Advanced Inference," introduces the analysis of variance (Chapter 10) and develops the techniques of simple and multiple regression and correlation (Chapters 11 and 12). Part Five, "Special Topics for Business and Economics," focuses on time series and forecasting (Chapter 13) and decision theory (Chapter 14).

This text assumes a knowledge of only college-entrance algebra, not of calculus. Certain additional skills (such as the use of the summation sign, the use of factorials, and the use of computer printouts) are introduced when needed, but the derivation of formulas is left to more advanced texts in mathematical statistics and, in part, to supplementary sections in the *Student Workbook* that accompanies this text. This approach has been dictated in part by space considerations: the sheer volume of important statistical techniques that are in common use and, thus, make a legitimate claim on the inevitably limited number of pages available for a standard-sized text leaves little room for extended mathematical proofs. A pedagogical consideration, however, has played a role, too: the text aims to prepare students for their future careers, and most students in business-administration and economics programs do not go on to careers in theoretical mathematics, but they do take up jobs that often require a fine appreciation of the power and applicability of statistical methods. Above all else, they must be taught sound statistical reasoning so that they know which techniques can be used under which assumptions and how the results, properly interpreted, can aid their decision making. The applications approach, which is also very space-consuming, is considered crucial for achieving this goal.

Instructors who wish to teach a shorter course than the text's 14 chapters require may wish to omit Part Five and occasional optional sections (designated by an asterisk). Others who wish to teach a more thorough course or one of longer duration should note the availability of my *Statistics for Business and Economics,* 2nd ed. (Glenview, Ill.: Scott, Foresman and Co., 1988).

Supporting Materials

Three types of supporting materials are available: an *Instructor's Manual,* a *Student Workbook,* and a set of two personal-computer diskettes, called DAYSAL-2.

The *Instructor's Manual* contains responses to or solutions for the even-numbered end-of-chapter Questions and Problems in this text. It also contains numerous true/false questions, multiple-choice questions, and problems (with solutions). These questions and problems do not duplicate those of the *Student Workbook* that is described in the Preface to the Student.

The DAYSAL-2 disks, designed to be used with IBM microcomputers and compatible machines, contain numerous mathematical programs, graphics programs, and educational games and simulations, as well as 26 files of data on the U.S. economy. Jointly, these programs cover almost every aspect of this text. Note the descriptions of relevant programs that follow each chapter's Questions and Problems section.

One final note: The most important skill, perhaps, that students in statistics learn is *the ability to discern which particular technique is applicable to the solution of a given problem.* Yet, inevitably, a text that introduces each technique in the context of a problem (or that lists problems at the end of a given chapter) gives away the answer to this most difficult task of fitting the technique to the problem at hand. Instructors, therefore, are advised to consider the text and *Workbook* problems as exercises in technique and to use the *Instructor's Manual* problems (as well as the *Workbook* "Achievement Tests") as a means of testing the additional student skill just noted.

TO THE STUDENT

Statistics is an important subject. Consider the business executive who claims that a 20 percent reduction in product price will so stimulate sales as to bring about a significant gain in the firm's market share. Consider the economist who argues that a $20 billion tax cut will induce a rise in national income by a multiple, will bring about a significant drop in the unemployment rate, and will actually increase government revenue. Decision makers about to act on the basis of such beliefs may, however, have second thoughts: Might the quantity sold rise so little that the firm's dollar market share actually *drops?* Might the tax cut fail to end the recession and produce soaring deficits instead of revenue gains? In quandaries such as these, decision makers can use statistical methods to help them choose a proper course of action because these methods can help them *test* the accuracy of mere opinions and introduce precision where ambiguity reigns supreme. A proper statistical test, for example, might conclude that the 20 percent price reduction will raise quantity sold by only 10 percent and will actually reduce the firm's market share by 5 percent. Similarly, after gathering and analyzing data, the economist might discover that the $20 billion tax cut will raise national income by $60 billion and will lower the unemployment rate by half a percentage point but will raise the government deficit by $15 billion. Those who master the theoretical knowledge of statistics are, thus, able to wield a powerful set of tools capable of unearthing the likely truth about matters merely hypothesized and of, thus, facilitating wise decision making in the face of uncertainty.

Statistics is also an exciting subject. This fact is evidenced throughout this book by a continuous link between theory and applications. Although the discipline of statistics is crucial for all sciences, this introductory book, being designed for students of business administration and economics, draws examples primarily from these two fields. Again and again, this book shows how the careful collection, effective presentation, and proper analysis of numerical information enables people to draw important inferences from inevitably limited information and, thus, to build windows that afford an occasional glance into the world of the unknown. The front endpapers of this book contain a partial list of the examples found throughout this text and its supplements.

Students will find a number of aids to the study of the material presented in this book. Some of these aids have been built into the text: the figures made self-contained by carefully worded captions, the formulas highlighted in boxed spaces and followed by numbered "Practice Problems" or "Applications," the free-standing "Analytical Examples" and "Close-Ups" that give additional real-world applications, the chapter "Summaries," the end-of-chapter listings of "Key Terms" (boldfaced in the text), the end-of-chapter "Questions and Problems," and the "Selected Readings." In addition, a "Glossary" of all key terms and answers to odd-numbered Questions and Problems appear at the end of the book. Finally, a "Glossary of Symbols" appears in the back endpapers. A separate *Student Workbook* is also available. For each chapter, the *Student Workbook* contains true/false questions, multiple-choice questions, and numerous problems, along with answers to all of these. For selected chapters, the *Workbook* also contains "Supplementary Topics" that expand upon the materials presented in the text. For each of the five parts of the text, the *Workbook* contains a major "Achievement Test" along with all the answers.

Students who have access to IBM microcomputers or compatible machines may wish to acquire DAYSAL-2, a set of diskettes that contain numerous mathematical programs, graphics programs, and educational games and simulations, as well as 26 files of data on the U.S. economy. Jointly, these programs cover almost every aspect of this text and can, thus, aid mightily in the number crunching that the application of formulas inevitably requires. Note the description of relevant programs that follow each chapter's Questions and Problems section.

Another kind of aid that is built into the text may appear like a hindrance at first. Many important concepts appear under more than one name, and many symbols take on more than one form. For example, a particular kind of correlation may be described as both *nonsense correlation* and *spurious correlation,* and a particular concept may be symbolized as both $s_{Y \cdot X_1 X_2}$ and $s_{Y \cdot 12}$. Both forms are given, not to make life hard for you (or because the author couldn't make up his mind), but because both versions are in common use. A statistics textbook should make the student aware of the fact that different books or journals may describe a concept in different ways.

Acknowledgments

I would like to express my sincere gratitude to many who have helped me in the creation of this text. Many reviewers took the time to examine at least a part of the project and gave me good advice:

Richard E. Beckwith	*California State University at Sacramento*
Harry Benham	*University of Oklahoma*
Stephen G. Cecchetti	*New York University*
Georgio Canarella	*California State University at Los Angeles*
John S. Gafar	*Long Island University*
Tarsaim L. Goyal	*University of the District of Columbia*
Paul W. Guy	*California State University at Chico*
Stephen Grubaugh	*Bentley College*
Hannah Hiles	*Wesleyan University*
Peter Hoefer	*Pace University*
Patrick R. Huntley	*University of Arkansas*
Mel Jameson	*University of Rochester*
Denis F. Johnston	*Georgetown University*
Allen Lee	*Northeastern University*
Robert C. Marshall	*Duke University*
Peter H. Michael	*California State University at Sacramento*
Keith T. Poole	*Carnegie-Mellon University*
Harry O. Posten	*University of Connecticut*
Lawrence B. Pulley	*Brandeis University*
David P. Rogers	*Purdue University*
John D. Scholl	*Rio Grande College*
Saul Schwartz	*Tufts University*
C. B. Sowell	*Berea College*
Leonard W. Swanson	*Northwestern University*
Stanley A. Taylor	*California State University at Sacramento*
Willbann D. Terpening	*University of Notre Dame*
Bruce Vermeulen	*Colby College*

I am equally grateful to acquiring editor George Lobell, developmental editor Bruce Kaplan, project editor Ann-Marie Buesing, designer Jacqueline Kolb, and layout artist Cathy Wacaser. They have guided this project through the long process of production and have created, as most will agree, a beautiful book. Permission to use selected materials from other books is gratefully acknowledged on the copyright page.

Heinz Kohler
Amherst College

CONTENTS

Chapter 1   THE NATURE OF STATISTICS   1

Descriptive Statistics   1

Analytical Statistics   2

Statistics—a Universal Guide to the Unknown   3

Collecting Data—the Statistician's Raw Material   4
Some Basic Concepts   Population vs. Sample
Qualitative vs. Quantitative Variables   Types of Data
Surveys vs. Experiments

Census Taking Versus Sampling   10
The Reasons for Sampling   Types of Samples   The Simple Random Sample   The Systematic Random
Sample   The Stratified Random Sample   The Clustered Random Sample

Errors in Survey Data   19
Random Error vs. Systematic Error   How Bias Can Enter Surveys

   Analytical Example 1.1   How Accountants Save Money by Sampling   26
   Close-Up 1.1   The 1970 Draft Lottery Fiasco   27
   Close-Up 1.2   Dubious Sampling: The Literary Digest Case   27
   Biography 1.1   Adolphe Quetelet   28

Part One

DESCRIPTIVE STATISTICS   37

Chapter 2   THE PRESENTATION OF DATA: TABLES AND GRAPHS   38

Tables   38
The Absolute Frequency Distribution   The Relative Frequency Distribution   The Cumulative Frequency
Distribution   A Final Note

Graphs   47
The Frequency Histogram   The Frequency Polygon and the Frequency Curve   Ogives

Other Types of Graphs   56
Bar and Column Charts   Time-Series Line Graphs   Maps, Pie Charts, and Pictograms

Analytical Example 2.1   Deciphering Secret Codes   60
Analytical Example 2.2   Quality Control in Manufacturing   63
Analytical Example 2.3   Deciding Authorship   64
Close-Up 2.1   How to Lie with Statistics   65

Chapter 3   THE PRESENTATION OF DATA: SUMMARY MEASURES   75

Summary Measures of Central Tendency   77
The Arithmetic Mean   The Median   The Mode   Other Measures

Summary Measures of Dispersion   90
Distance Measures of Dispersion   Average Deviation Measures of Dispersion   The Special Significance of the Standard Deviation

The Proportion   101

Analytical Example 3.1   Standard Scores   101
Analytical Example 3.2   Control Charts   102
Biography 3.1   Pafnuty Chebyshev   103

Part Two

ANALYTICAL STATISTICS: THE FOUNDATIONS OF INFERENCE   113

Chapter 4   THE THEORY OF PROBABILITY   114

Basic Concepts   114
The Random Experiment   The Sample Space   Random Events and Their Relationships

The Meaning of Probability   122
Objective Probability   Subjective Probability

Counting Techniques   129
Factorials   Permutations   Combinations

The Laws of Probability   134
The Addition Law   The Multiplication Law   Famous Applications of the Probability Laws

Application 1.   Striving for Reliability: The Case of the Apollo Mission   144
Application 2.   Probability in Court   146
Application 3.   The Paradox of the Chevalier de Méré   147

Probability Laws and Tree Diagrams

Revising Probabilities: Bayes's Theorem   150

Practice Problem 1:   The Crooked Die   152
Practice Problem 2:   Drilling for Oil   153
Practice Problem 3:   Screening Job Applicants   153

Analytical Example 4.1   The ESP Mystery   154
Biography 4.1   Abraham de Moivre   154
Biography 4.2   Pierre de Fermat   155
Biography 4.3   Blaise Pascal   155

Chapter 5   PROBABILITY DISTRIBUTIONS: DISCRETE VARIABLES   165

Basic Concepts   166
The Random Variable   The Probability Distribution

Summary Measures for the Probability Distribution   169
The Arithmetic Mean of a Random Variable   The Variance of a Random Variable   The Standard Deviation of a
Random Variable

The Binomial Probability Distribution   172
The Bernoulli Process

    Practice Problem 1:   Tossing Three Coins   173
    Practice Problem 2:   Sampling Three Items for Quality   175

The Binomial Formula

    Practice Problem 3:   Coin Tossing   177
    Practice Problem 4:   Defective Items   178
    Practice Problem 5:   The Production Process   178

The Binomial Probability Distribution Family   Binomial Probability Tables

    Practice Problem 6:   Producing Defective Items   181
    Practice Problem 7:   Punching Data Cards   182
    Practice Problem 8:   Consumers' Product Preferences   183

Summary Measures for the Binomial Probability Distribution

    Analytical Example 5.1   Budgeting Research and Development   185
    Analytical Example 5.2   Acceptance Sampling Plans   186
    Biography 5.1   James Bernoulli   187

Chapter 6   PROBABILITY DISTRIBUTIONS: CONTINUOUS VARIABLES   193

The Probability Density Function   194

The Normal Probability Distribution   196
A Graphical Exposition   The Formula   The Standard Normal Curve   The Standard Normal Tables

    Practice Problem 1:   Finding Probabilities Concerning the Lifetime of Aircraft Engines   202
    Practice Problem 2:   Designing an Aircraft Cockpit   206

Approximating Discrete Probability Distributions by the Normal Probability Distribution

    Practice Problem 3:   Invoice Errors   209

    Analytical Example 6.1   Honest Weights and the Normal Curve of Error   210
    Analytical Example 6.2   The Decision to Seed Hurricanes   211
    Biography 6.1   Carl Friedrich Gauss   212

Part Three

ANALYTICAL STATISTICS: BASIC INFERENCE   219

Chapter 7   SAMPLING DISTRIBUTIONS AND ESTIMATION   220

Estimators and Estimates   220

The Concept of the Sampling Distribution   221

Summary Measures of the Sampling Distribution   225

The Sampling Distribution vs. Its Parent Population   225
The Shapes of the Distributions and the Central Limit Theorem

    Practice Problem 1:   A Sample Mean Below 45   226
    Practice Problem 2:   A Sample Mean Between 37 and 41.5   227

Mathematical Relationships Among Summary Measures

The Choice of a Good Estimator   232
The Criterion of Unbiasedness   The Criterion of Efficiency   The Criterion of Consistency   A Final Note:
Mean Squared Error

Making Point Estimates   236

    Practice Problem 3:   Profit Survey of Fast-Food Outlets   236

The Nature of Confidence Intervals   237

Making Interval Estimates of the Population Mean or Proportion: Large Samples   241

    Practice Problem 4:   Estimating Mean Family Income   242
    Practice Problem 5:   Estimating Mean Weight Per Animal   242

Making Interval Estimates of the Population Mean and Proportion: Small Samples   243

    Practice Problem 6:   Estimating Mean Managerial Salaries   243
    Practice Problem 7:   Estimating Mean Waiting Times   244

Student's t Distribution   The t Distribution Table

    Practice Problem 8:   Estimating the Mean Time Lost from Machinery Breakdowns   247

Making Interval Estimates for the Difference Between Two Population Means   248
Large and Independent Samples

    Practice Problem 9:   Comparing the Mean Lives of Two Types of Tires   249

Large Matched-Pairs Sample

    Practice Problem 10:   Comparing the Effectiveness of Two Drugs   250

Small and Independent Samples

    Practice Problem 11:   Comparing the Effectiveness of Two Pesticides   252

Small Matched-Pairs Sample

Making Interval Estimates for the Difference Between Two Population Proportions   253

    Analytical Example 7.1   Matched-Pairs Sampling in the Chemical Industry   254
    Biography 7.1   William S. Gosset   255

Chapter 8   HYPOTHESIS TESTING: CLASSICAL TECHNIQUES   261

Step 1: Formulating Two Opposing Hypotheses   262
Hypotheses About a Population Mean   Hypotheses About the Difference Between Two Population Means
Hypotheses About a Population Proportion (or the Difference Between Two Such Proportions)

Step 2: Selecting a Test Statistic   264

Step 3: Deriving a Decision Rule   264
The Level of Significance   Three Illustrations

Step 4: Selecting a Sample, Computing the Test Statistic, and Confronting It with the
Decision Rule   270
The Possibility of Error   The Trade-Off Between a Type I Error and a Type II Error

A Historical Perspective   274

Large-Sample Tests   274
Tests of a Population Mean

    Practice Problem 1:   The Thickness of Aluminum Sheets   274
    Practice Problem 2:   The Tensile Strength of Steel Rods   275
    Practice Problem 3:   Additive in Aviation Fuel   277

Tests of a Population Proportion

    Practice Problem 4:   The Weights of Drug Dosages   278

Tests of the Difference Between Two Population Means: Independent Samples

    Practice Problem 5:   The Lifetimes of Aircraft Radios   280

Tests of the Difference Between Two Population Means: Matched-Pairs Samples

    Practice Problem 6:   Smoking and Heart Disease   281

Tests of the Difference Between Two Population Proportions: Independent Samples

    Practice Problem 7:   Sex and TV-Program Preference   283

Small-Sample Tests   284
Tests of a Population Mean

    Practice Problem 8:   The Length of Long-Distance Calls   284
    Practice Problem 9:   The Weight of Raisin Boxes   285

Tests of the Difference Between Two Population Means: Independent Samples

    Practice Problem 10:   The Cavity-Fighting Ability of Two Toothpastes   286

Tests of the Difference Between Two Population Means: Matched-Pairs Sampling

    Practice Problem 11:   Setting Retail Prices   288

Criticisms of Hypothesis Testing   288
Serious Violations of Assumptions   Rigid Interpretation of the "Acceptance" Criterion   Nonpublication of
Nonsignificant Results

Analytical Example 8.1   Antitrust Pork Barrel   289
Analytical Example 8.2   The Never-Ending Search for New and Better Drugs   291
Analytical Example 8.3   Do Nursing Homes Discriminate Against the Poor?   291
Close-Up 8.1   The Trial of the Pyx   292
Biography 8.1   Ronald A. Fisher   293
Biography 8.2   Jerzy Neyman   294
Biography 8.3   Egon S. Pearson   295

Chapter 9   HYPOTHESIS TESTING: THE CHI-SQUARE TECHNIQUE   302

Testing the Alleged Independence of Two Qualitative Variables   302
Statistical Independence Reviewed   A Contingency Table Test

        Practice Problem 1:   Sex and Brand Preferences   311

Making Inferences About More than Two Population Proportions   312

        Case 1:   Are Population Proportions Equal to Each Other?   313
        Case 2:   Are Population Proportions Equal to a Given Set of (Unequal) Values?   314

Making Inferences About a Population Variance   315
Probability Intervals for the Sample Variance   Confidence Intervals for the Population Variance   Testing
Hypotheses About the Population Variance

        Practice Problem 2:   A Bank's Waiting-Line Policy   317
        Practice Problem 3:   Altimeter Readouts   319
        Practice Problem 4:   The Diameters of Fuel-Tank Lids   320

Conducting Goodness-of-Fit Tests   321
Goodness of Fit to the Binomial Distribution

        Practice Problem 5:   Housing-Code Violations   322

Goodness of Fit to the Normal Distribution

        Practice Problem 6:   Trading in the Futures Market   324

Goodness of Fit to Any Specified Distribution

        Practice Problem 7:   Television Viewing Habits   326

        Analytical Example 9.1   Were Mendel's Data Fudged?   328
        Biography 9.1   Karl Pearson   329

||| Part Four

ANALYTICAL STATISTICS: ADVANCED INFERENCE   341

Chapter 10   ANALYSIS OF VARIANCE   342

Alternative Versions of ANOVA   345

One-Way ANOVA   346
Variation Among Columns: Explained by Treatments   Variation Within Columns: Due to Error   The ANOVA
Table   The F Distribution

Practice Problem 1:   The Effect of Three Toothpastes on Cavity Development   352

Discriminating Among Different Population Means   353
Establishing Confidence Intervals for Individual Population Means   Establishing Confidence Intervals for the Difference Between Two Population Means   Tukey's *HSD* Test

Chapter 11   SIMPLE REGRESSION AND CORRELATION   364

Simple Regression Analysis   365
Basic Concepts

Practice Problem 1:   A Preliminary Graphical Test of the Association Between Education and Income 369

Finding the Regression Line: The Use of Eyesight   Finding the Regression Line: The Method of Least Squares

Practice Problem 2:   Using the Least-Squares Method to Determine the Estimated Regression Line for Education and Income   373

The Estimated vs. the True Regression Line   Estimating the Average Value of Y, Given X

Practice Problem 3:   Estimating the Average Value of Income, Given Education   378
Practice Problem 4:   Establishing a Prediction Interval for the Average Value of Income, Given Education   382
Practice Problem 5:   Establishing a Prediction Interval for the Average Wheat Output, Given Fertilizer Input   384

Predicting an Individual Value of Y, Given X

Practice Problem 6:   Establishing a Prediction Interval for an Individual Value of Income, Given Education   385
Practice Problem 7:   Establishing a Prediction Interval for an Individual Value of Wheat Output, Given Fertilizer Input   387

Making Inferences About the True Regression Coefficients

Practice Problem 8:   Establishing a Confidence Interval for β in the Case of Education and Income 388
Practice Problem 9:   Establishing a Confidence Interval for α in the Case of Education and Income 389
Practice Problem 10:   A Two-Tailed Test of β in the Case of Education and Income   389
Practice Problem 11:   An Upper-Tailed Test of β in the Case of Education and Income   390

Simple Correlation Analysis   391
The Coefficient of Determination

Practice Problem 12:   Calculating the Sample Coefficient of Determination (r^2) for Education and Income   394

The Coefficient of Correlation

Practice Problem 13:   Calculating Pearson's Sample Coefficient of Correlation (r) for Education and Income   395

An Important Extension   396
Testing β with Analysis of Variance

Practice Problem 14:   An ANOVA Test of β in the Case of Education and Income   396

Analytical Example 11.1   Evaluating a New Food Product   397
Analytical Example 11.2   Of Bulls, Bears, and Beauty   399
Analytical Example 11.3   The Inflation Rate vs. Its Predictability   400
Analytical Example 11.4   Predicting Changes in the GNP   401
Analytical Example 11.5   Snowfall and Unemployment   402
Close-Up 11.1   Smoking and Health   402
Biography 11.1   Francis Galton   403

Chapter 12   MULTIPLE REGRESSION AND CORRELATION   414

Linear Multiple Regression: Two Explanatory Variables   415
The Regression Plane   Least Squares Revisited

        Practice Problem 1:   Regressing Income on Education and Job Experience   419

The Sample Standard Error of the Estimate of Y

        Practice Problem 2:   Calculating the Sample Standard Error of the Estimate of Income   421

Making Inferences

        Practice Problem 3:   Estimating Average Income, Given Education and Job Experience   424
        Practice Problem 4:   Estimating Individual Income, Given Education and Job Experience   424
        Practice Problem 5:   Establishing Confidence Intervals for β_1 and β_2 in the Regression of Income on
                               Education and Job Experience   425
        Practice Problem 6:   Using t Values for Practice Problem 5   426
        Practice Problem 7:   Using p Values for Practice Problem 5   427
        Practice Problem 8:   An Overall Significance Test of the Regression of Income on Education and Job
                               Experience   427

Multiple Correlation Analysis   428
The Coefficient of Multiple Determination

        Practice Problem 9:   Calculating the Sample Coefficient of Multiple Determination for the Regression of
                               Income on Education and Job Experience   429

The Coefficient of Multiple Correlation   The Adjusted Coefficient of Multiple Determination

        Practice Problem 10:   Adjusting the Sample Coefficient of Multiple Determination for the Regression of
                                Income on Education and Job Experience   430

Partial Correlation

        Practice Problem 11:   Calculating the Sample Coefficient of Partial Determination to Show the
                                Correlation of Income with Job Experience, When the Effect of Education is
                                Present but Constant   432

**Linear Multiple Regression and Correlation with Three or More Explanatory Variables
(Computer Applications)**   433
The Estimated Multiple Regression Equation   The Standard Error of the Estimate of Y Predicting the Average
or an Individual Value of Y Confidence Intervals for the True Regression Coefficients   Hypothesis Tests About
True Regression Coefficients   Testing the Overall Significance of the Regression   The Sample Coefficient of
Multiple Determination

Examples

1   Housing Prices and Proposition 13   437
2   The Price of Heroin and the Incidence of Crime   438
3   Revenue Sharing and the Flypaper Effect   439

Biography 12.1.   George Snedecor   440

Part Five

SPECIAL TOPICS FOR BUSINESS AND ECONOMICS   449

Chapter 13   TIME SERIES AND FORECASTING   450

The Composition of a Time Series   541
The Trend   Cyclical Fluctuations   Seasonal Fluctuations   Irregular Variations

Forecasting: Extrapolating Trends   457
Moving Averages

Practice Problem 1:   Constructing a 3-Year Moving-Averages Series for the Avionics Sales of Butler
                      Aviation   458
Practice Problem 2:   Constructing an 8-year Moving-Averages Series for the Avionics Sales of Butler
                      Aviation   460

Exponential Smoothing

Practice Problem 3:   Predicting Future Avionics Sales Using Single-Parameter Exponential Smoothing
                      463
Practice Problem 4:   Predicting Future Avionics Sales Using Two-Parameter Exponential Smoothing
                      466

Fitting Regression Lines

Practice Problem 5:   Letting a Linear Regression Line Approximate the Trend in Avionics Sales   470

Analytical Example 13.1   The Role of Time Series in Social Experiments   472
Analytical Example 13.2   The Slutsky-Yule Effect   473
Close-Up 13.1   The Dangers of Extrapolation   474

Chapter 14   DECISION THEORY   479

Basic Concepts   480
Actions, Events, Payoffs   The Payoff Table   The Decision Tree

Decision Making Without Probabilities   482
Maximin (or Minimax)   Maximax (or Minimin)   Minimax Regret   Criticisms

Decision Making with Probabilities: Prior Analysis   489
Maximum Likelihood   Expected Monetary Value   Expected Opportunity Loss   Expected Utility

Decision Making with Probabilities: Posterior Analysis   500
Gathering New Information   Applying Bayes's Theorem   Finding the Optimal Strategy

The Value of Information   505
The Expected Value of Perfect Information   The Expected Value of Sample Information   The Efficiency of Sample Information

Preposterior Analysis   508
Extensive-Form Analysis   Normal-Form Analysis   Pros and Cons

Sequential Sampling   512

The Great Controversy: Classical vs. Bayesian Statistics   513
An Example   The Unresolved Issue

Analytical Example 14.1   Decision Analysis Comes of Age   515
Analytical Example 14.2   The Decision to Seed Hurricanes—A Second Look   516
Analytical Example 14.3   The Value of Perfect Information: The Case of the U.S. Cattle Industry   517
Biography 14.1   Thomas Bayes   518
Biography 14.2   Abraham Wald   519

APPENDIX TABLES   *T–1*

A Random Numbers   *T–2*
B The Greek Alphabet   *T–3*
C Binomial Probabilities for Individual Values of x *T–3*
D Binomial Probabilities for Cumulative Values of x *T–8*
E Standard Normal Curve Areas   *T–14*
F Critical Normal Deviate Values   *T–15*
G Student t Distributions   *T–16*
H Chi-Square Distributions   *T–17*
I F Distributions   *T–18*
J Values of q_{α} in Tukey's *HSD* Test   *T–21*

GLOSSARY   *G–1*

ANSWERS TO ODD-NUMBERED QUESTIONS AND PROBLEMS   *A–1*

SUBJECT INDEX   *I–1*

NAME INDEX   *I–5*

ESSENTIALS OF STATISTICS

CHAPTER ONE

The Nature of Statistics

Most people associate the term *statistics* with masses of numbers or, perhaps, with the tables and graphs that display them and with the averages and similar measures that summarize them. This mental image is reinforced daily as people encounter an abundance of numerical information in newspapers, magazines, and on television screens: on the prices of bonds and stocks, on the performance of businesses and sports teams, on the movements of exchange rates and commodity futures, on the rates of unemployment and inflation, on the incidence of poverty and disease, on accidents, crime, and the weather—the list goes on. It is not surprising, therefore, that people who think about statistics not in the plural sense of the word, but in the singular (that is, as a field of study) imagine it to deal with the collection and presentation of numbers. As a matter of fact, this widely held view quite accurately depicts the original concern of the discipline as well as the continuing concern of one of its modern branches.

DESCRIPTIVE STATISTICS

Early statisticians were typically interested in collecting and displaying information useful to the *state*—and that is how *statistics* got its name. They collected, for example, data on births and deaths to aid officials in charge of the military draft, they collected data on diseases to aid others concerned with the public health, and they collected data on exports, imports, incomes, and expenditures in order to facilitate the collection of taxes. A look at present-day government publications, such as the *Statistical Abstract of the United States* or the *Economic Report of the President,* suggests that the tradition of statisticians-aiding-government has certainly not died out.

Accordingly, Part One of this book deals with **descriptive statistics,** a branch of the discipline that is concerned with developing and utilizing techniques for the effective presentation of numerical information. Later in this chapter, we will see how statistical data can be *collected* in a variety of ways, involving complete or sample surveys or even controlled experiments. The branch of descriptive statistics, however, focuses on the subsequent task of organizing and condensing data, usually with the help of tables, graphs, or numerical summary measures. Unlike a mere listing of masses of raw data, this type of presentation alone makes the data understandable and often reveals patterns otherwise hidden in them.

The importance of descriptive statistics is illustrated vividly by the examples found in Part One. As will be shown, the effective presentation of data can lead a production engineer to discover the secret behind the recent breakdowns of motors produced by a firm, can help a historian unravel the mystery of a disputed authorship, and can aid spies in the cracking of a secret code.

Useful as it may be, descriptive statistics, however, comprises only a small part of the modern discipline of statistics (and of this book). Contrary to the common view that identifies descriptive statistics with the entire field of study, another branch of the discipline is nowadays considerably more important.

ANALYTICAL STATISTICS

Modern statisticians direct most of their effort not toward the collection and presentation but toward the analysis of numerical information. They focus on applying reason to data in order to draw sensible conclusions from them; their chief concern is with the making of reasonable inferences, from the limited information that is available, about matters that are not known. Accordingly, Parts Two through Four of this book deal exclusively with **analytical statistics,** a branch of the discipline that is concerned with developing and utilizing techniques for properly analyzing numerical information. Because analyzing data means drawing inferences from them, this branch of statistics is also referred to as **inferential statistics.**

Sometimes a general truth is inferred from particular instances; at other times, statisticians reverse the process and draw conclusions about the particular from a knowledge of the general. The drawing of inferences about an unknown whole from a known part is **inductive reasoning.** Inductive reasoning is at work, for example, when a statistician concludes, with a 95-percent chance of being right and a 5-percent chance of being wrong, that between 1 and 3 percent of a firm's total output is defective because 2 percent of an output sample did in fact not meet quality standards. On the other hand, the drawing of inferences about an unknown part from a known whole is **deductive reasoning.** Deductive reasoning is at work, for example, when a statistician concludes that a particular unit of a firm's output (namely, a unit produced at plant 7, during the night shift, and with the help of components supplied by firm X) has 5 chances in 100 of being included in a quality test because (a) 5 percent of the firm's total output meets the above criteria and (b) the portion of output to be tested is being selected from total output by a random process and is, therefore, likely to reflect the characteristics of total output. As a matter of fact, deductive reasoning and inductive reasoning are complementary. Before statisticians can safely generalize from the part to the whole, they must study how the part has been generated from the whole.

The importance of analytical statistics is illustrated by the rich array of examples found throughout this book and its supplements. Statistical techniques help firms screen job applicants, budget research and development expenditures, determine the quality of raw materials received or of output produced, and decide whether sales personnel are better motivated by salary or commission, whether one shipping company makes faster deliveries than another, whether one production method is less time-consuming than another, and even whether people prefer brand A to brand B. Statistical techniques can, similarly, help firms choose the best one among several product designs, leasing arrangements, oil-drilling sites, fertilizer types, and advertising media. And statistics can tell firms precisely how the quantity of their product that is demanded relates to the product's price, to the prices of substitutes and complements, to consumer income, and, perhaps, even to the consumer's sex.

Government officials are equally avid users of what this book has to teach. Thus, analytical statistics plays an important role in assuring the reliability of space missions and of more mundane airport lighting systems. And statistical techniques help answer questions such as: Do motorcycle helmets really reduce accident fatalities? Do nursing homes discriminate against Medicaid recipients? Do the boxes of raisins marketed by this firm truly contain 15 ounces as claimed? Do the firms in this state meet antipollution standards? Have the rents in this city risen over the past year? Are wages in New York higher than in Chicago? Do different IRS offices perform work of equal quality? Should this hurricane be seeded to reduce its destructiveness? Is that defendant guilty? Where should the political candidate concentrate further campaign efforts? Do entry barriers explain the degree of dominance achieved by a few firms in this industry? What will happen to tax revenues, the level of employment, and the rate of inflation next year? How are the incomes of people related to their education, job experience, age, sex, or even race? This list, too, can be expanded at will.

All of the previous examples have one thing in common. They illustrate dramatically how analytical statistics can facilitate decision making in the face of uncertainty. Indeed, one can argue that facilitating decision making under uncertainty is the main purpose of the entire discipline.

STATISTICS—A UNIVERSAL GUIDE TO THE UNKNOWN

According to one widely held view, **statistics** is best defined as a branch of mathematics that is concerned with facilitating wise decision making in the face of uncertainty and that, therefore, develops and utilizes techniques for the careful collection, effective presentation, and proper analysis of numerical information. This definition clearly incorporates both descriptive and analytical statistics. Jointly, the two branches help decision makers extract the maximum usefulness from limited information. On the one hand, tables, graphs, and summary calculations highlight patterns otherwise buried in unorganized data. On the other hand, proper inferences provide reasonable estimates of things unknown, along with clearly stated probabilities of such estimates being right or wrong.

Note: By failing to specify who the decision makers are, the above definition quite correctly suggests the universal applicability of what statistics has to offer. As the earlier examples have shown, modern statistical techniques routinely guide business executives, as well as government economic-policy makers, in making reasonable decisions in the face of uncertainty. It may even be true, as some have argued, that modern statistics is particularly useful in the fields of business and economics because controlled experiments can rarely be performed in these fields. Therefore, relationships between variables (as between advertising expenditures and sales or between interest rates and investment) that are of interest to economic decision makers can only be established by applying statistical techniques to raw data that reflect the complicated and simultaneous interaction of large numbers of variables. Nevertheless, the same techniques are just as useful, and are just as frequently applied, outside the fields of business and economics. Like mathematics in general, statistics is a universal type of language that is regularly used by all sciences. The drawing of valid inferences from limited information is just as important to historians and psychologists, to geneticists and medical researchers, to astronomers and engineers as it is to business executives and economists. Of this universal use of statistics we shall see examples as well. Among others, questions such as these will be posed and answered: How can one decipher a secret code? How can one resolve the issue of a disputed authorship? Were Mendel's data fudged? Does smoking cause heart disease? Is toothpaste A better than toothpaste B? Is ESP real?

COLLECTING DATA—THE STATISTICIAN'S RAW MATERIAL

Statistical work cannot be performed in a vacuum. Before all else, such work requires the acquisition of a crucial type of raw material: information relevant to the subject matter under study. This section introduces a number of basic concepts employed in data gathering and then discusses the major methods of accomplishing that task.

Some Basic Concepts

A statistical investigation invariably focuses on people or things with characteristics in which someone is interested. The persons or objects possessing the characteristics that interest the statistician are referred to as **elementary units.** Thus, someone who wanted to learn about the racial composition of a firm's labor force would quickly identify the individual employees of that firm as the elementary units, but someone concerned about the amount of credit extended by that firm might view the individual credit accounts issued by it as the elementary units to be investigated. Even the flash cubes produced by the firm, the light bulbs installed in its plants, or the boxes of cereal shipped by one of its divisions could be regarded as elementary units—provided someone was interested in discovering, respectively, the percentage of defective flash cubes produced, the lifetimes of light bulbs used, or the content weights of boxes of cereal sold. A complete listing of all elementary units relevant to a statistical investigation is called a **frame.**

Any one elementary unit may possess one or more characteristics that interest the statistician. An investigator may, indeed, be interested only in the race of each employee, but it would be just as possible to observe, in addition, each employee's sex or salary. The characteristics of elementary units are themselves called **variables,** presumably because observations about these characteristics are likely to vary from one elementary unit to the next.

Any single observation about a specified characteristic of interest is called a **datum;** it is the basic unit of the statistician's raw material. Any collection of observations about one or more characteristics of interest, for one or more elementary units, is called a **data set.** A data set is said to be **univariate, bivariate,** or **multivariate** depending on whether it contains information on one variable only, on two variables, or on more than two.

An example can help us assimilate these new concepts. Consider a statistician who was hired to evaluate charges of racial and sex discrimination in a firm. Using the firm's personnel records, the statistician might record observations about various character- istics (such as race, sex, job title, and the like) associated with each one of the firm's employees. The resultant frame and data set is shown as Table 1.1. Any one entry in column 1 is an elementary unit; all the entries in the shaded portion of that column jointly represent the frame. The headings of columns 2–6 designate the variables; any one entry in these columns is a datum; all entries in these columns together constitute the entire (multivariate) data set. Indeed, we can learn more than this from Table 1.1.

Population versus Sample

Statisticians call the set of all possible observations about a specified characteristic of interest a **population** or **universe.** As Table 1.1 illustrates, it is possible to draw several such statistical populations from a given frame. Because there are five vari- ables—the headings of columns 2–6—our table contains five populations. The shaded entries in column 3, for example, make up the population of employee sexes; those in

Table 1.1   Selected Characteristics of All the Full-Time Employees of Mountain Aviation, Inc.; July 1, 1987

This table contains a statistical frame and the multivariate data set derived from it. The table illustrates the meaning of a number of basic statistical concepts. Thus, column 1 lists 9 *elementary units* that jointly constitute the *frame* (shaded). The headings of columns 2–6 show various characteristics of the elementary units that are called *variables* and that can be qualitative (race, sex, job title) or quantitative (years of service, annual salary). All possible observations about a given variable constitute a statistical *population*—the shaded entries in column 3 and in column 6 are two examples of popula- tions; any single observation is a *datum;* any subset of a population or of the frame is a *sample.*

	Employees (1)	Race (2)	Sex (3)	Job Title (4)	Years of Service (5)	Annual Salary (6)	
Qualitative variable							Quantitative variable
	Abel	White	Male	Pilot	2	$17,000	
	Cruz	White	Male	Chief mechanic	10	30,000	
	Dunn	Black	Male	Chief pilot	23	35,000	←Datum
	Hill	Black	Female	Secretary	5	7,000	
Elementary unit	King	White	Male	Janitor	8	8,500	
	Otis	White	Male	Grounds keeper	10	10,000	
	West	Black	Male	Mechanic	2	18,000	Sample of
	Wolf	White	Female	Pilot	7	18,000	← employee
	Zorn	White	Female	Mechanic	7	20,000	salaries
	↑ Frame		↑ Population of employee sexes			↑ Population of employee salaries	

column 6 make up the population of employee salaries; and so on for columns 2, 4, and 5.

It should be noted that a statistical population consists of *all possible* observations about a variable. It is irrelevant how many observations, if any, are actually being made. Because they correspond to *all* the employees of our firm, the entries in column 6 make up the population of employee salaries in our hypothetical study, regardless of whether anyone actually gathers these data.

In a firm of only nine employees, it is, of course, easy to observe the entire salary population (as we assumed), but imagine the difficulty of such an undertaking if our statistician attempted to carry out a similar study for the entire aviation industry or even for the entire labor force of the United States! Under such circumstances, statisticians often make observations concerning only selected elementary units. They observe only *n* such units out of the larger number *N* that exist. Naturally, they end up with a subset of all the possible observations about the characteristic of interest. Such a subset of a statistical population, or of the frame from which it is derived, is called a **sample.** The boxed entries in column 6 of Table 1.1, for instance, make up one of many possible samples of employee salaries—namely, the sample based on observing the salary characteristics of only West, Wolf, and Zorn. These three names themselves can, in turn, be viewed as a sample of the frame.

But note: What constitutes a population or a sample of it depends entirely on the context in which the question arises. If the goal is to study salaries only at Mountain Aviation, Inc., the data in shaded column 6 of Table 1.1 do, indeed, as a group, make up the relevant population. If the goal were to study salaries in the entire aviation industry, however, the identical column 6 data, even as a group, would constitute only a sample of the now much larger population of salaries throughout the industry.

Qualitative versus Quantitative Variables

Table 1.1 teaches us something else: Any given characteristic of interest to the statistician can differ in kind or in degree among various elementary units. A variable that is normally not expressed numerically (because it differs in kind rather than degree among elementary units) is called a **qualitative variable.** Table 1.1 contains three qualitative variables: race, sex, and job title. Qualitative variables can, in turn, be *dichotomous* or *multinomial.* Observations about a **dichotomous qualitative variable** can be made only in two categories: for example, male or female, employed or unemployed, correct or incorrect, defective or satisfactory, elected or defeated, absent or present. Observations about a **multinomial qualitative variable** can be made in more than two categories; consider job titles, colors, languages, religions, or types of businesses.

On the other hand, a variable that is normally expressed numerically (because it differs in degree rather than kind among elementary units) is called a **quantitative variable.** Table 1.1 contains two of them: years of service and annual salary. Quantitative variables can, in turn, be *discrete* or *continuous.* Observations about a **discrete quantitative variable** can assume values only at specific points on a scale of values, with gaps between them. Such data differ from each other by clearly defined steps. Consider observing the number of children in families, of employees in firms, of students in classes, of rooms in houses, of cars in stock, of cows in pastures. Invariably, the individual data will be disconnected from each other by gaps on the scale of values;

in the above instances, they will look like 1, 2, 3, . . . and 49; never like 3.28 or 20.13. But note: the gaps representing impossible values need not span the entire space between *whole* numbers. Stock prices, for example, are customarily reported in eighths of dollars (that is, in units of $0.125). These discrete figures can take on values of 67⅛, 67⅜, 67⅜ (in dollars per share) but cannot take on values between these. (The quoting of prices by eighths is a throwback to the old pirate days and the Spanish gold "pieces of eight.")

Observations about a **continuous quantitative variable** can, in contrast, assume values at all points on a scale of values, with no breaks between possible values. Consider height, temperature, time, volume, or weight. Weight, for instance, might be reported as 7 pounds or 8 pounds but also as 7.3 pounds or even 7.3425 pounds, depending entirely on the sensitivity of the measuring instrument involved. No matter how close two values are to each other, it is always possible for a more precise device to find another value between them.

Note: The distinction between qualitative and quantitative variables is visually obvious in Table 1.1. The observations about one type of variable are recorded in words; those about the other type in numbers. Yet that distinction can easily be blurred. Quantitative variables can be converted into seemingly qualitative ones, and the opposite is also true. Thus, a statistician could replace the column 5 and 6 data by words, such as *low, intermediate,* or *high,* although, presumably, nobody would in this way wish to give up the more precise information presently recorded in Table 1.1. On the other hand, it is very common to *code* observations about qualitative variables with the help of numbers. Thus, a statistician might turn the verbal entries of Table 1.1 into *numbers* by recording "white" as 1 and "black" as 2 in column 2, by recording "male" as 0 and "female" as 1 in column 3, and by assigning numbers between 0 and 6 to the seven job titles in column 4. Nevertheless, the distinction between qualitative and quantitative variables, although then hidden, would remain. Even when they are encoded numerically, one cannot perform arithmetic operations with qualitative data, while one can perform them with quantitative ones. Thus, it would make no sense to report the "sum of races" in our firm as 12 (using the code just noted), but one can report the sum of annual salaries as $163,500.

Types of Data

The preceding discussion makes clear why statisticians, who inevitably work with numbers, must be keenly aware of what, if anything, is measured by these numbers. The assignment of numbers to characteristics that are being observed—which is **measurement**—can yield four types of data of increasing sophistication. It can produce nominal, ordinal, interval, or ratio data, and different statistical concepts and techniques are appropriately applied to each type.

The weakest level of measurement produces **nominal data,** which are numbers that merely *name* or label differences in kind and, thus, can serve the purpose of classifying observations about qualitative variables into mutually exclusive groups. As noted, "male" might be numbered as 0, "female" as 1, but alternative labels of "male" = 100 and "female" = 17 would serve as well. Other examples of creating nominal data are classifying defective units of a product as 1, satisfactory units as 2; labeling rooms on first, second, or third floors by numbers in the 100s, 200s, or 300s, respectively; or designating rooms on the north or south side of buildings by even or odd last digits in room numbers: 102, 104, 106 for first-floor rooms to the north; 301, 303, 305 for third-floor rooms facing south. As these examples confirm, it never makes sense to add, subtract, multiply, or divide nominal data, but one can *count* them: The presence of five 17s, by one of the above codes, denotes the presence of five females; 25 odd room numbers mean there are 25 rooms facing south.

The next level of measurement produces **ordinal data,** which are numbers that by their size *order* or rank observations on the basis of importance, while intervals between the numbers, or ratios of such numbers, are meaningless. Assessments of a product as *superb, average,* or *poor* might be recorded as 2, 1, 0, as 250, 10, 2, or even as 10, 9, 4.5—the important thing is that larger numbers denote a more favorable assessment, or a higher ranking, while smaller ones do the opposite. Yet ordinal data make no statement about *how much* more or less favorable one assessment is compared to another. A 2 is deemed better than a 1 but not necessarily twice as good; a 250 is deemed better than a 10 but not necessarily 25 times as good; a 4.5 is deemed worse than a 9 but not necessarily half as good; and that is all. Thus, an *ordinal* coding of "professor," "associate professor," and "assistant professor" as 3, 2, and 1 does imply that, in some sense, professors are viewed as more important than assistant professors, while a *nominal* coding of male = 100, female = 17 does not imply the superiority of males over females any more than a coding of male = 0, female = 1 denotes the opposite. In either case, arithmetic operations with the data are out of the question.

Somewhat more sophisticated are **interval data** that permit at least addition and subtraction. Interval data are numbers that by their size rank observations in order of importance and between which *intervals* or distances are comparable, while their ratios are meaningless. Because these data do not possess an intrinsically meaningful origin, their measurement starts from an arbitrarily located zero point and utilizes an equally arbitrary unit distance for expressing intervals between numbers. Commonly used temperature scales are examples of interval data. The Celsius scale places zero at the water-freezing point; the Fahrenheit scale places it far below that. Within the context of either scale, the unit distance (degree of temperature) has a consistent meaning: each degree Celsius equals 1/100 of the distance between water's freezing and boiling points; each degree Fahrenheit equals 1/180 of that distance. Note that the zero point, being arbitrarily located, does not denote the absence of the characteristic being measured. Unlike 0° on the absolute (Kelvin) temperature scale that is familiar to astronauts, neither 0°C nor 0°F expresses a complete absence of temperature. Note also that any ratio of interval data is meaningless: 90°F is *not* twice as hot as 45°F. Indeed, the ratio of the corresponding Celsius figures (32.2° and 7.2°) does not equal 2:1 but well over 4:1. This discussion brings us to the most sophisticated types of numbers of all.

The most useful types of data, such as those in columns 5 and 6 of Table 1.1, are **ratio data,** which are numbers that by their size rank observations in order of importance and between which intervals as well as *ratios* are meaningful. All types of arithmetic operations can be performed with such data because these types of numbers have a natural or true zero point that denotes the complete absence of the characteristic they measure and makes the ratio of any two such numbers independent of the unit of measurement (unlike in the Fahrenheit-Celsius example). Thus, the measurement of salaries, age, distance, height, volume, or weight produces ratio data. For example, because it is meaningful to rank salary data and to say that a salary of $15,000 is larger than one of $10,000, which is larger than one of $5,000, salary data give the kind of information provided by ordinal data. Yet in addition, because it is meaningful to compare intervals between salary data and to say that the distance between $15,000 and $10,000 equals the distance between $10,000 and $5,000, these data also give the type of information provided by interval data. Further, these data are ratio data because we can safely describe $15,000 as three times as much money as $5,000; even a change in the unit of measurement, as from dollars to cents or from dollars to francs (at an exchange rate of, say, 6 francs to the dollar) does not change this conclusion: 1,500,000 cents still is three times as much money as 500,000 cents; 90,000 francs still is three times as much as 30,000 francs. The fact that ratios of numbers convey meaningful information is the advantage of ratio data over interval data. No wonder that statisticians, when they have a choice, prefer ratio data to interval data, interval data to ordinal data, and ordinal data to nominal data.

Having discussed the major types of data, we introduce, finally, the two basic methods of generating them.

Surveys versus Experiments

New data can be generated either by conducting a survey or by performing an experiment. A **survey,** or **observational study,** is the collection of data from elementary units without exercising any particular control over factors that may make these units different from one another and that may, therefore, affect the characteristic of interest being observed. A characteristic such as the annual salary of workers, for example, is simply being observed and recorded for different workers without regard to factors, such as education, work experience, or length of service, that make workers different from one another and that may be responsible for differences in their salaries. In contrast, an **experiment** involves the collection of data from elementary units, while exercising control over some or all factors that may make these units different from one another and that may, therefore, affect the characteristic of interest being observed. Thus, a firm may divide its 40 new employees into two similar groups of equal size (with the help of some random device) and then administer an obligatory special training program to one of the groups only. If the 20 employees who went through the program exhibited superior productivity later on, the training program might justifiably be credited because other factors that could account for this result (such as group differences in age, motivation, or prior work experience) were eliminated by the random division of the original group of 40 newcomers. Note: If the same data on superior productivity had been generated from a group of volunteers, the data would have to be viewed as mere observational ones because of the lack of special controls. Without the random assignment, some systematic difference might have emerged between the two groups of workers even prior to the training program. If only more mature, more motivated, or more experienced workers volunteered for the training, they would probably have shown greater productivity even without the program. Thus, the higher productivity later on could not be credited to that program with absolute confidence. Experimental data tend to be stronger than survey data.

Unfortunately, most data in business and economics, and in many other fields, are generated not by experiments but by surveys because it is often impossible, or extremely costly, to exercise proper experimental controls. One cannot easily divide the country's labor force (or even a segment of it) at random into three groups and then subject each group to different tax rates in order to study the effect of taxes on labor supply. One cannot simply divide the country's newly born (or even a segment of them) at random into two groups and then subject only one of the groups to a lifetime of smoking in order to study the effect of smoking on health. Nevertheless, with proper statistical techniques, one can learn a great deal about such matters even from survey data. Accordingly, the remainder of this chapter will first deal with surveys (this being the most common method of generating business and economic data); it will then point out the errors likely to be associated with survey procedures.

CENSUS TAKING VERSUS SAMPLING

Two types of surveys exist: complete and partial ones. In a complete survey, or **census,** observations about one or more characteristics of interest are being made for every elementary unit that exists. Accordingly, the resultant data set consists of at least one statistical population but possibly several of them, as in columns 2–6 of Table 1.1. But note: When the number of elementary units is very large, complete success in observing all of them is likely to elude the census takers. Nevertheless, any reasonably successful attempt to collect data on an entire statistical population is referred to as a census.

The precise procedure of data collection may vary. It may involve *direct observation* of some ongoing activity by an investigator who might record, say, the net weights of boxes of cereal as they are being filled by a machine or the numbers and brands of such boxes as they are being picked up by customers from a supermarket shelf. A census may instead involve a *personal or telephone interview* in which the investigator asks questions printed on a carefully prepared list and records the verbal answers given by people. Finally, a census may proceed by *self-enumeration,* as when people, having read a set of instructions, make written replies on questionnaires that they received in the mail, on the street corner, or, perhaps, when purchasing an appliance. More often than not, however, these same procedures are applied to a different type of survey. In a partial or **sample survey,** observations about one or more characteristics of interest are being made for only a subset of all existing elementary units. The resultant data set consists of one or more samples, as noted in column 6 of Table 1.1, and it is hoped that the sample data mirror the population data in the sense that valid inferences can be drawn from any given sample about the corresponding, but unknown, statistical population. There are good reasons why sample surveys are often undertaken in place of censuses.

The Reasons for Sampling

In one way or another, all the reasons typically cited for preferring sampling to census taking have something to do with reducing the cost of getting a given type of information or with increasing the quantity or quality of information received with a given cost.

First, the cost of collecting and processing data is obviously lower the fewer are the elementary units that have to be contacted. This becomes a crucial consideration whenever the number of relevant elementary units is large. The U.S. Bureau of Labor Statistics, for example, gathers its monthly unemployment data from a mere 60,000 households, not by questioning the more than 100 million members of the labor force.

Yet—and this is the crucial point—the information so received is, nevertheless, a reliable indicator of what would be known if a complete census were taken. When first encountered, this statement is hard to believe, but consider an analogy. Suppose we wanted to determine the proportion of green and yellow peas in a sackful of them. If the peas were thoroughly mixed, we could surely take a mere cupful from the sack and get our answer by counting the peas in that cup. The result would be almost exactly the same as when all the peas in the sack were counted. And the same would continue to be true if we had not a sackful, but a truckload of peas, or even an entire trainload of them. Even then a sample of a single cupful would answer our question, *provided* the peas were still thoroughly mixed. In the same way, one can count the employed (green peas) and the unemployed (yellow peas) in the labor force, provided the sample is properly selected (contains a thorough mixture of the employed and unemployed). Thus, desired information can, indeed, be gained at a fraction of what a census would cost. For the same reason, business firms that are interested in consumer preferences concerning old or new products never survey all existing consumers but only a tiny percentage of them. Firms even answer questions about their *internal* operations by sampling whenever a census would be too unwieldy and costly. Consider a bank that wishes to ascertain the percentage of errors made when crediting monthly interest to some 3 *million* savings accounts or when billing some 5 *million* credit accounts. The clerical cost of conducting a census of all accounts would be enormous.

Second, a census is sometimes physically impossible (that is, infinitely costly), as when the number of elementary units is infinitely large or when some of them are totally inaccessible. Any *process* that is expected to operate indefinitely under identical conditions, for example, generates an infinite number of outcomes. Thus, a census can never observe all the outcomes of this process. It can never record, for example, all the defective memory chips likely to be produced by a new and ongoing production process or all the effects ever to be produced by a new drug. In such situations, sampling is inevitable. The same is true when some elementary units are practically impossible to contact, such as airplanes presumed to have crashed in the ocean or in remote mountain regions or people with addresses unknown or temporarily hospitalized in intensive care. Third, a census is senseless (that is, again infinitely costly) whenever the acquisition of the desired information destroys the elementary units of interest. Consider questions about the lifetime of batteries or about the quality of flash cubes produced by a firm. If every battery or flash cube were tested, all of the output would be used up and the answers to the original questions would be useless. Fourth, a census is senseless whenever it produces information that comes too late. Consider political opinion polling undertaken prior to an election. A census of many millions of registered voters (or even repeated censuses in the face of rapidly fluctuating preferences) would take too long to yield results; only sampling can provide the desired information in time. Fifth, for a given cost, sampling can provide *more detailed* information than a census. Consider how many more questions the Bureau of Labor Statistics could ask those 60,000 households if it spent the same amount of money, or even a quarter of the money, that it would have to spend on a more-than-100-million-person census of the entire labor force. Sixth, sampling can provide *more accurate* data than a census. This seems paradoxical but is true because fewer statistical workers are needed, and they can be better trained and more effectively supervised. Thus, for a given cost, higher quality information is received.

Types of Samples

Different types of samples are obtained depending on the method by which elementary units are selected for observation. These include convenience samples, judgment samples, and various forms of random samples. When expediency is the primary consideration and only the most easily accessible elementary units are chosen for observation, the resultant subset of the frame, or of an associated population, constitutes a **convenience sample.** It is unlikely that a convenience sample of a statistical population is representative of it in the sense that valid inferences can be drawn from the sample about the population. The selection procedure based on convenience almost assures the opposite. The boxed salary figures in column 6 of Table 1.1 provide an example. For purposes of illustrating the concept of a sample, it was simply *convenient* to combine the last three numbers in that column, but no serious thought was given about their representativeness of the entire column 6 salary population. In the same way, a member of Congress who judges constituent attitudes on the basis of mail received is taking a convenience sample; so is the aide who telephones constituents about their opinions—without regard to the opinions of those who happen not to answer the phone, who do not have a phone, or who have unlisted telephone numbers.

A somewhat more sophisticated type of sample emerges when personal judgment, presumably based on prior experience, plays a major role in selecting elementary units for observation. Because the "expert" judgment is believed to make the sample representative of the whole, such a subset of the frame, or of an associated population, is called a **judgment sample.**

Making such a judgment can be next to impossible, however, especially when the elementary units are heterogeneous and when the desired sample is small. Consider, for example, selecting a "representative" group of 10 persons from among 1,000 people about whom we already know a great deal. If there were 400 males and 600 females, we could select 4 males and 6 females for our sample, and in this sense our sample would be a miniature of the population. But what about all the other characteristics people have, such as age, education, income, and race? If half the males and two thirds of the females were over 50 years of age, we could, of course, make sure that two of our 4 males were under 50 and two of them were over 50 and that two of our 6 females were under and four of them over that age. If the population also contained 20 percent blacks among males and 30 percent blacks among females, however, we couldn't also arrange for 0.8 black males and 1.8 black females to enter our sample. We might choose 1 black male and 2 black females, of course, but already our sample would have ceased being a miniature of the whole. The more characteristics we considered, the worse our problem would get. The more we tried, the more our judgment sample would look like a convenience sample, as if it were determined solely by our personal whim. Nevertheless, "expert" judgment is oftentimes used to select samples.

A good example of such judgment at work is the construction of the monthly Consumer Price Index. Someone decides, on the basis of personal judgment, which prices, among literally billions of prices, are to be sampled and what weights are to be assigned to them. Determining the CPI involves complex decisions on which stores in which geographic areas are to be surveyed on which days concerning the prices of which products. A price charged many customers in a popular store is more important than one charged few customers in a nearly abandoned one; prices in Hadley, Massachusetts, affect fewer people than prices in Chicago; more people shop on Saturdays than on Mondays (and then take advantage of weekend specials); and automobile tires, fruit, and meat somehow *seem* more important than horseshoes, tea, and matches. On all these matters and more, "expert" judgment is brought to bear. When it is wrong, as is easily possible, the sample ends up unrepresentative of the associated whole.

Most important, because it avoids the problem of unrepresentativeness, is the **random sample,** or **probability sample,** which is a subset of a frame, or of an associated population, chosen by a random process that gives each unit of the frame or associated population a known positive (but not necessarily equal) chance to be selected. If properly executed, the random selection process allows no discretion to the investigator as to which particular units in the frame or population enter the sample. As a result, such a sample tends to maximize our chances of making valid inferences about the totality from which it is drawn. Because random samples are so important, we must look at them in some detail. Four major types of random samples exist. We will consider each of them in turn, always assuming that the sample is to be taken (a) from a *finite* frame or population (containing a countable number of units) and (b) without replacement (so that any given unit can enter the sample only once).

The Simple Random Sample

A **simple random sample** is a subset of a frame, or of an associated population, chosen in such a fashion that every possible subset of like size has an equal chance of being selected. This procedure implies that each individual unit of the frame or population has an equal chance of being selected, but the converse is not true (giving each individual unit an equal chance of being selected does not assure that every possible subset of like size has an equal chance). Consider a frame consisting of six elementary units, A, B, C, D, E, and F. Suppose we divided these units into two categories, a subset of larger units (A, D, E) and a subset of small units (B, C, F). One *could* decide to select one of these two subsets as the sample, if a sample of three were desired, and the actual sample could then be selected by simply tossing a coin. Each of the two listed samples would have an equal chance of being selected—namely, 1 out of 2, depending on whether heads or tails appears. The fact that each of these samples would have an equal chance of being selected would imply that each individual unit contained in either one of these two samples, and, therefore, each unit in the frame, would also have an equal chance (1 out of 2) of being selected. Yet this procedure would not satisfy the above prescription for selecting a simple random sample. There would be some possible samples of size three—namely, combinations of large *and* small elementary units (such as A, B, C or C, E, F)—that would have a *zero* chance of being selected. In order to select a simple random sample of three from the above list of six elementary units, each conceivable subset of three must be given an equal chance of being selected. There are 20 such conceivable subsets[1]:

ABC	ACD	(ADE)	AEF	BCD	BDE	BEF	CDE	CEF	DEF
ABD	ACE	ADF		BCE	BDF		CDF		
ABE	ACF			(BCF)					
ABF									

---

[1] The number of different samples of size n that can be drawn without replacement from a list of N elements equals $\dfrac{N!}{(N-n)!\,n!}$ which, in this example, equals $\dfrac{6!}{(6-3)!3!} = \dfrac{6!}{3!3!} = \dfrac{6 \cdot 5 \cdot 4 \cdot 3 \cdot 2 \cdot 1}{(3 \cdot 2 \cdot 1)(3 \cdot 2 \cdot 1)} = \dfrac{720}{36}$ = 20. The expression $N!$ is pronounced "N factorial"; its meaning is obvious from the example. But note: By definition, $1! = 1$ and $0! = 1$. Factorials are discussed further in Chapter 4.

In order for the sample to qualify as a simple random one, all 20 of these subsets, and not just the encircled ones, must have an equal chance (here 1 in 20) of being selected.

A simple random sample of a city's residents, obviously, cannot be obtained by interviewing people on Main Street at 11 A.M. on a Monday. Samples including people at work or in hospitals, for example, would have no chance of being selected. Indeed, the most common procedure of obtaining a simple random sample involves a lottery process applied to the relevant frame or population. Consider the alphabetical listing, in Table 1.2, of the 100 largest multinational companies headquartered in the United States. For someone who wanted to investigate characteristics of these companies, such as their foreign and domestic sales, their profits, or their assets, this listing represents the frame. If a census were to be undertaken, relevant data could be collected for every one of the companies, and these would represent the statistical populations of sales, profits, or assets. Yet how could one acquire simple random samples of such populations corresponding to, say, a mere 10 of these companies?

One procedure is akin to a raffle. One records each of the 100 company names on a separate slip of paper, places the slips of paper in a container, mixes thoroughly, and draws 10 slips. The firms named on the slips drawn arc thcn investigated with respect to the variables of interest. This procedure is, however, troublesome when there is a large number of elements because it is difficult to mix them properly. (Close-Up 1.1, "The 1970 Draft Lottery Fiasco," provides a vivid illustration of this fact.)

A more common procedure nowadays makes use of a **random-numbers table;** that is, a listing of numbers generated by a random process in such a way that each possible digit is equally likely to precede or follow any other one. In such a table, there is no discernible pattern in the way the digits appear. Suppose we wanted to construct our own table of five-digit random numbers. All we would have to do is write the numbers from 0 to 9 separately on 10 slips of paper, mix the slips in a bowl, and pull out one. We would have the first digit for our first number. After returning the slip we pulled to the bowl, we would once more mix the 10 slips, pull out one, and thus get the second digit; then, repeating the process, we would determine the third, fourth, and fifth digits. We would, thus, have created our first five-digit random number, and if we cared to we could create 500 of them, such as those listed in Appendix Table A. Actually, the numbers listed there come from a list of 1 million random numbers generated by a computer at the Rand Corporation. Because the computer, like an electronic roulette wheel, followed the selection principle outlined above, each value between 00000 and 99999 had an equal chance of appearing at each of the 500 locations given in the table.

It is easy to select a simple random sample of the multinational companies with the help of our random-numbers table. For this purpose, the 100 company names have already been coded by two-digit numbers ranging from 00 to 99. (These numbers are a good example of nominal data.) We decide which 10 of the companies are to be in the sample simply by entering the random-numbers table at random and reading off two-digit numbers in any predetermined, systematic way. The first 10 two-digit numbers encountered that correspond to previously unused company codes tell us what the sample is. Let us elaborate.

Many random-numbers tables, unlike Appendix Table A, consist of many pages. It is a good idea to begin the process of selecting a simple random sample by first selecting a page of the table and then selecting a number on that page, at random. Any nonpatterned procedure will do. For example, we might simply open the 400-page

Table 1.2 The 100 Largest U.S. Multinationals

This table shows the names of those 100 U.S.-based multinational companies that had the highest foreign revenues in 1981. Such revenues ranged from $75.8 billion (Exxon) to $1.1 billion (Ingersoll-Rand). Company names have been coded with two-digit numbers from 00 to 99.

00 Allied Corporation	50 Honeywell
01 American Brands	51 IBM
02 American Cyanamid	52 Ingersoll-Rand
03 American Express	53 International Harvester
04 American Home Products	54 International Telephone & Telegraph
05 American International Group	55 Irving Bank
06 Atlantic Richfield	56 Johnson and Johnson
07 Avon Products	57 K mart
08 Bank America	58 Litton Industries
09 Bankers Trust New York	59 Manufacturers Hanover
10 Beatrice Foods	60 Marine Midland Banks
11 Bendix	61 Merck
12 Bristol-Myers	62 Minnesota Mining & Manufacturing
13 Burroughs	63 Mobil
14 Caterpillar Tractors	64 Monsanto
15 Chase Manhattan	65 Morgan, J. P.
16 Chemical New York	66 Motorola
17 Chrysler	67 Nabisco Brands
18 Citicorp	68 NCR
19 Coca-Cola	69 Occidental Petroleum
20 Colgate-Palmolive	70 Pan American
21 Consolidated Foods	71 Pepsi Co.
22 Continental Group	72 Pfizer
23 Continental Illinois	73 Phibro-Salomon
24 CPC International	74 Philip Morris
25 Dart and Kraft	75 Procter and Gamble
26 Deere	76 Procter and Gamble
27 Digital Equipment	77 Ralston Purina
28 Dow Chemical	78 RCA
29 Du Pont de Nemours	79 R. J. Reynolds Industries
30 Eastman Kodak	80 Safeway Stores
31 Exxon	81 Scott Paper
32 Firestone	82 Sears, Roebuck
33 First Chicago	83 Singer
34 First National Boston	84 Sperry
35 Fluor	85 Standard Oil California
36 Ford Motor	86 Standard Oil Indiana
37 General Electric	87 Sun Co.
38 General Foods	88 Tenneco
39 General Motors	89 Texaco
40 General Telephone and Electric	90 Texas Instruments
41 Getty Oil	91 Trans World
42 Gillette	92 TRW
43 Goodyear	93 Union Carbide
44 Grace, W. R.	94 Union Oil California
45 Gulf and Western Industries	95 United Technologies
46 Gulf Oil	96 Warner-Lambert
47 Halliburton	97 Westinghouse Electric
48 Heinz, H. J.	98 Woolworth, F. W.
49 Hewlett-Packard	99 Xerox

Source: Forbes, July 5, 1982, pp. 126–28.

Rand Corporation table at random and select page 377. (This page has, in fact, been reproduced as Appendix Table A.) Closing our eyes, we might then place a finger on the chosen page and thus determine number 00717 (row 24, column 5) as our starting number. Actually though, we should not even go about finding this number until we have decided how we will use it and how we will then proceed to find other numbers. If, as in our case, the data consist of two digits, while the random numbers contain five, we must decide how we will extract two-digit numbers from our chosen page. We could read only the first two digits of each five-digit number, we could read the last two digits only, or even the second and last digits in each number. A multitude of possibilities exist. And having found our first *two-digit* number, we can go about finding additional ones (up to the desired sample size) in any systematic fashion we like: by proceeding along our initial row to the right or to the left, by moving along our initial column down or up, even by reading numbers diagonally. We can follow any self-imposed plan we like, as long as we never read a given location twice.

Suppose, for example, that we had decided to select our sample of 10 companies by reading the last two digits of the five-digit random numbers and by proceeding down along (possibly successive) columns. Since our initial number was the 24th one in the fifth column of Appendix Table A, our simple random sample would quickly be determined as follows:

007⑰ Chrysler
107⑨⑦ Westinghouse Electric
739㉟ Fluor
53497 omitted (97 used above)
692⑭ Caterpillar Tractors
747⑭ Philip Morris
945⑧② Sears, Roebuck
151⓪⑦ Avon Products
807㊤ Johnson and Johnson
894㉗ Digital Equipment
43335 omitted (35 used above)
830⓪① American Brands

Two things should be noted: First, if one encounters a given value a second time during the above procedure (as happened with numbers 97 and 35 in our case), one skips to the next number. The same is recommended when the value encountered has no counterpart in the frame or population being sampled (as when one is choosing from a list of 60 companies only and one encounters numbers between 61 and 99 in the random-numbers table).

Second, there are those who argue that the above procedure for finding a starting position (our number 00717) may not be sufficiently random for *frequent* users of a random-numbers table. This may be so because a given book has the tendency to open to the same page repeatedly (and, thus, a frequent user might again and again select page 377), while a person placing a finger on a page has a tendency to choose a number toward the center of the page (as we did). In order to guard against these dangers, one can, of course, select a starting position in any number of truly random ways. For example, if we were working with an eight-page table, we might select a page by tossing a coin three times. There are two possible outcomes for the first toss, heads (H) or tails (T). There are two possible follow-ups for each of these two contingencies on the second toss, making for sequences of HH, HT, TH, or TT. And there are two possible follow-ups for each of these four contingencies on the third toss, making for eight conceivable sequences for the three tosses. These sequences might be linked to the choice of a page in the random-numbers table, as shown in the accompanying table.

Possible Sequence	Page Number Picked
H H H	1
H H T	2
H T H	3
H T T	4
T H H	5
T H T	6
T T H	7
T T T	8

Having thus found the page by tossing a coin three times, we might determine the starting position by reading two digits at a time along successive rows and letting the first usable set of two digits point us to a row and the second such set to a column. (Can you see why, in Appendix Table A, this procedure would have us start with row 27 and column 01; that is, with number 20545?) The possibilities of random selection are truly limitless.

The Systematic Random Sample

Another commonly used type of sample, the **systematic random sample,** is a subset of a frame, or of an associated population, chosen by randomly selecting one of the first k elements and then including every kth element thereafter. If this procedure is employed, k is determined by dividing population size, N, by desired sample size, n. Consider again Table 1.2. To select a systematic random sample of, say, five companies, one would wish to include every 20th company on the list (there being a total of 100 firms); hence $k = 20$. The procedure, however, requires that the starting point (the first company included) be chosen at random. By drawing a number between 00 and 19 out of a hat, or by using a table of random numbers, we might decide that company number 09 is to be selected as the starting point. Then our entire sample would consist of the companies coded as 09, 29, 49, 69, 89. Had our initial choice been 01 or 17, the sample would have been different, of course—namely, 01, 21, 41, 61, 81 or 17, 37, 57, 77, 97, respectively.

The Stratified Random Sample

Sometimes the frame or population to be sampled is known to contain two or more mutually exclusive and clearly distinguishable subgroups or *strata* that differ greatly from one another with respect to some characteristic of interest, while the elements within each stratum are fairly homogeneous. In such circumstances, one can select a

stratified random sample, which is a subset of a frame, or of an associated population, chosen by taking separate (simple or systematic) random samples from every stratum in the frame or population, often in such a way that the sizes of the separate samples vary with the importance of the different strata. If we knew that 10 of the 100 companies listed in Table 1.2 accounted for 70 percent of the 100 companies' sales, while the other 90 companies accounted for the remaining 30 percent, and if sales were the characteristic of interest to us, we might wish to ensure that the 10 giants were not missed by our sampling procedure (as might well happen if a simple or systematic random sample were taken). We could divide our list into two strata (10 giants and 90 dwarfs) and then create our overall sample by selecting some firms from each of these two groups. For a 10-percent overall sample, we might select 8 firms from the giant stratum and 2 firms from the dwarf stratum, and these 10 firms could be expected to account for more than half of all sales.

This type of procedure is, in fact, quite common. Some examples from the business world are summarized in Analytical Example 1.1, "How Accountants Save Money by Sampling." Other examples of stratified sampling include a study of banking (wherein strata were formed to ensure representation of very large banks that may be few in number but that hold a high percentage of all deposits), a study of affirmative-action programs (concerned with the representation of all ethnic or racial groups), and a study of teacher qualifications (designed to ensure the proportional representation of elementary-school, high-school, and college teachers). Stratified random sampling has also become crucially important for pollsters who wish to predict the outcome of U.S. presidential elections. It is not good enough to estimate correctly which candidate will win the popular vote in the nation as a whole. Such a candidate can still lose the election because the president is elected by the Electoral College in which representatives traditionally vote as a block *by state,* and electoral votes are rationed unequally to each state in accordance with its representation in Congress. To be successful, a candidate must, therefore, get popular vote margins in the large states. Hence, a pollster must treat each state as a separate stratum, must determine the winner in each state, and must combine the results in the same proportion as the electoral votes.

The Clustered Random Sample

Finally, there are occasions when the frame or population to be sampled is naturally subdivided into *clusters* on the basis of physical accessibility. In such circumstances, one can select a **clustered random sample,** which is a subset of a frame, or of an associated population, chosen by taking separate censuses in a randomly chosen subset of geographically distinct clusters. Someone who wanted to sample the residents or shops of a city, for example, might divide the city into blocks, randomly select a few of these (by any of the methods previously discussed), and then interview every resident or shop owner within the chosen blocks. Because of the geographic proximity of those interviewed, this procedure would save considerable transportation expenses and time in comparison to a citywide simple random sample of individual residents or shop owners (who would almost certainly be located in a multitude of different places). Acceptance sampling by business firms is likely to follow a similar route. Imagine a firm that had just received a shipment of 1 million coffee cups. Its warehouse would be filled, perhaps, with 10,000 sealed cartons, each containing 100 cups. An inspector who wanted to determine the quality of the cups by taking a 10,000-cup sample could take a simple or systematic random sample of them, but might then have to open and unpack nearly all of the 10,000 cartons in order to find the individual cups selected. It would be much easier to regard each carton as a geographic cluster, to select 100 cartons randomly, and to inspect every single cup in these cartons only. Some 9,900 of the cartons would not have to be opened at all.

Note: The procedure described here is also called *single-stage* cluster sampling. On occasion, this type of sampling is replaced by the more complex *multistage* cluster sampling. An example of the latter might be a nationwide household survey conducted as follows: first, a subset of states (primary clusters) is chosen at random; second, a subset of cities (secondary clusters) is randomly chosen within the previously selected states; third, a subset of city blocks (tertiary clusters) is randomly selected within the cities chosen before. As in the earlier example, censuses occur in the city blocks finally selected.

Note also that the random selection of clusters need not necessarily involve simple or systematic random sampling. By an appropriately designed type of stratified sampling, different clusters might, for example, be given unequal chances of selection, depending, perhaps, on the proportions of relevant units (such as people, shops, or coffee cups) that are located in different clusters (such as cities, city blocks, or cartons).

ERRORS IN SURVEY DATA

We have noted how data can be acquired by taking a census or by engaging in various forms of sampling. Inevitably, all survey data are subject to error; errors can arise from innumerable, often unexpected, sources. At best, errors obscure the truth only slightly; at worst, they can reduce the value of a survey below zero, for nothing is more unfortunate than "knowing" something that isn't so. Errors can be generated during the planning stage of a survey; more errors can be created during the later stages of data recording and processing. Indeed, it is a hopeless task to list all the ways in which surveys can go wrong, and we won't attempt it. It is possible, however, to raise our awareness of the problem by focusing on two broad categories of error.

Random Error versus Systematic Error

Suppose we were interested in the average annual salary received by the employees of a particular firm, such as those of Mountain Aviation, Inc. discussed earlier. If we had none of the information given in Table 1.1, except the listing of employee names, we could take a random sample of this frame, interview the employees so selected about their salaries, and let the average of these salaries serve as an estimate of the average salary of all employees. (The choice of such a sampling procedure would be all the more likely, of course, the greater was the number of employees of the firm.) A random sample of three might consist of Cruz, Hill, and West, and we could estimate the average annual salary of all employees by the average of the Cruz, Hill, and West salaries listed in column 6; that is, as $18,333 per year. Yet it is clear that this result depends entirely on the particular elementary units that happened to be selected for our sample. One can, in fact, select 84 different samples of three from among our nine hypothetical employees (see footnote 1), and almost every one of these 84 samples would provide us with a different average-salary estimate, if the above procedure were followed. The sample consisting of the salaries of West, Wolf, and Zorn—see the box in column 6 of Table 1.1—for example, would give us an average-salary estimate of $18,667 per year, but a sample consisting of the salaries of Hill, Otis, and Zorn would yield an estimate of $12,333 per year. All these estimates differ from the average-annual-salary figure a census would reveal, for the average of *all* the column 6 data is $18,167 per year. But note: The individual-sample estimates are just as likely to depart from the census figure in one direction as the other. In the long run, if one took repeated samples and averaged the resulting estimates of average salary made by all possible samples of three, the same census-derived figure would be found.

Thus, the first major type of error of which statisticians must be aware is called **random error,** or **chance error,** or **sampling error,** and equals the difference between the value of a variable obtained by taking a single random sample and the value obtained by taking a census (or by averaging the results of all possible random samples of like size). This definition is summarized in the left half of Figure 1.1.

This type of error, as one of its names suggests, is associated only with sample surveys. It arises from the operation of chance that determines which particular units of the population happen to be included in the sample. This error can, of course, be positive or negative, tiny or huge, but it can always be reduced by increasing the size or number of random samples taken (and it is zero in a census). Most important, as later chapters will show, the size of this error can itself be estimated (and reported alongside the observed data). As a result, statisticians can state, for example, that the average annual salary in a firm equals $18,333 and that there are 95 chances in 100 that a census would reveal a number within $205 (plus or minus) of the stated one.

Unfortunately, random errors are the least of our problems. Even if we had the complete census information given in column 6 of Table 1.1, we could not be *certain* that the average annual salary in our firm was $18,167. The true value might well equal $20,167, as noted on the right-hand side of Figure 1.1. This surprising statement is true because every type of survey (sampling or census taking) could produce incorrect data if it was subject to some fundamental flaw in the manner in which observations were made or recorded. Such a flaw might affect every single measurement in the same way, systematically pushing each one either below or above its true value.

Consider the story of the sales clerk who measured yards of cloth with an old, stretched-out cloth tape, thus causing all the recorded lengths to be lower than the true lengths. A similar predicament could befall the survey taker. Unbeknownst to anyone, an incorrectly programmed computer might, for example, deduct $2,000 from every true salary figure entered in it, and it might thus print out an incorrect set of column 6 data (in the case of a census) or an incorrect subset of them (in the case of sampling). The data, mind you, would look precisely like those in Table 1.1, but they would be wrong, nevertheless.

The people who are being interviewed can be at fault, too. Consider the story of the butcher who weighed each piece of meat with a thumb on the scales, thus causing all the recorded weights to be higher than the true weights. An analogous misfortune could befall the survey taker when "shopping" for information. Like the scheming butcher, the employees being interviewed about their salaries might not tell the truth. They might understate their salaries if they suspected the information was being channeled to the tax collector; they might overstate their salaries if they thought the information was about to be used to compute a 20-percent raise.

This sort of problem is the basis for defining the second major type of error that statisticians must consider. It is called **systematic error,** or **bias,** or **nonsampling error,** and equals the difference between the value of a variable obtained by taking a

Figure 1.1   Random Error and Systematic Error

This graph illustrates how values obtained by data gatherers can differ from true values as a result of random error (in the case of random sampling) or of systematic error (in the case of all types of surveys). Either type of error can be positive or negative.

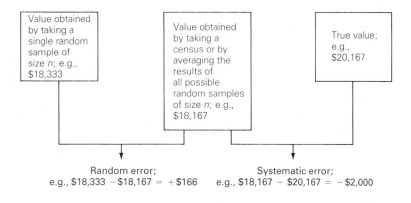

Value obtained by taking a single random sample of size *n*; e.g., $18,333

Value obtained by taking a census or by averaging the results of all possible random samples of size *n*; e.g., $18,167

True value; e.g., $20,167

Random error;
e.g., $18,333 − $18,167 = +$166

Systematic error;
e.g., $18,167 − $20,167 = −$2,000

census (or by averaging the results of all possible random samples of a given size) and the true value. This definition is summarized in the right half of Figure 1.1.

Unfortunately, bias can be hard to detect, and its size—unlike that of random error—cannot be estimated. For this reason, statisticians who seek to discover the truth must become aware of the major sources of bias and try their best to neutralize them. To that issue we now turn.

How Bias Can Enter Surveys

Bias can enter a census or sampling procedure at the planning stage, during the data-collection stage, or even while data are being processed.

In the planning stage. If they are not careful, statisticians can literally build systematic error into the very design of their surveys. This can happen, for example, as a result of **selection bias,** a systematic tendency to favor the inclusion in a survey of selected elementary units with particular characteristics, while excluding other units with other characteristics. As a result of selection bias, any data that are eventually collected are bound to overrepresent the former characteristics. The use of a faulty frame or the selection of a *nonrandom* sample can easily result in a selection bias.

Consider what would happen in our quest to determine the average salary earned in a firm if we planned our survey on the basis of a list of those employees who were seen walking through the firm's gates last Friday around 5 P.M. Regardless of whether we were planning a census or a random sample, the accuracy of our survey results would be compromised because we would not be working the proper frame (such as an up-to-date listing of all employees culled from the firm's personnel records). Many employees—such as those on vacation, those working part-time, those who are ill, and, perhaps, executives working late—would have no chance of being included in our survey. It would be hazardous to assume that their salaries would, on average, equal those of the people on our list. We would be studying an *actual* population (salaries of employees who were seen leaving the firm last Friday around 5 P.M.) that was almost certainly quite different from our *target* population (salaries of all employees). Unfortunately, the use of faulty frames is quite common; sometimes, for lack of better alternatives, it is even unavoidable. Examples include the firm that surveys its actual customers (instead of its potential customers) to assess the prospects of a new product and the political pollster who surveys a group of people whose names were taken from the telephone directory or from car-registration rolls or even from a list of *eligible* voters in order to assess the chances of a candidate (when the people who really matter are those *registered* voters who are sure to vote on election day).

As noted, selection bias can also result from the nonrandomness of sampling, even though a perfect frame is used. The use of any convenience or judgment sample is automatically suspect, but even an attempted random sample can turn out to be nonrandom in fact. Consider the attempted selection of a systematic random sample from a perfect frame that lists elementary units with some type of rhythm. If the frame were a 50-year listing of monthly stock prices, and we selected every 12th number (or some multiple thereof), we would end up with a set of data all of which would pertain to the same month—say December; these selected data could easily differ in some systematic way from all the excluded ones. Or consider an alphabetical listing of names. If most of the members of some ethnic group had names that began with certain letters (such as Irish names that begin with *Mc* or *O'*), a systematic random sample could easily miss every one of these people, and this would be disastrous if the survey were designed to study ethnic questions. The same sort of problem can occur in clustered random sampling. Because people with similar characteristics (such as education level, ethnic grouping, or income) tend to live close to each other, it is not impossible for cluster sampling to miss a whole range of backgrounds altogether, and this omission could seriously bias the survey results.

Another common type of systematic error is **response bias,** a tendency for answers to survey questions to be wrong in some systematic way. Nothing, probably, contributes more to this problem than the faulty design of questionnaires. As students who take exams know too well, it is not easy to ask good questions that elicit the answers really sought. Consider the frequently told story of the student who answered (c) when confronted with the following question:

4. Wind-eroded rocks are most commonly found
   - a. on mountain tops.
   - b. on river bottoms.
   - c. in fertile valleys.
   - d. in outer space.

The teacher had hoped for (a), but the student argued that hardly any people ever *were* on mountain tops, river bottoms, or in outer space; hence, if such rocks were to be *found,* it would most likely happen in (heavily populated) fertile valleys.

Nothing seems easier than to formulate similarly inept questions on survey questionnaires. When this happens, people cannot be expected to give reliable answers; data gatherers who imagine themselves measuring something may, in fact, end up measuring something else or nothing at all. When asked about their income, for instance, people give different answers depending on whether they suspect that their answer will affect their taxes, their pension, or their bank credit. (A good questionnaire, therefore, contains instructions that motivate respondents to tell the truth by providing them with reasons for the survey). The use of clearly defined concepts is crucial, too. It is not obvious what it means to live in poverty or in a substandard home or in a crime-ridden neighborhood. Such terms, therefore, should not be placed on a questionnaire unless they are carefully defined. Nor can people be expected to know what to do when asked to list their home town (what about soldiers or students?), their employment status (is a retired person *unemployed?*), their trips abroad (does Canada qualify?), their weight (with or without clothes? at morning or at night?), their age (last birthday or nearest birthday?), or their spending plans for next year (next *calendar* year or the next 12 months?).

Most worrisome of all, however, is the leading question that is likely to lead the respondent, inexorably, to a particular, predictable answer. Leading questions can be obvious or extremely subtle. Here are some obvious ones:

1. As usual, today's food in the dining hall was rotten.   Yes   No
2. Should we waste further billions to send people into outer space, while there is poverty on earth?   Yes   No
3. In light of their $3 billion profit, should we reject management's skimpy wage offer?   Yes   No
4. Does the name *Apple* come to mind when thinking about high-quality computers?   Yes   No

One can safely bet that most answers would be Yes, No, Yes, Yes. Such questions are sure to invite response bias. There are, however, more subtle ways to ask leading questions.

Response bias can result from the *sequencing* of different questions in such a way that people answer one question within a frame of reference provided by a (possibly unrelated) earlier question. Consider this:

10. Which of the following is responsible for making "Fruit and Fibre" such a superior cereal?
    a. It is fruitier   c. It is sweeter.
    b. It is nuttier.   d. It is crunchier.
11. Which of the following do you consider superior?
    a. Fruit and Fibre.   c. Granola.
    b. Bran Chex.   d. Rice Krispies.

In light of question 10, most people will answer question 11 by choosing (a).

Faulty questionnaire design can, finally, be responsible for **nonresponse bias,** a systematic tendency for selected elementary units with particular characteristics not to contribute data in a survey while other such units, with other characteristics, do. In the presence of this problem, even a census based on a perfect frame, or a perfectly selected random sample, will fail. They will yield faulty conclusions because the data actually collected will in fact constitute a convenience sample (for example, of the most strongly opinionated people among all the people that were supposed to be in the survey). Questionnaire features that contribute to nonresponse bias include a physically unattractive design; hard-to-read print; questions that are boring, unclear, or long and involved; an excessive number of questions; bad sequencing of questions (so that respondents are forced to jump back and forth from topic to topic), and, in the case of multiple-choice questions, the specification of answers that are not mutually exclusive or are excessively restricted to particular points of view, while omitting other possible views. Experience shows that high-income people and low-income people (unlike middle-income people) tend not to respond to surveys; it is easy to see how the exclusion of either group is apt to bias survey results.

Close-Up 1.2, "Dubious Sampling: The Literary Digest Case," gives an all too vivid picture of a survey gone wrong even at the planning stage.

In the collection stage. Even when the designers of surveys successfully avoid the emergence of bias by the use of proper frames, the careful selection of random samples, the meticulous creation (and pretesting) of questionnaires, and the like, those who execute surveys can still allow bias to enter the process.

Selection bias is apt to enter a survey when interviewers are instructed to select, within broad guidelines, the particular individuals they will question. Interviewers are routinely allowed to select their interviewees in the case of judgment sampling. If it is known that the population of a city contains 10 percent black men, 20 percent black women, 30 percent white men, and 40 percent white women, for example, an interviewer who is to question 100 persons may be given *quotas* of 10 black men, 20 black women, 30 white men, and 40 white women in order to assure that the questions are posed to a proper cross section of the population. Given the quotas, the interviewer, however, may be free to select particular individuals. Naturally, interviewers will select such people in some convenient way; perhaps, they will choose interviewees by standing at the nearest street corner (provided they can find people whose appearance is pleasant enough and whose behavior is conventional) or by knocking on doors (provided the neighborhood is attractive enough, there are no dogs in the front yard, and the time of day is right). This procedure will, of course, miss all sorts of people, such as those on vacation, those in hospitals or prisons, those at work, or those residing in unpleasant parts of town, in trailer camps, or in dormitories. Indeed, experience shows that quota sampling invariably leads to the selection of a disproportionate number of highly educated, high-income individuals. (Other types of quotas lead to similarly biased selections. When told to select 60 people over 50 years of age and 40 people under 50 years, the typical interviewer, understandably, selects lots of people who look like 70, lots of people who look like 20, and hardly anyone who is close to 50 years old). Indeed, selection bias can occur even in the execution stage of a properly selected *random* sample. Suppose a random-numbers table were used to select, from a complete list of a city's residents, a simple random sample of 100 persons (in the same way as a sample of 10 multinational firms was selected earlier in this chapter). Such a procedure would charge an interviewer with finding 100 specifically named individuals, no matter how hard the task. Yet actual interviewers have been known to substitute other persons for the randomly selected ones if the latter were hard to catch during (the interviewer's) working hours or lived in dangerous or out-of-the-way places. Naturally, such a failure to follow directions on the part of the interviewer dooms the entire survey by turning a probability sample into a convenience sample. Imagine the biased results in a survey about cosmetics use if the interviewer substituted grandmothers for all the teenagers because the latter were never around. Imagine the misleading results if elderly or working people or vacationers were excluded from surveys that sought to ascertain the demand, respectively, for household help or frozen food or leisure-time products.

Response bias can arise for a number of reasons during data collection. Both interviewers and respondents can be at fault. Interviewers, often inadvertently, may solicit a particular "acceptable" answer to a question by their dress and choice of words (which may betray their social class) or by their tone of voice and demeanor (a gesture of disapproval or surprise at some answers will almost surely affect other answers). Interviewers can, of course, also make systematic mistakes when recording answers—for example, by consistently categorizing part-time income or work as full-time. And interviewers have been known to fake answers completely, as happened during the 1980 census in New York City when drug addicts on methadone maintenance worked for the Census Bureau in ghetto areas. (Some individuals just wanted to get the work done fast; the lives of others had been threatened by unwilling respondents.) Even the timing of a survey can create response bias. What do you think would happen to all the answers if a survey on plant safety were conducted precisely one day after a rare but major accident or if a survey on hours recently worked were conducted shortly after a major holiday? Interviewers must not, however, receive all the blame.

Even when questionnaires are designed well, respondents often give false answers for a variety of reasons. Respondents may simply not know the answer but may give one anyway to conceal their ignorance. They may give whatever answer they think will please the interviewer. (It is well known that white and black interviewers get different answers from identical respondents when asking identical questions on racial issues.) Respondents may give distorted answers in line with some current fad toward optimism or pessimism. And they may tell deliberate falsehoods to mislead competitors or impress the interviewer. (People are particularly likely to boast of successes and hide failures on so-called "prestige questions" concerning, say, their knowledge of current events or famous people, their reading of books or level of education, their grade-point average or wealth, and even their brushing of teeth and taking of baths.)

The data-collection stage also generates its share of *nonresponse bias*. Respondents often refuse to answer questions they consider too personal, such as those about their age, health, and sexual habits, about their income, or about illegal and immoral activities. Indeed, many respondents refuse to answer *any* questions, particularly in surveys conducted by phone or mail. (The 1980 Census, claimed the city of New York, failed to count some 800,000 residents who refused to be counted; Chicago, Detroit, Philadelphia, and other cities made similar allegations.)

In the processing stage. The emergence of bias during data collection can conceivably be minimized by the careful training and supervision of interviewers and by a variety of techniques designed to forestall incorrect or refused responses. (These techniques include prominent and favorable coverage by the news media about surveys to be undertaken, personal special-delivery letters to respondents on the stationery of well-known organizations or over the signature of respected individuals, credible promises of anonymity or of rewards, and more.) Nevertheless, bias can enter even at the data-processing stage. Multitudes of people who code, edit, keypunch, tabulate, print, and otherwise manipulate data have multitudes of opportunities for making non-canceling errors. Those who code answers to *open-ended* questions (which elicit answers in people's own words) may consistently miscategorize them; editors may consistently introduce high values (or low ones) when encountering incomplete or illegible responses; they may fail to eliminate **outliers,** or "wild values" (maverick responses that are not believable because they differ greatly from the majority of observed values); keypunchers may consistently misread 7s as 9s; they may ignore all the decimal points, thus turning every 8.1 into an 81—the list goes on.

1.1 How Accountants Save Money by Sampling

Business firms and other organizations used to spend a great deal of time and money on censuses to determine their accounts receivable (or payable) and to take physical inventories of equipment, raw materials, goods in process, and the like. Recent examples have shown, however, that census taking is often unnecessary and that relatively small samples of the relevant populations, if carefully drawn, can produce high-quality results, along with dramatic savings in time and money.

Case 1. The Chesapeake and Ohio Railroad faced the problem of allocating total freight and passenger revenues among several carriers when freight or passengers had traveled over lines owned by different companies, a common occurrence. In one experiment, it examined the waybills for interline freight shipments that had occurred over a six-month period. (A *waybill* is a document describing the nature of goods shipped, their routing and the total freight charges.) As the accompanying table shows, nearly 23,000 waybills were examined, and the amount that the shippers of freight owed the Chesapeake and Ohio (rather than other railroads) was determined to be $64,651.

	Census	Sample
Number of waybills examined	22,984	2,072
Cost of examination	$5,000	$1,000
Amount found due Chesapeake and Ohio Railroad	$64,651	$64,568

At a much lower cost, a sample was also taken. It was a stratified sample, containing only 1 percent of waybills totaling $5 or less, 10 percent of waybills totaling between $5.01 and $10; 20 percent of waybills totaling between $10.01 and $20; 50 percent of waybills totaling between $20.01 and $40; and 100 percent of waybills in excess of $40. The particular waybills included in the sample were selected with the help of a random-numbers table on the basis of the last two digits of their serial numbers.

The result was amazing: The amount due was determined within $83 or 1⁄10 of 1 percent of the much costlier census figure. Similar accuracy was achieved when the railroad studied five months' worth of interline passenger tickets. A census of the 14,109 tickets revealed the amount due the railroad to be $212,164. A 5 percent sample of these tickets yielded a figure of $212,063. The $101 difference was even less than 1⁄10 of 1 percent of the census figure.

Case 2. Airlines face the same problem of settling accounts with each other concerning passengers traveling on several different airlines during a given trip. Each major airline picks up hundreds of thousands of tickets per month that were issued by and initially paid to other airlines. Clerks used to figure out the revenue allocation for each ticket, but this process could be extremely cumbersome and costly. (At one time, there existed 60 possible fares between Chicago and New York!) Three airlines (Northwest, United, and TWA), therefore, conducted a four-month test similar to the one by the Chesapeake and Ohio described above. A census was taken, then a stratified sample. (The strata used were first-class, coach, military, and other types of tickets. Individual tickets were pulled on the basis of the last digit in their serial numbers with the help of a random-numbers table.) The difference in the census and sample results came to less than $700 per $1 million of tickets. After introducing sampling procedures, one major carrier cut clerical expenses by $75,000 per year, the industry as a whole saved $500,000 per year.

Case 3. One firm, Minneapolis-Honeywell, used to take 100-percent inventories of its goods in process. Then it conducted two experiments. First, it took a simple random sample of 500 items out of 5,000 items in one department. The estimated size of the inventory came within 1⁄10 of 1 percent of the full count. Second, it took a stratified random sample, dividing work-in-process lots from all departments into those of high value ($500 or more) and those of low value. All of the former were counted; only 10 percent of the latter were counted. The sample of 4,200 out of 40,000 items yielded an estimate within 8⁄10 of 1 percent of the full count.

Conclusion. Sampling instead of census taking for accounting purposes is now widespread. Firms use it to estimate their bad debts (by estimating the total of accounts owed for more than 60 days) and to gauge the accuracy of their clerical workers (by contacting a sample of customers concerning their billings); the U.S. Air Force uses it when taking a worldwide audit of its motor-vehicle pool. The list could easily be lengthened.

Sources: John Neter, "How Accountants Save Money by Sampling," in Judith M. Tanur, et al., eds., *Statistics: A Guide to the Unknown* (San Francisco: Holden Day, 1972), pp. 203–11; Theodore J. Sielaff, *Statistics in Action: Readings in Business and Economic Statistics* (San Jose, Calif.: Lansford Press, 1963), pp. 20–21 and 26–27.

||| CLOSE-UPS

1.1 The 1970 Draft Lottery Fiasco

During both World Wars I and II, it became necessary to establish an order in which men were to be drafted into the U.S. military. In 1917, accordingly, 10,500 black capsules, containing numbers previously assigned to eligible men, were drawn from a glass fishbowl. In 1940, a similar procedure was adopted to draw 9,000 numbers, but there were criticisms: The small, wooden paddle (which was made from a piece of rafter traceable to Independence Hall in Philadelphia) used to stir the capsules in the bowl would not reach deep enough into the bowl. It also broke open some of the capsules, impeding the mixing process further. In the end, the numbers drawn looked like anything but random ones; they were concentrated in certain clusters of hundreds, apparently reflecting the fact that the numbers had been poured into the bowl in lots of 100 each. The lesson was clear: thorough physical mixing of capsules in a bowl is difficult.

Apparently the lesson had been forgotten when the 1970 draft lottery was instituted. Some 366 capsules,

containing all the possible birthdates in a year, were poured into a bowl but not stirred. (There had been some mixing during the process of inserting dated slips of paper into the capsules). Then capsules were drawn out, the order of their withdrawal determining the priorities for the draft. The observed sequence strongly reflected the order in which the capsules were created (one month at a time), with late-in-the-year birthdays (that were encapsulated last) being drawn first and early-in-the-year birthdays (that were encapsulated first) being drawn last and, therefore, less subject to the draft. Several young men filed suit in federal court seeking to have the 1970 lottery voided on the basis of the apparent lack of randomization. Note: A 1971 draft lottery, in response to widespread criticism of the 1970 lottery, made use of random-numbers tables, discussed in the text.

Source: Stephen E. Fienberg, "Randomization and Social Affairs: The 1970 Draft Lottery," *Science,* January 22, 1971, p. 255–61.

1.2 Dubious Sampling: The Literary Digest Case

A classic case of both selection bias and nonresponse bias occurred in 1936. The *Literary Digest* magazine, which had correctly predicted the winner in every U.S. presidential election since 1916, predicted confidently a comfortable victory for Alfred M. Landon, the Republican candidate, over Franklin D. Roosevelt, the Democratic candidate, by a margin of 57 to 43. Yet Roosevelt won by a landslide never before seen in U.S. history, receiving 62 percent of the votes cast. What had gone wrong?

First, there was selection bias. The *Digest* had mailed questionnaires to 10 million people whose

names had been taken from various lists such as its own subscribers, telephone directories, and automobile-registration rolls. During the Great Depression, higher-income people typically voted Republican, and these people were well represented in the *Digest's* sample. On the other hand, lower-income people, who heavily favored the Democrats, were underrepresented because a lower percentage could afford magazine subscriptions, telephones, and automobiles.

Second, there was nonresponse bias. Only 2.4 million of the 10 million questionnaires were mailed back. Although this made the survey the largest sample ever taken, more educated people are more likely to respond to mail questionnaires than less educated ones. The former, again, tended to favor Republicans; the latter, the Democrats. Thus, a much larger percentage of the nonrespondents than of the respondents were for Roosevelt. This bias reinforced the selection bias.

Note: The *Digest* never survived the debacle and folded shortly thereafter. At the same time, George Gallup was setting up his survey organization, and he correctly forecast the Roosevelt victory from a mere sample of 50,000 people. Yet in 1948, using another dubious procedure (a form of judgment sampling, called *quota sampling*), Gallup's organization (along with Crossley's and Roper's) incorrectly predicted the victory of Thomas Dewey over Harry S Truman.

Sources: Frederick Mosteller, et al., *The Pre-Election Polls of 1948* (New York: Social Science Research Council, 1949); Mildred B. Parten, *Surveys, Polls, and Samples* (New York: Harper, 1950); Frederick F. Stephan and Philip J. McCarthy, *Sampling Opinions* (New York: Wiley, 1958); and George H. Gallup, *The Sophisticated Poll-Watcher's Guide* (Princeton: Opinion Research, 1976). See also Maurice C. Bryson, "The *Literary Digest* Poll: Making a Statistical Myth," *The American Statistician,* November 1976, pp. 184–85, which lays the entire blame for the debacle on nonresponse bias.

BIOGRAPHY

1.1 Adolphe Quetelet

Lambert Adolphe Jacques Quetelet (1796–1874) was born in Ghent, Belgium. He had an early interest in the fine arts (he painted, wrote poems, and even produced an opera), but this interest was soon overshadowed by his attraction to mathematics. His was the first doctoral dissertation at the newly established University of Ghent, and it was widely acclaimed as an original contribution to analytical geometry. The dissertation resulted in his election, at age 24, to the Brussels Académie Royale des Sciences et des Belles-Lettres (in which he soon became the dominant spirit) and to a position of teaching mathematics, physics, and astronomy at the Brussels Athenaeum. He was a great teacher (students and visitors from all of Europe crowded his lectures), and he was a prodigious writer (producing a vast array of essays and books and editing a leading journal). Yet the seemingly undefatigable energy that he poured into his career also changed its orientation. His enthusiasm for astronomy that led, eventually, to the building of an observatory in Brussels, and his directorship of it, also brought him in contact with illustrious French mathematicians, such as Fourier, Laplace (whose Biography appears in Chapter 4 of the *Student Workbook*), and Poisson. Their interest in probability theory and its applications to social phenomena ex-

cited Quetelet. His subsequent active encouragement of the collection of empirical social data led to the first national census in Belgium and Holland (in 1829), the formation (in 1834) of the Statistical Society of London (now named the Royal Statistical Society), and the organization (in 1841) of the Belgian Central Statistical Commission, a central agency responsible for collecting statistics. As president of the latter, Quetelet did much to inspire the creation of statistical bureaus all over Europe and labored unstintingly to promote internationally uniform methods and terminology in data collection and presentation. Under his leadership, the first of a long series of International Statistical Congresses was held in Brussels in 1853.

Although there had been forerunners in England, France, and Germany, Quetelet earned the honor of being called the father of modern statistics by the publication of his *Sur l'Homme et le Développement de Ses Facultés* in 1835. In this book, Quetelet noted how social phenomena (such as crime or suicides) reproduced themselves with amazing regularity; he argued that such regularities were discoverable only by statistical techniques and, even more important, could also be linked to causes with the help of such techniques. Unlike earlier writers who had given a theological interpretation to social regularities (seeing in them evidence of a divine presence), Quetelet pointed to social conditions as causes and suggested that legislation could also ameliorate their effects (such as crime or suicides). As he put it[1]:

> The constancy with which the same crimes repeat themselves every year with the same frequency and provoke the same punishment in the same ratios, is one of the most curious facts we learn from the statistics of the courts. . . . And every year the numbers have confirmed my prevision in a way that I can even say: there is a tribute man pays more regularly than those owed to nature or to the Treasury; the tribute paid to crime! Sad condition of human race! We can tell beforehand how many will stain their hands with the blood of their fellow-creatures, how many will be forgers, how many poisoners, almost as one can foretell the number of births and deaths.

[1]From *Sur l'Homme . . .,* as cited in *Dictionary of Scientific Biography* vol. XI (New York: Charles Scribner's, 1975), p. 237.

Quetelet believed, however, that masses of numbers had to be studied before one could reach any reliable conclusions about causes[2]:

> It seems to me that *that which relates to the human species, considered en masse, is of the order of physical facts;* the greater the number of individuals, the more the influence of the individual will is effaced, being replaced by the series of general facts that depend on the general causes according to which society exists and maintains itself. These are the causes we seek to grasp, and when we do know them, we shall be able to ascertain their effects in social matters, just as we ascertain effects from causes in the physical sciences.

[2]*Recherches sur le Penchant au Crime aux Différens Âges,* 2nd ed. (Brussels: Hayez, 1833), pp. 80–81.

Summary

1. The academic discipline of statistics is a branch of mathematics and is itself divided into two branches. *Descriptive statistics* is concerned with developing and utilizing techniques for the effective presentation of numerical information.

2. *Analytical statistics,* in contrast, is concerned with developing and utilizing techniques for properly analyzing numerical information. Because analyzing data means drawing inferences from them, this branch of statistics is also referred to as *inferential statistics.* Inferences can be made by inductive or deductive reasoning. The degree of their validity can be indicated by attaching measures of reliability to them (for example, a 95-percent chance of being right, a 5-percent chance of being wrong).

3. By developing and utilizing techniques for the careful collection, effective presentation, and proper analysis of numerical information, statistics facilitates wise decision making in the face of uncertainty. Statistics can be viewed as a universal guide to the unknown.

4. The process of data collection employs a number of basic concepts. Thus, a statistical investigation focuses on persons or objects, which are called *elementary units;* these elementary units possess characteristics of interest, called *variables;* observations about them are called *data.* The set of all possible observations about a specified characteristic of interest is called a *population;* a subset of it (or of the frame from which the population is derived) is referred to as a *sample.* Variables themselves can be qualitative or quantitative; the assignment of numbers to observations about them, which is called *measurement,* can yield data of varying quality, such as nominal, ordinal, interval, and ratio data. Whatever their type, new data can be generated either by conducting a *survey* or by performing an *experiment.* A survey is the collection of data from elementary units without exercising any particular control over factors that may make these units different from one another and that may, therefore, affect the characteristic of interest being observed. In an experiment, in contrast, such control is exercised.

5. Surveys are the most common method of generating data in business and economics. In a complete survey, or *census,* observations are made about one or more characteristics of interest for every elementary unit that exists—by direct observation, personal or telephone interview, or self-enumeration. In a partial or *sample survey,* such observations are made only for a subset of existing elementary units. A number of reasons for preferring sampling to census taking can be cited; in one way

or another, these reasons have something to do with reducing the cost of getting a given type of information (or with raising the quantity or quality of information received at a given cost). Samples of different quality are obtained depending on the method by which elementary units are selected for observation; convenience samples, judgment samples, and probability samples are the major types. Among the latter, which are the most important, are simple random samples, systematic random samples, stratified random samples, and clustered random samples.

6. Whether collected by census or sample, all survey data are subject to error. Values obtained by data gatherers can differ from true values as a result of *random error* (in the case of random sampling) or of *systematic error* (in the case of all types of surveys). The size of the former can be estimated, but not so the size of the latter. Systematic error or bias can enter surveys at the planning stage, during the data-collection stage, and even while data are being processed, and such error can take the form of selection bias, response bias, or nonresponse bias.

Key Terms

analytical statistics
bias
bivariate data set
census
chance error
clustered random sample
continuous quantitative variable
convenience sample
data set
datum
deductive reasoning
descriptive statistics
dichotomous qualitative variable
discrete quantitative variable
elementary units
experiment
frame
inductive reasoning

inferential statistics
interval data
judgment sample
measurement
multinomial qualitative variable
multivariate data set
nominal data
nonresponse bias
nonsampling error
observational study
ordinal data
outliers
population
probability sample
qualitative variable
quantitative variable
random error

random-numbers table
random sample
ratio data
response bias
sample
sample survey
sampling error
selection bias
simple random sample
statistics
stratified random sample
survey
systematic error
systematic random sample
univariate data set
universe
variables

Questions and Problems

1. The accompanying table represents the 1985 mutual funds "honor roll" published by *Forbes* magazine (September 16, 1985, pp. 80–81). The twenty-three companies finished in the top half of the pack in both bull (up) and bear (down) markets; they also delivered good performance in the long run (over almost a nine-year period).
   a. How many *variables* are listed for these 23 mutual funds?
   b. List the qualitative variables, then the quantitative ones.
   c. Separate the variables into dichotomous/multinomial and discrete/continuous, respectively.

2. In Table 1.A, identify
   a. the elementary units.
   b. the population of managers.
   c. the population of net assets.
3. Classify the following variables, first, as qualitative or quantitative and, second, as dichotomous/multinomial or discrete/continuous:
   a. number of telephone calls made by someone during a day.
   b. dollar figures listed on a sheet of paper.
   c. sexes of corporate executives.
   d. running times of participants in a race.
   e. employment/unemployment status of workers.
   f. types of hair coloring sold in a drugstore.

Table 1.A

Fund (1)	Performance UP (2)	Performance DOWN (3)	Manager (4)	Net Assets 6/30/85 (millions) (5)	Annual Expenses per $100 Assets (6)	Portfolio Strategy (7)	Average Annual Total Return (8)
01 Acorn Fund	B	A	Ralph Wanger	$ 269	$0.85	Small Co Growth	20.6%
02 Amcap Fund	B	A	Marjorie Fisher	1,209	0.69	Growth	21.5
03 American Capital Pace Fund	B	A+	John Doney	1,521	0.62	Aggressive Growth	27.5
04 American Capital Venture Fund	B	A	Steve Hayward	438	0.72	Aggressive Growth	24.9
05 Claremont Capital Corp.	B	A+	Erik Bergstrom	44	1.03	Growth	23.1
06 Evergreen Fund	A+	B	Stephen Lieber	323	1.10	Growth	26.4
07 Explorer Fund	A	B	Frank Wisneski	355	1.00	Small Co Growth	20.7
08 Fidelity Destiny Fund	B	A	George Vanderheiden	764	0.70	Growth	22.1
09 Fidelity Magellan Fund	A+	A	Peter Lynch	2,879	1.12	Growth	33.3
10 Growth Fund of America	B	A	James Rothenberg	606	0.68	Growth	22.3
11 Janus Fund	B	A	Thomas Bailey	412	1.06	Aggressive Growth	20.9
12 Loomis-Sayles Capital Development Fund	A	B	Kenneth Heebner	169	0.70	Growth	22.7
13 Mass Cap Development Fund	A+	B	William S. Harris	785	0.75	Growth	23.8
14 NEL Growth Fund	A	B	Kenneth Heebner	216	0.75	Growth	22.2
15 Nicholas Fund	B	A+	Albert Nicholas	462	0.82	Growth	25.4
16 Over-the-Counter Securities Fund	B	A	Binkley Shorts	114	1.28	Small Co Growth	23.4
17 Pennsylvania Mutual Fund	B	A	Charles Royce	292	1.18	Small Co Growth	22.6
18 Pioneer II	B	A	David Tripple	1,834	0.71	Growth and Income	20.7
19 Scudder Development Fund	B	A	Edmund Swanberg	254	1.34	Small Co Growth	20.8
20 Sigma Venture Shares	A	B	Richard King	68	0.99	Small Co Growth	22.7
21 Tudor Fund	A	B	Melville Straus	130	1.59	Aggressive Growth	21.5
22 Twentieth Century Select	A+	B	Fund Has Four Managers	1,076	1.01	Growth	28.0
23 United Vanguard Fund	B	B	Hank Herrmann	409	1.04	Growth	21.0

4. Classify the following variables, first, as qualitative or quantitative and, second, as dichotomous/ multinomial or discrete/continuous:
   a. the weight lost by a dieter.
   b. the types of skills found among a firm's employees.
   c. the attendance record of students in a class.
   d. the ages of applicants for a marriage license.

e. the types of cars seen in a parking lot.

f. the pressure required to fracture a casting.

5. Classify the following variables as qualitative or quantitative:

a. a firm's average cost of production.

b. a town's tax rate.

c. the religious affiliations of a firm's employees.

d. the national unemployment rate.

e. the brands of gasoline for sale in a city.

f. a listing of the states in which 50 firms achieved their highest sales.

6. Classify the following variables as qualitative or quantitative:

a. a list of foreign exchange rates.

b. the number of black executives in an industry.

c. the depth of tread remaining on aircraft tires after 1,000 landings.

d. the Dow-Jones Industrial Average.

e. the political-party affiliations of a firm's employees.

f. the types of sports practiced by a group of people.

7. Classify the variables in problem 5 also as dichotomous/multinomial or discrete/continuous.

8. Classify the variables in problem 6 also as dichotomous/multinomial or discrete/continuous.

9. Classify numbers describing the following as nominal, ordinal, interval, or ratio data:

a. the location of voters by district.

b. the ages of employees.

c. the order in which cars finish a race.

d. models of computers.

e. white blood cells found in a cubic centimeter.

f. the colors of new cars.

10. Classify numbers describing the following as nominal, ordinal, interval or ratio data:

a. ratings of colleges.

b. temperature readings at the airport.

c. the daily receipts of a supermarket.

d. consumer brand preferences concerning types of coffee.

e. army ranks.

f. a corporate hierarchy from president to janitor.

g. calendar years.

11. Among numbers describing the following, which are nominal data?

a. distances traveled

b. student I.D. numbers

c. net assets

d. room numbers

e. sound levels inside different airplanes

f. drivers' ratings of handling characteristics of cars

12. Are there interval data in the listing of problem 11? If so, explain.

13. Consider the following situations and determine whether a census or a sample would be more appropriate:

a. A personnel director, who wants to improve labor relations, seeks to ascertain the attitudes of the firm's employees.

b. A pharmaceutical firm wants to ascertain the side effects, if any, of a drug that has proven effective against some types of cancer.

c. NASA wants to check the quality of all space-shuttle components.

14. Consider the following situations and determine whether a census or a sample would be more appropriate:

a. A manufacturer of matches has discovered that a certain percentage of production always seems to be defective and wants to ascertain what the percentage is.

b. A retail hardware store wants to determine the value of its inventory.

c. Someone wants to predict the outcome of a planned vote on the establishment of no-smoking zones in all public places.

15. If you had to do it, how would you go about measuring the popularity of next fall's new TV shows?

16. Which type of sample was actually taken in each of the following situations?

a. A newscaster reports having taken a random sample of city residents by interviewing people coming out of a supermarket.

b. Between 10 A.M. and 4 P.M., Mondays to Fridays, a pollster calls every 100th number in a city's telephone directory and asks whomever answers to rate various types of detergents. (No effort is made to contact people who do not answer the phone.)

c. The Dow-Jones Industrials stock index is constructed on the basis of observing the performance of a mere 30 stocks selected by experts as representative of all industrial stocks.

17. Which type of sample was actually taken in each of the following situations?

a. The state tourism bureau creates a profile of the typical out-of-state tourist by interviewing one evening all out-of-state guests at a single hotel chosen at random.

b. In order to ascertain how well a new product is doing, a pollster checks at known trendsetters among retail stores.

c. Using a random-numbers table, a quality inspector selects 20 from among 1000 numbered boxes containing paper napkins produced during a day, then inspects all the napkins in these boxes.

18. Which of the following are probability samples? How would you characterize the others?
    a. Using a table of random numbers, an interviewer selects 10 shopping malls from a list of 200, then interviews the owners of every store in these 10 malls.
    b. A farmer throws a pair of dice, comes up with a seven, and then, starting with tree number 7 inspects every peach on every twelfth tree in an orchard to ascertain frost damage.
    c. A manufacturer sends a questionnaire to all new-car purchasers. The results, based on a 20 percent response, are then tabulated.
    d. Representatives of the mayor interview three blacks and seven whites prior to a town meeting concerning a proposed ordinance to close bars at 10 P.M. The proportions chosen reflect the racial makeup of the town.

19. Consider all the probability samples you found when answering problems 16–18. What type were they?

20. How many different samples of three might one select from four units (A, B, C, D), and what probability of selection must each of these have to make the sample a simple random sample?

21. Consider Table 1.2. How many different simple random samples of size $n = 5$ can be drawn without replacement from the list of multinationals? Of size $n = 50$?

22. Consider Table 1.A given in problem 1. How many different simple random samples of size $n = 4$ can be drawn without replacement from the list of mutual funds? Of size $n = 10$?

23. Consider Table 1.1. How many different simple random samples of size $n = 3$ can be drawn without replacement from the list of employees? Of size $n = 1$?

24. Start at the bottom of column 7 of Appendix Table A. By using the first two digits of numbers and reading upwards, select a simple random sample of 10 multinationals from Table 1.2. What are their names?

25. Repeat the exercise given in problem 24, but assume that there are only 50 such multinationals; namely, the ones numbered 00 through 49. What are the names of sampled firms now?

26. Appendix Table A has 50 rows and 10 columns of five-digit numbers. Starting at the upper left-hand corner of the table, read two digits at a time along the first row (and possibly along subsequent

rows), letting the first usable set of two digits point you to a row and the second usable set to a column where random-number selection is to start. What is your starting number?

27. Given the starting number found in problem 26, proceed to the right along successive rows, reading the first and last digits of the 5-digit numbers in order to select a sample of six mutual funds from Table 1.A. What is your sample?

28. Given the starting number found in problem 26, proceed downward along successive columns, reading the second and third digits of the 5-digit numbers in order to select a sample of four mutual funds from Table 1.A. What is your sample?

29. Repeat problem 27, but move upward along successive columns.

30. Repeat problem 28, but proceed diagonally to the right and down. If you run out of numbers, read successive diagonals below the initial one.

31. What would your sample be if, in problem 27, you selected six multinationals from Table 1.2?

32. What would your sample be if, in problem 28, you selected four multinationals from Table 1.2?

33. Use the table of random numbers to select a simple random sample of three doctors from among those listed in the yellow pages of your local telephone directory.

34. Select a systematic random sample of five firms from Table 1.2. Assume the first number selected is 07. List the names.

35. Select a systematic random sample of four firms from Table 1.2. What is the value of k? Find the starting point for your sample by reading the last row of Appendix Table A and picking the first usable number between 00 and k. List your sample.

36. Select a systematic random sample of four firms from Table 1.A, using the procedure of problem 35. If necessary, read successive rows above the last one.

37. How might one select a systematic random sample of 200 from among 10,000 bills?

38. How might one take a 10-percent systematic random sample of a firm's 200 employees?

39. How big a sample would one have to take from each stratum if a population was stratified in such a way that each stratum contained completely homogeneous elements and if one desired information equivalent to a census?

40. Imagine that the 100 firms listed in Table 1.2 could meaningfully be subdivided into 10 strata, the members of which were, respectively, the firms numbered 00–09, 10–19, 20–29, and so on. Select

a stratified random sample of one firm from each group, assuming that you picked the following random numbers to represent the last digits of firms to be picked from the strata: 0   8   2   3   3   5   9   5   0   6.   List the names.

41. Once again consider Table 1.2, but now imagine that each successive five firms were a meaningful cluster. Select a sample of two from the 20 clusters in which to take censuses. Assume you assigned numbers 00 to 19 to the clusters, then picked the random numbers 17 and 05 from a table. List the firms to be sampled.

42. What kinds of errors are likely to occur as a result of each of the following:

    a. judging consumer preferences for a product by interviewing people stopped in cars at a red light.

    b. calling 100 people listed in the phone book to rate a new TV show.

    c. taking a random sample of 10 pages of this book to estimate the fractions of the book devoted to text, boxed examples, biographies, summaries, questions, and the like.

    d. taking a random sample of 10 sentences from the population of sentences found on the second page of Chapter 1 in order to estimate the length of sentences in this book.

    e. asking people how many phone calls they have made (or bars of soap they have used) in the past 24 months.

43. An economist randomly samples some small trucking firms, some medium-sized firms, and some large ones and finds their average annual insurance expenses to be $5,503 per truck. A census shows the figure to be $4,973. What kind of sample was taken, what kind of error was made?

44. Once again consider Table 1.A, but now imagine that each successive two firms and the last three firms were a meaningful cluster. Select a sample of two from the 11 clusters in which to take censuses. Assume you picked clusters 5 and 9 at random. List your sample.

45. What kinds of errors are likely to occur as a result of each of the following:

    a. consistently keypunching a wrong column on census cards used as computer input.

    b. the inability to interview a family selected as part of a simple random sample.

    c. having only 21 percent of questionnaires returned from a mail survey.

    d. judging public opinion from a radio talk show.

    e. asking people about their age rather than their birthday.

Selected Readings

American Statistical Association and Institute of Mathematical Statistics. *Careers in Statistics.* Washington, D.C.: 1974.

Cochran, William G. *Sampling Techniques,* 3rd ed. New York: Wiley, 1977. A lucid and clear account of sample survey techniques.

Moore, D. S. *Statistics: Concepts and Controversy.* San Francisco: Freeman, 1979. On how data are collected; the focus is on ideas and their impact on everyday life.

Morgenstern, Oskar. *On the Accuracy of Economic Observations,* 2nd ed. Princeton: Princeton University Press, 1963. A classic book about error in economic statistics.

Mosteller, Frederick, et al., eds. *Statistics By Example.* Reading, Mass.: Addison-Wesley, 1973. A good introduction to the range of issues dealt with by modern statistics: vol. 1, *Exploring Data;* vol. 2, *Weighing Chances,* vol. 3, *Detecting Patterns;* vol. 4, *Finding Models.*

Sielaff, Theodore J. *Statistics in Action: Readings in Business and Economic Statistics.* San Jose, Calif.: Lansford Press, 1963. Parts II–IV contain a dozen case studies on the use of sampling techniques.

Sommers, Albert T. *The U.S. Economy Demystified.* Lexington, Mass.: D.C. Heath, 1985. A superb discussion of the U.S. statistical system and the inevitable uncertainty of data.

Tanur, Judith M., et al., eds. *Statistics: A Guide to the Unknown,* 2nd ed. San Francisco: Holden-Day, 1978. A superb volume of 46 essays that describe important applications of statistics in many fields of endeavor. Prepared by a joint committee of the American Statistical Association and the National Council of Teachers of Mathematics.

U.S. Department of Commerce, Office of Federal Statistical Policy and Standards. *Revolution in United States Government Statistics, 1926–1976.* Washington, D.C.: U.S. Government Printing Office, 1978. A review of 200 years of federal government statistics gathering, with particular emphasis on such recent developments as probability sampling and the use of computers.

Computer Programs

The DAYSAL-2 personal-computer diskettes that accompany this book contain four programs of interest to this chapter:

1. Random Numbers lets you generate your own random-numbers tables.

2. Secret Code illustrates how random numbers can help create secret codes. You can write and decipher your own messages.

17. Random Walk uses random numbers to illustrate the random-walk hypothesis about stock prices or futures prices.

18. Cointoss uses random numbers to simulate the tossing of a coin and determines the ever-changing empirical probability of getting heads.

Note: Programs 17 and 18 require graphics capability.

PART ONE

DESCRIPTIVE STATISTICS:

2 The Presentation of Data: Tables and Graphs
3 The Presentation of Data: Summary Measures

CHAPTER TWO

The Presentation of Data: Tables and Graphs

After collecting their data, descriptive statisticians turn to an equally important task: the effective presentation of data. This task is particularly crucial when the data collection is large. No human mind is capable of grasping the meaning of any considerable quantity of numerical data, unless their mass is somehow reduced to relatively few convenient categories or is condensed with the help of some kind of visual aid. Yet it is not easy to cast masses of raw data into a form that is not only succinct or visually attractive, but also successful in weeding out the irrelevant and in then conveying all the relevant information in an accurate and unambiguous manner. This chapter discusses the two common procedures for making data collections intelligible: the construction of tables and the drawing of graphs.

TABLES

The construction of a good statistical table is an art. It involves a great deal more than simply presenting data in rows and columns. Imagine for example, that a census had been taken of the 100 multinational companies listed in Table 1.2 on page 15 and that for each of them data had been collected on the combined profit from domestic and foreign operations, as shown in Table 2.1. Even though Table 2.1 contains a mere 100 numbers, it is pretty confusing. A listing of these data in order of ascending or descending magnitude would certainly be a major improvement. Such an **ordered array,** in this case going from the smallest to the largest number, is provided by Table 2.2. More likely than not, however, most users would not be interested in the detail contained in either table and would prefer a considerable condensation of the data. The table maker obliges by carefully dividing the range of available data into nonoverlapping classes, noting how many observations fall into each, and constructing a table on that basis.

Table 2.1 A Population of Profits

This table shows 1981 combined profits from domestic and foreign operations, in millions of dollars, of the 100 largest U.S.-based multinationals listed in Table 1.2. Successive columns of numbers correspond to the alphabetical listing of companies in the earlier table.

1,071	724	457	2,060	722	499	81	−353	115	108
784	447	283	5,423	2,276	3,308	579	708	133	165
197	600	405	258	312	430	1,169	524	592	489
835	448	254	119	772	60	2,433	535	145	1,177
960	1,159	473	119	803	1,576	754	1,369	749	791
441	412	863	370	512	97	348	2,762	2,380	458
1,671	215	522	−1,060	1,231	838	339	668	1,992	580
403	−476	536	1,652	1,181	220	627	409	1,281	729
445	531	916	532	338	519	437	98	1,851	257
188	918	1,467	333	580	252	2,227	1,567	2,310	598

Source: Forbes, July 5, 1982, pp. 126-128.

The Absolute Frequency Distribution

It is a matter of personal judgment how many such collectively exhaustive and mutually exclusive groupings of data, called **classes,** best serve the goal of clarity. Obviously, this judgment will vary with the nature of the raw data in question and, thus, with the possible number of different qualitative observations or the possible range that quantitative observations can assume. If data pertain to a *qualitative* variable (such as types of businesses), the number of qualitative categories (such as proprietorships, partnerships, or corporations) may well point to the most appropriate number of classes, but even then alternatives exist. (One could combine observations about proprietorships and partnerships, for example, or one may wish to list data separately for small, medium-sized, and large corporations.) Data pertaining to *discrete quantitative* variables

Table 2.2 The Profit Population Rearranged into an Ordered Array

When the data of Table 2.1 are turned into an *ordered array* from the lowest to the highest number, better comprehension is the result. It is instantly clear, for example, that profits ranged from −1,060 million to +5,423 million. (These extremes referred to Ford and Exxon, respectively. The other two losers were Chrysler and Pan Am.)

−1,060	119	257	405	458	535	708	835	1,181	1,992
−476	133	258	409	473	536	722	838	1,231	2,060
−353	145	283	412	489	579	724	863	1,281	2,227
60	165	312	430	499	580	729	916	1,369	2,276
81	188	333	437	512	580	749	918	1,467	2,310
97	197	338	441	519	592	754	960	1,567	2,380
98	215	339	445	522	598	772	1,071	1,576	2,433
108	220	348	447	524	600	784	1,159	1,652	2,762
115	252	370	448	531	627	791	1,169	1,671	3,308
119	254	403	457	532	668	803	1,177	1,851	5,423

Source: Table 2.1.

(such as the number of television sets owned by households) can, similarly, be classified on the basis of the number of discrete possibilities (households owning zero television sets can be counted separately from those owning one set, two sets, and so on), but once more different classifications are possible. (One could combine the counts of households owning two or more sets, for example.) Data concerning *continuous quantitative* variables (such as the tonnage produced by firms in the steel industry), finally, can be classified in any convenient fashion. Whenever the choice on the matter is wide open, most practitioners recommend between 5 and 20 classes; a smaller number sacrifices too much detail, a larger one retains too much of it.[1]

In the case of quantitative data, it is also considered desirable to make the **class width**—the difference between the numerical lower and upper limit of a class—the same for all classes because this fact facilitates comparisons between classes, as well as the calculation of summary values (to be discussed in Chapter 3). Having decided on the desired *number* of classes, one can quickly approximate an appropriate class *width* by dividing the difference between the largest and smallest observation in one's data set by the desired number of classes.

If we tentatively decided to condense the Table 2.2 data into seven classes of equal width (a number, perhaps, suggested by Sturgess's rule noted in footnote 1), we would calculate

$$\text{approximate class width} = \frac{\text{largest value} - \text{smallest value in data set}}{\text{desired number of classes}}$$

$$= \frac{5{,}423 - (-1{,}060)}{7} = 926.1.$$

In order to set up nonoverlapping classes for the entire range of data, we might round this number to 900 and arrive at the following:

1st class:	−1,100 to under −200
2nd class:	−200 to under 700
3rd class:	700 to under 1,600
4th class:	1,600 to under 2,500
5th class:	2,500 to under 3,400
6th class:	3,400 to under 4,300
7th class:	4,300 to under 5,200
8th class:	5,200 to under 6,100

We would end up with eight classes of equal width and would have satisfied both of the recommendations (concerning the desirable number of classes and the equality of

---

[1]Some statisticians like to use *Sturgess's rule* to determine the desirable number of classes, k. According to this rule, $k = 1 + 3.3 \log n$, where n equals the size of the data set. In our example, with $n = 100$ and $\log 100 = 2$, this comes to $k = 1 + 3.3(2) = 7.6$, which might be rounded to 7 or 8. As will be shown, it is rarely desirable, however, to follow such a fixed rule blindly.

widths). We would then be able to place each of the Table 2.2 data into one, and only one, of these eight categories. (The use of the term *under* in the previous classification avoids ambiguity by clearly distinguishing between the upper limit of one class and the lower limit of the class above it. Thus, an observation of 699 would be placed in the second class, as would a 699.99. Yet an observation of 700 would clearly belong to the third class, a fact that would not be obvious if the term *under* were missing.)

Yet our creation of classes would hardly be an appropriate one in light of the distribution of values found in Table 2.2. We would end up with 3 observations in the 1st class, then lump fully 57 observations into the 2nd class, 27 more into the 3rd class, and place a mere 13 observations into the last five classes (none of them into classes 6 and 7). This is where good judgment must come to play.

Noting that most observations fall into the range between 0 and 2,500, we might divide that range into five equal classes and then place two larger classes at either end of the distribution, as column 1 of Table 2.3 illustrates. Then we would be ready to note the total number of observations that fall into each of our classes, a procedure that would be easy with the help of Table 2.2 and would yield the numbers in column 3 of Table 2.3. (Note: If we had worked with unordered raw data, such as those in Table 2.1, we would have first placed a tally mark in the appropriate row of column 2 for every value found in Table 2.1, while checking off the Table 2.1 data so that none of them were counted twice. At the end of this process, we would also have found the values in column 3 of Table 2.3 by counting the tally marks.)

As Table 2.3 indicates, the absolute number of observations that fall into a given class is referred to as the **absolute class frequency.** In turn, the tabular summary of a data set showing the absolute class frequencies in each of several collectively

Table 2.3 An Absolute Frequency Distribution of Profits of the 100 Largest U.S.-Based Multinationals in 1981

An absolute frequency distribution, such as columns 1 and 3, can highlight meaningful patterns hidden in raw data (such as Tables 2.1 or 2.2). Note: The sum of absolute class frequencies necessarily equals the total number of observations.

Class (size of profit in millions of dollars) (1)	Absolute Class Frequency (number of companies in class[a]) (2)	(3)
− 1,500 to under 0		3
0 to under 500		41
500 to under 1,000		32
1,000 to under 1,500		9
1,500 to under 2,000		6
2,000 to under 2,500		6
2,500 to under 5,500		3
		Total 100

[a]Strictly speaking, each class frequency refers to the number of *profit figures* from the profit population of Table 2.1 that fit into the designated class, but one can just as well talk about the number of companies because each profit figure is linked to a particular (different) company.
Source: Table 2.1.

exhaustive and mutually exclusive classes is called the data set's **absolute frequency distribution.** The column 1 and 3 data as a group are an example. Consider how this presentation tells us at a glance that three among the 100 largest U.S. multinationals made losses in 1981, that 24 of them made profits of $1 billion or more; while most of them (41 and 32, respectively) earned profits between 0 and under $500 million or between $500 million and under $1 billion. Such succinct information could hardly have been derived, equally rapidly, from the raw data of Table 2.1 and would have been even less possible to derive if the raw data had been more massive in number, as often is the case (except in textbooks).

Note: Although it is rarely done, the usefulness for later analysis of a summary such as Table 2.3 could be vastly improved by the addition of another column, showing the *sum* of data pertaining to each class. Thus, the sum of losses made by the 3 companies in class 1 might be given as $-$$1,889 million (as Table 2.2 attests), the sum of profits of the 41 companies in class 2 might be given as $12,057 million, and similar totals (of $21,901 million, $11,105 million, $10,309 million, $13,686 million, and $11,493 million) might be given for the other five classes. This sort of information could, for example, be used to calculate precise class averages (such as $-$$1,889 million divided by 3, or $-$$629.67 million for class 1) and would obviate the need for estimating them. Such estimates are often made by assuming (perhaps incorrectly) that the average pertaining to each class equals the midpoint of the class interval (such as $-$$750 million for class 1). The listing of class sums is particularly crucial for **open-ended classes** that have only one stated end point, the upper or the lower limit (examples are "under 0" or "above $2,500 million"). In that case, users do not even have the option of approximating the class average by the class midpoint. Because this desirable practice (of listing class sums) is rarely followed, anyone who prepares an absolute frequency distribution, such as Table 2.3, can render subsequent table users a great favor by selecting classes in such a fashion that the average of values in a class comes close to the class midpoint.[2]

The Relative Frequency Distribution

The frequency distribution discussed so far in this chapter is termed an absolute one because it is based on counting absolute *numbers* of observations for each class. One can, however, also determine the *fractions* or *proportions* of all observations that fall into various classes; any such proportion is a **relative class frequency** and is the ratio of the number of observations in a particular class to the total number of observations made. Accordingly, the tabular summary of a data set showing the relative class frequencies in each of several collectively exhaustive and mutually exclusive classes is called the data set's **relative frequency distribution.** Table 2.4 illustrates how a relative frequency distribution is derived from an absolute one, but as the table caption cautions, it is an accident that the total number of observations equals 100. As a result of this accident, the column 2 and column 3 entries look very similar in this case. In all cases, however, proportions can be read as *percentages* by mentally multiplying them by 100. Thus, .03(100) = 3 percent of the multinationals in question made losses,

---

[2]The sets of class midpoints and precise class averages for the seven classes in Table 2.3 are $-$750 and $-$629.67; 250 and 294; 750 and 684; 1,250 and 1,234; 1,750 and 1,718; 2,250 and 2,281; and 4,000 and 3,831.

Table 2.4   Deriving the Relative Frequency Distribution of Profits of the 100 Largest U.S.-Based Multinationals in 1981

An absolute frequency distribution such as columns 1 and 2, is easily converted into a relative one—columns 1 and 3—by successively dividing each absolute frequency—such as the number 3 in row 1 of column 2—by the total number of observations (here $N = 100$) to yield the corresponding relative frequency (such as .03). *Caution:* It is an accident that the total number of observations equals 100. Only because of that fact do the relative frequencies look so similar to the absolute ones. They would not if the column 2 total added to a number other than 100 (as a look at Table 2.5 can confirm). It is no accident, however, that the sum of relative frequencies equals 1.00; such is always necessarily the case.

Class (size of profit in millions of dollars) (1)	Absolute Class Frequency (number of companies in class) (2)	Relative Class Frequency (proportion of all companies in class) (3)
−1,500 to under 0	3	.03
0 to under 500	41	.41
500 to under 1,000	32	.32
1,000 to under 1,500	9	.09
1,500 to under 2,000	6	.06
2,000 to under 2,500	6	.06
2,500 to under 5,500	3	.03
Totals	100	1.00

Source: Table 2.1.

.41(100) = 41 percent earned profits between 0 and under 500 million dollars, and so on.

Relative frequency distributions are particularly useful for comparisons of populations that are similar in type but vastly different in size. A most dramatic illustration of this fact is provided by Analytical Example 2.1, "Deciphering Secret Codes," wherein a small population of letters from the words of a secret message is compared with the vast population of letters in all the words of the English language. For a more mundane example, consider Table 2.5, which shows hypothetical frequency distributions of the racial and sexual composition of the labor force in a single firm and in the entire industry. It would be absurd to argue, on the basis of absolute numbers, that Mountain Aviation, Inc. was discriminating against blacks or females in hiring because it had hired so few of them compared to the industry as a whole. Relative frequencies alone tell a meaningful story here: 33 percent of the employees of Mountain Aviation, Inc. are black, while only 20 percent of the entire industry's employees are—note the first two entries in columns 4 and 5. On the other hand, 33 percent of that firm's employees are also female, while the industrywide percentage is 49. [Add the first and third entries in columns 4 and 5 for this comparison.]

Note: As a comparison of the column 1 entries in Table 2.5 with those in Table 2.4 confirms, absolute and relative frequency distributions can be produced for qualitative variables as well as quantitative ones. Such is not the case, however, for another type of distribution.

Table 2.5   Absolute and Relative Frequency Distributions of Selected Characteristics of Aviation Industry Employees

As these hypothetical data indicate, comparisons of absolute numbers—columns 2 vs. 3—can sometimes be less insightful than comparisons of relative numbers—columns 4 vs. 5.

Class (employee race and sex) (1)	Absolute Class Frequencies (number of employees in class)		Relative Class Frequencies (proportion of all employees in class)	
	Mountain Aviation, Inc. (2)	Entire Aviation Industry (3)	Mountain Aviation, Inc. (4)	Entire Aviation Industry (5)
Black female	1	69,000	.111	.100
Black male	2	69,000	.222	.100
White female	2	269,000	.222	.390
White male	4	283,000	.444	.410
Totals	9	690,000	1.000	1.000

Table 2.6   Deriving the LE Type of Cumulative Frequency Distribution of Profits of the 100 Largest U.S.-Based Multinationals in 1981

An absolute frequency distribution—columns 1 and 2—is easily converted into a cumulative absolute one—columns 1 and 3—just as a relative frequency distribution—columns 1 and 4—can easily become a cumulative relative one—columns 1 and 5. Note how column 2 has been turned into column 3 and how 4 has been turned into 5. When cumulation proceeds from lesser to greater classes, a "less-than-or-equal," or LE, type of cumulative frequency distribution emerges. Each member of such a distribution tells us the number or proportion of observations that are less than, or at most equal to, the *upper* limit of the class in question. Note: While it is an accident that the largest cumulative absolute frequency equals 100 in this example, the largest cumulative relative frequency will always equal 1.00.

Class (size of profit in millions of dollars) (1)	Absolute Class Frequency (number of companies in class) (2)	Cumulative Absolute Class Frequency (number of companies in class or lower ones) (3)	Relative Class Frequency (proportion of all companies in class) (4)	Cumulative Relative Class Frequency (proportion of all companies in class or lower ones) (5)
		Starting Point		Starting Point
−1,500 to under 0	3	3	.03	.03
0 to under 500	41	3 + 41 = 44	.41	.03 + .41 = .44
500 to under 1,000	32	44 + 32 = 76	.32	.44 + .32 = .76
1,000 to under 1,500	9	76 + 9 = 85	.09	.76 + .09 = .85
1,500 to under 2,000	6	85 + 6 = 91	.06	.85 + .06 = .91
2,000 to under 2,500	6	91 + 6 = 97	.06	.91 + .06 = .97
2,500 to under 5,500	3	97 + 3 = 100	.03	.97 + .03 = 1.00

Source: Table 2.4.

The Cumulative Frequency Distribution

Only when variables are quantitative can one determine a **cumulative class frequency** as the sum of the (absolute or relative) class frequencies for all classes up to and including the class in question, beginning at either end of the frequency distribution. If the cumulation process proceeds from lesser to greater classes, the result is a "less-than-or-equal," or LE, type of distribution because the frequencies counted pertain to the class in question or lower ones. If cumulation proceeds from greater to lesser classes, a "more-than-or-equal," or ME, type of distribution emerges because the frequencies counted pertain to the class in question or higher ones. Accordingly, a **cumulative frequency distribution** is a tabular summary of a data set showing for each of several collectively exhaustive and mutually exclusive classes the absolute number or proportion of observations that are less than or equal to the upper limits of the classes in question (LE type) or that are more than or equal to their lower limits (ME type). All this is illustrated in Tables 2.6 and 2.7, with the help of the Table 2.4 data.

Consider Table 2.6. Note how the third (encircled) entry in column 3 tells us that 76 companies had profits that were less than $1,000 million (and at most equal to the upper class limit of just under $1,000 million) while the fourth (encircled) entry in column 5 tells us that 85 *percent* of the companies had profits that were less than $1.5 billion.

Now consider Table 2.7. Note how the fourth (encircled) entry in column 3 tells us that 24 companies had profits that were more than, or at least equal to, $1 billion, while the third (encircled) entry in column 5 tells us that 56 *percent* of the companies had profits of at least $500 million.

Table 2.7   Deriving the ME Type of Cumulative Frequency Distribution of Profits of the 100 Largest U.S.-Based Multinationals, in 1981

An absolute frequency distribution—columns 1 and 2—is easily converted into a cumulative one—columns 1 and 3. The same is true of a relative frequency distribution; note how column 4 has been turned into column 5. When cumulation proceeds from greater to lesser classes, a "more-than-or equal," or ME, type of cumulative frequency distribution emerges. Each member of such a distribution tells us the number or proportion of observations that are more than, or at least equal to, the *lower* limit of the class in question.

Class (size of profit in millions of dollars) (1)	Absolute Class Frequency (number of companies in class) (2)	Cumulative Absolute Class Frequency (number of companies in class or higher ones) (3)	Relative Class Frequency (proportion of all companies in class) (4)	Cumulative Relative Class Frequency (proportion of all companies in class or higher ones) (5)
−1,500 to under 0	3	97 + 3 = 100	.03	.97 + .03 = 1.00
0 to under 500	41	56 + 41 = 97	.41	.56 + .41 = .97
500 to under 1,000	32	24 + 32 = 56	.32	.24 + .32 = .56
1,000 to under 1,500	9	15 + 9 = 24	.09	.15 + .09 = .24
1,500 to under 2,000	6	9 + 6 = 15	.06	.09 + .06 = .15
2,000 to under 2,500	6	3 + 6 = 9	.06	.03 + .06 = .09
2,500 to under 5,500	3	3	.03	.03
		Starting Point		Starting Point

Source: Table 2.4.

A Final Note

The preceding sections have focused on how statisticians condense masses of raw data in order to highlight the most important information contained in them. We have paid no particular attention, however, to the proper construction of the tables themselves that are to convey the condensed data to their ultimate users. Good table making requires that attention be paid to a number of important factors. Table 2.8 can serve as an example of what it takes to make a table publishable. The overriding goal must be to help the user.

Number and title. When published along with other tables, the table should have a number so it can be referred to easily. Often, as in this case, it is convenient to use a chapter number for the purpose; Table 2.8 then refers to the eighth table in Chapter 2. Sometimes it is useful to use page numbers instead; Table 361a then refers to the first table on p. 361. The table number must be followed by a title that is as short as possible but also complete. It must focus on what, where, and when: in this case, sales of household washing machines (what), in Boston and Massachusetts (where), during 1976–85 (when). Consider how useless the entire table would be if any one of these ingredients were left out of the title.

Footnotes. Additional information can be extremely valuable to a user and can often easily be provided in a footnote (designated by letter or symbol to avoid confusion with numbers in the body of the table). When the information contained in the table is based on a sample survey, details about the size and type of sample can be given, along with the size of the sampling error. (A later chapter will note how the sampling error is calculated.) Information useful in determining the likely extent of bias can also be added, such as the exact wording of the question that was asked or the rate of response to a mail questionnaire. (Anyone noting the low response rate in footnote *a* is likely to suspect a significant nonresponse bias.) In other instances, other types of information may be equally useful. Anyone who wanted to publish a table based on columns 1, 4, and 5 of Table 2.5, for example, would serve the user immensely by footnoting the absolute totals (9 and 690,000) to which the relative frequencies refer. Any user could then easily reconstruct columns 2 and 3.

Decimals and rounding. A decision must inevitably be made on how many decimals to retain in the numbers appearing in the table. Should an absolute value of 600.493 be reported as such, as 600.49, or even as 600, with all the decimals dropped? Should a proportion of .389855 be reported as such or would .39 do just as well? An advantage of many decimals in proportions is that the user can more accurately recalculate absolute numbers, provided the total on which the proportions are based has been provided. Thus, a reader who is given the column 3 total of 690,000 in Table 2.5 in a footnote, but is not given the other column 3 values, can calculate the third entry of 269,000 in that column precisely when multiplying the given total by a column 5 proportion of .389855. Yet the calculation would yield 269,100 if a proportion of .39 were used. Still, this less precise calculation would lie within a few hundredths of 1 percent of the correct one. More often than not, the retention of many decimals serves no good purpose (in that it adds little information) and may, in fact, do harm (by giving the published data a totally unwarranted air of precision). Having decided on a reasonable

Table 2.8   Sales of Household Washing Machines in Boston and Massachusetts, 1976–1985[a]

| Year | Thousands of Units Sold | | Year | Thousands of Units Sold | |
	Boston	Massachusetts		Boston	Massachusetts
1976	60	300	1981	61	295
1977	51	255	1982	67	345
1978	42	210	1983	72	360
1979	37	200	1984	75	375
1980	59	245	1985	79	395

[a]Based on questionnaire mailed to a 5 percent simple random sample of retail outlets. ("How many household washing machines did you sell in each of the stated years?") Sampling error: xxx. Sample size: Boston 50, Massachusetts 500. Response: Boston 21, Massachusetts 101. Massachusetts data include Boston.

Source: Hypothetical data.

number of decimals to retain, some systematic procedure for rounding must be employed. Typically, decimals from 1 to 4 are eliminated by rounding down (a 5.42 becomes a 5.4); those between 6 and 9 are eliminated by rounding up (a 5.47 becomes a 5.5); and an arbitrary rule is followed with the decimal 5, such as always rounding it to the *even* possibility (a 5.45 becomes a 5.4 and not a 5.5, but a 5.35 also becomes a 5.4 and not a 5.3). This arbitrary rule serves the purpose of avoiding bias from rounding up more often than down. Finally, note that numbers rounded to zero, as from .03, are listed as 0.0, while an original observation of zero is recorded as 0.0e (e = exact).

GRAPHS

Graphs, undoubtedly, are the most popular way of presenting statistical data. In one way or another, the information condensed into any type of table can also be presented with the help of a graph.

The Frequency Histogram

Consider graphing the information contained in Table 2.3. A graphical portrayal of an absolute or relative frequency distribution of a continuous quantitative variable is called a **frequency histogram.** As Figure 2.1 indicates, a frequency histogram is drawn on a set of rectilinear coordinates and consists of a series of contiguous rectangles (also referred to as bars or columns) placed on the horizontal axis so their bases correspond to classes of possibly different widths and their *areas* to class frequencies.

It is easy to see how the various classes of profit from column 1 of Table 2.3 can be represented on the horizontal axis by starting at the lower limit of the first class and going to the upper limit of the last one. (Hence, the first rectangle of a histogram need not start at the zero point.) The widths of the various classes can be measured in terms of any convenient standard. If, as in Table 2.3, most classes span intervals of $500 million, that range might just as well be used as the unit of measurement. Note the tick

Figure 2.1   The Frequency Histogram

A frequency histogram is a graphical portrayal of a frequency distribution for continuous quantitative data. While classes are shown horizontally, frequency densities (here the absolute numbers of observations per $500 million class range) are plotted vertically. As a result, the *area* of each rectangle, rather than its height, equals the frequency (here the absolute number of observations) for the corresponding class. Notice how the area of the first rectangle equals *three* $500 million units (base) times *one* observation per $500 million range (height), or *three* observations altogether. Similarly, the areas of the second or last rectangles (and, therefore, the absolute frequencies of the second and last classes) equal, respectively, $1 \times 41 = 41$ or $6 \times \frac{1}{2} = 3$ (just as noted in Table 2.9).

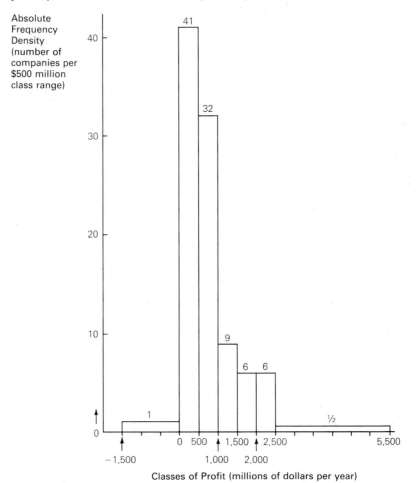

marks on the horizontal axis of Figure 2.1 that clearly show the first class to be three units wide, classes 2 to 6 to be one unit wide, and the last class to be six units wide.

It is much more difficult to determine what to measure on the vertical axis. Vertical lines are to be erected at the class limits to form the sides of rectangles the *areas* of which are to represent class frequencies, but how long should these lines be? A reminder from basic geometry quickly solves the puzzle. The area of any rectangle equals base times height. Height, therefore, equals area divided by base—or, in our case, class frequency (area) divided by class width (base). This ratio of class frequency to class width is also called **frequency density** and is, inevitably, what must be measured on the vertical axis if the *areas* of histogram rectangles are to correspond to class frequencies. These relationships are summarized in Table 2.9.

Note how the column 4 data—and not the column 2 data—have been plotted on the vertical axis of Figure 2.1. The first class, for example, spans an interval of $1,500 million; it is *three* times as wide as the standard unit of measurement employed. Hence, the absolute class frequency of 3 is properly represented by the first rectangle's *area* if that rectangle's height is plotted as the frequency density of 1. In this large class, one is likely to find only one company per $500 million range. Similarly, the last class spans $3,000 million (or $3 billion); it is *six* units wide. Therefore, the absolute frequency of 3 is properly represented by the last rectangle's area only if the rectangle's height equals 1/2.

There is, of course, good reason for all this adjustment. Failure to draw the heights of all histogram columns with respect to the same standard (such as a $500 million range) would make these columns not comparable and would present a false visual impression. In our case, drawing the first and last rectangles at a height of 3 [as the first and last entries of column 2, Table 2.9, might suggest] would make these two classes of profit seem much more important than they really are. Per $500 million range, these two "countries" are sparsely populated, indeed.

Table 2.9   Preparing to Draw a Frequency Histogram

The data of columns 2 through 4 shown here become, respectively, the area, base, and height of the corresponding histogram (Figure 2.1). Note how the concept of frequency density is analogous to the familiar one of human population density. Just as one commonly speaks of so many people living per square mile in Texas, Michigan, or New York, the column 4 data tell us of so many companies "living" per $500 million range in the "countries" of column 1. The total "populations" of these "countries" are shown in column 2; the sizes of these "countries" are given in column 3.

Basic Data			Absolute Frequency Density
Class Interval (size of profit in millions of dollars) (1)	Absolute Class Frequency (number of companies in class) (2)	Class Width (number of $500 million units in class) (3)	(number of companies per $500 million class range) (4) = (2)/(3)
−1,500 to under 0	3	3	3:3 = 1
0 to under 500	41	1	41:1 = 41
500 to under 1,000	32	1	32:1 = 32
1,000 to under 1,500	9	1	9:1 = 9
1,500 to under 2,000	6	1	6:1 = 6
2,000 to under 2,500	6	1	6:1 = 6
2,500 to under 5,500	3	6	3:6 = ½
	Area	= Base	× Height

Source: Table 2.3.

Figure 2.2   Common Types of Histograms

This graph pictures some of the relative frequency distributions that will be discussed further in Chapters 5 and 6. Histograms for relative frequency distributions such as these are always constructed in such a way that the total area covered by a given set of columns equals unity. Therefore, the proportion of the total area taken by any one column (which is indicated throughout) can be readily converted into the percentage of observations falling into the class above which this column has been erected. Note: As was true of the absolute frequency densities in Figure 2.1, the relative frequency densities measured here are expressed with respect to some unit that is defined on the horizontal axis. This unit equals .004 feet in panel (a), $500 million per year in panel (b), 5 miles per gallon in panel (c), 1 year in panel (d), and 2 inches in panel (e). In all cases, multiplying a column's vertical distance, or frequency density, by the number of such units on which the column stands yields its share of the total area of the histogram. The numbers shown inside the columns refer to these areas.

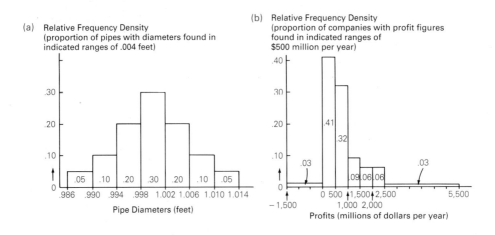

Histograms for relative frequency distributions. Histograms for relative frequency distributions are constructed on the same principles as those for absolute distributions. A histogram constructed on the basis of columns 1 and 3 of Table 2.4, for example, would have the same shape as that shown in Figure 2.1; however, the labeling on the vertical axis (and, therefore, the tops of the columns) would change to reflect the fact that *proportions* of companies, rather than absolute numbers of the companies, were being measured per 500-million dollar range.

Common types of histograms. All sets of quantitative data can be fitted into some form of frequency distribution that is likely to yield one or another of a relatively few common types of histogram. Figure 2.2 shows some of the more common shapes of histograms. Histogram (a) pictures the relative frequency distribution of the actual diameters of supposedly 1-foot-wide water pipes produced during a day. This histogram is symmetrical about the interval from .998 to 1.002 feet, with the histogram columns to the left and right becoming progressively shorter. The larger the divergence of the actual diameter, in either direction, from the desired diameter of 1 foot, the smaller is the proportion of such pipes produced. Thus, 30 percent of all pipes produced fit in the class on which the central column stands, 20-percent each fit in the two adjacent classes, 10-percent each into the next two, and 5-percent each into the two extreme classes (making a total of 100 percent). This distribution follows from the fact that the area of the central column takes up .3 times the entire histogram area; the areas of the two adjacent columns each take up .2 times the total area, and so on, as the numbers inside the columns indicate. These numbers must add to 1.00. This type of histogram describes a *normal distribution* that is typical of populations of physical measurements.

Histogram (b) once more shows the profits of multinationals from Figure 2.1. This time the proportion of the total area of the histogram taken by each column is indicated (which is not to be confused with the column heights given in Figure 2.1). Again, these proportions add to 1.00, and they tell us this: while 41 percent of the companies fall in the 0 to $500 million category and only 3 percent of them are found below that range, 56 percent of companies are above that range, which gives the distribution a long "tail" to the right. This shape corresponds to a general class of *skewed* distributions in which there are relatively few small observations concentrated in a narrow range below a dominant class, while relatively more large observations are spread over a wide range of values above that class.

Histogram (c) pictures gasoline mileages of a year's new-car models and represents an opposite type of skewed distribution. In this case, the smaller values are widely spread out below a relatively important range, but there are hardly any observations above it. This distribution is skewed to the left.

Histogram (d) depicts the *exponential distribution*. Such a picture is typical of populations that exhibit changes over time, such as the lifetimes of machines. A relatively large proportion of machines last one year, for example, but the proportion of machines that exceeds that number by one, two, three, and more years gets rapidly smaller and smaller.

Histogram (e), finally, depicts the *rectangular* or *uniform distribution*. For a construction job, lumber may have to be purchased at lengths of 8 feet, while pieces of all sizes from 14 inches to 8 feet are needed. As a result, lots of scrap is produced, and every one of the scrap pieces measures less than 14 inches. About equal proportions of them are found in each of the classes between 0 and 14 inches.

Vivid illustrations of the usefulness of histograms are provided by Analytical Examples 2.2, "Quality Control in Manufacturing," and 2.3, "Deciding Authorship."

The Frequency Polygon and the Frequency Curve

An alternative graphical way of portraying an absolute or relative frequency distribution is provided by the **frequency polygon.** To draw this many-sided figure, the same set of coordinates is used as for the histogram, but this time the **class mark,** or midpoint of each class width, is identified (as the average of the two class limits) and a dot is positioned above it at a height equal to the (absolute or relative) frequency density. The dots are then connected by straight lines. To complete the polygon, straight lines are also drawn from the dots above the first and last class marks, respectively, to points on the horizontal axis that lie one-half the length of the standard unit of measurement below the lowest or above the highest class limit. Figure 2.3 indicates how histogram (a) of Figure 2.2 can be converted into a polygon.

The shape of a histogram can also be approximated by a smoothed **frequency curve.** This approximation can be accomplished by mathematical or graphical techniques and typically serves the purpose of removing irregularities as a result of sampling error from a histogram depicting information gathered in a sample survey. The frequency curve can then be viewed as an estimate of the unknown histogram that would emerge if census information were graphed with many classes of tiny widths. Figure 2.4 on page 54 indicates how histogram (b) of Figure 2.2 might be converted into a smoothed frequency curve.

Ogives

Any cumulative frequency distribution, such as those noted in Table 2.6 or 2.7, can be shown graphically as well. The resulting curve is called an **ogive** (pronounced ōjive).

In the case of a less-than-or-equal type of cumulative frequency distribution, each cumulative class frequency is plotted vertically above the *upper* limit of the corresponding class (which, in turn, is measured horizontally). Panel (a) of Figure 2.5 on page 55, for example, shows an ogive based on columns 1 and 3 of Table 2.6. The encircled entry in column 3 now appears as point *b:* 76 companies had profits of less than $1,000 million ($1 billion).

In the case of a more-than-or-equal type of cumulative frequency distribution, each cumulative class frequency is plotted vertically above the *lower* limit of the corresponding class. Panel (b) of Figure 2.5, for example, shows an ogive based on columns 1 and

Figure 2.3 Turning a Histogram into a Polygon

Panel (a) depicts the histogram of pipe diameters (now dashed) that was first depicted in Figure 2.2. Fat dots *b* through *h* have been placed over the midpoints of every class width at heights equal to that of the histogram columns (and thus measuring the relative frequency densities). Two additional dots, *a* and *i*, have been placed on the horizontal axis, one-half the length of the .004 foot standard unit of measurement, below the lowest and above the highest class limit, respectively. The polygon is created by connecting all these dots *(a* through *i)* by straight lines. Panel (b) shows the polygon standing alone, but class marks have replaced class limits on the horizontal axis.

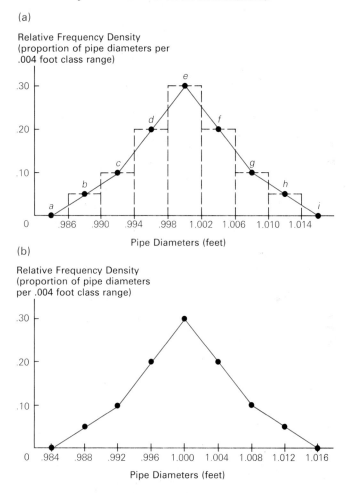

3 of Table 2.7. The encircled entry in column 3 now appears as point *e:* 24 companies had profits of $1,000 million ($1 billion) or more.

Note: In the case of discrete quantitative data, an ogive is typically drawn in stair-step fashion, the horizontal portions of the line being placed in the gaps between possible values. Panels (c) and (d) of Figure 5.1 on p. 168 provide examples.

Figure 2.4   **Turning a Histogram into a Frequency Curve**

Panel (a) depicts the histogram of miles per gallon (now dashed) that was first noted in Figure 2.2. A smooth curve is drawn to approximate the shape of this histogram. This frequency curve is shown separately in panel (b).

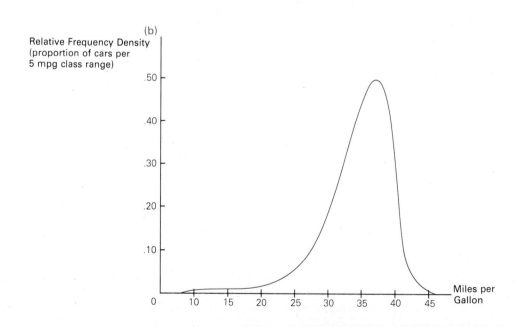

Figure 2.5 Two Types of Ogives

Just as there are two types of cumulative frequency distributions, there are two types of graphs to portray them. Panel (a) is based on columns 1 and 3 of Table 2.6 and represents a less-than type ogive. As did Table 2.6, it tells us the number of companies with 1981 profit less than a certain figure. Thus point *a* indicates that 0 companies earned less than − $1,500 million, point *b* shows that 76 companies earned less than $1,000 million, and point *c* tells us that 100 companies earned less than $5,500 million. Panel (b) is based on columns 1 and 3 of Table 2.7 and represents a more-than-or-equal type ogive. As did Table 2.7, it gives us the number of companies with 1981 profit of a certain figure or more. Point *d,* for example, indicates that all 100 companies earned − $1,500 million or more, point *e* shows that 24 companies earned $1,000 million or more, and point *f* tells us that 0 companies earned $5,500 million or more. All this can, of course, be shown for relative frequency distributions as well.

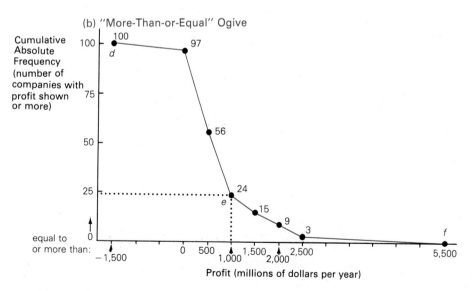

OTHER TYPES OF GRAPHS*

Bar and Column Charts

Bar charts and **column charts** are particularly effective in presenting data about qualitative or discrete quantitative variables. Such data are portrayed by a series of noncontiguous horizontal bars or vertical columns, the lengths of which are proportional to the values that are to be depicted. Many variations of these types of charts exist; Figure 2.6, which is based on Table 2.5, shows just some of them.

Time-Series Line Graphs

Some data, such as those of Table 2.8, that are linked with specific points in time or intervals of time can be presented in bar or column charts but are also suited to being plotted on a rectilinear coordinate surface. Successive points are then joined by a continuous line, yielding a **time-series line graph.** On page 58 panel (a) of Figure 2.7 provides the simplest example; panel (b) provides a slightly more complex one. Panel (a) is called an *arithmetic* chart because it uses the regular arithmetic scale on both axes. It has a disadvantage, however, when, as in this case, two series of numbers are being compared that move within rather different ranges of values. The series within the higher range of values (in this case, statewide sales) will appear to be considerably more volatile, but this volatility can be an illusion. The illusion can be removed by using a *semilogarithmic* chart, as in panel (b). The horizontal scale remains arithmetic, the vertical one becomes logarithmic. As a result, equal percentage changes on two series of data (as between Boston sales and statewide sales until 1978 and again from 1983 onwards) produce lines of equal slope; the illusion disappears.

Maps, Pie Charts, and Pictograms

A great number of other devices can be used to display data visually. Some of them are shown on page 59 in Figure 2.8. Panel (a) pictures a **statistical map,** which portrays data for areal units, such as world regions, nations, states, counties, or even census tracts, by crosshatching or shading these units in different ways on a geographic map. Such a map can be very effective, as in panel (a).

Note how anyone with the proper geographical knowledge can tell instantly, for example, that 1972 per capita personal income reached $5,000 or more in only 6 among the lower 48 states of the United States (namely, in California, Nevada, Illinois, New York, New Jersey, and Connecticut), while such per capita income in Maine, Louisiana, and many other states was below $4,000. Other statistical maps are, however, less easy to read—particularly three-dimensional ones that attempt, for example, to depict the extent of urban sprawl or air pollution by drawing pictures of buildings or billowing chimneys on top of a geographic map.

Panel (b) is a typical **pie chart,** a segmented circle that is amenable to portraying divisions of some aggregate, provided there are not too many divisions. The circular pie is cut into slices by radii in such a way that the central angles (and, therefore, the circumference arcs and sector areas) are proportional to the sizes of the divisions

---

*Optional section.

Figure 2.6 Bar and Column Charts

Panel (a) is a bar chart; panels (b) through (d) are various forms of column charts (an ordinary column chart, a component-column chart, and a grouped-column chart). Note that the identical information is conveyed by panels (a) through (c); panel (d), unlike the other panels, portrays relative rather than absolute frequencies.

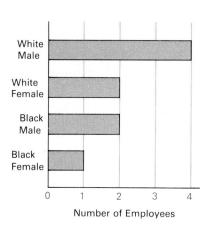

(a) Mountain Aviation, Inc.
Composition of Employees by
Race and Sex, 7/1/85

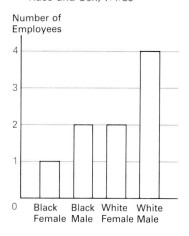

(b) Mountain Aviation, Inc.
Composition of Employees by
Race and Sex, 7/1/85

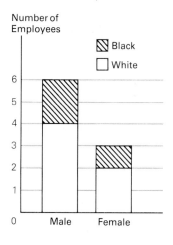

(c) Mountain Aviation, Inc.
Composition of Employees by
Race and Sex, 7/1/85

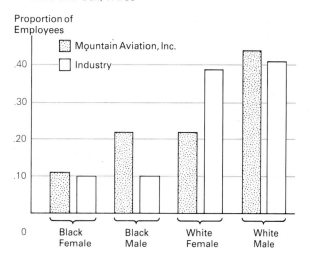

(d) Mountain Aviation, Inc. vs. Aviation Industry:
Composition of Employees by
Race and Sex, 7/1/85

Source: Table 2.5.

Figure 2.7   Time-Series Line Graphs

Line graphs can be drawn on a regular arithmetic scale on both axes, as in panel (a). They can also be drawn on a semilogarithmic scale, as in panel (b). In the latter case, equal percentage changes in different series of data produce lines of equal slope. Note: A logarithmic scale does not have a zero point because the logarithm of zero equals minus infinity. Such a scale is calibrated in log values in various "cycles" from 1–10, 10–100, 100–1,000 and so on (or, in the other direction, from 1–10, .1 to 1, .01 to .1, and so forth).

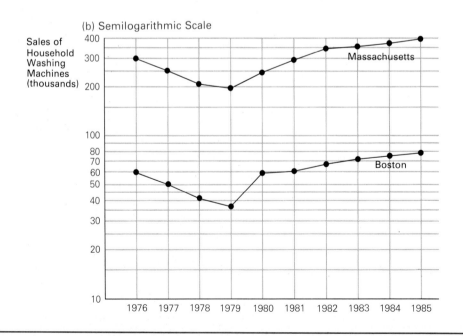

Figure 2.8 Maps, Pie Charts, and Pictograms

These are just three examples of a near-infinite variety of graphical devices commonly used to portray statistical data.

(a) Per Capita Personal Incomes, 1972

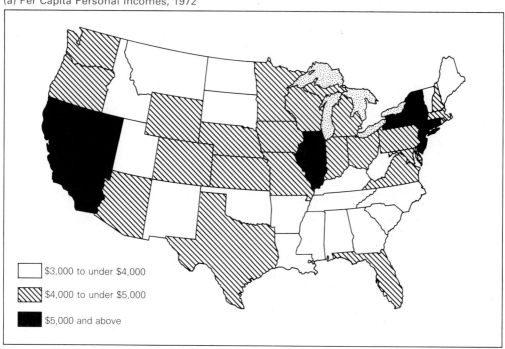

□ $3,000 to under $4,000

▨ $4,000 to under $5,000

■ $5,000 and above

(b) The U.S. Analgesic Market, 1981

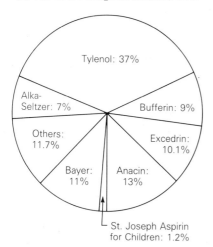

(c) Employment at City Motors, 1950 vs. 1985

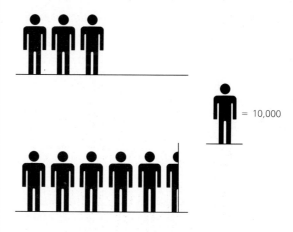

= 10,000

Sources: panel (a): U.S. Bureau of the Census, *Statistical Abstract of the United States: 1973* (Washington, D.C.: U.S. Government Printing Office, 1973), p. 326; panel (b): Paine Webber Mitchell Hutchins, Inc.; panel (c): hypothetical data.

being displayed.[3] The pie chart shown in panel (b) of Figure 2.8 demonstrates how Tylenol dominated the 1981 U.S. market for over-the-counter pain relievers, which at the time gave aspirin makers a headache that even extra-strength Tylenol couldn't cure.

A pictorial chart or **pictogram** depicts data with the help of symbols, such as those in panel (c). All kinds of symbols can be used, from stars to stacks of silver coins, from tanks to trees, from crosses to castles, from smiling faces to billowing smokestacks. Typically, each symbol represents a definite and uniform value, as is the case here, but sometimes the size of the pictorial symbol is made proportional to the values that are to be portrayed. Making the size of the symbol proportional to values is tricky business, however, and can easily lead to confusion. How, for instances, does one double the size of a three-dimensional symbol, such as a box? By doubling each length, one inevitably quadruples the areas portrayed, and one raises the volume eightfold. (For an illustration of this problem see Close-Up 2.1, "How to Lie with Statistics.")

---

[3]Suppose the aggregate in question were industry sales of $132 million, of which firm X sold $48.84 million, or 37 percent. This market share would be represented in a pie chart by a central angle of $.37(360°)$ = $133.2°$, since a circle has a total of 360 degrees and $133.2°$ equal 37 percent of this total, as the accompanying graph illustrates.

2.1 Deciphering Secret Codes

Cryptanalysis, the scientific study of converting ciphers or codes to which the key is not known into plain text, provides a dramatic illustration of how frequency distributions shed light on masses of raw data. Consider the secret message (presumed to be written in English) contained in Table A.

Table A

IMEUX	ANEMO	MNPKN	EONUC	MOMXB
MJDAM	VNNES	NSXXW	MVSPM	RPMSN
MAMQK	SXNES	NNEMG	SPMMV	AUIMA
CGNEM	DPRPM	SNUPI	DNERM	PNSDV
KVSXD	MVSCX	MPDLE	NONES	NSWUV
LNEMO	MSPMX	DBMXD	CMPNG	SVANE
MTKPO	KDNUB	ESTTD	VMOON	ESNNU
OMRKP	MNEMO	MPDLE	NOLUJ	MPVWM
VNOSP	MDVON	DNKNM	ASWUV	LWMVA
MPDJD	VLNEM	DPZKO	NTUIM	POBPU
WNEMR	UVOMV	NUBNE	MLUJM	PVMA

To decipher the message, cryptanalysis first establishes the absolute frequency distribution of the letters in the secret message. Table B provides a count of the number of time each letter occurs and also shows the implied relative frequency distribution. Next, the relative frequency distribution is compared with that of the letters in a normal English language text, shown in Table C (small letters are used in the list of normal

Table B Frequency Distribution of 274 Letters in Coded Message

Letter	Absolute Frequency (number)	Relative Frequency (percent)	Letter	Absolute Frequency (number)	Relative Frequency (percent)
A	9	3.3	N	36	13.1
B	5	1.8	O	16	5.8
C	4	1.5	P	21	7.7
D	16	5.8	Q	1	.4
E	19	6.9	R	5	1.8
F	0	0	S	19	6.9
G	3	1.1	T	4	1.5
H	0	0	U	14	5.1
I	4	1.5	V	18	6.6
J	4	1.5	W	6	2.2
K	8	2.9	X	9	3.3
L	7	2.5	Y	0	0
M	45	16.4	Z	1	.4
			Totals	274	100.0

Table C Frequency Distribution of 200 Letters of a Normal English-Language Text

Letter	Absolute Frequency (number)	Relative Frequency (percent)	Letter	Absolute Frequency (number)	Relative Frequency (percent)
a	16	8	n	14	7
b	3	1.5	o	16	8
c	6	3	p	4	2
d	8	4	q	.5	.25
e	26	13	r	13	6.5
f	4	2	s	12	6
g	3	1.5	t	18	9
h	12	6	u	6	3
i	13	6.5	v	2	1
j	1	.5	w	3	1.5
k	1	.5	x	1	.5
l	7	3.5	y	4	2
m	6	3	z	.5	.25
			Totals	200	100.0

Source: David Kahn, *The Codebreakers: The Story of Secret Writing* (New York: Macmillan, 1967), p. 100.

English letters to avoid confusion with the capital letters in the coded message). To make the comparison, the letters of normal English and of the secret message are arranged in Table D in order of decreasing relative frequency.

Although one can hardly expect an immediate letter-for-letter matching in Table D, the juxtaposition can be very helpful in breaking the code. After some trying about, some letters can be identified (for example, *M* as *e* and *N* as *t*), then portions of words, and finally, the whole message. (Note: A good codebreaker looks not only at general letter frequencies, but also considers the preferred associations of one letter with other letters, the order of frequency of the most common doubles, of initial letters, of final letters, of the most frequent one-letter words, and much more.) The key for this particular message is a simple one: the word *SCRAMBLED* is used for the first nine letters of the alphabet, and the remaining code letters (now excluding the letters in the word *SCRAMBLED*) are listed in reverse alphabetical sequence. The first of the following pair of lines shows the plain text and the second line shows the coded text.

```
a b c d e f g h i j k l m n o p q r s t u v w x y z
S C R A M B L E D Z Y X W V U T Q P O N K J I H G F
```

As an analysis of Table D can show, even this simple table correctly identifies the meanings of five code letters *(M, N, A, X,* and *B)*, and it provides strong hints in the case of a dozen others, the meanings of which are found within a line or two of the code letter. (Thus a *V* is not an *i*, but an *n;* a *D* is not an *r*, but an *i*, and so on). In fact, all but two of the alphabet's 26 letters are decipherable by looking within four lines of the code letter in Table D. The decoded message is an excerpt from the Declaration of Independence.

We hold these truths to be self-evident, that all men are created equal, that they are endowed by their Creator with certain unalienable rights, that among these are life, liberty and the pursuit of happiness, that to secure these rights, governments are instituted among men, deriving their just powers from the consent of the governed.

Table D   Comparative Relative Frequency Distributions of Plain and Coded English

Plain English		Coded English		Plain English		Coded English	
Letter	Relative Frequency	Letter	Relative Frequency	Letter	Relative Frequency	Letter	Relative Frequency
e	13	M	16.4	u	3	W	2.2
t	9	N	13.1	f	2	B	1.8
a	8	P	7.7	p	2	R	1.8
o	8	E	6.9	y	2	C	1.5
n	7	S	6.9	b	1.5	I	1.5
i	6.5	V	6.6	g	1.5	J	1.5
r	6.5	D	5.8	w	1.5	T	1.5
h	6	O	5.8	v	1	G	1.1
s	6	U	5.1	j	.5	Q	.4
d	4	A	3.3	k	.5	Z	.4
l	3.5	X	3.3	x	.5	F	0
c	3	K	2.9	q	.25	H	0
m	3	L	2.5	z	.25	Y	0
				Total	100.0	Total	100.0

2.2 Quality Control in Manufacturing

A manufacturer of small motors was in trouble. Recently produced motors were breaking down at an unexpected rate. A close look at the manufacturing process was in order. The investigation centered on steel rods that might be too loose in their bearings.

The investigator noted that quality inspectors had been instructed to inspect the inside diameters of the rods and to reject all those with measurements of .9995 centimeters or less. The inspection records of 500 such rods were requisitioned and the physical measurements were plotted as a histogram. The investigator expected to find a symmetrical histogram, such as histogram (a) in Figure 2.2 because in the past, errors in either direction from the standard had been equally frequent and smaller errors had been more frequent than larger errors. Yet the investigator found the histogram reproduced in the accompanying figure, showing absolute frequencies of 10, 30, 0, 80, 60, 100, 90, 60, 40, 20, and 10 for the eleven classes of diame-

ters. (Note: Because all the classes are of equal size, absolute frequencies and absolute frequency densities coincide in this case.) The unusual gap just below the lower tolerance limit and the unusually tall column just above it were too obvious to be ignored.

As it turned out, inspectors had felt pressure to keep scrappage low (because any rod thrown out meant labor, materials, and other costs were wasted in its manufacture). As a result, they had misclassified borderline defective rods (with diameters, say, of .9993 centimeters) as borderline acceptable (as if they had diameters of .9995 centimeters). No wonder that too few rods appeared in class 3 and too many in class 4. When their inspection procedures were corrected, not 40, but 105 rods had to be scrapped from the next batch of 500.

The high number of defective rods itself was subsequently traced to a faulty machine setting and corrected. The breakdowns of the motors disappeared.

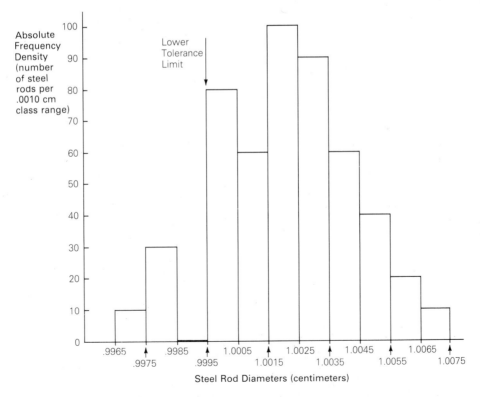

Source: W. Edwards Deming, "Making Things Right," in Judith M. Tanur et al., eds., *Statistics: A Guide to the Unknown* (San Francisco: Holden-Day, 1972), pp. 229–31.

2.3 Deciding Authorship

On occasion, an important question arises concerning the authorship of a piece of writing. A well-known case concerned the authorship of the Paulines, a set of Christian religious writings possibly written by St. Paul. A second famous case was the Shakespeare-Bacon-Marlowe controversy over who wrote certain plays traditionally attributed to Shakespeare. And then there was the matter of the *Federalist* papers, some of which were claimed simultaneously by Alexander Hamilton and James Madison. Let us consider the latter case to show how descriptive statistics can play an important role in solving the puzzle simply by counting something in a systematic way.

After most of them had appeared anonymously in various newspapers, some 85 essays written by Alexander Hamilton, John Jay, and James Madison ap-

(a) Occurrence of *by*   (b) Occurrence of *to*   (c) Occurrence of *upon*

peared in book form in 1788. *The Federalist,* as the book was called, has ever since been a leading source of information concerning the intent of the framers of the U.S. Constitution, but between 1807 and 1818 a variety of lists appeared with conflicting information about which authors wrote which papers. In particular, 12 papers were claimed by both Hamilton and Madison after they had become bitter political enemies. Historians have tried in vain to decide authorship on the basis of political content. This line of inquiry led nowhere partly because both authors were at the time writing these papers as lawyers' briefs in favor of ratifying the Constitution (and not every argument put forward was necessarily their own), partly because the authors' political opinions changed over time (and one couldn't infer their earlier positions from later ones). Nor could average sentence length be used to differentiate the disputed papers. It was 34.5 words in 51 undisputed Hamilton papers and 34.6 words in 14 undisputed Madison papers—the authors wrote in similar styles.

Statisticians were able, however, to employ as discriminators *rates of word use.* They ignored contextual words, such as *law* and *liberty,* that were likely to appear in all the papers to a great extent, simply because

of the topic discussed. They focused instead on filler words, such as *by, to,* and *upon,* that were unrelated to the topic. Three sets of histograms depicting rates of word use are shown in panels (a)–(c) of the accompanying figure. Focus on the tallest columns in panel (a): In 18 of 48 undisputed Hamilton papers, the word *by* occurred between 7 and 9 times per 1,000 words; yet in 16 of 50 undisputed Madison papers, it occurred between 11 and 13 times per 1,000 words, just as it did in 4 of 12 disputed papers. Compared to Hamilton's, the entire Madison histogram is displaced to the right (as is that of the disputed papers). The more frequent use of *by* suggested Madison authorship.

Similarly, the *less* frequent use of the words *to* and *upon* in Madison's undisputed papers and in the disputed papers lead to the same conclusion. Indeed, the three histogram sets shown here, along with 27 other sets like these, provided overwhelming evidence in favor of *Madison* being the author of each one of the disputed papers.

Source: Frederick Mosteller and David L. Wallace, "Deciding Authorship," in Judith M. Tanur et al., eds., *Statistics: A Guide to the Unknown* (San Francisco: Holden-Day, 1972), pp. 164–75.

||| CLOSE-UP

2.1 How to Lie with Statistics

Benjamin Disraeli, the British Prime Minister from 1874–80, purportedly said: "There are three kinds of lies: ordinary lies, damned lies, and statistics." Regardless of whether this tale is true, it is a fact that one can lie with statistics. Such distortion may happen accidentally, out of ignorance; it may also happen deliberately, out of malice, in order to defraud the gullible. The ways of doing so are legion. Consider just some of the vices illustrated in the accompanying graphs.

Figure A on page 66 graphs the identical data about Boston washing-machine sales (Table 2.8) in three ways. Every one of the graphs is technically correct, but consider the different impressions one can make just by stretching the horizontal or vertical axis.

Figure B is a worse case, although commonly encountered. By not labeling the vertical axis at all, the

desired impression is achieved, yet the graphs are totally meaningless. Does the panel (b) change from 1980–81, for example, depict an annual increase of 50¢ or of $5 million?

Finally, consider Figure C on page 67. Panel (a) pictures a 3.5-fold increase in the number of a firm's employees with the help of two human figures, the second of which is 3.5 times as tall as the first. Note the false impression this type of pictogram is likely to give. Compare it with the alternative version given in panel (b).[1]

---

[1]For further discussion of graphical fallacies, see Darrell Huff and Irving Geis, *How to Lie with Statistics* (New York: W. W. Norton, 1954), which inspired this example.

Figure A

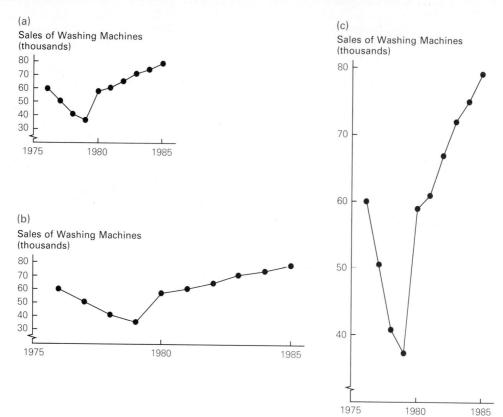

(a)
Sales of Washing Machines
(thousands)

(b)
Sales of Washing Machines
(thousands)

(c)
Sales of Washing Machines
(thousands)

Figure B

(a) Breakfast Cereal, Brand X

Energy Release ⟶

Time of
Eating

1 Minute
Later

2 Minutes
Later

(b) Ad Agency Business

Figure C

(a)

10,000   35,000

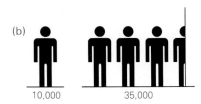

(b)

10,000   35,000

Summary

1. This chapter discusses the two most common procedures for making masses of data intelligible: the construction of tables and the drawing of graphs. Any considerable quantity of unorganized numerical data is likely to be confusing; even an ordered array is an improvement. Greater clarity still is provided by dividing the range of available data into a relatively small number of collectively exhaustive and mutually exclusive classes, noting how many observations fall into each, and constructing a table on that basis. If desired, such an absolute frequency distribution can easily be converted into a relative one that shows the proportions of all observations in the chosen classes (rather than the absolute number of observations). In the case of quantitative data, further insight can be gained from the construction of cumulative frequency distributions (of the LE or ME type). Prior to the publication of any such tabular summary, however, a number of factors concerning table title, footnotes, and the like should be designed to provide maximum help to the user.

2. Some of the most important devices for presenting absolute or relative frequency distributions of continuous quantitative data graphically include frequency histograms, frequency polygons, and frequency curves. Cumulative frequency distributions are, in turn, graphed as ogives.

3. Data about qualitative or discrete quantitative variables are most effectively presented in bar or column charts. Other key types of graphs include time-series line graphs, statistical maps, pie charts, and pictograms.

Key Terms

absolute class frequency
absolute frequency distribution
bar charts
classes
class mark
class width
column charts
cumulative class frequency

cumulative frequency distribution
frequency curve
frequency density
frequency histogram
frequency polygon
ogive
open-ended classes
ordered array

pictogram
pie chart
relative class frequency
relative frequency distribution
statistical map
time-series line graph

Questions and Problems

The computer programs noted at the end of this chapter can be used to work many of the subsequent problems.

1. An automated filling machine filled 80 cans with supposedly 16 ounces of peas each. Upon inspection, the actual net weights, however, were found to be those listed in Table 2.A. Present these data as an ordered array.
2. Table 2.B lists the miles (in thousands) driven by the 40 sales representatives of a publishing firm during the past 12 months. Present these data as an ordered array.
3. Table 2.C lists the average salaries of public school teachers during the 1984–85 school year (*The New York Times,* August 31, 1985, p. 15). Present these data as an ordered array.
4. Table 2.D lists the price-earnings ratios of selected stocks. Present these data as an ordered array.

Table 2.A

15.83	15.39	15.93	15.98	15.98	15.85	16.00	15.98	16.04	15.79
16.01	15.91	16.21	15.89	15.77	16.05	15.77	15.96	15.75	16.37
16.24	15.71	15.84	16.09	15.75	16.38	16.43	16.49	16.60	16.63
16.42	16.01	16.17	15.91	16.13	15.82	15.91	16.27	15.92	15.92
15.33	15.90	16.01	16.08	15.57	16.12	16.11	16.81	16.65	16.68
15.44	15.51	15.92	15.72	16.25	15.90	15.53	16.00	16.29	15.69
16.88	16.10	16.20	15.94	15.93	15.73	16.47	16.07	16.00	16.53
16.31	15.62	15.81	15.99	16.40	15.84	15.95	16.19	15.74	15.90

Table 2.B.

42	17	33	31	21	16	58	51	38	12
32	58	19	40	18	8	25	13	42	38
18	11	41	45	18	36	29	48	29	47
60	27	16	9	11	40	50	10	53	34

Table 2.C

Alabama ... $20,209	Kentucky ... $21,486	North Dakota ... $19,900
Alaska ... 39,751	Louisiana ... 19,690	Ohio ... 22,737
Arizona ... 23,380	Maine ... 18,329	Oklahoma ... 18,954
Arkansas ... 19,180	Maryland ... 25,861	Oregon ... 24,378
California ... 26,300	Massachusetts ... 24,110	Pennsylvania ... 24,435
Colorado ... 24,456	Michigan ... 28,401	Rhode Island ... 27,384
Connecticut ... 24,520	Minnesota ... 25,920	South Carolina ... 19,800
Delaware ... 22,924	Mississippi ... 15,971	South Dakota ... 17,356
D.C. ... 28,621	Missouri ... 20,452	Tennessee ... 22,141
Florida ... 21,057	Montana ... 21,705	Texas ... 22,600
Georgia ... 20,607	Nebraska ... 20,153	Utah ... 21,307
Hawaii ... 24,628	Nevada ... 22,520	Vermont ... 19,014
Idaho ... 19,700	New Hampshire ... 18,577	Virginia ... 21,536
Illinois ... 25,829	New Jersey ... 25,125	Washington ... 25,610
Indiana ... 23,089	New Mexico ... 22,064	West Virginia ... 19,563
Iowa ... 20,934	New York ... 29,000	Wisconsin ... 24,577
Kansas ... 21,208	North Carolina ... 20,691	Wyoming ... 26,709
		U.S. ... 23,546

Table 2.D

28	9	11	8	14	14	10	4	13
13	7	8	13	12	57	15	8	19
13	9	35	18	14	9	15	8	48
10	21	7	19	9	26	29	6	17
14	23	22	10	9	11	13	24	37

5. Following Sturgess's rule, what is the recommended number of classes, k, for a frequency distribution with $n = 80$ observations? For $n = 400$?

6. How many classes would Sturgess's rule recommend for the data sets given in problems 1 through 4? Explain.

7. Comment on the following two arrangements for dividing the net weights of vegetable cans into classes.

Arrangement 1:	Arrangement 2:
15.3 to under 15.5	15.3 to 15.5
15.6 to under 15.8	15.5 to 15.7
15.8 to under 16.0	15.7 to 15.9
and so forth	and so forth

8. Contrary to Sturgess's rule, create an absolute frequency distribution from the Table 2.B data, starting at 8,000 miles and using
a. classes of 4,000 miles.
b. classes of 28,000 miles.

9. Contrary to Sturgess's rule, create an absolute frequency distribution from the Table 2.A data, starting at 15.3 ounces and using
a. classes of 0.1 ounce.
b. classes of 0.8 ounce.

10. Follow Sturgess's rule and create an absolute and relative frequency distribution for the Table 2.B

data. Start the first class with the minimum value in the data set.

11. Divide the Table 2.A data set into 0.2-ounce classes, beginning at 15.3 ounces. Construct an absolute and relative frequency distribution.

12. Follow Sturgess's rule and create an absolute and relative frequency distribution for the Table 2.D data. Start the first class with a value of 4.

13. Follow Sturgess's rule and create an absolute and relative frequency distribution for the Table 2.C data and a minimum value of $15,500. Round the class width suggested by Sturgess's rule upwards to the nearest $500.

14. Consider your answer to problem 10. Use it to create the cumulative absolute frequency distribution and the cumulative relative frequency distribution, each of the "less-than-or-equal" type.

15. Consider your answer to problem 11. Use it to create the cumulative absolute frequency distribution and the cumulative relative frequency distribution, each of the "less-than-or-equal" type.

16. Consider your answer to problem 12. Use it to create the cumulative absolute frequency distribution and the cumulative relative frequency distribution, each of the "more-than-or-equal" type.

17. Consider your answer to problem 13. Use it to create the cumulative absolute frequency distribution and the cumulative relative frequency distribution, each of the "more-than-or-equal" type.

18. Consider Table 2.E concerning the households in a suburban development.
a. Fill in the missing values (denoted by dots).
b. What percentage of these households has at least one TV set? Has at most three sets? Has three or four sets?

Table 2.E

			Cumulative Relative Class Frequency	
Class (number of TV sets per household) (1)	Absolute Class Frequency (number of households) (2)	Relative Class Frequency (proportion of all households in class) (3)	"less than or equal" (proportion of all households in class or lower ones) (4)	"more than or equal" (proportion of all households in class or higher ones) (5)
0	60	•	•	•
1	•	.28	•	•
2	•	•	.72	•
3	•	•	•	•
4	24	•	•	•
5	•	•	•	.04
Totals	300	•		

Table 2.F

Loans and Investments (1)	Relative Frequency (2)	Absolute Frequency (3)
Commercial loans	.274	
Other loans	.492	
U.S. gov't securities	.152	
Other securities	.082	
Total	1.000	

19. In December of 1984, U.S. commercial banks held loans and investments valued at $1,713.6 billion (*Economic Report of the President,* February 1985, p. 307). The relative class frequencies of four categories of these assets are shown in column 2 of Table 2.F. Calculate the absolute class frequencies that are missing in column 3.

20. Based on your answer to problem 8a, draw the two frequency histograms (absolute and relative).

21. Depict your answers to problem 9 as absolute frequency histograms.

22. Depict your problem 10 answers as histograms.

23. Consider whether each of the following variables (measured on the horizontal axis) is likely to be described by a histogram approximating a frequency distribution that is normal (symmetrically bell-shaped), skewed to the right, skewed to the left, exponential, or rectangular:
a. family money incomes in the United States.
b. random numbers.
c. dollar values of mail-order-house sales.
d. height of boats on the Mississippi.
e. ages of pinball machine players.
f. ages of people with false teeth or bald heads.
g. last digits of all New York City telephone numbers.
h. grades of all students at midterm.
i. ages of children in elementary school.

24. Consider whether each of the following variables (measured on the horizontal axis) is likely to be described by a histogram approximating a frequency distribution that is normal (symmetrically bell-shaped), skewed to the right, skewed to the left, exponential or rectangular:
a. the number of printing errors found on a book's pages.
b. the number of minutes people have to wait at the bank, doctor's office, post office, supermarket checkout counter.
c. the number of days people spend in a hospital.
d. the size of errors made by a supermarket checkout clerk.
e. the number of days it takes people to pay their credit-card bills.
f. response times of a fire or police department.
g. the yields per acre of U.S. wheat farmers.
h. the ages at which famous authors wrote their best books.
i. the mileages driven in a year by all cars registered in the United States.

25. A firm was using two machines to produce metal parts, but customers complained about erratic changes in thickness of parts contained in different shipments. The firm's statistician developed the accompanying histogram from a random sample of a week's output. What does it tell you?

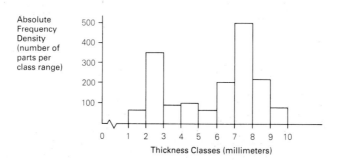

26. Consider the cumulative absolute frequency distribution developed in problem 14. Plot these data as a "less-than" ogive.

27. Consider the absolute frequency distribution developed in problem 11. Plot it as
  a. an absolute frequency histogram.
  b. a frequency polygon.
  c. a frequency curve.

28. Consider the cumulative relative frequency distribution developed in problem 16. Plot these data as a "more-than-or-equal" ogive.

29. Consider the cumulative absolute frequency distribution developed in problem 15. With appropriate changes, plot these data as a "more-than-or-equal" ogive.

30. Consider the first of the two cumulative relative frequency distributions developed in problem 18. Plot the corresponding ogive.

31. Graph Table 2.G as an *ordinary horizontal bar chart.*

Table 2.G **Net Import Reliance of Selected Minerals in the United States, 1981**

Mineral	Import as Percent of Consumption
Chromium	90
Cobalt	91
Columbium	100
Fluorspar	85
Iron ore	28
Manganese	98
Nickel	72
Tin	80
Tungsten	52
Vanadium	42
Zinc	67

Source: American Iron and Steel Institute, *Steel and America: An Annual Report* (Washington, D.C.: May 1982), p. 31.

32. Graph Table 2.H as a *duo-directional* horizontal bar chart, with negative values graphed toward the left and positive values toward the right of a vertical zero line.

33. Graph Table 2.I as a *component bar chart.*

34. From Table 2.J create a *grouped bar chart,* titled "Foreign Direct Investment in the United States—Percent Distribution by Area: 1970 and 1983."

Table 2.H **Percentage Change in U.S. GNP (in 1972 dollars)**

1978	+4.8%
1979	+3.2
1980	−0.2
1981	+2.0
1982	−1.8

Source: Federal Reserve Bulletin, various issues.

Table 2.I **State and Local Government Taxes in the United States—Percent Distribution by Type, 1965–1982, Selected Years**

Year	Property	Individual Income	Sales and Gross Receipts	Other
1965	44.1	8.0	33.4	14.5
1970	39.2	12.5	34.9	13.4
1975	36.4	15.2	35.2	13.2
1979	31.6	18.0	36.1	14.3
1980	30.7	18.8	35.8	14.7
1981	30.7	19.0	35.2	15.1
1982	30.8	19.1	35.2	14.9

Source: Statistical Abstract of the United States, 1985, p. 265.

Table 2.J **Foreign Direct Investment in the United States (in millions of dollars)**

	1970	1983
United Kingdom	$ 4,127	$ 32,512
Netherlands	2,151	28,817
Canada	3,117	11,115
Japan	229	11,145
Germany	680	10,482
Switzerland	1,545	7,132
Other Europe	1,051	13,747
Other Areas	370	20,573
Total	13,270	135,523

Source: Statistical Abstract of the United States, 1985, p. 804.

35. From 2.K create a *grouped bar chart.*

Table 2.K   Retail Store Sales in the United States by Kind of Business: 1982 and 1983 (billions of dollars)

	1982	1983
Food Stores	$249.3	$259.4
Automotive Dealers	182.6	221.7
General Merchandise Group	132.6	143.0
Gasoline Service Stations	103.5	103.1
Eating and Drinking Places	104.7	115.7
Building Material and Hardware	51.3	59.9
Apparel and Accessory Stores	51.4	54.0
Furniture, Furnishings, and Equipment Stores	46.1	51.8
Drug and Proprietary Stores	36.0	38.8
Liquor Stores	19.4	19.7
Mail-order Houses	4.2	4.4

Source: Statistical Abstract of the United States, 1985, p. 782.

36. From Table 2.L create an *ordinary column chart.*

Table 2.L   Health Expenditures in the United States as Percent of GNP, 1970–83

1970	7.6
1975	8.6
1980	9.4
1981	9.7
1982	10.5
1983	10.8

Source: Statistical Abstract of the United States, 1985, p. 96.

37. From Table 2.M create an *ordinary column chart.*

Table 2.M   Annual Percentage Change in U.S. Consumer Price Index, 1970–1984

1970	5.9	1977	6.5
1971	4.3	1978	7.7
1972	3.3	1979	11.3
1973	6.2	1980	13.5
1974	11.0	1981	10.4
1975	9.1	1982	6.1
1976	5.8	1983	3.2
		1984	4.2

Source: Statistical Abstract of the United States, 1985, p. 467.

38. From Table 2.N create an *ordinary column chart.*

Table 2.N   Business Failure Rate in the United States, 1970–1983 (per 10,000 firms)

1970	44	1977	28
1971	42	1978	24
1972	38	1979	28
1973	36	1980	42
1974	38	1981	61
1975	43	1982	88
1976	35	1983	110

Source: Statistical Abstract of the United States, 1985, p. 519.

39. Graph Table 2.O as a *component-column chart.*

Table 2.O   U.S. Business Inventories (in billions of 1972 dollars)

Year	Total	Farm	Nonfarm
1961	$171.8	$33.2	$138.6
1971	267.0	39.2	227.8
1981	348.7	44.2	304.5

Source: Economic Report of the President, February 1982, p. 253.

40. U.S. federal receipts in 1970 were distributed as follows: individual income taxes 47 percent, social insurance taxes and contributions 23 percent, corporate income taxes 17 percent, other sources 13 percent. The corresponding 1983 figures were 48.1 percent, 34.8 percent, 6.2 percent, 10.9 percent. Illustrate these data by means of a *component-column chart.*

41. Graph Table 2.P as a *grouped-column chart.*

Table 2.P   Personnel Problems at City Motors, 1986 (hypothetical data)

	Atlanta Plant	Chicago Plant
Average daily percent absent	15	4
Average daily percent tardy	10	1

42. Graph Table 2.Q as a *duo-directional column chart.*

43. Graph Table 2.R as a *pie chart.*

Table 2.Q   U.S. Air Carriers Net Operating Income (in millions of dollars), 1975–83

1975	$129	1980	$ −6
1976	576	1981	−264
1977	654	1982	−750
1978	995	1983	−176
1979	123		

Source: Statistical Abstract of the United States, 1985, p. 614.

Table 2.R   Car-Rental Market Shares, March 1982

Hertz	39.7%
Avis	23.4
National	17.0
Budget	14.5
Others	5.4
Total	**100.0%**

Source: Hertz Corporation.

44. Draw a *time-series line graph* depicting the percent return on stockholders' equity in the U.S. iron and steel industry from 1975 to 1984: 9.8, 7.8, 0.1, 7.3, 6.5, 9.6, 13.4, −15.0, −19.7, −1.3 (American Iron and Steel Institute, *An Annual Report,* May 1985).

45. Graph Table 2.S as a pictogram.

Table 2.S   Fewer Farmers Feeding More People

Year	Population (of the United States in millions)	
	Farm	Total
1940	31	132
1970	10	203

Source: U.S. Bureau of the Census, *Statistical Abstract of the United States, 1973,* pp. 5 and 584.

Selected Readings

Brainerd, B. *Weighing Evidence in Language and Literature: A Statistical Approach.* Toronto: University of Toronto Press, 1974. Very readable.

Hall, Ray O. *Handbook of Tabular Presentation: How to Design and Edit Statistical Tables.* New York: Ronald Press, 1946. A style manual and case book.

Huff, Darrell, and Irving Geis. *How to Lie with Statistics.* New York: W. W. Norton, 1954. A discussion of graphical fallacies. (See also Close-Up 2.1.)

Kjetsaa, Geir. "The Battle of the Quiet Don." *Computers and the Humanities.* Pergamon Press, 1977, vol. 11, pp. 341–46. A statistical study of disputed authorship of *The Quiet Don.* Was it written by Kryukov or Sholokhov (who received the 1965 Nobel Prize for Literature for it)?

Mosteller, Frederick, and David L. Wallace. *Inference and Disputed Authorship: The Federalist.* Reading, Mass.: Addison-Wesley, 1964. A more extensive discussion, including historical details and a variety of alternative analyses, of Analytical Example 2.3., "Deciding Authorship."

Pratt, Fletcher. *Secret and Urgent: The Story of Codes and Ciphers.* Indianapolis: Bobbs-Merrill, 1939. Contains a discussion of the Shakespeare-Bacon controversy noted in Analytical Example 2.3, "Deciding Authorship."

Smith, Laurence Dwight. *Cryptography: The Science of Secret Writing* (New York: W. W. Norton, 1943). More about codes and ciphers, including word-frequency listings for English, French, German, Italian, and Spanish.

Computer Programs

The DAYSAL-2 personal-computer diskettes that accompany this book contain four programs of interest to this chapter:

2. Secret Code illustrates how random numbers can help create secret codes. Lets you write and decipher your own secret messages.

3. Frequency Distributions From Raw Data lets you enter a set of unorganized, raw data; then produces an ordered array plus six types of frequency distributions: absolute, relative, cumulative absolute LE-type and ME-type, cumulative relative LE-type and ME-type. You can specify (a) only the desired number of classes, (b) only the desired class width, or (c) the precise limits of all classes. The program also produces six horizontal histograms, like column 2 of Table 2.3, corresponding to the six frequency distributions just noted, as well as statistical summary measures, to be discussed in Chapter 3.

4. Summary Measures From Grouped Data takes an absolute frequency distribution that you supply and produces, among other things, five other frequency distributions: cumulative absolute LE-type and ME-type, relative, cumulative relative LE-type and ME-type.

19. General Graphics lets you draw line graphs, bar charts, column charts, histograms like Figures 2.1 and 2.2, and pie charts. Numerous designs are available; for example, up to 18 bars with as many as 4 subdivisions and 3 different widths, or pies with up to 10 slices, 3 colors, and 7 patterns.

CHAPTER THREE

The Presentation of Data: Summary Measures

As Chapter 2 has shown, properly constructed tables and graphs can be of considerable help in making otherwise confusing masses of data intelligible and in revealing the secrets hidden in such unorganized data. Yet an even more radical approach to condensing collections of data is the calculation of *arithmetic summary measures* that are designed to express the most salient features of data sets in the most compact fashion imaginable. Data about quantitative variables can be neatly summarized in three ways: by measures of central tendency, measures of dispersion, and measures of shape.

Summary **measures of central tendency** (or **location**) are values around which observations tend to cluster and that describe the location of what in some sense might be called the "center" of a data set. Consider the two frequency curves in panel (a) of Figure 3.1. They describe two hypothetical data sets on the sizes of life-insurance policies taken out by women and men. It is immediately obvious that men in general take out larger policies than women do, in spite of the fact that some women take out policies for as much as $100,000, while some men take out policies for as little as $50,000. If there were to be a single and "typical" measure of the size of women's policies, it would be $50,000; such a measure for men would be $100,000. Each of these would be a measure of central tendency because it would locate, in this case on the horizontal axis, the general center of the respective data sets. What, however, are the centers of the asymmetrical data sets in panel (c)? The answer is less obvious and, as will be shown depends very much on how the term *center* is defined.

Summary **measures of dispersion** (or **variability**) are numbers that indicate the spread or scatter of observations; they show the extent to which individual values in a data set differ from one another and, hence, differ from their central location. The two data sets on life-insurance policies that are visually summarized in panel (a) of Figure 3.1 do not differ from each other with respect to variability. Note, for example, how

Figure 3.1   Summary Measures and Frequency Curves

Different types of summary measures describe different aspects of frequency curves, such as the curves shown here. Measures of *central tendency* locate the center of a data set, as at 50 (women) or 100 (men) in panel (a). Measures of *dispersion* focus on the spread of data around their center, as on the ranges from 0 to 100 (women) or 50 to 150 (men) in panel (a) or from 9.7 to 10.3 (pounds of cookies) or 9.9 to 10.1 (pounds of sugar) in panel (b). Measures of *shape* describe how symmetrical frequency curves are—as are the curves in panels (a) and (b)—or how asymmetrical they are—as are the curves in panel (c). Measures of shape also describe how peaked (sugar) or how flat (cookies) frequency curves are.

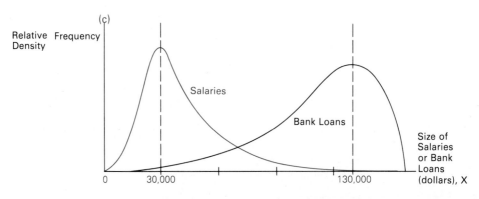

both women's and men's policies vary ± $50,000 from their respective centers. Women's policies cover a $100,000 range (from 0 to $100,000); men's policies also cover a $100,000 range (from $50,000 to $150,000). But now consider the two frequency curves in panel (b), which describe two data sets on the actual net weights of supposedly 10-pound cartons of cookies and of sugar. Even though both data sets have the identical central tendency (of 10 pounds per carton), they differ significantly as far as data dispersion is concerned. The sugar-carton net weights are tightly packed around the center of the distribution. All the weights are found within the 0.2-pound range from 9.9 to 10.1 pounds; that is, within 1 percent of the stated weight. The net weights of cookie cartons, on the other hand, are more widely spread; they range over 0.6 pounds; that is, within ± 3 percent of the desired weight. (This greater dispersion may occur in the case of the cookie-carton weights because their final weight must be determined by adding or subtracting an entire cookie rather than a few tiny grains of sugar.) Clearly, the two net-weight data sets could quite nicely be described by some measure of central tendency *along with* appropriate measures of dispersion.

Summary **measures of shape** are numbers that indicate either the degree of asymmetry or the degree of peakedness in a frequency distribution. Note how the four distributions pictured in panels (a) and (b) of Figure 3.1 are symmetrical about the dashed vertical lines; the left and right halves of the distributions in each case are mirror images of each other. This symmetry need not be present, however, as is indicated in panel (c). The salaries received by the employees in an industry might cluster near a dominant value of, say, $30,000. None being below zero, the spread below $30,000 would be limited, while the spread above that number may be extreme. As a result, the frequency curve comes to look like the side view of a child's slide. Asymmetry may be found in the other direction as well: the sizes of loans made by a bank may be concentrated near $130,000, there being none much larger because of the bank's lending limit, perhaps. Yet there may be many scattered loans of much lower amount. Measures of shape try to capture this type of asymmetry, as well as the flatness or peakedness of frequency curves such as those depicted graphically in panel (b) of Figure 3.1.

Finally, in the case of qualitative data, the only summary measure available is the **proportion,** a number that describes the frequency of observations in a particular category as a fraction of all observations made. (As the Chapter 2 discussion of pie charts indicates, proportions can, however, be calculated for quantitative data as well.)

In the following sections, the most important types of summary measures will be discussed in turn, but note: any summary measure can be calculated either for a statistical population or for a sample. If it is based on population data, the summary measure is called a **parameter;** it is referred to as a **statistic** if it is based on sample data only. In addition, the calculations can proceed from sets of original, raw, or ungrouped data, or they can be based, with some loss of accuracy as a result, on previously processed or grouped data, such as those data found in a frequency distribution.

SUMMARY MEASURES OF CENTRAL TENDENCY

The most important summary measures of central tendency are the following: (1) the *mean,* (2) the *median,* and (3) the *mode.* Because many people refer to all of them using the same word—*averages*—it is important to keep their meanings distinct so that you can use each summary measure of central tendency in an appropriate way.

The Arithmetic Mean

Most often encountered and most widely known among measures of central tendency is, no doubt, the *mean*—more precisely called the *arithmetic mean* (and in everyday usage simply referred to as the *average*).

Calculation from ungrouped data. When ungrouped data are available, the **arithmetic mean** can be calculated with precision by adding all the values observed and dividing their sum by the number of observations. If the annual profits of five firms (in millions of dollars) were 2, 2, 4, 7, and 15, the arithmetic mean would equal

$$\frac{2 + 2 + 4 + 7 + 15}{5} = \frac{30}{5} = 6.$$

This number (6) would be the *population mean* if the frame of interest contained only five firms (say, all the makers of airplanes in the United States or all the makers of beer in Detroit); the number would be a *sample mean* if it referred to five firms among a much larger group of interest (say, five among dozens of airplane manufacturers in the world or five among hundreds of breweries in the United States). The procedure just followed can also be expressed symbolically, as in Box 3.A.

Symbolic expression. By tradition, the small or the capital letter X is used to represent an observed value. Different values are distinguished by subscripts 1, 2, 3, . . . and so on, yielding the series $X_1, X_2, X_3, \ldots$ and so on, for a set of data. The symbol X_1 is pronounced "X sub one," and, in the example discussed before, $X_1 = 2$. Similarly, $X_2 = 2, X_3 = 4, X_4 = 7$, and $X_5 = 15$. Also by tradition, the total number of observations (here 5) is referred to as N in the case of a population and as n in the case of a sample. Finally, *parameters* are always symbolized by Greek letters, *statistics* by Roman ones (Appendix Table B lists the Greek alphabet). The population mean, in particular, is represented by μ (which is the lowercase Greek m, is pronounced "mu," and stands for *mean*); the sample mean, in turn, is represented by capital $\overline{X}$ (pronounced "X bar"). Thus, the calculation of the arithmetic mean from ungrouped data can be symbolized by the formulas in Box 3.A.

3.A Arithmetic Mean from Ungrouped Data

For a population:

$$\mu = \frac{X_1 + X_2 + X_3 + \ldots X_N}{N}$$

For a sample:

$$\overline{X} = \frac{X_1 + X_2 + X_3 + \ldots + X_n}{n}$$

where the X's are observed population (or sample) values, N is the number of observations in the population, and n is the number of observations in the sample.

A considerably more succinct version of these formulas can, however, be developed with the help of the *summation sign,* Σ (which is the uppercase Greek *S,* is pronounced "sigma," and stands for *sum*). Thus, we can write

$$X_1 + X_2 + X_3 + \ldots + X_N \text{ as } \sum_{i=1}^{N} X_i$$

which means "the sum of all the individual values of X with subscripts ranging from 1 to N." Similarly, if it were desirable to sum not all of the values, but only some of them, one could indicate this fact by appropriate subscripts and superscripts connected to the summation sign. Thus,

$$\sum_{i=1}^{3} X_i$$

would indicate that only the values of X with subscripts from 1 to 3 should be added, such as $X_1 + X_2 + X_3$. Analogously,

$$\sum_{i=5}^{8} X_i$$

would equal $X_5 + X_6 + X_7 + X_8$.

As it turns out, however, all the summations in this book, with a single exception that will be duly noted, involve the summing of *all* the values in question (of *all* the individual values of X_i between 1 and N, for example). It is, therefore, possible to simplify the summation notation by dropping the subscripts and superscripts connected to the summation sign. Thus, we will write

$$\frac{X_1 + X_2 + X_3 + \ldots + X_N}{N} \text{ not as } \frac{\sum_{i=1}^{N} X_i}{N} \text{ but simply as } \frac{\Sigma X}{N}.$$

And we will, similarly, write

$$\frac{X_1 + X_2 + X_3 + \ldots + X_n}{n} \text{ not as } \frac{\sum_{i=1}^{n} X_i}{n} \text{ but as } \frac{\Sigma X}{n}.$$

In each case, the invisible message attached to the summation sign will be the same: *all* the available values are to be summed. The context in which the simplified notation appears will indicate what these values are. (In the case of ungrouped population data, the values will range from 1 to N, for example; in the case of sample data, from 1 to n.)[1]

---

[1] An extended discussion of the rules of summation appears in Chapter 3 of the *Student Workbook* that accompanies this text.

Thus, the formulas in Box 3.A can be simplified to those in Box 3.B.

3.B Arithmetic Mean from Ungrouped Data—Alternative Formulation

For a population:

$$\mu = \frac{\Sigma X}{N}$$

For a sample:

$$\overline{X} = \frac{\Sigma X}{n}$$

where ΣX is the sum of all observed population (or sample) values, N is the number of observations in the population, and n is the number of observations in the sample.

Calculation from grouped data. Sometimes raw data are unavailable, and the mean, if it is to be calculated at all, must be calculated from a frequency distribution. At other times, raw data are available but are so massive in number that it is preferable to use grouped data to speed up the calculations. Calculating the mean from grouped data can be accomplished only by assuming (perhaps incorrectly) that the observations falling into a given class are equally spaced within it and are, therefore, on the average, equal to the midpoint of the class interval. (In the case of open-ended classes, the procedure must be abandoned or a wild estimate of the midpoint must be made.) Each midpoint is then multiplied by the absolute class frequency, and the sum of these products is divided by the population size, N (or the sample size, n). If each midpoint is represented by X, and the corresponding absolute class frequency by f, the procedure can be described by the formula in Box 3.C.

3.C Arithmetic Mean from Grouped Data

For a population:

$$\mu = \frac{\Sigma fX}{N}$$

For a sample:

$$\overline{X} = \frac{\Sigma fX}{n}$$

where ΣfX is the sum of all class-frequency (f) times class-midpoint (X) products, N is the number of observations in the population, and n is the number of observations in the sample.

Consider, for example, calculating the arithmetic mean of the 1981 profits of the 100 largest U.S.-based multinationals. If the original data were available, such as those

Table 3.1 Approximation of the Arithmetic Mean from a Frequency Distribution

When the arithmetic mean is to be calculated from a frequency distribution, and (as is usually the case) the actual sums of values in each class are not known, an approximation can still be derived by weighting each class midpoint, X, by the corresponding class frequency, f, and dividing the sum of the weighted products, ΣfX, by the population size, N (or the sample size, n).

Class (size of profit in millions of dollars)	Absolute Class Frequency (number of companies in class) f	Class Midpoint X	fX	Calculation
− 1,500 to under 0	3	− 750	− 2,250	
0 to under 500	41	250	10,250	$\mu = \dfrac{\Sigma fX}{N} =$
500 to under 1,000	32	750	24,000	
1,000 to under 1,500	9	1,250	11,250	$\dfrac{79,250}{100} =$
1,500 to under 2,000	6	1,750	10,500	
2,000 to under 2,500	6	2,250	13,500	$792.5 million
2,500 to under 5,500	3	4,000	12,000	
Totals:	$\Sigma f = N = 100$		$\Sigma fX = 79,250$	

Source: Table 2.3.

in Table 2.2 on page 39, one could sum them, divide by $N = 100$, and find the answer to be $786.62 million. Suppose, however, that only the grouped data of Table 2.3 (page 41) were available. The calculation would then have to proceed as shown in Table 3.1. The result—$792.5 million—is only an approximation but a close one; it is off the true value by less than 1 percent. (When discussing the construction of absolute frequency distributions in Chapter 2, it was noted how helpful it would be for subsequent users if the maker of the distribution chose class intervals with midpoints close to the average of the observations falling into the class or even reported the sums of observations in each class. These assertions are now vindicated. If each class midpoint equaled the average of all the observations in the class, each fX would equal the precise sum of class observations, and the mean calculated on that basis would be precise also. The same would, of course, be true if the precise sums of class values were provided in the first place.)

The nature of the mean. Knowing how to calculate the mean is one thing; knowing what it tells us is another. The mean is best viewed as a point of balance in a data set, very much like the fulcrum of a seesaw. In the case of a seesaw, or teeterboard, the number of (equal-sized) weights on one side of the fulcrum multiplied by their respective distances from it must equal the number of such weights times their distances on the other side in order for the board to be horizontal. Like the fulcrum, the arithmetic mean similarly balances the number of observations on one side of the mean times their respective deviations from this mean with the number of observa-

Figure 3.2   The Nature of the Arithmetic Mean

The mean of a data set can be compared to the fulcrum of a seesaw. The mean occupies a central position in the sense that the sum of negative and positive deviations of individual observations from the mean equals zero. Thus, for the data set represented by the blocks in panel (a), the mean equals 6, and the sum of deviation equals $-4 - 4 - 2 + 1 + 9 = 0$. For the data set represented by the blocks in panel (b), the mean equals 8, and the sum of deviations equals $-6 - 6 - 4 - 1 + 17 = 0$. Note how the arithmetic mean responds to the change in any single observation (as when 15 becomes 25) and how it is quite possible for the vast majority of all observations to be on one side of the mean—as in panel (b).

tions times their deviations on the other side. This role of the mean is illustrated in Figure 3.2.

Panel (a) illustrates the relationship between the five hypothetical profit figures noted above and the mean calculated from them. The five original observations are represented by the numbered blocks on top of the teeterboard; the $6 million mean is represented by the numbered fulcrum. The deviations from the mean are shown on the side of the board. Note how one observation of $4 million lies $2 million below the mean, while two observations of $2 million lie $4 million below the mean; hence, the negative deviations add to $- \$2$ million $- \$4$ million $- \$4$ million $= - \$10$ million. They are perfectly matched by the positive deviations from the mean of the $7 million and $15 million profit figures, the deviations of which come to $+ \$1$ million $+ \$9$ million $= + \$10$ million.

Panel (b) shows how the arithmetic mean instantly responds to a change in even a single observation (such as a change from $15 million to $25 million). Indeed, the mean is so responsive to extreme values that it can easily lie above (or below) the vast majority of individual observations! (Contrary to what most people think, it is, there-fore, quite possible for *most* people to weigh more than the average, be less intelligent than the average, or even be shorter than the average.) The mean always remains a *central* value, however, in the sense that the sum of deviations of all observations from the mean equals zero.[2] Nevertheless, and quite understandably, many people are confused when confronted with situations, such as that in panel (b), in which the vast majority of observations are below (or above) the average. This situation is not what *average* denotes to them; they prefer to use an entirely different concept of central tendency, a concept to which we now turn.

The Median

Another important measure of central tendency, quite different in nature from the mean, is the **median**—the value in an ordered array of data above and below which an equal number of observations can be found. This *middle value*, too, can be calculated precisely from ungrouped data, or it can be approximated from grouped ones.

Calculation from ungrouped data. For the profit figures (in millions of dollars) of 2, 2, 4, 7, and 15, the median equals 4 because it is located at the center of the array. Two values lie above it; two values lie below it—their sizes do not matter. Thus, if the array had been 2, 2, 4, 7, and 25, as in panel (b) of Figure 3.2, the median would still have been 4; and so it would have been for the array of 4, 4, 4, 4, and 25. But note: when there is an even number of observations, as in the series 2, 2, 4, 7, 15, and 25, there would seem to be two central values (of 4 and 7, in this example). In such a case, the two central values are averaged, and that average (here 5.5) is designated as the median. In our example, as the definition requires, three firms would earn less and three firms would earn more than the median (the values of 2, 2, and 4 would lie below 5.5, and those of 7, 15, and 25 would lie above it). Can you see with the help of Table 2.2 on page 39 that the median 1981 profit of the 100 multinationals equals $533.5 million, the average of the $532 million and $535 million values appearing in the center of the ordered array?

Symbolic expression. Denoting the median by M (the Greek *capital* mu) for a population and by m for a sample, we can express the definition of median symbolically by the formulas in Box 3.D.

3.D Median from Ungrouped Data

For a population:

$$M = X_{\frac{N+1}{2}} \text{ in an ordered array}$$

For a sample:

$$m = X_{\frac{n+1}{2}} \text{ in an ordered array}$$

where X is an observed population (or sample) value, N is the number of observations in the population, and n is the number of observations in the sample.

---

[2]A mathematical proof of this assertion appears in Chapter 3 of the *Student Workbook* that accompanies this text.

The median, these formulas tell us, always equals the value of the middle observation in an ordered array. In our sample of five profit figures, it equals the value of

$$X_{\frac{5 + 1}{2}} = X_{\frac{6}{2}} = X_3.$$

The value of the third observation is 4. Similarly, in our population of 100 multinational-company profits, the median equals the value of

$$X_{\frac{100 + 1}{2}} = X_{50.5};$$

that is, the arithmetic mean of $X_{50} = 532$ and $X_{51} = 535$, or 533.5.

Calculating the median from grouped data. Just as the arithmetic mean can be approximated from grouped data, so can the median. Consider again Table 3.1. It is fairly easy to find the class that contains the median and is, therefore, called the **median class.** Because the size of the population equals 100, the median is the average of the 50th and 51st observation in an ordered array of the population. There being 3 + 41 = 44 observations in the first two classes, the median cannot lie in either of them, but it is bound to be found among the next 32 observations in the third class. Indeed, the median must be the average of the sixth and seventh observations in the median class (which would correspond to the 50th and 51st observations among all the data). If one imagined the 32 observations in the third or median class to be equally spaced within the $500 million width of that class, each observation would be 15.625 units apart from the next one. If one further assumed that the first observation within the class was one half this distance above the lower class limit (and that the last observation within the class was one half this distance below the upper limit), the sixth and seventh observations would, respectively, be found at 585.9375 and 601.5625, making their arithmetic mean equal to $M = \$593.75$ million. Other interpolation procedures are possible also, but note again how the value determined in this way is bound to be an approximation only (this time within roughly 11 percent) of the correct figure of $533.5 million determined above.

This procedure, too, can be expressed by formulas, as shown in Box 3.E on the opposite page. In Box 3.E, L designates the lower limit of the median class; N or n is the size of the population or sample; F is the sum of frequencies up to, but not including, the median class; f is the frequency of the median class; and w is the width of the median class. In our case, this formula produced a median estimate of

$$M = 500 + \frac{(100/2) - 44}{32}(500) = 500 + \frac{6}{32}(500) = 593.75.$$

Median vs. mean. Although both median and mean are often referred to as *averages,* they are such in very different senses of the term. The arithmetic mean is an average of the observed values; the median is whatever value happens to be found at the average of all the positions in an ordered array. [It is, for example, the value found at position $(1 + 100) \div 2 = 50.5$ among 100 ordered numbers]. As a result, the mean can be algebraically manipulated in many ways; the median cannot. For example, if we know the means and the sizes of two populations, we can calculate their combined

3.E Median from Grouped Data

For a population:

$$M = L + \frac{(N/2) - F}{f}\, w$$

For a sample:

$$m = L + \frac{(n/2) - F}{f}\, w$$

where L is the lower limit of the median class, f is its absolute frequency, and w is its width, while F is the sum of frequencies up to (but not including) the median class, N is the number of observations in the population, and n is the number of observations in the sample.

mean; yet we cannot calculate the combined median from a knowledge of the medians and sizes of two populations. Consider the two populations, A and B:

```
A: 10   17   39
B:  6   20   31   57   82   97   99
```

The arithmetic means are 22 and 56, respectively; the medians are 17 and 57. One can easily calculate the combined mean as $[(3 \times 22) + (7 \times 56)] \div 10 = 45.8$, knowing only the sizes of the two populations and their means. One *cannot* calculate the combined median by a similar procedure, such as $[(3 \times 17) + (7 \times 57)] \div 10 = 45$. The combined median is, in fact, not 45, but the average of the fifth and sixth value in an ordered array of the combined populations, or $(31 + 39) \div 2 = 35$.

In spite of this disadvantage, the median is, on the other hand, a better measure of central tendency than the mean when the data set contains a few extreme values, high or low. Consider these annual income figures of the families living on a certain street:

```
$15,000      $19,000
$17,000      $19,000
$17,000      $301,000
$18,000
```

The arithmetic mean of these incomes equals $58,000, which certainly does not look like a very good summary of neighborhood incomes. The median of $18,000, on the other hand, is much more representative. Indeed, the median is often called "democratic" because in a sense it gives each value, regardless of its size, an equal "vote" in determining the central location. In the above example, the three values below $18,000 have one vote each, the three values above $18,000 have one vote each, so $18,000 is the "winner." Unlike in the case of the arithmetic mean, the sizes of the individual values below or above the median are irrelevant for determining the median. [The same issue is visually illustrated in Figure 3.2. Under the influence of one extreme observation, the mean changes from panel (a) to panel (b), and it hardly seems representative of the panel (b) data. Yet the median in both panels equals 4.]

The Mode

A third measure of central tendency—the **mode**—is defined as the most frequently occurring value in a set of data. In both panels (a) and (b) of Figure 3.2, for example, the mode equals 2. Just as the expression to be "à la mode" means to be in fashion, the mode is the most "fashionable" value in a data set. Yet, as the examples of Figure 3.2 suggest, the mode can easily be found at either extreme of a data set and can be quite atypical of the majority of observations. Sometimes a population contains two modes or even more; consider the profit figures of $119 million and $580 million in Table 2.2 on page 39. In practice, the mode is seldom used for describing ungrouped data; much more frequent is its designation for grouped data.

Calculation from grouped data. In the case of grouped data, the class containing the mode is the **modal class** and is the class with the highest frequency density. Consider Table 3.2. The modal class clearly is the second one, but there exist several alternative ways of approximating the mode. One possibility is simply to designate the midpoint of the modal class, here $250 million, as the mode. Another possibility proceeds from the assumption that the mode will be closer to the adjacent class that has the greater frequency density (here closer to the third class than the first one). A favorite formula for finding the mode (*Mo* for population, *mo* for sample) based on the second alternative just described is given in Box 3.F and is applied in Table 3.2.

3.F Mode from Grouped Data

For a population or a sample:

$$Mo \text{ or } mo = L + \frac{d_1}{d_1 + d_2} w$$

where L is the lower limit of the modal class, w is its width, and d_1 and d_2, respectively, are the differences between the modal class frequency *density* and that of the preceding or following class.

In Box 3.F, L designates the lower limit of the modal class; d_1 and d_2 are the respective differences between frequency densities (*not* absolute frequencies) of the modal class and that of the preceding (or following) class; and w is the modal class width.

The same procedure can also be shown graphically, as Figure 3.3 illustrates. Note how the formula gives the same result as the graph:

$$\text{For panel (a), } Mo = 3 + \frac{2}{2 + 6}(2) = 3.5.$$

$$\text{For panel (b), } Mo = 3 + \frac{2}{2 + 2}(2) = 4.$$

$$\text{For panel (c), } Mo = 3 + \frac{6}{6 + 2}(2) = 4.5.$$

Table 3.2 Approximation of the Mode from a Frequency Distribution

The mode can be estimated from a frequency distribution in several ways, one of which is illustrated here. This method assumes that the mode is found in the modal class but is also "attracted" to the two adjacent classes by forces that are proportional to the frequency densities of these classes. But note: the mode is highly unstable. It is likely to change with every change in the grouping of data.

Class (size of profit in millions of dollars)	Absolute Frequency Density (number of companies per $500 million class range)	Calculation
−1,500 to under 0	1	
0 to under 500	41	$Mo = L + \dfrac{d_1}{d_1 + d_2} \cdot w =$
500 to under 1,000	32	
1,000 to under 1,500	9	$0 + \dfrac{41 - 1}{(41 - 1) + (41 - 32)} (500) =$
1,500 to under 2,000	6	
2,000 to under 2,500	6	$\dfrac{40}{49} (500) = \$408.16$ million
2,500 to under 5,500	½	

Source: Table 2.9.

Figure 3.3 Approximating the Mode from a Histogram

The graphical procedure for finding the mode that is shown here corresponds to the formula given in Box 3.F. In each case, the second class represents the modal class, but the mode equals the midpoint of that class only in panel (b) where both adjacent classes have equal frequency densities. Thus, the two adjacent classes exert equal pull on the midpoint of the modal class. In contrast, the mode in panel (a) is more "attracted" to the preceding class (which has a higher frequency density than the following one), while in panel (c) the mode is more "attracted" to the following class (because it has a higher frequency density than the preceding one). Thus, the mode equals 3.5 in panel (a), 4.0 in panel (b), and 4.5 in panel (c).

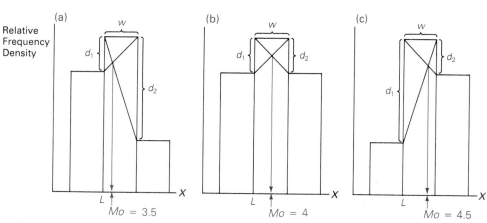

The mode and the frequency curve. The mode can also be found on a smooth frequency curve as the value lying underneath its highest point. However, if two or more different values in a data set occur with the highest frequency, or almost that, two or more modes exist, and the data set is said to have a **multimodal frequency distribution.** This distribution shows up as two or more peaks on a frequency curve. Such a situation usually arises because the population or sample in question contains two or more groups that by some criterion are fairly homogeneous internally but differ significantly from each other. For purposes of statistical analysis, it is often wiser to study such groups separately. Consider Figure 3.4 on page 89. Panels (a) and (b) give two examples of the **bimodal frequency distribution,** so called because it contains two modes. Panel (a) portrays the frequency distribution of the heights of all the production workers in a large firm. Two modes are visible, at 5′2″ and 5′11″, but they reflect nothing more mysterious than the fact that the firm has an equal number of female and male workers (and women tend to be shorter). To illustrate why it is often wiser to study such groupings separately, imagine that someone compared the mean heights of production workers now and 50 years ago (when almost all the workers were male). The apparent conclusion that workers had become shorter would be false, a fact that would become obvious if female workers of 50 years ago were compared with female workers now, while male workers then and now were also compared separately. The presence of more than one mode, therefore, is a warning signal to the investigator that suggests the utmost of caution.

Panel (b) presents another example. It shows the frequency distribution of the high-school grade-point averages of newly hired production workers. Humps in the distribution occur at 2.1 and 3.3; although the frequency of the former is below that of the latter (which, strictly speaking, makes 3.3 the mode), one still refers to this distribution as bimodal. The reason might be that two groups of workers were hired, a large one of apparently highly qualified ones, a small one of seemingly unqualified ones. Yet the latter group—hired, perhaps, under affirmative-action rules—may be given special training and be expected, eventually, to perform as well as the other on the production line. Once again, the presence of these heterogeneous groups suggests to an investigator (who might wish to compare early and later work performance) that these two groups should be studied separately. The mean performance at work of the entire group may hardly change over time; the mean performance at work of the small special-training group may improve dramatically over time.

Panel (c), finally, gives an example of an even more heterogeneous statistical population. The city's residents may, for example, fall into three distinct groups: unskilled laborers, skilled workers, and professionals. Their respective modal incomes of $8,000, $20,000, and $34,000 per year may produce the three-peaked frequency curve when all of them are included in a single data set.

Other Measures

Additional but less frequently used measures of central tendency include (1) the *weighted mean,* (2) the *geometric mean,* and (3) the *harmonic mean.* We will here illustrate the meaning of the first of these; the second and third are discussed in Chapter 3 of the *Student Workbook* that accompanies this text.

The **weighted mean, μ_w** or $\overline{X}_w$, is calculated when different observations must be given unequal weights in accordance with their unequal relative importance. The weighted mean equals the sum of the products of observed values and their respective

Figure 3.4 Multimodal Frequency Distributions

Multipeaked frequency curves illustrate the presence of two or more modes and strongly suggest the existence of some nonhomogeneous factor in the underlying data set. Under such circumstances, any single measure of central location is likely to be misleading; it is advisable to describe the two or more underlying data sets separately.

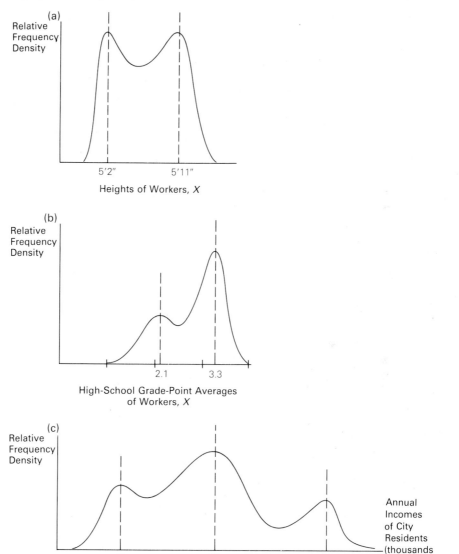

weights, divided by the sum of weights. If a firm pays hourly wages of $5, $10, and $15 to different groups of workers, it would be unwise to conclude that its workers were earning, on the average, $10 per hour, unless equal numbers of workers were found in the three categories. If, however, there were 100 unskilled workers earning $5, 50 semiskilled workers earning \$10, and 10 skilled workers earning \$15, these numbers of workers should be used as weights to count \$5 one hundred times, \$10 fifty times, and \$15 ten times. Thus, the formulas for the unweighted arithmetic mean noted in Box 3.B are adjusted to read like those in Box 3.G.

3.G Weighted Mean from Ungrouped Data

For a population:

$$\mu_w = \frac{\Sigma wX}{\Sigma w}$$

For a sample:

$$\overline{X}_w = \frac{\Sigma wX}{\Sigma w}$$

where ΣwX is the sum of all weight (w) times observed-value (X) products, while Σw equals N (the number of observations in the population) or n (the number of observations in the sample).

In our case, this procedure yields a weighted mean,

$$\mu_w = \frac{(100 \times \$5) + (50 \times \$10) + (10 \times \$15)}{100 + 50 + 10} = \frac{\$1,150}{160} = \$7.19,$$

a far cry from the unweighted average hourly wage of \$10.

Note: The calculation of the arithmetic mean from grouped data, in fact, produces a weighted mean; the class frequencies, f, then serve the function of weights, w. The sum of these weights, Σf, is simply designated as the total number of observations, N or n. Thus, the formulas in Box 3.C are a special case of those in Box 3.G.

SUMMARY MEASURES OF DISPERSION

A knowledge of the dispersion or variability of data around their center of location is crucial in many situations. Consider someone about to decide which of two types of business to enter. Both types may promise the same median income, yet the incomes observed in one type of activity may be scattered widely about the median while the incomes in the other type may fall within a relatively narrow range of the median. Depending on their attitudes toward risk (a matter that is discussed at length in Chapter 14), different people will evaluate these two business prospects differently. Risk seekers will welcome a high variability of income; at the risk of getting a very low income, this high variability of income offers them a chance to get an exceptionally high income.

Figure 3.5 The Importance of Variability

Even though two populations may have identical medians, the spread of observations around them may differ significantly. This spread of observations is likely to be a crucial factor in decision making. Thus, a risk-averse individual may prefer the Type A Business (that promises a less varied income), while a risk-seeking person may choose to enter the Type B Business (just because it offers a chance of fairly large incomes, along with the danger of extremely low incomes, of course).

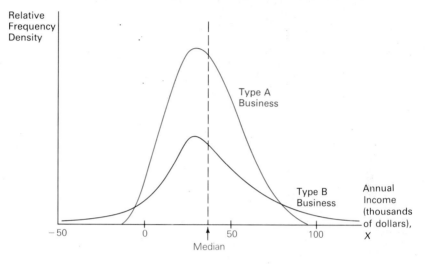

Risk-averse people, on the other hand, will prefer the activity with a low variability of income; it offers them a middle-level income with near-certainty and offers them no chance for very low or very high income. The importance of variability is reviewed in Figure 3.5.

Dispersion is measured in two ways: as distances between selected observations or as average deviations of individual observations from a central value. We consider these possibilities in turn.

Distance Measures of Dispersion

The most popular distance measures of dispersion include (1) the (overall) range and (2) various interfractile ranges.

The range. The overall range is usually called just the **range** and is the difference between the largest and smallest observation in a set of ungrouped data. It is the difference between the upper limit of the largest class and the lower limit of the smallest class for grouped data (and, therefore, cannot be determined at all if grouped data have open-ended classes). Thus, the range in our set of ungrouped profit figures listed in Table 2.2 on page 39 is $5,423 million − (−$1,060 million) = $6,483 million. The 1981 profits of U.S. multinationals, this figure tells us, were spread out over a vast range of almost $6.5 billion. Yet the major disadvantage of this measure is obvious: it ignores all the values except the two extremes. These extremes may be atypical not only among all the values in the data set, but even among the higher and lower ones—which surely is the case here. As another look at Table 2.2 confirms, most profits are found in the much narrower $2.5 billion-wide range between $0 and $2.5 billion. This problem is even worse when the range is calculated from the corresponding grouped data, found in Table 3.1. Here the range appears to equal $5,500 million − (−$1,500 million) = $7,000 million.

Interfractile ranges. A definite improvement over using the overall range is to use various **interfractile ranges,** which measure differences between two values called *fractiles.* A **fractile** is a value below which a specified proportion of all values is found. The same information can also be given as a **percentile,** which is a value below which a specified *percentage* of all values is found. Thus, the median is the .50 fractile (or the 50th percentile) because half (or 50 percent) of all observations lie below it. After the median, the most common fractiles or percentiles used are the **quartiles,** which divide the total of observations into four quarters, each of which contains .25 (or 25 percent) of the observed values. The .25 fractile (or 25th percentile) is called the **first quartile,** and a quarter (or 25 percent) of all observations lie below it. The median itself is the **second quartile,** and the .75 fractile (or 75th percentile) is referred to as the **third quartile.** Can you see in Table 2.2. on page 39 that the first, second, and third quartiles equal $335.5 million, $533.5 million, and $939 million, respectively? (In each case, the average of two adjacent values has been taken to ensure that 25, 50, and 75 percent of all observations lie below the quartile in question.)

Most popular among interfractile ranges is the **interquartile range,** which is the difference between the third and first quartiles and, hence, is the range containing the middle 50 percent of all values observed.

Although they are seldom used, other interfractile ranges can, of course, also be constructed, such as those between various **deciles**—the values in a data set that divide the total of observations into 10 parts rather than into 4 parts, as quartiles do, or into 100 parts, as percentiles do). Thus, the range between the 1st and 9th deciles (or the .10 and .90 fractiles or the 10th and 90th percentiles) contains the middle 80 percent of all values observed.

Can you see in Table 2.2 on p. 39 that the interquartile range equals $939 million − $335.5 million = $603.5 million? Can you see that the .10 to .90 interfractile range equals $1,802.5 million?

Figure 3.6 explains how fractile values can be read from a graph of a less-than type ogive.

Average Deviation Measures of Dispersion

All the distance measures of dispersion share a common disadvantage: they do not take into account all observations. Average deviation measures of dispersion—most important, (1) the mean absolute deviation, (2) the variance, and (3) the standard deviation—do take all observations into account.

The mean absolute deviation. To accomplish the feat of not ignoring any observed value, one might be tempted to calculate how much each individual observation, X, deviates from some central value, such as μ or $\overline{X}$, and then to combine these deviations

Figure 3.6 Reading Fractiles from an Ogive

Fractile values can be approximated from an ogive. The first quartile is about $336 million (below *b*), and the third quartile is $939 million (below *c*); thus, the interquartile range is about $603 million wide. By the same token, the .10 to .90 interfractile range is seen to equal about $1,803 million (going from below *a* to below *d* on the horizontal axis).

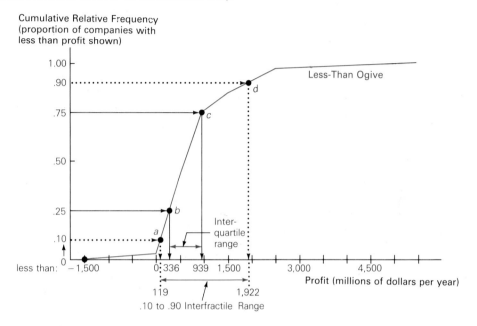

Source: Panel (a) of Figure 2.5.

---

by averaging them. However, averaging deviations from the arithmetic mean necessarily produces a result of zero, as a review of Figure 3.2 (the seesaws) can quickly remind us. Consider the five numbers 2, 2, 4, 7, and 15. Their arithmetic mean equals 6. Subtracting this mean from each value yields deviations of -4, -4, -2, $+1$, and $+9$, as panel (a) of Figure 3.2 shows. These deviations necessarily average to 0. The difficulty is overcome by realizing that the objective is merely to find the average distance of all the data from their center; for this purpose, it is irrelevant whether the individual datalie above or below the center. Hence, positive and negative deviations can be treated alike; one can ignore all the above plus and minus signs, and one can average all the *absolute* values of the deviations of individual observations from the mean (a fact that is indicated symbolically by vertical lines before and after each value). This procedure yields the **mean absolute deviation, MAD,** here equal to $[(|4| + |4| + |2| + |1| + |9|)/5] = (20/5) = 4$. Can you confirm that this measure of dispersion comes to $566.70 million for the profit data of Table 2.2 (the arithmetic mean of which came to $786.62 million)? Symbolically, the mean absolute deviation can be expressed as shown in Box 3.H (with vertical lines denoting absolute value).

3.H Mean Absolute Deviation

From Ungrouped Data
For a population:

$$MAD = \frac{\Sigma |X - \mu|}{N}$$

For a sample:

$$MAD = \frac{\Sigma |X - \overline{X}|}{n}$$

where $\Sigma |X - \mu|$ is the sum of the absolute differences between each observed population value, X, and the population mean, μ, while N is the number of observations in the population, and where $\Sigma |X - \overline{X}|$ is the sum of the absolute differences between each observed sample value, X, and the sample mean, $\overline{X}$, while n is the number of observations in the sample.

From Grouped Data
Denoting absolute class frequencies by f and class midpoints by X, substitute $\Sigma f |X - \mu|$ or $\Sigma f |X - \overline{X}|$ for the numerators given here.

Note: Occasionally, absolute deviations from the median rather than from the mean are calculated; in which case μ is replaced by M, and $\overline{X}$ is replaced by m.

The variance. A much more important measure of dispersion, however, which is calculated by averaging the *squares* of the individual deviations from the mean, is the **mean of squared deviations,** or the **variance.**[3] Using the five values from panel (a) of Figure 3.2, the variance is calculated in Table 3.3.

When the variance is calculated from a statistical population, it is symbolized by σ^2 (the letter σ is the lowercase Greek *sigma* and σ^2 is pronounced "sigma squared"). When the variance refers to sample data, it is denoted by s^2 instead (in line with the common procedure of labeling parameters by Greek letters and statistics by Roman ones). The variance is calculated as shown in Box 3.I.

Note that this formulation contains only one surprise: the sample variance is obtained by dividing the sum of the squared deviations of individual sample values, X, from the sample mean, $\overline{X}$, by $n - 1$ rather than by the sample size, n. The reason, to be explained in Chapter 7, is that this procedure makes the sample variance, s^2, a more accurate estimator of the usually unknown population variance, σ^2.

---

[3]The arithmetic mean has the property of minimizing the sum of the squared deviations of all the individual observations from itself, as shown in Chapter 3 of the *Student Workbook* that accompanies this text.

Table 3.3 Calculating the Variance for a Population of $N = 5$ Values

The population variance is calculated as the arithmetic mean of the individually squared deviations of each observed value from the population mean. Note how the process of squaring deviations (a) eliminates negative values (and thus produces a measure of dispersion that focuses on the existence of deviations rather than their direction) and (b) emphasizes large deviations more than small ones (a 9 squared counts not 9 times, but 81 times as much as a 1 squared).

Observed Values X	Population Mean $\mu = \dfrac{\Sigma X}{N}$	Deviations $X - \mu$	Squared Deviations $(X - \mu)^2$
2	6	−4	16
2	6	−4	16
4	6	−2	4
7	6	+1	1
15	6	+9	81

$\Sigma X = 30$ Sum of squared deviations $= \Sigma (X - \mu)^2 = 118$

$$\text{Mean of squared deviations} = \text{variance} = \frac{\Sigma (X - \mu)^2}{N} = \frac{118}{5} = 23.6$$

3.1 Variance from Ungrouped Data

For a population:

$$\sigma^2 = \frac{\Sigma(X - \mu)^2}{N}$$

For a sample:

$$s^2 = \frac{\Sigma(X - \overline{X})^2}{n - 1}$$

where $\Sigma(X - \mu)^2$ is the sum of squared deviations between each population value, X, and the population mean, μ, with N being the number of observations in the population, while $\Sigma(X - \overline{X})^2$ is the sum of squared deviations between each sample value, X, and the sample mean, $\overline{X}$, with n being the number of observations in the sample.

The variance, too, can be estimated from grouped data by the now familiar procedure of substituting class midpoints for actually observed values, X, and replacing the numerators in Box 3.I by $\Sigma f(X - \mu)^2$ and $\Sigma f(X - \overline{X})^2$, respectively, as shown in Box 3.J.

3.J Variance from Grouped Data

For a population:

$$\sigma^2 = \frac{\Sigma f(X - \mu)^2}{N}$$

For a sample:

$$s^2 = \frac{\Sigma f(X - \overline{X})^2}{n - 1}$$

where absolute class frequencies are denoted by f, class midpoints of grouped population (or sample) values by X, the population (or sample) mean by μ (or $\overline{X}$), and the number of observations in the population (or sample) by N (or n).

Practical problems. Unfortunately, there are two practical problems associated with the use of the variance concept. First, the variance tends to be a large number compared to the observations the spread of which it is to describe. When the original observations, as in Table 2.2 on p. 39 are at most equal to a few billions, their variance can equal many hundreds of billions. Second and worse yet, the variance, being a squared number, is not expressed in the same units as the observed values themselves. Thus, the variance of the Table 2.2 profit population is not expressed in hundreds of billions of *dollars*, but in hundreds of billions of *squared* dollars; in fact, it equals 704.4 billion squared dollars (and who knows the meaning of that?). But there is good news as well: both of the conceptual difficulties just cited can be overcome in one fell swoop by working with the square root of the variance, a concept to which we turn.

The standard deviation. The positive square root of the variance is called the **standard deviation** and is, without doubt, the most important measure of dispersion. It falls in the same range of magnitude as, and is expressed in the same units as, the observations themselves. Thus, the standard deviation to the data in Table 3.3 equals $\sqrt{23.6} = 4.86$; the standard deviation for the Table 2.2 profit data equals $\sqrt{704,408.34}$ million squared dollars $= \$839.29$ million. The definition of the standard deviation is expressed symbolically in Box 3.K.

Once again, the standard deviation is estimated from grouped data by substituting class midpoints for actually observed values, X, and replacing the numerators under the square roots in Box 3.K by $\Sigma f(X - \mu)^2$ and $\Sigma f(X - \overline{X})^2$, respectively, as shown in Box 3.L.

Shortcut calculations of variance and standard deviation. Unless the variance or standard deviation is to be calculated by a computer, it is helpful to use a shortcut procedure that is mathematically equivalent to the boxed formulas encountered in the book

3.K Standard Deviation from Ungrouped Data

For a population:

$$\sigma = \sqrt{\frac{\Sigma(X - \mu)^2}{N}}$$

For a sample:

$$s = \sqrt{\frac{\Sigma(X - \overline{X})^2}{n - 1}}$$

where $\Sigma(X - \mu)^2$ is the sum of squared deviations between each population value, X, and the population mean, μ, with N being the number of observations in the population, while $\Sigma(X - \overline{X})^2$ is the sum of squared deviations between each sample value, X, and the sample mean, $\overline{X}$, with n being the number of observations in the sample.

3.L Standard Deviation from Grouped Data

For a population:

$$\sigma = \sqrt{\frac{\Sigma f(X - \mu)^2}{N}}$$

For a sample:

$$s = \sqrt{\frac{\Sigma f(X - \overline{X})^2}{n - 1}}$$

where absolute class frequencies are denoted by f, class midpoints of grouped population (or sample) values by X, the population (or sample) mean by μ (or $\overline{X}$), and the number of observations in the population (or sample) by N (or n).

so far.[4] The shortcut formulas for the variance appear in Box 3.M; the standard deviation is, of course, always the square root of the variance. A numerical example, based on the data in Table 3.3, is given in Table 3.4.

---

[4]Consider this proof for the population variance from ungrouped data: $\sigma^2 = \dfrac{\Sigma(X - \mu)^2}{N} = \dfrac{\Sigma(X^2 - 2X\mu + \mu^2)}{N} = \dfrac{\Sigma X^2 - 2\mu\Sigma X + N\mu^2}{N} = \dfrac{\Sigma X^2}{N} - 2\mu^2 + \mu^2 = \dfrac{\Sigma X^2 - N\mu^2}{N}$

Table 3.4   Calculating the Variance and Standard Deviation for a Population of Ungrouped Data of N = 5 Values—Shortcut Method

This shortcut method of calculating the variance and standard deviation is particularly recommended whenever the calculation proceeds with pencil and paper or by simple calculator.

Observed Values, X	X^2	
2	4	$\mu = \dfrac{\Sigma X}{N} = \dfrac{30}{5} = 6$
2	4	
4	16	$\sigma^2 = \dfrac{\Sigma X^2 - N\mu^2}{N} = \dfrac{298 - 5(6)^2}{5} = \dfrac{298 - 180}{5}$
7	49	
15	225	$= 23.6$
$\Sigma X = 30$	$\Sigma X^2 = 298$	$\sigma = \sqrt{23.6} = 4.86$

3.M Variance from Ungrouped Data—Shortcut Method

For a population:

$$\sigma^2 = \frac{\Sigma X^2 - N\mu^2}{N}$$

For a sample:

$$s^2 = \frac{\Sigma X^2 - n\overline{X}^2}{n - 1}$$

where ΣX^2 is the sum of squared population (or sample) values, μ^2 is the squared population mean and $\overline{X}^2$ the squared sample mean, N is the number of observations in the population, and n is the number of observations in the sample.

A shortcut method is equally helpful when making approximations from grouped data, such as those in Table 3.1. The formulas appear in Box 3.N, and Table 3.5 gives an example of how to use the formulas.

3.N Variance from Grouped Data—Shortcut Method

For a population:

$$\sigma^2 = \frac{\Sigma f X^2 - N\mu^2}{N}$$

For a sample:

$$s^2 = \frac{\Sigma f X^2 - n\overline{X}^2}{n - 1}$$

where $\Sigma f X^2$ is the sum of absolute-class-frequency (f) times squared-class-midpoint (X) products, μ^2 is the squared population mean and $\overline{X}^2$ the squared sample mean, N is the number of observations in the population and n is the number of observations in the sample.

Table 3.5 Approximating the Variance and Standard Deviation for a Population from Grouped Data—Shortcut Method

This table illustrates how closely one can approximate the true variance and standard deviation of ungrouped data (such as those in Table 2.2) with the help of the corresponding grouped data (shown here). Compare the approximations calculated at the bottom of this table with the precise calculations from the underlying ungrouped data:

The *variance from ungrouped data* is

$$\sigma^2 = \frac{132,317,936 - 61,877,102}{100} = \$704,408.34 \text{ million squared.}$$

The *standard deviation from ungrouped data* is $\sigma = \$839.29$ million.

Class (size of profit in millions of dollars)	Absolute Class Frequency (number of companies in class), f	Class Midpoint, X	fX	X^2	fX^2
−1,500 to under 0	3	−750	−2,250	562,500	1,687,500
0 to under 500	41	250	10,250	62,500	2,562,500
500 to under 1,000	32	750	24,000	562,500	18,000,000
1,000 to under 1,500	9	1,250	11,250	1,562,500	14,062,500
1,500 to under 2,000	6	1,750	10,500	3,062,500	18,375,000
2,000 to under 2,500	6	2,250	13,500	5,062,500	30,375,000
2,500 to under 5,500	3	4,000	12,000	16,000,000	48,000,000
Totals:	$N = 100$		$\Sigma fX = 79,250$		$\Sigma fX^2 = 133,062,500$

$$\mu = \frac{\Sigma fX}{N} = \frac{79,250}{100} = \$792.5 \text{ million}$$

$$\sigma^2 = \frac{\Sigma fX^2 - N\mu^2}{N} = \frac{133,062,500 - 100(792.5)^2}{100}$$

$$= \frac{133,062,500 - 62,805,625}{100} = \frac{70,256,875}{100}$$

$$= \$702,568.75 \text{ million squared}$$

$$\sigma = \sqrt{\$702,568.75 \text{ million squared}} = \$838.19 \text{ million}$$

Source: Tables 2.2 and 3.1.

The Special Significance of the Standard Deviation

Together with the arithmetic mean, the standard deviation is, perhaps, the most significant parameter of a statistical population. Consider just two of its many uses: describing the normal frequency distribution and comparing the degree of dispersion of different data sets.

Describing the normal frequency distribution. When a population is correctly described by the perfectly symmetrical *normal frequency distribution* shown in the graph on p. 102 (and to be discussed at length in Chapter 6), one can predict the precise percentage of observations falling within ranges from the mean defined by any desired number of standard deviations. Some 68.3 percent of all observations will then lie within the range of $\mu \pm 1\sigma$; that is, within one standard deviation of the mean. Similarly, some 95.4 percent of the population will lie within $\mu \pm 2\sigma$, and 99.7 percent will lie within $\mu \pm 3\sigma$. Thus, if the mean height of all male production workers in a large firm were $\mu = 5'8''$, if the standard deviation was $\sigma = 2$ inches, and if the heights were distributed like a normal curve, we could confidently assert that 68.3 percent of these men were between 5′6″ and 5′ 10″ tall or that 99.7 percent of them had heights between 5′2″ and 6′2″.

It is difficult to overemphasize the significance of these facts. Think of it, a normally distributed population can be fully described by a mere two numbers—the parameters μ and σ. From these two numbers, the entire frequency distribution can be reconstructed. Analytical Examples 3.1, "Standard Scores," and 3.2, "Control Charts," illustrate just two of a multitude of applications. Indeed, as the discussion of *Chebyshev's theorem* in Biography 3.1 at the end of the chapter attests, the standard deviation can tell us important things about all types of populations, even those that are not normally distributed.

Comparing the degree of dispersion of different data sets. Sometimes the ratio of the standard deviation to the arithmetic mean is used as an indicator of *relative* dispersion. This ratio is called the **coefficient of variation** and is symbolized by V (or v), as shown in Box 3.O.

3.0 Coefficient of Variation

For a population:

$$V = \frac{\sigma}{\mu}$$

For a sample:

$$v = \frac{s}{\overline{X}}$$

where σ is the population standard deviation, μ the population mean, s the sample standard deviation, and $\overline{X}$ the sample mean.

For the profit population of Table 2.2, the coefficient of variation comes to

$$V = \frac{\$839.29 \text{ million}}{\$786.62 \text{ million}} = 1.067.$$

Put differently, the standard deviation equals 106.7 percent of the mean.

Note that the coefficient of variation, unlike all the measures of *absolute* dispersion, is expressed as a pure number without any units. As a result, the coefficient of variation can be used to compare the relative dispersion of two or more distributions that are expressed in different units. For example, if the employees of a firm have a mean height of 66 inches (with a standard deviation of 4 inches), but a mean weight of 150 pounds (with a standard deviation of 30 pounds), one can say that heights are less varied than weights because

$$\text{for heights, } V = \frac{4''}{66''} = .06,$$

$$\text{but for weights, } V = \frac{30 \text{ lb.}}{150 \text{ lb.}} = .20.$$

THE PROPORTION

For qualitative data, the *proportion* of observations falling into a given category is the only summary measure available. It is calculated as the number of observations in the category divided by the total of all observations. Thus, if 138 of 200 firms sampled are single proprietorships, if 18 are partnerships, and if the rest are corporations, one could designate the first group as the modal type of firm, but one could hardly calculate the arithmetic mean, the median, or the standard deviation of firm types. It would be meaningful, however, to state that the *proportion* of single proprietorships and partnerships among all types of firms equaled (138 + 18)/200, or .78. By the same token, corporations could be reported as making up a proportion of .22. Naturally, these numbers, like all proportions, can instantly be converted to percentages, here of 78 and 22 percent.

When referring to a population, the proportion is denoted by π (which is the lowercase Greek p, is pronounced "pī," and stands for *proportion* and *not*—here—for the geometric constant 3.1416, which equals the ratio of the circumference to the diameter of a circle). When referring to a sample, the proportion is represented by the capital Roman letter P.

3.1 Standard Scores

As this chapter indicates, one of the significant features of the standard deviation is the fact that it can be used to describe the precise percentage of observations falling within various ranges from the mean, provided the frequency distribution fits the so-called *normal distribution,* such as the one first noted in panel (a) of Figure 2.2 on page 39. Consider the accompanying figure. This graph shows a hypothetical normal distribution of aptitude test scores (such as the familiar SAT scores) with a mean score of $\mu = 500$ and a standard deviation of $\sigma = 100$. As a result of the normal distribution of the population of scores, 68.3 percent of all scores lie between 400 and 600 (that is, within 1 standard deviation of the mean), 95.4 percent of all scores lie between 300 and 700 (within 2 stan-

dard deviations of the mean), and 99.7 percent of all scores lie between 200 and 800 (within 3 standard deviations of the mean). As will be discussed at length in Chapter 6, it is also true that 68.3 percent of the *area* under a normal curve lies within 1 standard deviation of the mean, while 95.4 and 99.7 percent of the area, respectively, lie within 2 and 3 standard deviations.

In order to make possible easy comparisons among normal distributions of different types of populations (be they test scores, heights, weights, or wages), one can convert raw data into **standard scores** that indicate the numbers of standard deviations between particular observations and the mean of all observations in a data set. As is indicated underneath the graph, the standard score, z, equals the difference between an

Relative Frequency Density

68.3%

$\mu = 500$
$\sigma = 100$

95.4%

99.7%

$X =$ Aptitude Test Scores

200   300   400   500   600   700   800
Mean

-3 -2 -1 0 $+1$ $+2$ $+3$ $z = \dfrac{X-\mu}{\sigma} =$ Standard Score

observed value, X, and the mean, μ, divided by the standard deviation, σ; thus, a z score always shows how many standard deviations *away from the mean* a corresponding X score is located.

In this example, with $\mu = 500$ and $\sigma = 100$, someone scoring $X = 300$ has a standard score of $z = -2$ because $(300 - 500) \div 100 = -2$. Such a score lies below 50 percent of all scores (the area under the right half of the bell-shaped curve) and also below an additional $(95.4/2) = 47.7$ percent of all scores (the

area under the left side of the bell between the mean and $X = 300$). Thus, a score of $X = 300$ lies below 97.7 percent of all scores.

Can you see, similarly, why a score of 800 would correspond to a standard score of $+3$ and would lie above 99.85 percent of all scores? Why someone with standard scores of $+.5$ on the weight distribution and -1 on the wage distribution would be heavier and less well paid than the average person whose weight and wages were plotted?

3.2 Control Charts

Managers of businesses know that there exists an inherent amount of variability in any data series, whether the series describes the sugar content of bottled drinks, the extent of clerical errors, the size of inventories, of labor costs, or advertising expenses, or even the loss rate of old customers. **Control charts** are graphical devices, such as those shown in the accompanying figure, that highlight the average performance of data series and the dispersion around this average. (They were introduced in 1924 by Walter A. Shewhart of the Bell Telephone Laboratories.) The average performance of the past is viewed as a standard with which to compare current performance, and the typical dispersion of the past is used to set allowable limits for performance now. When plotting new data, it becomes immediately obvious whether they fall within the range of the expected or whether there are

undesirable trends that need the immediate attention of the manager.

Note how all the data series in the figure, except panels (c) and (e), are proceeding within normal limits of, say, ± 2 standard deviations from the average observed over some past period. Managers would waste their time if they worried about the recent decrease observed in series (a) or the recent increase in series (f). Both these changes are still within the upper and lower control limits of $\mu \pm 2\sigma$. Managers would be well advised, however, to pay special attention to series (c) and (e) that have broken through the limits and are, thus, "out of control." Managers might even consider checking up on series (b) that seems to be exhibiting a lengthy undesirable trend.

Note: For our purposes it is unimportant how the original data of the various series are measured—

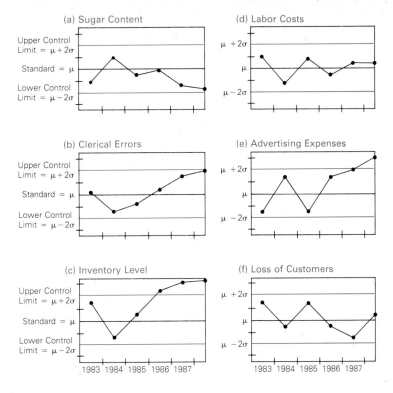

(a) Sugar Content

Upper Control Limit = $\mu + 2\sigma$

Standard = μ

Lower Control Limit = $\mu - 2\sigma$

(b) Clerical Errors

Upper Control Limit = $\mu + 2\sigma$

Standard = μ

Lower Control Limit = $\mu - 2\sigma$

(c) Inventory Level

Upper Control Limit = $\mu + 2\sigma$

Standard = μ

Lower Control Limit = $\mu - 2\sigma$

1983 1984 1985 1986 1987

(d) Labor Costs

$\mu + 2\sigma$

μ

$\mu - 2\sigma$

(e) Advertising Expenses

$\mu + 2\sigma$

μ

$\mu - 2\sigma$

(f) Loss of Customers

$\mu + 2\sigma$

μ

$\mu - 2\sigma$

1983 1984 1985 1986 1987

whether, for example, sugar content is measured in milligrams per bottle or milligrams per cubic inch and whether advertising expenses are expressed in dollars per million bottles sold or dollars per year. Important

only is the fact that the data of each series, however originally measured, can be converted into "standard scores" defined by their own means and standard deviations.

BIOGRAPHY

3.1 Pafnuty Chebyshev

Pafnuty Lvovich Chebyshev (1821–1894) was born in Okatovo, Russia. His parents, who belonged to the gentry, had him privately tutored; he quickly became fascinated by mathematics and eventually studied mathematics and physics at Moscow University. Even as a student, he won a silver medal for a now famous paper on calculating the roots of equations. It was only the first of many brilliant papers that he wrote while teaching mathematics at St. Petersburg University and pursuing a keen interest in mechanical engineering. (Among

other things, he contributed significantly to ballistics, which gave rise to various innovations in artillery, and he invented a calculating machine.) Always, he stressed the unity of theory and practice, saying[1]:

Mathematical sciences have attracted especial attention since the greatest antiquity; they are attracting still more interest at present because of their influence on industry and arts. The agreement of theory and practice brings most

[1]*Dictionary of Scientific Biography,* vol. 3 (New York: Charles Scribner's, 1971), p. 226.

beneficial results; and it is not exclusively the practical side that gains; the sciences are advancing under its influence as it discovers new objects of study for them, new aspects to exploit in subjects long familiar.

Chebyshev typically worked toward the effective solution of problems by establishing *algorithms* (methods of computation) that gave either an exact numerical answer or an approximation correct within precisely defined limits. A most important example of this approach in the field of statistics is his formulation of what is now called **Chebyshev's theorem.** It states that "regardless of the shape of a population's frequency distribution, the proportion of observations falling within k standard deviations of the mean is at least

$$1 - \frac{1}{k^2}$$

given that k equals 1 or more." Thus, we can make the predictions listed in the accompanying table.

Notice how this theorem is vindicated by the profit data of Table 2.2 on page 39. The mean profit equals $786.62 million; the associated standard deviation equals $839.29 million. Can at least 75 percent of the 100 profit figures be found within $\mu \pm 2\sigma$—that is,

within the range of $-\$891.96$ million and $+\$2,465.2$ million? Can at least 89 percent be found within the range of $-\$1,731.25$ million and $+\$3,304.49$ million? Can at least 94 percent be found within the range of $-\$2,570.54$ million and $+\$4,143.78$ million? As a look at Table 2.2 attests, the answers are yes in all cases. In fact, the answers provided by Chebyshev's theorem are most conservative; the actual percentages of observations found within these ranges are 96, 98, and 99, respectively, in this example.

For practical purposes, many a frequency distribution that is only slightly skewed can be treated like a perfectly symmetrical one, and the higher percentages applicable to a normal curve (and shown in Analytical Example 3.1, "Standard Scores") can be applied to estimate the proportions of observations falling within specified distances from the mean.

Chebyshev's theorem, however, demonstrates well a radical change brought about by him: He was the first mathematician to insist on absolute accuracy in limit theorems. In the words of A. N. Kolmogorov, another eminent Russian mathematician, "he always aspired to estimate exactly in the form of inequalities absolutely valid under any number of tests the possible deviations from limit regularities."[2]

---

[2]Ibid., p. 231.

If the number of standard deviations, k, equals	then the proportion of all observations lying within the range of $\mu \pm k\sigma$ equals at least
1	$1 - (1/1^2) = 0$
2	$1 - (1/2^2) = 0.75$
3	$1 - (1/3^2) = 0.89$
4	$1 - (1/4^2) = 0.94$

Summary

1. This chapter discusses an approach to condensing collections of data that is even more radical than the construction of tables and graphs: the calculation of arithmetic summary measures. The most important summary measures are the measures of central tendency, of dispersion, and of shape for quantitative variables and the proportion for qualitative ones.

2. Measures of central tendency or location are values around which observations tend to cluster and that describe the location of the "center" of a data set. They can be calculated with precision from ungrouped data and can be approximated from grouped data. The most commonly used measures of central tendency are the mean, the median, and the mode; the mean comes in several versions, including the unweighted mean and weighted mean (among others).

3. Measures of dispersion or variability are numbers that indicate the spread or scatter of observations; they show the extent to which individual values in a data set differ from one another and, hence, differ from their central location. Dispersion can be measured as distances between selected observations or as average deviations of individual observations from a central value. The most important distance measures of dispersion include the (overall) range and a variety of interfractile ranges, usu-

ally defined by quartiles, deciles, or percentiles. The most important average deviation measures of dispersion include the mean absolute deviation, the variance, and the standard deviation. There are a number of important uses for the standard deviation (and its derivative coefficient of variation).

4. The only summary measure available for qualitative data is the proportion, a number that describes the frequency of observations in a particular category as a fraction of all observations made. Even this humble measure can lead to important insights that are hidden in raw data.

Key Terms

arithmetic mean (μ or $\overline{X}$)
bimodal frequency distribution
Chebyshev's theorem
coefficient of variation (V or v)
control charts
deciles
first quartile
fractile
interfractile ranges
interquartile range
mean absolute deviation *(MAD)*
mean of squared deviations
measures of central tendency (or location)
measures of dispersion (or variability)
measures of shape
median (**M** or m)

median class
modal class
mode *(Mo* or *mo)*
multimodal frequency distribution
parameter
percentile
proportion (π or P)
quartiles
range
second quartile
standard deviation (σ or s)
standard scores *(z)*
statistic
third quartile
variance (σ^2 or s^2)
weighted mean (μ_w or $\overline{X}_w$)

Questions and Problems

The computer programs at the end of this chapter can be used to work many of the subsequent problems.

1. Consider the accompanying sample data on the 1981 research-and-development spending of fifteen U.S. firms.
   a. Calculate the mean, median, and mode for each of the listings in Table 3.A on page 106.
   b. Would your answers change, if these were population data? If so, how?
2. Consider columns 5 and 8 of Table 1.A that accompanies problem 1 of Chapter 1. For each data set, calculate the sample mean, median, and mode.

3. Consider Table 2.A that accompanies problem 1 of Chapter 2. Calculate the sample mean, median, and mode.
4. Consider Table 2.B that accompanies problem 2 of Chapter 2. Calculate the sample mean, median, and mode.
5. Consider Table 2.C that accompanies problem 3 of Chapter 2.
   a. Calculate the population mean, median, and mode.

Table 3.A   Fifteen U.S. Companies: Research-and-Development Spending in 1981

In Total Dollars (millions) (1)		In Percent of Sales (2)		In Dollars per Employee (3)	
1. General Motors..	$2,250	1. Telesciences......... 22.1		1. Cray Research........	$15,060
2. Ford Motor......	1,718	2. Kulicke & Soffa...... 18.9		2. Amdahl..............	14,851
3. AT&T...........	1,686	3. Computer Consoles.. 17.8		3. Auto-Trol	
4. IBM.............	1,612	4. Auto-Trol		Technology...........	14,760
5. Boeing.........	844	Technology.......... 17.2		4. Telesciences.........	11,130
6. General Electric .	814	5. Amdahl.............. 17.0		5. Computer Consoles...	10,677
7. United		6. Cray Research........ 16.0		6. Applied Materials.....	9,722
Technologies	736	7. Floating Point		7. Intergraph...........	9,393
8. Du Pont.........	631	Systems.............. 15.3		8. Onyx & IMI..........	9,039
9. Exxon...........	630	8. Dysan 15.2		9. Apple Computer......	8,532
10. Eastman Kodak ..	615	9. Intel................. 14.8		10. Merck................	8,462
11. Xerox...........	526	10. Applied Materials..... 14.4		11. Floating Point	
12. ITT.............	503	11. Cordis.............. 14.0		Systems	8,418
13. Dow Chemical...	404	12. Intergraph.......... 13.1		12. Boeing...............	8,357
14. Honeywell.......	369	13. Teradyne 12.7		13. International Flavors &	
15. Hewlett-Packard..	347	14. Genrad............. 12.6		Fragrances	8,297
		15. Anderson Jacobson... 11.8		14. Cado Systems.........	8,226
				15. Eli Lilly..............	8,210

Source: Business Week, July 5, 1982, p. 54.

b. Why does your answer for the mean differ from that given in the table (that lists an average of $23,546)?

6. Consider the absolute frequency distribution found in the answer to problem 11 of Chapter 2 at the back of the book (Table A.6). Calculate the sample mean, median, and mode. Show your calculations.

7. Consider the absolute frequency distribution found in the answer to problem 13 of Chapter 2 at the back of the book (Table A.7). Calculate the population mean, median, and mode.

8. Consider the absolute frequency distribution found in Table 3.B. Calculate the sample mean, median, and mode.

9. Consider the absolute frequency distribution found in Table 3.C. Calculate the sample mean, median, and mode.

10. Consider the absolute frequency distribution found in Table 3.D. Calculate the population mean, median, and mode.

Table 3.B   Sample of Orders Received By Aircraft-Parts Supplier

Price Range	Absolute Class Frequency
$ 0 to under $50	200
$ 50 to under $100	700
$ 100 to under $200	1,953
$ 200 to under $500	2,311
$ 500 to under $1,000	582
$1,000 to under $10,000	320

Table 3.C   Sample of Household Monthly Telephone Bills

Size of Bill	Absolute Class Frequency
$ 0 to under $10	11
$ 10 to under $30	67
$ 30 to under $50	129
$ 50 to under $75	732
$ 75 to under $150	82
$150 to under $500	11
$500 to under $1,000	3

Table 3.D Age Distribution of an Aircraft Fleet

Age Class (months)	Absolute Class Frequency
11 to under 24	513
24 to under 48	1,607
48 to under 80	192
80 to under 150	33
150 to under 240	12
240 to under 480	2

11. Reconsider problem 7. Determine the mode graphically by drawing an appropriate histogram.

12. Reconsider problem 9. If you wanted to determine the mode graphically, what kind of calculations would you have to perform prior to drawing the appropriate histogram?

13. Invent a set of 10 figures on family incomes that are *simultaneously* consistent with each of the following statements: (i) On the average, the family income in my neighborhood equals $35,000 per year. (ii) On the average, the family income in my neighborhood equals $10,000 per year. (iii) On the average, the family income in my neighborhood equals $2,000 per year.

14. Invent a set of 10 figures on the tar content of cigarettes that are *simultaneously* consistent with each of the following statements: (i) Brand A cigarettes contain 40 percent less tar than the arithmetic average of all the other brands. (ii) Eighty percent of all brands of cigarettes contain less tar than Brand A.

15. In each case determine the measure of central tendency that is most appropriate: (i) The manager of a clothing store wants to order those sizes of different types of clothing that are most likely to promote low markup, high volume. (ii) An economist wants to compare the average family income in the 50 states. (iii) A manager wants to distribute a profit-sharing fund equally among all the workers. (iv) The personnel director of a firm argues that there are as many people with above-average intelligence as with below-average intelligence. (v) The personnel director's assistant argues that it is possible, in the presence of a single genius (or moron) for most people to have below-average (or above-average) intelligence. (vi) A reporter wants to know the price most commonly paid for houses. (vii) At a New York conference of 20 scientists (of whom 18 are from New York and one each is from Sweden and Japan), someone wants to determine the average distance of the conference site from the participants' homes. (viii) A manager who wants to calculate the total of accounts receivable asks for the number of accounts and the average balance.

16. Consider Table 1.A that accompanies problem 1 of Chapter 1. Calculate a weighted mean of the column 8 data, using the column 5 data as weights. Why does your answer differ from the answer in problem 2 of these questions and problems?

17. The U.S. labor-force and unemployment data listed in Table 3.E were reported for December, 1981. Calculate the weighted average unemployment rate for the entire U.S. labor force; compare it with an unweighted arithmetic average. Show your calculations.

18. In 1981, the five largest industrial corporations outside the United States reported the data listed in Table 3.F on the following page. Calculate the weighted average profit rate; compare it with the unweighted average. Show your calculations.

Table 3.E

	Labor Force (millions)	Unemployment Rate (percent)
Both sexes, 16–19 years	8.62	21.5
Males, 20 years and over	57.51	7.9
Females, 20 years and over	42.91	7.4

Source: Economic Report of the President (Washington, D.C.: U.S. Government Printing Office, 1982), pp. 268–69.

Table 3.F

	Stockholders' Equity (billions of dollars)	Profit as Percent of Equity (percent)
Royal Dutch Shell	25.2	14.4
British Petroleum	14.8	13.9
ENI	2.2	17.5
Unilever	4.8	16.5
Française des Pétroles	3.7	4.7

Source: Fortune, August 23, 1982, p. 183.

19. From the data in Table 3.G, calculate the weighted mean rate of return from public electric utilities in Canada between 1967 and 1981. Compare your answer with an unweighted mean.

Table 3.G

Year	Rate of Return (percent)	Total Invested Capital ($ millions)
1967	3.66	12,081.67
1968	3.71	13,025.41
1969	3.84	14,390.25
1970	3.74	16,406.42
1971	3.62	18,266.16
1972	3.51	20,057.25
1973	3.68	23,027.77
1974	3.19	30,277.61
1975	2.32	39,361.00
1976	3.00	46,523.68
1977	3.01	52,066.63
1978	3.94	62,785.06
1979	3.95	74,715.24
1980	3.69	85,349.86
1981	3.79	99,643.62

Source: Glenn P. Jenkins, "Public Utility Finance and Economic Waste," Canadian Journal of Economics, August 1985, p. 493.

20. From the data in Table 3.H on the opposite page, calculate the weighted mean unemployment rate

of the nine listed countries in 1982. Compare your answer with an unweighted mean.

21. Consider the answer at the back of this book for Chapter 2's problem 1. Then determine
   a. the range of the data.
   b. the first and third quartiles.
   c. the second and seventh deciles.
   Finally, determine *graphically*
   d. the median and the 87th percentile. [For this purpose, convert the column 3 relative frequency distribution of Chapter 2's answer to problem 11 into a *cumulative* relative frequency distribution and plot an appropriate ogive.]

22. Consider columns 5 and 8 of Table 1.A that accompanies problem 1 of Chapter 1. For each data set, calculate
   a. the range.
   b. the interquartile range.

23. Consider Table 3.A. For each of the three data sets, calculate
   a. the range.
   b. the interquartile range.

24. Calculate (a) the range and (b) the interquartile range for the data sets in Chapter 2's Tables 2.B and 2.C.

25. Calculate (a) the range and (b) the interquartile range for the percentage data sets in Tables 3.F, 3.G, and 3.H.

26. Consider the first and third of the three sets of sample data in Table 3.A and determine in each case
   a. the mean absolute deviation.
   b. the variance.
   c. the standard deviation.

27. Consider the second of the three sets of figures in Table 3.A. Treat it as a sample and determine

Table 3.H

Country	Percent Unemployed	Civilian Labor Force (millions)
Australia	7.1	6.9
Canada	11.0	11.9
France	8.5	22.9
Great Britain	12.3	25.7
Italy	4.8	21.4
Japan	2.4	57.0
Sweden	3.1	4.4
United States	9.7	110.2
West Germany	5.9	26.6

Source: Statistical Abstract of the United States 1984, p. 872.

   a. the mean absolute deviation.
   b. the variance (long and short method).
   c. the standard deviation.
   In each case, show your calculations.
28. Consider columns 5 and 8 of Table 1.A that accompanies problem 1 of Chapter 1. For each of these sets of sample data, determine
   a. the mean absolute deviation.
   b. the variance.
   c. the standard deviation.
29. Consider the data sets in Chapter 2's Tables 2.A, 2.B, and 2.C. In each case, determine
   a. the mean absolute deviation.
   b. the variance.
   c. the standard deviation.
30. Consider the percentage data sets in Tables 3.F, 3.G, and 3.H. In each case, determine
   a. the mean absolute deviation.
   b. the variance.
   c. the standard deviation.
31. Consider the answer at the back of the book to Chapter 2's problem 11 (Table A.6). From the absolute frequency distribution shown there, estimate the variance and standard deviation, treating the data as sample data and using the shortcut method. Show your calculations.
32. Consider the answer at the back of the book to Chapter 2's problem 13 (Table A.7). From the absolute frequency distribution shown there, estimate
   a. the mean absolute deviation.
   b. the population variance.
   c. the population standard deviation.
   Would any of your answers change if these were sample data? If so, how?

33. Consider the sample data in Table 3.B. Use them to estimate
   a. the mean absolute deviation.
   b. the variance.
   c. the standard deviation.
   Would any of your answers change if these were population data? If so, how?
34. Consider the sample data in Table 3.C. Use them to estimate
   a. the mean absolute deviation.
   b. the variance.
   c. the standard deviation.
   Would any of your answers change if these were population data? If so, how?
35. Consider the population data in Table 3.D. Use them to estimate
   a. the mean absolute deviation.
   b. the variance.
   c. the standard deviation.
   Would any of your answers change if these were sample data? If so, how?
36. A computer firm offers a paid leave of absence to its engineers who wish to get an advanced degree. However, applicants are tested and their aptitude test scores must indicate a superb chance of success (as defined by z-scores of $+2$ or better). In one test, applicants A through F earned raw scores of 500, 631, 760, 438, 598, and 720. The mean score of all 200 applicants was 520, the standard deviation was 60. Who among the six will go back to school?
37. The quality controller of a vegetable-canning firm is constructing a control chart. The production process fills cans of peas with an average of $\mu =$

16.02763 ounces, the standard deviation being σ = .3168132 ounces. Any can filled with $\mu \pm 2\sigma$ is considered acceptable. A sample reveals net weights of 15.33, 15.53, 15.77, 15.94, 16.05, 16.24, 16.43, 16.63, 16.88. Is the production process "out of control"?

38. Make a list of 10 numbers such that each number is a 1 or a 9, while the standard deviation of the set is as small (or as large) as possible.

39. The manager of a firm announced: "The light bulbs produced by our firm have a mean lifetime of 1,500 hours, with a standard deviation of 150 hours; thus, it follows that almost all of our light bulbs last between 1,200 and 1,800 hours." What was the manager assuming? What if the assumption did not hold?

40. Consider Table 2.K in Chapter 2's problem 35. In which of the two years did the sample of retail businesses show the greater variability of sales?

41. Consider again the three alternative listings of the top R&D spenders given in Table 3.A. By which of

the three criteria does R&D spending show the greatest (or smallest) variability? Show your calculations.

42. Consider Analytical Example 2.1, "Deciphering Secret Codes." Determine the proportions of vowels and consonants in 200 letters of a normal English language text.

43. Consider Table 1.A that accompanies Chapter 1's problem 1. What proportions of the sampled mutual funds pursued a portfolio strategy of "aggressive growth" or "small company growth"?

44. Consider the frequency curve in Figure A. What is the *minimum* proportion of observations falling in the crosshatched area? In the dotted area?

45. Reconsider the ordered array of Table A.2 that constitutes the answer to Chapter 2's problem 1, along with your answers to problems 3 and 29 of the current chapter. What percentage of the data fall within ± 2.5 standard deviations of the mean? Is this number consistent with Chebyshev's theorem?

Figure A

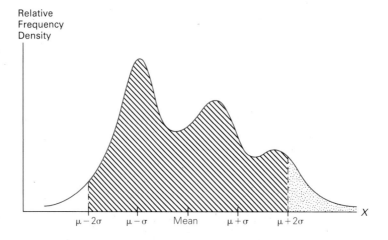

Selected Readings

Campbell, Stephen K. *Flaws and Fallacies in Statistical Thinking*. Englewood Cliffs, N.J.: Prentice-Hall. 1974. An excellent and amusing discussion aimed at helping nonstatisticians recognize (intentional and unintentional) abuses of statistical tools and gain the ability to judge the quality of statistical evidence.

Kohler, Heinz. *Statistics for Business and Economics*, 2nd ed. Glenview, Ill.: Scott, Foresman and Co., 1988. Chapter 4 contains a detailed discussion of summary measures of shape (skewness and kurtosis).

Computer Programs

The DAYSAL-2 personal-computer diskettes that accompany this book contain three programs of interest to this chapter:

3. Frequency Distributions from Raw Data calculates all major summary measures as a by-product of setting up frequency distributions.

4. Summary Measures from Grouped Data takes an absolute frequency distribution that you supply and computes all major summary measures, along with other types of frequency distributions.

5. Summary Measures from Raw Data calculates all major summary measures of central tendency, dispersion, and shape.

PART TWO

Analytical Statistics:
The Foundations of Inference

4   The Theory of Probability
5   Probability Distributions: Discrete Variables
6   Probability Distributions: Continuous Variables

CHAPTER FOUR

The Theory of Probability

Every day, each of us makes decisions the outcomes of which lie in the realm of the unknown, but we do not make them blindly. Figuring the *chances* of various possible outcomes helps us make the right choices. Without even knowing it, we rely on a special kind of calculus—"the calculus of the likelihood of specific occurrences"— which is what the **theory of probability** has been called. Consider how we cancel the planned picnic or move the graduation indoors when the chance of rain seems too high; how we proceed with the wedding or introduce the new product when success seems more likely than failure. Wherever we turn, we are weighing the chances. How likely is it, we ask, that the price hike will reduce sales, that the new process will raise productivity, that the merger will raise profit, that the inspection will turn up defective parts, that the tax cut will put an end to the recession? What are the chances, we ask, for life on other planets, for nuclear war in this century, for our candidate's election, for this horse to win, for this item to be on the quiz? The list goes on.

As we noted in Chapter 1, statistics as a discipline can be viewed as a guide to the unknown. Is it any wonder that all of analytical statistics is built upon the theory of probability? As the Latin origin of the word suggests (*probabilis* means "likely" or "like truth"), the theory helps us find the likely truth when we cannot know it with certainty. But we must be patient. The content of the current chapter may seem to be remote from business and economics applications; yet it will turn out to be the very foundation for the development, in later chapters, of a host of statistical techniques that are routinely and successfully used to solve problems arising in these very fields.

BASIC CONCEPTS

Probability theorists employ a number of important concepts; crucial among them are (1) the *random experiment,* (2) the *sample space,* and (3) *random events.* We discuss each of these concepts in turn.

The Random Experiment

Any activity that will result in one and only one of several well-defined outcomes but that does not allow us to tell in advance which of these will prevail in any particular instance is called a **random experiment.** The tossing of a coin, for example, is such an experiment. Assuming that tosses with such unlikely outcomes as the coin landing on its edge or disappearing in a hole do not count, the *type* of outcome is known with certainty: it must be heads or tails. Yet the *actual* outcome associated with a given toss is bound to vary from one trial to the next and, therefore, cannot be predicted with certainty. The rolling of a die is another such random experiment; so is the drawing of a card from a deck of cards or the spinning of the wheel in the game of roulette. These examples come most easily to mind because probability was first studied some 300 years ago in connection with gambling. However, the concept of probability is just as applicable to matters outside the gambling situation. Pulling selected items from the production line for inspection or selecting invoices for an audit represent other examples of random experiments (with such possible alternative outcomes as defective or satisfactory in the first case and correct or erroneous in the other). Even investing a billion dollars in a new type of car can be viewed as a random experiment (leading to profit, loss, or even bankruptcy), and so can the placing of a coin into a vending machine (leading to the desired coffee, to money returned, to an infuriating neither, or even to a quite unwelcome chicken soup).

The Sample Space

Any one of a random experiment's possible outcomes, the occurrence of which rules out the occurrence of all the alternative outcomes, is called a **basic outcome.** In turn, a listing of all the basic outcomes of a random experiment is generally referred to as the **sample space,** although other terms, such as **outcome space** and **probability space,** are occasionally used as well. The basic outcomes that constitute the sample space can be represented symbolically, pictorially, and in other ways still; let us consider some of them.

Tossing a single coin once. The single toss of a single coin involves a sample space *(S)* of two basic outcomes only, heads (H) and tails (T). This sample space might be represented symbolically as

$$S = \{H, T\}.$$

This sample space can also be shown pictorially, as in panel (a) of Figure 4.1 on page 116. In any single toss, one (and only one) of the two outcomes shown must occur.

Rolling a single die once. The single roll of a single die involves a slightly larger sample space because six basic outcomes exist. Symbolically,

$$S = \{1, 2, 3, 4, 5, 6\}.$$

This sample space is shown pictorially in panel (b) of Figure 4.1.

Drawing a single card once. Larger still is the sample space describing the possible consequences of randomly selecting a single card from a standard deck of 52 playing cards. Such a deck has four colored *suits*—namely, black clubs, red diamonds, red

Figure 4.1   Sample Spaces for Simple Experiments

The sample spaces of simple random experiments can be represented pictorially.

(a) Tossing a Single Coin Once

(b) Rolling a Single Die Once

(c) Drawing a Single Card Once

	Clubs	Diamonds	Hearts	Spades
King	♣ K	♦ K	♥ K	♠ K
Queen	♣ Q	♦ Q	♥ Q	♠ Q
Jack	♣ J	♦ J	♥ J	♠ J
10	♣ 10	♦ 10	♥ 10	♠ 10
9	♣ 9	♦ 9	♥ 9	♠ 9
8	♣ 8	♦ 8	♥ 8	♠ 8
7	♣ 7	♦ 7	♥ 7	♠ 7
6	♣ 6	♦ 6	♥ 6	♠ 6
5	♣ 5	♦ 5	♥ 5	♠ 5
4	♣ 4	♦ 4	♥ 4	♠ 4
3	♣ 3	♦ 3	♥ 3	♠ 3
Deuce	♣ 2	♦ 2	♥ 2	♠ 2
Ace	♣ A	♦ A	♥ A	♠ A

(d) Spinning a Roulette Wheel Once

---

hearts, and black spades; each suit has 13 *denominations,* from ace to 10 plus three face cards (jack, queen, and king). Symbolically,

$$S = \{\text{ace of clubs, deuce of clubs, 3 of clubs, . . ., king of spades}\}.$$

Pictorially, we can represent the same set of alternative outcomes as in panel (c) of Figure 4.1.

Spinning a roulette wheel once. Also large is the set of basic outcomes from spinning a roulette wheel. The Monte Carlo wheel that is shown in panel (d) of Figure 4.1 has 37 numbers, ranging from 0 to 36. Apart from the zero, half of them are black, half red. Symbolically,

$$S = \{\text{red 0, red 1, black 2, . . ., red 36}\}.$$

Note: The Monte Carlo wheel shown here should not be confused with the American wheel that adds a double zero or even a triple zero to the numbers shown (a version of which appears in problem 1 at the end of this chapter).

Tossing a single coin twice. More complicated experiments, such as tossing a coin twice or tossing two coins at a time, can be depicted similarly, although various other ways are often preferred. Among the favorite devices for defining the sample space of random experiments involving multiple steps is the **tree diagram,** so called because it reminds one of the branches of a tree. Consider panel (a) of Figure 4.2. The tree diagram shown there pictures the two possible outcomes of the first toss of a coin: heads (H_1) or tails (T_1), where the subscripts denote the number of the toss. For either contingency, two further possibilities exist on the second toss: H_2 or T_2. Thus, one can

Figure 4.2 Sample Spaces for Multiple-Step Experiments

Sample spaces for multiple-step experiments are typically represented by the devices shown here.

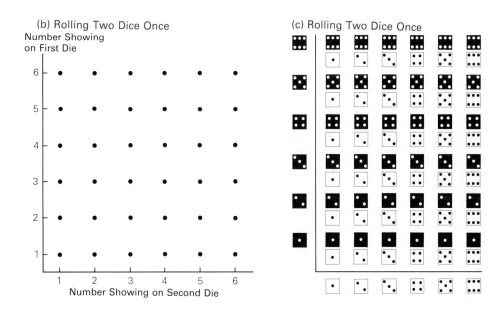

follow the various paths along the branches from the starting position prior to the first toss and trace out the four basic outcomes of the entire two-stage experiment, and these outcomes are conveniently summarized in the box to the right of the tree diagram. Note how the identical diagram could be used to identify the possible consequences of tossing two coins at once. One would only have to replace "First Toss" by "First Coin" and "Second Toss" by "Second Coin" to see the perfect analogy. Naturally, more complex trees can be drawn, too, in order to find, say, the sample space associated with tossing a coin three times or even with tossing it more often than that.

Rolling two dice once. Finally, consider panels (b) and (c) of Figure 4.2. They depict other ways still of determining the sample space in multiple-step experiments. In this case, both panels tell us immediately that there are 36 basic outcomes of an experiment that rolls two dice at once (or, for that matter, that rolls a single die twice).

Random Events and Their Relationships

We have now seen that any random experiment will yield one of two or more possible outcomes and that these basic outcomes, as a group, make up the sample space. Any subset of this sample space is called a **random event,** of which there are two types.

Simple vs. composite events. Any single one of a random experiment's basic outcomes is called a **simple event.** Thus, the sample space for experiment (a) in Figure 4.1 consists of two simple events: the appearance of a head and the appearance of a tail. The sample spaces for experiments (b) through (d) are, similarly, made up of 6, 52, and 37 simple events, respectively.

Finally, in Figure 4.2, we find four simple events in panel (a) and 36 simple events each in panels (b) and (c), although each of these events (unlike those in Figure 4.1) inextricably links together two elements, such as "head on the first toss" with "head on the second toss" or "6 on the black die" with "1 on the white die." Some people distinguish basic outcomes or simple events that are *univariate* (such as those in Figure 4.1 where each outcome is defined by the appearance of a single element) from those that are *bivariate* (such as those in Figure 4.2 where each outcome is defined by the joint appearance of two elements), or, perhaps, even *multivariate* (not shown). On occasion, basic outcomes or simple events are also referred to as **elementary events,** or **sample points**—the latter name, perhaps, being suggested by the type of graphical representation of the sample space shown in panel (b) of Figure 4.2.

It is often useful, however, to consider subsets of the sample space that contain more than one basic outcome or simple event. Any such combination of two or more basic outcomes is called a **composite event.** Consider the event "getting an even number" when rolling a single die once. Consider the event "getting a face card" when drawing a single card from a deck. Consider, finally, "getting a sum of 10 or more" when rolling two dice. Such composite events are pictured by the shaded areas in Figure 4.3.

Mutually exclusive events. Different random events that have no basic outcomes in common are called **mutually exclusive,** or **disjoint,** or **incompatible events.** Such events cannot occur at the same time; the occurrence of one automatically precludes the occurrence of the other. As an example, consider "getting an even number" and "getting an odd number" when rolling a single die once. If one of these composite

Figure 4.3 Composite Events

The shaded subsets of the three boxed sample spaces shown here illustrate the nature of composite events. As the shaded portion of panel (a) shows, "getting an even number" when rolling a single die once is a composite event that consists of three basic outcomes or simple events (getting a 2, a 4, or a 6). The shaded portion of panel (b) shows, similarly, that "getting a face card" when drawing a single card from a deck is a composite event that consists of 12 basic outcomes or simple events (getting a jack, a queen, or a king from any one of the four suits). Finally, panel (c) tells us that "getting a sum of 10 or more" on a single roll of two dice is a composite event that consists of six simple (but bivariate) events; namely, those depicted by the six combinations of two faces of dice in the shaded area.

(a) Getting an Even Number

(b) Getting a Face Card

	Clubs	Diamonds	Hearts	Spades
King	♣ K	♦ K	♥ K	♠ K
Queen	♣ Q	♦ Q	♥ Q	♠ Q
Jack	♣ J	♦ J	♥ J	♠ J
10	♣ 10	♦ 10	♥ 10	♠ 10
9	♣ 9	♦ 9	♥ 9	♠ 9
8	♣ 8	♦ 8	♥ 8	♠ 8
7	♣ 7	♦ 7	♥ 7	♠ 7
6	♣ 6	♦ 6	♥ 6	♠ 6
5	♣ 5	♦ 5	♥ 5	♠ 5
4	♣ 4	♦ 4	♥ 4	♠ 4
3	♣ 3	♦ 3	♥ 3	♠ 3
Deuce	♣ 2	♦ 2	♥ 2	♠ 2
Ace	♣ A	♦ A	♥ A	♠ A

(c) Getting a Sum of 10 or More

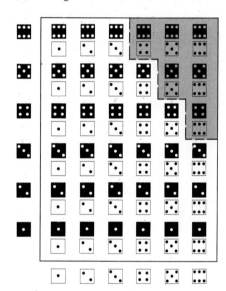

events occurs, the other one cannot possibly occur. The event in question need not, however, be a composite one. The same random experiment can yield *simple* events that are mutually exclusive as well, such as "getting a 6" (when rolling a single die) and "getting a 4." Equally incompatible is "getting the king of hearts" and "getting the ace of hearts" when drawing a single card from a deck or "getting a defective part" and "getting a perfect part" when pulling an item off the production line. In all these cases, it is one *or* the other.

As another look at Figure 4.3 can confirm, mutually exclusive events always occupy entirely different parts of the sample space. Note how, in panel (a), the even numbers are found in some areas (the shaded ones) and the odd ones are found in other areas (the unshaded ones). Equally distinct parts of the sample space are occupied by the six face and the four face in panel (a), or by the king of hearts and the ace of hearts in panel (b).

Collectively exhaustive events. Different random events that jointly contain all the basic outcomes in the sample space are called **collectively exhaustive events.** When the appropriate random experiment is conducted, one of these events is bound to occur. Again consider "getting an even number" and "getting an odd number" when rolling a single die once. These two events, as we saw, are mutually exclusive, but they are collectively exhaustive as well. Although they cannot occur at the same time, one of them is bound to occur. This characteristic is shown visually in panel (a) of Figure 4.3 by the fact that the shaded areas (even numbers) and the unshaded areas (odd numbers) jointly cover the entire sample space represented by the box.

But note: Not every set of mutually exclusive events is collectively exhaustive. In the same experiment, the events "getting a six" and "getting a four" are mutually exclusive, but they hardly exhaust the sample space. On the other hand, not every set of collectively exhaustive events need consist of mutually exclusive ones, either. Consider the three events of "getting a six," "getting an even number," and "getting an odd number." In our experiment, these events are surely collectively exhaustive; they cover the entire sample space in panel (a). Yet these events are not mutually exclusive because "getting a six" is a simple event that is already included in the composite event of "getting an even number."

Complementary events. There exists one relationship, however, that always describes events that are both mutually exclusive and collectively exhaustive at the same time. Two random events such that precisely all those basic outcomes that are not contained in one are contained in the other are called **complementary events.** Once more, Figure 4.3 provides perfect examples if we contrast, in each panel, the event pictured by the shaded area with that described by the unshaded one. Thus, in panel (a), one event is "getting an even number"; its complement of "not getting an even number" is, of course, equivalent to "getting an odd number." Together, these two mutually exclusive events exhaust the sample space. Similarly, in panel (b), one event is "getting a face card" (shaded area); its complement is "not getting a face card" (unshaded area). Finally, in panel (c), one event is "getting a sum of 10 or more"; its complement is "not getting a sum of 10 or more" or "getting a sum of 9 or less." Again, the two mutually exclusive events precisely exhaust the sample space.

Compatible events. Just as there are incompatible events that have no basic outcomes in common, there are **compatible events** that have at least some basic outcomes in common. Such events can occur at the same time; the occurrence of one does not rule out the occurrence of the other. The two events, "getting a face card" and "getting a heart," provide an example. This situation is depicted in Figure 4.4. The shaded area in panel (a) depicts the 12 simple events associated with the composite event of "getting a face card"; the shaded area in panel (b) pictures the 13 simple events described by the composite event "getting a heart."

Sometimes one is interested in the occurrence of *either* of two such composite events; for this purpose one defines the **union of two events** as all the basic outcomes contained in either one random event *or* another or, conceivably, in both. Such a union of two events is illustrated by the shaded area in panel (c).

On other occasions, one might be interested only in those occurrences that two composite events have in common—in our example, all the possibilities of getting a

Figure 4.4   Union and Intersection

Two composite events, such as those pictured by the shaded areas in panels (a) and (b), can be considered together either by focusing on their union—panel (c)—or on their intersection—panel (d). The union contains all the simple events found in *either one* of the composite events—found, that is, in one of the composite events only *or* in the other only *or* in both at the same time. The key word describing such a situation is *or*. In contrast, the intersection contains only the simple events found in *both* of the composite events—found, that is, in one of the composite events *and* also in the other. The key word for this situation is *and*.

(a) Getting a Face Card

	Clubs	Diamonds	Hearts	Spades
King	♣ K	♦ K	♥ K	♠ K
Queen	♣ Q	♦ Q	♥ Q	♠ Q
Jack	♣ J	♦ J	♥ J	♠ J
10	♣ 10	♦ 10	♥ 10	♠ 10
9	♣ 9	♦ 9	♥ 9	♠ 9
8	♣ 8	♦ 8	♥ 8	♠ 8
7	♣ 7	♦ 7	♥ 7	♠ 7
6	♣ 6	♦ 6	♥ 6	♠ 6
5	♣ 5	♦ 5	♥ 5	♠ 5
4	♣ 4	♦ 4	♥ 4	♠ 4
3	♣ 3	♦ 3	♥ 3	♠ 3
Deuce	♣ 2	♦ 2	♥ 2	♠ 2
Ace	♣ A	♦ A	♥ A	♠ A

(shaded: King, Queen, Jack rows — all suits)

(b) Getting a Heart

	Clubs	Diamonds	Hearts	Spades
King	♣ K	♦ K	♥ K	♠ K
Queen	♣ Q	♦ Q	♥ Q	♠ Q
Jack	♣ J	♦ J	♥ J	♠ J
10	♣ 10	♦ 10	♥ 10	♠ 10
9	♣ 9	♦ 9	♥ 9	♠ 9
8	♣ 8	♦ 8	♥ 8	♠ 8
7	♣ 7	♦ 7	♥ 7	♠ 7
6	♣ 6	♦ 6	♥ 6	♠ 6
5	♣ 5	♦ 5	♥ 5	♠ 5
4	♣ 4	♦ 4	♥ 4	♠ 4
3	♣ 3	♦ 3	♥ 3	♠ 3
Deuce	♣ 2	♦ 2	♥ 2	♠ 2
Ace	♣ A	♦ A	♥ A	♠ A

(shaded: Hearts column — all ranks)

(c) Union: Face Card *or* Heart

	Clubs	Diamonds	Hearts	Spades
King	♣ K	♦ K	♥ K	♠ K
Queen	♣ Q	♦ Q	♥ Q	♠ Q
Jack	♣ J	♦ J	♥ J	♠ J
10	♣ 10	♦ 10	♥ 10	♠ 10
9	♣ 9	♦ 9	♥ 9	♠ 9
8	♣ 8	♦ 8	♥ 8	♠ 8
7	♣ 7	♦ 7	♥ 7	♠ 7
6	♣ 6	♦ 6	♥ 6	♠ 6
5	♣ 5	♦ 5	♥ 5	♠ 5
4	♣ 4	♦ 4	♥ 4	♠ 4
3	♣ 3	♦ 3	♥ 3	♠ 3
Deuce	♣ 2	♦ 2	♥ 2	♠ 2
Ace	♣ A	♦ A	♥ A	♠ A

(shaded: King, Queen, Jack rows — all suits; plus Hearts column — all ranks)

(d) Intersection: Face Card *and* Heart

	Clubs	Diamonds	Hearts	Spades
King	♣ K	♦ K	♥ K	♠ K
Queen	♣ Q	♦ Q	♥ Q	♠ Q
Jack	♣ J	♦ J	♥ J	♠ J
10	♣ 10	♦ 10	♥ 10	♠ 10
9	♣ 9	♦ 9	♥ 9	♠ 9
8	♣ 8	♦ 8	♥ 8	♠ 8
7	♣ 7	♦ 7	♥ 7	♠ 7
6	♣ 6	♦ 6	♥ 6	♠ 6
5	♣ 5	♦ 5	♥ 5	♠ 5
4	♣ 4	♦ 4	♥ 4	♠ 4
3	♣ 3	♦ 3	♥ 3	♠ 3
Deuce	♣ 2	♦ 2	♥ 2	♠ 2
Ace	♣ A	♦ A	♥ A	♠ A

(shaded: Hearts column, King, Queen, Jack rows only)

face card and heart simultaneously. For this purpose, one defines the **intersection of two events** as all the basic outcomes simultaneously contained in both one random event *and* another. Such an intersection is illustrated by the shaded area in panel (d).

Summary. All the foregoing event relationships can be nicely summarized with the help of **Venn diagrams,** which are graphical devices, such as those in Figure 4.5, that depict sample spaces and random events symbolically (and that are named after their originator, John Venn). In each panel, the box represents the sample space containing all the basic outcomes of a random experiment. Particular events are depicted by circles or sections within the sample space. Thus, in panel (a), events A and B are mutually exclusive because, when performing the random experiment of "running the economy for a year," it is impossible to find inflation simultaneously at 0–5 percent and also at 10–15 percent. On the other hand, these two events are not collectively exhaustive; other outcomes, such as inflation above 5 but below 10 percent or inflation above 15 percent, are possible also.

Panel (b), on the other hand, depicts a complete *partition* of the sample space. When performing the random experiment of "running a firm for a year," one of the four noted events is bound to occur.

Panel (c) depicts the sample space for the random experiment of "checking the quality of items produced." Only two events are possible; they are complementary. Such events will be described symbolically as A vs. *not* A, or A vs. $\overline{A}$, in this book; another symbol traditionally used for the complement of A is A'.

Panels (d) and (e) once more relate to the experiment of "running the economy for a year." Note that the two events cited do not exhaust all the possibilities; therefore, they do not cover all of the sample space. On the other hand, the two events may or may not happen simultaneously. To the extent that they do, the two circles overlap. Thus, in panel (d), any outcome that fits *either* event A *or* B (and, perhaps, even both) is found in the crosshatched area, and such a *union* of the two events will be described symbolically as A *or* B in this book. (Other writers also use the designation A ∪ B, where ∪ stands for "union.") Finally, the smaller number of outcomes that fit *both* event A *and* B is found in the smaller crosshatched area of overlap in panel (e), and such an intersection of the two events will be described symbolically as A *and* B throughout this book. (Other writers also use the designation A ∩ B, where ∩ stands for "intersection.")

THE MEANING OF PROBABILITY

There are two types of random experiments, those that can be repeated over and over again under (essentially) identical conditions and those that are unique and cannot be repeated because the conditions surrounding them are forever changing. Correspondingly, there are two types of probability numbers, although both seek to measure the degree of likelihood that a particular event will occur. A numerical measure of chance that estimates the likelihood of a specific occurrence in a repeatable random experiment is called an **objective probability;** a corresponding value for a unique random experiment is called a **subjective probability.** We will discuss each of these in turn.

Figure 4.5 Venn Diagrams

Venn diagrams depict symbolically various relationships among random events.

Examples:

(a) Mutually Exclusive, but Not Collectively Exhaustive Events

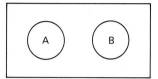

Event A = Inflation between
    0 and 5% per year
Event B = Inflation between
    10 and 15% per year

(b) Mutually Exclusive and Collectively Exhaustive Events

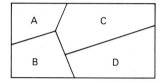

Event A: Loss
Event B: Profit between $0 and
    under $10 million
Event C: Profit between $10 million
    and under $100 million
Event D: Profit of $100 million or more

(c) Complementary Events: A vs. *Not* A

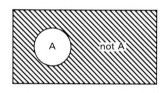

Event A: Defective items
Event not-A: Satisfactory items

(d) Union of Two Events: A *or* B

Event A: Inflation between
    0 and 5% per year
Event B: Unemployment between
    5 and 10% of the
    labor force

(e) Intersection of Two Events: A *and* B

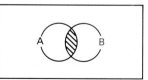

Event A: Inflation between
    0 and 5% per year
Event B: Unemployment between
    5 and 10% of the
    labor force

Objective Probability

An objective probability value can be determined in one of two ways: theoretically, by conducting repeated *thought* experiments, or empirically, by conducting repeated *actual* experiments.

The theoretical approach. The theoretical approach to determining probability is also called the *classical* approach because it was used by the originators of probability theory, such as Pierre Simon de Laplace.[1] This approach relies entirely on abstract reasoning; it does not resort to performing actual experiments because logic is deemed sufficient for providing all the answers. As Laplace saw it, if a hypothetical experiment could have N equally likely basic outcomes, and if n of these outcomes were favorable to event A, then the probability of this event, $p(A)$, equaled the ratio of n to N.

4.A The Probability of Event A: Classical Approach

$$p(A) = \frac{\text{number of equally likely basic outcomes favorable to the occurrence of event A}}{\text{number of equally likely basic outcomes possible}} = \frac{n}{N}$$

The application of this definition inevitably produces a number between 0 and 1, as Figure 4.6 illustrates. But note: while probability will usually be measured in this fashion throughout this book, there are alternative ways of expressing it—for example, in terms of percentages, chances, or odds. Thus, a probability of 7/10 or .7 might be expressed as a probability of 70 percent or as 70 chances in 100 or as odds of 70:30 for or even as odds of 30:70 against.

The classical formula for measuring probability is easily applied to games of chance, such as those pictured in Figures 4.1 to 4.3. Regardless of whether the event is a simple or composite one, the juxtaposition of the number of favorable basic outcomes to that of all possible basic outcomes immediately produces a probability value. Consider Figure 4.1. The probability of getting a head when tossing a fair coin is easily seen as $p(H) = 1/2$; that of getting a 6 when rolling a balanced die is $p(6) = 1/6$; and that of getting the king of hearts when randomly drawing a card from a well-shuffled deck is $p(\text{king of hearts}) = 1/52$, always by reference to the Box 4.A formula. Similarly, the probability of getting a 15 when drawing a card is

$$p(15) = \frac{0}{52} = 0$$

(because there is no such card), while that of getting a black card is

$$p(\text{black}) = \frac{26}{52} = \frac{1}{2}$$

In contrast, the probability of getting black when spinning the Monte Carlo wheel is

$$p(\text{black}) = \frac{18}{37}, \text{ or somewhat less than } \frac{1}{2}.$$

[1] A biographical sketch of Laplace (1749–1847) is contained in Chapter 4 of the *Student Workbook* that accompanies this text.

Figure 4.6   Measuring Probability

Probability as defined in Box 4.A is inevitably measured on a scale from 0 to 1. Because there can be no negative basic outcomes of a random experiment, the lowest possible number of basic outcomes favorable to an event is zero, and such an occurrence would make the probability of that event zero as well. [As a look at Box 4.A will confirm, when $n = 0$ and N is positive, $p(A) = (n/N)$ must be zero.] On the other hand, the highest possible number of basic outcomes favorable to an event equals the total number of basic outcomes that are possible, and such an occurrence would make the probability of that event equal to one. [As Box 4.A confirms, when $n = N$, their ratio $p(A) = (n/N)$ must equal one.]

The same procedure of contrasting favorable basic outcomes with all possible ones is applied just as easily to the slightly more complex games of Figures 4.2 to 4.4.

You should be able to see from Figure 4.2 that getting two heads when tossing a coin twice carries a probability of

$$p(2\text{H}) = \frac{1}{4}$$

that getting two heads *or* two tails in the same game has a probability of

$$p(2\text{H } or \text{ 2T}) = \frac{2}{4} = \frac{1}{2}$$

and that getting two fives when rolling two dice has a probability of

$$p(2 \text{ fives}) = \frac{1}{36}.$$

You should be able to confirm from Figure 4.3 that getting an even number when rolling a die once carries a probability of

$$p(\text{even number}) = \frac{3}{6} = \frac{1}{2}$$

that getting a face card when drawing one card has a probability of

$$p(\text{face card}) = \frac{12}{52} = \frac{3}{13}$$

and that getting a sum of 10 or more when rolling two dice once has a probability of

$$p(\text{sum of 10 or more}) = \frac{6}{36} = \frac{1}{6}.$$

Finally, it should be evident from Figure 4.4 that getting a face card *or* a heart carries a probability of

$$p(\text{face card } or \text{ heart}) = \frac{22}{52} = \frac{11}{26}$$

while getting a face card *and* a heart has a probability of only

$$p(\text{face card } and \text{ heart}) = \frac{3}{52}.$$

How easy! The classical approach allows the determination of probability values without ever *actually* tossing a coin, rolling a die, and the like. Yet there are those who have trouble with this definition of probability: the definition seems circular, for by specifying *equally likely* basic outcomes it employs the notion of probability in order to define probability! In addition, there exist repeatable random experiments (such as tossing a biased coin, rolling a loaded die, or sampling the quality of items coming off the production line) for the various possible outcomes of which no amount of arm-chair reasoning can ever discover probability values. In such cases, numerical assignments require *experience*.

The empirical approach. In order to counter the criticisms of the classical approach, many prefer to think of probability values as derived from experience, the more of it the better. They suggest that the probability of an event be considered equal to the relative frequency with which it has actually been observed in the past over the course of a large number of experiments.

4.B The Probability of Event A: Empirical Approach

$$p(\text{A}) = \frac{\begin{array}{c}\text{number of times event A did occur in the past}\\\text{during a large number of experiments}\end{array}}{\begin{array}{c}\text{maximum number of times event A could have}\\\text{occurred during these experiments}\end{array}} = \frac{k}{M}$$

The probability value is still a ratio between 0 and 1, but it is now based on empirical data rather than on *a priori* reasoning. Thus, the probability of getting a head with a *biased* coin might be established by experiment as $p(\text{H}) = (9/10)$ and that of finding a defective item on the production line as $p(\text{D}) = (3/100)$, both of these being values that could not be derived by abstract reasoning alone.

Yet here, too, controversy exists. What is a "large number" of experiments? Do there have to be a thousand of them, or even a million, or must M, as some purists

Table 4.1   Determining Probability Empirically: Tossing a Coin

When determining by actual experiment the probability of getting a head when tossing a coin, the computed value differs depending on the number of trials.

Toss Number	Result	Cumulative Number of Heads, k	Cumulative Number of Tosses, M	Probability of Head Implied by Past Experience $p(\text{H}) = \dfrac{k}{M}$
1	T	0	1	0
2	T	0	2	0
3	H	1	3	.33
4	H	2	4	.50
5	H	3	5	.60
•	•	•	•	•
•	•	•	•	•
•	•	•	•	•
10	T	6	10	.60
•	•	•	•	•
•	•	•	•	•
•	•	•	•	•
20	T	10	20	.50
•	•	•	•	•
•	•	•	•	•
•	•	•	•	•
30	H	14	30	.47
•	•	•	•	•
•	•	•	•	•
•	•	•	•	•
40	T	20	40	.50
•	•	•	•	•
•	•	•	•	•
•	•	•	•	•
50	H	26	50	.52

suggest, approach infinity? The latter would certainly not be practical, but it is true that the probability of an event, calculated by the empirical formula in Box 4.B, will vary with the number of experiments performed and will hardly be very meaningful if that number is small. To appreciate this fact, consider tossing a (presumably unbiased) coin to determine *empirically* the probability of getting a head.

At the time of this writing, such an experiment was performed by the author. Fifty tosses were made; some of the results are shown in Table 4.1, and all of them are graphed in Figure 4.7 on page 128. Note how the probability measure so determined varies with the number of experiments performed. What number of trials, therefore, reveals the "true" probability? This question is particularly acute in situations in which the empirical approach is the only one available.

Note: Contrary to popular belief (among nonstatisticians), and contrary to what Figure 4.7 might suggest, it is not *certain* that the empirical probability value will cease to fluctuate after a "large enough" number of trials (and will thus reveal which number can be relied upon). Even if the experiment described here were continued for 20,000

Figure 4.7   Determining Probability: The Classical versus the Empirical Approach

The classical approach to probability determines the probability of getting a head when tossing a fair coin by reasoning alone, and the value equals .5. The empirical approach requires actual experiments; its measurement varies with the number of trials. Using the empirical approach raises the question of what number of trials can be relied upon to reveal the "true" probability—a question that is particularly acute when the alternative, classical approach is unavailable.

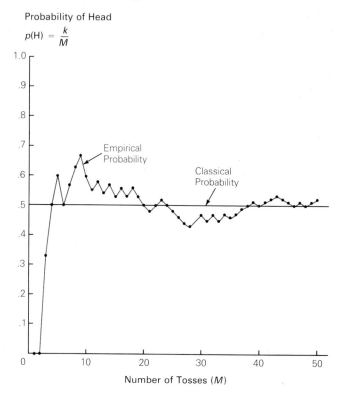

tosses, the probability value may well rise again to the high of .67 observed after the ninth toss or may even fall by an equivalent amount below the classical .5. However, as the **law of large numbers** suggests, the probability of a significant deviation of an empirically determined probability value from a theoretically determined one is smaller the larger is the number of repetitions of the random experiment in question.

Subjective Probability

Considerably more controversial than objectively determined probability values are subjective ones. Such numbers reflect a purely personal degree of belief in the likelihood of the occurrence of some event. They reflect the hunches people have, their "feelings in the bone," and these vary over time and from one person to the next.

Consider how people speculate about the likelihood of a recession this year, of General Motors being in the red next fall, of research finding a source of abundant energy, or of oil being found at a given site. "There are 3 chances out of 10," they might say, yet the named experiments have never been performed before and will never be repeated. Such a probability value of .3 is based neither on logical deduction nor on repeated observations of actual events—that is, historical experience. In the view of some probability theorists, therefore, such values are totally worthless. Yet it is a fact that subjective estimates of the chances for various outcomes of unique experiments play an important role in millions of decisions every day. This fact will be noted in the last section of this chapter and, at greater length, in the last chapter of this book.

COUNTING TECHNIQUES

All the random experiments discussed so far in this chapter have been easy ones in the sense that the numbers of favorable outcomes and the numbers of possible outcomes were invariably small, and counting them presented no obstacle to the calculation of probability. Such simplicity is not always present, however. Some random experiments involve truly huge sample spaces; in such cases, the determination of probabilities would be severely hampered without special counting techniques. In order to be prepared for such occasions, we must learn a special way of counting and become familiar with the concepts of (1) factorials, (2) permutations, and (3) combinations.

Factorials

Suppose we had to know in how many different ways n distinct items can be sequenced. If the number of such items were small enough and involved, say, only the three letters *A, B,* and *C,* we could, of course, do the counting rather quickly, perhaps with the help of the tree diagram in Figure 4.8 on page 130. If we started out with the three letters in our hands, we would have three alternative choices for placing a letter in the first position: *A, B,* or *C.* Having made that decision, only two choices would remain for the second position. (If the first choice were *A,* for example, only *B* or *C* would be available for the second position.) And having made the second choice, our hands would be tied; only one possibility would remain for the third position. (If the first choice were *A* and the second *B,* for example, we would *have* to place *C* in the third place.) By following the various branches of the tree diagram, we would discover the six possible sequences shown in the box and, thus, the answer to our question: 3 distinct items can be sequenced in 6 possible ways.

Now consider this: our answer can also be found by multiplying together the number of choices for the first, second, and third positions; that is, our 6 possible ways equal $3 \cdot 2 \cdot 1$. Such a product of a series of positive whole numbers that descends from a given number n down to 1 is called a **factorial product** or simply a **factorial.** A factorial always tells us in how many different ways n distinct items can be sequenced. In our example, where $n = 3$, the answer is written as 3! and pronounced "three factorial." The more general case is given in Box 4.C.

4.C The Factorial

$$n! = n \cdot (n-1) \cdot (n-2) \cdot \ldots \cdot 3 \cdot 2 \cdot 1$$

where 0! and 1! are defined to equal 1.

Figure 4.8   Alternative Sequences of Three Distinct Items

As this tree diagram indicates, three distinct items can be sequenced in six possible ways.

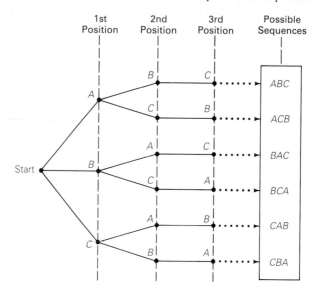

Consider an advertising agency that wants to place 6 different items in a sales brochure. In how many different ways can they be sequenced? The answer is 6! = 6 · 5 · 4 · 3 · 2 · 1 = 720. Yet, even if there were only 10 items, the answer would already be 10! = 3,628,800; for 20 items, it would exceed 2.4 quintillion. (A *quintillion* is a 1 followed by 18 zeros.) Can you imagine in how many different ways one could sequence the thousands of items in the Sears, Roebuck catalogue?

As these examples show, factorials of rather small numbers can take the place of extremely large ordinary numbers.

Permutations

The previous section has in fact discussed special cases of **permutations,** which are distinguishable ordered arrangements of items all of which have been drawn from a given group of items. Thus, the box in Figure 4.8 contains 6 permutations of the letters *A, B, C* because (a) it shows 6 distinct alternative sequences of these letters and (b) all of these sequences have been created out of the same group of letters: *A, B,* and *C.* Nevertheless, the letter-sequencing example of Figure 4.8 constitutes a special case of permutations for two reasons: First, the example involves a selection of n items out of n, but one could select fewer than n. Second, in this example, selections are made out of a group of *clearly distinguishable* items, but all of the elements of some groups may not be this distinct from one another. Both of these matters are discussed in the remainder of this section.

Selecting x *out of* n *distinct items.* Our letter-sequencing example involved finding the permutations for all the items in existence—that is, they involved finding all the different ways of arranging n distinct items into groups of n. Oftentimes, however, one is interested in the possible number of ordered sequences of only x items out of n, where x is smaller than n. In how many different ways, we may ask, can one sequence $x = 3$ items taken out of $n = 10$ distinct items? The answer cannot be $n!$ because that number also counts the ways of sequencing the last $(n - x) = 7$ items in the sequences of which we are not now interested. Thus, the $(n - x)!$ portion of the expression $n!$ must be eliminated; this can most easily be done by dividing $n!$ by $(n - x)!$. The resultant formula for the possible number of permutations, P, when x items at a time are taken out of n distinct items is given in Box 4.D.

4.D Permutations for x at a Time out of n Distinct Items (with no repetitions allowed among the x items)

$$P^n_x = \frac{n!}{(n - x)!} \text{ where } x \le n$$

Applying the new formula in Box 4.D to the selection of 3 items out of 10, we find the answer as

$$P^{10}_3 = \frac{10!}{(10 - 3)!} = \frac{10!}{7!} = \frac{10 \cdot 9 \cdot 8 \cdot 7!}{7!} = 10 \cdot 9 \cdot 8 = 720.$$

Out of a total of 10 items, one can create 720 different ordered sequences of 3 items. For example, a firm that wants to place three new managers in three of its ten plants can do so in 720 different ways!

Similarly, applying the Box 4.D formula to the selection of $x = 2$ letters from $n = 3$ distinct letters (A, B, and C),

$$P^3_2 = \frac{3!}{(3 - 2)!} = \frac{3 \cdot 2 \cdot 1}{1} = 6.$$

These six sequences are *AB, AC, BA, BC, CA,* and *CB.*

Note: If one wishes, one can apply the Box 4.D formula to the special case of $x = n$, but one must realize that $n - x$ then equals 0 and that $0! = 1$ *by definition.* Thus, if $x = n$, the entire formula for P^n_n collapses to $n!$, which returns us to the formula given in Box 4.C.

Selecting n *out of* n *nondistinct items.* In our letter-sequencing example of Figure 4.8, the elements contained in the given group of items (namely, the letters A, B, C) were *distinct* in the sense that every one of the $n = 3$ items in the group differed from every other item in the group. Yet the elements contained in a given group of items can also be *nondistinct* in the sense that x_1 out of n items are of one kind, x_2 are of another kind, and so on for k different kinds, such that $x_1 + x_2 + \ldots + x_k = n$.

(Consider the group A, A, A, B, B, C, C, C, C, C, for example.) In such a case, one cannot distinguish a sequence such as the one printed here from another one in which only the first and second A's have exchanged places. In how many distinct ways, we may ask, can one sequence the above 10 nondistinct items? The answer cannot be $n!$ because of all the unrecognizable ways in which one can internally rearrange the $x_1 = 3$ As, the $x_2 = 2$ B's, and the $x_k = 5$ C's in our example. These rearrangement possibilities, however, number $x_1!$ and $x_2!$ and $x_k!$, respectively, and we can get our answer by dividing $n!$ by the product of all these factorials, as shown in Box 4.E.

4.E Permutations for n Out of n Items, Given k Distinct Types of Items and k < n

$$P^n_{x_1, x_2, \ldots x_k} = \frac{n!}{x_1! \, x_2! \cdot \ldots \cdot x_k!}$$

where x_1 items are of one kind, x_2 items are of a second kind, ... and x_k items are of a kth kind and where $x_1 + x_2 + \ldots + x_k = n$.

When applying this formula, we discover that the above 10 nondistinct items can be put into 2,520 distinct sequences because

$$P^{10}_{3,2,5} = \frac{10!}{3!2!5!} = 2,520.$$

What if our initial group contained only two kinds of items, such as $x = 2$ A's and $(n - x) = 1$ B? From this group of $n = 3$ (consisting of A, A, B), we could form only 3 distinct permutations:

$$P^n_{x, n-x} = \frac{n!}{x!(n - x)!} = \frac{3!}{2!(2 - 1)!} = 3.$$

These permutations are AAB, ABA, and BAA.

Combinations

Sometimes one is interested not in permutations, but in **combinations,** which are different selections of items such that possible alternative sequences among the components of any one such selection are considered immaterial. Thus, the boxed list in Figure 4.8 shows six *permutations* but represents a single *combination* of the letters A, B, C. If one doesn't care about the order of the items, there is only one way to select three items out of three. While the sequence of the elements in the selected group matters in the case of permutations (ABC is not considered the same as ACB, for example), sequence does not matter in the least for combinations (combination ABC is considered the same as combination ACB because each grouping contains the same elements). Sequence does not matter in the case of card hands or of committee members or of samples of all sorts. The fact that a hand of cards contains the king of hearts, the ace of diamonds, and the 2, 8, and 10 of clubs is all that matters. The identical elements listed in a different order —ace of diamonds, 2 of clubs, king of hearts, 8 and 10 of clubs—are viewed as the same hand. They represent a different permutation but the same combination.

When calculating probabilities, one is often interested in this question: in how many different ways can one select a given number of items (say $x = 5$) from a group of distinct items (say, $n = 10$) without any concern whatsoever for the sequence in which the selected items appear? The answer cannot be $n!/(n - x)!$ precisely because the permutation formula counts separately all the different orderings of the group of x items—orderings that are now irrelevant to us. In order to count these once only, one can divide $n!$ by $x!$ as well (for $x!$ shows all the different and now irrelevant ways in which x items can be sequenced), which yields the formula in Box 4.F.

4.F Combinations for x at a Time out of n Distinct Items (with no repetitions allowed among the x items)

$$C_x^n = \frac{n!}{x! \, (n - x)!}, \text{ where } x \le n$$

Note how this formula collapses to 1 for the limiting case of $x = n$:

$$C_n^n = \frac{n!}{n! \, (n - n)!} = \frac{n!}{n! \, 0!} = \frac{n!}{n! \, (1)} = 1$$

When the order of selection is irrelevant, there exists only one way to select n items out of n items.

In how many ways, however, can one select 5 cards from a deck of 52 when the order of selection does not matter? Our formula gives us the answer almost instantly:

$$C_5^{52} = \frac{52!}{5!(52 - 5)!} = \frac{52!}{5! \, 47!} = \frac{52 \cdot 51 \cdot 50 \cdot 49 \cdot 48 \cdot 47!}{5 \cdot 4 \cdot 3 \cdot 2 \cdot 1 \cdot 47!}$$

$$= \frac{52 \cdot 51 \cdot 50 \cdot 49 \cdot 48}{5 \cdot 4 \cdot 3 \cdot 2 \cdot 1} = \frac{311,875,200}{120} = 2,598,960.$$

One can select 2,598,960 different hands of 5 cards from a deck of a mere 52 cards. Imagine how long it would take you to find this answer with the help of a tree diagram such as the one in Figure 4.8. Now ask yourself this: what is the probability of picking the top 5 cards from a deck of 52 well-shuffled cards and ending up with a *straight* (an ordered denominational sequence of 5 cards in which the deuce counts as the lowest and the ace as the highest card acceptable)? Although the lowest possible card in a 5-card straight can be a 2 (or deuce), it can belong to any one of the four different suits. The 2 can be clubs, diamonds, hearts, or spades, and the same possibilities exist for the next four cards (which would be 3, 4, 5, and 6). Thus there can be $4 \cdot 4 \cdot 4 \cdot 4 \cdot 4 = 1,024$ different straights starting with 2. Yet there are nine possible low cards— namely, all the numbers between 2 and 10. (A 5-card straight starting with 10 would reach to the end of the series, going 10, jack, queen, king, and ace.) Hence there can be $9 \cdot 1,024 = 9,216$ such straights altogether. Contrast this number with the entire sample space of 5-card sets, and you find the (low) probability of a 5-card straight:

$$p(\text{5-card straight}) = \frac{9,216}{2,598,960} = .003546.$$

Analytical Example 4.1, "The ESP Mystery," provides an interesting case in which special counting techniques were applied.

THE LAWS OF PROBABILITY

Deriving probability values for single events is one thing; manipulating these values to determine the combined probability for the union or intersection of several events is another. Such manipulation is often useful in that it obviates the necessity for repeated counting of favorable and possible basic outcomes once some probability values have been derived in this way. Just as a person who has separately counted the fingers on the left hand and on the right hand can determine the number of fingers on both hands by simply adding $5 + 5$, rather than by looking at both hands simultaneously and counting again, so a person who has access to some probability values can often combine them to gain further knowledge. We will consider two important laws for doing just that: (1) the addition law and (2) the multiplication law. Both were first stated by Abraham de Moivre (see Biography 4.1 at the end of the chapter).

The Addition Law

The **addition law** is a law of probability theory that is used to compute the probability for the occurrence of a *union* of two or more events. Such an occurrence was first illustrated (for the case of two events) in panel (c) of Figure 4.4 and then was shown symbolically in panel (d) of Figure 4.5. For the time being, we will continue to focus on two events, but, as will be shown later, the law can easily be extended to a larger number of events. Regardless of the *number* of events, a general addition law can be formulated that applies to all *types* of events; a special and simpler law can be stated for mutually exclusive events.

The general law. The general addition law for two events is given in Box 4.G, and it is easy to interpret. The law states that the probability of event A *or* B happening equals the probability of A alone plus the probability of B alone minus the probability of A *and* B happening at the same time. All this is most easily understood by reviewing panel (d) of Figure 4.5. The probability of A alone, $p(A)$, is the probability of getting the basic outcomes contained in the black circle, labeled A. The probability of B alone, $p(B)$, is the probability of getting the basic outcomes contained in the colored circle, labeled B. If one now determined the probability of either A *or* B happening by adding the separate probabilities of the two events, the probabilities of all the basic outcomes contained in the intersection of the two events and shown separately by the cross-hatched area in panel (e) would be counted twice: once as a part of circle A and again as a part of circle B. To eliminate the double counting, the probabilities of the basic outcomes found in this overlapping section are deducted once; hence, the *negative* term $p(A$ *and* $B)$ in Box 4.G. As a whole, the formula gives us precisely the probability of all the basic outcomes inside the crosshatched area of panel (d).

$$p(A \text{ } or \text{ } B) = p(A) + p(B) - p(A \text{ } and \text{ } B)$$

Consider the concrete example of Figure 4.4. The probability of event A (getting a face card) equals

$$p(A) = \frac{12}{52}.$$

This calculation requires nothing more difficult than applying the classical formula: counting the basic outcomes favorable to this event, of which there are 12 in the shaded area of panel (a), and relating the count to all the possible basic outcomes, of which there are 52 in the entire boxed sample space. Similarly, in panel (b), we can find the probability of event B (getting a heart) as

$$p(B) = \frac{13}{52}.$$

What is the probability of the union of these events, of getting *either* a face card *or* a heart? Does it equal

$$\frac{12}{52} + \frac{13}{52} = \frac{25}{52}?$$

The answer is no, as a simple count of favorable to possible basic outcomes in panel (c) reveals. The probability of getting a face card *or* a heart, $p(A \text{ } or \text{ } B)$, equals only 22/52, the number of panel (c) favorable basic outcomes (shaded) divided by the number of all possible basic outcomes (shaded plus unshaded). The difference of 3/52 between the incorrect result of 25/52 and the correct one of 22/52 is easily accounted for. It arises from the double counting of the jack, queen, and king of hearts when the probabilities implied by the shaded areas of panels (a) and (b) are thoughtlessly added together. The error can be eliminated by deducting the probability of getting a face card *and* a heart, $p(A \text{ } and \text{ } B)$, as the formula demands. That value, of course, equals 3/52, as panel (d) so vividly illustrates. Thus,

$$
\begin{array}{ccccccc}
p(A) & + & p(B) & - & p(A \text{ } and \text{ } B) & = & p(A \text{ } or \text{ } B) \\
p(\text{face card}) & + & p(\text{heart}) & - & p(\text{face card } and \text{ heart}) & = & p(\text{face card } or \text{ heart}) \\
\frac{12}{52} & + & \frac{13}{52} & - & \frac{3}{52} & = & \frac{22}{52}
\end{array}
$$

If we know the probabilities implied by panels (a), (b), and (d) in Figure 4.4, we can calculate the probability implied by panel (c).

Note: If there were three events, we would simply expand the formula to read

$$p(A) + p(B) + p(C) - p(A \text{ and } B) - p(A \text{ and } C) - p(B \text{ and } C)$$
$$+ p(A \text{ and } B \text{ and } C) = p(A \text{ or } B \text{ or } C).$$

Further expansion is carried out similarly.

The special law. The general addition law can be used to calculate the probability for a union of all types of events, but it reduces to a simpler form for mutually exclusive events. The reason is simple: In the case of mutually exclusive events, such as those symbolized in panel (a) of Figure 4.5, there is no intersection, hence no double counting, hence no need for the negative term in our formula. The formula in Box 4.G can still be used, but the expression $p(A \text{ and } B)$ will equal zero; hence, we can just as well use the simpler version in Box 4.H.

4.H Special Addition Law for Mutually Exclusive Events

$$p(A \text{ or } B) = p(A) + p(B)$$

Consider calculating the probability of getting a face card (event A) *or* getting an ace (event B) for the experiment of drawing one card at random from a well-shuffled deck. As a quick look at panel (a) of Figure 4.4 confirms, the separate probabilities of these mutually exclusive events equal

$$p(A) = \frac{12}{52} \text{ and } p(B) = \frac{4}{52}$$

and the probability of their union is

$$p(A) + p(B) = p(A \text{ or } B)$$
$$p(\text{face card}) + p(\text{ace}) = p(\text{face card } or \text{ ace})$$
$$\frac{12}{52} + \frac{4}{52} = \frac{16}{52}$$

Naturally, the probability for any union of mutually exclusive events that are also collectively exhaustive must equal 1, and this fact, too, can easily be confirmed. Consider panel (c) of Figure 4.3. In the experiment shown there, the probability of getting a sum of 10 or more (event A) *or* getting a sum of 9 or less (event B) exhausts the sample space. Accordingly, our special formula tells us that

$$p(A) + p(B) = p(A \text{ or } B)$$
$$p(\text{sum of 10 or more}) + p(\text{sum of 9 or less}) = p(\text{sum of} \geq 10 \text{ or sum of} \leq 9)$$
$$\frac{6}{36} + \frac{30}{36} = 1$$

Thus, the probability for a union of complementary events always equals 1. When event B is complementary to event A, one can write the above also as

$$p(A) + p(\overline{A}) = p(A \text{ or } \overline{A}) = 1.$$

An awareness of this truism is often helpful when it is desired to calculate the probability of one event, say A, but when it is easier to calculate that of its complement, *not* A = $\overline{A}$. In that case $p(A)$ can simply be found as $1 - p(\overline{A})$. Similarly, of course, $p(\overline{A})$ can always be found by $1 - p(A)$.

Again note that the special formula in Box 4.H can easily be expanded to any number of mutually exclusive events. For example,

$$p(A \text{ or } B \text{ or } C \text{ or } D) = p(A) + p(B) + p(C) + p(D)$$

in a case such as the one illustrated in panel (b) of Figure 4.5.

The Multiplication Law

The **multiplication law** is a law of probability theory that is used to compute the probability for an *intersection* of two or more events. Such an occurrence was first illustrated (for the case of two events) in panel (d) of Figure 4.4 and then shown symbolically in panel (e) of Figure 4.5. For the time being, we will once again continue to focus on two events, but, as will be shown later, this law, too, can easily be extended to a larger number of events. As was true for the addition law, a general multiplication law can be formulated that applies to all types of events; a special and simpler one can be stated for a more restricted group of events. Before we can understand these laws, however, we must become familiar with a number of new concepts.

Unconditional probability. Consider an appliance dealer who has been promoting a certain product by means of a major television advertising campaign. In order to evaluate the effectiveness of the campaign, let us suppose, all of the 600 customers who visit the dealership during a certain period are asked whether they remember the TV ad; records are kept of their answers as well as their possible purchase of the product in question. The results of this survey are given in Table 4.2.

Table 4.2 **Summary of Customer Survey**

These hypothetical figures represent numbers of customers classified in two ways: whether they purchased a certain product and whether they remembered a TV ad promoting it.

TV Ad / Product	Remembered, R	Not Remembered, $\overline{R}$	Total
Purchased, P	120	60	180
Not Purchased, $\overline{P}$	80	340	420
Total	200	400	600

Now consider the following four events: the purchase of the product, P; its non-purchase, $\bar{P}$; remembering the TV ad, R; not remembering it, $\bar{R}$. Clearly, a person can calculate regular empirical probability values for these four events by using the numbers in the shaded last column and shaded last row of Table 4.2:

$$p(P) = \frac{180}{600} = .30 \qquad p(R) = \frac{200}{600} = .33$$

$$p(\bar{P}) = \frac{420}{600} = .70 \qquad p(\bar{R}) = \frac{400}{600} = .67$$

Experience thus reveals a probability of .30 that a customer buys the product, a probability of .70 that a customer doesn't, a probability of .33 that a customer remembers the ad, and a probability of .67 that a customer doesn't remember the ad. Because each of these values measures the likelihood that a particular event occurs, regardless of whether another event occurs, each of these values is called an **unconditional probability.** Thus .30 equals the probability of purchase by a customer without regard to the question of whether that customer does or does not remember the ad. Similarly, the value of .67 equals the probability of not remembering by a customer without regard to the question of whether that customer does or does not purchase the product. Because these unconditional probabilities are calculated from the margins of our frequency table, they are also referred to as **marginal probabilities.**

Conditional probability. It will come as no surprise that the concept of unconditional probability has a twin—namely, that of **conditional probability,** which measures the likelihood that a particular event occurs, given the fact that another event has already occurred or is certain to occur. For two events, A and B, such a probability is always denoted by $p(A|B)$ or $p(B|A)$, which is read as "the probability of A, given B" or "the probability of B, given A"—the vertical line standing for "given."

By focusing, successively, on the values in the first row only, then on those in the second row only, next on those in the first column only, and finally on those in the second column only, we can calculate the following conditional probabilities.

1. From the first row, we can calculate the probability of remembering, given the fact that a purchase has occurred, and also the probability of not remembering, given that a purchase has occurred:

$$p(R|P) = \frac{120}{180} = .67$$

$$p(\bar{R}|P) = \frac{60}{180} = .33$$

2. From the second row, we can calculate the probability of remembering, given the fact that no purchase has occurred, and also the probability of not remembering, given that no purchase has occurred:

$$p(R|\overline{P}) = \frac{80}{420} = .19$$

$$p(\overline{R}|\overline{P}) = \frac{340}{420} = .81$$

3. From the first column, we can calculate the probability of purchase, given the fact that the ad is remembered, and also the probability of no purchase, given that the ad is remembered:

$$p(P|R) = \frac{120}{200} = .60$$

$$p(\overline{P}|R) = \frac{80}{200} = .40$$

4. From the second column, we can calculate the probability of purchase, given the fact that the ad is not remembered, and also the probability of no purchase, given that the ad is not remembered:

$$p(P|\overline{R}) = \frac{60}{400} = .15$$

$$p(\overline{P}|\overline{R}) = \frac{340}{400} = .85$$

Note: It is no accident that each pair of these conditional probabilities sums to 1.00, just as the two pairs of unconditional probabilities given in the previous section do. The reason is that the events in question are in each case mutually exclusive and collectively exhaustive (either there was a purchase or there was no purchase, either the ad was remembered or it was not remembered), and the probabilities of such events always add to 1.

Joint probability. We must, finally, turn to the concept of **joint probability,** a measure of the likelihood of the simultaneous occurrence of two or more events. Four such values can be calculated from Table 4.2 because the events P, $\overline{P}$, R, and $\overline{R}$ intersect in the four ways shown in the unshaded portion of our table. It is customary to present this information in a *joint probability table,* such as Table 4.3. Note that each probability value shown therein simply equals the number of relevant observations given in Table 4.2, divided by the total number of observations (which was 600). The relative frequencies so computed are, of course, nothing else but probability numbers determined by the empirical approach (Box 4.B). The numbers within the colored box in Table 4.3 are the joint probabilities, and they are symbolized as follows:

$$p(P \text{ and } R) = .20 \qquad p(P \text{ and } \overline{R}) = .10$$
$$p(\overline{P} \text{ and } R) = .13 \qquad p(\overline{P} \text{ and } \overline{R}) = .57$$

Table 4.3   A Joint Probability Table

The joint probabilities for the designated events appear inside the colored box; the remaining values are the marginal probabilities, which equal the respective sums of the joint probabilities in the various rows and columns.

Product / TV Ad	Remembered, R	Not Remembered, $\bar{\text{R}}$	Total
Purchased, P	$\dfrac{120}{600} = .20$	$\dfrac{60}{600} = .10$	$\dfrac{180}{600} = .30$
Not Purchased, $\bar{\text{P}}$	$\dfrac{80}{600} = .13$	$\dfrac{340}{600} = .57$	$\dfrac{420}{600} = .70$
Total	$\dfrac{200}{600} = .33$	$\dfrac{400}{600} = .67$	$\dfrac{600}{600} = 1.00$

Source: Table 4.2

Thus, we are told, the probability of finding a customer who purchases the product and also remembers the ad is .20, and the probability of finding a customer who does neither is .57. Again, it is no accident that the sum of the four joint probabilities of the four mutually exclusive and collectively exhaustive joint events equals one. (The special addition law applies.)

Two things should be noted: First, the margins of the joint probability table once again show the marginal or unconditional probabilities computed earlier. Second, by dividing any given joint probability by the marginal probability of its row (or column), the conditional probability for the given row (or column) event can be calculated. Thus,

$$\frac{.20}{.30} = .67, \text{ which means } \frac{p(\text{P } and \text{ R})}{p(\text{P})} = p(\text{R}|\text{P}).$$

The joint probability of purchase *and* remembering, this expression tells us, when divided by the unconditional probability of purchase, equals the conditional probability of remembering, given the fact of purchase. Similarly, it is true that

$$\frac{.20}{.33} = .60, \text{ which means } \frac{p(\text{P } and \text{ R})}{p(\text{R})} = p(\text{P}|\text{R}).$$

By making similar calculations based on the other three joint probabilities in Table 4.3, you can confirm the remaining six conditional probability values given in the earlier section on conditional probability.

The general law. The relationships just discovered between joint, unconditional, and conditional probabilities can be rewritten for any two events, A and B, as shown in Box 4.I, the general multiplication law.

The law states that the joint probability of two events happening at the same time equals the unconditional probability of one event times the conditional probability of the other event, given that the first event has already occurred (or is certain to occur).

4.1 General Multiplication Law

$$1.\ p(A\ and\ B) = p(A) \cdot p(B|A)$$

and also

$$2.\ p(A\ and\ B) = p(B) \cdot p(A|B)$$

Consider again the example of Figure 4.4. The unconditional probability of event A (getting a face card) is

$$p(A) = \frac{12}{52}.$$

The conditional probability of event B (getting a heart, assuming a face card is being picked) is

$$p(B|A) = \frac{3}{12}$$

because there are 3 face cards with hearts among a total of 12 face cards. Version 1 of our formula tells us that the joint probability should be

$$p(A\ and\ B) = p(A) \cdot p(B|A) = \frac{12}{52} \cdot \left(\frac{3}{12}\right) = \frac{3}{52}$$

which is precisely what panel (d) of Figure 4.4 suggests (there are 3 basic outcomes in the shaded intersection and 52 basic outcomes in the entire sample-space box).

Note that the same result could have been reached by the use of version 2 of the formula: the unconditional probability of event B (getting a heart) is

$$p(B) = \frac{13}{52}.$$

The conditional probability of event A (getting a face card, assuming a heart is being picked) is

$$p(A|B) = \frac{3}{13}$$

because there are 3 face cards with hearts among a total of 13 cards with hearts. Thus, version 2 of our formula tells us that

$$p(A\ and\ B) = p(B) \cdot p(A|B) = \frac{13}{52} \cdot \frac{3}{13} = \frac{3}{52}.$$

Note also that the general multiplication law can be extended to more than two events such that (for three events)

$$p(A \text{ and } B \text{ and } C) = p(A) \cdot p(B|A) \cdot p(C|A \text{ and } B).$$

If the probability of a person seeing a TV ad is .3 (event A) and if the probability of such a person walking into the advertising dealership's showroom is .1 (event B, given A) and if the probability of such a visitor making a purchase is .5 (event C, given A *and* B), the joint probability of someone seeing the ad, visiting the showroom, and making a purchase (event A *and* B *and* C) is .3(.1)(.5) = .015. Obviously, this type of calculation gets more complicated as the number of events increases.

Fortunately, a special and simpler version of the multiplication law can be applied whenever the events in question are independent rather than dependent.

Dependent versus independent events. Two random events are said to be **dependent events** when the probability of one event is affected by the occurrence of the other event. In contrast, two events are called **independent events** when the probability of one event is *not* affected by the occurrence of the other event.

A good example of dependent events is the drawing of more than one card from a deck *without replacing* the cards drawn. Prior to the first draw, the probability of getting the king of hearts is 1/52, and so is the probability of getting the ace of diamonds or the queen of spades. Now imagine drawing the king of hearts on the first try *and setting it aside.* Notice how all the probabilities will now differ because the deck is reduced to 51 cards, and the king of hearts is gone. The subsequent probabilities are zero for the king of hearts and 1/51 for all the other cards, and so it goes until, when only one card is left, the probability of drawing it equals 1, at which point the probability is zero for drawing any of the other cards already drawn.

Consider, in contrast, the same experiment, with the immediate replacement of any card drawn. The probability of drawing any one card will then remain unaffected by any of the preceding events; it will always equal 1/52. Even if the king of hearts were drawn on the first try, after replacing it and reshuffling the deck, the probability of drawing it on the second try would again be 1/52, just as for any other card. In this case, different drawings are independent events.

Note: While one can, if one wishes, set aside the king of hearts and draw again, it is physically impossible to set aside a head (or tail) and then toss a coin again; hence, different tosses of a coin are always independent events. Nevertheless, many gamblers refuse to believe it. After observing a series of heads, they bet on a tail coming up next because, as they put it, "it is the tail's turn" or "the tail's chances are now mature." In fact, this is nonsense. The probability of getting a head (or a tail) remains 1/2 for any toss of a fair coin, even when 37 heads have just been tossed in a row. The coin has no conscience or memory and doesn't care or know whose turn is next.

As is true throughout this chapter, the many examples from the world of gambling have innumerable analogies in the world of business and economics. Thus, an inspector who randomly picks selected items out of a warehouse in order to check quality (and who wants to determine the probability of the existence of defective items within the population of all those produced) must be aware of whether successive pickings are dependent events or independent ones. If each item inspected is set aside, as is usually the case, dependent events are occurring; if each item is replaced before the next one is picked, the events are independent.

It is easy to test for the dependence or independence of any two events, such as the remembering of an ad and the purchase of a product or the drawing of a king and a red card from a deck. Three simple steps are involved:

1. *Determine the unconditional probability for one event in the pair.* For example, note in Table 4.3 that the unconditional probability of remembering is $p(R) = .33$, or note in panel (c) of Figure 4.1 that the unconditional probability of drawing a king from a deck is $p(\text{king}) = 4/52$.
2. *Determine the conditional probability for the same event, given the occurrence of the second event in the pair.* For example, note that the conditional probability of remembering, given purchase, is $p(R \mid P) = .67$, or note that the conditional probability of drawing a king, given that a red card is drawn, equals $p(\text{king}|\text{red}) = 2/26$ (there are 2 kings among 26 red cards).
3. *Compare the unconditional with the conditional probability.* If the values differ, the events are *statistically dependent;* if they do not differ, the events are *statistically independent.* In our case, $.33 \neq .67$; remembering and purchase are dependent events. On the other hand, $4/52 = 2/26$; getting a king and getting a red card are independent events.

Symbolically, dependence implies $p(A) \neq p(A \mid B)$, while independence implies $p(A) = p(A \mid B)$. This discovery brings us directly to the simplification of the general multiplication law.

The special law. The general multiplication law can be used to calculate the probability for an intersection of all types of events, but it reduces to an easier form for independent events. The reason is simple: in the case of independent events, the unconditional and the conditional probabilities are identical; hence, the conditional probabilities, included in Box 4.I, can be replaced by the unconditional probabilities, as shown in Box 4.J.

4.J Special Multiplication Law for Independent Events

$$p(A \text{ and } B) = p(A) \cdot p(B)$$

In the case of independent events, the joint probability of A *and* B (or of any number of such events) simply equals the product of the unconditional probabilities of these events. For example, as we noted above, drawing a king and drawing a red card from a deck are independent events. Hence, their joint probability

$$p(\text{king and red card}) = \frac{4}{52}\left(\frac{26}{52}\right) = \frac{1}{26}.$$

This probability is, indeed, obvious by looking at panel(c) of Figure 4.1: there are 2 red kings in the deck of 52 cards.

Or consider the quality inspector who selects 3 items (like kings) from a group of 100 (like a deck of cards). If the inspected item is replaced after each selection and the whole lot is remixed, what is the probability that the inspector will find 3 defective items if (unbeknownst to the inspector) there are only 10 defective items in the lot?

The answer is simple: the unconditional probability of finding a defective item, $p(D)$, equals $(10/100) = .1$. Because replacement makes the events of finding a defective item on the first, second, and third try (D_1, D_2, D_3) independent ones, the special multiplication law applies: $p(D_1$ *and* D_2 *and* $D_3) = p(D_1) \cdot p(D_2) \cdot p(D_3) = .1(.1)(.1) = .001$. There is just one such chance in 1,000 tries.

FAMOUS APPLICATIONS OF THE PROBABILITY LAWS

When properly applied, the laws of probability can tremendously enhance people's ability to make proper decisions in the face of uncertainty. This fact is vividly illustrated by Application 1, "Striving for Reliability: The Case of the Apollo Mission." On the other hand, disastrous consequences await those who have failed to do their homework. Consider the mix-up of the general multiplication law (that should have been applied) with the special multiplication law (that was applied), as reported in Application 2, "Probability in Court." Or study the equally famous mix-up of the special multiplication law (that should have been applied) with the special addition law (that was applied), a case reported in Application 3, "The Paradox of the Chevalier de Méré." That case was solved jointly by two great mathematicians, Pierre de Fermat and Blaise Pascal (see Biographies 4.2 and 4.3 at the end of the chapter).

Application 1: Striving for Reliability—The Case of the Apollo Mission*

The Apollo mission that was to land Americans on the moon required the creation of a complex artificial world to sustain three astronauts in a hostile environment for two weeks. The failure of any component of the Apollo system could lead to failure of the entire mission. The reliability problem was the greatest challenge facing the program. Because *reliability* of components is closely linked with the *probability* of their success or failure, the entire issue had to be dealt with by probability theory. First, the mission planners developed a mathematical model of the Apollo system showing the location in various subsystems of 2 million functional parts, miles of wiring, and thousands of joints. Such a simplified model is given in panel (a) of Figure A; an even more simplified model appears in panel (b).

Second, mission planners estimated the reliability of each subsystem from experimental data and then used the laws of probability to predict the reliability of the Apollo system as a whole.

Figure A is a typical example of a *series system:* if any component fails, the entire system fails (just as, in the case of old-fashioned strings of Christmas tree lights, the entire tree was darkened whenever a single bulb failed). Assuming that the components operate independently and that the individual probability-of-success values are $p(A)$, $p(B)$, $p(C)$, $p(D)$, and $p(E)$, respectively, the reliability of the entire system, $p(S)$, can be calculated by the special multiplication law for independent events (see Box

---

*Adapted from Gerald J. Lieberman, "Striving for Reliability," in Judith M. Tanur et al., eds., *Statistics: A Guide to the Unknown* (San Francisco: Holden-Day, 1972), pp. 400–406.

Figure A

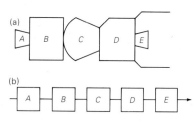

(a)

(b)

A Main Engine
B Service Propulsion System
C Command Service Module
D Lunar Excursion Module (LEM)
E LEM Engine

Figure B

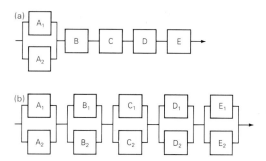

(a)

(b)

4.J) as $p(S) = p(A) \cdot p(B) \cdot p(C) \cdot p(D) \cdot p(E)$. Thus, if each component has a reliability of .95, the probability of success for the entire system shown in Figure A is given by $p(S) = (.95)^5 = .774$, which is not too encouraging. If the performance of each subsystem is not independent of that of the others (and, say, the successful performance of A portends the successful performance of E or if the failure of B puts a high load on other subsystems and causes the failure of D), the *general* multiplication law must, of course, be used to calculate $p(S)$.

The Apollo planners were also aware, however, of the implications of *parallel configurations*—of replacing, say, the large main engine A by two smaller engines, A_1 and A_2. If, then, failure occurs only if *both* A_1 and A_2 fail (or only if an intersection of two events occurs), reliability might be increased. Such a possible new engine configuration is shown in panel (a) of Figure B.

Under this scheme, the probability of success for the engine function now equals the probability of success for engine A_1 *or* A_2; that is,

$$p(A) = p(A_1 \text{ or } A_2).$$

According to the general addition law, the above equation can also be written as:

$$p(A) = p(A_1) + p(A_2) - p(A_1 \text{ and } A_2).$$

Assuming again independence of engines A_1 and A_2, the latter term can be replaced by $p(A_1) \cdot p(A_2)$ in accordance with the special multiplication law, yielding

$$p(A) = p(A_1) + p(A_2) - [p(A_1) \cdot p(A_2)].$$

Assuming the same probability of success for the smaller engine as for the larger one—that is, $p(A_1) = .95$ and $p(A_2) = .95$—the success probability for the engine function becomes

$$p(A) = .95 + .95 - (.95)^2 = 1.9 - .9025 = .9975,$$

which is a great improvement over the original .95. Together with the other components, this engine setup raises the probability of success for the entire mission from .774 to

$$p(S) = .9975 \, (.95)^4 = .8125.$$

Indeed, one can do better than that! Consider parallel configurations for all of the subsystems, as shown in panel (b) of Figure B. Assuming probabilities of success of .95 for all the components, and applying the same reasoning, one can raise the probabilities of success for all the other subsystems to .9975 as well.

Thus, the probability of success for the entire system to sustain the three astronauts, shown in panel (b) of Figure B, becomes

$$p(S) = (.9975)^5 = .9876,$$

which is a dramatic improvement over the original .774. Further improvements of the same type can also be made, of course—for example, by the substitution of two engines for A_1 and of two more engines for A_2. Nor is this idle speculation: During the Apollo 13 launch, the center engine of the second stage of the Saturn rocket failed, yet a satisfactory earth orbit was achieved with the remaining *four* rocket motors.

And note: the same principle of reliability control applies in the case of other and possibly less complex systems, be they automobiles, household appliances, power supplies, or telephones. Indeed, auto manufacturers introduced the five-year unconditional guarantee after examining the reliability of the affected components and determining the overall reliability of their cars in the fashion indicated here.

Application 2: Probability in Court*

On many an occasion, probability theory has had its day in court. Two famous cases stand out.

The Trujillo Case. In 1946, a police officer was shot to death. Joseph Trujillo was charged with the murder. The prosecutor argued that 11 matching fibers were found that apparently had been transferred between the victim's and the defendant's clothing. The probability for one accidental fiber match was given as .1; hence, that of 11 matches was calculated by the special multiplication law as $(.1)^{11}$ or a tiny .00000000001. More than that! The prosecutor also noted five matching characteristics between the bullet that had killed the victim and the pistol that belonged to the defendant.

1. Both were of .38 caliber, a probability given as .2.
2. Rifling was right-handed, a probability given as .5.
3. Both pistol and bullet showed five lands and grooves, a probability given as .2.

---

*Adapted from Darrell Huff, *How to Take a Chance* (New York: Norton, 1959), pp. 138–40; *Time,* January 8, 1965, p. 42, and April 26, 1968, p. 41; and W. Fairley and Frederick Mosteller, "A Conversation About Collins," *University of Chicago Law Review,* 1974.

4. Both pistol and bullet showed the chamber out of line with the barrel, a probability given as .2.
5. The pistol was not, and the bullet had not come from, an automatic revolver, a probability given as .5.

Again, the prosecutor multiplied together the estimated probabilities to show the small joint probability of all these events and, hence, the tiny likelihood that someone else could have been the killer.

Trujillo was convicted and executed. No thought, apparently, was given to whether the stated events were independent (and the special multiplication law was even applicable). Some 18 years later, however, a similar case had quite a different outcome.

The Collins Case. In 1964, an elderly woman was mugged in an alley. Shortly afterward, a witness saw a blonde girl, her ponytail flying, run out of the alley, get into a yellow car driven by a bearded black man with a moustache, and speed away. Eventually, the police arrested Janet and Malcolm Collins, a married couple who fit the physical description given above and who also owned a yellow Lincoln.

Once again, the prosecutor invoked the laws of probability. He suggested to the jury probability values:

1. for being a blonde of $p = .33$.
2. for wearing a ponytail of $p = .10$.
3. for being an interracial couple of $p = .001$.
4. for owning a yellow car of $p = .10$.
5. for wearing a moustache of $p = .25$.
6. for being a black man wearing a beard of $p = .10$.

He multiplied the numbers to produce a joint probability of .00000008 that the Collinses could have been duplicated on the morning of the crime. The jury, overwhelmed by the logic of it all, found them guilty beyond a reasonable doubt. Malcolm Collins received a sentence of "one year to life"; Janet Collins received a sentence of "not less than one year."

Yet, a few years later, the California Supreme Court overturned the verdict. There was no evidence, it said, that the individual probability values were even roughly accurate. Most important, they were not shown to be independent of one another, as they must be to satisfy the probability law that was applied. The fact that the black man had a beard, said the court, was not independent of the fact that the couple was interracial, for example. Indeed, the court was able to show that there was a 41 percent chance that at least one other couple in the area might have had the admittedly unusual characteristics listed above.

Application 3: The Paradox of the Chevalier de Méré*

A good example of what happens when the special law of multiplication is confused with the special law of addition comes to us from 17th-century France. A nobleman, Antoine Gombauld, also known as Chevalier de Méré, thought he had found a surefire way of winning in games of dice. For a single roll of a die, he reasoned (correctly), the

---

*Adapted from Darrell Huff, *How to Take a Chance* (New York: Norton, 1959), pp. 63–67.

probability of getting a 6 to come up was 1/6. Hence, he argued (incorrectly), the probability of at least one 6 to come up was 2/6 for two rolls of the die, 3/6 for three rolls (making for even odds), and 4/6 for four rolls (turning the odds in favor of this event). Thus, he was prepared to gamble that he would get at least one 6 in four rolls of a die; as he saw it, he would win two wagers out of three in the long run. Gamble he did, and, amazingly, he became quite rich.

Then, for reasons unknown, the Chevalier changed his strategy, betting that he could get at least one 12 in 24 rolls of two dice. His reasoning was the same. As he saw it, the probability of getting a 12 in a single roll of two dice was 1/36. So far, he was right, as panels (b) and (c) of Figure 4.2 attest. Hence, he concluded (incorrectly), the probability of getting a 12 in 24 such rolls was 24/36 or, once more, two out of three. Yet this time, his luck seemed to have run out; he lost more often than he won.

Perplexed, he turned for help to the mathematician and philosopher Blaise Pascal (see Biography 4.3) who, in turn, consulted with a friend, Pierre de Fermat (see Biography 4.2). Their joint inquiry into this *problem of points,* as it is called, gave birth to the theory of probability and quickly solved the Chevalier's paradox. The gambler's application of the special addition law for mutually exclusive events (see Box 4.H), argued the two, was a mistake. The events in question were not mutually exclusive at all: one could conceivably get a 6 on every one of four rolls of a die; one could get a 12 on every one of 24 rolls of two dice as well. Indeed, by the logic of our gambling friend, the probability of getting a 6 on *six* rolls of a die was 6/6; that is, such an event would occur with certainty!

The issue, argued Pascal and de Fermat, called for the use of the special *multiplication* law for independent events (see Box 4.J), but they found it easier first to calculate the probability for the gambler's failure rather than his success: the probability of *not* getting a 6 on a single roll, they figured, was 5/6. Hence, the probability of not getting a 6 on four such rolls was

$$\frac{5}{6} \cdot \frac{5}{6} \cdot \frac{5}{6} \cdot \frac{5}{6} = \frac{625}{1296}.$$

Thus, the first strategy had a probability of *success* of

$$1 - \frac{625}{1296} = \frac{671}{1296}$$

or about .52. This was less than the gambler's imagined 2/3, but high enough to make him rich.

By the same reasoning, the probability of *not* getting a 12 on a single roll of two dice was 35/36. Hence the probability of not getting a 12 on 24 such rolls was $(35/36)^{24}$ or about .509. Thus, the second strategy had a probability of success of $1 - (\approx .509)$, or about .491. No wonder the gambler's luck ran out.

Probability Laws and Tree Diagrams

Both addition and multiplication laws can be illustrated with the help of tree diagrams, such as those in Figure 4.9.

Both panels (a) and (b) illustrate the survey summarized in Table 4.2 but do so in slightly different ways. The first fork in panel (a) pictures the possibilities of a customer making or not making a purchase and lists the associated unconditional probabilities of these events along the two branches. Because a customer must do one or the other,

Figure 4.9 Probability Laws and Tree Diagrams

The two equivalent versions of the general multiplication law that helps us find the joint probability of two events by multiplying the unconditional probability of one event with the conditional probability of the other event, given the first event, are illustrated in panels (a) and (b). The special addition law is seen at work, too, although indirectly: the sum of probabilities at each fork and the sums of joint probabilities always equal 1, a requirement for mutually exclusive and collectively exhaustive events. Panel (c), finally, illustrates an application of the special multiplication law to the determination of joint probabilities for three independent events.

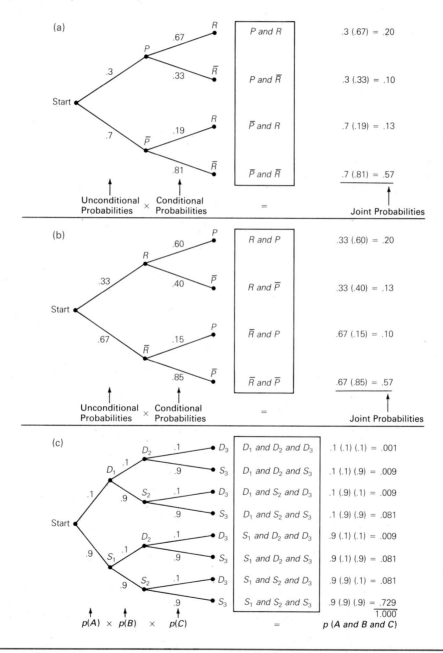

the special additon law applies; the sum of .3 and .7 equals 1, implying certainty that one *or* the other of these events will occur. If a purchase is made, two mutually exclusive and collectively exhaustive possibilities exist once again at P: the customer does or does not remember seeing the ad. The conditional probabilities (*given* that a purchase has been made) equal .67 and .33 (as was noted in an earlier section) and, once more, add to 1. If no purchase is made, the same two possibilities exist at $\bar{P}$, but the conditional probabilities are .19 and .81; once more, they add to 1. If the numbers along any path to the ultimate outcome are multiplied, joint probabilities are found, just as the general multiplication law suggests. Note that the joint probabilities of the four possible outcomes again add to 1 and that each equals the value previously calculated (and given in Table 4.3).

Panel (b) reaches the identical conclusion, but it does so in a different way, by employing the alternative version of the multiplication law. The choice between remembering and not remembering the ad is considered first; that between purchase and no purchase is considered second. The same joint probabilities result (but note that they are listed in a different order).

Note also how easy it is to recognize all the events in question as *dependent* ones. The unconditional probability of purchase, given as .3 in panel (a), differs from the conditional probabilities of purchase, given remembering (or not remembering) shown as .6 (or .15) in panel (b). So purchase and remembering (or not remembering) are dependent events. Similarly, the unconditional probability of not remembering, given as .67 in panel (b), differs from the conditional probabilities of not remembering, given purchase (or no purchase) shown as .33 (or .81) in panel (a). Not remembering and purchase (or no purchase) are dependent events as well.

Panel (c) pictures the case of quality inspection *with replacement* in which 3 items are picked from a lot of 100 of which 10 percent are defective. Thus, the panel simultaneously illustrates a case of independent events and a case of more than two events. The fact of independence is visible in the tree diagram because the probabilities at every fork remain the same (.1 for picking a defective item; .9 for picking a satisfactory one). Thus $p(A)$ symbolizes the unconditional probability of finding a defective or satisfactory item on the first trial; $p(B)$ symbolizes the corresponding conditional probabilities on the second trial, given the result of the first trial, but these probabilities equal the unconditional ones. Similarly, $p(C)$ symbolizes the conditional probabilities on the third trial, given the results of the first and second trials, but these probabilities, too, equal the unconditional ones, because the three trials are independent events when replacement occurs. Note how the tree diagram shows clearly why the 8 outcomes of the completed experiments are not equally likely. While the joint probability of finding three defective items in a row is a tiny .001, that of finding three satisfactory items in a row is a respectable .729.

REVISING PROBABILITIES: BAYES'S THEOREM*

It was noted earlier that *subjective* probability figures, although commonly used, are also rather controversial. In this section, we consider how such values can be confirmed or revised in the light of empirical evidence. The idea goes back to an 18th-century British clergyman, the Rev. Thomas Bayes (Biography 14.1), who stumbled on-

*Optional section.

to probability theory when searching for a mathematical proof of the existence of God. Bayes pictured people as possessing, prior to any empirical investigation, an initial subjective estimate of the likelihood of an event; he called this value a **prior probability.** He then imagined an investigation to take place and to yield certain results. He developed what is now called **Bayes's theorem:** a formula for revising an initial subjective probability value on the basis of results obtained by an empirical investigation. The end result is a new probability value, a prior probability modified on the basis of new information and called a **posterior probability.** The sequence of probability revision can be summarized in the following sketch:

| Prior Probability | + | New Information | + | Application of Bayes's Theorem | → | Posterior Probability |

If we denote the probability of an event as $p(E)$, the probability of its complement as $p(\overline{E})$, and the result of an investigation as R, Bayes suggested that the posterior probability of event E could be calculated by the formula in Box 4.K.

4.K Bayes's Theorem

$$p(E|R) = \frac{p(E) \cdot p(R|E)}{p(E) \cdot p(R|E) + p(\overline{E}) \cdot p(R|\overline{E})}$$

Two things should be noted. First, the prior probability of the event, $p(E)$, is an unconditional one. The posterior probability, $p(E|R)$, is a conditional one, given the empirical result, R. Second, the entire formula can be rewritten to take the form of the expression given in Box 4.L. As a quick glance at the general multiplication law (Box 4.I) confirms, the numerator in Bayes's formula is nothing else but the joint probability of the given event and the empirical result, or $p(E \ and \ R)$. The denominator, therefore, is the *sum* of the joint probability of the given event and the observed result, or $p(E \ and \ R)$, and the joint probability of the given event's complement and the observed result, or $p(\overline{E} \ and \ R)$. This sum, as Table 4.3 confirms, is nothing else but the unconditional probability of the result, or $p(R)$. Thus, Bayes's theorem can also be written as in Box 4.L.

4.L Bayes's Theorem Rewritten

$$p(E|R) = \frac{p(E \ and \ R)}{p(R)}$$

The Bayesian posterior probability of an event is the ratio of the joint probability of this event and the empirical result to the unconditional probability of this result. The following examples illustrate the Bayesian approach.

Practice Problem 1

The Crooked Die

Consider being faced with three dice, one of which is known to be crooked. The crooked die, let us suppose, contains a lead weight that makes a 6 appear on half the tosses; yet the dice all look alike. If you know all of the above and now select one of the dice, what is the probability that it is the crooked one? Let us follow the revision sequence sketched above.

Step 1. Prior Probabilities. Letting C stand for "crooked" and $\overline{C}$ for "not crooked," the *prior* probabilities that you subjectively and arbitrarily select (because you are considering one die out of three) are the following:

$$p(C) = 1/3;$$
$$p(\overline{C}) = 2/3.$$

Step 2. New Information. Now comes the empirical investigation: you roll the die, and its face shows a 6. We have our result, and we can state the two conditional probabilities of this result, given each of the two possible events, easily enough:

$$p(6|C) = 1/2 \text{ (information given before)};$$
$$p(6|\overline{C}) = 1/6 \text{ (classical probability value)}.$$

Step 3. Applying the Theorem. We now apply Bayes's theorem from Box 4.K:

$$p(C|6) = \frac{p(C)p(6|C)}{p(C)p(6|C) + p(\overline{C})p(6|\overline{C})}$$

$$= \frac{\dfrac{1}{3} \cdot \dfrac{1}{2}}{\left(\dfrac{1}{3} \cdot \dfrac{1}{2}\right) + \left(\dfrac{2}{3} \cdot \dfrac{1}{6}\right)}$$

$$= \frac{\dfrac{1}{6}}{\dfrac{1}{6} + \dfrac{2}{18}}$$

$$= \frac{\dfrac{1}{6}}{\dfrac{5}{18}} = \frac{3}{5}.$$

Step 4. Posterior Probability. In light of the new information, the probability that the selected die is the crooked one is revised upward from the prior value of 1/3 (when nothing was known about the particular die) to a posterior value of 3/5 (when it was known that this die showed a 6 when rolled once). In light of the evidence, crookedness is more likely, but it is not certain, of course. (The 6 face could have been rolled by any die.) |||

||| Practice Problem 2

Drilling for Oil

Consider an oil company about to drill for oil. The company geologists might state subjective probabilities for finding oil at the given site as follows:

$$p(\text{oil}) = .5$$
$$p(\text{no oil}) = .5$$

They drill for 500 feet, find no oil yet, but sample the soil. From past experience, they know the conditional probabilities of finding this type of soil, given oil or no oil:

$$p(\text{this type soil} \mid \text{oil}) = .2$$
$$p(\text{this type soil} \mid \text{no oil}) = .8$$

What is the posterior probability for finding oil, according to Bayes?

$$p(\text{oil} \mid \text{this type soil}) = \frac{.5(.2)}{[.5(.2)] + [.5(.8)]}$$
$$= \frac{.1}{.1 + .4} = \frac{.1}{.5} = .2$$

The new information reduces the probability of an oil find from .5 to .2. |||

||| Practice Problem 3

Screening Job Applicants

An employment agency administers a placement test to an equal number of men and women. Thus, the prior probability is .5 that any randomly selected test paper was written by a man.

The tests are graded. Some 60 percent of the tests score a C, and 30 percent of the tests score a C *and* are written by a man.

Given this new information, what is the likelihood that a randomly selected paper with a score of C was written by a man? The alternative version of Bayes's theorem can easily be applied:

$$p(\text{man} \mid \text{given score C}) = \frac{p(\text{man } and \text{ score C})}{p(\text{score C})}$$
$$= \frac{.3}{.6} = .5$$

The posterior probability that a C paper was written by a man is .5, which equals the prior probability of .5 for any paper. There is no reason for revising the prior probability value. |||

ANALYTICAL EXAMPLE

4.1 The ESP Mystery

In 1882, the Society for Psychical Research was founded in London to study paranormal (odd and unexplained) kinds of events. Mind reading, fortune-telling, and similar forms of alleged extrasensory perception (ESP) have in particular been subjected to experimental investigation. Is the phenomenon real?

Consider a typical card-guessing experiment. One person randomly deals 10 cards in one room, concentrating on their color (red or black). Another person, who knows that half the cards are red and half black, sits in another room and building. At the very time when the cards are dealt (watches have been synchronized), this person writes down the order of colors dealt. If, for example, 8 of the 10 cards are correctly identified, does this support the claim of ESP?

Probability theory can help. In how many ways could a person who was *guessing* assign red to 5 of the guesses (and black to 5)? This is the number of combinations of 10 items, taken 5 at a time, or

$$C_5^{10} = \frac{10!}{5!(10-5)!} = \frac{10 \cdot 9 \cdot 8 \cdot 7 \cdot 6}{5 \cdot 4 \cdot 3 \cdot 2 \cdot 1}$$
$$= \frac{30{,}240}{120} = 252.$$

If these 252 outcomes are equally likely, what is the probability of identifying correctly 8 of the 10 cards? The number of ways to identify correctly 4 of the 5 red cards is $C_4^5 = 5$; and for each of these ways, 4 black cards can be identified in $C_4^5 = 5$ ways as well. So the number of ways to identify 8 cards is $5 \times 5 = 25$. Hence, the probability of identifying correctly 8 of the cards is 25/252 or about .1. There is almost a 10 percent chance of getting the observed result by pure luck.

Many other experiments, however, have produced remarkable odds in favor of the existence of ESP. In one experiment, 558 correct identifications were made of 1,850 cards showing 5 types of symbols (cross, star, circle, square, wave). Guessing might have produced 370 correct identifications; the probability of achieving the actual result by accident was determined to be $1/10^{22}$—that is, once in 10 thousand billion billion times.

Sources: Adapted from J. Leroy Folks, *Ideas of Statistics* (New York: Wiley, 1981), pp. 86 and 87; and Darrell Huff, *How to Take a Chance* (New York: Norton, 1959), pp. 125–29.

BIOGRAPHIES

4.1 Abraham de Moivre

Abraham de Moivre (1667–1754) was born at Vitry, France, where his father was a surgeon. De Moivre studied mathematics and physics in Paris, but in 1685, after the Edict of Nantes was revoked, he was imprisoned for being a Protestant. When released three years later, he emigrated to England to escape religious persecution. He never returned to France and never published anything in French.

By all accounts, he was a mathematical genius, and he was in constant touch (at the Royal Society) with the leading thinkers of his day, including Isaac Newton who became a close friend. Yet de Moivre never suc-

ceeded in obtaining a university appointment. He eked out a living by tutoring the sons of nobility and by advising gamblers and speculators. This unwelcome fate was posterity's gain, for his successful solution of the problems he met in his consulting practice led to his writing of two great books. His text on probability, *The Doctrine of Chances,* emanated from an article first published in Latin in 1711 and was published posthumously in its final and third edition in 1756. It is notable (among many other contributions) for the origin of the general laws of addition and multiplication of probabilities (discussed in the present chapter of this book), for the origin of the binomial distribution law (discussed in Chapter 5), and for the origin of the formula for the normal curve (discussed in

Chapter 6), which de Moivre discovered in 1733. De Moivre's other book, *A Treatise of Annuities on Lives,* was published in 1752. It was highly original and laid foundations for the mathematics of life insurance.

De Moivre took great pains to free the science of probability from its connection with gambling and also to establish a connection between probability and theology. He says in *The Doctrine of Chances:*[1]

And thus in all cases it will be found, that although Chance produces irregularities, still the Odds will be infinitely great, that in process of Time, those Irregularities will bear no proportion to the recurrency of that Order which naturally results from Original Design. . . . Again, as it is thus

[1] Abraham de Moivre, *The Doctrine of Chances,* 3rd ed. (London: 1756; reprinted New York: Chelsea, 1967), pp. 251–54.

demonstrable that there are, in the constitution of things, certain Laws according to which Events happen, it is no less evident from Observation, that these Laws serve to wise, useful and beneficent purposes, to preserve the stedfast Order of the Universe, to propagate the several Species of Beings, and furnish to the sentient Kind such degrees of happiness as are suited to their State. . . . Yet there are Writers, of a Class indeed very different from that of *James Bernoulli,* who insinuate as if the *Doctrine of Probabilities* could have no place in any serious Enquiry; and that studies of this kind, trivial and easy as they be, rather disqualify a man for reasoning on every other subject. Let the Reader chuse.

Source: International Encyclopedia of Statistics, vol. 1 (New York: Free Press, 1978), pp. 601–4.

4.2 Pierre de Fermat

Pierre de Fermat (1601–1665) was born at Beaumont-de-Lomagne, France, where his father was a leather merchant. Even though he practiced law, when the courts went into recess he studied literature and mathematics. His knowledge of the chief European languages and the literature of Europe was wide; he even wrote verses in Latin, French, and Spanish. His main contribution, though, was to mathematics—in particular, the theory

of probability. Many of his original ideas have been preserved in a vast correspondence with scientists all over Europe. Among the many correspondents was Blaise Pascal (Biography 4.3) whose name is always linked with Fermat as one of the joint discoverers of the laws of probability. Application 3: "The Paradox of the Chevalier de Méré," shows, perhaps, the most famous instance of this collaboration.

Source: F. N. David, *Games, Gods, and Gambling* (New York: Hafner, 1962), Chapters 8, 9, and Appendix 4.

4.3 Blaise Pascal

Blaise Pascal (1623–1662) was born at Clermont-Ferrand, France. He was a youngster of precocious mental ability and, from an early age, his father devoted himself almost entirely to developing the boy's reasoning powers. At age 14, the elder Pascal introduced Blaise into Mersenne's famous academy where musicians, mathematicians, and natural scientists met regularly. By age 16, Blaise wrote his famous *Essaie pour les Coniques,* which drew acclaim from many for the boy's astonishing mathematical powers. From projective geometry, he turned to mechanical computation, and two years later he invented a calculating machine. Still at a young

age, he made a number of lasting contributions to mathematics (some of which are preserved in his correspondence with de Fermat—see Biography 4.2).

Unfortunately, Pascal's intellectual work was cut short by poor health and his desire to devote himself to prayer and self-denial. After his father's death, Pascal vacillated between a powerful entanglement with Jansenism, an anti-Jesuit religious cult, and a dissolute life (in which he quickly lost the moderate fortune he inherited but also met the Chevalier de Méré noted in Application 3).

Today, strangely enough, Pascal is best known for Pascal's Triangle, although, contrary to his own claims in his *Traité du Triangle Arithmétique,* he was surely not its originator. The triangle represents an ancient system of quickly answering questions such as these: In a family of 10 children, what is the likelihood that exactly 3 will be girls? When tossing a coin 10 times, what is the probability that exactly 3 heads will appear? When spinning the roulette wheel 10 times (and not counting the zero) what is the chance for getting red precisely 3 times?

The answer, of course, is the same each time, and it can be found with the help of a tree diagram that notes all the possible $2^{10} = 1,024$ outcomes and allows one to count the ones that meet the specifications. Yet less laboriously, one can create the triangle in question and shown in the accompanying figure.

To begin with, note how the triangle is constructed, without trying to fathom its meaning. To construct it, one would begin by writing down two 1's in the first row, side by side. Next, in the second row, below to the left and below to the right, one would put two more 1's and would then fill in the gap between with the *sum* of the two numbers above the gap. These numbers in row 1 are 1's, so one would write 2. Next, one would put 1's at the extremes of the third line and would then fill in the two gaps with 3's (once more the *sum* of the numbers above each gap, which are 2 and 1). And so it goes. By the time one would reach the 10th line, one would have constructed Pascal's triangle as printed here. But note: there is no limit to the number of rows one could construct. Can you see why an 11th row would contain entries of 1, 11 (the sum of 1 + 10), 55 (the sum of 10 + 45), 165, 330, 462, 462, 330, 165, 55, 11, and 1?

What does it all mean? First, as the accompanying table lists, the *number of each row* represents the number of trials of some experiment with two possible types of outcome per trial. Row 1, for example, represents *one* trial, such as tossing a coin once. Row 2 represents *two* trials, such as tossing a coin twice. Row 10 represents *ten* trials, such as tossing a coin ten times.

Second, the *sum of the actual entries in each row* represents the number of possible outcomes. The numbers in row 1 sum to 2; there are 2 possible outcomes of one trial (heads or tails, if a coin is tossed once). The numbers in row 2 sum to 4; there are 4 possible outcomes of two trials (HH, HT, TH, and TT if a coin is tossed twice). And the sum of entries in row 10 tells us that there are 1,024 possible outcomes in 10 trials (as when a coin is tossed ten times).

Third, *each individual number in a given row* represents the possible number of times some particular outcome (such as x heads) will occur. The first number in a given row tells us how often n particular outcomes (such as n heads) will occur in n trials; the second number tells us how often $n - 1$ particular outcomes (such as $n - 1$ heads) will occur in n trials, and so on. In row 1, for example, the first 1 might denote 1 possibility of getting $n = 1$ heads in 1 trial; the second 1 might denote 1 possibility of getting $n - 1 = 0$ heads in 1 trial. In row 2, similarly, the first 1 might denote 1 possibility of getting $n = 2$ heads in 2 trials; the 2 might denote 2 possibilities of getting $n - 1 = 1$ head in 2 trials, and the last 1 might denote 1 possibility of getting $n - 2 = 0$ heads in 2 trials. Given the 4 possible outcomes, these numbers imply, of course, probabilities of 1/4, 2/4, and 1/4, respectively. Thus row 10 gives, for example, the probabilities for a sequence of 10 children, 10 tosses of a coin, 10 spins of the wheel at roulette, or 10 of any series of equal chances. Note how the numbers in row 10 add to 1,024. These are all the *possible* outcomes. We can

```
              1   1
            1   2   1
          1   3   3   1
        1   4   6   4   1
      1   5  10  10   5   1
    1   6  15  20  15   6   1
  1   7  21  35  35  21   7   1
1   8  28  56  70  56  28   8   1
1  9  36  84 126 126  84  36   9   1
1 10 45 120 210 252 210 120  45  10  1
```

Row Number = Number of Trials, n	Number of Possible Outcomes, 2^n
1	2
2	4
3	8
4	16
5	32
6	64
7	128
8	256
9	512
10	1,024

read the numbers in the row, from left to right. The first one is a 1. There is a probability of 1 in 1,024, it tells us, of having 10 girls out of 10 children (or 10 heads in 10 tosses or 10 reds in 10 spins of the wheel).

We turn to the second number, a 10. There is a probability of 10 in 1,024, it says, of having 9 girls out of 10 children (or 9 heads or 9 reds). And there is a probability of 45 in 1,024, the third number says, of having 8 girls out of 10 (or 8 heads or 8 reds). And so it goes until the final number in row 10 gives us the probability of 0 girls (or heads or reds) out of 10: it is 1 in 1,024.

Thus, we have our answer: there are 120 chances in 1,024 for each of the 3-in-10 events listed.

Note: Can you see that row 2 tells us the same thing as the tree diagram in panel (a) of Figure 4.2? There are four possibilities, it says, the sum of the numbers in line 2. There is 1 chance in 4 of getting 2 heads, there are 2 chances in 4 of getting 1 head, there is 1 chance in 4 of getting 0 heads. Pascal's triangle, it turns out, is another way (and a quicker one) of getting some of the results a tree diagram might provide.

Sources: F. N. David, *Games, Gods, and Gambling* (New York: Hafner, 1962), Chapters 8, 9, and Appendix 4; Darrell Huff, *How to Take a Chance* (New York: Norton, 1959), pp. 170–73; *Dictionary of Scientific Biography,* vol. 10 (New York: Scribner's, 1974), pp. 330–42.

Summary

1. The theory of probability helps us figure the likelihood of specific occurrences. Among the important concepts that the theory employs are those of the random experiment, the sample space, and the random event. Any activity that will result in one and only one of several well-defined outcomes, but that does not allow us to tell in advance which of these will prevail in any particular instance, is called a *random experiment.* Any one of a random experiment's possible outcomes, the occurrence of which rules out the occurrence of all the alternative outcomes, is called a *basic outcome;* a listing of all basic outcomes constitutes the *sample space.* Any subset of the sample space is a *random event;* it can be simple (containing one basic outcome) or composite (containing more than one). It is crucial to distinguish between mutually exclusive, collectively exhaustive, complementary, and compatible events as well as to distinguish the union from the intersection of two events.

2. Two types of random experiments exist: those that can be repeated over and over again and those that are unique. Accordingly, there are two numerical measures of chance that estimate the likelihood that a particular event will occur. Such a measure for a repeatable random experiment is called an *objective* probability; a corresponding value for a unique random experiment is called a *subjective* probability. The objective probability can be determined theoretically (by conducting repeated thought experiments) as the ratio of the number of equally likely basic outcomes favorable to an event A to the possible number of such outcomes.

An objective probability can also be determined empirically (by conducting repeated actual experiments) as the ratio of the number of times an event A did occur in the past to the maximum number of times event A could have occurred. In either case, the resultant ratio must lie between 0 and 1. The same is true for a subjective probability value that, in contrast, is derived neither by logical deduction nor from repeated experiments but reflects solely a personal degree of belief in the likelihood of an occurrence.

3. When sample spaces are extremely large, special counting techniques are needed to determine the numbers of favorable and possible outcomes of random experiments (and, thus, probability values). Such counting techniques employ the concepts of factorials, permutations, and combinations.

4. Deriving probability values for single events is one thing; manipulating these values to determine the combined probability for the union or intersection of several events is another. The procedure for computing the probability for the occurrence of a *union* of two or more events is summarized by the *addition law.* A general version applies to all types of events; a special law applies to mutually exclusive events only. The procedure for computing the probability for the occurrence of an *intersection* of two or more events is summarized by the *multiplication law.* A general version applies to all types of events; a special law applies to independent events only. A full understanding of the general multiplication law requires an understanding of the concepts of unconditional, conditional, and joint probability; any application of the special law presupposes our ability to distinguish dependent from independent events. There are a number of famous applications of the probability laws of addition and multiplication, and these laws can be illustrated with the help of tree diagrams.

5. An important extension of the traditional calculus of probability is provided by *Bayes's theorem,* a formula for revising a subjective *prior* probability value on the basis of results obtained by an empirical investigation and for, thus, obtaining a *posterior* probability value.

Key Terms

addition law
basic outcome
Bayes's theorem
collectively exhaustive events
combinations
compatible events
complementary events
composite events
conditional probability
dependent events
disjoint events
elementary events
factorial

factorial product
incompatible events
independent events
intersection of two events
joint probability
law of large numbers
marginal probabilities
multiplication law
mutually exclusive events
objective probability
outcome space
permutations
posterior probability

prior probability
probability space
random event
random experiment
sample points
sample space
simple event
subjective probability
theory of probability
tree diagram
unconditional probability
union of two events
Venn diagrams

Questions and Problems

The computer programs noted at the end of this chapter can be used to work many of the subsequent problems.

1. a. Determine the number of basic outcomes on the *American-style* roulette wheel shown in Figure A. Are they univariate?

   b. Determine the number of basic outcomes when tossing three coins once. Are they univariate? Find your answer with the help of a tree diagram.

2. With the help of a tree diagram, determine the sample space for a random experiment that takes

Figure A

3 balls from an urn containing 1 red ball, 1 green ball, and 1 white ball, and do so *with* replacement.

3. In the four sample spaces given in Figures B, C, D, and E (shown at the top of the opposite page), show the favorable outcomes for the events noted below by shading the appropriate subsets of the sample spaces. (i) Shade the subset of Figure B that corresponds to getting a sum of 5 or 6 when rolling two dice once. (ii) Shade the subset of Figure C that corresponds to getting a sum of 2 when rolling two dice once. (iii) Shade the subset of Figure D that corresponds to drawing a black king from a deck. (iv) Shade the subset of Figure E that corresponds to drawing a black card or an ace from a deck.

4. Classify each of the following random events as simple or composite: (i) Getting a 3 when rolling a single die once. (ii) Getting an odd number when rolling a single die once. (iii) Getting a sum of 5 or 6 when rolling two dice once. (iv) Getting a sum of 2 when rolling two dice once. (v) Getting a 7 when drawing one card from a deck.

Figure B

Figure C

Figure D

	Clubs	Diamonds	Hearts	Spades
King	♣ K	♦ K	♥ K	♠ K
Queen	♣ Q	♦ Q	♥ Q	♠ Q
Jack	♣ J	♦ J	♥ J	♠ J
10	♣ 10	♦ 10	♥ 10	♠ 10
9	♣ 9	♦ 9	♥ 9	♠ 9
8	♣ 8	♦ 8	♥ 8	♠ 8
7	♣ 7	♦ 7	♥ 7	♠ 7
6	♣ 6	♦ 6	♥ 6	♠ 6
5	♣ 5	♦ 5	♥ 5	♠ 5
4	♣ 4	♦ 4	♥ 4	♠ 4
3	♣ 3	♦ 3	♥ 3	♠ 3
Deuce	♣ 2	♦ 2	♥ 2	♠ 2
Ace	♣ A	♦ A	♥ A	♠ A

Figure E

	Clubs	Diamonds	Hearts	Spades
King	♣ K	♦ K	♥ K	♠ K
Queen	♣ Q	♦ Q	♥ Q	♠ Q
Jack	♣ J	♦ J	♥ J	♠ J
10	♣ 10	♦ 10	♥ 10	♠ 10
9	♣ 9	♦ 9	♥ 9	♠ 9
8	♣ 8	♦ 8	♥ 8	♠ 8
7	♣ 7	♦ 7	♥ 7	♠ 7
6	♣ 6	♦ 6	♥ 6	♠ 6
5	♣ 5	♦ 5	♥ 5	♠ 5
4	♣ 4	♦ 4	♥ 4	♠ 4
3	♣ 3	♦ 3	♥ 3	♠ 3
Deuce	♣ 2	♦ 2	♥ 2	♠ 2
Ace	♣ A	♦ A	♥ A	♠ A

5. Classify each of the following sets of events in two ways: first, as mutually exclusive or compatible and, second, as collectively exhaustive or not collectively exhaustive: (i) Money supply growth of 5 percent per year; money supply growth of 9 percent per year. (ii) Sales under $5 million per year; sales between $5 million and under $10 million per year; sales of $10 million per year or more. (iii) Profits between $1 and $7 million per year; profit of $5 million per year. (iv) Defective parts below 7 percent of output; defective parts above 2 percent of output; defective parts equal to 5 percent of output. (v) Real GNP growth of 3 percent per year; inflation of 7 percent per year; unemployment of 8 percent of the labor force.

6. Consider the sample space $S = \{1, 2, 3, 4, 5, 6, 7, 8, 9, 10\}$, as well as these random events:

$A = \{2, 4, 6, 8, 10\}$ $D = \{3, 4, 5, 6, 7\}$

$B = \{1, 2, 9, 10\}$ $E = \{7, 8, 9\}$

$C = \{1, 3, 5, 7, 9\}$ $F = \{2, 3, 4, 5\}$

Are the following sets of events mutually exclusive, collectively exhaustive, both, or neither?

A, B	B, C	C, E	A, D, F
A, C	B, D	C, F	B, D, E
A, D	B, E	D, E	C, E, F
A, E	B, F	D, F	
A, F	C, D	E, F	

7. State the complements of each of the following events: (i) Finding defective items in a group of items. (ii) Inflation above 10 percent per year. (iii) Real GNP growth of 3 percent per year or more. (iv) Drawing an ace from a deck of cards. (v) Finding no defective items in a group of items.

8. Identify each of the following descriptions as either the union or the intersection of events: (i) Drawing the king of hearts from a deck of cards. (ii) Drawing a black king from a deck of cards. (iii) Drawing a black card or an ace from a deck of cards. (iv) Drawing an ace or a deuce from a deck of cards. (v) Drawing a red face card from a deck of cards.

9. A questionnaire is sent to 500 homes. If we define encountering a brick house as event A, encountering a house over 40 years of age as event B, and encountering an all-electric house as event C, are the following unions, intersections, or complements of events: A and B; $\overline{A}$; $\overline{B}$ and C; A or C; B or $\overline{C}$.

10. In Figures F and G, depict your answers to 8 (iv) and 8 (v) graphically.

11. Classify each of the following as an objective or subjective probability statement (and if the former, as a classical or empirical one): (i) There is one chance in two for a recession this fall. (ii) There is 1 chance in 100 that we will meet a car coming up the other side of this hill. (iii) The probability is 1/52 that one picks the king of spades when taking one card from a deck. (iv) There is no chance for a meltdown of this nuclear plant. (v) The odds are 20 to 1 in favor of sales going up because of our product's new name. (vi) Past experience indicates that the odds against engine-mount failure on this type of airplane are a million to one. (vii) There is one chance in three that the next U.S. President will be a Democrat. (viii) In 600 tosses of a die, you can expect to get a 3 face 100 times. (ix) The probability is .0001 that radicals will get hold of nuclear weapons. (x) The probability of picking an even-denomination card from a deck equals 5/13.

12. Calculate the following (objective) probabilities:
   a. Getting red when spinning the American roulette wheel (Figure A).
   b. Getting two heads and one tail in the random experiment described in problem 1b.
   c. Getting two red balls in the random experiment described in problem 2.
   d. For each of the four events described in problem 3.
   e. For each of the five events described in problem 4.
   f. For each of the five events described in problem 8.

Figure F

	Clubs	Diamonds	Hearts	Spades
King	♣ K	♦ K	♥ K	♠ K
Queen	♣ Q	♦ Q	♥ Q	♠ Q
Jack	♣ J	♦ J	♥ J	♠ J
10	♣ 10	♦ 10	♥ 10	♠ 10
9	♣ 9	♦ 9	♥ 9	♠ 9
8	♣ 8	♦ 8	♥ 8	♠ 8
7	♣ 7	♦ 7	♥ 7	♠ 7
6	♣ 6	♦ 6	♥ 6	♠ 6
5	♣ 5	♦ 5	♥ 5	♠ 5
4	♣ 4	♦ 4	♥ 4	♠ 4
3	♣ 3	♦ 3	♥ 3	♠ 3
Deuce	♣ 2	♦ 2	♥ 2	♠ 2
Ace	♣ A	♦ A	♥ A	♠ A

Figure G

	Clubs	Diamonds	Hearts	Spades
King	♣ K	♦ K	♥ K	♠ K
Queen	♣ Q	♦ Q	♥ Q	♠ Q
Jack	♣ J	♦ J	♥ J	♠ J
10	♣ 10	♦ 10	♥ 10	♠ 10
9	♣ 9	♦ 9	♥ 9	♠ 9
8	♣ 8	♦ 8	♥ 8	♠ 8
7	♣ 7	♦ 7	♥ 7	♠ 7
6	♣ 6	♦ 6	♥ 6	♠ 6
5	♣ 5	♦ 5	♥ 5	♠ 5
4	♣ 4	♦ 4	♥ 4	♠ 4
3	♣ 3	♦ 3	♥ 3	♠ 3
Deuce	♣ 2	♦ 2	♥ 2	♠ 2
Ace	♣ A	♦ A	♥ A	♠ A

13. Comment on the following bet someone suggests to you: "Let's toss two dimes at a time. Three outcomes are possible: 2 heads, 2 tails, or one of each. If two heads show or two tails, you win; if there is one tail and one head, I win. Because you have two chances out of three to win and I have only one, I will pay you 50¢ whenever you win, but you must pay me $1 whenever I win. In the long run, neither one of us will win or lose."

14. Comment on this reasoning by a sales clerk: "From past experience, I know that 1 customer out of 20 buys my product. I have just had 19 customers who didn't buy. The next one is certain to buy."

15. **a.** Consider Figure A.13 at the back of the book. If the data still pertain, what is the probability that a randomly selected car renter is an Avis customer?

    b. Consider Table A.11 at the back of the book. What is the probability that a randomly selected can of peas will have a net weight in the 15.5 to under 15.7 oz. range?

16. Select a single letter at random from the English alphabet. What is the probability of it being a vowel? What if you had selected this letter from among the alphabet's first 10 letters?

17. Consider Table 3.B that accompanies Chapter 3's problem 8. If an order is randomly selected, what is the probability of its value being under $100? Between $100 and under $500? Above $500?

18. Compute factorials for the following numbers: 3, 11, 19, 25.

19. For each of the following sets, compute the number of permutations of x at a time out of n distinct items, with no repetitions allowed.

n	x
2	1
6	4
15	2
25	7
25	32

20. Create a table entitled "Number of Permutations (x at a time out of n distinct items)" that lists values of n from 0 to 10 in the first column, values of x from 0 to 10 in the first row, and the values of P_x^n in the body of the table.

21. A ship has five flags each of green, red, and yellow. If it uses all 15 flags for each, how many different messages can it send? What if it had ten flags, of which two each were green and red, while three each were blue and yellow?

22. For each of the following sets, compute the number of combinations of x at a time out of n distinct items, with no repetitions allowed.

n	x
2	2
7	3
13	3
22	10
30	35

23. Create a table entitled "Number of Combinations (x at a time out of n distinct items)" that lists values of n from 0 to 50 in the first column, values of x from 0 to 10 in the first row, and the values of C_x^n in the body of the table.

24. In the Massachusetts Megabuck Lottery, six numbers out of 1 to 36 must be picked (in any order, with no repetitions allowed). How many different tickets would one have to buy in order to be *certain* to win? Given the fact that the jackpot usually varies between $3 million and $10 million, and a ticket costs only $1, why doesn't anyone do it?

25. If a restaurant serves 3 salads, 5 entrees, 4 vegetables, 6 drinks, and 7 desserts, in how many different ways can one get 2 entrees, 2 vegetables, and 2 desserts? If a complete dinner includes two of everything, how many different dinner combinations are being offered?

26. How many different combinations of 3 cards can be taken from a deck of 52 cards? How many permutations? Show your calculations.

27. How many different committees of 3 males and 2 females can be formed from a group of 10 males and 10 females?

28. If each participant in a 10-player tournament must play every other participant, how many plays must be made?

29. If 13 cards are taken from a deck, what is the probability that they include the king of spades and the king of hearts?

30. If 5 cards are taken from a deck, what is the probability of getting a *royal* flush (a sequence from 10 to king plus ace of the same suit)? Of getting *any* straight flush (a sequence of 5 cards from the same suit, starting with any number between 2 and 10)?

31. If you think it applies, use the general or special addition law to determine each of the following probabilities.

    a. The probability of getting a sum of 5 *or* 6 when rolling two dice.

    b. The probability of getting a black card *or* an ace from a deck of cards.

c. The probability of inflation *or* recession, if that of inflation is .8, that of recession is .2, and that of both at the same time is .1.

d. The probability for a city experiencing, before the end of this century, a natural disaster such as an earthquake, a flood, or a tornado, if the separate probabilities are .1, .4, and .7, respectively.

32. A market survey shows that 60 percent of consumers like Brand A, 40 percent like Brand B, 15 percent like both. What is the probability of someone chosen at random liking A *or* B?

33. A market survey yields the results given in Table 4.A. What is the probability of someone chosen at random liking A *or* B *or* C:

Table 4.A   Percent of Consumers Who Like Brand

A	20
B	10
C	5
A and B	8
A and C	2
B and C	7
All three	5

34. From the information in Table 4.B, construct a joint probability table and clearly identify all the joint and marginal probabilities.

35. Calculate all the conditional probabilities associated with problem 34.

36. Calculate the six conditional probabilities associated with the joint probabilities of .10, .13, and .57 that are given in Table 4.3.

37. From the information given in Tables 4.C to 4.F, calculate the following probabilities and identify them as conditional, joint, or marginal: (i) The probability of an inspector tagging a satisfactory item as unacceptable. (ii) The probability of a female guest ordering dish A. (iii) The probability of a random person being a nonsmoker with heart disease. (iv) The probability of a person with heart disease being a nonsmoker. (v) The probability of a smoker being free of heart disease. (vi) The probability of a loan being under $500 and also in excess of 30 days old. (vii) The probability of a loan being above $500 in size.

38. Still using Tables 4.C to 4.F, calculate the following probabilities and identify them as conditional, joint, or marginal: (i) The probability of an item being defective. (ii) The probability of a defective

Table 4.B   Number of Workers Whose . . .

Score on Hiring Test Was . . .	Performance on the Job During First Year Was . . .		Total
	Good, G	Bad, B	
High, H	240	60	300
Low, L	10	90	100
Total	250	150	400

Table 4.C   Number of Items . . .

Inspected and . . .	Tagged as Acceptable, A	Tagged as Unacceptable, U	Total
In Fact Satisfactory, S	600	30	630
In Fact Defective, D	35	35	70
Total	635	65	700

Table 4.D Number of Restaurant Patrons Who . . .

	Ordered Dish A	Ordered Dish B	Total
Were Male, M	20	180	200
Were Female, F	90	10	100
Total	110	190	300

Table 4.E Number of Persons Who . . .

	Had Heart Disease, H	Were Free of Heart Disease, F	Total
Were Smokers, S	300	200	500
Were Nonsmokers, N	100	300	400
Total	400	500	900

Table 4.F Number of Loans . . .

	30 Days Old or Less, L	In Excess of 30 Days Old, E	Total
Of up to $500 Inclusive, U	700	1,000	1,700
Above $500, A	1,400	2,000	3,400
Total	2,100	3,000	5,100

item being tagged acceptable. (iii) The probability of a B order having come from a male. (iv) The probability of a B order. (v) The probability of heart disease. (vi) The probability of a loan being above $500 and also in excess of 30 days old.

39. Apply the general multiplication law, according to which $p(A \ and \ B) = p(A) \cdot p(B|A)$, to the intersection of the two events given in the first row and second column of each of Tables 4.B to 4.F. Confirm your results by direct calculation from the table data.

40. Rework problem 39, but with respect to the intersection of the two events given in the second row and first column of Tables 4.B to 4.F.

41. Consider Tables 4.C to 4.F and, using an appropriate formula from the calculus of probability determine $p(S \ or \ A), p(M \ or \ A), p(N \ or \ H), p(U \ or \ L)$.

42. Once again find the answers to problem 41, but this time do so without a formula, simply by inspecting the table entries in question.

43. Are the events noted in problem 39 dependent or independent? Why?

44. Are the events noted in problem 40 dependent or independent? Why?

45. A new movie is released. The producing company judges the probability of a smashing success at $p = .3$. It also knows that a certain reviewer has liked 75 percent of all movies that became greatly successful and has disliked 90 percent of all movies that later turned out to be ghastly failures. If the company followed the Bayesian approach, how would it revise its probability-of-success figure if it learned that the reviewer praised the movie? What if the reviewer criticized it severely?

Selected Readings

David, F. N. *Games, Gods, and Gambling*. New York: Hafner, 1962. On the origin and history of probability from the earliest times to the Newtonian era.

Feller, William. *An Introduction to Probability Theory and Its Applications*. New York: Wiley, 1968–1971, 2 volumes. A classic text.

Heron House Editors. *The Odds on Virtually Everything*. New York: Putnam's, 1980. A popular book that uses empirical data to determine probabilities for practically anything, anywhere, any time.

Huff, Darrell, *How to Take a Chance*. New York: W. W. Norton, 1959. An amusing but well-informed book on probability and its many applications.

Mises, Richard von. *Probability, Statistics and Truth*. London: Allen and Unwin, 1961. A classic statement of the objectivist view of probability theory that defines probability as the relative frequency of the observed attribute that would be found if the observations were indefinitely continued.

Savage, Leonard J. *The Foundations of Statistics*. New York: Dover, 1972. A classic statement arguing that subjective probability alone is essential to decision making and rejecting the von Mises view noted above.

Schlaifer, Robert. *Probability and Statistics for Business Decisions*. New York: McGraw-Hill, 1959. One of the first books emphasizing the Bayesian approach to decision making that blends subjective probabilities with objective ones.

Computer Programs

The DAYSAL-2 personal-computer diskettes that accompany this book contain two programs of interest to this chapter:

6. Factorials, Permutations, Combinations lets you make the calculations called for by Boxes 4.C–4.F, as well as others that involve permutations or combinations which do allow repetition among the x items picked at a time.

18. Cointoss, if you have graphics capability, lets you perform the kind of experiment illustrated in Figure 4.7.

CHAPTER FIVE

Probability Distributions:
Discrete Variables

In an earlier chapter, we defined *variables* as characteristics possessed by persons or objects, called elementary units, in which a statistician is interested. Such variables, we noted, can be *qualitative* or *quantitative*. Qualitative variables differ in kind rather than in degree among elementary units and are normally not described numerically. Quantitative variables, on the other hand, differ in degree rather than in kind among elementary units, and these variables are expressed numerically. Consider conducting a random experiment such as tossing three coins or taking a sample of five employees of a firm. If one were interested only in qualitative characteristics, such as the types of faces showing after the coins had been tossed or the races of the selected employees, one might record the experimental outcomes in words, such as "head, tail, head" or "black, white, black, white, white." If, on the other hand, one were interested in quantitative characteristics such as the numbers (or proportions) of heads showing in different trials of the experiment or the numbers (or proportions) of black employees found in different samples, one would record the outcomes numerically; perhaps, as 0, 1, 2, or 3 heads in a toss, as 0, 1, 2, 3, 4, or 5 blacks in a sample, or as the proportions such outcomes represent among all possible outcomes (1/3 heads, 2/5 blacks, and so on). This chapter and the next will deal exclusively with such quantitative variables. The present chapter, moreover, will be concerned only with *discrete* quantitative variables that can assume values only at specific points on a scale of values, with inevitable gaps between. The possible observations about such variables are countable and (with some exceptions) are finite. (The number of heads showing when coins are tossed or the number of black employees in firms are examples of such discrete quantitative variables.) The next chapter will focus on *continuous* quantitative variables that can assume values at all points on a scale of values, with no breaks between possible values.

The possible observations about such variables are not countable and are infinite in number. (Unlike the numbers of black employees, their ages, heights, or weights provide examples of such continuous quantitative variables.)

BASIC CONCEPTS

Before proceeding, we must become familiar with two important new concepts—namely, *random variable* and *probability distribution*.

The Random Variable

It is customary to refer to any quantitative variable the numerical value of which is determined by a random experiment (and, thus, by chance) as **random variable,** or **chance variable.** The number of heads observed when tossing three coins is such a variable; chance determines whether the number is 0, 1, 2, or 3. The number of points showing on the top face of a die that is rolled once is a random variable, too; so is the difference or *point spread* observed between the top faces of two dice that are rolled at the same time. The serial number of a lottery ticket about to be drawn from among 100,000 tickets in a bowl is a random variable as well.

As is usual in the field of statistics, this new concept is, however, applicable far beyond the world of games and gambling. Imagine yourself about to draw a random sample from among a list of employees of a firm in order to determine, say, their mean salary (or the proportion of blacks on the payroll). Before the sample is taken, the sample mean (or proportion) can also be viewed as a random variable because chance will determine which of many possible samples will actually be selected and, therefore, which of many possible sample means or proportions will actually be observed. Innumerable other variables that are of interest to business executives and economists are random variables, too. These range widely from tomorrow's closing price of a given stock to the number of customers arriving during the next 10 minutes at the supermarket checkout counter to the future rate of return on an investment undertaken today.

The Probability Distribution

Each of the possible numerical values of a random variable will occur, however, with a certain probability; a systematic listing of each possible value of a random variable with the associated likelihood of its occurrence is called the random variable's **probability distribution.** Table 5.1 shows two such distributions for discrete random variables. Part A gives the possible values, *x*, of the point spread when two dice are rolled at the same time, along with the associated probability, *p,* that the actual value of the random variable, *R,* emerging from such a random experiment equals any given possible value, *x.* The stated probabilities can be confirmed by referring to panel (b) of Figure 4.2 on page 117. Naturally, the sum of the separate probabilities of the six mutually exclusive and collectively exhaustive events equals 1. Part B shows another discrete probability distribution (the data are assumed); this distribution concerns the

Table 5.1 Two Discrete Probability Distributions

As these examples indicate, a probability distribution is similar to a relative frequency distribution. A probability distribution provides a probability value, p, for each value, x, of a random variable, R.

(A) Random Variable, R, = Point Spread from Rolling Two Dice		(B) Random Variable, R, = Hourly Credit-Verification Requests	
Possible Values, x	Probability, $p(R = x)$	Possible Values, x	Probability, $p(R = x)$
0	6/36	0	.05
1	10/36	1	.10
2	8/36	2	.15
3	6/36	3	.17
4	4/36	4	.19
5	2/36	5	.09
	$\Sigma p = (36/36) = 1.00$	6	.08
		7	.06
		8	.06
		9	.02
		10	.02
		11	.01
			$\Sigma p = 1.00$

hourly credit-verification requests received by a credit bureau that never receives more than 11 such requests per hour.[1] Once again, the separate probabilities of all the mutually exclusive and collectively exhaustive events sum to 1.

The information given in Table 5.1 can be presented in graphical form as well, as Figure 5.1 demonstrates. Panels (a) and (b) depict the data of Table 5.1 directly. Panels (c) and (d), on the other hand, indicate how any regular probability distribution that shows probabilities for individual values of a random variable can be converted into a **cumulative probability distribution** that shows the probabilities of a random variable being less than or equal to any given possible value. Nothing more complicated is involved than adding the probability of any given value of the random variable to the probabilities associated with all the preceding values.

A final note: Often it is impossible to express even a discrete probability distribution in a table or a graph because too many possible values of the random variable exist. When that happens, the distribution can be expressed only symbolically. Take the case of drawing one lottery ticket from among 100,000 tickets (or of sampling one item for every 100,000 items sitting in a warehouse or moving along a production line.) It would be tedious and unnecessary to construct a table with 100,000 rows, stating that

[1]Throughout this chapter, we follow the simplified summation notation that was first explained on p. 79 of Chapter 3. Without exception, whenever we encounter the Σ sign, we sum *all* the available values of the variables following that sign.

Figure 5.1   Regular and Cumulative Probability Distributions of Discrete Random Variables

Stick diagrams, such as those in panels (a) and (b), are graphical devices that illustrate discrete probability distributions, such as those given in Table 5.1. As comparisons of panel (a) with (c) and of panel (b) with (d) indicate, any regular probability distribution can easily be converted into a cumulative one by adding the probability of any given value of the random variable to that of all the preceding values. While the probability of a point spread of 2 equals 8/36, for example, that of a point spread of 2 or less equals 6/36 + 10/36 + 8/36 = 24/36.

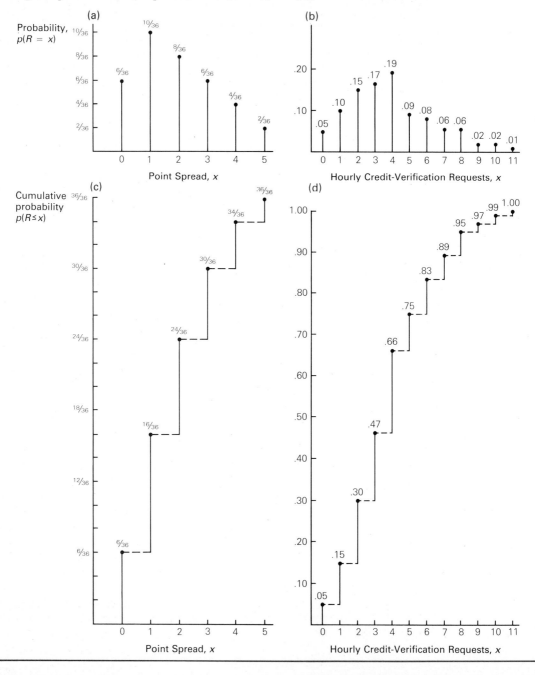

$$p(R = \text{serial number 1}) = \frac{1}{100,000}$$

$$p(R = \text{serial number 2}) = \frac{1}{100,000}$$

$$\vdots \qquad\qquad \vdots$$

$$p(R = \text{serial number 100,000}) = \frac{1}{100,000}$$

It is much easier simply to designate the actual number selected by R and possible serial numbers by x and to write

$$p(R = x) = \frac{1}{100,000}, \text{ where } x = 1, 2, \ldots, 100,000.$$

We will employ this notation throughout. It tells us succinctly that the probability is 1/100,000 for the random variable R to be equal to any specific possible number, x, where the latter can be any integer between 1 and 100,000.

SUMMARY MEASURES FOR THE PROBABILITY DISTRIBUTION

A gambler contemplating the information given in Part A of Table 5.1 [and graphed in panel (a) of Figure 5.1] might wonder what the *average* point spread would be for a large number of rolls. The manager of our credit bureau contemplating the information given in Part B of Table 5.1 [and graphed in panel (b) of Figure 5.1] might similarly wonder about the average hourly credit-verification requests over the long haul. For purposes such as these, it is useful to calculate summary measures. A probability distribution can most easily be summarized (and compared with other such distributions) by using measures of location and spread, like those introduced in Chapter 3. Consider calculating a random variable's arithmetic mean, variance, and standard deviation.

The Arithmetic Mean of a Random Variable

The most important measure of location for a probability distribution—the (weighted) arithmetic mean of the random variable's possible values—is calculated by weighting each of these possible values by its associated probability. This mean of the random variable R is symbolized by μ_R and is also called the **expected value, E_R,** of the random variable because it is the value of the random variable one can expect to find on the average by numerous repetitions of the random experiment that generates the variable's actual value.

Consider, for instance, Part A of Table 5.1. If the experiment of rolling two dice were repeated over and over again, one could expect a point spread of 0 in 6 out of 36 cases, a point spread of 1 in 10 out of 36 cases, and so on. The (weighted) average point spread for a large number of rolls can be calculated as shown in Table 5.2.

The same procedure can be applied to calculate the expected number of hourly credit-verification requests from the data in Part B of Table 5.1. The calculation is shown in Table 5.3.

Table 5.2   Expected Value of Point Spread from Rolling Two Dice

Calculating the expected value of a random variable is equivalent to calculating the weighted arithmetic mean from grouped data, as in Table 3.1 on page 81. Our present x corresponds to the earlier class midpoint, X, and our present p corresponds to the earlier class frequency, f. Thus, the earlier formula (equivalent to Box 3.C), $\mu = \dfrac{\Sigma f \cdot X}{\Sigma f}$, becomes $\mu = \dfrac{\Sigma p \cdot x}{\Sigma p}$, but (as the second column shows) $\Sigma p = 1$ for a probability distribution, making μ simply equal to the sum of the last column, or $\Sigma p \cdot x$.

Possible Value of Point Spread, x	Probability, $p(R = x)$	Weighted Value, $p(R = x) \cdot x$	Calculation
0	6/36	0	
1	10/36	10/36	
2	8/36	16/36	
3	6/36	18/36	$\mu_R = E_R = \dfrac{70}{36} = 1.9444$
4	4/36	16/36	
5	2/36	10/36	
	$\Sigma p = (36/36) = 1$	$\Sigma p \cdot x = (70/36)$	

Source: Table 5.1.

Table 5.3   Expected Number of Hourly Credit-Verification Requests

Possible Number of Requests, x	Probability, $p(R = x)$	Weighted Value, $p(R = x) \cdot x$
0	.05	0
1	.10	.10
2	.15	.30
3	.17	.51
4	.19	.76
5	.09	.45
6	.08	.48
7	.06	.42
8	.06	.48
9	.02	.18
10	.02	.20
11	.01	.11
	1.00	$\mu_R = 3.99$

Source: Table 5.1.

The Variance of a Random Variable

The variance of a random variable, denoted by σ_R^2, is also calculated by using probability weights. As the variance formula demands (see Box 3.J on page 96), these weights are, of course, attached to the squared deviations of the random variable's possible values from its expected value. This procedure is illustrated for our two examples in Tables 5.4 and 5.5.

Table 5.4   Variance of Point Spread from Rolling Two Dice

Calculating the variance of a random variable is equivalent to calculating the variance from grouped data, by the formula given in Box 3.J on p. 96: $\sigma^2 = \dfrac{\Sigma f(X - \mu)^2}{N}$. However, as was true when calculating the arithmetic mean (Table 5.2), the probability values, p, now replace the earlier class frequencies, f, while the sum of these weights (second column below), which corresponds to the earlier $N = \Sigma f$, now equals 1. Thus, the variance simply equals the sum of the last column. Note: the same result can be reached by utilizing the corresponding shortcut method given in Box 3.N on p. 98. Can you show this to be true?

Possible Value of Point Spread, x	Probability, $p(R = x)$	Deviation, $x - \mu_R$	Squared Deviation, $(x - \mu_R)^2$	Weighted Value, $p(R = x) \cdot (x - \mu_R)^2$
0	(6/36) = .1667	−1.9444	3.7807	.6302
1	(10/36) = .2778	−.9444	.8919	.2478
2	(8/36) = .2222	.0556	.0031	.0007
3	(6/36) = .1667	1.0556	1.1143	.1858
4	(4/36) = .1111	2.0556	4.2255	.4695
5	(2/36) = .0556	3.0556	9.3367	.5191
	$\Sigma p = 1.0000$			$\sigma_R^2 = 2.0531$

Source: Tables 5.1 and 5.2.

Table 5.5   Variance of Number of Hourly Credit-Verification Requests

Possible Number of Requests, x	Probability, $p(R = x)$	Deviation, $x - \mu_R$	Squared Deviation, $(x - \mu_R)^2$	Weighted Value, $p(R = x) \cdot (x - \mu_R)^2$
0	.05	−3.99	15.9201	.7960
1	.10	−2.99	8.9401	.8940
2	.15	−1.99	3.9601	.5940
3	.17	−.99	.9801	.1666
4	.19	.01	.0001	.0000
5	.09	1.01	1.0201	.0918
6	.08	2.01	4.0401	.3232
7	.06	3.01	9.0601	.5436
8	.06	4.01	16.0801	.9648
9	.02	5.01	25.1001	.5020
10	.02	6.01	36.1201	.7224
11	.01	7.01	49.1401	.4914
	1.00			$\sigma_R^2 = 6.0898$

Source: Tables 5.1 and 5.3

The Standard Deviation of a Random Variable

As always, the standard deviation equals the square root of the variance. Thus, in our two examples, $\sigma_R = 1.4329$ for case (A) and $\sigma_R = 2.4678$ for case (B).

The formulas for the (weighted) arithmetic mean, the variance, and the standard deviation of a random variable, R, are summarized in Box 5.A.

5.A Summary Measures for the Probability Distribution of a Random Variable, R

$$\mu_R = E_R = \Sigma p(R = x) \cdot x$$
$$\sigma_R^2 = \Sigma p(R = x) \cdot (x - \mu_R)^2$$
$$\sigma_R = \sqrt{\sigma_R^2}$$

where μ_R is the arithmetic mean, σ_R^2 is the variance, and σ_R is the standard deviation of the probability distribution of random variable, R, while $p(R = x)$ is the probability of the random variable equaling x and x is an observed value of that variable.

THE BINOMIAL PROBABILITY DISTRIBUTION

Different types of random experiments give rise to different types of probability distributions; the **binomial probability distribution** is the most important one for discrete variables.[2] It shows the probabilities associated with possible values of a discrete random variable that are generated by a type of experiment called a *Bernoulli process.*

The Bernoulli Process

A **Bernoulli process** (named after James Bernoulli, discussed in Biography 5.1 at the end of the chapter) is a sequence of n identical trials of a random experiment such that each trial (a) produces one of two possible complementary outcomes that are conventionally called *success* and *failure* and (b) is independent of any other trial so that the probability of success or of failure is constant from trial to trial. The number of successes achieved in a Bernoulli process is the **binomial random variable.**

A sequence of coin tosses provides a perfect example. Each toss necessarily produces one of two possible outcomes: heads or tails. Quite arbitrarily, one of these outcomes is called a "success," the other one a "failure"; it does not matter which is called which. No matter how many tosses are made, the probability of success (for example, of getting a head) is precisely 1/2; that of failure (for example, of getting a tail) remains a constant 1/2 as well. The same situation arises in a series of die tosses, provided one classifies the outcomes as "odd" and "even" numbers of dots. Each toss must lead to one of these two outcomes, and each of these outcomes has an unchanging probability of 3/6. But note: the Bernoulli process requires that the success and failure probabilities be constant from one trial to the next; it does not require these

---

[2]Other such distributions for discrete variables, not discussed in this book, include the *hypergeometric* probablity distribution and the *Poisson* probability distribution. See the *Selected Readings* at the end of the chapter.

two probabilities to be *equal* to each other, as happens to be the case in the examples just cited. A constant success probability of .1 and a constant failure probability of .9 satisfies the definition as well.

Interestingly, the conditions defining the Bernoulli process are met by a multitude of random experiments all of which lead to only one of two outcomes. Consider how a new baby must be male or female, how a new product must be liked or disliked, how a new drug must be effective or ineffective. Consider how a salesperson will sell or not sell, how a loan will be granted or denied, how a bid will be won or lost, how the quality of a product will be acceptable or unacceptable, how an account will be correct or in error, how an exam will be passed or failed, how a patient will be alive or dead, how an organ will be infected or not infected . . . the list goes on. But note: Whenever the random experiment involves sampling, successive elementary units must be selected *after replacement* of the units previously sampled in order to meet condition (b) and, thus, keep the probabilities of success and failure from changing as the sequence of trials proceeds. Consider the case of acceptance sampling. If there are 10 defective items in a lot of 100, and a sample of 20 items is to be taken, the probability of finding a defective item is 10/100 for the first item sampled. If it is set aside rather than replaced, the corresponding probability is 10/99 or 9/99 for the second item sampled (depending on whether the first item taken was satisfactory or defective). And the probability continues to change for the next items selected until it might lie between 10/81 and 0/81 for the 20th item (depending on how many defectives were found in the meantime). Only by replacing each item after inspection and remixing the lot can the probability in question be kept unchanged (and can the requirement of the Bernoulli process be strictly met). For practical purposes, sampling without replacement can, however, often be viewed as closely resembling a Bernoulli process, provided the sample is small relative to the population and constitutes, say, 5 percent or less of the population. In such a case, the probabilities involved change from sampling one unit to the next, but the changes are negligible. (If one took a sample of only 2 out of 100 or only 20 out of 1,000—and did so without replacement—the second item's probability of being defective could hardly be distinguished from that of the first item, as the above numbers indicate.)

‖ Practice Problem 1

Tossing Three Coins

To illustrate how a binomial probability distribution is generated, let us first consider a random experiment of tossing three coins, while defining the binomial random variable of interest as the number of heads appearing. Panel (a) of Figure 5.2 uses a probability tree diagram of the type employed in Chapter 4 to determine the possible outcomes of such an experiment. The 8 possible outcomes are shown in the box, wherein H and T refer to head or tail showing and the subscripts, 1, 2, and 3, refer to the first, second, and third coins, respectively. The random variable of interest (number of heads) is given in the next column; the probability of each of these events is calculated by the special multiplication law using the probabilities given along the branches of the tree diagram. Since we are not interested in which particular coin shows a head, but only in the total number of heads, we can combine all the outcomes with the same total and derive the following binomial probability distribution:

Figure 5.2 Bernoulli Trials

These two diagrams illustrate the Bernoulli process with an example in panel (a) concerning the tossing of 3 coins and an example in panel (b) concerning the sampling of 3 items to determine quality. In each case, one of two outcomes must occur: a head or a tail must appear in one case; the sampled item must be defective or satisfactory in the other. In both examples, each trial is independent of all others, and the probabilities involved are constant throughout the experiment. As a comparison of panels (a) and (b) shows, the probabilities of success or failure need not, however, be equal to each other.

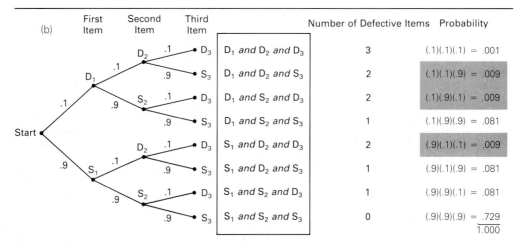

$$p(R = 3 \text{ heads}) = .125$$
$$p(R = 2 \text{ heads}) = .375$$
$$p(R = 1 \text{ head}) = .375$$
$$p(R = 0 \text{ heads}) = \underline{.125}$$
$$1.000$$

Note, for example, how 2 heads can occur in 3 different ways: via the H_1, H_2, T_3 route (with a probability of .125) *or* via the H_1, T_2, H_3 route (with a like probability) *or* via the T_1, H_2, H_3 route (once again with a like probability). Because these three ways of achieving the 2-heads outcome are mutually exclusive events, we can apply the special addition law (Box 4.H on p. 136) and say that the probability of getting 2 heads equals .125 + .125 + .125 = .375. |||

||| Practice Problem 2

Sampling Three Items for Quality

In order to confirm that the Bernoulli process requires the constancy rather than the equality of probabilities from trial to trial, let us reconsider the case of acceptance sampling first illustrated in panel (c) of Figure 4.9 on p. 149 and now reproduced in panel (b) of Figure 5.2. It is assumed that 3 items are picked from a lot of 100 items of which 10 percent are defective. If each sampled item is replaced before the next one is picked (which we assume), the strict conditions of the Bernoulli process are met. The probability of "success" (finding a defective item, say) remains a constant .1 from one trial to the next, and the probability of "failure" (finding a satisfactory item) equals a constant .9. As the box in panel (b) tells us, once again 8 possible outcomes exist, but, unlike in the coin-tossing experiment, these outcomes are not equally likely. Once again, however, several outcomes can be combined because they involve the same number of defective items, and when this is done (and the special addition law is applied), the following binomial probability distribution is found:

$$P(R = 3 \text{ defectives}) = .001$$
$$p(R = 2 \text{ defectives}) = .027$$
$$p(R = 1 \text{ defective}) = .243$$
$$p(R = 0 \text{ defectives}) = \underline{.729}$$
$$1.000 \quad |||$$

The Binomial Formula

Luckily, we need not always draw complicated tree diagrams to find the types of probabilities just calculated. They can also be determined more directly with the help of the formula given in Box 5.B.

5.B The Binomial Formula

$$p(R = x|n,\pi) = c_x^n \cdot \pi^x \cdot (1 - \pi)^{n-x},$$

where $x = 0, 1, 2, \ldots, n$ is the number of successes in n trials, $c_x^n = \dfrac{n!}{x! \, (n - x)!} = $ the binomial coefficient, and π is the probability of success in any one trial.

This **binomial formula** gives us the probability of x successes in n trials of a random experiment that satisfies the conditions of a Bernoulli process. Although the formula looks complicated, it is fairly easy to interpret:

1. The left-hand side of the equation symbolizes the probability of x successes, given the occurrence of n trials and given a probability of success of π in each of these trials. (As the box indicates, x cannot exceed n: one cannot succeed more often than one tries.)

2. The first term on the right-hand side, or c_x^n, is called the **binomial coefficient.** It tells us how many permutations of x successes (and, therefore, $n - x$ failures) can be achieved in n trials. The answer is indicated in Box 5.B as

$$c_x^n = \frac{n!}{x!\,(n - x)!}$$

It tells us, for example, how many sequences exist that contain $x = 2$ successess in $n = 3$ trials. The answer is 3 sequences:

$$c_2^3 = \frac{3!}{2!(3 - 2)!} = \frac{3 \cdot 2 \cdot 1}{2 \cdot 1(1)} = 3.$$

Indeed, as another look at Figure 5.2 confirms, three instances of 2 successes in 3 trials appear in the column labeled "Number of Heads" and also in the column labeled "Number of Defective Items." The kind of information given in these columns, and laboriously derived in Figure 5.2, can thus be gained instantly by calculating the binomial coefficient.

Pascal's triangle, given in Biography 4.3 on page 155, is another device for generating binomial coefficients. Consider how each row stands for a different value of n and how each successive entry in a given row stands for the number of outcomes with 0, 1, 2, . . . , n successes. Thus, the third row represents $n = 3$ trials and the numbers therein tell us to expect 1 outcome with 0 successes, 3 outcomes with 1 success, 3 outcomes with 2 successes, and 1 outcome with 3 successes—a total of 8 possible outcomes. This result corresponds precisely to the answers given in the "Number of Heads" and "Number of Defective Items" columns of Figure 5.2.[3]

3. The next term in the formula, or π^x, represents the probability of success, π, in any one trial of the experiment, such as that of getting a head (.5) or finding a defective item (.1), raised to the xth power, where x is the number of successes.[4]

---

[3]The binomial coefficient, denoted by a lowercase c, is often confused with the combinatorial symbol, denoted by a capital C, but fortunately it does not matter because both are equal to the same ratio. Compare Box 5.B with Box 4.F. The binomial coefficient, however, does not measure combinations. Rather, it measures *permutations* involving *nondistinct* items (Box 4.E). It so happens that in a Bernoulli process any n trials contain precisely *two* subsets of like items (namely, x successes and $n - x$ failures); hence, the Box 4.E formula simplifies to $P^n_{x,\,n-x} = \dfrac{n!}{x!(n - x)!}$ and, thus, gives precisely the same result (now simply denoted as c_x^n) that one obtains when counting *combinations* of x items out of n distinct items (usually denoted by C_x^n). A look at Figure 5.2 can quickly establish the truth of the foregoing. Note, for example, that there are $c_2^3 = 3$ sequences of 2 heads and 1 tail: HHT, HTH, and THH. Because order matters in the case of permutations, these are 3 *permutations*. Because order does not matter in the case of combinations, these results represent a single *combination*. Then why does the combinatorial formula also give the correct answer? It is pure accident, and one cannot see this in Figure 5.2 because the combinatorial $C_2^3 = 3$ tells us a totally different thing, such as the number of combinations of 2 items that one can take out of 3 *distinct* items. If those 3 items are A, B, C, the combinations in question are $AB = BA$, $AC = CA$, and $BC = CB$.

[4]The symbol π is commonly used because the probability in question in fact equals a population proportion; for example, the ratio of defective items to all items in the population. Thus, π must not be confused here with the physical constant that is the ratio of the circumference to the diameter of the circle.

This term provides us with the joint probability of x successes, calculated by using the special multiplication law for independent events (Box 4.J on page 143). In each of the three shaded rows of panel (b) in Figure 5.2 (which correspond to the three outcomes with 2 successes), π^x appears as $(.1)(.1)$, which is, of course, the same thing as $\pi^x = (.1)^2$.

4. The last term in the formula, or $(1 - \pi)^{n-x}$, represents the probability of failure, $1 - \pi$, in any one trial of the experiment, such as that of getting a tail (.5) or finding a satisfactory item (.9), taken to the $(n - x)$th power, where $n - x$ is the number of failures. This term provides us with the joint probability of $n - x$ failures, once more calculated using the special multiplication law. In each of the three shaded rows of panel (b) in Figure 5.2 (which correspond to the three outcomes with 2 successes and, therefore, 1 failure), $(1 - \pi)^{n-x}$ appears as $(.9)^{3-2} = (.9)^1 = .9$.

5. When the binomial formula *combines* the last two terms and calculates $\pi^x \cdot (1 - \pi)^{n-x}$, it is in fact calculating (once more by using the special multiplication law for independent events) the joint probability of x successes *and* $n - x$ failures in a specific sequence of experimental outcomes. Thus, in panel (b) of Figure 5.2, the joint probability of a sequence of 2 successes *and* 1 failure equals $(.1)^2 (.9)^{3-2} = (.1)(.1)(.9) = .009$, precisely as each one of the shaded rows indicates.

6. Because, however, there exist $c_2^3 = 3$ different ways of getting the type of sequence just discussed (2 successes and 1 failure in 3 trials) and because these 3 outcomes are mutually exclusive events, the special addition law (Box 4.H on page 136) can be applied in order to calculate the overall probability of getting 2 successes in 3 trials of this experiment *somehow*—that is, regardless of whether this result occurs in the fashion shown in row 2 or 3 or 5 of panel (b), Figure 5.2. This overall probability equals .009 (first shaded row) + .009 (second shaded row) + .009 (third shaded row) = .027 (which is the very result for the probability of 2 defective items that was calculated above). Instead of adding $\pi^x(1 - \pi)^{n-x} = .009$ three times, one can, however, simply multiply the term by 3, which is precisely what the binomial coefficient, c_x^n, in the binomial formula accomplishes.

We can test the accuracy of the binomial formula by pretending to seek with its help some of the answers already provided in Figure 5.2.

Practice Problem 3

Coin Tossing
What is the probability of getting 3 heads when tossing 3 coins?

$$p(R = 3 | n = 3, \pi = .5) = c_3^3 \cdot (.5)^3 (1 - .5)^{3-3}$$

$$= \frac{3!}{3!(3-3)!} (.125)(.5)^0 = 1(.125) 1 = .125.$$

This result is precisely the same as that found in row 1 of panel (a). |||

||| Practice Problem 4

Defective Items

What instead is the probability of finding 0 defective items when sampling 3 items under the conditions of panel (b)?

$$p(R = 0|n = 3, \pi = .1) = c_0^3 (.1)^0 (1 - .1)^{3-0}$$

$$= \frac{3!}{0!(3 - 0)!} (1)(.9)^3$$

$$= 1(1)(.9)^3$$

$$= .729$$

Once more, the result is confirmed by that found in the last line of panel (b). |||

Note: While the binomial formula tells us directly the probability of x successes in n trials, it tells us indirectly the probability of the complementary event. Thus, the probability of *not* getting 3 heads in the Practice Problem 3 example equals $1 - .125 = .875$; that of finding *some* defective items in the Practice Problem 4 example equals $1 - .729 = .271$.

||| Practice Problem 5

The Production Process

To show how an entire binomial probability distribution can rapidly be determined by using the binomial formula (without resort to the type of tree diagrams seen in Figure 5.2), consider another problem: a manufacturer uses a production process that produces 20 percent defective items. A defect in any one item is, however, independent of possible defects in other items. (The problem is *not* a matter of some machine becoming occasionally maladjusted and turning out a whole series of defective items until a correction is made. If that were the case, the appearance of one defective item would significantly affect the probability of finding another such item.) If under these conditions 5 units are produced, what is the probability of 0, 1, 2, 3, 4, and 5 of them being defective? The answers are quickly found by a systematic application of the binomial formula, as Table 5.6 attests. |||

The Binomial Probability Distribution Family

A look at the binomial formula in Box 5.B clearly indicates that the probability for any given number of successes varies with two factors: the number of trials, n, and the probability of success in any one trial, π. Each different combination of n and π thus produces a different binomial probability distribution, even though the same formula is applied to derive the distribution. It is customary to refer to any specific binomial probability distribution, such as that given in Table 5.6, as *a member of the binomial distribution family,* it being understood that other members can be found by varying

Table 5.6 Determining a Binomial Probability Distribution for $n = 5$ and $\pi = .2$ Using the Binomial Formula

The entries in the first column together with those in the last column represent a binomial probability distribution. Its mathematical determination, using the three terms in the binomial formula, is illustrated with the help of the three middle columns. The second rows tells us, for example, that there exist 5 different ways of achieving 1 success (along with 4 failures) in the experiment in question and that the probability of each of these sequences equals (.2) multiplied by (.8)(.8)(.8)(.8). Thus, the probability of getting any one of these 5 sequences is five times the former product, or .4096.

Number of "Successes" (defective items found), x	$C_x^n = \dfrac{n!}{x!\,(n-x)!}$	π^x	$(1-\pi)^{n-x}$	$p(R = x\mid n,\ \pi)$
0	$\dfrac{5!}{0!(5-0)!} = 1$	$(.2)^0 = 1$	$(.8)^{5-0} = .32768$	$1(1)(.32768) = .32768$
1	$\dfrac{5!}{1!(5-1)!} = 5$	$(.2)^1 = .2$	$(.8)^{5-1} = .4096$	$5(.2)(.4096) = .4096$
2	$\dfrac{5!}{2!(5-2)!} = 10$	$(.2)^2 = .04$	$(.8)^{5-2} = .512$	$10(.04)(.512) = .2048$
3	$\dfrac{5!}{3!(5-3)!} = 10$	$(.2)^3 = .008$	$(.8)^{5-3} = .64$	$10(.008)(.64) = .0512$
4	$\dfrac{5!}{4!(5-4)!} = 5$	$(.2)^4 = .0016$	$(.8)^{5-4} = .8$	$5(.0016)(.8) = .0064$
5	$\dfrac{5!}{5!(5-5)!} = 1$	$(.2)^5 = .00032$	$(.8)^{5-5} = 1$	$1(.00032)(1) = .00032$
				1.00000

the values of n or π. Figure 5.3, on page 180, depicts nine members of this family, each of them determined by the procedure illustrated in Table 5.6. Indeed, the distribution calculated in Table 5.6 is graphed in panel (d). Three things should be noted. First, given any number of trials, n, the probability of a high proportion of successes rises with the probability of achieving success in a given trial. Consider, for example, panels (a) through (c), all of which involve two trials of a random experiment. The probability of achieving two successes in two trials rises from .04 to .25 and to .64 as the probability of success in a single trial, π, rises from .2 to .5 and then to .8. Similar observations can be made for panels (d) through (f) and panels (g) through (i), respectively.

Second, any binomial probability distribution with a success probability of $\pi = .5$ is perfectly symmetrical regardless of the value of n, as panels (b), (e), and (h) indicate. On the other hand, any such distribution with $\pi < .5$ or $\pi > .5$ is skewed to the right

Figure 5.3   Nine Members of the Binomial-Probability-Distribution Family

In accordance with the binomial formula, each different combination of *n* (number of trials) and *π* (success probability in a given trial) gives birth to a different member of the binomial probability distribution family. Each of the nine distributions shown here has been calculated by the procedure illustrated in Table 5.6, with the results rounded to two decimals.

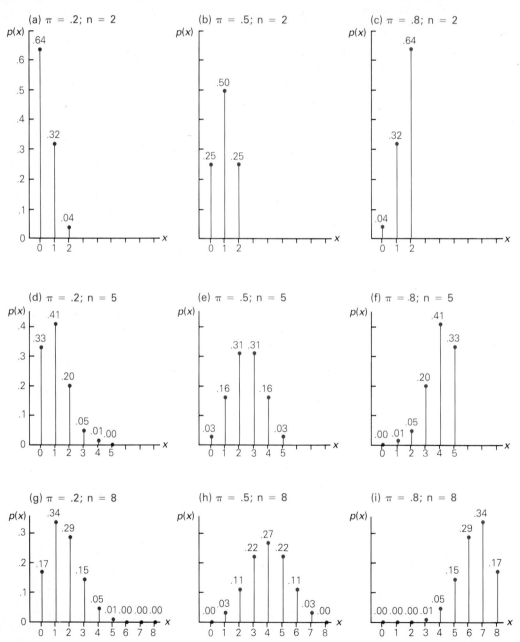

or left, respectively. This is illustrated by panels (a), (d), and (g) on the one hand and panels (c), (f), and (i) on the other. Although Figure 5.3 does not show it, the skewness gets more pronounced the closer π gets to 0 or 1, respectively.

Third, given any value of $\pi \neq .5$ and, therefore, the presence of skewness, the skewness becomes less pronounced as n rises. For high values of n, all such binomial probability distributions approach symmetry, as a progressive comparison of panel (a) with (d) and (g) or of panel (c) with (f) and (i) indicates. As will be noted in the next chapter, this tendency of the binomial probability distribution to take on a bell-shaped form as n increases allows one to approximate such distributions by a normal probability distribution.

Binomial Probability Tables

Undoubtedly, the use of the binomial formula for determining binomial probabilities is less cumbersome than drawing tree diagrams. However, even this mathematical approach can become awkward and can require much time and effort—especially when large factorials are involved and the formula requires us to raise numbers to large powers. Under such circumstances, it would be convenient to have the probabilities already computed for us and listed in tables similar to those of square roots and logarithms. Luckily, such **binomial probability tables** do exist; they list binomial probabilities for various combinations of possible values of n and π. Two types of binomial probability tables exist; one for individual values of x (such as Appendix Table C) and another for cumulative values of x (such as Appendix Table D).

Binomial probabilities for individual values of x. Probabilities for achieving x numbers of successes in a Bernoulli process, given various combinations of n and π, are listed in Appendix Table C. Each value in the body of the table was calculated by the formula in Box 5.B. This particular table is set up for values of n ranging from 1 to 20 and for selected probabilities of success in any one trial, π, between .05 and .50. The table can, however, still be used to determine binomial probabilities for π in excess of .50, as will be shown presently.

To become familiar with Appendix Table C consider the following experiments, always assuming that they meet the conditions of a Bernoulli process.

||| Practice Problem 6

Producing Defective Items

a. A manufacturer uses a production process that produces 20 percent defective items. If 5 units are produced, what is the probability of none of them being defective? Of 3 units being defective? Of all 5 being defective? Looking at Appendix Table C in the section for $n = 5$ and under the column for $\pi = .20$, the desired probabilities can be found instantly: the probability for $x = 0$ is seen to be .3277, that for $x = 3$ is .0512, and that for $x = 5$ is .0003. Note: These answers correspond precisely to those calculated directly in the corresponding rows of Table 5.6.

b. Pursing (a) further, what is the probability of at most 1 unit being defective? Of 2 or 3 units being defective? Of at least 3 units being defective? The special addition law for mutually exclusive events can be applied to the numbers given in Appendix Table C. The probability of at most 1 unit being defective equals the sum of the separate probabilities of 0 or 1 unit being defective; that is, .3277 + .4096 = .7373. By similar reasoning, the probability of 2 or 3 units being defective is found as .2048 + .0512 = .256, while that of at least 3 units being defective equals the sum of the separate probabilities of 3 and 4 and 5 units being defective; that is, .0512 + .0064 + .0003 = .0579. ‖‖

‖‖ Practice Problem 7

Punching Data Cards

a. Now consider a case in which the probability of success in any one trial, π, exceeds .50, the highest value found in Appendix Table C. Take a data processor who punches cards correctly 95 percent of the time. If 20 cards are punched, what is the probability that precisely 16 cards are correct?

The Appendix Table C section with $n = 20$ seems to be of no help because the row containing $x = 16$ contains no value for $\pi = .95$. Yet the desired answer can still be found by rephrasing the question in terms of failure rather than success. Our data processor fails to punch cards correctly 5 percent of the time. If we now read π as the probability of failure on any one trial, we can find, in the $\pi = .05$ column, a probability of getting precisely 4 incorrect cards as .0133. This implies a probability of getting 16 correct cards of .0133 as well.

b. Finally, pursing (a) further, what is the probability of getting at least 4 incorrect cards? Of getting at most 3 incorrect cards? Here the answer requires once again the addition of various probabilities. Getting at least 4 incorrect cards involves adding the probabilities of 4 incorrect cards, of 5 incorrect cards, of 6 incorrect cards, . . . , and, finally, of 20 incorrect cards. As the $\pi = .05$ column in the $n = 20$ section of Appendix Table C tells us, this amounts to .0133 + .0022 + .0003 = .0158. Similarly, the probability of getting at most 3 incorrect cards is found as the sum of the probabilities of getting 0, 1, 2, and 3 incorrect cards; that is, .3585 + .3774 + .1887 + .0596 = .9842. It is no accident that our two answers sum to 1; the events described are complementary. ‖‖

Binomial probabilities for cumulative values of x. Practice Problem 7b illustrates why it has been found desirable to construct tables of binomial probabilities for cumulative values of x as well as for individual values of x: while it is easy enough to calculate cumulative probabilities when n is small (we had to sum only four separate probability values in Problem 7b), this process can become extremely tedious for large values of n. Appendix Table D is such a binomial probability table for cumulative values of x. It is set up for values of n ranging from 1 to 20, as well as for $n = 50$ and $n = 100$. Once again, selected probability-of-success values between .05 and .50 are given. Unlike in Appendix Table C, the values in the body of Appendix Table D now give the probabilities not of x successes, but of x *or fewer* successes.

||| Practice Problem 8

Consumers' Product Preferences

Consider using Appendix Table D to answer questions about a population of consumers 30 percent of whom favor and 70 percent of whom dislike a new product. If 100 persons are sampled, what are the following probabilities?

    a. What is the probability of finding 8 or fewer consumers who favor the product? As a look at the $\pi = .30$ column in the $n = 100$ section of Appendix Table D tells us, the probability is 0.

    b. What is the probability of finding precisely 40 consumers who favor the product? The answer, in the $\pi = .30$ column and the $n = 100$ section of Appendix Table D, is the difference between the probabilities of finding 40 or fewer such consumers (.9875) and finding 39 or fewer such consumers (.9790), or .0085.

    c. What is the probability of finding fewer than 20 consumers who favor the product? The answer is equivalent to finding 19 or fewer such consumers, or .0089.

    d. What is the probability of finding 26 such consumers or more? The answer is equivalent to 1 minus the probability of finding 25 or fewer such consumers, or $1 - .1631 = .8369$.

    e. What is the probability of finding more than 35 such consumers? The answer is equivalent to 1 minus the probability of finding 35 or fewer such consumers, or $1 - .8839 = .1161$.

    f. What is the probability of finding between 20 and 40 such consumers, inclusive? The answer is equivalent to the probability of finding 40 or fewer such consumers (.9875) minus that of finding 19 or fewer of them (.0089), or .9786. |||

It should be noted that a cumulative probability table provides no information that could not also be gathered from the equivalent section of an ordinary probability table; the only (and not inconsiderable) advantage of the cumulative table is that it frees the user from the need to add or subtract large numbers of probability values. Analytical Examples 5.1, "Budgeting Research and Development," and 5.2, "Acceptance Sampling Plans," provide two cases in point.

Summary Measures for the Binomial Probability Distribution

Earlier in this chapter (see Box 5.A), we noted that the arithmetic mean or expected value of *any* random variable, R, can be calculated by a simplified formula, $E_R = \Sigma p \cdot x$. For a *binomial* random variable, an even simpler approach can be used.

    The expected value. As Box 5.C states, the expected value of a binomial random variable always equals the number of trials of the experiment multiplied by the probability of success in any one trial; that is, $E_R = n \cdot \pi$. This equality is shown in Table 5.7 on page 184, wherein the expected value of a binomial random variable is calculated in both ways, using the binomial probability distribution of Table 5.6 as a starting point.

Table 5.7 Determining the Expected Value of a Binomial Random Variable Given $n = 5$ and $\pi = .2$

The expected value of a binomial random variable can be calculated using the formula in Box 5.A, or it can be calculated more simply using Box 5.C as the number of trials of the experiment, n, multiplied by the probability of success in any one trial, π.

Number of Successes, x	Probability, $p(R = x)$	Weighted Value, $p(R = x) \cdot x$
0	.32768	0
1	.4096	.4096
2	.2048	.4096
3	.0512	.1536
4	.0064	.0256
5	.00032	.0016
		$E_R = \Sigma p \cdot x$ $= 1.0000$

Alternative Calculation: $E_R = n \cdot \pi = 5(.2) = 1$

Source: Table 5.6

The variance and standard deviation. We noted earlier (see Box 5.A) that the variance of *any* random variable, R, can be calculated by a simplified formula as well:

$$\sigma_R^2 = \Sigma p(x - \mu_R)^2 \tag{5.1}$$

—the standard deviation being, of course, the square root of this expression. For a *binomial* random variable an even simpler formula exists: the variance of a binomial random variable always equals the number of trials of the experiment multipled by the probability of success in any one trial, multiplied by the probability of failure in any one trial; that is,

$$\sigma_R^2 = n \cdot \pi(1 - \pi). \text{ (See Box 5.C.)} \tag{5.2}$$

5.C Summary Measures for the Binomial Probability Distribution of Random Variable, R.

$$\mu_R = E_R = n \cdot \pi$$
$$\sigma_R^2 = n \cdot \pi(1 - \pi)^{\bullet}$$
$$\sigma_R = \sqrt{\sigma_R^2}$$

where μ_R is the arithmetic mean, σ_R^2 is the variance, and σ_R is the standard deviation of the probability distribution of binomial random variable, R, while n is the number of trials and π is the probability of success in 1 trial.

Table 5.8 Determining the Variance of a Binomial Random Variable Given $n = 5$ and $\pi = .2$

The variance of a binomial random variable can be calculated using the formula in Box 5.A, or it can be calculated more simply using Box 5.C as the number of trials of the experiment, n, multiplied by the probability of success in any one trial, π, multiplied by the probability of failure in any one trial, $(1 - \pi)$. The standard deviation is, of course, the square root of this expression; in this case $\sigma_R = \sqrt{.8} = .8944$.

Number of Successes, x	Probability, $p(R = x)$	Deviation, $x - \mu_R$	Squared Deviation, $(x - \mu_R)^2$	Weighted Value, $p(R = x)(x - \mu_R)^2$
0	.32768	−1	1	.32768
1	.4096	0	0	0
2	.2048	1	1	.2048
3	.0512	2	4	.2048
4	.0064	3	9	.0576
5	.00032	4	16	.00512
				$\sigma_R^2 = .8$

Alternative Calculation: $\sigma_R^2 = n \cdot \pi(1 - \pi) = 5(.2)(.8) = .8$

Source: Table 5.6 and Table 5.7.

The equivalence of equations 5.1 and 5.2 is demonstrated clearly in Table 5.8, wherein the variance of a binomial random variable is calculated both ways, again using the binomial probability distribution of Table 5.6 as a starting point.[5]

---

[5]Chapter 5 of the *Student Workbook* that accompanies this text contains a number of mathematical derivations concerning the binomial probability distribution.

ANALYTICAL EXAMPLES

5.1 Budgeting Research and Development

Firms and governments throughout the world spend large sums of money on basic or applied research and on product development. In the 1950s, an approach called *parallel-path strategy* was developed by the Rand Corporation to help achieve research-and-development (R&D) goals at minimum cost. This strategy involves n teams working independently of one another trying to achieve some goal, such as the development of a new product. Clearly, each team needs an annual budget of, say, $10 million and even then has a less than perfect chance of, say, merely 5 percent of succeeding. The more teams that are put to work, the greater is the chance of at least one of them having a success, but the greater must the overall budget be. This is where probability theory comes in; it helps answer questions such as: how much money must be spent by a sponsoring firm or government agency if it wants to have a 90 percent chance of at least one success? Because the activity in question involves a Bernoulli process (one of two outcomes, independent teams), the binomial probability distribution can be used to establish the probability of at least one success by n teams.

$$p(R \geq 1) = 1 - p(R = 0)$$
$$= 1 - c_0^n \cdot \pi^0 \cdot (1 - \pi)^{n-0}$$
$$= 1 - (1 - \pi)^n.$$

Given $\pi = .05$, we get $p(R \geq 1) = 1 - .95^n$

If the agency were content with only a 90 percent chance of success, this expression would become

$$p(R \geq 1) = 1 - .95^n = .90,$$

hence,

$$.95^n = .10.$$

The value of n implied by this expression is about 45, implying (at $10 million a team) an annual budget of $450 million. The accompanying table shows other budgets required for other acceptable probabilities of achieving at least one success, given a .05 success probability of any one team.

Note how certainty of at least one success would require an infinite budget, but quite feasible budgets can produce near-certainty. (In 1981, in spite of a recession, General Motors spent $2,250 million on R&D, which was the highest total spent by any U.S. company. Next came Ford with $1,718 million, AT&T with $1,686 million and IBM with $1,612 million. Firms such as Boeing, General Electric, United Technologies, DuPont, Exxon, Eastman Kodak, Xerox, and ITT each spent between $500 million and $1 billion.)

Desired Probability of at Least 1 Success	Number of R&D Teams Needed	Budget per Team (millions of dollars per year)	Total Required Budget
.50	14	10	140
.60	18	10	180
.70	24	10	240
.80	32	10	320
.90	45	10	450
.95	59	10	590
.99	90	10	900
1.00	∞	10	∞

Sources: Richard R. Nelson, "Uncertainty, Learning, and the Economics of Parallel Research and Development Efforts," *Review of Economics and Statistics,* November 1961, pp. 351–64; Moshe Ben Horim and Haim Levy, *Statistics* (New York: Random House, 1981), pp. 246–48; *Business Week,* July 5, 1982, p. 54.

5.2 Acceptance Sampling Plans

Many firms and government agencies regularly receive shipments of goods from outside suppliers, and they want to accept these shipments only if they contain a reasonably low percentage of defective items. Thus, they take random samples of each shipment to determine the percentage of defective items. To make a decision on whether to accept or reject a shipment, however, they must have a *sampling plan;* that is, they must decide (a) on the desirable sample size and (b) on the maximum number of defective units they are willing to find, while still accepting the shipment. The binomial probability distribution can be extremely helpful here.

Consider the U.S. Air Force purchasing, say, electronic navigation equipment. If it sampled 5 such items in each shipment of 100 units and followed the self-imposed rule of accepting the entire shipment only if 0 defective units were found, what would it in fact be doing? The binomial formula provides a quick answer: With $n = 5$ and $x = 0$, the Air Force would under this sampling plan nevertheless be accepting shipments containing 5 percent defective items ($\pi = .05$) 77 percent of the time, it would be accepting shipments containing 10 percent defective items 59 percent of the time, and it would even accept shipments with 30 percent defective items 17 percent of the time. (These numbers can be confirmed quickly in Appendix Table C, section $n = 5$, $x = 0$, for values of π of .05, .10, and .30.)

If such a situation were deemed undesirable, matters could be improved by increasing the sample size to, say, $n = 20$. A sample plan of $n = 20$ and $x = 0$, would cause acceptance of lots with 5, 10, or 30 percent defectives only 36, 12, and 0 percent of the time.

Naturally, sampling plans need not be based on $x = 0$; one can decide to accept lots that contain $x = 3$ defective items or fewer of them. As Appendix Table D shows, shipments with 5 percent defective items ($\pi = .05$) will be accepted 98 percent of the time when samples of 20 are taken and findings of 3 or fewer defective items are deemed acceptable. On the other hand, shipments with 10 or 30 percent de-

fective items will then be accepted only 87 or 11 percent of the time.

Can you show why someone who wanted to accept shipments with 10 percent defectives at most 3 percent of the time, but who was willing to accept all shipments the samples of which contained up to 1 defective item, would have to choose a sample size of $n = 50$? (Hint: Note in Appendix Table D how the cumulative probability numbers in the $x \leq 1$ row and the $\pi = .10$ column steadily decline as n increases.)

BIOGRAPHY

5.1 James Bernoulli

James Bernoulli (1654–1705), also known as Jacques or Jakob, was born in Basel, Switzerland, into a family that had fled Antwerp a few decades earlier because of its Protestant faith and that was to produce nine mathematicians of first rank within three generations. James himself was made to study theology and completed these studies, but, against his father's will, he soon turned to mathematics and astronomy. In travels through Holland, France, and England, he familiarized himself with the state of these sciences, and he returned with theories of his own concerning comets and gravity. Before long, he was appointed professor of mathematics at the University of Basel. Inspired by the work of Leibniz concerning the infinitesimal calculus, he used it to tackle an abundance of mathematical problems (especially in astronomy and mechanics) and to describe the properties of many important curves. Indeed, he became so fascinated by the logarithmic spiral that he wrote[1]:

Because our wonderful curve always in its changes remains constantly the same and identical in type, it can be regarded as the symbol of fortitude and constancy in adversity: or even of the resurrection of our flesh after various changes and at length after death itself. Indeed, if it were the habit to imitate Archimedes today, I would order this spiral to be inscribed on my tomb with the epitaph

Eadem mutata resurgo (Although changed, I will resurrect).

Such was done. Yet, much more important than Bernoulli's description of the logarithmic spiral was his work on probability. Having been born in the very year in which de Fermat and Pascal (see Biographies 4.2 and 4.3) exchanged the famous letters that developed the fundamental principles of probability, James Bernoulli became the first to devote an entire book to the subject. His *Ars Conjectandi* (The Art of Conjecturing) was written in Latin and published posthumously in 1713. It contains, among other things, a systematic presentation of the theory of permutations and combinations, applications of probability theory to contemporary games (including the *jeu de paume,* a forerunner of tennis), and the development of the binomial probability distribution (discussed in this chapter). Bernoulli's great work also takes up a great many philosophical ideas concerning the very nature of probability (objective vs. subjective) and is interspersed with delightful examples of wit[2]:

Even as the finite encloses an infinite series
And in the unlimited limits appear,
So the soul of immensity dwells in minutia
And in narrowest limits no limits inhere.
What joy to discern the minute in infinity!
The vast to perceive in the small, what divinity!

[1]As cited by F. N. David, *Games, Gods, and Gambling* (New York: Hafner, 1962), pp. 138–39.

[2]*Dictionary of Scientific Biography*, vol. 2 (New York: Scribner's, 1970), p. 50.
Source: International Encyclopedia of Statistics, vol. 1 (New York: The Free Press, 1978), pp. 18–19.

Summary

1. Any quantitative variable the numerical value of which is determined by a random experiment (and, thus, by chance) is called a *random variable*. A systematic listing of each possible value of a random variable with the associated likelihood of its occurrence is called the random variable's *probability distribution*. This chapter focuses on probability distributions of discrete random variables (that can assume values only at specific points on a scale of values, with inevitable gaps between values).

2. Any probability distribution can be summarized by calculating the random variable's arithmetic mean (or expected value), along with its variance and standard deviation. (Appropriate formulas are given in Box 5.A.)

3. The *binomial probability distribution* is the most important probability distribution for discrete variables. It shows the probabilities associated with possible values of a random variable that are generated by a *Bernoulli process*. Such a process is a sequence of n identical trials of a random experiment such that each trial (a) produces one of two possible complementary outcomes that are conventionally called *success* and *failure* and (b) is independent of any other trial so that the probability of success or of failure is constant from trial to trial. The number of successes achieved in a Bernoulli process is called the *binomial random variable;* the probabilities for different values of it can be calculated with the help of the *binomial formula* (Box 5.B). The formula shows that the probability for any given number of successes varies with the number of trials, n, and with the probability of success in any one trial, π. A different binomial probability distribution can, therefore, be derived for each combination of n and π; many such distributions have been tabulated in *binomial probability tables* for both individual and cumulative values of the binomial random variable. Simplified formulas exist for determining a binomial random variable's summary measures of location and spread (Box 5.C).

Key Terms

Bernoulli process
binomial coefficient (c_x^n)
binomial formula
binomial probability distribution

binomial probability tables
binomial random variable
chance variable
cumulative probability distribution

expected value, E_R
probability distribution
random variable

Questions and Problems

The computer programs noted at the end of the chapter can be used to work many of the subsequent problems.

1. Consider a canasta deck; it consists of four jokers plus two ordinary decks of 52 playing cards. Cards are given point values as follows: red 3 = 100 points; joker = 50 points; ace or deuce = 20 points; 8 through king = 10 points; 4 through 7 = 5 points; black 3 = 5 points. Determine the probability distribution for a random variable, R, that equals the point value of the first card drawn from a shuffled canasta deck.

2. For the random variable described in problem 1, determine

a. the expected value.
b. the variance.
c. the standard deviation.

3. Could the following represent probability distributions? If so, write out the distribution and determine its summary measures (mean, variance, standard deviation).

a. $p(x) = \dfrac{x}{10}$, where $x = 1, 2, 3,$ or 4.

b. $p(x) = .7x$, where $x = -1, 0, 1,$ or 2.

4. Could the following represent probability distributions? If so, write out the distribution and determine its summary measures (mean, variance, standard deviation).

a. $p(x) = \dfrac{x^2}{15}$, where $x = 0, 1, 2,$ or 3.

b. $p(x) = \dfrac{1}{x}$, where $x = 1, 2$ or 3.

5. Could the following represent probability distributions? If so, write out the distribution and determine its summary measures (mean, variance, standard deviation).

a. $p(x) = (10 - x)/50$, where $x = -1, 0, 5,$ or 6.
b. $p(x) = (10 - x)/40$, where $x = 0, 1, 2, 3,$ or 4.

6. Determine the probability distribution for a random variable, R, that equals the number of points showing when a regular six-sided die is rolled once.

7. For the random variable described in problem 6, determine
a. the expected value.
b. the variance.
c. the standard deviation.

8. Referring to Table 5.4, show why the sum of weighted values in the last column of the table equals the result yielded by using the shortcut method of determining variance given in Box 3.N on page 98.

9. The Kentfield Hardware store is about to place an order for the *expected number* of lawn mowers demanded, based on the probability distribution in Table 5.A. How many will be ordered? What is the variance and standard deviation of the number demanded?

Table 5.A

Number Demanded	Probability
0	.05
1	.10
2	.20
3	.30
4	.20
5	.10
6	.05

10. In a certain hour each Monday, the Metropolitan Transport Authority noted that the data in Table 5.B applied. What would be the expected revenue if the Authority instituted a toll of $1 per car?

Table 5.B

Cars Crossing Throgs Neck Bridge	Absolute Frequency of Observation (Mondays)
100	20
120	35
140	16
150	7

11. A state auto inspection station collected the following data during the past year. If the observed pattern continued to hold, what would be the station's expected daily revenue if it instituted a $10 fee per car?

Table 5.C

Cars Coming For Inspection	Absolute Frequency of Observation (Days)
20	79
25	121
32	19
40	25
47	39
51	30

12. An office service company has collected the accompanying data. The manager wants to know
a. the probability of fewer than 30 service calls on a given day.
b. the expected company revenue per day if the pattern observed in the past continues and all service requests are answered (at a $50 fee per call).

Table 5.D

Requests For Servicing Broken-Down Xerox Machines	Absolute Frequency of Observation (Days)
5	276
9	59
13	30
25	36
31	38
52	21

13. The Massachusetts government has collected the following data over the past 50 years. If the pattern continues next year, what is the expected number of trees that will be defoliated by gypsy moths? What is the standard deviation?

Table 5.E

Number of Trees Defoliated by Gypsy Moths in Year	Probability
4,000 to under 7,000	.05
7,000 to under 10,000	.07
10,000 to under 13,000	.11
13,000 to under 16,000	.54
16,000 to under 19,000	.11
19,000 to under 22,000	.07
22,000 to under 25,000	.05

14. The Massachusetts sales tax is supposed to collect 5 cents on the dollar, but, on the cents part of any purchase, sales tax is collected according to the accompanying schedule. What is the percentage of tax actually collected? In your answer, make use of a probability distribution.

Table 5.F

Cents	Tax
1–29	1¢
30–49	2¢
50–69	3¢
70–89	4¢
90–99	5¢

15. An automobile manufacturer has fitted all cars with the identical pollution-control device (designed to meet government standards), yet experience shows that 5 percent of cars tested perform below government pollution standards. Assume that 20 cars coming off the assembly line during a given month are selected at random and that the condition of a Bernoulli process are met. What is the probability distribution for the numbers, x, of below-standard cars found in the sample?

16. Determine the expected value, variance, and standard deviation of the random variable noted in problem 15.

17. Reconsider problem 15. What is the probability that a government inspector who tests 20 cars a month in the above fashion will unjustly accuse the manufacturer of producing more than 5 percent of all cars below standard?

18. A family wants ten children: five boys and five girls. Assuming there is an equal chance for getting a child of either sex on any one try, what is the probability of this happening? Make your determination in two ways: first with Pascal's triangle on page 156, then with the help of Appendix Table C.

19. Grover C. Jones of Peterson, W. Virginia, in fact had 15 children—all boys. What is the probability of this happening?

20. Determine the expected value, variance, and standard deviation of the random variable in problems 18 and 19.

21. An employer has noted a 10-percent annual quit rate of the firm's employees. Assume that three employees are sampled at random and that the conditions of a Bernoulli process are met. With the help of a tree diagram, determine the probability distribution for the numbers, x, of employees in the sample who are likely to quit next year. Check your result in Appendix Table C.

22. Assuming that the conditions of a Bernoulli process are met, determine the following:
   a. A test is administered 10 times; the probability of getting the correct answer is .9. Use an appropriate formula to figure the likelihood of getting precisely 9 correct answers.
   b. What is the probability of having exactly 3 boys in a family of 10 children (assuming that boys and girls are equally likely and different births are independent events)? Find your answer with the help of a table and check it by using Pascal's Triangle on page 156.

23. A car-rental firm rents only compact and medium-sized cars; experience shows that three persons out of four prefer compacts. Considering the next 15 requests, what is the probability distribution (a) for the numbers of medium-sized cars requested? (b) For the numbers of compact cars requested? (You may assume that the conditions of a Bernoulli process apply.)

24. A manufacturer is sampling (with replacement) incoming shipments of 100 parts produced by other firms. (a) If (unbeknownst to the manufacturer) 5 percent of the parts are defective, what is the probability distribution for discovering defective units for a sample of 5 parts? (b) What is the probability distribution if in fact 40 percent of all units are defective and 5 of them are sampled? (c) What if 10 percent are defective and 3 units are sampled? (d) What if 60 percent are defective and 3 units are

sampled? (You may assume that the conditions of a Bernoulli process apply.)

25. A polling organization randomly samples 100 consumers from among the U.S. population about the new design of a product. If in fact 40 percent of all consumers favor the new design, what is the likelihood that among the sampled individuals the design will be favored by (i) 20 or fewer? (ii) Precisely 30? (iii) Fewer than 40? (iv) More than 45? (v) 50 or more? (vi) Between 30 and 40, inclusive? (You may assume that the conditions of a Bernoulli process apply.)

26. A new drug in nationwide use seems to be effective 40 percent of the time. What are the chances that a random sample of 100 patients using the drug will show success (i) in at most 20 patients? (ii) In exactly 30 patients? (iii) In at least 40 patients? (iv) In fewer than 50 patients? (v) In more than 50 patients? (You may assume that the conditions of a Bernoulli process apply.)

27. A new disease affects 20 percent of the population (or a new "bug" affects 20 percent of all units produced by a production process). If the conditions of the Bernoulli process are met, what is the likelihood that in a random sample of 5, (i) all are affected? (ii) None are affected? (iii) At least 1 is affected? (iv) At least 3 are affected? (v) Precisely 2 are affected? (vi) Between 2 and 4, inclusive, are affected?

28. A fair coin is tossed 16 times. What is the probability of getting (i) exactly 2 heads? (ii) Exactly 8 heads? (iii) Exactly 1 tail? (iv) Exactly 14 tails? [Compare your answer to (iv) with that given to (i) and explain.]

29. A lopsided coin (that gives heads a 70 percent chance of coming up) is tossed 50 times. What is the probability of getting (i) 44 or more heads? (ii) Exactly 40 heads? (iii) More than 37 heads? (iv) 31 or fewer heads? (v) Fewer than 28 heads? (vi) Between 30 and 40 heads, inclusive?

30. Twenty radar transponders are sitting on the bench at an avionics repair shop; of these, nine are inoperative, but this is not obvious to a casual observer. A thief steals five transponders. Assuming that the conditions of a Bernoulli process apply, determine the probability that the thief got (i) five good transponders, (ii) three bad transponders, (iii) only bad transponders, (iv) at least two good transponders.

31. What are the expected value, the variance, and the standard deviation of the binomial random variable (the number of affected units) in problem 27 above?

32. A bottling company checks out empty returned bottles before cleaning and refilling them. Some 10 percent are chipped and discarded. If the conditions of a Bernoulli process apply and 18 bottles are sampled,
  a. What is the probability of 10 or fewer being chipped?
  b. What are the distribution's summary measures?

33. A market survey shows that 30 percent of all families own a Polaroid camera. If the conditions of a Bernoulli process apply and 100 families are sampled,
  a. What is the probability that 20 or fewer have such a camera?
  b. What is the probability that precisely 31 have a camera?
  c. What is the probability that 47 or more have a camera?
  d. What are the distribution's summary measures?

34. According to a well-known private firm, 90 percent of its parcels are delivered within 2 days. If the conditions of a Bernoulli process apply and 14 parcels (that were shipped at different times) are sampled,
  a. What is the probability that all 14 arrive within 2 days?
  b. What is the probability that none arrive within 2 days?
  c. What is the probability that precisely 8 arrive within 2 days?
  d. What are the summary measures for the probability distribution of a random variable that denotes *late* deliveries?

35. According to a well-known stockbroker, 95 percent of all purchase orders are executed within 15 minutes of request. If the conditions of a Bernoulli process apply and 7 such orders are sampled, determine these probabilities: (i) that all are executed within 15 minutes; (ii) that none are executed within 15 minutes; (iii) that at least 5 are so executed; (iv) that at most 6 are so executed.

36. A random sample of 3 units is taken from a group of 10 in which 4 units are defective. With the help of the appropriate formulas, determine the probability distribution for the numbers of defective units found, along with summary measures, if sampling occurs with replacement. Compare your result with that given in the appropriate probability table.

37. A Xerox machine fails to print on every page. On average, 1 percent of the pages are blank. If 100 copies are run off, what is the probability of none of them being blank? Of one being blank?

38. Is the following statement *true* or *false*? Explain. A store permits returns of purchases within 10 days and knows from experience that the chance of return is 10 percent. If 18 units are sold to different customers, the probability of having at most 5 items returned equals .9936.

39. Are the following statements *true* or *false*? Explain.
   a. If for a binomial probability distribution the mean is 4 and the variance is .8, the probability of success in any one trial equals. .5.
   b. The number of trials, *n*, that determines the probability distribution referred to in (a) equals 5.
   c. The entire probability distribution implied by (a), for 0-5 successes, respectively, equals .0003, .0064, .0512, .2048, .4096, and .3277.

40. Is the following statement *true* or *false*? Explain. The probability is .1310 that at least 30 correct answers will be given in a 50-question true/false test, provided the answers are given at random by a totally unprepared student.

41. Is the following statement *true* or *false*? Explain. The probability of getting 4 heads in 6 tosses of a balanced coin is .2344.

42. Is the following statement *true* or *false*? Explain. The probability is .2430 that oil will be found in precisely one of three holes being drilled (in different locations), although 90 out of 100 such holes drilled are dry.

43. Is the following statement *true* or *false*? Explain. The probability is .4114 that fewer than 4 students among any 20 being picked at random from the very large number of entering freshmen in the U.S. will drop out of college before graduation, provided that 20 percent of all entering freshmen drop out.

44. Is the following statement *true* or *false*? Explain. The probability is .0729 of finding precisely 2 books that have never been borrowed when taking a random sample of 5 books from a large library in which 10 percent of all books have never been borrowed.

45. Is the following statement *true* or *false*? Explain. The probability is .1511 that 9 acceptable items will be found in as sample of 10 picked at random from a very large annual output of a plant that produces 90 percent acceptable units.

Selected Readings

Feller, William. *An Introduction to Probability Theory and Its Applications,* 2 vols. New York: Wiley, 1950–1966. By far the best textbook on the theory and application of discrete probability distributions.

Kohler, Heinz. *Statistics for Business and Economics,* 2nd ed. Glenview, Ill.: Scott, Foresman and Co., 1988. Chapter 6 contains a detailed discussion of two other discrete probability distributions, the hypergeometric distribution and the Poisson distribution.

Computer Programs

The DAYSAL-2 personal-computer diskettes that accompany this book contain two programs of interest to this chapter:

7. *Probability Distributions*, lets you compute binomial, hypergeometric, and Poisson probabilities, along with the summary measures of the respective probability distributions.

16. *Expected Monetary Values,* although designed for another purpose, can be used to compute the mean of probability distributions, such as that in Table 5.3.

CHAPTER SIX

Probability Distributions: Continuous Variables

Chapter 5 focused on quantitative random variables that are *discrete*—that can assume values only at specific points on a scale of values, with inevitable gaps between. As was noted, many everyday processes generate values of such variables, and the likelihood of occurrence of specific values can be gauged with the help of one or another of certain common probability distributions, such as the binomial distribution. The present chapter extends the earlier discussion to quantitative random variables that are *continuous* in nature—that can assume values at all points on a scale of values, with no breaks between possible values. Consider characteristics measured in units of money, time, distance, or weight, to name just a few of the possibilities. The possible observations about such variables are infinite in number.

The profit per dollar of sales earned by a firm is a case in point. It might be -10 cents or $+23$ cents or even $+41.37895$ cents. The random experiment of operating a firm for a year can generate *any* value for this variable—positive, zero, or negative. Or contemplate the possible measurements concerning the lifetime of new appliances, of car batteries, of light bulbs, or of tires, the flight time of a plane, the completion time of a task, the reaction time after a stimulus. Consider distance measurements of the drilling depths of oil wells, of inches of rainfall, of the length of rods of steel or of ears of corn, of the thickness of tablets, of the height of weather satellites. Imagine measuring the weight of apples harvested per tree, of cereal boxes filled, of ingots of metal produced. In all these cases and a million more, the observations actually made will be among an infinite number of conceivable ones. As a result, one cannot *list* all the possible values of continuous random variables; nor can one list the associated probabilities for each one of the infinite number of conceivable values. When continuous random variables are involved, it is common practice instead to associate probabilities with *ranges* of values along the continuum of possible values that the random variable might take on.

THE PROBABILITY DENSITY FUNCTION

Panel (a) of Figure 6.1 represents one way of picturing the probability distribution of a continuous random variable. The graph is a relative frequency histogram of the type first introduced in Chapter 2. The continuous random variable in this example is the length of service, as of December 31 of a given year, of the 49,000 employees of a firm. For any one employee, this period could clearly be measured by any positive number: as 1 year, 5.2 years, or even as 7.13946 years. The possible values of this variable have been grouped on the horizontal axis in 5-year intervals. The height of any column standing on top of any one of these equal-sized intervals represents the relative frequency density, telling us simultaneously the proportion of all employees whose service length falls within the given range and the probability that any one employee, randomly chosen, will have a service length within that range. Naturally, the sum of these proportions (and of these probabilities) equals 1.

Thus, we can tell instantly that the probability equals .3 that a randomly chosen employee has been with the firm between 10 and 15 years: $p(R = 10 \text{ to } 15 \text{ years}) = .3$. Yet we cannot tell the probability of our random variable lying in the range from 11 to 13 years because that interval is completely contained within the 10- to 15-year

Figure 6.1   A Continuous Probability Distribution

The probability distribution of a continuous random variable can be represented by a relative frequency histogram, as in panel (a); it can also be approximated by a smoothed frequency curve, such as the color line in panel (b), which is called a *probability density function.*

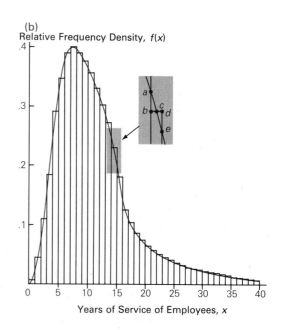

Figure 6.2 Two Probability Density Functions

The probability that a continuous random variable (such as years of employee service) takes on values within any given range of possible values (such as 11 to 13 years) is represented by the ratio of two areas: the area (here crosshatched) under the portion of the probability density function covering the range of interest and the entire area under the function (which is taken to equal 1). In this example, the probability that a randomly chosen employee has seen between 11 and 13 years of service is larger, therefore, in Firm A, than in Firm B.

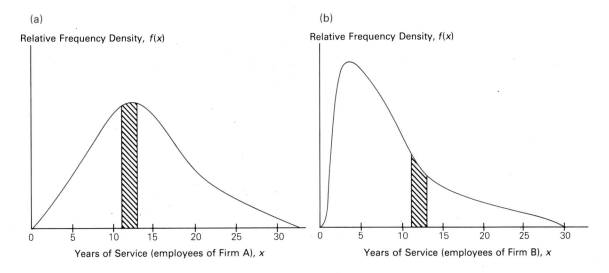

range in our histogram. We can imagine, however, constructing from our data a series of alternative histograms with ever narrower years-of-service intervals. If we did, the columns would get ever more numerous and also narrower until a line along their tops approached the smooth frequency curve given in panel (b). This curve is called a **probability density function.** It describes the probability distribution of a continuous random variable, R, and is denoted by $f(x)$, pronounced "f of x," where x represents all the possible values of this variable. Note that the total area under the probability density fucntion, just as that under the relative frequency histogram, equals 1. As the inset in panel (b) indicates, the smooth frequency curve adds some areas, such as *abc,* to the total histogram area, but it cuts off equivalent areas, such as *cde.* Given the probability density function, the probability that a continuous random variable takes on values within any given range is, thus, represented by the area under the portion of this curve covering this range.

Consider Figure 6.2, which shows probability density functions for the years of service of employees in two different firms, A and B. (The total area underneath either function equals 1). One can instantly tell that the probability of finding an employee with 11 to 13 years of service equals whatever proportion the shaded area constitutes of the total area underneath the curve; this probability is clearly larger in Firm A than in Firm B. The total area equals 1 in both cases, but the crosshatched area is larger in panel (a).

Note: in the case of continuous random variables, it is customary never to consider the probability of occurrence of a single value, such as 11.92374 years of service, to be anything but zero. This probability would have to equal the area above a *point* on the horizontal axis up to the probability density function, and, given the fact that a point has no width, this area above the point equals zero. Indeed, given an infinite number of possible values, the probability of one specific value is practically nil. When calculating cumulative probabilities for continuous random variables, it is, therefore, deemed irrelevant to make a distinction between the probability of the random variable being "smaller than x" and being "smaller than or equal to x":

$$p(R < x) = p(R \leq x) \text{ because } p(R = x) = 0.$$

Similarly, no distinction is made between the probability of a continuous random variable being "larger than x" and being "larger than or equal to x":

$$p(R > x) = p(R \geq x) \text{ because } p(R = x) = 0.$$

Figure 6.2 pictures two rather differently shaped probability density functions; as one can expect, an infinite variety exists, each one describing the probabilities associated with different types of continuous random variables. Three types of probability density functions, however, are encountered more often than most. These are (1) the *normal* probability distribution, (2) the *exponential* probability distribution, and (3) the *uniform* probability distribution. Their general shapes were noted in panels (a), (d), and (e) of Figure 2.2 on pp. 50–51; this chapter will focus on the first of these.[1]

THE NORMAL PROBABILITY DISTRIBUTION

The **normal probability distribution** is a probability density function that (a) is single-peaked above the random variable's mean, median, and mode (all of which are equal to one another), (b) is perfectly symmetrical about this central value (and, thus, said to be "bell-shaped"), and (c) is characterized by tails extending indefinitely in both directions from the center, approaching but never touching the horizontal axis (which implies a positive probability for finding values of the random variable within any range from minus infinity to plus infinity).[2] This probability distribution was first described in 1733 by Abraham de Moivre (Biography 4.1), and it was popularized by Adolphe Quetelet (Biography 1.1), who used it to discuss "the average man" (l'homme moyen), and by Carl Friedrich Gauss (see Biography 6.1 at the end of the chapter), who used it to describe errors of measurement in astronomy. Indeed, many measurements of natural phenomena—ranging from the heights, weights, or IQs of people to the distances, volumes, or speeds of heavenly bodies—have frequency distributions that closely resemble the normal distribution. As we shall see in Chapter 7, this fact makes the normal distribution very important for making inferences about unknown populations from known samples.

---

[1]Those interested in the other two should note the *Selected Readings* at the end of this chapter.

[2]Chapter 6 of the *Student Workbook* that accompanies this text considers the mathematical properties of the normal distribution in detail.

Figure 6.3   Three Members of the Normal Probability Distribution Family

Members of the normal probability distribution family, called *normal curves* for short, differ from one another only by the values of the distribution's mean (μ_R) and standard deviation (σ_R). The value of μ_R positions the distribution on the horizontal axis; that of σ_R determines its spread and, thus, its appearance as peaked, when σ_R is small, or as flat, when σ_R is large. In every case, the normal probability density function has points of inflection precisely 1 standard deviation below and above the mean and then approaches, respectively, minus and plus infinity on the horizontal axis (which, naturally, the graph can only suggest but not show). As a practical matter, the height of every normal curve is near 0 within 3 standard deviations of the mean.

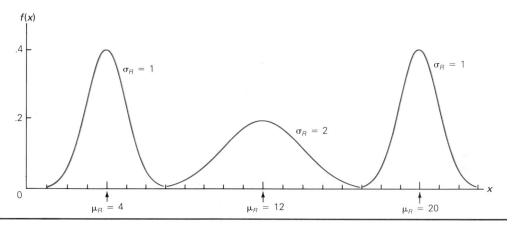

A Graphical Exposition

Figure 6.3 shows three members of the normal probability distribution family. They differ from one another only by the magnitudes of the mean, μ_R, and standard deviation, σ_R, of the random variable in question. Note that every one of the three probability density functions shown in the graph exhibits the three characteristics listed before, although it is, of course, impractical in a graph to show how the tails of the functions extend indefinitely in both directions without ever touching the axes. Analytical Example 6.1, "Honest Weights and the Normal Curve of Error," provides additional illustrations.

The Formula

The precise height of the normal probability density function, $f(x)$, for different values, x, of the normal random variable, R, can be determined with the help of the formula given in Box 6.A. This rather formidable expression is the equation for the types of curves seen in Figure 6.3; as usual, it becomes less formidable upon closer inspection.

6.A The Normal Probability Density Function

$$f(x) = \frac{1}{\sigma_R\sqrt{2\pi}} \cdot e^{-\frac{1}{2}\left(\frac{x - \mu_R}{\sigma_R}\right)^2}$$

where $f(x)$ is a function of observed value, x, while μ_R is the mean and σ_R is the standard deviation of the probability distribution of random variable, R, and $\pi = 3.14159$, $e = 2.71828$.

The mean and standard deviation of the normal random variable's probability distribution appear as μ_R and σ_R, and they appear alongside two of the most famous constants in mathematics: π is in this case not the population proportion, but the ratio of the circumference to the diameter of the circle; it equals 3.14159. The constant e, is the base of the natural logarithms; it equals 2.71828.

Thus, given μ_R and σ_R, one can calculate the height of the probability density function $f(x)$ for each x; by integrating that function over a specified range of x, one can calculate the type of crosshatched area shown in Figure 6.2 and, thus, determine the probability of finding values of the random variable within that range of x. In order to facilitate the determination of such probabilities, statisticians have found a way of converting normal curves with differing shapes and positions into a single *standard* normal curve and of then finding probabilities in appropriately prepared tables.

The Standard Normal Curve

The **standard normal curve** is a normal probability density function with a mean of 0 and a standard deviation of 1. As Figure 6.4 illustrates, any normal curve—whatever the value of μ_R and σ_R—can quickly be transformed into this standard curve by plotting the value of μ_R as 0 and expressing values above and below this mean in units of σ_R. The deviation from the mean, μ_R, of any given value, x, of a normally distributed random variable, when measured in units of standard deviations, σ_R, is called the **standard normal deviate** (see Box 6.B), and it is symbolized by z. (In education statistics, z is often referred to as the *standard score,* as was noted in Analytical Example 3.1).

6.B The Standard Normal Deviate

$$z = \frac{x - \mu_R}{\sigma_R}$$

where x is an observed value of random variable, R, while μ_R is the mean and σ_R is the standard deviation of its probability distribution.

Thus, if $\mu_R = 4$ and $\sigma_R = 1$, a value of $x = 2$ becomes $z = -2$, indicating (as Figure 6.4 shows) that this value of the random variable lies 2 standard deviations below the mean. Similarly, when $\mu_R = 12$ and $\sigma_R = 2$, a value of $x = 14$ becomes $z = +1$, indicating a distance of 1 standard deviation *above* the mean. By the same procedure (not shown in Figure 6.4), a value of $x = 22.5$, given $\mu_R = 20$ and $\sigma_R = 1$, would be plotted as $z = +2.5$ (but as $z = +5.25$, given $\mu_R = 12$ and $\sigma_R = 2$).

The Standard Normal Tables

When x values have been transformed into z values by setting $\mu_R = 0$ and $\sigma_R = 1$, the Box 6.A formula simplifies to that given in Box 6.C.

Figure 6.4   Constructing the Standard Normal Curve

The three different normal curves of Figure 6.3, reproduced here in the top panel, can be converted into an identical *standard* normal curve, shown in the bottom panel. Each actual value, *x*, of the normal random variable is translated into a standardized value, *z*, by first calculating the deviation of *x* from the mean and then expressing this deviation in terms of standard deviation units. This procedure is equivalent to setting μ equal to zero and treating the distance of 1σ as equal to one.

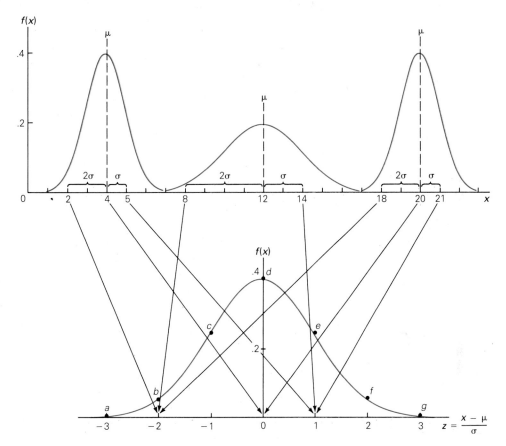

6.C The Standard Normal Probability Density Function

$$f(z) = \frac{1}{\sqrt{2\pi}} \cdot e^{-\frac{1}{2}z^2}$$

where $\pi = 3.14159$, $e = 2.71828$, $z = \dfrac{x - \mu_R}{\sigma_R}$ and $-\infty \le z \le +\infty$.

Based on this simplified formula, two types of tables can be prepared: tables of ordinates and tables of cumulative relative frequencies.[3] We will discuss the nature of these tables first and will then show how helpful they can be in typical business and economics applications.

Tables of ordinates. One set of these tables shows the *height* of the standard normal probability density function for different values of z. Thus, the height of the curve corresponding to z values of 0, 1, 2, 3, and 4 is shown in the accompanying table.

As was noted earlier, strictly speaking, the normal curve touches the horizontal axis only at minus and plus infinity, but for all practical purposes it can be considered to do so within 3 standard deviations of the mean (or when $z = 3$).

z	Height of Curve	Corresponding Points in Figure 6.4
0	.39894	d
1	.24197	c and e
2	.05399	b and f
3	.00443	a and g
4	.00013	—

Tables of cumulative relative frequencies. A second set of tables typically shows the *area* under the standard normal probability density function between $\mu = 0$ and alternative *positive* values of z. (Because of the symmetry of the normal curve, these tables can also be interpreted to show the noted area between $\mu = 0$ and alternative *negative* values of z.) This type of table, as we shall see presently, is of much greater importance than a table of ordinates, and it has been reproduced, for selected values of z, as Appendix Table E. Note that the value of z, up to the first decimal, is shown in the leftmost column of that table, while the second decimal of z is found in one of the 10 entries in the top row. By *adding together* any given row heading with any given column heading, one obtains a z value with two decimal places. Thus, a row heading of .5 and a column heading of .00 corresponds to a z value of .5 + .00 = .50. At the intersection of the .5 row and the .00 column, one finds, accordingly, the area under the standard normal curve between the mean of 0 and a z value of .50, and this area value equals .1915. This means that there is a probability of .1915 that the value of a normally distributed random variable lies within half a standard deviation above the mean. Given the curve's symmetry, there exists, of course, an equal probability of .1915 that such a value lies within half a standard deviation *below* the mean. Thus, there exists a probability of .1915 + .1915 = .383 that such a value lies somewhere between half a standard deviation below to half a standard deviation above the mean.

Moving along the $z = .5$ row to the column headed .01, .02, .03, and so on, we can find, similarly, the probability of any value of the random variable lying $z = .51$, .52, .53 standard deviations above the mean. The probabilities involved equal .1950,

---

[3]These tables were first prepared in 1799 by Chrétien Kramp, a French mathematician and physicist (who also invented the factorial notation—n!—that was discussed in Chapter 4). The most recent tables were issued in 1953 by the U.S. National Bureau of Standards, as noted in Appendix Table E.

Figure 6.5   Finding Areas Under the Normal Curve

With the help of Appendix Table E, one can determine areas under the normal curve and, thus, the probability of finding values of a normally distributed random variable within specified numbers of standard deviations below or above the mean. Because the area under the normal curve between the mean and 1 standard deviation below it covers .3413 of the entire area under the curve, as represented by the shaded portion in panel (a), and because the same is true for the range of values between the mean and 1 standard deviation above it, as shown in panel (b), .3413 + .3413 = .6826 of the total area is found within the range of $\mu \pm 1\sigma$; thus .6826 is the probability of finding a value of the random variable within that range. Similarly, the probability is .9544 of finding a value of the normal random variable within $\mu \pm 2\sigma$.

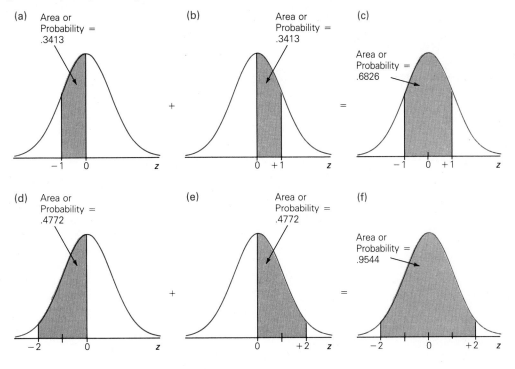

.1985, .2019, respectively. Hence, the probability of finding a value of a normally distributed random variable somewhere between .53 standard deviations below the mean and .53 standard deviations above it equals .2019 + .2019 = .4038.

Figure 6.5 shows, similarly, the likelihood of encountering values of such a random variable within 1 or 2 standard deviations below or above the mean. The graph clearly implies that the probability of finding a value of a normally distributed random variable that lies more than 2 standard deviations either below or above its mean equals a tiny 1 − .9544 = .0456. Yet the probability remains positive even for large values of z. (According to the calculations by the National Bureau of Standards, noted in Appendix Table E, the probability of finding a value that lies more than 3 standard deviations either below or above the mean equals 1 − .9972 = .0028; that for a value beyond 8 standard deviations above or below the mean equals a tiny .00000 00000 00001.)

Note: The graphical exposition in Figure 6.5 illustrates the manner in which more complicated questions can also be answered. For example, the probability of finding a value of this normal random variable lying between 2 standard deviations below the mean and 1 standard deviation above it can easily be found by combining panels (d) and (b) as .4772 + .3413 = .8185. Similarly, the probability of finding a value that lies more than 1 standard deviation below the mean *or* more than 2 standard deviations above it can be found by using panels (a) and (e) and combining the unshaded area in the left half of panel (a)—that is, .5000 − .3413 = .1587—with the unshaded area in the right half of panel (e)—that is, .5000 − .4772 = .0228; the result is .1815. (Naturally, the area under half the normal curve always equals .5000.) Analytical Example 6.2, "The Decision to Seed Hurricanes," provides a vivid illustration of why one may want to know normal-curve probabilities. Other examples are discussed in the next sections.

||| Practice Problem 1

Finding Probabilites Concerning the Lifetime of Aircraft Engines

Suppose that a manufacturer of aircraft engines knows their lifetimes to be a normally distributed random variable with a mean of 2,000 hours and a standard deviation of 100 hours. Here are some of the questions of interest to the manufacturer and to potential users (along with their answers).

a. What is the probability that a randomly chosen engine has a lifetime between 2,000 and 2,075 hours? The problem amounts to finding an area under the normal curve between the mean and a value above it. It helps to sketch the problem, as in Figure A, to convert x values to z values, and to find the answer in Appendix Table E. Because μ = 2,000 hours and σ = 100 hours, the x value of 2,075 hours equals a z value of 2,000 − 2,075 divided by 100, or of .75. The area between μ = 0 and z = .75 (shaded in Figure A) can be found in Appendix Table E at the intersection of the .7 row and .05 column as .2734. This number is also the probability sought.

Figure A

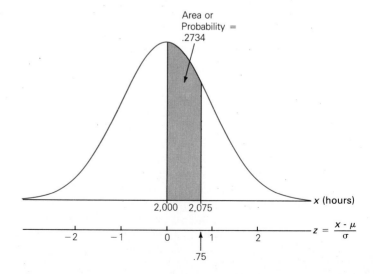

Area or
Probability =
.2734

$z = \dfrac{x - \mu}{\sigma}$

b. What is the probability that a randomly chosen engine has a lifetime between 1,880 and 2,000 hours? The problem amounts to finding an area under the normal curve between the mean and a value below it. It is, therefore, solved by a procedure analogous to (a) above, as shown by Figure B. Because $\mu = 2,000$ hours and $\sigma = 100$ hours, the x value of 1,880 hours equals a z value of -1.20. The area between $\mu = 0$ and $z = -1.20$ (shaded in Figure B) can be found in Appendix Table E at the intersection of the 1.2 row and .00 column as .3849. This number is also the probability sought.

Figure B

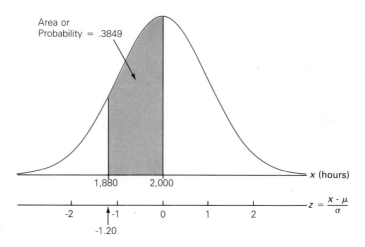

c. What is the probability that a randomly chosen engine has a lifetime between 1,950 and 2,150 hours? The problem amounts to finding an area under the normal curve overlapping the mean, as shown in Figure C. Because $\mu = 2,000$ hours and $\sigma = 100$ hours, the x value of 1,950 hours (or 2,150 hours) equals a z value of $-.50$ (or $+1.50$). The area between $\mu = 0$ and $z = -.50$ (cross-

Figure C

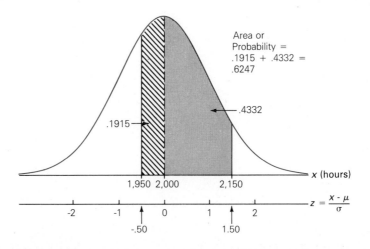

hatched) and that between $\mu = 0$ and $z = +1.50$ (shaded) can be found in Appendix Table E at the intersection of the .5 row and .00 column (and of the 1.5 row and .00 column) as .1915 (and .4332). Thus, the probability sought equals the sum of the crosshatched and shaded areas, or .1915 + .4332 = .6247.

d. What is the probability that a randomly chosen engine has a lifetime above 2,170 hours? The problem amounts to finding an area under the upper tail of the normal curve, as shown in Figure D. Because $\mu = 2,000$ hours and $\sigma = 100$ hours, the x value of 2,170 hours equals a z value of $+1.70$. The entire area to the right of μ always equals .5, that between $\mu = 0$ and $z = +1.70$ (dotted in Figure D) can be found in Appendix Table E at the intersection of the 1.7 row and .00 column as .4554. Thus, the probability sought (which corresponds to the shaded upper-tail area) equals .5000 − .4554, or .0446.

Figure D

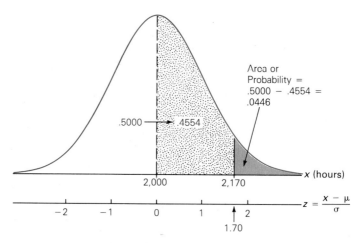

e. What is the probability that a randomly chosen engine has a lifetime below 1,840 hours? The problem amounts to finding an area under the lower tail of the normal curve, as shown in Figure E. Because $\mu = 2,000$ hours and $\sigma = 100$

Figure E

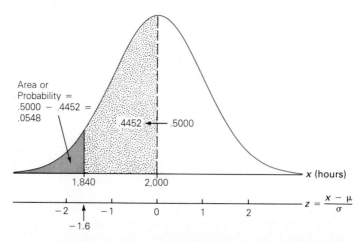

hours, the x value of 1,840 hours equals a z value of -1.60. The entire area to the left of μ always equals .5; the area between $\mu = 0$ and $z = -1.60$ (dotted in Figure E) can be found in Appendix Table E at the intersection of the 1.6 row and .00 column as .4452. Thus, the probability sought (which corresponds to the shaded lower-tail area) equals .5000 − .4452, or .0548.

f. What is the probability that a randomly chosen engine has a lifetime between 2,071 and 2,103 hours? The problem amounts to finding an area under the normal curve between two values both of which lie above the mean, as shown in Figure F. Because $\mu = 2,000$ hours and $\sigma = 100$ hours, the x values of 2,071 and 2,103 hours equal z values, respectively, of $+.71$ and $+1.03$. The area between $\mu = 0$ and $z = +1.03$ (the dotted plus the shaded areas in Figure F) can be found in Appendix Table E at the intersection of the 1.0 row and .03 column as .3485. Similarly, the area between $\mu = 0$ and $z = +.71$ (the dotted area in Figure F) can be found at the intersection of the .7 row and .01 column as .2612. Thus, the probability sought (which corresponds to the shaded area only) equals .3485 − .2612, or .0873.

Figure F

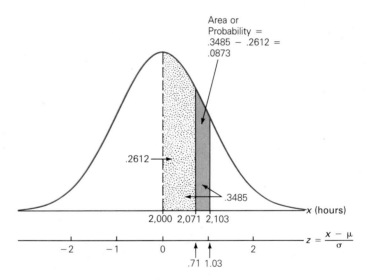

g. What is the probability that a randomly chosen engine has a lifetime between 1,849 and 1,923 hours? The problem amounts to finding an area under the normal curve between two values both of which lie below the mean, as shown in Figure G. Because $\mu = 2,000$ hours and $\sigma = 100$ hours, the x values of 1,849 and 1,923 hours equal z values, respectively, of -1.51 and $-.77$. The area between $\mu = 0$ and $z = -1.51$ (the dotted plus the shaded areas in Figure G) can be found in Appendix Table E at the intersection of the 1.5 row and .01 column as .4345. Similarly, the area between $\mu = 0$ and $z = -.77$ (the dotted area in Figure G) can be found at the intersection of the .7 row and .07 column as .2794. Thus, the probability sought (which corresponds to the shaded area only) equals .4345 − .2794, or .1551.

Figure G

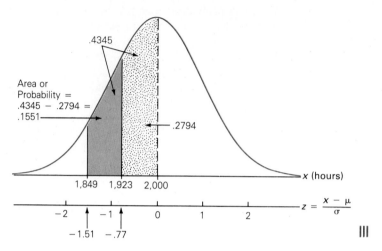

Area or
Probability =
.4345 − .2794 =
.1551

.4345

.2794

1,849   1,923   2,000

x (hours)

−2   −1   0   1   2

$z = \dfrac{x - \mu}{\sigma}$

−1.51   −.77

III

III Practice Problem 2

Designing an Aircraft Cockpit

A manufacturer of aircraft (or, for that matter, of any product) is likely to be very much concerned about the ability of potential users to use the product. When 60 percent of all pilots cannot reach the rudder pedals when seated in the pilot's seat or find themselves unable to tune navigation equipment because it is nicely out of reach, something is wrong. Suppose a manufacturer knows that the lengths of pilots' legs and arms are normally distributed random variables with mean of 33 and 28 inches, respectively, and standard deviations of 2 inches in both cases.

a. If the manufacturer wants to design a cockpit such that precisely 90 percent of all pilots can reach the rudder pedals with their feet while seated (and can, thus, fly this particular airplane), what is the desired distance between seat and pedals? The question requires us to find the 10th percentile—that value of the random variable (length of pilots' legs) below which lies 10 percent of the area under the normal curve. Because the 10th percentile lies below the mean and 50 percent of the entire area under the normal curve lies below the mean, the 10th percentile corresponds to that negative value of *z* which places 50 − 10 = 40 percent of the area between itself and the mean, as shown in Figure H. Accordingly, we can search the body of Appendix Table E for an entry of .4000, and such is found between a *z* of −1.28 (area: .3997) and a *z* of −1.29 (area: .4015). Interpolation yields a *z* value of −1.2817 for the 10th percentile. Just as *x* values can be converted into *z* values, so the reverse is also true; a *z* of −1.2817 corresponds to 30.44 inches. This means that 10 percent of all pilots cannot reach as far as 30.44 inches—which is, therefore, the distance sought because it assures that 90 percent of all pilots easily can reach the rudder pedals.

Figure H

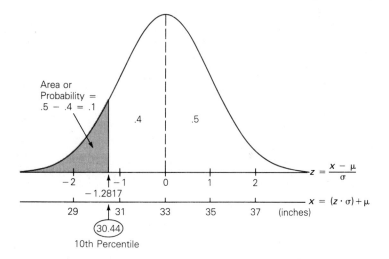

Area or
Probability =
.5 − .4 = .1

.4   .5

$z = \dfrac{x - \mu}{\sigma}$

$x = (z \cdot \sigma) + \mu$

(inches)

-2 -1 0 1 2
-1.2817

29   31   33   35   37

(30.44)
10th Percentile

b. If it were true that precisely 60 percent of all pilots could not reach the navigation equipment when seated, what must have been the distance between the pilot's seat and the relevant knobs? The question requires us to find the 60th percentile—that value of the random variable (length of pilots' arms) below which lies 60 percent of the area under the normal curve. Because the 60th percentile lies above the mean, but 50 percent of the entire area under the normal curve lies below it, the 60th percentile corresponds to that positive value of z which places 10 percent of the area between itself and the mean, as shown in Figure I. Accordingly, we can search the body of Appendix Table E for an entry of .1000, and such is found between a z of .25 (area .0987) and a z of .26

Figure I

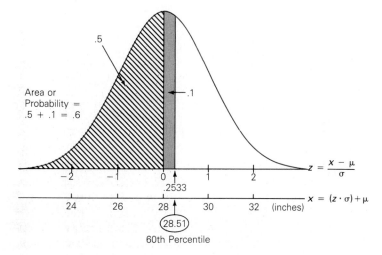

.5

.1

Area or
Probability =
.5 + .1 = .6

$z = \dfrac{x - \mu}{\sigma}$

$x = (z \cdot \sigma) + \mu$

(inches)

-2 -1 0 1 2
.2533

24   26   28   30   32

(28.51)
60th Percentile

(area .1026). Interpolation yields a *z* value of .2533 for the 60th percentile. Again, *z* values can be converted into *x* values; in this case, the 60th percentile corresponds to 28.51 inches. Such a distance would account for only 40 percent of all pilots being able to reach the navigation-equipment knobs.[4] |||

Approximating Discrete Probability Distributions by the Normal Probability Distribution

Under certain circumstances, and in spite of the fact that it is a continuous one, the normal probability distribution can be used to approximate various discrete probability distributions. This use of the normal probability distribution is very convenient when the calculation of precise probabilities is a cumbersome procedure.

Approximating binomial probabilities. Binomial probability tables typically do not extend to large values of *n* (numbers of trials of a random experiment), or they may not cover a given value of π (the probability of success in one trial). Appendix Table C, for example, contains no section for *n* = 50, nor does the more detailed table from which it is derived. Appendix Table C, as well as its source, similarly contains no column for π = .255. If one were interested in establishing the probabilities for an experiment involving these numbers and did not have a computer, one could still do so, but one would have to go through the type of cumbersome calculations illustrated in Table 5.6 on page 179, using the binomial formula given in Box 5.B. Luckily, there are circumstances when the normal probability distribution gives almost identical results. Such results can be obtained whenever the value of *n* is large or whenever that of π is close to .5. It is easy to see why this similarity in results occurs by having another look at Figure 5.3 on page 180: regardless of the value of π, as *n* increases, a binomial probability distribution becomes less and less skewed (and, thus, looks more and more like the symmetrical normal probability distribution). Furthermore, whenever π = .5, any binomial probability distribution is perfectly symmetrical regardless of the value of *n*.

It is, of course, a matter of arbitrary judgment which combination of *n* and π makes a binomial probability distribution "sufficiently" symmetrical to justify the use of the normal curve as an approximation to it. According to one widely used rule of thumb, the normal-curve approximation can be used without undue loss of accuracy whenever $n \cdot \pi$ *as well as* $n(1 - \pi)$ equals or exceeds 5.

Once the decision to substitute normal for binomial probabilities has been made, the relevant probabilities can be discovered quickly. Because the probability of any one *x* value, such as 11 successes in so many trials, is theoretically zero for a continuous distribution, one converts the discrete value of *x* into a small *range* of *x* by subtracting and adding half the distance between the discrete values to any one value of *x*. This so-called *continuity correction factor* turns a discrete 11 into a range of 10.5 to 11.5, for example, and it turns a discrete "11 or larger" into a range of "10.5 or more," while

---

[4]The concerns noted in Practice Problem 2 are far from being merely academic exercises. In 1984, the Navy, for example, issued new flight-training standards that stipulated a pilot's maximum "sitting height" at 40.5 inches, maximum "buttock-to-leg length" at 48 inches, and maximum "functional reach" at 29 inches—all because of the design of new Navy fighter planes. The standards excluded 73 percent of all college-age women and 13 percent of all college-age men from naval flying. (*The New York Times,* November 25, 1984, p. 49.)

replacing a discrete "11 or fewer" by an "11.5 or less." This correction having been made, the new continuous x value is transformed into a z value by the standard formula, while noting (Box 5.C) that the binomial $\mu = n \cdot \pi$ and $\sigma = \sqrt{n \cdot \pi(1 - \pi)}$. Hence,

$$z = \frac{x - \mu}{\sigma} = \frac{x - (n\pi)}{\sqrt{n \cdot \pi(1 - \pi)}}.$$

Given values of z, the desired probabilities can be found in Appendix Table E.

||| Practice Problem 3

Invoice Errors

Consider a company that has experienced 25.5 percent errors in its invoices. It takes a sample (with replacement) of 50 invoices. What is the probability that exactly 11 of the sample invoices are in error? One could calculate the precise probability by the binomial formula:

$$p(R = 11|50, .255) = \frac{50!}{11!(50 - 11)!} (.255)^{11}(.745)^{39} = .1144.$$

One can also note that $\mu = n \cdot \pi = 50(.255) = 12.75$, while $\sigma = \sqrt{n \cdot \pi(1 - \pi)} = \sqrt{50(.255)(.745)} = \sqrt{9.49875} = 3.082$. Accordingly, the problem can be pictured as shown in Figure J. The binomial value of 11 appears as the range from 10.5 to 11.5, and the corresponding z values equal $-.73$ and $-.41$. Accordingly, Appendix Table E gives the desired probability as the area (the dotted plus the shaded areas in Figure J) corresponding to $z = -.73$ (or .2673) minus the area (dotted) corresponding to $z = -.41$ (or .1591)—that is, as .1082. This value lies within 5.4 percent of the correct one.

Figure J

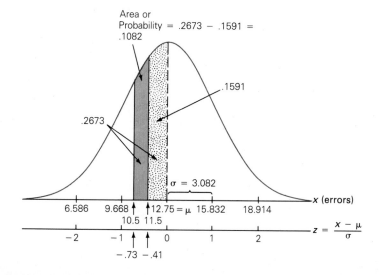

By the same procedure, one can quickly estimate the probability of finding, say 13 or fewer errors, as shown in Figure K. The binomial value of 13 or fewer appears as "13.5 or less." The corresponding z value equals $+ .24$. Accordingly, Appendix Table E gives the probability of the shaded area as .0948. The desired probability (cross-hatched plus shaded areas), therefore, is approximated by .5948.

Figure K

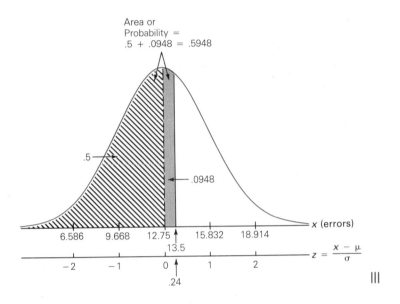

ANALYTICAL EXAMPLES

6.1 Honest Weights and the Normal Curve of Error

Since time immemorial, merchants and their customers have had problems with accurate measurement. Whether they were weighing bags of spices, live sheep, or flasks of olive oil, whether they were measuring off lengths of cloth or lengths of pastureland, they were soon aware that each trial of measurement could easily produce a different result. Even today, with infinitely more refined measurement techniques, repeated measurement of the same thing tends to produce different results each time. Each result is contaminated by a different *random error*. (See Chapter 1 for a more detailed discussion of this concept.) These errors are as likely to be positive as negative; they are also more likely to be small than large (that is, they are more likely to be close to zero than farther away from it). In short, errors of careful measurement can

be described by the normal curve. This fact was clearly stated in 1809 by Gauss (Biography 6.1) and, some eight decades later, led Francis Galton (Biography 11.1) to exclaim[1]: "I know of scarce anything so apt to impress the imagination as the wonderful form of cosmic order expressed by the Law of Frequency of Error."

Consider, for example, how it is decided that a pound of sugar you buy really is a pound. In 1875, a Treaty of the Meter was signed in Paris.[2] Among other things, the signatories agreed to define a certain object

[1]Francis Galton, *Natural Inheritance* (London: Macmillan, 1888).
[2]See David Freedman, Robert Pisani, and Roger Purves, *Statistics* (New York: W. W. Norton, 1978), Chapter 6.

Figure A

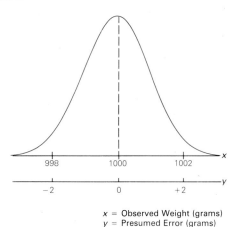

x = Observed Weight (grams)
y = Presumed Error (grams)

Figure B

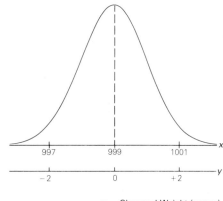

x = Observed Weight (grams)
y = Presumed Error (grams)

as the International Prototype Kilogram and to determine all other weights in relation to it. (The object in question is made of platinum-iridium and is held, under standard conditions of air pressure and temperature, at the International Bureau of Weights and Measures in Paris.) Each signatory nation owns a copy of the original Kilogram; that of the United States (number 20) is held at the National Bureau of Standards in Washington. With its help, similar weights in each state are calibrated, and these are used to check merchant scales periodically. In the end, the amount of sugar you buy is called 1 pound because its weight equals .4539237 of the original kilogram.

Yet even at the National Bureau of Standards, weighing the same object repeatedly may produce results such as these: 999.9231 grams, 999.9752 grams, 999.9603 grams, 1000.0598 grams, 1000.1003 grams, 999.9501 grams, 1000.0001 grams, and so on. Slight variations in the position of weights or imperceptible amounts of play in the balance mechanism will yield

ever new results, but the probability distribution of these results may well fit a normal curve perfectly. If the distribution of results looks like Figure A, the object may well be deemed to weigh 1 kilogram; if the distribution looks like Figure B, the object may well be deemed to weigh less than 1 kilogram.

It is clear from the previous discussion why the normal curve has also been called the *normal curve of error.* If the mean result of many measurements is defined as the correct weight, the mean translates into an error of zero, while positive and negative errors (associated with results in excess of or below the mean) are normally distributed around the mean. Note: repeated comparisons of the U.S. Kilogram (K_{20}) with the Paris prototype suggest that the U.S. kilogram is 19 parts per billion lighter than the original.[3]

[3]Mort LaBrecque, "After 185 Years, Physicists Are Still Weighing the Kilogram," *Popular Science,* May 1984, pp. 96–99 and 170.

6.2 The Decision to Seed Hurricanes

In 1961, it was suggested that seeding hurricanes with silver iodide would mitigate their destructive force. The extent of that force was evidenced when Hurricane Betsy in 1965 and Hurricane Camille in 1969 each caused $1.5 billion in property damage. Also in 1969, strong evidence for the effectiveness of the seeding procedure was obtained when Hurricane

Debbie was seeded twice and reductions in peak wind speed of 31 and 15 percent, respectively, were observed as a result. The U.S. Department of Commerce subsequently sponsored a study of the behavior of hurricanes; it was found that in a "typical" hurricane of 100 miles per hour the probability distribution of various percentage changes in maximum sustained

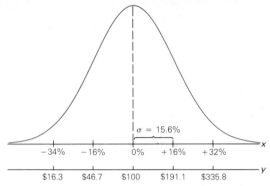

$\sigma = 15.6\%$

-34% -16% 0% +16% +32% →x

$16.3 $46.7 $100 $191.1 $335.8 →y

x = Percentage Change in Maximum Sustained Surface Wind
Within 12 Hours Before Landfall, Given a 100mph Hurricane
y = Associated Property Damage (millions of dollars)

surface winds within 12 hours before landfall could be described by a normal curve with a mean of 0 and a standard deviation of 15.6 percent. Such a curve is shown in the accompanying figure; typical property damage totals are also indicated.

Thus, a hurricane the 100-mph peak sustained surface wind of which intensifies 32 percent within the 12 hours prior to landfall can be expected to destroy property well in excess of $300 million, but the damage will only be about $15 million if the winds decrease by 34 percent. (Note: The probabilities for the two events can be determined by using the normal probabilities table. In this example, +32 percent

equals a z score of $32 - 0$, divided by 15.6, or 2.05; -34 percent equals a z score of -2.18. The associated probabilities, according to Appendix Table E, for $z \geq 2.05$ or $z \leq -2.18$ equal $.5 - .4798 = .0202$ and $.5 - .4854 = .0146$, respectively.) Clearly, the kind of information embodied in the accompanying normal curve can become a crucial input in the decision to engage in, or refrain from engaging in, costly hurricane seeding.

Source: R. A. Howard, J. E. Matheson, D. W. North, "The Decision to Seed Hurricanes," *Science,* June 16, 1972, pp. 1191–1202.

BIOGRAPHY

6.1 Carl Friedrich Gauss

Carl Friedrich Gauss (1777–1855) was born in Brunswick, Germany, where he grew up in humble circumstances (his father was a bricklayer and gardener; his mother had had no schooling and could not even write). Yet Gauss grew up to become one of the greatest mathematicians of all time. His precocity was extraordinary and probably unequaled. Before he was three, while watching his father's wage calculations, he detected an error and announced the correct result. At school, the speed and accuracy of his mental calculations astonished his beginning teacher in arithmetic; before long, Gauss mastered the best available textbooks and went beyond his teachers. Gauss's amazing powers came to the attention of the Duke of Brunswick who became his patron, sending him to the University of Göttingen. While a student there, Gauss discovered that a regular polygon of 17 sides is amenable to straightedge-and-compass construction; such a feat had eluded mathematicians for 2,000 years. Shortly thereafter, in 1801,

Gauss published *Disquisitiones Arithmeticae,* a masterpiece that took the theory of numbers far beyond its earlier state and established Gauss as a mathematical genius of the first rank, equal to Archimedes and Newton. One great work was followed by another in 1809, the *Theoria Motus Corporum Coelestium,* in which Gauss developed methods for calculating the orbits of heavenly bodies. It was here that he made first use of two of the most common tools of all sciences today, the theory of least squares (to be discussed in Chapter 11) and the normal curve (discussed in this chapter). Although Gauss acknowledged the priority of Laplace with respect to the normal curve,[1] it soon came to be known as the *Gaussean curve,* and it is still known as such in many countries today. (It was only in 1893 that Karl Pearson, Biography 9.1, introduced the term *normal* curve, and this is somewhat unfortunate because it creates the false impression that other-than-normal probability density functions are somehow abnormal or rare.) Gauss, apparently, was the first to view the bell-shaped curve as a model for depicting random error, as noted in Analytical Example 6.1. Writing in the context of his book on celestial mechanics, he said[2]:

> The investigation of an orbit having, strictly speaking, the *maximum* probability, will depend upon a knowledge of the law according to which

the probability of errors decreases as the errors increase in magnitude: but that depends upon so many vague and doubtful considerations—physiological included—which cannot be subjected to calculation, that it is scarcely, and indeed less than scarcely, possible to assign properly a law of this kind in any case of practical astronomy. Nevertheless, an investigation of the connection between this law and the most probable orbit, which we will undertake in its utmost generality, is not to be regarded as by any means a barren speculation. Let $\phi\Delta$ be the probability to be assigned to each error Δ. Now although we cannot precisely assign the form of this function, we can at least affirm that it should be a maximum for $\Delta = 0$, equal, generally, for equal opposite values of Δ, and should vanish, if, for Δ is taken the greatest error, or a value greater than the greatest error.

By age 30, Gauss was director of the Göttingen observatory, where he stayed for the rest of his life. He contributed mightily to astronomy but also to every branch of mathematics and physics. Hundreds of his publications testify to a man who combined in a most unusual way a pure mathematician's interest in abstract ideas and logical rigor, a theoretical physicist's interest in the creation of mathematical models of the physical world, an astronomer's talent for keen observation, and an experimentalist's skill in the application and invention of methods of measurement. (Among other things Gauss invented the heliotrope, the magnetometer, the photometer, and the telegraph.)

---

[1] A biographical sketch of Laplace can be found in Chapter 4 of the *Student Workbook* that accompanies this text.
[2] Cited in Helen M. Walker, *Studies in the History of Statistical Method* (Baltimore: Williams and Wilkins, 1929), p. 23.

Summary

1. This chapter focuses on probability distributions of *continuous* random variables, which can assume values at all points on a scale of values. Naturally, one cannot list all the conceivable values of such variables and then associate their infinite number with an equally infinite number of probabilities. Instead, one associates probabilities of continuous random variables with various *ranges* of their values, and one measures these probabilities as appropriate areas under a smooth frequency curve that approximates a relative frequency histogram and is called a *probability density function.* An infinite variety of continuous probability distributions exist, although the nor-

mal, exponential, and uniform ones are most frequently encountered.

2. A *normal probability distribution* is a probability density function that is (a) single-peaked above the random variable's mean, median, and mode, (b) perfectly symmetrical about this central value, and (c) characterized by tails extending indefinitely in both directions from the center, approaching but never touching the horizontal axis. The height of the normal probability density function for different values of the normal random variable can be determined by formula (Box 6.A), and probabilities for specified ranges of values of this variable can be found by integrating that function over

these ranges. It is common practice, however, to convert all members of the normal probability distribution family into a *standard normal curve* (with a mean of zero and a standard deviation of 1); probabilities for this curve are readily available in appropriately prepared tables (such as Appendix Table E). In certain circumstances, normal probabilities can be used to approximate various discrete probability distributions. This is illustrated with respect to Chapter 5's binomial probability distribution.

Key Terms

normal probability distribution
probability density function

standard normal curve
standard normal deviate

Questions and Problems

The computer program at the end of this chapter can be used to work many of the subsequent problems.

1. The time required by a bank teller to cash a check has a mean of 30 seconds and a standard deviation of 10 seconds. Find the times representing the tenth and seventy-fifth percentiles. (Assume that you are dealing with a normally distributed random variable.)

2. An architect wishes to design doors so that 95 percent of all people have at least a 1-inch clearance when passing through. If people have a mean height of 66 inches with a standard deviation of 4 inches, how high must the doors be? (Assume that people's heights are a normally distributed random variable.)

3. The Stanford-Binet IQ test has a mean of 100 and a standard deviation of 16. How likely is it that a randomly chosen person gets a score of at least 140? (Assume that IQs and test scores are normally distributed random variables.)

4. The output of workers in a plant is a normally distributed random variable. The plant's manager suggests that workers switch from hourly pay ($5.16 per hour) to a piece-work plan. If workers on the average produce 30 pieces per hour and the standard deviation is 5 pieces, which pay per piece would assure workers their present hourly pay at least 80 percent of the time? If the workers were paid 25 cents per piece, what percentage of the time would they earn between $4.50 and $5.50 per hour?

5. A company administers a test to all of its employees. The mean score is 500, and the standard deviation is 100. If the workers with the 25 percent highest scores are to be given special training, what is the lowest acceptable score for entrance into the training program? Furthermore, if the average time required to finish the test equals 60 minutes, with a standard deviation of 12 minutes, when should the exam be terminated so that 95 percent of the workers have completed all parts of the test? (Assume that test scores as well as finishing times are normally distributed random variables.)

6. A firm uses 2,000 light bulbs; their lifetime has a mean of 500 hours and a standard deviation of 50 hours. How often must the bulbs be replaced if all of them are to be replaced at once and at most 1 percent of them are to burn out between replacements? (Assume that the lifetime in question is a normally distributed random variable.)

7. The city miles-per-gallon (mpg) ratings of cars have a mean of 25.9 and a standard deviation of 2.45. If an automobile manufacturer wants to build a car the mpg ratings of which improve upon 99 percent of existing cars, what must the new car's mgp rating be? (Assume that the mpg rating is a normally distributed random variable.)

8. The shelf life of a battery has a mean of 525 days and a standard deviation of 50 days. If the battery has been on the shelf for 647 days, what is the probability of it being dead? (Assume that the shelf life of the battery is a normally distributed random variable.)

9. The time required to install a new aircraft engine is a normally distributed random variable with a

mean of 20 hours and a standard deviation of 1 hour. What is the probability that the next installation takes
a. between 20 and 21.5 hours?
b. between 18 and 20 hours?
c. between 19 and 22 hours?
d. over 23 hours?
e. at most 16.1 hours?
f. between 21 and 22 hours?
g. between 17 and 18 hours?
h. at most 23.7 hours?
i. more than 18.3 hours?

10. A manufacturer has deveoped a new type of automobile tire. The marketing department believes that the mileage guarantee offered to customers will be the crucial factor in winning consumer acceptance. Naturally, management wants to know the probability distribution of the lifetimes of these tires. Accordingly, 400 tires are tested. Engineers report that lifetimes are normally distributed with a mean of 36,000 miles and a standard deviation of 4,800 miles.
a. How many of the tested tires had a lifetime between 31,200 and 40,800 miles?
b. How many lasted at most 45,600 miles?
c. How many lasted at least 26,400 miles?
d. How many lasted between 26,400 and 31,200 miles?
e. Within what limits were the lifetimes of the 20 percent worst tires?
f. What mileage guarantee should the firm offer if it wants to make no more than 10 percent of all tires eligible?

11. The weekly number of checks cleared by a bank is a normally distributed random variable with a mean of 122,000 checks cleared and a standard deviation of 10,000 checks. In what proportion of weeks will the bank have to clear more than 140,000 checks?

12. The number of gallons of effluent processed at a sewage treatment plant is a normally distributed random variable with a mean of 50,000 gallons per day and a standard deviation of 7,000 gallons per day. The chief engineer wants to know: In what proportion of days will the plant be called upon to process in excess of 60,000 gallons (the plant's designed capacity)?

13. A state's turnpike authority finds its annual toll revenues to be normally distributed with a mean of $700,000 and a standard deviation of $50,000. The manager wants to know
a. the probability that next year's operating expenses of $680,000 will be covered.

b. the probability that these expenses will be covered in the next two years (assuming that the size of one year's revenue is independent of the size of the previous year's).
c. the size of revenue in the best 25 percent of years.

14. In a recent year, the mathematics SAT scores were normally distributed with a mean of 500 for men and of 460 for women; both had a standard deviation of 100.
a. Estimate the percentages of men and women, respectively, with a score of 600 or more. Of 300 or less.
b. Now you are to pick one man at random from the group and guess his score; if you are correct within 50 points, you get $100. What score do you pick if you want to win? What is your chance of winning?

15. Each year Amherst College accepts 700 students. From historical experience it is known that the number from this group who choose to come is a normally distributed random variable with a mean of 375 and a standard deviation of 40.
a. Calculate the probability that more than 400 students will choose to come.
b. Calculate the probability that between 350 and 400 will choose to come.
c. Calculate the probability that fewer than 350 will choose to come.

16. College Board scores are normally distributed with standard deviations of 100 points for both left- and right-handed students. However, the mean math score for left-handed students is 525 whereas it is 500 for right-handed students. In the population as a whole, 10 percent of all students are left-handed. What fraction of students with math scores over 700 are left-handed?

17. A machine makes parts the lengths of which are normally distributed with a mean of 3 inches and a standard deviation of .15 inches. The acceptable range of lengths is 2.87 to 3.10 inches. What proportion of output is *not* acceptable?

18. A quality inspector tests the strength of aircraft control cables. Their breaking pressure is a normally distributed random variable with a mean of 500 lb. and a standard deviation of 50 lb. Some 100 cables ruptured under 475 lb. pressure or less. How many withstood pressure of 560 lb. or more?

19. A quality inspector tests the strength of window panes. Their breaking pressure is a normally distributed random variable with a mean of 80 mph wind velocity and a standard deviation of 7 mph. Some 50 panes broke under simulated wind velocities of 73 mph and less. How many withstood the hurricane velocity of 100 mph or more?

20. There are x gas stations in a state. Their incomes are normally distributed with a mean of $29,000 per year and a standard deviation of $5,100 per year. Some 189 stations earn between $26,000 and $31,000 per year. How many gas stations exist in the state?

21. There are x psychiatrists in a city. Their incomes are normally distributed with a mean of $89,000 per year and a standard deviation of $19,000 per year. Three psychiatrists earn more than $127,000 per year. How many psychiatrists are there in the entire city?

22. A time-and-motion study of 100,000 workers shows that they can accomplish a task in 35 seconds on the average. Seventy-four percent of the workers need more than 32 seconds. What percentage needs fewer than 36 seconds?

23. A study of 50,000 coal miners shows that the daily output per worker equals 17.3 tons on the average. Some 89.25 percent produce 18.1 tons or fewer. What percentage produces more than 16 tons?

24. An airline has 105 seats available per flight. If it sells 115 tickets, the actual arrivals are a normally distributed random variable with a mean of 100 and a standard deviation of 5. What is the probability that the number of passengers showing up exceeds the number of available seats?

25. A cattle feedlot finds that the daily weight gain of animals is 1.5 lb on the average with a standard deviation of .3 lb. If weight gain is a normally distributed random variable, what is the probability that a randomly chosen animal gains in a day
   a. less than 1 lb.?
   b. less than 2 lb.?
   c. more than 1.4 lb.?
   d. between 1.6 and 1.7 lb.?

26. The rents charged in a city are a normally distributed random variable with a mean of $551 per month and a standard deviation of $62. The mayor wants to know: What percentage of households would be eligible for rent subsidies if all households paying more than $600 were made eligible.

27. A personnel manager has discovered that the hours of sickleave taken by employees during a year are a normally distributed random variable with a mean of 52 hours and a standard deviation of 10 hours. She considers the hours taken by A (63) and B (93) highly unusual. Do you agree?

28. The weights of fish in Long Island Sound are normally distributed with a mean of 37 lb. and a standard deviation of 8 lb. One man claims to have caught an 80-lb. fish. Do you believe it? Explain.

29. An efficiency expert knows that the time required by electricians to wire a certain type of house is a normally distributed random variable with a mean of 14 hours and a standard deviation of 1.3 hours. Evaluate these claims by two electricians:
   a. "I can wire a house like this in 11 hours."
   b. "It took me 17 hours of hard work to wire this house."

30. A city water department finds household water use to be a normally distributed random variable with a mean of 27 gallons and a standard deviation of 3 gallons per day.
   a. Find the probabilities that a randomly chosen household uses more than 25 gallons per day, between 20 and 25 gallons per day, fewer than 20 gallons per day.
   b. If the mayor wants to give a tax rebate to the 17-percent lowest water users, what should the gallons-per-day cutoff be?
   c. If the rebate changes the mean to 22 gallons and the standard deviation to 4 gallons per day, what should the 17-percent cutoff be?

31. The manager of a flower shop has a cash flow problem: 20 days of inflow are followed by 10 days of outflow, but the latter is a normally distributed random variable with a mean of $1,500 and a standard deviation of $400. If the manager wants only a 5-percent chance of running out of cash before the end of the outflow period, how much money should he have on hand at the beginning of that period?

32. The debt-asset ratios of all firms in an industry are normally distributed with a mean of 48 percent and a standard deviation of 9.9 percent. A particular firm's ratio is 37.2 percent. Do you find it unusual? Explain.

33. A bank fails. The FDIC finds it had *insider loans* (to directors and stockholders) of over $1.9 million. Given the fact that the size of such loans in the banking industry is a normally distributed random variable with a mean of $437,000 and a standard deviation of $129,000, is the situation unusual?

34. An investigation reveals that the local electric company has been overcharging customers. The overcharges are normally distributed with a mean of 700 kwh per month and a standard deviation of 200 kwh per month. What is the probability that a randomly chosen household was overcharged 800 kwh or more? A maximum of 500 kwh?

35. An unemployed worker sets out on a job search. The available jobs pay wages that are normally distributed with a mean of $400 per week and a standard deviation of $30 per week. What is the prob-

ability that the first job offered to the worker will pay $500 per week or more? $350 per week or less? Between $410 and $430 per week?

36. An insurance company finds that the life span of participants in its annuity program is a normally distributed random variable with a mean of 69 years and standard deviation of 8.1 years. What proportion of participants will get benefits beyond age 70? Age 75? Age 85?

37. The price-earnings ratio of all stocks is a normally distributed random variable with a mean of 9.6 and a standard deviation of 2.1. You suspect that some of the 10-percent lowest P/E ratio stocks are undervalued. You want to look at the companies involved. In order to make the selection, you should look at companies with a P/E ratio of x or less. What is x?

38. Since they are needed so often, find the z values corresponding to percentiles of 5, 10, 15, and so on up to 100 in increments of 5. Make your calculations to four decimals.

39. Experience shows that 25 percent of the people entering a store make a purchase.
  a. What is the probability distribution for purchases made for the next 20 customers? (Do *not* use binomial probability tables.)
  b. What is the probability that at most 3 of the next 20 customers will make a purchase?

40. In bridge tournaments players are assigned randomly to positions at the table (North, South, East, and West). Charles Goren once complained that he sat at the North position far more often than he should have. If Goren sat North in 250 out of 800 tournaments, was his complaint justified? What is

the probability that such a happening could have occurred by chance? (Hint: Use the normal approximation to the binomial distribution.)

41. The absentee rate in a class of 500 statistics students is 10 percent. With the help of an appropriate approximation, compute the probability that in the next class
  a. more than 400 will attend.
  b. fewer than 450 will attend.
  c. between 413 and 463 will attend.

42. A firm ships thermometers in boxes of 1,000. Typically, 10 percent are broken upon arrival. Using an approximation, determine the probability that, in a randomly chosen box,
  a. 100 to 130 thermometers are broken.
  b. 75 thermometers are broken.
  c. 200 thermometers are broken.
  d. none are broken.

43. A census of telephone bills has revealed that the conversation times charged in 10 percent of the bills exceed the actual conversation times found elsewhere in the company records. A random sample of 100 bills is taken. Using an appropriate approximation, determine the probability of discovering fewer than 5 such errors.

44. Experience shows that in 40 percent of IRS audits taxpayers do not owe additional taxes. If we take a sample of 1,000 audited taxpayers, what is the probability that between 300 and 600 owe nothing? (Hint: Use an appropriate approximation.)

45. Reconsider Practice Problem 1 of this chapter. What is the probability that a randomly chosen engine has a lifetime of less than 2,075 hours?

Selected Readings

Cohen, Daniel. *Intelligence, What Is It?* New York: M. Evans, 1974. Traces the testing of intelligence from the ideas of Galton (Biography 11.1) in the 1890s to lawsuits about IQ tests in the 1970s.

Kohler, Heinz. *Statistics for Business and Economics,* 2nd ed. Glenview, Ill.: Scott, Foresman and Co., 1988. Chapter 7 contains a detailed discussion of two other

continuous probability distributions, the exponential distribution and the uniform distribution.

Walker, Helen M. *Studies in the History of Statistical Method.* (Baltimore: Williams and Wilkins, 1929), Chapter 2. A superb discussion of the history of the normal curve.

Computer Program

The DAYSAL-2 personal-computer diskettes that accompany this book contain one program of interest to this chapter:

7. Probability Distributions, among other things, lets you find areas under the normal curve between any two z-values you specify.

PART THREE

Analytical Statistics: Basic Inference

7   Sampling Distributions and Estimation

8   Hypothesis Testing: Classical Techniques

9   Hypothesis Testing: The Chi-Square Technique

CHAPTER SEVEN

Sampling Distributions and Estimation

As was noted in Chapter 1, decision makers frequently gain vital information about a population not by taking a full-fledged census, but merely by taking a sample. Consider a manufacturer who wants to know the average time workers require to complete a given task, the average amount of fuel needed to ship a truckload a given distance, or the average age of those who use a given product. Consider questions about a product's market share, about the percentage of acceptable units in a shipment of parts, or about the proportion of people watching a TV show, favoring a tax hike, or opposing a labor contract. In all these cases and many more, the statistician is interested in learning something about a statistical population. In the absence of a census, the desired knowledge about such *parameters* as the population mean (μ), the population standard deviation (σ), or the population proportion (π) can be gained only by drawing a random sample from the population, calculating such *statistics* as the sample mean ($\overline{X}$), the sample standard deviation (s), or the sample proportion (P), and making inferences about the parameters from the statistics.[1] The process of inferring the values of unknown population parameters from known sample statistics is called **estimation.**

ESTIMATORS AND ESTIMATES

The type of sample statistics that is used to make inferences about a given type of population parameter is called the **estimator** of that parameter. A later section of this chapter will consider major criteria for choosing an appropriate estimator—that is, one the actual numerical value of which is likely to be close to the unknown value of the parameter of interest. Obviously, the control of error is of major importance

---

[1]The initial discussion of *parameters* and *statistics* can be reviewed in the introductory section of Chapter 3.

when making estimates. For this reason, random sampling, rather than convenience or judgment sampling, is a necessity; this type of sampling helps eliminate *systematic* error or bias. In addition, as more advanced texts show, *random* or sampling error can be reduced (at a cost) by constructing samples that are sufficiently large. (For a review of the concepts of error, see the Chapter 1 section on "Errors in Survey Data.")

When the estimate of a population parameter is expressed as a single numerical value, it is referred to as a **point estimate.** Thus, the average time workers require to complete a given task may be estimated at 32 minutes; a product's market share may be estimated at 70 percent. These are point estimates. Any estimate, of course, can be wrong; users of estimates are always eager to know how wrong any given estimate might be. That is why it is often deemed desirable to provide a range of values within which the unknown population parameter presumably lies; such a range is called an **interval estimate.** As will be shown below, depending on the size of the interval, one can attach varying degrees of confidence to it and state with, say, 95-percent confidence that a product's market share lies between 64 and 76 percent or state with, say, 80-percent confidence that it lies between 67 and 73 percent. This type of statement calls attention to the fact that any estimate can be wrong. Thus, a procedure that yields interval estimates with 95-percent confidence is expected to be correct 95 percent of the time but wrong 5 percent of the time.

A final note: the process of estimation begins by sampling an already existing, clearly identifiable population. It must not be confused with *forecasting,* to be discussed in Chapter 13, which seeks to make statements about parameters of *future* populations—as about the median incomes of Americans in the year 2000, about the proportion of consumers likely to favor a new product that is still on the drawing board, or about the average daily demand for electric power five years hence.

THE CONCEPT OF THE SAMPLING DISTRIBUTION

To understand the nature of statistical estimation, one needs a firm grasp of the concept of the sampling distribution because it provides an important link between the single sample that is typically taken and the population about which inferences are to be made. As a basis for illustrating the concept, let us consider the two statistical populations given in Table 7.1. Merely in order to keep our example simple, we imagine that there exist only $N = 5$ business executives in an industry. Their salaries and sexes are given in the table. With this complete information in hand, we can, of course, calculate such population summary measures as the mean annual salary (μ), along with its variance (σ^2) and standard deviation (σ), as well as, say, the proportion of females (π_F) in the executive ranks. The summary calculations are indicated underneath the table. (The formulas involved were discussed in Chapter 3.) The mean annual salary equals $40,000 (with a standard deviation of $9,508); there are 40-percent females among the executives.

Suppose, however, that this census-type information were not available to us and couldn't easily be acquired. (In reality, there might be 30,000 executives in the industry, and contacting them all would be too costly or time-consuming.) Suppose we, therefore, decided to *estimate* the four population parameters by taking a simple random sample of the executives. As an example, let us take a sample of size $n = 3$, and let us

Table 7.1   Hypothetical Populations of Executive Salaries and Sexes

Executive	Annual Salary (thousands of dollars)	Sex
A	39	M
B	41	F
C	25	F
D	55	M
E	40	M

Summary Measures

a. Salary

$$\mu = \frac{\Sigma X}{N} = \frac{200}{5} = 40$$

$$\sigma^2 = \frac{\Sigma(X - \mu)^2}{N} = \frac{452}{5} = 90.4$$

$$\sigma = \sqrt{90.4} = 9.508$$

b. Sex

$$\pi_F = \frac{2}{5} = .40$$

do so without replacement. (Sampling without replacement is also common practice; once an elementary unit has been chosen, there is typically no chance that it will be chosen again.) The sample we pick might consist of executives B, D, and E, and we could instantly calculate the appropriate sample summary measures (once more using the applicable formulas from Chapter 3). Thus, the summary measures for salary would be the following:

$$\overline{X} = \frac{\Sigma X}{n} = \frac{136}{3} = 45.333 \text{ (thousands of dollars)}$$

$$s^2 = \frac{\Sigma(X - \overline{X})^2}{n - 1} = \frac{140.666}{2} = 70.333 \text{ (thousands of squared dollars)}$$

$$s = \sqrt{70.333} = 8.386 \text{ (thousands of dollars)}$$

The summary measure for sex would be the following:

$$P_F = \frac{1}{3} = .333$$

From our sample, we would estimate the mean annual salary at $45,333 (with a standard deviation of $8,386); and we would presume that 33.3 percent of the executives are female.

We might also, however, be bothered by a disturbing thought: if the random-selection procedure had produced a different sample of $n = 3$, we would surely have calculated different summary measures for the sample and would, thus, have made different point estimates of the corresponding population parameters. Indeed, what the point estimates might have been (if we had selected different samples) can easily be visualized with the help of Table 7.2. The table lists all the possible simple random

Table 7.2 Possible Samples of $n = 3$ Taken Without Replacement from a Population of $N = 5$, Along with Summary Measures of $\overline{X}$, s^2, s and P_F

Sample Number	Units in Sample	(a) for Salary $\overline{X}$	(a) for Salary s^2	(a) for Salary s	(b) for Sex P_F	Probability p
(1)	ABC	35.000	76.000	8.718	.667	.1
(2)	ABD	45.000	76.000	8.718	.333	.1
(3)	ABE	40.000	1.000	1.000	.333	.1
(4)	ACD	39.667	225.333	15.011	.333	.1
(5)	ACE	34.667	70.333	8.386	.333	.1
(6)	ADE	44.667	80.333	8.963	0	.1
(7)	BCD	40.333	225.333	15.011	.667	.1
(8)	BCE	35.333	80.333	8.963	.667	.1
(9)	BDE	45.333	70.333	8.386	.333	.1
(10)	CDE	40.000	225.000	15.000	.333	.1

Source: Table 7.1.

samples of size $n = 3$ that we might have taken (without replacement) from among our population of size $N = 5$. (This ability to *list* all the possible samples is a lucky circumstance due to the utter simplicity of our example; it would not be practical to do so if we had considered a population of, say, 30,000 executives.) If we consider the order in which people enter a sample as irrelevant (so that a sample of B, D, E is treated as identical to a sample of E, D, B, and the like), we can apply the combinatorial formula of Box 4.F and determine that the possible number of samples of $n = 3$ out of a population of $N = 5$ equals

$$C_3^5 = \frac{5!}{3!\,(5-3)!} = \frac{5!}{3!\,2!} = \frac{5 \cdot 4}{2} = 10$$

These 10 samples are listed in Table 7.2; four summary measures (calculated on the basis of the corresponding data in Table 7.1) are given for each one. Note that the sample evaluated above now appears as sample number (9).

When we contemplate Table 7.2, we realize that every sample summary measure (such as $\overline{X}$, s^2, s, and P_E) is in fact a random variable *prior* to the selection of an actual sample. Before we picked sample number (9), the sample mean, $\overline{X}$, for instance, could have turned out to be any one of the values in the $\overline{X}$ column of Table 7.2, depending entirely on which elementary units happened to be included in our sample. Because we were about to engage in simple random sampling, each of the 10 possible samples given in Table 7.2 was equally likely to be picked; hence, each of the 10 sets of associated summary measures was equally likely to become the set of sample statistics ultimately revealed to us. The probablities involved are given in the last column of Table 7.2. For any given value that is listed only once in the $\overline{X}$ column (or s^2 column or s column or P_F column), there existed, prior to our sample selection, a proability of

.1 that it would become the sample statistic (because there was a probability of .1 that any one of the listed samples would actually be drawn). Similarly, the probability equaled two times .1, or .2, for any value listed twice, and so on.

This list of probabilities in the last column of Table 7.2 brings us to the concept of the **sampling distribution**—a probability distribution of a random variable that happens to be a summary measure based on sample data. The sampling distribution, thus, shows the likelihood of occurrence associated with all the possible values of a sample statistic (which values would be obtained when drawing all possible simple random samples of a given size from a population). This probability distribution, therefore, describes the manner in which sample characteristics vary from sample to sample. In Table 7.2, for example, the data in the $\overline{X}$ column, along with those in the probability column, represent the sampling distribution of the sample mean; similarly, the data in the s^2, the s, and the P_F columns, along with those in the probability column, represent the sampling distributions, respectively, of the sample variance, the sample standard deviation, and the sample proportion.

Note: to the extent that some values of the random variable occur more than once, the presentation of the sampling distribution can, of course, be simplified. The sampling distribution of the mean executive salary (or of the proportion of female executives) can, for example, be presented as shown in Table 7.3.

One way of interpreting Table 7.3 is: if simple random samples of $n = 3$ were to be selected, say, 1,000 times from the population of $N = 5$ given in Table 7.1, one could expect to calculate a $34,667 mean salary 100 times, a $40,000 mean salary 200 times, and so on. In addition, one could expect to find a 0 proportion of females 100 times, a proportion of .333 some 600 times, and a proportion of .667 in the remaining 300 times.

Table 7.3   Sampling Distributions of the Mean and of the Proportion

These sampling distributions, culled from Table 7.2, show that, prior to sample selection, there exists a probability of .1 that the sample statistic of the mean salary will turn out to be $34,667; the probability also equals .1 that $\overline{X}$ will be calculated as $35,000, and so on. Similarly, the probability is .1 that the proportion of females will be found to be 0, but .6 that it will be calculated as .333, and .3 that this sample statistic will equal .667.

$\overline{X}$ (thousands of dollars)	Probability	P_F	Probability
34.667	.1	0	.1
35.000	.1	.333	.6
35.333	.1	.667	.3
39.667	.1		1.0
40.000	.2		
40.333	.1		
44.667	.1		
45.000	.1		
45.333	.1		
	1.0		

Source: Table 7.2.

Table 7.4   Summary Measures of Two Sampling Distributions

With the help of the formulas presented in Box 5.A, sampling distributions of Table 7.3 can be summarized as shown here.

(a) Summary Measures of the Sampling Distribution of the Mean, $\overline{X}$	(b) Summary Measures of the Sampling Distribution of the Proportion, P_F
$\mu_{\overline{X}} = 40$	$\mu_P = .40$
$\sigma^2_{\overline{X}} = 15.067$	$\sigma^2_P = .04$
$\sigma_{\overline{X}} = 3.882$	$\sigma_P = .20$

SUMMARY MEASURES OF THE SAMPLING DISTRIBUTION

Any sampling distribution can, in turn, be described by summary measures of its own. Given the data in the $\overline{X}$, s^2, s, and P_F columns of Table 7.2 (or given a consolidated version of these data as shown in Table 7.3), we can calculate their means or expected values, their variances, and their standard deviations. Such calculations have in fact been performed, with the help of the formulas in Box 5.A, for the two sampling distributions of $\overline{X}$ and P_F. The results are given in Table 7.4.

Thus, $\mu_{\overline{X}}$ (pronounced "mu sub X bar") is the mean of all the possible sample salary means (or the expected value of the sampling distribution of the mean); μ_P is, similarly, the mean of all the possible sample proportions of female executives. The expressions $\sigma^2_{\overline{X}}$ and σ^2_P are the respective variances of the sampling distributions of the sample mean and of the sample proportion; their square roots (the standard deviations) are also referred to as the *standard errors* of the respective sample statistics. Thus, $\sigma_{\overline{X}}$ is the **standard error of the sample mean,** and σ_P is the **standard error of the sample proportion.**

A note of caution: it is important not to confuse $\sigma_{\overline{X}}$ with s. This mistake is easy to make because both are standard deviations. However, $\sigma_{\overline{X}}$ (the standard error of the sample mean from Table 7.4) measures the variability of *possible* $\overline{X}$ values that might be obtained—a variability a statistician is likely to contemplate in the planning stage of sampling. In contrast, s (the standard deviation of a particular sample result found in any one row of Table 7.2) measures the variability of X values around an *actual* $\overline{X}$ value observed; this variability is calculated at the final stage of sampling, after one of the many possible samples has actually been selected.

THE SAMPLING DISTRIBUTION VERSUS ITS PARENT POPULATION

The answer to one question is of crucial importance to statistical estimation—that is, the answer is crucial to inferring the values of unknown population parameters from known sample statistics: what is the relationship between a sampling distribution and its parent population? We can answer the question in two ways: by focusing on the general shape of the two frequency distributions (of possible sample statistics on the one hand and of population values on the other) or by focusing on the mathematical relationships between summary measures. We will do both.

The Shapes of the Distributions and the Central Limit Theorem

The sampling distribution of the sample mean. Let us first consider the sampling distribution of the sample mean, $\overline{X}$, and how its shape relates to the associated population values. Two of the most important theorems in all of statistics describe this relationship.

> *Theorem 1:* If $\overline{X}$ is the mean of a random sample taken from a population and if the population values are normally distributed, the sampling distribution of $\overline{X}$ is also normally distributed, regardless of sample size.

The above discussion, although not precisely applicable, at least hints at the validity of this theorem. Note how the salary population values in Table 7.1 are symmetrically distributed around the population mean of $\mu = 40$; then notice how the sampling distribution of the mean in Table 7.3 is also symmetrically distributed around its mean of $\mu_{\overline{x}} = 40$. To be sure, a symmetrical distribution is not the same thing as a precisely normal distribution. Nevertheless, we can consider the Table 7.1 salary distribution *approximately* normal.[2]

Given a population of X values that is normally distributed in the strict sense, rather than only approximately so, and given an associated sampling distribution of $\overline{X}$ that is—in accordance with the above theorem—therefore also normal, it is easy to find the probability that the mean of any given random sample will be within any given interval. All one needs to do is transform the range of $\overline{X}$ values into the corresponding z values and use Appendix Table E to establish the appropriate areas under the standard normal curve. The following problems apply to a normally distributed sampling distribution of $\overline{X}$ in which $\mu_{\overline{x}} = 40$ and $\sigma_{\overline{x}} = 4.655$. (These values might have been derived like those in Table 7.4 but from a population, unlike that in Table 7.1, that was very large and normally distributed in the strict sense. Thus, Theorem 1 permits the stated assertion concerning the normal shape of the sampling distribution of $\overline{X}$.)

Practice Problem 1

A Sample Mean Below 45
What is the probability that a sample mean will lie below 45? The area in question lies below $z = \dfrac{45 - 40}{4.655} = 1.07$. Thus, the probability equals .8577, as shown in Figure A by the sum of the shaded plus crosshatched areas.

---

[2]First, 60 percent of these population values lie within 1 standard deviation of the mean (within the range of 30.492 to 49.508) as opposed to 68.26 percent in the case of the normal distribution. Second, 100 percent of the population values lie within 2 standard deviations of the mean (within the range of 20.984 and 59.016) as opposed to 95.44 percent in the case of the normal distribution. Third, various measures of shape (skewness, kurtosis) that are not discussed in this book yield magnitudes very similar to a normal distribution. Similar statements can be made for the Table 7.3 sampling distribution of the mean.

Figure A

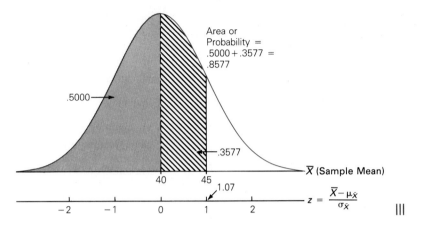

Area or
Probability =
.5000 + .3577 =
.8577

.5000

.3577

$\overline{X}$ (Sample Mean)

$z = \dfrac{\overline{X} - \mu_{\overline{X}}}{\sigma_{\overline{X}}}$

|||

Practice Problem 2

A Sample Mean Between 37 and 41.5

What is the probability that a sample mean will lie between 37 and 41.5? The area in question lies between $z = \dfrac{37 - 40}{4.655} = -.64$ and $z = \dfrac{41.5 - 40}{4.655} = .32$. Thus, the probability equals .2389 + .1255 = .3644 or the sum of the shaded plus crosshatched areas in Figure B.

Figure B

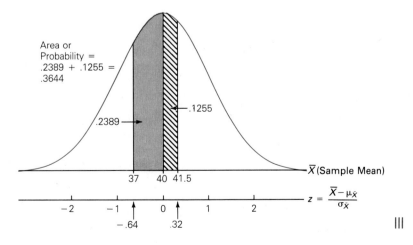

Area or
Probability =
.2389 + .1255 =
.3644

.1255

.2389

$\overline{X}$ (Sample Mean)

$z = \dfrac{\overline{X} - \mu_{\overline{X}}}{\sigma_{\overline{X}}}$

|||

Theorem 2: If $\overline{X}$ is the mean of a random sample taken from a population and if the population values are *not* normally distributed, the sampling distribution of $\overline{X}$ nevertheless approaches a normal distribution as sample size increases—provided sample size, n, remains small relative to population size, N; this approximation is near-perfect for samples of $n \geq 30$ but $n < .05N$.

This theorem is called the **central limit theorem** and is, perhaps, the most important one in the entire field of statistical inference. As long as we take random samples that are sufficiently large absolutely ($n \geq 30$) but that are fairly small relative to population size ($n < .05N$), the theorem allows us to infer population parameters from sample statistics without knowing the shape of the population distribution (which is precisely the type of knowledge that is often unavailable). In addition, as the next section shows, the theorem can be adapted for use with discrete as well as continuous random variables.

Figure 7.1 pictures a variety of population distributions that are anything but normal, and it clearly shows the tendency toward normality in the associated sampling distributions as sample size, n, increases.

The sampling distribution of the sample proportion. Under specified circumstances, the sampling distribution of the sample proportion, P, can also be treated as if it were normally distributed. Recall from Chapter 5 that the proportion of successes in a Bernoulli process is a random variable having a binomial distribution. As sample size increases, the binomial distribution, too, approaches normality (see Figure 5.3 on page 180). Also recall the Chapter 6 section, "Approximating Discrete Probability Distributions by the Normal Probability Distribution." According to the rule of thumb there discussed, the binomial distribution is extremely close to the normal distribution whenever $n \cdot \pi$ as well as $n(1 - \pi) \geq 5$. In this expression, n stands for sample size, and π stands for the probability of success in any one trial (which is nothing else but the population proportion, π). In our above example, involving the proportion of females among executives, $n = 3$ and $\pi = .4$; thus, the rule of thumb is not satisfied [$n \cdot \pi = 1.2$ and $n(1 - \pi) = 1.8$]. Indeed, the above sampling distribution of the sample proportion (Table 7.3) is not normally distributed. If the rule of thumb is satisfied, however, one can determine probabilities for various ranges of values of the sample proportion by means of the standard-normal-curve table, using the procedure illustrated in the aforementioned Chapter 6 section, and calculating $z = (P - \mu_P)/\sigma_P$. Thus, we can state the following adaptation of the central limit theorem:

Theorem 2A: If P is the proportion in a random sample taken from a population and if the population values are *not* normally distributed, the sampling distribution of P nevertheless approaches a normal distribution provided $n \cdot \pi$ as well as $n(1 - \pi) \geq 5$.

Mathematical Relationships Among Summary Measures

Concerning the sampling distribution of $\overline{X}$. The way in which the summary measures of the sampling distribution of the sample mean relate to those of the parent population is shown in Box 7.A. In addition, mathematical derivations are provided in Chapter 7 of the *Student Workbook* that accompanies this text.

Figure 7.1   Illustrations of the Central Limit Theorem

These graphs picture vividly the incredibly rapid tendency toward normality in sampling distributions as sample size increases, regardless of the shape of the associated population distributions—provided only that $n < .05N$. (The vertical dotted lines coincide with the means of the various distributions.)

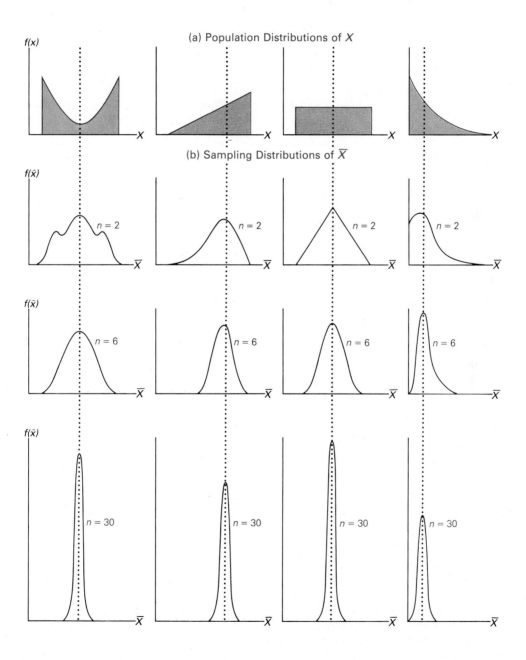

7.A Summary Measures of the Sampling Distribution of $\bar{X}$

$$\mu_{\bar{X}} = \mu$$

$$\left.\begin{array}{c} \sigma_{\bar{X}}^2 = \dfrac{\sigma^2}{n} \\[2mm] \text{and} \\[2mm] \sigma_{\bar{X}} = \dfrac{\sigma}{\sqrt{n}} \end{array}\right\}$$ when selections of sample elements are statistically *independent* events; typically referred to as "the large-population case," because $n < .05N$.

$$\left.\begin{array}{c} \sigma_{\bar{X}}^2 = \dfrac{\sigma^2}{n}\left(\dfrac{N-n}{N-1}\right) \\[3mm] \text{and} \\[3mm] \sigma_{\bar{X}} = \dfrac{\sigma}{\sqrt{n}}\sqrt{\dfrac{N-n}{N-1}} \end{array}\right\}$$ when selections of sample elements are statistically *dependent* events; typically referred to as "the small-population case," because $n \geq .05N$.

where $\mu_{\bar{X}}$ is the mean, $\sigma_{\bar{X}}^2$ is the variance, and $\sigma_{\bar{X}}$ is the standard deviation of the sampling distribution of the sample mean, $\bar{X}$, while μ, σ^2, and σ are population mean, variance, and standard deviation, respectively, and N or n are population or sample size.

Notice that the mean of all possible sample means, $\mu_{\bar{X}}$, always equals the population mean, μ. This fact was evident in the above example (Tables 7.1 and 7.4) that showed $\mu = \mu_{\bar{X}} = 40$. This fact is also shown by the vertical dotted lines in Figure 7.1 that coincide with the means of the various distributions shown there.

The relationship between the variance of the sampling distribution, $\sigma_{\bar{X}}^2$ (or the standard deviation, $\sigma_{\bar{X}}$), on the one hand and the corresponding population measures is somewhat more complicated. The relationship also differs depending on whether the selections of sample elements are statistically independent or dependent events. As was noted in Chapter 4, two events are statistically independent if the probability of the occurrence of one is not affected by the occurrence of the other; in the case of statistical dependence, the occurrence of one event does affect the probability of the occurrence of the other. Now consider the event of selecting an element of the population for inclusion in a sample. Selecting one element does not in the least affect the probability of selecting any other element if the population is infinite in size or if sampling occurs with replacement. Even if sampling occurs without replacement, selecting one element does not affect the probability of selecting any other element noticeably as long as population size, N, is large relative to the sample size, n. (By tradition, the latter condition is considered fulfilled whenever $n < .05N$.) Statistical independence in sample selection is, therefore, frequently referred to as "the large-population case."

On the other hand, selecting an element of the population for inclusion in a sample *does* noticeably affect the probability of selecting any other element when sampling occurs without replacement and when the population is small relative to the sample ($n \geq .05\ N$). Statistical dependence in sample selection is, therefore, typically referred to as "the small-population case."

As the formulas in Box 7.A attest, as long as $n > 1$, the variance (or standard deviation) of the sampling distribution of $\overline{X}$ is smaller than the corresponding population summary measure. (Tables 7.1 and 7.4 once more provide examples: $\sigma^2 = 90.4$, but $\sigma_{\overline{X}}^2 = 15.067$; $\sigma = 9.508$, but $\sigma_{\overline{x}} = 3.882$.) Note, in addition, that the variance (or standard deviation) of the sampling distribution of $\overline{X}$ is larger the larger is the population variance (or standard deviation), given sample size, n. On the other hand, the larger the sample, given the population variance (or standard deviation), the smaller is the variance (or standard deviation) of the sampling distribution.

Note: the summary data of the sampling distribution of $\overline{X}$ in Table 7.4 were derived with the help of the general formulas first presented in Chapter 5. We can now confirm these results with the Box 7.A formulas (small-population case), applied to the population summary data of Table 7.1.

$$\mu_{\overline{x}} = \mu = 40$$

$$\sigma_{\overline{x}}^2 = \frac{\sigma^2}{n}\left(\frac{N - n}{N - 1}\right) = \frac{90.4}{3}\left(\frac{5 - 3}{5 - 1}\right) = 30.133(.5) = 15.067$$

$$\sigma_{\overline{x}} = \frac{\sigma}{\sqrt{n}}\sqrt{\frac{N - n}{N - 1}} = \frac{9.508}{\sqrt{3}}\sqrt{\frac{5 - 3}{5 - 1}} = 5.489(.707) = 3.881$$

Except for rounding error, the results are identical.

Concerning the sampling distribution of P. The way in which the summary measures of the sampling distribution of the sample proportion relate to that of the parent population (namely, π) is shown in Box 7.B.

7.B Summary Measures of the Sampling Distribution of P

$$\mu_P = \pi$$
$$\sigma_P^2 = \frac{\pi(1 - \pi)}{n}$$
$$\sigma_P = \sqrt{\frac{\pi(1 - \pi)}{n}}$$

when selections of sample elements are statistically *independent* events; typically referred to as "the large-population case," because $n < .05N$.

$$\sigma_P^2 = \frac{\pi(1 - \pi)}{n}\left(\frac{N - n}{N - 1}\right)$$
$$\sigma_P = \sqrt{\frac{\pi(1 - \pi)}{n}}\sqrt{\frac{N - n}{N - 1}}$$

when selections of sample elements are statistically *dependent* events; typically referred to as "the small-population case," because $n \geq .05N$.

where μ_P is the mean, σ_P^2 is the variance, and σ_P is the standard deviation of the sampling distribution of the sample proportion, P, while π is the population proportion and N or n are population or sample size.

Once more, we can confirm our earlier results (Table 7.4) that were derived independently. When applying the Box 7.B formulas (small-population case) to the only population summary value available (Table 7.1):

$$\mu_P = \pi = .40$$

$$\sigma_P^2 = \frac{\pi(1 - \pi)}{n} \left(\frac{N - n}{N - 1} \right) = \frac{.4(.6)}{3} \left(\frac{5 - 3}{5 - 1} \right) = .08(.5) = .04$$

$$\sigma_P = \sqrt{\frac{\pi(1 - \pi)}{n}} \sqrt{\frac{N - n}{N - 1}} = \sqrt{\frac{.4(.6)}{3}} \sqrt{\frac{5 - 3}{5 - 1}} = .2828(.707) = .20$$

THE CHOICE OF A GOOD ESTIMATOR

The linkage between understanding sampling distributions and engaging in statistical estimation becomes evident the moment one considers the question of what makes a good estimator. Three major criteria are commonly employed: unbiasedness, efficiency, and consistency.

The Criterion of Unbiasedness

A sample statistic that, on the average, across many samples, takes on a value equal to the population parameter that is being estimated with its help is called an **unbiased estimator.** On the other hand, if repeated random samplings from a given population produce a statistic with a mean or expected value that differs from the target to be estimated, bias is said to exist, as illustrated in panel (a) of Figure 7.2. The target, or true value of the population parameter, is designated by T. The sampling distribution of the statistic that is being used as an estimator, E, is also shown. In the upper graph of panel (a), the mean of the sampling distribution, μ_E, equals the true parameter; the estimator, therefore, is said to be unbiased. In the lower graph of panel (a), the mean of the sampling distribution lies far above T; the estimator, therefore, contains upward bias. The extent of bias is always measured as $\mu_E - T$. (For an alternative discussion of the subject, see Figure 1.1 on page 21.)

We can now see why the sample mean, $\overline{X}$, is considered an unbiased estimate of the population mean, μ. As was noted in Box 7.A above, regardless of whether samling occurs from a large or small population, the mean of sample means, $\mu_{\overline{X}}$, always equals μ, and that equality satisfies the definition of unbiased estimator.

Similarly, because $\mu_P = \pi$ (Box 7.B), regardless of whether sampling occurs from a large or small population, the sample proportion, P, can always be viewed as an unbiased estimator of the population proportion, π.

At this point, we can also solve a riddle first noted in Chapter 3 (on page 94) when introducing the formula for the sample variance. Instead of being defined (in perfect analogy to the formula for the population variance) as the mean of the squared deviations of observed sample values from their mean, or as $\Sigma(X - \overline{X})^2/n$, the divisor was changed to $n - 1$ (see Box 3.I). The reason for this change in the divisor is this: the mean of the squared deviations of sample values from their mean turns out to be a

Figure 7.2 Criteria Employed in the Choice of Estimators

This set of graphs illustrates major criteria commonly used to define good estimators of unknown population parameters. Panel (a) shows that an estimator the expected value of which (μ_E) equals the true value of the population parameter *(T)* is *unbiased.* Panel (b) shows that, among several unbiased estimators, the one with the least variance is *efficient.* As panel (c) shows, an estimator that homes in on its target as sample size increases (so that bias and variance approach zero as sample size approaches infinity) is *consistent.*

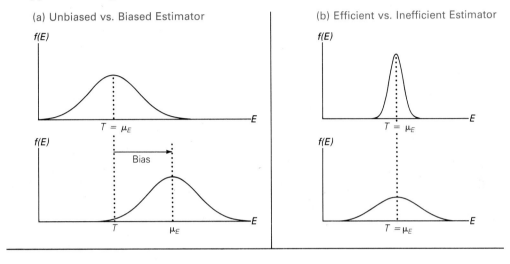

(a) Unbiased vs. Biased Estimator

(b) Efficient vs. Inefficient Estimator

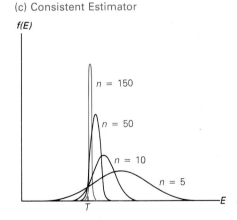

(c) Consistent Estimator

biased estimator of the population variance; it produces an underestimate of it. This result is intuitively reasonable because a part cannot be more varied than the whole from which it comes, but it can easily be less diverse. Therefore, in order to estimate correctly the population variance from sample data, one may want to inflate the above expression somewhat, and this inflation is achieved by using $(n - 1)$ as a divisor and defining the sample variance as $s^2 = \Sigma(X - \overline{X})^2/(n - 1)$. It turns out that s^2, so defined, is an unbiased estimator of σ^2, *provided* that selections of sample elements are statistically independent events. In the small-population case, however, s^2 (even with the divisor of $n - 1$) is a biased estimator of σ^2. (In our small-population example, the expected value of s^2, when calculated from Table 7.2, is found to be 113, which is a far cry from the Table 7.1 population variance of 90.4).

The sample standard deviation, s, finally, is a biased estimator of the population standard deviation, σ, regardless of whether sampling involves the large- or small-population case. (Again, to use our above small-population example, the expected value of s, when calculated from Table 7.2, turns out to be 9.816, while Table 7.1 shows σ to equal 9.508.)

The Criterion of Efficiency

As a look at the upper graph in panel (a) of Figure 7.2 confirms, even though an estimate may be right on target *on the average,* it can, nevertheless, be off on many occasions. That is why a statistician naturally comes to think of another criterion for the merit of an estimator: all else being equal, one would surely prefer an estimator with a small variance (with a distribution that is highly concentrated near the mean) over one with a large degree of variability. Accordingly, among all the available unbiased estimators, the one with the smallest variance for a given sample size is considered to be the **efficient estimator.** This criterion is illustrated in panel (b) of Figure 7.2. Note that both estimators in panel (b) have a sampling distribution the mean of which hits the target precisely. Yet the statistic with the sampling distribution shown in the upper graph of panel (b) is the efficient one because of its lower variability around the mean. A good example is provided by the sample median and sample mean, both of which are unbiased estimators of the mean of a normally distributed population. Yet in the normally distributed sampling distributions shown, the variance of the sample median is 57 percent larger than that of the sample mean (hence, the variance of the sample mean is only 64 percent as large as that of the sample median); this difference makes $\overline{X}$ the more efficient estimator of the two. Sometimes this relative efficiency is expressed by the ratio of variances:

$$\text{Relative efficiency of a given statistic} = \frac{\text{variance of efficient statistic}}{\text{variance of given statistic}}.$$

In the example just noted, this expression would amount to

$$\begin{aligned}\text{Relative efficiency of sample median} \atop \text{(as estimator of population mean)} &= \frac{\text{variance of sample mean}}{\text{variance of sample median}}\\[2mm] &= \frac{\sigma^2/n}{\left(\dfrac{\pi}{2}\right)\left(\dfrac{\sigma^2}{n}\right)} = \frac{2}{\pi} = .64\end{aligned}$$

(In this expression, π does refer to the well-known constant, not to the population proportion.) This relationship means that a sample median is considerably less likely than a sample mean to be a correct estimate of a population mean.

The Criterion of Consistency

If a large variance of the sampling distribution of the estimator seems undesirable, anything that can reduce the variance becomes of interest. When discussing the formulas in Box 7.A above, we noted one such possibility. At least so far as the sampling distribution of $\overline{X}$ is concerned, one can lower the variance by increasing sample size. In the large-population case, for example, given the population variance, σ^2, a large n reduces $\sigma_{\overline{X}}^2$ because the latter equals σ^2/n. Whenever the probability that a statistic will be close to the parameter being estimated gets ever larger (and, thus, approaches unity) as sample size increases, the statistic is called a **consistent estimator.** The sample mean and sample proportion are such consistent estimators of the corresponding population parameters. Panel (c) of Figure 7.2 illustrates a consistent estimator the sampling distribution of which increasingly concentrates on its target as sample size increases.

A Final Note: Mean Squared Error

More often than not, the search for a good estimator yields several prospects, none of which are perfect. One estimator may be unbiased but have a huge variance. Another may have a tiny variance but a huge bias. A third one may have some of both faults. This problem of choosing among imperfect estimators is illustrated in Figure 7.3. The estimator with distribution A has an expected value right on target but a large variance. The estimator with distribution C has a low variance but a large bias. The estimator with distribution B stands in the middle with respect to both criteria. Surely it would be foolish in this case to insist on no bias and to choose estimator A. Note how a trade-off could be made by switching from A to B, accepting some bias, but vastly reducing variability.

Statisticians do in fact choose the best estimator under such circumstances by calculating (and then minimizing) the **mean squared error,** which is the sum of an estimator's squared bias plus its variance, or $(\mu_E - T)^2 + \sigma_E^2$. In terms of Figure 7.3, this procedure picks the statistic with sampling distribution B as the best estimator.

Figure 7.3   Choosing Among Estimators

When all available estimators are imperfect, it is wise to choose the one with a combination of small bias and small variance (such as the one with distribution B).

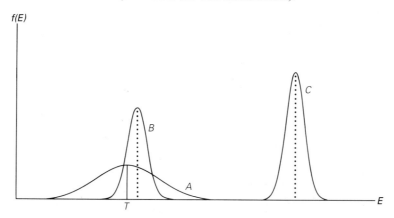

MAKING POINT ESTIMATES

Whenever the statistical population of interest is normally distributed or whenever the conditions of the central limit theorem are fulfilled [$n \geq 30$, but $n < .05N$, or $n \cdot \pi$ as well as $n(1 - \pi) \geq 5$], and regardless of whether sampling occurs from a large or a small population, sample mean and sample proportion are unbiased, efficient, and consistent estimators of the corresponding population parameters. In the large-population case, the same can be said about the sample variance. No wonder these statistics are routinely used as estimators under the stated circumstances. Indeed, because of a lack of better alternatives, statisticians employ the sample variance also as an estimator of the population variance in the small-population case (even though the estimate is then biased). For the same reason, and being equally aware of its bias under all circumstances, statisticians use the sample standard deviation as an estimator of that of the population. Let us consider an example of point estimation made on this basis.

||| Practice Problem 3

Profit Survey of Fast-Food Outlets

Suppose a franchiser of fast-food outlets wishes to gather information about the annual profits of current franchise holders. Past experience suggests that profits are normally distributed; a random sample of $n = 12$ outlets yields these profit data for the previous year:

$61,242	$10,404	$91,053	$48,912
$28,785	$76,326	$67,422	$40,056
$92,203	$96,803	$97,357	$92,233

a. What is a point estimate for the mean annual profit of all the franchised outlets? The sample mean equals

$$\bar{X} = \frac{\Sigma X}{n} = \frac{\$802,796}{12} = \$66,899.67.$$

This figure is the point estimate of the *population* mean as well because the sample mean, $\bar{X}$, is an unbiased, efficient, and consistent estimator of μ.

b. What is a point estimate for the standard deviation of the annual profit of all the franchised outlets?

$$s = \sqrt{\frac{\Sigma X^2 - n\bar{X}^2}{n-1}} = \sqrt{\frac{63,203,823,000 - 53,706,784,800}{11}}$$

$$= \sqrt{863,367,109} = \$29,383.10.$$

The sample's standard deviation serves as the estimate of the population's.

c. What is the proportion of profits, among all the franchisees, in excess of $50,000 (below $20,000)? Because these proportions in the sample are 8/12 and 1/12, respectively, and because the sample proportion, P, is an unbiased, efficient, and consistent estimator of π, these sample proportions are the estimates of the population proportions as well. |||

THE NATURE OF CONFIDENCE INTERVALS

Even though an unbiased point estimator will *on the average* take on a value equal to the parameter being estimated, any one estimate is unlikely to be on target. Yet, typically, only one estimate is made because only one sample is taken; hence, the typical point estimate is almost certain to lie below or above the true value of the parameter.

Take, for example, the sample mean, $\overline{X}$, as an estimator of the population mean, μ. As Table 7.2 so vividly demonstrates, only if sample number (3) or sample number (10) happens to be selected, will $\overline{X} = 40$, which is also the value of μ (see Table 7.1). In the other eight cases, $\overline{X} \neq \mu$, and the fact that $\mu_{\overline{X}} = \mu$ is then of little comfort, indeed. We must recognize that any parameter being estimated equals the statistic serving as a point estimate minus (negative or positive) sampling error, e. For instance,

$$\mu = \overline{X} - e.$$

In the previous example, if sample number 1 is selected, $\overline{X} = 35$, but a negative sampling error of $e = -5$ occurs; hence, $\mu = 35 - (-5) = 40$. If sample number 2 is selected, $\overline{X} = 45$, but a positive sampling error of $e = +5$ occurs; hence, $\mu = 45 - (+5) = 40$. And so it goes.

The inevitable uncertainty attached to any point estimate can be made explicit by presenting an interval estimate and stating, say, that $a \leq \mu \leq b$. Statisticians construct an interval within which the unknown population parameter presumably lies with the help of a point estimate (that serves as the interval center) and the standard error of the estimator (which, as we noted, is the standard deviation of the sampling distribution of the estimator). The limits of the interval, however, need not lie precisely 1 standard error below or above the point estimate that serves as the interval center. The limits can lie a fraction or a multiple of a standard error to either side of the point estimate. If we define a coefficient A as any (fractional or whole) positive value (such as .1, .5, 1, 2, and so on), we can make the following statement.

Upper and lower limits of interval estimate
= point estimate ± [A · (standard error of estimator)]

Consider, for example, estimating the population mean by using the sample mean. If sample number 4 in Table 7.2 had been selected, one could present $\overline{X} = 39.667$ as a point estimate of μ. Given $\sigma_{\overline{X}} = 3.882$ (Table 7.4), one could, however, also present various interval estimates for μ, each one being constructed such that μ falls within the range of $\overline{X} \pm (A \cdot \sigma_{\overline{X}})$, as follows:

1. If A is taken as .5, μ is estimated as lying between the limits of $39.667 \pm [.5(3.882)]$; hence, $37.726 \leq \mu \leq 41.608$.
2. If A is taken as 1.0, μ is estimated as lying between the limits of $39.667 \pm [1(3.882)]$; hence, $35.785 \leq \mu \leq 43.549$.
3. If A is taken as 2.0, μ is estimated as lying between the limits of $39.667 \pm [2(3.882)]$; hence, $31.903 \leq \mu \leq 47.431$.

And so it goes for other values of A. The fact that larger values of A produce wider intervals (less precise estimates) is no accident because the width of the interval has important implications for the degree of confidence with which one can state that the population parameter that is being estimated lies within the two limits, also called **confidence limits,** that define the interval. Let us see why.

Recall that a sampling distribution, such as that of the sample mean, is a normal distribution (a) when the underlying population values are normally distributed and (b) even when the population is not normal, provided only that the sample size is sufficiently large absolutely ($n \geq$ 30), but fairly small relative to population size ($n <$ .05N), to validate the central limit theorem. By choosing a proper sample size, a statistician, therefore, can *assure* that the sampling distribution of the estimator being used is approximately normal. As a result, the coefficient A in the previous expression can be treated as a z value, such that

Upper and lower limits of interval estimate
= point estimate ± [z · (standard error of estimator)].

Appendix Table E can then be employed to figure the areas under the normal curve lying within the ranges so calculated. Consider again the example employed in Practice Problems 1–2 earlier: a normally distributed sampling distribution of $\overline{X}$ with $\mu_{\overline{X}}$ = 40 and $\sigma_{\overline{X}}$ = 4.655. In order to make interval estimates of the population mean, we can view μ as falling in the range of $\overline{X} \pm (z \cdot \sigma_{\overline{X}})$. If we assume a random sample produces an $\overline{X}$ equal to 35, what can we conclude?

1. If z = .5, we can figure that 38.3 percent of all $\overline{X}$ values fall within the range of $\mu_{\overline{X}} \pm .5\sigma_{\overline{X}}$ because Appendix Table E tells us that the area under the standard normal curve between the center and z = .5 equals .1915; hence, the area between z = −.5 and z = +.5 equals 2(.1915) = .383. Because the center of the sampling distribution lies at $\mu_{\overline{X}}$ = μ, we can be 38.3 percent confident that our method of interval construction will produce an interval that will actually contain μ. In fact, we can calculate an interval that reaches from 32.6725 to 37.3275; this interval does not contain μ = $\mu_{\overline{X}}$ = 40.

2. If z = 1.0, we can similarly figure that 68.26 percent of all $\overline{X}$ values fall within the range of $\mu_{\overline{X}} \pm 1\sigma_{\overline{X}}$ [because 2(.3413) = .6826]. We can, thus, be 68.26 percent confident that our method of interval construction will produce an interval that will contain μ. In fact, we can calculate an interval that reaches from 30.345 to 39.655.

3. If z = 2.0, it is likely that 95.44 percent of all $\overline{X}$ values fall within the range of $\mu_{\overline{X}} \pm 2\sigma_{\overline{X}}$ [because 2(.4772) = .9544]. We can, thus, be 95.44 percent confident that our method of interval construction will produce an interval that will contain μ. In fact, we can calculate an interval that reaches from 25.69 to 44.31.

The foregoing discussion shows why the intervals in which estimates are stated are called **confidence intervals.** Given sample size, the level of confidence attached to an estimate varies directly with the value of z and, thus, with the interval width. Notice how a smaller z value means greater precision of an estimate (a narrower interval) but also implies a smaller degree of confidence in the estimate, while a larger z value means less precision (a wider interval) but implies a greater degree of confidence. The **confidence level,** or the percentage of interval estimates (obtained from repeated random samples, each of size n, taken from a given population) that can be expected to contain the actual value of the parameter being estimated, can clearly be set by choosing the value of z. Anyone who wishes to make an estimate with 68.26 percent confidence, for example, must choose the z value corresponding to $(.6826/2) = .3413$ in Appendix Table E (hence, $z = 1$), for this will assure that the resultant interval will cover $2(.3413) = .6826$ of the area under the normal curve. Appendix Table F, "Critical Normal Deviate Values," contains z values for the most commonly used confidence levels; others can be derived from Appendix Table E in the manner just shown.

What is meant by confidence level must, however, be considered with care. The three above statements, for example, do *not* imply probabilities of .383, .6826, or .9544 that the unknown population mean lies within the limits of the respective intervals. (The unknown parameter either does or does not lie within any given interval. Period.) The statements do mean the following:

> If all possible samples of a given size are selected from a given population and if all possible values of a given estimator are then calculated and a confidence interval of a given width is then constructed around each of these values, the percentage of resulting intervals that will contain the true population value is referred to as *the confidence level* of the interval.

This meaning of *confidence level* is illustrated in Figure 7.4 on page 240. It depicts the sampling distribution of $\overline{X}$ that was first postulated in Practice Problems 1–2 and was imagined to have been derived from a very large and normally distributed population. Even though this sampling distribution is known to us (the readers of this textbook), it would not, of course, be known to the statistician about to estimate μ by taking a single random sample from a population. If this statistician wanted to estimate μ with, say, an 80-percent confidence level, that statistician would have to select a z value such that 80 percent of all possible $\overline{X}$ values fall within the range of $\pm z\sigma_{\overline{X}}$ from the unknown μ. As Appendix Table F indicates, that z value is 1.28. Now let us imagine for a moment that our statistician knew $\sigma_{\overline{X}} = 4.655$. Then the 80-percent confidence interval could be set as lying between the limits of $\mu \pm 1.28(4.655)$ or $\mu \pm 6$ (rounded). As we learned from our simplified example of Table 7.2, however, many different random samples of a given size can be taken from a given population. Our statistician might select any one of them. If from, perhaps, thousands of possible samples sample number 1 is selected, the statistician might calculate $\overline{X} = 35$. He or she would then use this point estimate of μ to make an interval estimate of μ as the range of 35 ± 6—that is, between 29 and 41, as shown in Figure 7.4. We can see (but the statistician could not) that the point estimate of 35 is wrong (μ actually equals 40; hence, a sampling error of -5 was made); we can also see (and, again, the statistician would not know) that the interval estimate of 29 to 41 does contain the true value of $\mu = 40$.

But what if the statistician had selected a different sample that we might call sample number (2)? Then $\overline{X}$ would probably have been different, perhaps equal to 45 (also, unbeknownst to the statistician, a wrong point estimate of μ), and the interval estimate would have ranged from 39 to 51 (again, unbeknownst to the statistician, containing the true value of $\mu = 40$). A few of the many other possible sample results are indicated by the unlabeled fat dots in Figure 7.4, along with the confidence intervals our statistician would calculate. Notice that all but two $\overline{X}$ values—samples (3) and (10) being the exceptions—would be incorrect point estimates of μ, yet fully 8 of the 10 interval estimates would, nevertheless, contain the true value of μ—the two exceptions here being the interval constructed after taking sample numbers (5) and (9). And therein lies the meaning of the term "80-percent confidence interval": Any one confi-

Figure 7.4   Confidence Intervals

An interval estimate of a population mean (μ) can be constructed around a sample mean ($\overline{X}$) as $\overline{X} \pm z\sigma_{\overline{X}}$. Depending on the value of z, one can place different degrees of confidence in the method of estimation. In this example, $z = 1.28$, which implies that 80 percent of the $\overline{X}$ values (heavy dots) that might be found by repeated sampling from the same population lie within the range of $\mu_{\overline{X}} \pm 1.28$ $\sigma_{\overline{X}}$. Because $\mu = \mu_{\overline{X}}$, this procedure also implies that 80 percent of the intervals that might be constructed will contain μ. Any interval constructed by this method is, therefore, called an "80-percent confidence interval." By choosing smaller or larger z values, one can, similarly, construct narrower or wider intervals with smaller or higher confidence levels. What is illustrated here for population mean and sample mean holds equally for other parameters and other estimators.

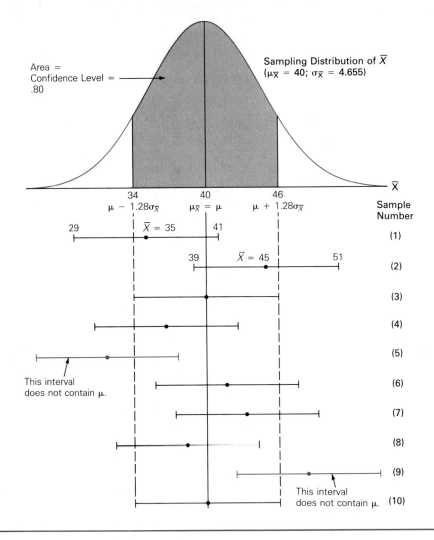

dence interval is either dead right or dead wrong (it does or does not contain μ, but the statistician does not know it). Yet the statistician can have 80-percent confidence in *the method employed.* If repeated sampling occurred, 80 percent of the intervals would be right and 20 percent would be wrong, as this example illustrates.

Notice again: There is *not* an 80-percent probability that μ lies within the confidence limits of the first interval. Unbeknownst to the statistician, there is certainty that μ is contained in that interval (just as there is certainty that it is not bracketed by the fifth interval). All that the statistician knows is that of, say, 1000 intervals, each constructed with a value of $z = 1.28$, some 800 will contain the unknown parameter, but the statistician has no way of identifying these particular 800 intervals. A stated confidence level gives the probability *prior to actual sample selection* that the parameter being estimated will lie within the interval being constructed. (Once a specific sample has been taken and a specific interval has been constructed, the parameter is certain to lie or not to lie within that interval.)

MAKING INTERVAL ESTIMATES OF THE POPULATION MEAN AND PROPORTION: LARGE SAMPLES

Let us consider a series of cases, focusing first on those involving (a) large samples [when estimating means, $n \geq 30$; when estimating proportions, $n \cdot \pi$ and also $n(1 - \pi) \geq 5$] and (b) normally distributed sampling distributions (either because the population values are themselves normally distributed or because the central limit theorem applies). Confidence intervals for the population mean (μ) or proportion (π) can be constructed as shown in Box 7.C.

7.C Upper and Lower Limits of Confidence Intervals for the Population Mean, μ, and Proportion, π (using the normal distribution)

$$\mu = \overline{X} \pm (z\sigma_{\overline{x}}) \text{ and } \pi = P \pm (z\sigma_P)$$

where $\overline{X}$ is the sample mean and P is the sample proportion, while $\sigma_{\overline{x}}$ and σ_P are the standard deviations of their respective sampling distributions.

Assumptions: Sampling distribution of $\overline{X}$ or P is normal because (i) population distribution is normal *or* (ii) the central-limit theorem applies [for the mean, $n \geq 30, n < .05N$; for the proportion, $n \cdot \pi \geq 5$ and $n(1 - \pi) \geq 5$].

The actual sample taken will clearly yield a value for $\overline{X}$ or P, and the value of z (and hence, the desired confidence level) can be determined at will. Once values for $\overline{X}$ or P and z are determined, we are left with the necessity to determine $\sigma_{\overline{x}}$ or σ_P. If only one sample is taken (and that is the usual case), these values cannot be obtained in the fashion indicated in Table 7.4 because the respective sampling distributions are not available. Boxes 7.A and 7.B provide an alternative approach.

According to Box 7.A, $\sigma_{\overline{x}}$ can be calculated if σ, the population standard deviation, is known, and on occasion such is the case. (The statistician obviously does know the size of the sample taken; hence, $\sigma_{\overline{x}}$ can be calculated, given σ, in one of the two ways shown in Box 7.A, depending on whether the large- or small-population case applies.) This information provides us with all the ingredients necessary for solving Practice Problem 4.

Note: According to Box 7.B, σ_P can be calculated if π, the population proportion, is known, but if that were the case, there would be no need to estimate it. This approach, therefore, leads nowhere.

What if σ, like π, is not known, all else being the same? In that case, statisticians estimate these values by substituting, in the Box 7.A and 7.B formulas, the sample standard deviation, s, and the sample proportion, P, for σ and π, respectively. This substitution provides us with all the ingredients for solving Practice Problem 5.

Practice Problem 4

Estimating Mean Family Income ($n \geq 30$, $n < .05N$, σ known)

Consider a large city. From prior surveys, it seems reasonable to assume that the city population's family incomes have a standard deviation of $4,000. A 95-percent confidence interval for the city's mean family income is to be constructed with the help of a simple random sample (without replacement) of 100 families. The sample mean income turns out to be $19,763.

The sample size $n = 100$ clearly is less than .05N; thus, the large-population case (Box 7.A) applies, and

$$\sigma_{\overline{X}} = \frac{\sigma}{\sqrt{n}} = \frac{\$4,000}{\sqrt{100}} = \$400.$$

A 95-percent confidence level requires $z = 1.96$ (see Appendix Table F). Hence, the limits of the confidence interval are

$$\mu = \overline{X} \pm (z\sigma_{\overline{X}}) = \$19,763 \pm [1.96 (\$400)] = \$19,763 \pm \$784.$$

Note: In order to simplify the wording, the expression, $x = a \pm b$, shall henceforth imply that x can take on any value in the interval from $(a - b)$ to $(a + b)$. Thus, the expression, $x = a \pm b$, shall be considered equivalent to $(a - b) \leq x \leq (a + b)$ and shall not be construed to imply that x can equal only $(a - b)$ or $(a + b)$. The expression, $\mu = \$19,763 \pm \$784$, therefore, implies that the 95-percent confidence interval goes from $18,979 to $20,547. Remember: This does not mean that there exists a 95-percent probability that the city population's mean income lies within this range. As Figure 7.4 showed, the true mean either does or does not lie within any given interval. But we can say that μ would on the average lie within 95 out of 100 such intervals, if we took repeated samples of $n = 100$ from the city population and repeatedly made interval estimates by this method. |||

Practice Problem 5

Estimating Mean Weight per Animal ($n \geq 30$, $n < .05N$, σ unknown)

Consider a cattle feed lot. It is about to ship 800 head of cattle by train. A 90-percent confidence interval for the entire shipment's mean weight per animal is to be constructed with the help of a simple random sample (without replacement) of 30 animals. The sample mean weight per animal turns out to be 1,301 pounds, with a sample standard deviation of 290 pounds.

The sample size of $n = 30$ is less than .05N; thus, the large-population case (Box 7.A) applies, and (in the absence of σ) one can estimate (using s):

$$\sigma_{\bar{x}} \cong \frac{s}{\sqrt{n}} = \frac{290 \text{ lbs.}}{\sqrt{30}} = 52.95 \text{ lbs.}$$

A 90-percent confidence level requires $z = 1.64$. Hence,

$$\mu = \bar{X} \pm (z\sigma_{\bar{x}}) = 1{,}301 \text{ lbs.} \pm [1.64(52.95 \text{ lbs.})] = 1{,}301 \text{ lbs.} \pm 86.84 \text{ lbs.}$$

The 90-percent confidence interval goes from 1,214.14 pounds to 1,387.84 pounds. |||

MAKING INTERVAL ESTIMATES OF THE POPULATION MEAN AND PROPORTION: SMALL SAMPLES

Now let us consider a series of cases involving small samples (when estimating means, $n < 30$; when estimating proportions, $n \cdot \pi$ or $n(1 - \pi) < 5$). Such cases are quite common for a variety of reasons, including: the high cost of sampling, the desire to get faster results, and the rarity of some phenomena (only two skyscrapers are produced per year, only one nuclear reactor accident occurs in a decade, and so on). Under such circumstances, it is not safe to invoke the central limit theorem. Yet it is still possible for the sampling distribution to be normal, provided only that the underlying population values are normally distributed. If such is known to be the case, and if the population standard deviation is also known, a confidence interval for a population mean can be estimated with the help of z values as in Practice Problem 4. One example is provided by Practice Problem 6. Practice Problem 7 is an example of the opposite case in which it is known that the underlying population values are *not* normally distributed but in which the population standard deviation is given. In this case, a confidence interval can still be estimated with the help of Chebyshev's theorem (see Biography 3.1).

||| Practice Problem 6

Estimating Mean Managerial Salaries ($n < 30$, $n \geq .05N$, population normal, σ known)

Consider a normally distributed population of 5 managerial salaries, but assume that only the fact of this normality along with $\sigma = \$11{,}400$ are known. An 80-percent confidence interval for the mean salary of all 5 managers is to be constructed with the help of a simple random sample (without replacement) of 3 managers. The sample mean turns out to be $\$48{,}333$.

The sample size of $n = 3$ exceeds .05N; thus, the small-population case (Box 7.A) applies, and

$$\sigma_{\bar{x}} = \frac{\sigma}{\sqrt{n}} \sqrt{\frac{N - n}{N - 1}} = \frac{\$11{,}400}{\sqrt{3}} \sqrt{\frac{5 - 3}{5 - 1}} = \$4{,}654.$$

An 80-percent confidence level requires $z = 1.28$. Hence,

$$\mu = \overline{X} \pm (z\sigma_{\overline{x}}) = \$48{,}333 \pm [1.28(\$4{,}654)] = \$48{,}333 \pm \$5{,}957.$$

The 80-percent confidence interval goes from $42,376 to $54,290. |||

||| Practice Problem 7

Estimating Mean Waiting Times ($n < 30$, $n < .05N$, population not normal, σ known)

Consider a hotel manager who wants to know the mean time guests have to wait for room service. The population of waiting times is known to be not normally distributed, with a standard deviation of 11 minutes. A 95-percent confidence interval of the mean waiting time for 1,000 orders is to be constructed with the help of a simple random sample (without replacement) of 20 orders. The sample mean turns out to be 32 minutes.

The sample size of $n = 20$ is less than $.05N$; thus, the large-population case (Box 7.A) applies, and

$$\sigma_{\overline{x}} = \frac{\sigma}{\sqrt{n}} = \frac{11 \text{ minutes}}{\sqrt{20}} = 2.46 \text{ minutes.}$$

According to Chebyshev's theorem, at least $[1 - (1/k^2)]$ 100 percent of all observations fall within k standard deviations of the mean. For a 95-percent confidence level, we need

$$95 = \left(1 - \frac{1}{k^2}\right)100, \text{ or } .95 = 1 - \frac{1}{k^2}.$$

This confidence level requires $k = 4.472$. Hence,

$$\mu = \overline{X} \pm (k\sigma_{\overline{x}}) = 32 \text{ minutes} \pm [4.472(2.46 \text{ minutes})]$$
$$= 32 \text{ minutes} \pm 11 \text{ minutes.}$$

The 95-percent confidence interval goes from 21 minutes to 43 minutes. |||

Note: When $n < .05N$, the wide interval inevitably derived for any reasonable confidence level when using Chebyshev's theorem can easily be escaped by raising the sample size to 30 or above. This increase in sample size validates the central limit theorem and allows the kind of estimation procedure shown in Practice Problem 4. Thus, if the sample result in Practice Problem 7 had been obtained from 30 orders, we could estimate

$$\mu = \overline{X} \pm (z\sigma_{\overline{x}}) = 32 \text{ minutes} \pm [1.96(2.46 \text{ minutes})]$$
$$= 32 \text{ minutes} \pm 4.82 \text{ minutes.}$$

This escape route (of raising n) is also recommended when the population is not normally distributed and when σ is not known. Then the procedure described in Practice Problem 5 can be used.

Student's t Distribution

What remains is the most important small-sample case of all: the presence of a (presumably) normally distributed population but the absence of any other information about this population. In this small-sample case (unlike in situations illustrated by Practice Problem 5), the standard errors of the mean and proportion, $\sigma_{\bar{x}}$ and σ_P, are only poorly estimated with the help of the sample standard deviations, s and P. As was first shown by William S. Gosset, who wrote under the name "Student" (see Biography 7.1 at the end of the chapter), under these circumstances, better interval estimates can be derived by using a probability density function somewhat different from the normal curve.

Gosset described a sampling distribution for a random variable, t, derived from a normally distributed population and defined in analogy to the standard normal deviate, z:

$$t = \frac{\overline{X} - \mu}{\dfrac{s}{\sqrt{n}}} \qquad z = \frac{X - \mu}{\sigma}.$$

Like the standard normal curve, Gosset's probability density function, now called **Student's t distribution,** is (a) single-peaked above the random variable's mean, median, and mode of zero, (b) perfectly symmetrical about this central value, and (c) characterized by tails extending indefinitely in both directions from the center, approaching but never touching the horizontal axis. The only difference is that the random variable is t rather than z; as a result, the distribution's variance does not equal 1 (as is true of the z distribution), but equals

$$\frac{n - 1}{(n - 1) - 2}.$$

This variance of t implies that a different t distribution exists for each sample size, n, and also that the t distribution approaches the z distribution as sample size increases. The t distribution for $n = \infty$ has a variance of 1 and is indistinguishable from the standard normal curve. All these characteristics of the t distribution are illustrated in Figure 7.5.

As Figure 7.5 shows, members of the t-distribution family are labeled not on the basis of sample size, n, but on the basis of **degrees of freedom,** $n - 1$, which equal the number of values that can be "freely chosen" when calculating a statistic. Consider the number of independent deviations, $X - \overline{X}$, of sample observations from the sample mean that enter the computation of the sample standard deviation, s. Once we have "freely chosen" $n - 1$ of these deviations, the remaining one is determined for us because the sum of all deviations necessarily equals 0. In a sample of $n = 3$, for

Figure 7.5   Student *t* Distributions

Different bell-shaped *t* distributions exist for different degrees of freedom, *d.f.,* defined as sample size minus one, $n - 1$. Unless the degrees of freedom are infinite (which makes the *t* distribution equal to the standard normal curve), the *t* distribution is flatter than the standard normal curve, and more of its area is found in the tails. Strictly speaking, each *t* distribution is a sampling distribution of the *t* statistic derived by taking a small sample from a normally distributed population. Nevertheless, in practice, statisticians use the *t* distribution to make small-sample inferences about all types of populations, provided only that these populations are not highly skewed.

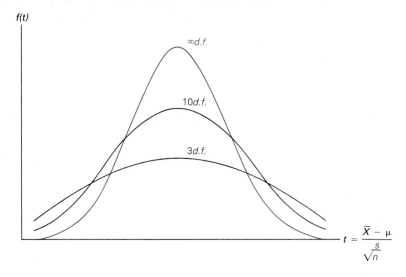

example, we have only $n - 1 = 2$ degrees of freedom. Take sample number (1) in Table 7.2. The individual sample observations (for sample units A, B, and C) are 39, 41, and 25 (see Table 7.1). The sample mean, therefore, equals $\overline{X} = 35$. Proceeding to compute *s*, once we have determined two deviations—say, $39 - 35 = +4$ and $41 - 35 = +6$—the last one has effectively been determined for us by the fact that all three deviations must sum to 0. The last one can be found logically as $0 - (+4) - (+6) = -10$; indeed, this value is correct $(25 - 35 = -10)$. For an earlier discussion of degrees of freedom (that does not mention the term) see the Chapter 4 section on "Factorials."

The *t* Distribution Table

Just as areas under the standard normal curve have been tabulated to make it easy for us to gauge various probabilities, so there exist tables for areas under various *t* distributions. Appendix Table G, "Student *t* Distributions," is a case in point. By tradition, the table shows the total area under a specified curve (defined by a given number of degrees of freedom) that lies *to the right of* a specified value of *t*. As is pictured in Appendix Table G, this upper-tail area is called α (the lowercase Greek letter *alpha*), and this *t* value is designated as t_α. Frequently, the applicable degrees of freedom *(d.f.)* are added to the α subscript, either in parentheses or following a comma: $t_{\alpha(d.f.)}$ or $t_{\alpha,d.f.}$

As a look at Appendix Table G will confirm, for 10 degrees of freedom and an upper-tail area of $\alpha = .1$, $t_{.1(10)} = 1.372$. In other words, .1 of the area under the t curve appropriate for a sample of $n = 11$ is associated with $t > 1.372$. Because of the curve's symmetry, .1 of the area under this curve is also associated with $t < -1.372$. Consequently, .8 of the area under this curve is associated with t values between -1.372 and $+1.372$; the probability for such t values equals .8. To construct an interval estimate with a confidence level of .8, or 80 percent, we must use $t = 1.372$.

By comparing this discussion to our earlier discussion, confidence intervals can be constructed with the help of Appendix Table G's t values, which are shown in Box 7.D below.

7.D Upper and Lower Limits of Confidence Interval for the Population Mean (using the t distribution)

$$\mu = \overline{X} \pm \left(t \frac{s}{\sqrt{n}} \right)$$

where $\overline{X}$ is the sample mean, s is the sample standard deviation, and n is sample size.

Assumption: $n < 30$, but population distribution is normal.

||| Practice Problem 8

Estimating the Mean Time Lost from Machinery Breakdowns ($n < 30$; $n < .05N$; population normal; σ unknown)

Consider a manufacturing plant. On many occasions during the course of a year, labor hours are lost because machinery breaks down. The manager (who knows from previous experience that the time-lost population is normally distributed) decides to keep track of the time lost for a week and calculates the mean time lost during 10 breakdowns as 820 hours, with a standard deviation of 510 hours. A 98-percent confidence interval is to be constructed for the mean time lost during all of the year's breakdowns.

Given a normal population, but an unknown σ and a small sample, the t distribution must be used. For sample size $n = 10$, there are 9 degrees of freedom, while the 98-percent confidence level corresponds to an upper-tail area of $\alpha = .01$. It is helpful to designate the corresponding t value with respect to α as well as with respect to the degrees of freedom—in this case as $t_{.01(9)}$. From Appendix Table G, we find $t_{.01(9)} = 2.821$. According to Box 7.D,

$$\mu = \overline{X} \pm \left(t \frac{s}{\sqrt{n}} \right) = 820 \text{ hours} \pm \left(2.821 \frac{510 \text{ hours}}{\sqrt{10}} \right)$$

$$= 820 \text{ hours} \pm 455 \text{ hours}.$$

The 98-percent confidence interval is disconcertingly wide; it goes from 365 hours to 1,275 hours. |||

MAKING INTERVAL ESTIMATES FOR THE DIFFERENCE BETWEEN TWO POPULATION MEANS

Oftentimes decision makers seek an estimate not of a single population mean, but of the difference between the means of two populations. Two methods of production may be available; on the average, which one requires the greater amount of labor time? Two drugs may be effective in reducing high blood pressure; which one, on the average, is more effective? Two types of raw material may be usable; which one, on the average, has the greater tensile strength? Which of two work rules is more conducive to raising average labor productivity? Which of two types of tires has the longer average life? To answer questions such as these, statisticians take samples from two populations, A and B, with means of μ_A and μ_B and estimate the difference, $\mu_A - \mu_B$. Once again, the proper estimating procedure depends on whether the samples are large or small, but the existence of two populations requires a further choice: whether to take independent or matched-pairs samples. We will consider all of these possibilities.

Large and Independent Samples

We first imagine taking two completely independent samples of sizes n_A and n_B from the two populations. These samples need not be of equal size; they will yield sample means $\overline{X}_A$ and $\overline{X}_B$; their difference, $d = \overline{X}_A - \overline{X}_B$, is an unbiased point estimator of $\mu_A - \mu_B$.

We could take such samples again and again, each time getting a different d. If we take large and independent samples of $n_A \geq 30$ and $n_B \geq 30$ (as we now assume), the sampling distribution of d will be normally distributed when $n < .05 N$ (because the central limit theorem then holds) and even when $n \geq .05 N$ provided the underlying populations are normally distributed. This sampling distribution will have a mean of $\mu_A - \mu_B$ and a standard deviation of

$$\sigma_d = \sqrt{\frac{\sigma_A^2}{n_A} + \frac{\sigma_B^2}{n_B}}$$

where σ_A^2 and σ_B^2 are the two population variances. Given large samples, we can also estimate the standard error of the difference between the means, σ_d, using the sample variances:

$$\sigma_d \cong s_d = \sqrt{\frac{s_A^2}{n_A} + \frac{s_B^2}{n_B}}$$

We can now adapt the earlier expression of a confidence interval for the population mean (Box 7.C) to establish a confidence interval for the difference between population means, $\mu_A - \mu_B$, as lying between the limits of $d \pm z\sigma_d$. These limits can also be written as shown in Box 7.E.

7.E. Upper and Lower Limits of Confidence Interval for the Difference Between Two Population Means (large and independent samples from large or normal populations)

$$\mu_A - \mu_B \cong (\overline{X}_A - \overline{X}_B) \pm \left[z \sqrt{\frac{s_A^2}{n_A} + \frac{s_B^2}{n_B}} \right]$$

where $\overline{X}$ is the sample mean, s is the sample standard deviation, and n is sample size.

Assumptions: $n_A \geq 30$ and $n_B \geq 30$ and sampling distribution of $\overline{X}_A - \overline{X}_B$ is normal because $n_A < .05N_A$ and $n_B < .05N_B$ *or* population distributions are normal.

Note: As will be shown, for *small* and independent samples, the same formula applies, except that t replaces z. In that case, $n_A < 30$ and $n_B < 30$, but the two populations are assumed to be normally distributed and to have equal variances.

Practice Problem 9

Comparing the Mean Lives of Two Types of Tires
Consider a tire manufacturer who wishes to estimate the difference between the mean lives of two types of tires, type A and type B, as a prelude to a major advertising campaign. A sample of 100 tires is taken from each production process, making $n_A = n_B = 100$. (As noted earlier, this equality of the two sample sizes is not a necessity.) The sample mean lifetimes equal $\overline{X}_A = 30,100$ miles and $\overline{X}_B = 25,200$ miles; the sample variances are $s_A^2 = 1,500,000$ miles squared and $s_B^2 = 2,400,000$ miles squared.

Clearly, the point estimate of the difference is $30,100 - 25,200 = 4,900$ miles, but the manufacturer desires a 99-percent confidence interval of the difference (requiring, as Appendix Table F indicates, a z value of 2.57). Using the Box 7.E formula,

$$\mu_A - \mu_B \cong 4,900 \pm \left[2.57 \sqrt{\frac{1,500,000}{100} + \frac{2,400,000}{100}} \right]$$

$$= 4,900 \pm (2.57 \sqrt{39,000}) = 4,900 \pm [2.57(197.48)]$$

$$= 4,900 \pm 507.52.$$

The 99-percent confidence interval goes from 4,392.48 miles to 5,407.52 miles. The tire from population A is clearly superior. |||

Large Matched-Pairs Sample

Now let us turn to an alternative approach to making interval estimates for the difference between two means. Consider taking a sample after each elementary unit in population A has been matched with a "twin" from population B so that any sample observation about a unit in population A automatically yields an associated observation about a unit in population B. This procedure is referred to as taking a **matched-pairs**

sample. Such matching is based on other factors besides the one under study that might also influence the characteristic being measured. For example, when testing the effectiveness of a new drug compared to a traditional one, each patient in an experimental group might be matched with a partner in a control group of the same age, weight, height, sex, occupation, medical history, lifestyle, and so on. The individual differences in response to the experimental stimulus between each pair are then used to estimate population differences.

The procedure can be summarized as follows: Let the sample observation for the *i*th pair equal X_{Ai} and X_{Bi}, depending on whether it refers to the partner from population A or B. Then the matched-pair difference equals $D_i = X_{Ai} - X_{Bi}$. From all the matched-pair differences involving n pairs, a mean and standard deviation can be calculated in the usual fashion as:

$$\overline{D} = \frac{\Sigma D_i}{n}$$

and

$$s_D = \sqrt{\frac{\Sigma D_i^2 - n\overline{D}^2}{n - 1}}$$

Now we can treat $\overline{D}$ and s_D equivalently to $\overline{X}$ and s in our earlier examples and, given our large sample ($n \geq 30$) and large population ($n < .05\ N$), we can assume that $\overline{D}$ is normally distributed with a mean of $\mu_A - \mu_B$ and a standard deviation of $s_D/\sqrt{n}$. Accordingly, we can establish the desired confidence interval in the fashion indicated in Box 7.F.

7.F Upper and Lower Limits of Confidence Interval for the Difference Between Two Population Means (large and matched-pairs sample from large or normal populations)

$$\mu_A - \mu_B = \overline{D} \pm \left(z\, \frac{s_D}{\sqrt{n}} \right)$$

where $\overline{D}$ is the mean of sample matched-pairs differences, s_D is the associated standard deviation, and n is the number of matched sample pairs.

Assumptions: $n \geq 30$ and sampling distribution of $\overline{D}$ is normal because $n < .05N$ *or* the populations are normally distributed.

Note: As will be shown, for a *small* and matched-pairs sample, the same formula applies, except that t replaces z. In that case, $n < 30$, but the populations are assumed to be normally distributed and to have equal variances.

Practice Problem 10

Comparing the Effectiveness of Two Drugs

Consider a drug manufacturer who wishes to compare the performance of a new drug for treating high blood pressure with that of one traditionally used, perhaps as a prelude to a marketing campaign. Sixty people are combined into 30 pairs of "near-twins."

In each pair, one person receives the new drug (A), and the other person receives the old drug (B). The individual percentage reductions in blood pressure are recorded, as shown in Table 7.6.

Table 7.6 Calculation of Matched-Pairs Differences of Effects on Patients Using New Drug (A) and Old Drug (B)

Pair i	Percentage Reduction in Blood Pressure, While Using Drug A, X_{Ai}	Percentage Reduction in Blood Pressure, While Using Drug B, X_{Bi}	Matched-Pair Difference, $D_i = X_{Ai} - X_{Bi}$	Squared Matched-Pair Difference, D_i^2
1	22	18	4	16
2	21	22	−1	1
3	26	19	7	49
4	23	22	1	1
5	21	22	−1	1
6	23	23	0	0
7	24	19	5	25
8	30	18	12	144
9	25	23	2	4
10	24	22	2	4
11	27	23	4	16
12	23	19	4	16
13	27	22	5	25
14	26	17	9	81
15	23	25	−2	4
16	25	17	8	64
17	21	19	2	4
18	23	23	0	0
19	25	15	10	100
20	28	25	3	9
21	28	16	12	144
22	30	18	12	144
23	29	16	13	169
24	28	20	8	64
25	21	20	1	1
26	30	18	12	144
27	24	25	−1	1
28	30	20	10	100
29	25	15	10	100
30	22	25	−3	9
			148	1,440

$$\overline{D} = \frac{\Sigma D_i}{n} = \frac{148}{30} = 4.933 \qquad s_D = \sqrt{\frac{\Sigma D_i^2 - n\overline{D}^2}{n-1}} = \sqrt{\frac{1,440 - 30(4.933)^2}{29}} = \sqrt{24.482} = 4.948$$

The matched-pair difference, D_i, is calculated for each pair, along with the mean and standard deviation of the 30-pair sample. For a 95-percent confidence interval, we use a z value of 1.96; hence, according to Box 7.F:

$$\mu_A - \mu_B = 4.933 \pm \left[1.96\left(\frac{4.948}{\sqrt{30}}\right) \right] = 4.933 \pm 1.771.$$

Thus, the 95-percent confidence interval goes from 3.162 percent to 6.704 percent. We can say with 95 percent confidence that drug A reduces blood pressure between 3.2 and 6.7 percent more than drug B. |||

Small and Independent Samples

Let us now consider taking independent samples from two (probably normal) populations when both sample sizes are small ($n_A < 30$ and $n_B < 30$). As we learned earlier, in the absence of any precise knowledge about population variances, the Student t distribution must be used. Essentially, the same procedure can be used as in the case of large samples, except that t_α replaces z. (Strictly speaking, use of the t distribution in this case requires an equality of the two unknown population variances, and these are often estimated by a weighted average of the two sample variances, but we leave this refinement to the next chapter.)

||| Practice Problem 11

Comparing the Effectiveness of Two Pesticides

Consider an agricultural experiment in which the effectiveness of a new pesticide compared to an old one is tested by measuring cotton production on 30 sample plots of land. Of the 30 plots, 15 use the new pesticide (A); the mean output is $\overline{X}_A = 1,575$ pounds per acre, with a standard deviation of $s_A = 250$ pounds. Another 15 plots use an old pesticide (B); the mean output is $\overline{X}_B = 1,450$ pounds per acre, with a standard deviation of $s_B = 300$ pounds.

Clearly, the point estimate of the difference between the population means is 125 pounds per acre, but suppose we desire to establish a 99-percent confidence interval for the difference between the population means. To find the appropriate t value, we consult Appendix Table G in the column for an upper-tail $\alpha = .005$ and in the row for 28 degrees of freedom, finding $t_{.005(28)} = 2.763$. [We use $(n_A - 1) + (n_B - 1)$ for the degrees of freedom.] Accordingly, our 99-percent confidence interval equals

$$\mu_A - \mu_B = \overline{X}_A - \overline{X}_B \pm \left[t\sqrt{\frac{s_A^2}{n_A} + \frac{s_B^2}{n_B}} \right]$$

$$= 125 \pm \left[2.763\sqrt{\frac{250^2}{15} + \frac{300^2}{15}} \right] = 125 \pm [2.763(100.830)]$$

$$= 125 \pm 278.59$$

The interval ranges from -153.59 pounds per acre to 403.59 pounds per acre, leaving open the possibility that the new pesticide is worse *or* better—hardly conclusive evidence in its favor. |||

Small Matched-Pairs Sample

In the case of small matched-pairs sampling, the formula in Box 7.F is applied, with t_α again replacing z (and $n - 1$ degrees of freedom being employed). Analytical Example 7.1, "Matched-Pairs Sampling in the Chemical Industry," provides an application of this procedure.

MAKING INTERVAL ESTIMATES FOR THE DIFFERENCE BETWEEN TWO POPULATION PROPORTIONS

Interval estimates not of a single population proportion, but of the difference between two such proportions are made by procedures similar to the ones described in the previous sections. We illustrate this fact with the case involving large and independent samples.

Imagine taking two large and independent samples of sizes $n_A = 250$ and $n_B = 350$ from two large populations of interest—say, cookies produced by process A and cookies produced by process B. Let the sample proportions of broken cookies equal $P_A = .08$ and $P_B = .10$; the difference, $d = P_A - P_B = -.02$, is the unbiased point estimate of the difference between the population proportions, $\pi_A - \pi_B$.

We can assume that the sampling distribution of d is normally distributed with a mean of $\pi_A - \pi_B$ and a standard deviation of

$$\sigma_d = \sqrt{\frac{\pi_A (1 - \pi_A)}{n_A} + \frac{\pi_B (1 - \pi_B)}{n_B}}$$

and that the latter can be approximated by

$$\sigma_d \cong s_d = \sqrt{\frac{P_A(1 - P_A)}{n_A} + \frac{P_B(1 - P_B)}{n_B}}.$$

We can now adapt the earlier expression of a confidence interval for the population proportion (Box 7.C) to establish a confidence interval for the difference between population proportions, $\pi_A - \pi_B$ as lying between the limits of $d \pm z\,\sigma_d$. These limits can also be written as shown in Box 7.G.

In our example and for a 95-percent confidence level, this interval comes to

$$\pi_A - \pi_B \cong -.02 \pm \left(1.96\sqrt{\frac{.08(.92)}{250} + \frac{.10(.90)}{350}}\right) = -.02 \pm [1.96(.0235)]$$

$$= -.02 \pm .046.$$

The 95-percent confidence interval ranges from $-.066$ to $+.026$, suggesting that production process A may be better *or* worse than process B (it may yield a smaller *or* larger proportion of broken cookies). Caution is advised: a true difference of zero in the proportion of broken cookies lies within the interval.

7.G Upper and Lower Limits of Confidence Interval for the Difference Between Two Population Proportions (large and independent samples from large populations)

$$\pi_A - \pi_B \cong (P_A - P_B) \pm \left(z \sqrt{\frac{P_A(1 - P_A)}{n_A} + \frac{P_B(1 - P_B)}{n_B}} \right)$$

where P is the sample proportion and n is sample size.
Assumptions: $n_A \geq 30$ and $n_B \geq 30$ and sampling distribution of $P_A - P_B$ is normal because $n\pi \geq 5$ and $n(1 - \pi) \geq 5$ for both samples.

||| ANALYTICAL EXAMPLE

7.1 Matched-Pairs Sampling in the Chemical Industry

Imperial Chemical Industries, a British firm, wanted to test the effect of a chlorinating agent on the abrasion resistance of a certain type of rubber. Ten pieces of rubber were cut in half; one of the two halves in each pair was chosen by the toss of a coin to be treated with the agent, the other one was not treated. Subsequently, the abrasion resistance of all the pieces was tested by a machine and the difference for each pair of halves was recorded, as indicated in the accompanying table.

The mean and standard deviation for the sample differences were computed as $\overline{D} = 1.27$ and $s_D = 1.1265$. The firm's statisticians then calculated a 98-percent confidence interval for the mean difference (using $t_{.01}$ for 9 degrees of freedom):

$$\mu_A - \mu_B = \overline{D} \pm \left(t \frac{s_D}{\sqrt{n}} \right)$$

$$= 1.27 \pm \left[2.821 \left(\frac{1.1265}{\sqrt{10}} \right) \right] = 1.27 \pm 1.005.$$

Thus, the mean difference in abrasion resistance of treated and untreated (otherwise identical) rubber pieces was an increase between .265 and 2.275, suggesting, with 98-percent confidence, a positive effect of such treatment.

Pair	Abrasion Resistance of Treated Minus that of Untreated Piece, D	Pair	Abrasion Resistance of Treated Minus that of Untreated Piece, D
1	2.6	6	1.2
2	3.1	7	2.2
3	$-.2$	8	1.1
4	1.7	9	$-.2$
5	.6	10	.6
			$\Sigma D = 12.7$

Source: Owen L. Davies, ed., *The Design and Analysis of Industrial Experiments* (London: Oliver and Boyd, 1960), p. 13.

7.1 William S. Gosset

William Sealy Gosset (1876–1937) was born in Canterbury, England, the latest descendant of an old Huguenot family that had left France after the revocation of the Edict of Nantes. He studied at Winchester, then at Oxford, where he focused on mathematics and natural sciences. Upon graduation, he joined Arthur Guinness and Son, a Dublin brewery, and he remained employed there throughout his life, ultimately becoming chief brewer at a new brewery in London.

Early on, Gosset saw a need for careful scientific analyses of a variety of processes, from barley production to yeast fermentation, all of which profoundly affected the quality of the brewery's final product, beer. His firm sent him to study under Karl Pearson (Biography 9.1) at the University College, London.

At the time, the theory of estimation based on large samples had been fully worked out, but Gosset noticed a void with respect to small-sample estimation theory. Small samples, however, were typical of Gosset's work at the brewery. So Gosset developed the theory himself. In a now famous 1908 paper, "The Probable Error of a Mean," he noted that s is an erratic estimator of σ when n is small; hence, customary measures of the precision of estimates were invalid for small samples and unknown σ. His paper presented the sampling distribution of a statistic now known as "Student's t" and introduced small-sample estimation by means of the t-distribution family. It is difficult to overestimate the importance of this achievement. It has proven fundamental to statistical inference as it exists today, not only in the realm of estimation, but also in hypothesis testing and the analysis of variance. Sir Ronald Fisher (Biography 8.1), who held great admiration for Gosset and shared with him a keen interest in agricultural experimentation, quite aptly called Gosset "the Faraday of statistics," for Gosset had a similar ability to grasp general principles and develop them further by applying them to practical ends.

All but one of Gosset's numerous papers were published under the pseudonym "Student" to protect the advances in his firm's quality control from nosy competitors. For many years, an air of romanticism surrounded the appearance of "Student's" papers, and only a few individuals knew his real identity, even for some time after his death.

Source: Dictionary of Scientific Biography (New York: Scribner's, 1972), pp. 476–77; *International Encyclopedia of Statistics,* vol. 1 (New York: Free Press, 1978), pp. 409–13.

Summary

1. The process of inferring the values of unknown population parameters from known sample statistics is referred to as *estimation.* The type of sample statistic that serves the purpose of making inferences about a given type of population parameter is called the *estimator* of that parameter. The estimate of a population parameter can be expressed as a single numerical value (called a *point estimate*) or as a range of values within which the unknown parameter presumably lies (called an *interval estimate*).

2. The concept of the *sampling distribution* is crucial to an understanding of the nature of estimation. Every sample summary measure is in fact a random variable prior to the selection of an actual sample, and a sampling distribution is simply a probability distribution of a random variable that happens to be a summary measure based on sample data. It shows the likelihood of occurrence associated with all the possible values of a sample statistic (which values would be obtained when drawing all possible simple random samples of a given size from a population).

3. Any sampling distribution can, in turn, be described by summary measures of its own, such as mean, variance, and standard deviation. The standard deviation of a sampling distribution is also referred to as the *standard error* of the sample statistic in question.

4. The relationship between a sampling distribution and its parent population is of great importance; it can be described in a general way by focusing on the general shapes of the two frequency distributions or more precisely by focusing on mathematical relationships among their summary measures. Two important theorems relate the general shape

of the sampling distribution of the mean to the population distribution:

a. If $\overline{X}$ is the mean of a random sample taken from a population and if the population values are normally distributed, the sampling distribution of $\overline{X}$ is also normally distributed, regardless of sample size.

b. If $\overline{X}$ is the mean of a random sample taken from a population and if the population values are *not* normally distributed, the sampling distribution of $\overline{X}$ nevertheless approaches a normal distribution as sample size increases provided sample size, n, remains small relative to population size, N; this approximation is near-perfect for samples of $n \geq 30$, but $n < .05\ N$ (*central limit theorem*).

Similar general relationships exist between other types of sampling distributions and their associated population distributions. More specific mathematical relationships are given in Boxes 7.A and 7.B.

5. Ideally, an estimate should be unbiased, efficient, and consistent. An *unbiased estimator* is a sample statistic that, on the average, across many samples, takes on a value equal to the population parameter that is being estimated with its help. An *efficient estimator* is that sample statistic, among all the available unbiased estimators, which has the smallest variance for a given sample size. A *consistent estimator* is a sample statistic such that the probability of its being close to the parameter being estimated gets ever larger (and, thus, approaches unity) as sample size increases. Whenever several possible estimators exist, but none of them is perfect by the above criteria, statisticians choose the best one by calculating (and then minimizing) *mean squared error,* the sum of an estimator's squared bias plus its variance.

6. Whenever the statistical population of interest is normally distributed, or whenever the conditions of the central limit theorem are fulfilled, sample mean and sample proportion are unbiased, effi-

cient, and consistent estimators of the corresponding population parameters. The same is true of the sample variance, provided sampling occurs under conditions of statistical independence—but not otherwise. The sample standard deviation is always a biased estimator of that of the population, but it is still used routinely for lack of a better alternative.

7. The typical point estimate, even if the estimator is unbiased, is almost certain to lie below or above the true value of the parameter being estimated. Statisticians, therefore, construct interval estimates centered on the point estimate and ranging below and above this value by A standard errors of the estimator, where A can equal any (fractional or whole) positive value. The width of the interval has important implications for the degree of confidence with which one can state that the population parameter lies within the two limits of the interval.

8. Confidence intervals for the mean or proportion can be constructed from large samples with the help of the normal distribution, and this procedure can be applied under a variety of conditions.

9. Confidence intervals for the mean or porportion can be constructed from small samples as well. Depending on circumstances, the procedure may rely on the normal distribution, may use Chebyshev's theorem, or may proceed from one of Student's t distributions. Tabular presentations of t values facilitate the use of these special types of sampling distributions.

10. Interval estimates can also be derived for the difference between two means, a parameter that frequently is of great interest to decision makers. The proper estimating procedure depends on whether the samples taken from the two populations are large or small but also on whether two independent samples are taken or a matched-pairs sample is taken.

11. Interval estimates can be constructed similarly for the difference between two proportions.

Key Terms

central limit theorem
confidence intervals
confidence level
confidence limits
consistent estimator
degrees of freedom, *d.f.*
efficient estimator
estimation
estimator

interval estimate
matched-pairs sample
mean squared error
point estimate
sampling distribution
standard error of the sample mean, $\sigma_{\overline{X}}$
standard error of the sample proportion, σ_P
Student's t distribution
unbiased estimator

Questions and Problems

The computer programs noted at the end of this chapter can be used to work many of the subsequent problems.

1. Public-health officials in charge of the safety of swimming pools and beaches have determined that the population of index numbers (that define bacterial contamination per gallon of water) is normally distributed with a mean of 150 and a standard deviation of 50. If a random sample of 25 gallons is taken, what is the probability that the sample mean index will
   a. lie between 140 and 160?
   b. exceed 140?
   c. lie below 130?
   d. lie between 160 and 170?
   e. exceed 155?

2. The manufacturer of batteries for aircraft emergency-locator transmitters claims that the lifetime of these batteries is normally distributed with a mean of 30 months and a standard deviation of 3 months. An aircraft manufacturer checks out 50 batteries and discovers a sample mean of only 29 months. What is the probability that the manufacturer's claim is true?

3. Various samples are drawn from normally distributed (infinite) populations with means and variances as shown below. In each case, determine the mean and standard deviation of the sampling distribution of $\overline{X}$.
   a. $n = 10;\ \mu = 30;\ \sigma^2 = 9$
   b. $n = 15;\ \mu = 50;\ \sigma^2 = 4$
   c. $n = 30;\ \mu = 100;\ \sigma^2 = 100$
   d. $n = 100;\ \mu = 400;\ \sigma^2 = 64$

4. In each of the four cases given in problem 3, determine the probability of observing a sample mean of $(\mu_{\overline{X}} - 1)$ or less.

5. A bank reports (accurately) that the population of its demand deposit balances is normally distributed with a mean of $1,200 and a standard deviation of $250. An auditor refuses to certify the bank's claim and takes a random sample of 36 account balances. He will certify the bank's report only if the sample mean lies within $50 of the alleged population mean. What is the probability for such a finding?

6. A manufacturer of strapping tape claims that the lengths of tape on the firm's rolls have a mean of 90.10 feet and a standard deviation of .14 feet. If you take a random sample of the firm's output and $n = 49$, what is the nature of the sampling distribution of the sample mean? What are the chances that your sample mean will be 90 feet or less?

7. Consider a group of persons with these ages: 34, 29, 22, 30, 25. If a simple random sample of 2 is to be taken *without replacement,* what is the sampling distribution for their mean age? What are the distribution's expected value, variance, and standard deviation?

8. Consider panel (b) of Figure 5.2 on page 174, which pictures the case of sampling 3 items *with replacement* from a large lot containing 10 percent defective items. Find the sampling distribution for the sample proportion of defectives; then determine its summary measures.

9. Review this chapter's Practice Problems 1 and 2. Then determine the probability that a sample mean will exceed 43. Show your answer graphically.

10. Construct *point* estimates in the following situations:
    a. At an automobile assembly plant, a simple random sample of 40 workers is taken. Of these, 21 are found to have been vaccinated against tetanus. Estimate the proportion of all the workers who are so vaccinated.
    b. The state fisheries department caught 1,000 fish in a lake, marked them, and returned them to the lake. A week later, it caught another 1,000 fish and found 200 of them marked. Estimate the proportion of all the fish in the lake so marked (and, hence, the size of the lake's fish population).
    c. The governor wants to finance a new prison by raising the state's sales tax from 3 percent to 4 percent. Of 900 residents randomly sampled, 810 oppose the idea. Estimate the proportion of all residents so opposed.

11. Construct *point* estimates in the following situations:
    a. A labor union is negotiating a new contract. The bargaining committee randomly samples the opinions of 100 workers of whom 61 favor the contract. Estimate the proportion of all workers in favor.
    b. After taking a random sample of 50 students, a professor discovers that 30 of them have no idea of the meaning of the term *c.i.f.* Estimate the proportion of all the students equally ignorant.
    c. A rental-car company takes a sample of three of its cars, finding annual repair costs of $510, $98, and $121. Estimate the annual repair cost per car for all the firm's cars.

12. A credit-card company has been billing its millions of customers on the last day of each month, receiving payment, on the average, 13 days later. It decides to experiment with the billing date, hoping to accelerate payments. A simple random sample of 500 customers is switched to a billing date in the middle of the month. The average length of time between billing and payment in the sample comes to 18.9 days, with a standard deviation of 5.3 days. Prepare a 98-percent confidence interval for the mean period between billing and payment for all customers if all of them were given the mid-month billing date.

13. The operator of a telephone exchange is aware that the population of telephone-call durations is normally distributed with a standard deviation of 4 minutes. A sample of 50 calls yields a mean duration of 9.1 minutes. Construct a 95-percent confidence interval for the mean duration of all calls.

14. In a given year, 1,000 apartments are vacated in a city; a random sample of 169 of them shows a mean repair cost of $171.32, with a standard deviation of $15.39. Construct a 90-percent confidence interval for the mean repair costs of all the vacated apartments.

15. The normally distributed population of stock investments by the 75 employees of a firm is known to have a standard deviation of $99. A random sample of 36 employees shows a mean investment of $736. Construct a 99.8-percent confidence interval for the mean investment of all the employees.

16. The personnel manager of a large corporation takes a random sample of $n = 100$ from $N = 600,000$ employees and determines their average annual sick leave at 9.8 days, with a sample standard deviation of 3.1 days. Determine a 90-percent confidence interval for the average annual sick leave of all employees.

17. The Internal Revenue Service is auditing the operators of some 13,000 private airports by taking a random sample of 100 of them. The IRS discovers an average error in reported taxable income of $14,750, with a sample standard deviation of $3,600. Determine a 95-percent confidence interval for the average error made by all the existing airports.

18. A real estate agent wants to determine a 99-percent confidence interval for the average home value in a metropolitan area. A random sample of 36 homes shows a mean value of $156,900 and a sample standard deviation of $40,333. Make the determination.

19. There are 200 gas stations in a city; an economist takes a random sample of 50 of them. Their average gasoline price is $1.339 per gallon, with a sample standard deviation of 23.1¢ per gallon. Determine an 80-percent confidence interval for the average price citywide, while assuming that the population distribution is normal.

20. The state police has taken a random sample of 300 cars on the turnpike and has calculated an average speed of 59.13 mph, with a sample standard deviation of 3.13 mph. Determine a 98-percent confidence interval for the average speed of all cars.

21. An airline has randomly sampled 81 of its thousands of annual flights and has discovered a sample mean of 13.9 unoccupied seats, with a standard deviation of 5.2 seats. Determine a 99.8-percent confidence interval for the mean number of unoccupied seats per flight on all of the airline's flights.

22. On the New York Stock Exchange in 1985, the average price of a share of stock in a random sample of 32 stocks was $52.07, with a standard deviation of $17.09. Determine a 90-percent confidence interval for the average price of all stocks.

23. The manager of United Parcel Service has taken a random sample of the week's 30,195 packages of $n = 100$ and has determined the average weight at 5.75 lb., with a sample standard deviation of 2.01 lb. Determine a 97.5-percent confidence interval for the average weight of all the week's packages.

24. The manager of a canning factory wants a 94.26-percent confidence interval for the average amount of green peas put into cans by a filling machine. A sample of 94 cans is taken from the ongoing production process; sample mean and standard deviation are 15.98 oz. and .21 oz., respectively. Make the determination.

25. Just before the Presidential elections, a Gallup poll of 1,500 voters showed the proportions of voters favoring the candidates indicated in Table 7.A. Construct 95-percent confidence intervals for the population proportion favoring the Democratic candidate. (The actual election outcomes were: Kennedy .501, Johnson .613, Humphrey .497, McGovern .382.)

Table 7.A

	Democratic		Republican	
1960	Kennedy	.51	Nixon	.49
1964	Johnson	.64	Goldwater	.36
1968	Humphrey	.50	Nixon	.50
1972	McGovern	.38	Nixon	.62

26. In a random sample of 100 households, 59 are found to prefer brand X. Construct a 98-percent confidence interval for the proportion of all households so inclined.

27. In a random sample of 100 taken from 55,000 fatal-traffic-accident records, alcohol is the apparent cause 39 times. Construct a 95-percent confidence interval for the proportion of all accidents so caused.

28. A lumber company ships 1 million pine boards. A random sample of 50 boards shows 17 being excessively warped. Construct a 95-percent confidence interval for the proportion of all boards so warped.

29. A random sample of 30 police officers taken from a 100-person police force shows that 21 believe their chief is doing a good job. Construct a 99-percent confidence interval for the proportion of all officers so inclined.

30. A random sample of 300 floppy disks taken from an ongoing production process shows 12 defective. Construct a 95-percent confidence interval for the proportion of defective disks in the firm's entire output.

31. A home-improvement contractor takes a random sample of 50 homes in a city and finds that 19 of them have vinyl siding. Construct a 77.76-percent confidence interval for the proportion of all city homes with such siding.

32. A mayoral study commission takes a sample of 500 apartments in the city (from among a total of 5,000) and finds 50 to be vacant. Construct a 98-percent confidence interval for the citywide proportion of vacant apartments.

33. The Federal Aviation Administration takes a sample of 50 planes landing at an airport where a total of 150 planes are landing. In the sample, 13 planes have illegal transponders (that have not been inspected and certified within the past 24 months as required). Determine a 98.8-percent confidence interval for the proportion of illegal transponders on all 150 planes.

34. A personnel manager is investigating why 1,900 employees quit last year. A random sample of 500 shows "boredom" to be the answer 225 times. Determine a 68.76-percent confidence interval for the proportion of all former employees who left the firm because of boredom. What would a 95-percent confidence interval be?

35. A bank claims that no more than 2 percent of its monthly customer statements are in error. An auditor doesn't believe the claim, selects 100 accounts randomly from the bank's 15,233 accounts, and

contacts the customers in question. Of these, 12 report an error in last month's statement and prove it. Construct a 98-percent confidence interval for the true proportion of erroneous bank statements that the bank mails out. What, precisely, is the *meaning* of this interval?

36. A manufacturer is aware that the lifetimes of batteries produced by the firm are normally distributed. A random sample of 10 batteries shows a mean lifetime of 6 hours with a standard deviation of 1 hour. Construct a 99-percent confidence interval for the mean lifetime of all batteries produced by the same process.

37. The daily gasoline consumption of a fleet of 35 taxis is known to be normally distributed with a population standard deviation of 3.7 gallons. One day, a random sample of 10 taxis is taken, and their mean gasoline consumption is found to have been 20 gallons. Construct a 96-percent confidence interval for the day's mean gasoline consumption by all the taxis.

38. An insurance company wants to estimate the average claim on its automobile collision policies. It believes the sizes of such claims to be normally distributed. It uses the last 21 claims as a sample and finds their average to equal $657, with a standard deviation of $310. Construct a 95-percent confidence interval for the average claim on all policies.

39. A pharmaceutical company deliberately infects 20 volunteers and then tests a new drug on them, finding a mean recovery time of 8 days, with a standard deviation of 2.5 days. Construct a 95-percent confidence interval for the mean recovery time of all potential users of this drug (i) on the assumption that recovery times are normally distributed and (ii) on the assumption that recovery times are definitely *not* normally distributed, but the population standard deviation equals 3 days.

40. A quality engineer is checking out the machinery that is supposed to put 20 oz. of liquid detergent into a container. A sample of 12 containers shows the mean amount dispensed to be 18.9 oz., the standard deviation being 3.1 oz. Construct a 90-percent confidence interval for the mean amount dispensed by the machine, assuming that the amounts dispensed are normally distributed.

41. Military analysts believe that the top speeds of the 5,000 Soviet submarines are normally distributed around an unknown mean that is of great interest to them. They do, however, have information on a random sample of five submarines. Their average top speed is 52 nautical miles per hour, the standard deviation is 2.3 knots. Construct a 99-percent confidence interval for the average top speed of all Soviet submarines.

42. The credit-card company noted in problem 12 takes a second random sample of 500 customers and switches their billing date to the twenty-fourth of the month. The average period between billing and payment in that sample comes to 7.5 days, with a standard deviation of 2.1 days. Construct a 99-percent confidence interval for the difference between the mean days until payment if all customers are switched to a billing date of the fifteenth or the twenty-fourth.

43. A Federal Aviation Administration flight-service specialist wants to determine the difference between the mean thunderstorm intensities (measured by peak gust velocities) for two regions of the country. The following data for a five-year period are obtained and used as a sample:

> Region A: 290 occurrences; sample mean-59 miles per hour; sample standard deviation-20 miles per hour.

> Region B: 333 occurrences; sample mean-71 miles per hour; sample standard deviation-30 miles per hour.

Construct a 99-percent confidence interval for the difference between the mean intensities.

44. A polling organization wants to estimate the difference between the proportion of urban and rural residents favoring agricultural price-support programs. It draws a simple random sample of 500 urban residents (250 favor the programs) and of 500 rural residents (400 favor the programs). Construct a 95 percent confidence interval for the difference between the proportions of those favoring the program.

45. An airline wants to estimate the difference between the proportion of passengers who carry only hand luggage on its New York-to-Chicago versus its New York-to-Miami flights. Two random samples of 50 passengers each show 34 versus 11 such passengers. Construct a 98-percent confidence interval for the difference between the proportions of hand-luggage-only passengers.

Selected Readings

Adams, William J. *The Life and Times of the Central Limit Theorem.* New York: Kaedmon, 1974.

Kohler, Heinz. *Statistics for Business and Economics,* 2nd. ed. Glenview, Ill.: Scott, Foresman and Co., 1988.

Chapter 8 contains discussions of finding optimal sample size (to control sampling error) and of Bayesian estimation.

Computer Programs

The DAYSAL-2 personal-computer diskettes that accompany this book contain three programs of interest to this chapter:

7. Probability Distributions, among other things, lets you find areas under the normal curve between any two z values you specify.

8. Interval Estimation lets you make interval estimates for a population mean, a population proportion, a difference between 2 population means—independent samples or matched pairs—and a difference between 2 population proportions. Estimates are made for large and small samples, using the normal distribution, the t-distribution, or Chebyshev's theorem, as appropriate.

16. Expected Monetary Values, although designed for another purpose, can be used to find the means of sampling distributions.

CHAPTER EIGHT

Hypothesis Testing: Classical Techniques

When people make decisions, they inevitably do so on the basis of beliefs they hold concerning the true state of the world. They carry in their minds a certain image of reality; they hold some things to be true, and others to be false, and they act accordingly. Thus, one government agency may ban cigarette ads because its officials have come to believe that cigarettes cause heart and lung disease; another one may refuse to license a new anticancer drug because no credible case has been made for its alleged effectiveness. A third agency yet may require the wearing of helmets on motorcycles because this precaution is believed to reduce fatality rates, and a fourth one may set out to stem the destruction of orchards by gypsy moths—not by means of the traditional spraying of pesticides, but by means of the introduction of intestinal (gypsy-moth) parasites, a procedure deemed much more effective in achieving the desired goal. Business executives, similarly, make crucial decisions every day because they hold certain beliefs: that a given type of filling machine puts at least one pound of detergent into a box, that a certain steel cable has a breaking strength of 5,000 pounds or more, that the average lifetime of a battery equals 100 hours, that a certain process yields shirt sleeves that are precisely 33 inches long (or capsules that contain precisely 100 milligrams of a drug), that the wages of construction workers in New York exceed those in Chicago by $8 per hour, that one type of fertilizer is more effective than another, that shipping company A has faster delivery times than company B, that the output of the east-coast plant contains fewer defective units than that of the west-coast plant, that one type of engine yields fewer exhaust emissions than another—the list goes on. In all these cases and a million more, people act on the basis of some belief about reality, a belief that first came into the world as a mere conjecture—perhaps a little more than an informed guess, a proposition tentatively advanced as possibly true, and therefore, called a **hypothesis.** Sooner or later, however, every hypothesis is confronted with evidence that either substantiates or refutes it, and in this way people's image of reality moves from much uncertainty to less.

Indeed, **hypothesis testing** (which is the subject of this chapter and the next two) is nothing else but a *systematic* approach to assessing beliefs about reality: it is confronting a belief (such as a tentative idea about the value of an unknown population parameter) with evidence (such as the value of a statistic computed from a random sample taken from the population in question) and then deciding, in light of this evidence, whether the initial belief (or hypothesis) can be maintained as reasonable or must be discarded as untenable. In this chapter, we will consider the testing of hypotheses about a population mean or proportion and about the difference between two such means or proportions. In every case, such testing involves four major steps.

STEP 1: FORMULATING TWO OPPOSING HYPOTHESES

The first step in hypothesis testing is always the formulation of two hypotheses that are mutually exclusive and also collectively exhaustive of the possible states of reality. Each of these complementary hypotheses is a proposition about a population parameter such that the truth of one hypothesis implies the falsity of the other. The first hypothesis in the set is called the **null hypothesis** and is symbolized by H_0; often, but not necessarily, it represents that proposition about an unknown population parameter which is tentatively assumed to be true. The second hypothesis is called the **alternative hypothesis** and is symbolized by H_A; often, but not necessarily, it is initially assumed to be false (but is later accepted as true if there is strong evidence against the null hypothesis). This procedure of formulating hypotheses is quite common and is similar to that in a criminal trial in which the defendant is assumed innocent until proven guilty. Being free of guilt is tentatively assumed to be true (a *null* hypothesis of *zero* guilt is made), but contrary evidence can reverse this judgment (an alternative hypothesis of guilt can be substituted).

Note: Which of two opposing propositions is called H_0 and which is called H_A is obviously quite arbitrary. Unlike defense lawyers, prosecutors, for example, view the assumption of innocence as something to be attacked and, ultimately, rejected in favor of a verdict of "guilty." In the same way, some researchers prefer to set up the null hypothesis in such a way that it *contradicts* what they suspect to be true. If they are right, the hypothesis-testing process leads to the acceptance of the alternative hypothesis, while the null hypothesis is rejected or "nullified" (which gives us another way yet to account for its name).

Hypotheses About a Population Mean

The opposing hypotheses about the value of a population mean, μ, are typically stated in one of three forms by reference to a specified value of the mean, μ_0.

Form 1	*Form 2*	*Form 3*
$H_0: \mu = \mu_0$	$H_0: \mu \geq \mu_0$	$H_0: \mu \leq \mu_0$
$H_A: \mu \neq \mu_0$	$H_A: \mu < \mu_0$	$H_A: \mu > \mu_0$

In Form 1, the null hypothesis is stated as an **exact hypothesis;** that is, as one that specifies a single value for the unknown parameter. The alternative hypothesis, on the other hand, is stated as an **inexact hypothesis,** or one that specifies a range of values for the unknown parameter. As examples, consider the null hypothesis that an industrial process yields metal parts with an average length, μ, of $\mu_0 = 5$ inches, drills holes with an average diameter of 1.2 inches, or makes shirt sleeves with an average length of 33 inches. The alternative hypothesis in each case suggests that the average is in fact larger or smaller than μ_0. Whenever an alternative hypothesis holds for deviations from the null hypothesis in either direction, it is called a **two-sided hypothesis.**

While the alternative hypothesis is always stated as an inexact one, the null hypothesis may be so stated (as in Forms 2 and 3), but even then it is customary to have it contain the equality sign. In Form 2, the null hypothesis claims that the parameter is greater than or equal to a specified value, while the alternative hypothesis claims it to be smaller than that value. As examples consider the null hypothesis that the average quantity of detergent, μ, put into a box by a filling machine equals or exceeds $\mu_0 = 1$ pound, that the average breaking strength of a certain type of cable is at least 5,000 pounds, or that the average lifetime of a certain type of battery is at least 100 hours. The alternative hypothesis in each case suggests that the average in fact falls short of μ_0. Whenever an alternative hypothesis holds for deviations from the null hypothesis in one direction only, it is called a **one-sided hypothesis.**

In Form 3, finally, the null hypothesis claims that the parameter is less than or equal to a specified value, while the alternative hypothesis claims it to be larger than that value. As examples, consider the null hypothesis that a shipping company's average delivery time, μ, equals or falls short of $\mu_0 = 3$ days, that the average drying time of a paint is at most 4 hours, or that an engine produces exhaust emissions of at most 17 parts per million. The (one-sided) alternative hypothesis in each case suggests that the average in fact exceeds μ_0.

Hypotheses About the Difference Between Two Population Means

The opposing hypotheses about the difference between two population means, μ_A and μ_B, are, similarly, stated in one of three forms:

Form I	*Form II*	*Form III*
$H_0: \mu_A = \mu_B$	$H_0: \mu_A \geq \mu_B$	$H_0: \mu_A \leq \mu_B$
$H_A: \mu_A \neq \mu_B$	$H_A: \mu_A < \mu_B$	$H_A: \mu_A > \mu_B$

Thus, the null hypothesis may claim (Form I) that two population means are the same (for example, that the average lifetimes of two types of tires are identical). Or it may claim (Form II) that one population mean equals or exceeds another (for example, that average construction-industry wages in New York are at least equal to, but possibly larger than, those in Chicago). Or it may claim (Form III) that one population mean equals or falls short of another (for example, that the average yield on acres using one type of fertilizer is at most equal to, but possibly less than, that on acres using a different type of fertilizer). In each case, the alternative hypothesis suggests the opposite.

Note: Often hypotheses about the difference between two population means are stated in the following ways that are mathematically equivalent to Forms I–III:

Form A	*Form B*	*Form C*
$H_0: \mu_A - \mu_B = 0$	$H_0: \mu_A - \mu_B \geq 0$	$H_0: \mu_A - \mu_B \leq 0$
$H_A: \mu_A - \mu_B \neq 0$	$H_A: \mu_A - \mu_B < 0$	$H_A: \mu_A - \mu_B > 0$

Here, the null hypothesis claims that the difference between the mean of population A and population B is precisely zero (Form A), is zero or positive (Form B), or is zero or negative (Form C). The alternative hypothesis sees this difference, respectively, as positive or negative (Form A), as negative (Form B), or as positive (Form C).

Hypotheses About a Population Proportion (or the Difference Between Two Such Proportions)

Hypotheses about a population proportion, π (such as the proportion of defective units put out by a production process), are formulated in a manner precisely analogous to those about a population mean. One simply substitutes π for μ in Forms 1 to 3 above, leaving all else unchanged, including subscripts.

Similarly, hypotheses about the difference between two population proportions, π_A and π_B (such as the proportions of cancer victims among smokers on the one hand and nonsmokers on the other), are formulated in analogy to those about the difference between population means. One simply substitutes π for μ in Forms I to III above (or in the equivalent Forms A to C), again leaving all else unchanged, subscripts included.

STEP 2: SELECTING A TEST STATISTIC

The second step in hypothesis testing is the selection of a **test statistic,** which is the type of statistic to be computed from a simple random sample taken from the population of interest and to be used for establishing the probable truth or falsity of the null hypothesis. Obvious choices are (1) the sample mean, $\overline{X}$, when the hypothesis test involves the population mean; (2) the difference between two sample means, $\overline{X}_A - \overline{X}_B$, when the test involves the difference between two population means; and (3) the sample proportion, P, or (4) the difference between two sample proportions, $P_A - P_B$, respectively, when the test is about the population proportion or the difference between two such proportions.

As we learned in Chapter 7, however, every sample statistic has a sampling distribution of its own, and such a distribution can often be approximated by the normal distribution for large samples [when $n \geq 30$ but $n < .05N$ or when $n \cdot \pi$ as well as $n(1 - \pi) \geq 5$] or by a Student t distribution for small samples (when the underlying population values are normally distributed or at least unimodal and fairly symmetrical). When a sample statistic's sampling distribution can be, thus, approximated, one can convert each of the four sample statistics noted above into a corresponding z value or t value by *dividing the difference between the sample statistic and the extreme value of the population parameter postulated in the null hypothesis by the standard error of the sample statistic.* Such z or t values are commonly used as test statistics, too.

STEP 3: DERIVING A DECISION RULE

Having formulated two opposing hypotheses and having selected the type of statistic with which to test them, the next step in hypothesis testing is the derivation of a **decision rule** that specifies in advance, for all possible values of the test statistic that might be computed from a sample, whether the null hypothesis should be accepted or whether it should be rejected in favor of the alternative hypothesis. At first thought, such a rule may seem superfluous. Isn't it bound to be obvious whether the computed value of the test statistic agrees with the null hypothesis or contradicts it? On reflection it becomes clear, however, why the matter is more complicated than might appear to be the case at first sight. While it is true enough that anyone can tell instantly whether the computed value of the test statistic does or does not agree with the null hypothesis, it is not obvious that a given divergence between the observed value and the hypothetical value automatically proves the null hypothesis to be false. What is proved by such a divergence is questionable because any sample statistic is a *random variable;* its value depends very much on the particular sample that happens to be selected from the population in question. Consider, for example, the population of metal parts produced by an industrial process. The average length of these parts may in fact be $\mu_0 =$ 5 inches, as required by the users of these parts. A quality inspector who does not know this fact may formulate two hypotheses as follows:

H_0: $\mu = 5$
H_A: $\mu \neq 5$

Now let a large sample of $n \geq 30$ be taken from this infinite population. Even though *we* know H_0 to be true, the chance factor operating in the sampling process can, nevertheless, provide the inspector with many possible sample means, as shown in Figure 8.1. All kinds of sample means might be found: $\overline{X} = 5$, $\overline{X} = 4.9$, $\overline{X} = 5.2$, and even $\overline{X} = 5.32$; yet all these results (many of which are apparently contradicting the null hypothesis) are consistent with the null hypothesis being true! So what is our quality inspector to do?

One escapes the dilemma by recognizing that even though all kinds of $\overline{X}$ values might be discovered when a simple random sample is taken from a population for

Figure 8.1   A Sampling Distribution of the Sample Mean

This graph shows the sampling distribution of the sample mean, $\overline{X}$, for a large simple random sample taken from a population of metal parts with a mean length of $\mu = 5$ inches. The sampling distribution is approximately normally distributed with a mean, $\mu_{\overline{X}}$, equal to the population mean of 5 and a standard deviation, $\sigma_{\overline{X}}$, of .1 inch. Thus, even though a null hypothesis of H_0: $\mu = 5$ is true, half of the possible sample statistics show $\overline{X} < 5$, and half show $\overline{X} > 5$. Thus, a finding of $\overline{X} \neq 5$ does not prove the correctness of the alternative hypothesis of H_A: $\mu \neq 5$.

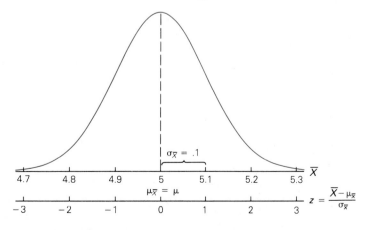

which the null hypothesis is true, not all of these values are equally probable. In Figure 8.1, for example, finding a sample mean between 5 and 5.1 is more likely than finding one between 5.1 and 5.2, and that is more likely still than finding one between 5.2 and 5.3. Given that H_0: $\mu = 5$ is true, finding a sample mean of 5.5 or larger, even though possible, would be a highly unusual result. Thus, one can proceed as follows:

If a particular sample result is very likely when H_0 is true (if, in our example, the probability of finding $\overline{X}$ in a given range is very high when H_0 is true), one ascribes any divergence between the observed and hypothesized values to chance factors operating during the sampling process, and one accepts the null hypothesis as true. If, on the other hand, a particular sample result is very unlikely to occur when H_0 is true (if, in our example, the probability of finding $\overline{X}$ in a given range is very low when H_0 is true), one ascribes a divergence between the observed and hypothesized values to nonchance factors (such as the fact that H_A is true), and one rejects the null hypothesis as false. In short, even seemingly contradictory sample results that are highly likely and nonsurprising when H_0 is true lead to the acceptance of the null hypothesis; others that are highly unusual and surprising when H_0 is true lead to the rejection of it.

The Level of Significance

The foregoing discussion of surprising and nonsurprising results makes a lot of sense but still leaves us with a crucial question: given that H_0 is true, when is the probability of observing a given sample statistic "high" (and the result, therefore, "nonsurprising") and when is that probability "low" (and the result "surprising")? The only possible answer is a *totally arbitrary* one. Whoever undertakes to test a hypothesis must make a subjective decision here; he or she must decide which value of the test statistic is unlikely enough when the null hypothesis is true that it can be viewed as sufficient evidence to reject the null hypothesis and embrace the alternative one. In the situation depicted in Figure 8.1, for example, one might decide to reject the null hypothesis for all sample results that differ by more than two standard errors of the sample mean (by more than $2\ \sigma_{\overline{X}}$) from the mean of the sampling distribution, $\mu_{\overline{X}}$ (and, therefore, from the population mean, μ, if the null hypothesis is true). Thus, values of $\overline{X}$ below 4.8 and above 5.2 (two ranges that correspond to all the values beyond $\pm\ 2z$) would be regarded as occurring with a sufficiently low probability (if H_0: $\mu = 5$ were true) to call H_0 false. As we can see in Appendix Table E, the (combined) area under the normal curve beyond $\pm\ 2z$ equals $1 - [2(.4772)] = .0456$. Thus, the decision rule just stated treats all sample results that differ from the null hypothesis—and that could occur by chance at most 4.56 percent of the time if H_0 were true—as sufficiently strong evidence to reject H_0. Any such sample result that leads to the rejection of H_0 is called a **statistically significant result.** On the other hand, the above decision rule treats all sample results that differ from the null hypothesis—but that could occur by chance more than 4.56 percent of the time if H_0 were true—as evidence too weak to reject H_0. Any such sample result that leads to the continued acceptance of H_0 is called a **not statistically significant result.** Such a result does not prove H_0 to be true, but it doesn't disprove it either. It does, however, make H_0 somewhat more credible than it was before the sample was taken.

Indeed, among all the sample results that are possible when the null hypothesis is true, the (arbitrary) maximum proportion that is considered sufficiently unusual to reject the null hypothesis is called the **significance level,** or **size of a hypothesis**

test, and it is symbolized by α (the lowercase Greek letter *alpha*). In the example, α equals .0456; more commonly used values are .10, .05, .025, and even lower ones. Thus, when the significance level, α, is set at .05, or 5 percent, only statistics that occur with a probability of .05 or less when H_0 is true are sufficient evidence to reject H_0.

Three Illustrations

The three panels of Figure 8.2 (p. 268) illustrate how easily a decision rule can be derived once the desired significance level of a hypothesis test has been specified. The three examples focus (a) on hypothesis tests about a population mean (stated in each of the three forms noted above) and (b) on the large-sample case of $n \geq 30$ but $n < .05N$ (so that the sampling distribution is represented by the normal curve). As will be shown later in this chapter, decision rules for other types of hypothesis tests (or for sampling distributions that are more accurately described by a t distribution) can easily be derived in an analogous fashion.

A two-tailed hypothesis test.   Panel (a) pictures the case in which the opposing hypotheses take Form 1. The null hypothesis claims, for example, that the lengths of metal parts produced average 5 inches; the alternative hypothesis denies it. The alternative hypothesis, thus, is a two-sided one. There is concern about departures from the 5-inch norm in *both* directions; very low as well as very high values of the test statistic ($\overline{X}$ or z) are to be viewed as evidence against H_0. If the desired significance level is set at $\alpha = .05$, then values of the test statistic that are so far below the hypothesized mean as to occur with a probability of at most $(\alpha/2) = .025$ when H_0 is true are to be viewed as evidence against H_0, and the same is true for values so far above the hypothesized mean as to occur with the same low probability. Jointly, the statistically significant values of the test statistic, thus, have a probability of $\alpha = .05$ of occurring when H_0 is true. By consulting Appendix Table E, one can easily determine the lower value of the test statistic (here of $-z_{\alpha/2}$ or $-z_{.025}$) and the upper value of the test statistic (here $+z_{\alpha/2}$ or $+z_{.025}$) that marks the border between two regions of values—namely, those that are to signal the acceptance of H_0 (and, thus, lie in the **acceptance region**) and those that are to signal the rejection of H_0 (and, thus, lie in the **rejection region**). The value of z below (or above) which .025 of the area under the standard normal curve lies is also the value that places $.500 - .025 = .475$ of that area between itself and the mean; therefore, as Appendix Table E shows, $-z_{.025} = -1.96$, while $+z_{.025} = +1.96$. If we assume that $\sigma_{\overline{X}} = .1$ (as in Figure 8.1), the corresponding values of $\overline{X}$ equal $5 - [1.96(.1)] = 4.804$ and $5 + [1.96(.1)] = 5.196$. Note: the value of a test statistic that in this way divides all possible values into an acceptance and rejection region is called a **critical value,** a **cutoff point,** or an **acceptance number.** As panel (a) of Figure 8.2 clearly indicates, the decision rule emerging in this example is: "Accept H_0 for any sample mean between 4.804 and 5.196 inches (or for any z value between -1.96 and $+1.96$); reject H_0 (and accept H_A) for values of the test statistic outside these limits."

A lower-tailed hypothesis test.   Panel (b) of Figure 8.2 pictures the case in which the opposing hypotheses take Form 2. The null hypothesis claims, for example, that the lifetime of a battery is at least 100 hours; the alternative hypothesis denies it. The alternative hypothesis is one-sided. There is concern about departures from the 100-

Figure 8.2   Two-Tailed vs. One-Tailed Hypothesis Tests

This set of graphs pictures sampling distributions of sample means (and their corresponding standard normal deviates) on the assumption that the stated null hypotheses are true. When the alternative hypothesis is two-sided, as in panel (a), any hypothesis test is said to be a **two-tailed test** because the null hypothesis is then rejected for values of the test statistic located in either tail of that statistic's sampling distribution. The rejection region is then split between the two tails of that distribution. When, on the other hand, the alternative hypothesis is one-sided, as in panels (b) and (c), any hypothesis test is said to be a **one-tailed test** because the null hypothesis is then rejected only for very low (or only for very high) values of the test statistic located entirely in one tail of the sampling distribution. The entire rejection region is then found in either the lower tail, making for a **lower-tailed test,** as in panel (b), or in the upper tail, making for an **upper-tailed test,** as in panel (c). Note that the critical values of the test statistics are encircled.

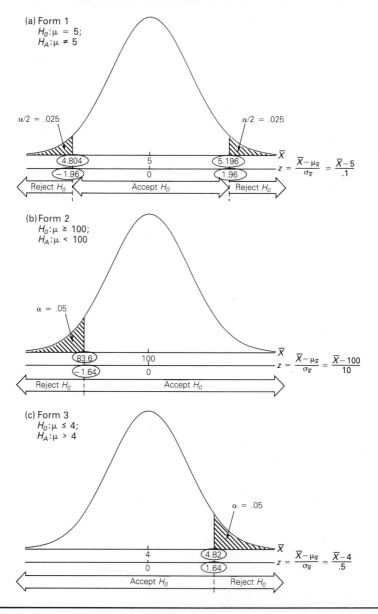

hour norm only in the lower direction (nobody will complain if the batteries in fact last longer than 100 hours); only very low values of the test statistic are to be viewed as evidence against H_0. If the desired significance level is again set at $\alpha = .05$, values of the test statistic so far below the hypothesized mean as to occur with a maximum probability of $\alpha = .05$ when H_0 is true must be viewed as evidence against H_0. According-ing to Appendix Table E, the critical z value of $-z_\alpha$ or $-z_{.05}$ equals -1.64 because this value places $.50 - .05 = .45$ of the area under the standard normal curve between itself and the mean above it (and thus, leaves .05 of that area in the lower tail of the sampling distribution). If we assume that $\sigma_{\overline{x}} = 10$ hours, the corresponding critical value of $\overline{X}$ equals 83.6 hours. Thus, the decision rule is: "Accept H_0 for any sample mean of 83.6 hours or more (or for z values of -1.64 and above); reject H_0 (and accept H_A) for all lower values of the test statistic."

An upper-tailed hypothesis test. Panel (c) of Figure 8.2 pictures the case in which the opposing hypotheses take Form 3. The null hypothesis claims, for example, that the drying time of a paint is at most 4 hours; the alternative hypothesis denies it. The alternative hypothesis is again one-sided, but this time the story differs from that in panel (b). Now there is concern about departures from the 4-hour norm only in the *upper* direction (nobody will complain if the paint in fact dries faster than claimed); only very *high* values of the test statistic are to be viewed as evidence against H_0. If the desired significance level is again set at $\alpha = .05$, values of the test statistic so far above the hypothesized mean as to occur with a maximum probability of $\alpha = .05$ when H_0 is true must be viewed as evidence against H_0. This time, the critical z value equals $+1.64$; if we know that $\sigma_{\overline{X}} = .5$ hour (because, say, $\sigma = 3$ hours and sample size is $n = 36$), the corresponding critical value of $\overline{X}$ equals 4.82 hours. Thus, the decision rule is "Accept H_0 for any sample mean of 4.82 hours or less (or for z values of 1.64 and below); reject H_0 (and accept H_A) for all higher values of the test statistic." Note: Part (b) of Appendix Table F provides critical z values for two-tailed and one-tailed hypothesis tests and for the most commonly used significance levels.

Decision rules summarized. The formulas in Box 8.A summarize the decision rules for large-sample hypothesis tests about a mean, which we have discussed so far. In Box 8.A, $\overline{X}$ and z are the computed values of the test statistics; their critical values are printed in black.

8.A Decision Rules for Large-Sample Hypothesis Tests About a Mean*

1. Two-tailed test, H_A: $\mu \neq \mu_0$
   Accept H_0 if $\mu_0 - (z_{\alpha/2} \cdot \sigma_{\overline{x}}) \leq \overline{X} \leq \mu_0 + (z_{\alpha/2} \cdot \sigma_{\overline{x}})$
   or if $-z_{\alpha/2} \leq z \leq z_{\alpha/2}$

2. Lower-tailed test, H_A: $\mu < \mu_0$
   Accept H_0 if $\overline{X} \geq \mu_0 - (z_\alpha \cdot \sigma_{\overline{x}})$
   or if $z \geq -z_\alpha$

3. Upper-tailed test, H_A: $\mu > \mu_0$
   Accept H_0 if $\overline{X} \leq \mu_0 + (z_\alpha \cdot \sigma_{\overline{x}})$
   or if $z \leq z_\alpha$

*Assuming $n \geq 30$ and $n < .05N$ or population is normal.

The decision rules discussed above can easily be extended to the small-sample case, provided the underlying population values are normally distributed. One simply substitutes Student t values for z in the expressions of Box 8.A. Similarly, one can derive analogous decision rules for hypothesis tests about a proportion (by substituting π for μ and P for $\overline{X}$) or about the difference between means or proportions (by substituting differences where now individual values appear).

STEP 4: SELECTING A SAMPLE, COMPUTING THE TEST STATISTIC, AND CONFRONTING IT WITH THE DECISION RULE

The final step in hypothesis testing involves (a) the selection of a simple random sample of arbitrary size, n, from the population of interest, (b) the computation of the actual (as opposed to critical) value of the test statistic (selected in Step 2) and (c) its confrontation with the decision rule (derived in Step 3). Note: as a later section of this chapter will show, even though sample size is arbitrary, some sample sizes may be preferred to others—in order to control, for example, the occurrence of various types of errors (also discussed below) that are inherent in the hypothesis-testing process.

If, for example, a sample of metal parts proved to have an average length of $\overline{X} = 5.1$ inches ($z = 1.00$), the null hypothesis stated in panel (a) of Figure 8.2 would be upheld. Similarly, if a sample of batteries turned out to have an average lifetime of $\overline{X} = 90$ hours ($z = -1.00$), the null hypothesis stated in panel (b) of Figure 8.2 would be upheld. If, however, a sample of paint canisters showed an average drying time of $\overline{X} = 4.9$ hours ($z = 1.80$), the null hypothesis given in panel (c) of Figure 8.2 would be rejected.

But note: given $\alpha = .05$, the latter conclusion (that the paint's average drying time exceeded the mere 4 hours claimed for it) would be incorrect in 5 out of 100 times in which this type of statistical procedure was employed. After all, the unusual sample result cited does occur with a 5 percent probability when the panel (c) null hypothesis is true!

The Possibility of Error

The hypothesis-testing procedure outlined in this chapter so far can lead to one of four results that are qualitatively quite different. As is noted in Table 8.1, the result of the procedure can be correct, but it can also be erroneous.

When the null hypothesis is true. When the null hypothesis is in fact true (and those who engage in hypothesis testing, of course, do not know this, or they would have no reason to undertake the test), two possible consequences can flow from a hypothesis test: the null hypothesis may be accepted (a correct result), or it may be rejected (an erroneous outcome). The latter possibility—the rejection of a true null hypothesis—is referred to as a **type I error,** or as the **error of rejection.** Such an error can be costly, indeed. It may lead to the interruption of the production process in order to adjust machinery that needs no adjusting; it may lead a firm to switch to a new supplier of raw materials although the performance of the old one was satisfactory; it may lead a government agency to condemn a firm for ignoring pollution standards even though

Table 8.1   Evaluating the Outcome of a Hypothesis Test

Depending on the true state of the world, the result reached in a hypothesis test can be correct—(a) and (d)—or erroneous—(b) and (c). The probabilities associated with these outcomes are given in the brackets.

True State of the World	Test Result	
	Null Hypothesis is Accepted	Null Hypothesis is Rejected
Null Hypothesis is in Fact True	(a) Correct result $[p = 1 - \alpha]$	(b) Type I error $[p = \alpha]$
Null Hypothesis is in Fact False	(c) Type II error $[p = \beta]$	(d) Correct result $[p = 1 - \beta]$

it has done no such thing. Committing a type I error is equivalent to condemning a defendant in a criminal trial even though he or she is innocent. But note: the logic of hypothesis testing *guarantees* that this type of error occurs sometimes. Picture in your mind a sampling distribution, such as Figure 8.1, that shows all the possible values of a test statistic, given that the null hypothesis is true. Then recall what happens when we select a significance level for a hypothesis test. We arbitrarily place some of these possible values in the rejection region; we interpret the occurrence of those values that have a low probability of occurring when H_0 is true as evidence that H_0 is false. Hence, in the long run, as we use a given procedure again and again, we are bound to reject a true null hypothesis the proportion of times we have set as the significance level. In short, the probability of making a type I error—of rejecting a null hypothesis that is true—is equal to α, the significance level of the hypothesis test, which is why it is also referred to as the **α risk.**

$$p \text{ (rejecting } H_0 | H_0 \text{ is true)} = \alpha = \text{significance level of the hypothesis test}$$

It follows that the complementary probability of avoiding a type I error when the null hypothesis is true equals $1 - \alpha$, which is also called the **confidence level of the hypothesis test.**

$$p \text{ (accepting } H_0 | H_0 \text{ is true)} = 1 - \alpha = \text{confidence level of the hypothesis test}$$

When the null hypothesis is false.   When the null hypothesis is in fact false, a hypothesis test once again leads to one of two possible consequences: the null hypothesis may be accepted (an erroneous outcome), or it may be rejected (a correct result). The former possibility—the acceptance of a false null hypothesis—is referred to as a **type II error,** or as the **error of acceptance.** Such an error, too, has its costs. It may lead to uninterrupted production at a time when the machinery needs to be adjusted badly (because this fact is not recognized); it may lead a firm to continue using a supplier whose performance is unsatisfactory (because this fact is not recognized); it may lead a government agency to do nothing about a firm's violation of pollution standards (because this fact is not recognized). Committing a type II error is equivalent to acquitting a defendant in a criminal trial even though he or she is guilty. And just as in the field of criminal justice it is deemed less serious to let a guilty person go than to convict an innocent one, it is common practice to phrase the null hypothesis in such a way that a type I error is more serious than a type II error.

The probability of committing a type II error is symbolized by β and is referred to as the **β risk.**

$$p \text{ (accepting } H_0|H_0 \text{ is false)} = \beta$$

In turn, the complementary probability of avoiding a type II error when the null hypothesis is false equals $1 - \beta$, which is also called the **power of the hypothesis test.**

$$p \text{ (rejecting } H_0|H_0 \text{ is false)} = 1 - \beta = \text{power of the hypothesis test}$$

The Trade-Off Between a Type I Error and a Type II Error

Whoever undertakes to test a hypothesis knows neither before sampling nor afterwards whether a type I or type II error has been committed. It is possible, however, to control the probability of having a given error occur. Unfortunately, given sample size, n, anything that reduces α automatically raises β; the reverse is also true, as can easily be seen with the help of Figure 8.3. Panel (a) of Figure 8.3 is a copy of panel (c) of Figure 8.2. It shows the sampling distribution of the mean drying time of paint when the null hypothesis ("average drying time equals at most 4 hours") is in fact true. If a significance level of $\alpha = .05$ is selected and if, as in the earlier example, $\sigma_{\overline{x}} = .5$, the critical values are $\overline{X} = 4.82$ hours and $z = 1.64$, respectively. The probability of making a type I error, thus, appears as the crosshatched area in panel (a); it equals the area under the upper tail of the sampling distribution prevailing if H_0 is true.

Panel (b) of Figure 8.3 shows the sampling distribution when the null hypothesis is in fact false (and the average drying time equals 5.8 hours). The critical value of $\overline{X} = 4.82$ (that was established on the basis of assuming H_0 to be true) now is shown to be lower than the mean of the true panel (b) sampling distribution; a corresponding z value is -1.96. To the left of this value (and, thus, in the acceptance region), we find .025 of the area under the standard normal curve. This crosshatched area in panel (b) gives us the probability of making a type II error. For any given sample, the selection of α, which leads to the decision rule, thus, ultimately determines β.

Two things should be noted about type I and type II errors:

1. A lower α (which would move the dashed borderline between the acceptance and rejection region to the right) would clearly raise β. A higher α (which would move that borderline left) would lower β. As we shall see in a later section of this chapter, only a larger sample size can reduce α and β at the same time (or can reduce one of the two without raising the other).

2. Unscrupulous investigators clearly could manipulate the size of α (and, thus, the position of the borderline between the acceptance and rejection regions) so as to accommodate *any* observed value of a test statistic. Someone eager to have a null hypothesis accepted could make α extremely small (at the cost of having β very large). Someone eager to have a null hypothesis rejected could make α

Figure 8.3   The Trade-Off Between α and β

Someone about to test a null hypothesis obviously does not know whether it is true or false and whether, therefore, the sampling distribution looks like that shown in panel (a) or like that in panel (b). By assuming the former and selecting a value for α, a decision rule is derived; this decision rule automatically determines the value of β if H_0 is in fact false. Given sample size, any decrease in α raises β (as the dashed vertical line moves right). The opposite is also true; any increase in α lowers β (as that line moves left). This result again is analogous to our experience with the criminal justice system: anything that lowers the probability, α, of convicting an innocent person (such as a rule against self-incrimination) also raises the probability, β, of acquitting a guilty one. Anything that lowers the probability, β, of acquitting a guilty person (such as a rule that allows split juries to convict) thereby raises the probability, α, of convicting an innocent one.

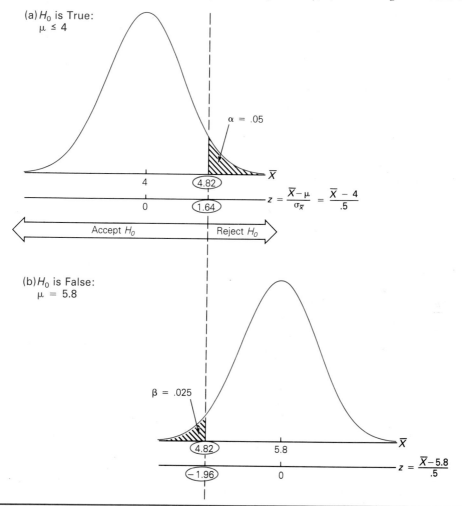

extremely large (which would make β very small). Obviously, such behavior would make a mockery of scientific inquiry. The only way to avoid it is to select α *before* the investigator has seen the sample data. Step 3 must be performed before Step 4.

A HISTORICAL PERSPECTIVE

Although the construction of formal hypothesis tests is a 20th-century innovation in statistical theory, certain ideas underlying this theory have a long history, going back at least eight centuries, probably to the reign of England's King Henry II (1154–89). Close-Up 8.1, "The Trial of the Pyx," tells the story. The modern theory was developed in England, too, but not without a lot of controversy, as is evidenced in Biographies 8.1 to 8.3 at the end of the chapter. The following sections show applications of the modern theory.

LARGE-SAMPLE TESTS

Let us now apply the four-stage procedure introduced above to typical testing situations. We begin by assuming that the samples to be taken are large ($n \geq 30$ but $n < .05N$ for tests involving one or two means; $n \cdot \pi$ and $n(1 - \pi) \geq 5$ for tests involving one or two proportions); therefore, we can assume that the test statistic's sampling distribution is approximated by the normal curve.

Tests of a Population Mean

We first consider each of the three types of hypothesis tests of a single population mean.

Practice Problem 1

The Thickness of Aluminum Sheets (a two-tailed test)
An aircraft manufacturer needs aluminum sheets that are .01 inches thick—no more, no less. The firm's quality inspector is to test the quality of supplies accordingly by measuring thickness in a simple random sample of incoming sheets.

Step 1: Formulating two opposing hypotheses.

H_0: μ = .01 inch
H_A: $\mu \neq$ .01 inch

Step 2: Selecting a test statistic. The inspector selects the standard normal deviate for the sample mean thickness of the sheets,

$$z = \frac{\overline{X} - \mu_0}{\sigma_{\overline{X}}}.$$

Step 3: Deriving a decision rule. The inspector selects a significance level of $\alpha = .05$, which (according to Appendix Table F) implies critical values of $\pm z_{\alpha/2} = \pm 1.96$ (this being a two-tailed test). Thus, the decision rule must be: "Accept H_0 if $-1.96 \leq z \leq +1.96$." The critical values are encircled in Figure A.

Figure A

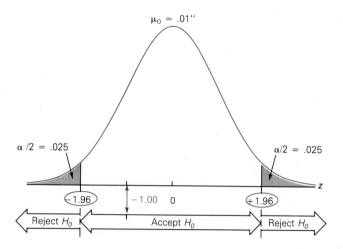

Step 4: Selecting a sample, computing the test statistic, and confronting it with the decision rule. After taking a sample of $n = 100$, the inspector finds a sample mean thickness of $\overline{X} = .009$ and a sample standard deviation of $s = .01$. Given an infinite population (the sheets come from an ongoing production process), the large-population case applies; hence $\sigma_{\overline{X}}$ is estimated as

$$\sigma_{\overline{X}} \cong \frac{s}{\sqrt{n}} = \frac{.01}{\sqrt{100}} = \frac{.01}{10} = .001.$$

Accordingly, the computed value of the test statistic equals

$$z = \frac{\overline{X} - \mu_0}{\sigma_{\overline{X}}} = \frac{.009 - .01}{.001} = -1.00.$$

This value corresponds to the colored arrow in Figure A; it suggests that the null hypothesis should be *accepted*. At the 5 percent significance level, the sample result is not statistically significant. The observed divergence from the desired average .01-inch standard of thickness is more likely due to chance factors at work during the sampling process than due to a faulty production process that systematically puts out sheets with an average thickness other than .01 inch. |||

||| Practice Problem 2

The Tensile Strength of Steel Rods (a lower-tailed test)

For purposes of mounting engines on aircraft, a manufacturer needs specialty steel rods with a tensile strength of at least 5,000 pounds. The firm's quality inspector is to test incoming supplies accordingly by figuring the tensile strength evidenced in a simple random sample of these rods. (This test is performed by noting the force at which the rods become distorted.)

Step 1: Formulating two opposing hypotheses.

H_0: $\mu \geq$ 5,000 pounds
H_A: $\mu <$ 5,000 pounds

Step 2: Selecting a test statistic. The inspector selects the standard normal deviate for the sample mean tensile strength of rods,

$$z = \frac{\overline{X} - \mu_0}{\sigma_{\overline{X}}}.$$

Step 3: Deriving a decision rule. The inspector selects a significance level of $\alpha =$ .01, which (according to Appendix Table F) implies a critical value of $-z_\alpha = -2.33$ (this being a lower-tailed test). Thus, the decision rule must be: "Accept H_0 if $z \geq -2.33$." The critical value is encircled in Figure B.

Figure B

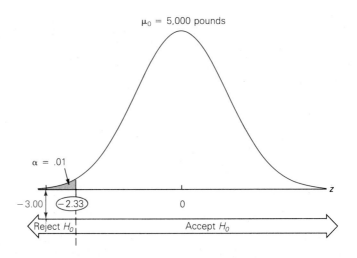

Step 4: Selecting a sample, computing the test statistic, and confronting it with the decision rule. After taking a sample of 64 rods, the inspector finds that the rods become distorted by a mean strength of $\overline{X} =$ 4,700 pounds, the sample standard deviation being $s =$ 800 pounds. Once again, the large-population case applies, so $\sigma_{\overline{X}}$ is esti mated as

$$\sigma_{\overline{X}} \cong \frac{s}{\sqrt{n}} = \frac{800}{\sqrt{64}} = \frac{800}{8} = 100 \text{ pounds.}$$

Accordingly, the computed value of the test statistic equals

$$z = \frac{\overline{X} - \mu_0}{\sigma_{\overline{X}}} = \frac{4,700 - 5,000}{100} = -3.00.$$

This value corresponds to the colored arrow in Figure B; it suggests that the null hypothesis should be *rejected*. At the 1-percent significance level, the sample result is statistically significant. The observed divergence from the 5,000-pound standard is unlikely to be due to chance factors operating during sampling; it is more likely to be the result of a production process that puts out low-quality (and, therefore, unusable) rods. ‖

‖‖ Practice Problem 3

Additive in Aviation Fuel (an upper-tailed test)
An airline wants to test a manufacturer's claim that 1 ounce of additive will significantly increase the performance of aviation fuel. The airline's records show that the fuel exclusive of the additive has delivered an average of 6 minutes of flight time per gallon (mpg). Future flights are to be flown with the additive; a random sample is to be evaluated by the airline's statistician who has reason to assume that the minutes-per-gallon population will be normally distributed.
    Step 1: Formulating two opposing hypotheses.

H_0: $\mu \leq 6$ minutes per gallon
H_A: $\mu > 6$ minutes per gallon

    Step 2: Selecting a test statistic. The statistician selects the sample mean minutes per gallon, $\overline{X}$.
    Step 3: Deriving a decision rule. The statistician selects a significance level of $\alpha = .10$, which (according to Appendix Table F) implies a critical value of $+z_\alpha = 1.28$ (this being an upper-tailed test). Thus, the decision rule must be: "Accept H_0 if $\overline{X} \leq \mu_0 + (1.28\ \sigma_{\overline{x}})$." Assuming the population standard deviation, σ, is unknown, the precise magnitude of this critical $\overline{X}$ cannot be determined until after the sample has been taken and $\sigma_{\overline{x}} = (\sigma/\sqrt{n})$ can be estimated as $s/\sqrt{n}$. The critical value of 6.1 given in Figure C is in this case found in Step 4.

Figure C

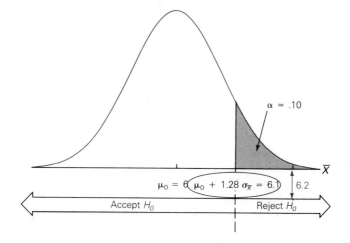

$\mu_0 = 6$ $\mu_0 + 1.28\ \sigma_{\overline{x}} = 6.1$ 6.2

$\alpha = .10$

Accept H_0 Reject H_0

Step 4: Selecting a sample, computing the test statistic, and confronting it with the decision rule. After 100 flights have been flown with the additive, the statistician takes a sample of $n = 36$ and finds the mean minutes per gallon to be $\overline{X} = 6.2$, with a sample standard deviation of $s = .6$ minutes. Because the small-population case applies $(n \geq .05N)$, the statistician estimates $\sigma_{\overline{X}}$ as

$$\sigma_{\overline{X}} \cong \frac{s}{\sqrt{n}} \sqrt{\frac{N-n}{N-1}} = \frac{.6}{\sqrt{36}} \sqrt{\frac{100-36}{100-1}} = .1(.804) = .0804 \text{ minutes per gallon.}$$

Accordingly, the critical value of $\overline{X}$ equals $6 + [1.28(.0804)] = 6.1$ minutes per gallon. This result suggests that the null hypothesis should be *rejected*. At the 10 percent significance level, the sample result is statistically significant. The additive does stretch the flight time squeezed out of a gallon. |||

Note: This is a good place to observe that *statistical* significance is not necessarily the same thing as *practical* significance. The statistical significance of this test simply tells us that the divergence of the sample result (an average 6.2 minutes of flight time per gallon) from the previous experience (an average 6 minutes per gallon) is probably not merely the result of the sampling process, but quite likely attributable to the additive. Yet the practical significance of all this may be nil: it may be cheaper, for example, to buy the added 3.3 percent of flight time by paying for extra fuel than by paying for an expensive additive. (This example also highlights the fact that the term *statistically significant,* being a technical term, must not be confused with such everyday terms as *important* or *impressive.*)

Tests of a Population Proportion

Hypothesis tests concerning a single population proportion are conducted analogously to those about a mean; Practice Problem 4 is a two-tailed test.[1]

||| Practice Problem 4

The Weights of Drug Dosages (a two-tailed test)
A hospital administrator needs to know whether it is true that 90 percent of the drug dosages prepared by a machine weigh precisely 100 milligrams. A random sample of dosages from among thousands prepared during a week is to be taken to evaluate the claim.

Step 1: Formulating two opposing hypotheses.

H_0: $\pi = .9$
H_A: $\pi \neq .9$

---

[1] For an illustration of a lower-tailed and an upper-tailed test of a population proportion, see problems 5 and 13 in Chapter 8 of the *Student Workbook* that accompanies this text.

Step 2: Selecting a test statistic. The evaluator selects the standard normal deviate for the sample proportion,

$$z = \frac{P - \pi_0}{\sigma_P}.$$

Step 3: Deriving a decision rule. A significance level of .05 is set; hence, critical values are $\pm z_{\alpha/2} = \pm 1.96$ (this being a two-tailed test). Thus, the decision rule must be: "Accept H_0 if $-1.96 \leq z \leq +1.96$." The critical values are encircled in Figure D.

Figure D

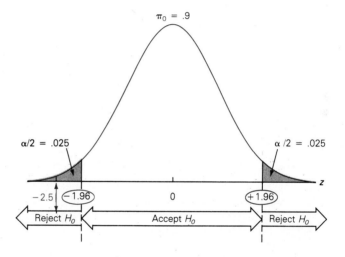

$\pi_0 = .9$

$\alpha/2 = .025$ $\alpha/2 = .025$

-2.5 -1.96 0 $+1.96$ z

Reject H_0 Accept H_0 Reject H_0

Step 4: Selecting a sample, computing the test statistic, and confronting it with the decision rule. After taking a sample of $n = 100$, it is found that the sample proportion of precise 100-milligram dosages equals $P = .8$. For the large-population case, σ_P is, thus, estimated as

$$\sigma_P \cong \sqrt{\frac{P(1 - P)}{n}} = \sqrt{\frac{.8(.2)}{100}} = .04.$$

Accordingly, the computed value of the test statistic equals

$$z = \frac{P - \pi_0}{\sigma_P} = \frac{.8 - .9}{.04} = -2.5.$$

This value corresponds to the colored arrow in Figure D; it suggests that the null hypothesis should be *rejected*. At the 5-percent significance level, the sample result is statistically significant. The observed divergence between the hypothesized population proportion and the sample proportion is unlikely to be the result of pure chance. |||

Tests of the Difference Between Two Population Means: Independent Samples

We first consider hypothesis tests of the difference between means, while employing *independent samples* such that the elements making up the sample taken from population A are chosen independently of the elements making up the sample taken from population B.[2]

The Lifetimes of Aircraft Radios (a two-tailed test)

An airline wants to test the claim that the mean lifetimes of two types of aircraft radios are identical. It installs 800 radios of each type in its current fleet and later selects two simple random samples of radios of each type.

Step 1: Formulating two opposing hypotheses.

H_0: $\mu_A - \mu_B = 0$
H_A: $\mu_A - \mu_B \neq 0$

Step 2: Selecting a test statistic. The firm's statistician selects

$$z = \frac{d}{\sigma_d} = \frac{\overline{X}_A - \overline{X}_B}{\sigma_{\overline{X}_A - \overline{X}_B}}.$$

Step 3: Deriving a decision rule. The statistician selects a significance level of $\alpha = .10$, which implies critical values of $\pm z_{\alpha/2} = \pm 1.64$ (this being a two-tailed test). Thus, the decision rule must be: "Accept H_0 if $-1.64 \leq z \leq +1.64$." The critical values are encircled in Figure E.

Figure E

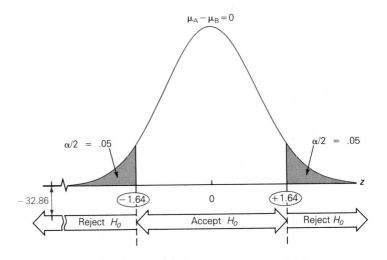

$\mu_A - \mu_B = 0$

$\alpha/2 = .05$ $\alpha/2 = .05$

-32.86 $\left(-1.64\right)$ 0 $\left(+1.64\right)$ z

Reject H_0 Accept H_0 Reject H_0

---

[2]Practice Problem 5 here illustrates a two-tailed test; for an illustration of a lower-tailed independent-samples test of the difference between two population means, see problem 6 or problem 14 in Chapter 8 of the *Student Workbook* that accompanies this text.

Step 4: Selecting the samples, computing the test statistic, and confronting it with the decision rule. After taking two samples of $n = 36$ each, the statistician finds mean lifetimes and sample standard deviations of $\overline{X}_A = 4,120$ hours, $s_A = 80$ hours and $\overline{X}_B = 4,910$ hours, $s_B = 120$ hours. The large-population case applies ($n \leq .05N$); σ_d is estimated as

$$\sigma_d \cong \sqrt{\frac{s_A^2}{n_A} + \frac{s_B^2}{n_B}} = \sqrt{177.78 + 400} = 24.04.$$

Accordingly, the computed value of the test statistic equals

$$z = \frac{d}{\sigma_d} = \frac{4,120 - 4,910}{24.04} = -32.86.$$

This value corresponds to the colored arrow in Figure E; it suggests that the null hypothesis should be *rejected*. At the 10-percent significance level, the sample result is statistically significant. It is very likely that radio B has a longer lifetime than radio A. |||

Tests of the Difference Between Two Population Means: Matched-Pairs Samples

We now turn to hypothesis tests of the difference between means, while employing *matched-pairs samples,* taken after each elementary unit in population A has been matched with a "twin" from population B.[3]

||| Practice Problem 6

Smoking and Heart Disease (a two-tailed test)

A medical researcher wants to test whether there exists no connection between smoking and heart disease because the mean age at which heart disease is first detected is the same for smokers (A) and nonsmokers (B). Some 100 smokers are matched with 100 nonsmokers according to age, lifestyle, medical history, occupation, sex, and so on.

Step 1: Formulating two opposing hypotheses.

H_0: $\mu_A - \mu_B = 0$
H_A: $\mu_A - \mu_B \neq 0$

Step 2: Selecting a test statistic. The researcher selects

$$z = \frac{\overline{D}}{s_D / \sqrt{n}}$$

---

[3]Practice Problem 6 here illustrates a two-tailed test; for an illustration of a lower-tailed matched-pairs-samples test of the difference between two population means, see problem 7 in Chapter 8 of the *Student Workbook* that accompanies this text.

Step 3: Deriving a decision rule. The researcher selects a significance level of $\alpha = .05$, which implies critical values of $\pm z_{\alpha/2} = \pm 1.96$ (this being a two-tailed test). Thus, the decision rule must be: "Accept H_0 if $-1.96 \leq z \leq +1.96$." The critical values are encircled in Figure F.

Figure F

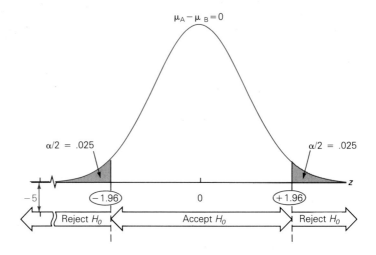

Step 4: Selecting the sample, computing the test statistic, and confronting it with the decision rule. After taking the sample, the researcher finds the mean age of the onset of heart disease to be 6 years earlier for smokers than for their "twin" nonsmokers; thus, $\overline{D} = -6$ years. The sample standard deviation of $s_D = 12$ years. Thus, the computed value of the test statistic equals

$$z = \frac{\overline{D}}{s_D/\sqrt{n}} = \frac{-6}{12/\sqrt{100}} = -5.$$

This value corresponds to the colored arrow in Figure F; it suggests that the null hypothesis should be *rejected.* At the 5-percent significance level, the sample result is statistically significant. Because the test statistic is negative, we conclude, furthermore, that smoking *hastens* the onset of heart disease. (If z had been in the right-hand rejection region, it would indicate that smoking delays heart disease.) |||

Tests of the Difference Between Two Population Proportions: Independent Samples

We, finally, consider large-sample hypothesis tests of the difference between proportions, again employing independent samples.[4]

---

[4]Practice Problem 7 here illustrates a two-tailed test; for an illustration of a lower-tailed independent-samples test of the difference between two population proportions, see problem 8 of the *Student Workbook* that accompanies this text.

||| Practice Problem 7

Sex and TV-Program Preference (a two-tailed test)

A television station wants to know whether there is any difference in the proportion of males (A) or females (B) who favor a given program over another during a certain time slot. Two simple random samples of 100 each are to be taken from among the city's adult population.

Step 1: Formulating two opposing hypotheses.

$$H_0: \pi_A - \pi_B = 0$$
$$H_A: \pi_A - \pi_B \neq 0$$

Step 2: Selecting a test statistic. The station manager selects

$$z = \frac{d}{\sigma_d} = \frac{P_A - P_B}{\sigma_{P_A - P_B}}.$$

Step 3: Deriving a decision rule. The station manager selects a significance level of $\alpha = .05$, which implies critical values of $\pm z_{\alpha/2} = \pm 1.96$ (this being a two-tailed test). Thus, the decision rule must be: "Accept H_0 if $-1.96 \leq z \leq +1.96$." The critical values are encircled in Figure G.

Figure G

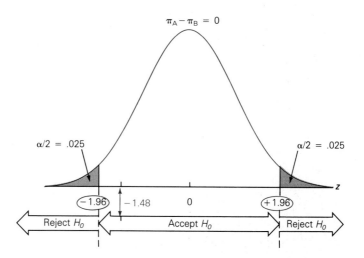

Step 4: Selecting the samples, computing the test statistic, and confronting it with the decision rule. After taking the samples, the manager finds proportions of $P_A = .6$ and $P_B = .7$ and then calculates the pooled estimator of the population proportion as

$$P_P = \frac{100(.6) + 100(.7)}{200} = .65.$$

Accordingly, σ_d is estimated as

$$\sigma_d \cong \sqrt{P_P(1 - P_P)\left(\frac{1}{n_A} + \frac{1}{n_B}\right)} = \sqrt{.65(.35)\left(\frac{1}{100} + \frac{1}{100}\right)} = .0675.$$

Thus, the value of the test statistic is computed as

$$z = \frac{d}{\sigma_d} = \frac{.6 - .7}{.0675} = -1.48.$$

This value corresponds to the colored arrow in Figure G; it suggests that the null hypothesis should be *accepted*. At the 5-percent significance level, the sample result is not statistically significant. It is quite likely that the proportions of males and females are the same.

Note: Analytical Example 8.1, "Antitrust Pork Barrel," provides another instance of the use of this procedure. |||

SMALL-SAMPLE TESTS

We now turn to cases where samples are small ($n < 30$), but the test statistic's sampling distribution can be approximated by a t distribution (because the underlying population values are approximately normally distributed and because, when two populations are involved, their standard deviations, though unknown, are known to be equal). Having discussed large-sample tests in such detail, we can restrict ourselves here to fewer examples, and we can also omit the graphical illustrations. (All graphs would be analogous to those given in the previous section, except that t distributions for specified degrees of freedom would replace the normal curves.)

Tests of a Population Mean

We first consider two tests concerning a single population mean.[5]

||| Practice Problem 8

The Length of Long-Distance Calls (a two-tailed test)
A telephone company executive wants to know whether it is still true (as in the past) that the average length of evening long-distance telephone calls equals 18.1 minutes. A simple random sample of 25 of an evening's calls is to be used to find the answer.

Step 1: Formulating two opposing hypotheses.

H_0: μ = 18.1 minutes
H_A: $\mu \neq$ 18.1 minutes

---

[5]Practice Problems 8 and 9 illustrate a two-tailed test and a lower-tailed test; for an illustration of an upper-tailed small-sample test of a population mean, see problem 9 or problem 15 in Chapter 8 of the *Student Workbook* that accompanies this text.

Step 2: Selecting a test statistic. The company statistician selects

$$t = \frac{\overline{X} - \mu_0}{s/\sqrt{n}},$$

given that the population of telephone-call durations is likely to be normally distributed and that the large-population case applies ($n < .05N$ because of the large number of calls made).

Step 3: Deriving a decision rule. The statistician selects a significance level of $\alpha = .05$, which (according to Appendix Table G) implies critical values of $\pm t_{\alpha/2} = \pm t_{.025,24} = \pm 2.064$ (this being a two-tailed test and there being $n - 1 = 24$ degrees of freedom). Thus, the decision rule must be: "Accept H_0 if $-2.064 \leq t \leq +2.064$."

Step 4: Selecting a sample, computing the test statistic, and confronting it with the decision rule. After taking the sample, the statistician finds a sample mean duration of calls of $\overline{X} = 17.2$ minutes and a sample standard deviation of $s = 4$ minutes. Thus, the value of the test statistic is computed as

$$t = \frac{\overline{X} - \mu_0}{s/\sqrt{n}} = \frac{17.2 - 18.1}{4/\sqrt{25}} = -1.125.$$

Accordingly, the null hypothesis should be *accepted*. At the 5-percent level of significance, the sample result (a somewhat smaller duration of the calls than noted in the past) is not statistically significant. The observed divergence is likely due to chance factors. ‖‖

‖‖ Practice Problem 9

The Weight of Raisin Boxes (a lower-tailed test)
A government agency has received a consumer complaint that boxes allegedly containing 15 ounces of raisins actually contain less. A simple random sample of 10 such boxes is to be selected from various stores to clarify the issue.

Step 1: Formulating two opposing hypotheses.

H_0: $\mu \geq 15$ ounces
H_A: $\mu < 15$ ounces

Step 2: Selecting a test statistic. The government statistician selects

$$t = \frac{\overline{X} - \mu_0}{s/\sqrt{n}},$$

given the likelihood of a normal distribution of the population of net weights and the large-population case ($n < .05N$ because of the large number of such boxes being produced and sold).

Step 3: Deriving a decision rule. The government desires a significance level of $\alpha = .01$, which (according to Appendix Table G) implies a critical value of $-t_{\alpha} = -t_{.01,9} = -2.821$ (this being a lower-tailed test and there being $n - 1 = 9$ degrees of freedom). Thus, the decision rule must be: "Accept H_0 if $t \geq -2.821$."

Step 4: Selecting a sample, computing the test statistic, and confronting it with the decision rule. After taking the sample, the statistician finds a sample mean weight of $\overline{X} = 13.5$ ounces and a sample standard deviation of $s = 1$ ounce. Thus, the value of the test statistic is computed as

$$t = \frac{\overline{X} - \mu_0}{s/\sqrt{n}} = \frac{13.5 - 15}{1/\sqrt{10}} = -4.743.$$

Accordingly, the null hypothesis should be *rejected.* At the 1-percent level of significance, the sample result in satistically significant. The observed divergence of the mean sample weight from the advertised weight is unlikely to be the result of chance factors operating in the sampling process; it is more likely due to the fact that the population of raisin boxes being marketed has a mean net weight below 15 ounces, as the alternative hypothesis claims. |||

Tests of the Difference Between Two Population Means: Independent Samples

We will consider one example of hypothesis tests concerning two means that involve independent samples.[6]

||| Practice Problem 10

The Cavity-Fighting Ability of Two Toothpastes (a two-tailed test)
The American Dental Association wants to determine whether there is a difference in the cavity-fighting ability of two toothpastes, A and B. A simple random sample of 21 users of each type is taken, and the mean number of cavities over a decade is counted.

Step 1: Formulating two opposing hypotheses.

H_0: $\mu_A - \mu_B = 0$
H_A: $\mu_A - \mu_B \neq 0$

Step 2: Selecting a test statistic. The investigator selects

$$t = \frac{\overline{X}_A - \overline{X}_B}{\sqrt{s_P^2\left(\frac{1}{n_A} + \frac{1}{n_B}\right)}},$$

given that the two populations of numbers of cavities are likely to be normally distributed and that the population variances are likely to be the same. The expression s_P^2

---

[6]Practice Problem 10 here illustrates a two-tailed test; for an illustration of an upper-tailed independent-samples test of the difference between two population means, see problem 11 in Chapter 8 of the *Student Workbook* that accompanies this text.

inside the square root is, in fact, an estimate of these equal population variances derived by pooling the information contained in the two sample variances. This **pooled variance,** s_P^2, is the weighted average of the two sample variances, where the weights equal the degrees of freedom associated with each variance:

$$s_P^2 = \frac{(n_A - 1)s_A^2 - (n_B - 1)s_B^2}{n_A + n_B - 2}$$

Step 3: Deriving a decision rule. The investigator selects a significance level of $\alpha = .01$, which (according to Appendix Table G) implies critical values of $\pm t_{\alpha/2} = \pm t_{.005,40} = \pm 2.704$ (this being a two-tailed test and there being $n_A + n_B - 2 = 40$ degrees of freedom). Thus, the decision rule must be: "Accept H_0 if $-2.704 \leq t \leq +2.704$."

Step 4: Selecting the samples, computing the test statistic, and confronting it with the decision rule. After taking the samples, the investigator finds sample mean numbers of cavities (and associated sample standard deviations) of $\overline{X}_A = 27$ and $\overline{X}_B = 23$ (with $s_A = 6$ and $s_B = 2$). Thus, the pooled variance is

$$s_P^2 = \frac{(n_A - 1)s_A^2 + (n_B - 1)s_B^2}{n_A + n_B - 2}$$
$$= \frac{20(6)^2 + 20(2)^2}{40} = 20,$$

while the value of the test statistic is computed as

$$t = \frac{\overline{X}_A - \overline{X}_B}{\sqrt{s_P^2\left(\dfrac{1}{n_A} + \dfrac{1}{n_B}\right)}} = \frac{27 - 23}{\sqrt{20\left(\dfrac{1}{21} + \dfrac{1}{21}\right)}} = 2.898.$$

Accordingly, the null hypothesis should be *rejected.* At the 1-percent significance level, the sample result (fewer cavities with toothpaste B) is statistically significant. The observed divergence of the sample difference from the hypothesized zero population difference is unlikely to be due to chance; toothpaste B does a better job.

Note: Analytical Example 8.2, "The Never-Ending Search for New and Better Drugs," provides another instance of the use of this procedure. So does Analytical Example 8.3, "Do Nursing Homes Discriminate Against the Poor?" except that this example involves an upper-tailed test. |||

Tests of the Difference Between Two Population Means: Matched-Pairs Sampling

We consider, finally, a hypothesis test concerning two means that makes use of matched-pairs sampling.[7]

---

[7] Practice Problem 11 illustrates a two-tailed test; for an illustration of an upper-tailed, matched-pairs-sample test, see problem 12 or problem 16 in Chapter 8 of the *Student Workbook* that accompanies this text.

||| Practice Problem 11

Setting Retail Prices (a two-tailed test)

A manufacturer markets two models of a product, A and B. The manufacturer has suggested to retailers nationwide that the retail prices of the two models should be kept the same. The manufacturer wishes to test whether this suggestion is being followed by sampling 10 retail outlets, all of which sell both models.

Step 1: Formulating two opposing hypotheses.

$$H_0: \mu_A - \mu_B = 0$$
$$H_A: \mu_A - \mu_B \neq 0$$

Step 2: Selecting a test statistic. The statistician selects

$$t = \frac{\overline{D}}{s_D/\sqrt{n}}$$

given that each retail outlet in the sample provides two price observations (and, thus, the desired difference) and that the population of price differences in all of the nation's retail outlets is likely to be normally distributed.

Step 3: Deriving a decision rule. The statistician selects a significance level of $\alpha = .05$, which (according to Appendix Table G) implies critical values of $\pm t_{\alpha/2} = \pm t_{.025,9} = \pm 2.262$ (this being a two-tailed test and there being $n - 1 = 9$ degrees of freedom). Thus, the decision rule must be: "Accept H_0 if $-2.262 \leq t \leq +2.262$."

Step 4: Selecting the sample, computing the test statistic, and confronting it with the decision rule. After taking the sample, the company statistician finds an average difference in model A vs. B prices of $\overline{D} = \$18.23$ (along with a standard deviation of $s_D = \$1.25$). Thus, the computed value of the test statistic is

$$t = \frac{\overline{D}}{s_D/\sqrt{n}} = \frac{18.23}{1.25/\sqrt{10}} = 46.12.$$

Accordingly, the null hypothesis should be *rejected.* At the 5 percent significance level, the sample result (that a large price difference is being maintained) is statistically significant. Model A is being sold for considerably more than model B; the observed sample difference is not a fluke inherent in the sampling process. |||

CRITICISMS OF HYPOTHESIS TESTING

The testing procedures discussed in this chapter are widely used, yet there is considerable controversy about their usefulness. Without trying to present an exhaustive list, we shall consider three of the criticisms that are commonly encountered.

Serious Violations of Assumptions

Some critics charge that many hypothesis tests are performed even though the assumptions that would validate the procedure used are not met. There might be a difference, for example, between the statistical (target) population about which inferences are to be made and the population that is actually being sampled. (This difference is likely because a perfect frame is usually unavailable for any nontrivial population.) In addition, an intended simple random sample can in fact turn out to be a nonprobability sample. (This result is likely because high nonresponse rates are common.) Or, perhaps, the populations being sampled may not be normally distributed, or they may not have equal variances (requirements for using the t statistic in two-sample tests). In situations such as these, the testing procedure will grind out nonsense.

Rigid Interpretation of the "Acceptance" Criterion

While it is meaningful to *reject* a null hypothesis as a result of uncovering contrary evidence that cannot reasonably be attributed to chance, it is not meaningful, critics claim, to *accept* a null hypothesis in the absence of such contrary evidence. Yet these words are commonly used. It would be much wiser, critics urge, to talk of "failure to reject" rather than "acceptance of" a null hypothesis. All the things we know ultimately are things that haven't yet been proven false and are things concerning which a wait-and-see attitude is most appropriate. All that we know is at best *tentatively* accepted. Any rigid interpretation of the acceptance criterion (as if it assured us of the truth with *certainty*), critics conclude, is to be avoided at all cost.

Note: this criticism is clearly semantic and can be met by agreeing to interpret "acceptance" loosely, as a token word for "failure to reject," except, perhaps, in cases where the probability β of a Type II error is known to be extremely small. Admittedly, it is often difficult to know this probability.

Nonpublication of Nonsignificant Results

Results of hypothesis tests that are statistically nonsignificant (and, thus, nonsurprising) are unlikely to be published. This situation has serious consequences, critics note. Consider a null hypothesis that is in fact true (unbeknownst to researchers) and that is tested independently by many, always at the $\alpha = .05$ level of significance. Each of these investigators has only 5 chances out of 100 of (incorrectly) finding statistical significance (and falsely rejecting H_0). Yet the chance that *at least one* among the many will reach such a result is much higher. If there are 10 such investigators, the probability that at least one will falsely reject H_0 equals $1 - (.95)^{10} = .401$—and that person's (wrong) results *will* be published!

ANALYTICAL EXAMPLES

8.1 Antitrust Pork Barrel

It has long been asserted that federal regulatory agencies are significantly impaired in their tasks by their dependence on Congress. This impairment is said to occur because each member of Congress inevitably seeks to further the provincial interests of citizens in the home district, the welfare of which may depend disproportionately on a few key industries. When those industries are threatened in any way by regulatory agencies, members of Congress who have power over these agencies will naturally deflect their actions. A group of researchers decided to test whether the case-bringing activity of the Federal Trade Commission (FTC) during the 1961–79 period was biased in favor of firms that operated in the districts of members of those congressional committees (three in the Senate and five in the House) that have important budgetary and oversight powers with respect to the FTC.

Although the FTC is empowered to initiate antitrust investigations on its own, almost 90 percent of its investigations are begun as a result of cases brought to it by the public. Ultimately, these investigations lead to a consent decree (by which the accused party promises to reform) or to a formal complaint; the latter, in turn, leads to dismissal for insufficient evidence of a violation or to a cease-and-desist order. The investigators formulated the null hypothesis that there was a zero difference between (a) the proportion of dismissals to cases brought affecting districts with representatives on FTC-relevant committees (π_A) and (b) this proportion in districts without such representatives (π_B). Accordingly, the alternative hypothesis supported the pork-barrel thesis:

$$H_0: \pi_A - \pi_B = 0$$
$$H_A: \pi_A - \pi_B \neq 0$$

Using

$$z = \frac{P_A - P_B}{\sigma_d}$$

as a test statistic and choosing the 10 percent significance level, the critical values for accepting H_0 were $\pm z_{\alpha/2} = \pm z_{.05} = \pm 1.64$. Selected sample results and computed values of z are shown in the accompanying table.

There was no reason to reject the null hypothesis in cases 1 and 2 but plenty of reason to do so for cases 3–5 because of the statistically significant z values. After reviewing similar data for the 1970–79 period (as well as for the ratio of dismissals to *formal complaints*), the investigators concluded that there was considerable support for the thesis that members of certain congressional committees that have important oversight and budgetary powers with respect to the FTC do deflect commission decisions in favor of firms in their home districts.

Congressional Committee	Ratio of Dismissals to Cases Brought (1961–69)		z
	Within Committee Members' Districts, P_A	Outside Committee Members' Districts, P_B	
(1) Senate Committee on Interior and Insular Affairs	$\frac{17}{285} = .0596$	$\frac{148}{2,190} = .0675$	$-.50$
(2) Senate Committee on Commerce, Science, and Transportation	$\frac{32}{570} = .0561$	$\frac{133}{1,905} = .0698$	-1.14
(3) Senate Subcommittee on Antitrust and Monopoly of the Senate Judiciary Committee	$\frac{60}{638} = .0940$	$\frac{105}{1,837} = .0572$	3.22
(4) House Subcommittee on Independent Offices of the House Appropriations Committee	$\frac{14}{87} = .1609$	$\frac{151}{2,388} = .0632$	3.59
(5) All Five Relevant House Subcommittees	$\frac{84}{1,104} = .0761$	$\frac{81}{1,371} = .0591$	1.69

Source: Roger L. Faith, Donald R. Leavens, Robert D. Tollison, "Antitrust Pork Barrel," *Journal of Law and Economics,* October 1982, pp. 329–42. Table copyright © 1982 by the University of Chicago. All rights reserved. Reprinted by permission of the University of Chicago Press.

8.2 The Never-Ending Search for New and Better Drugs

Pharmaceutical companies are engaged in a never-ending search for new and better drugs. There exists a virtually infinite supply of substances that might possibly be effective against one disease or another. In the United States, perhaps as many as 175,000 compounds are evaluated in research labs each year; only 20 of these eventually make it to the drug store. Sometimes success comes fairly soon (the remarkable antibiotic Aureomycin was found after screening "merely" 4,000 soil samples); at other times, success eludes researchers for decades. The search for an anticancer drug is a case in point. In all cases, however, the hypothesis-testing procedures introduced in this chapter are put to work.

In one actual case, researchers implanted cancer cells in 9 mice, then treated 3 mice with a new chemical compound to test whether it would retard the growth of cancer tumors. After a certain period, all the tumors were removed and weighed. It was expected that the compound, if effective, would have caused the tumor weights of the treated mice (experimental group) to be less than those of the untreated ones (control group). The results are shown in the accompanying table.

Was the .53-gram reduction in the average tumor weight of the experimental group significant enough to warrant further tests with the compound? A two-sided hypothesis test employing a known standard error of the difference between the two population means of $\sigma_d = .35$ grams and a significance level of $\alpha = .001$ would conclude in the negative.

Step 1:

$$H_0: \mu_A - \mu_B = 0$$
$$H_A: \mu_A - \mu_B \neq 0$$

Tumor Weights (grams)	
Experimental Group	Control Group
.96	1.29
1.59	1.60
1.14	2.27
	1.31
	1.88
	2.21
$\overline{X}_A = 1.23$	$\overline{X}_B = 1.76$

Step 2:

$$t = \frac{\overline{X}_A - \overline{X}_B}{\sqrt{\sigma_d^2 \left(\frac{1}{n_A} + \frac{1}{n_B}\right)}}$$

Step 3: Critical values for 7 degrees of freedom:

$$\pm t_{\alpha/2} = \pm t_{.0005(7)} = \pm 5.408.$$

Decision rule: "Accept H_0 if $-5.408 \leq t \leq +5.408$."

Step 4:

$$t = \frac{1.23 - 1.76}{.247} = -2.15$$

Source: Charles W. Dunnett, "Drug Screening: The Never-Ending Search for New and Better Drugs," in Judith M. Tanur et al., eds., *Statistics: A Guide to the Unknown* (San Francisco: Holden-Day, 1972), pp. 23–33.

8.3 Do Nursing Homes Discriminate Against the Poor?

For-profit nursing homes set their own fees for private patients but are dictated lower fees, called *reimbursements,* for patients supported by government agencies. Provided that the cost of care is the same for both types of patients, it seems reasonable to assume that such nursing homes will give preference in admissions to those patients (such as private ones) who bring in the greatest revenue and will discriminate in admissions against other patients (such as the poor who are supported by Medicaid) who bring in less revenue. A group of researchers investigated the 1969 practices of 18 southern California nursing homes to test this proposition. Of these, $n_A = 11$ facilities (group A) claimed not to discriminate in admissions, the other $n_B = 7$ facilities (group B) clearly preferred non-Medicaid patients. The mean percentage of Medicaid-supported patient days (and the standard deviations) were found to be $\overline{X}_A = 75.4$ ($s_A = 16.3$) and

$\bar{X}_B = 40.4$ ($s_B = 30.8$). A hypothesis test was performed as follows:

Step 1:

$H_0: \mu_A - \mu_B \leq 0$

$H_A: \mu_A - \mu_B > 0$

Step 2:

$$t = \frac{\bar{X}_A - \bar{X}_B}{\sqrt{s_P^2\left(\dfrac{1}{n_A} + \dfrac{1}{n_B}\right)}}$$

Step 3: $\alpha = .05$; given 16 degrees of freedom, $t_\alpha = t_{.05(16)} = 1.746$. The decision rule is: "Accept H_0 if $t \leq 1.746$."

Step 4:

$$s_P^2 = \frac{(n_A - 1)s_A^2 + (n_B - 1)s_B^2}{n_A + n_B - 2}$$

$$= \frac{(10)(16.3)^2 + (6)(30.8)^2}{16} = 521.8$$

$$t = \frac{\bar{X}_A - \bar{X}_B}{\sqrt{s_P^2\left(\dfrac{1}{n_A} + \dfrac{1}{n_B}\right)}}$$

$$= \frac{75.4 - 40.4}{\sqrt{521.8\left(\dfrac{1}{11} + \dfrac{1}{7}\right)}}$$

$$= \frac{35}{11.04}$$

$$= 3.17$$

Accordingly, H_0 was rejected. At the 5 percent level of significance, the alternative hypothesis (that group A facilities, on the average, had more Medicaid-supported patient days than group B facilities did) was accepted.

Source: John S. Greenlees, John M. Marshall, Donald E. Yett, "Nursing Home Admission Policies Under Reimbursement," *The Bell Journal of Economics,* Spring 1982, pp. 93–106.

||| CLOSE-UP

8.1 The Trial of the Pyx

The trial of the Pyx is an ancient ceremony at the Royal Mint in London. The avowed purpose of the ceremony is to ascertain whether the coinage issued by the mint meets the Crown's specifications. During a period of time, coins are taken at random from those minted and placed in a locked box, called *the Pyx.* At irregular intervals, perhaps as frequently as once each year but usually less often, a trial of the Pyx is declared, and a jury of members of the Worshipful Company of Goldsmiths assembles. At this time, the Pyx is opened, and the contents are counted, weighed, and assayed. The results are then compared with a set standard that includes an allowable tolerance above and below a given number. After a successful trial, a banquet is held in celebration.

From the statistician's point of view, the trial is a marvelous example of a hypothesis test: it involves the drawing of a random sample, a null hypothesis to be tested (the standard), a two-sided alternative (the Crown is worried about excessive deviations either above or below the standard), a test statistic (the total weight of the coins in the box), and a critical region (the allowable tolerance).

To be sure, there is also much to criticize from the point of view of modern theory: the way the sampled coins are selected has always been ill defined (leaving open the possibility of serious bias). The effect of sample size on the probability of a type I error (the sample size traditionally chosen makes the probability of type I error practically zero) was long unrecognized, as was the fact that the power of the test to detect a false null hypothesis is extremely low for parameter values near the one specified in the null hypothesis.

Source: Stephen M. Stigler, "Eight Centuries of Sampling Inspection: The Trial of the Pyx," *Journal of the American Statistical Association,* September 1977, pp. 493–500.

8.1 Ronald A. Fisher

Ronald Aylmer Fisher (1890–1962) was born in London, England, and studied mathematics and physics at Cambridge University. He spent his early years indecisively moving about—working at an investment house, on a farm in Canada, as a teacher in British public schools. He developed an interest in biometrics, however, and this interest led him, in 1919, to join the staff of the world-famous agricultural experiment station at Rothamsted. There he was charged with sorting and reassessing a 66-year accumulation of data on field trials and weather records—and in the process, he became one of the century's leading statisticians. Early on, he published the epochal *Statistical Methods for Research Workers* (1925), a book that was destined to go through 14 editions in many languages and to become the "bible" for researchers throughout the world. This book was followed by two equally influential works, *The Genetical Theory of Natural Selection* (1930), a book that reconciled Darwinian evolution with Mendelian genetics, and *The Design of Experiments* (1935); these book established Fisher as a top-ranking geneticist as well as statistician. Indeed, prior to his move to Australia late in life, Fisher held long-term positions as professor of eugenics, first at University College, London, and later at Cambridge University.

The works just cited, however, hardly begin to tell the story of Fisher's productivity. Over the course of half a century, he published one paper every two months, and most of them broke new ground! Thus, it is difficult to decide which of his many contributions are the most praiseworthy, and it is surely impossible in a few paragraphs to show how thoroughly the land of statistics is crisscrossed with the footprints of this prolific scholar. He is the pioneer in experimentation who revolutionized the field with randomized blocks, Latin squares, factorial designs, and confounding. Then he is the man who pushed ahead the theory of estimation (and introduced concepts of unbiasedness, consistency, efficiency, and more) and who made correlation, regression, and analysis of variance (and covariance) what they are today. It was Fisher who erected upon the work of William S. Gosset (Biography 7.1) a comprehensive theory of hypothesis testing based on small samples.

It is not surprising that honors and prizes were heaped upon Fisher throughout his life (he was even knighted in 1952); it is unfortunate that Fisher also became involved in an extraordinarily long-lasting and acrimonious battle with other statisticians concerning the nature of estimation and hypothesis testing. Fisher's views will be summarized here; those of his opponents will be outlined in Biographies 8.2 and 8.3. In his theory of estimation, Fisher introduced the concept of the **fiducial interval,** a range of values that can be *trusted,* with a specified probability, to contain the value of some parameter. Such an interval might take the form of $\mu = \overline{X} \pm (z\sigma_{\overline{X}})$; Fisher would establish this range *after* a sample had been taken and $\overline{X}$ had been calculated. Given the specific sample result, he would, thus, make a probability statement about the parameter in the form of the fiducial interval and would tell us that we could trust the parameter to lie within that range (the Latin *fiducia* means *trust*)—with a specified probability, of course, depending on the value of z. (This view is generally rejected by statisticians today, a fact already implied by the Chapter 7 discussion of *confidence intervals* and to be explained further in Biographies 8.2 and 8.3.)

In subtle but important ways, Fisher's view of hypothesis testing also differed from that of other statisticians (and from that found in the body of Chapter 8). Fisher viewed hypothesis testing as a form of data analysis and opinion formation engaged in by someone in a research situation. This person would

1. hypothesize the value of some parameter,
2. select a test statistic (such as z or t) the distribution of which was completely known if the hypothesized parameter value was true,
3. take a random sample from the population of interest,
4. calculate the value of the test statistic,
5. calculate the probability, also called the p **value,** or **observed significance level,** that a value of the test statistic as extreme as or more extreme than the one actually observed could have occurred by chance, assuming that the hypothesized parameter value is true, and

6. use the following arbitrary cutoff points to judge the plausibility of the hypothesized parameter value:

   a. If the p value is $\leq .01$, the person would render a strong judgment against the hypothesized parameter value.

   b. If the p value is $\leq .05$ but $> .01$, the person would render a weak judgment against the hypothesized parameter value.

   c. If the p value is $> .05$, the person would judge in favor of the hypothesized parameter value.

Figure 8.1 in the chapter can be used to illustrate the Fisher approach. The hypothesized value might be $\mu = 5$; the test statistic might be z, and its distribution would be the normal curve shown. The random sample might yield $\overline{X} = 5.1$, from which the value of the test statistic would be calculated as $z = 1$. Thus, Fisher would figure the p value (from Appendix Table E) as $.5000 - .3413 = .1587$, and he would form the opinion that the hypothesized parameter value was reasonable.

This Fisherian view of hypothesis testing, too, is generally rejected by statisticians today, a fact already implied by this chapter's discussion and to be explained further in Biographies 8.2 and 8.3. Nevertheless, Fisher's work greatly helped the development of the modern theory of hypothesis testing, and his scientific achievements are in any case so varied and so crucial that nothing can dim their luster or reduce his ranking as one of the greatest scientists of this century.

Sources: Dictionary of Scientific Biography, vol. 5 (New York: Scribner's, 1972), pp. 7–11; *International Encyclopedia of Statistics,* vol. 1 (New York: Free Press, 1978), pp. 352–58. For more information on Fisher, see *Encyclopedia of Statistical Sciences* (New York: Wiley-Interscience, 1982–86), vol. 3, pp. 103–10.

8.2 Jerzy Neyman

Jerzy Neyman (1894–1981) was born in Benderey, Bessarabia. He studied at the University of Warsaw and, in the 1920s, achieved world renown by extending sampling theory to sampling from finite populations without replacement and carrying out complex stratified sampling schemes for the Polish government. In the 1930s, he joined the faculty at University College, London, where Ronald A. Fisher (Biography 8.1) and Egon S. Pearson (Biography 8.3) had just filled two positions created to replace the one long held by Karl Pearson (Biography 9.1). Neyman, who soon was to move on to a long and distinguished career at the University of California, Berkeley, shared many interests with Fisher—agricultural experimentation, weather-modification experiments, genetics, astronomy, medical diagnosis—yet the two ended up on opposites sides when the famous and bitter controversy arose concerning the nature of estimation and hypothesis testing. For years, however, Neyman collaborated fruitfully with Egon S. Pearson; jointly the two men were immortalized by the Neyman-Pearson theory of estimation and hypothesis testing that is now generally accepted.

Neyman and Pearson introduced the concept of *confidence intervals* into the theory of estimation at about the same time that Fisher wrote about *fiducial intervals,* and for a time the two concepts lived amiably side by side, appearing to be two names for the same thing. Eventually, however, it became clear that these were different concepts, indeed. A 95-percent fiducial interval à la Fisher, for example, would claim a 95-percent probability that a given parameter lay within the interval constructed around a sample statistic already calculated. Unlike Fisher, Neyman-Pearson would establish the interval *before* the sample was taken and before any statistic was calculated. A 95-percent confidence interval à la Neyman-Pearson only claims that use of their formula in the long run produces intervals such that 95 out of 100 of them contain the parameter, while any actual interval, constructed *after* sampling, was *certain* either to contain or to exclude the parameter. (This important concept of confidence levels has already been discussed in Chapter 7 in connection with Figure 7.4 which can be found on page 240).

For additional information on Neyman, see *Encyclopedia of Statistical Sciences* (New York: Wiley-Interscience, 1982–86), vol. 6, pp. 215–23.

8.3 Egon S. Pearson

Egon Sharpe Pearson (1895–1980) was born in London, England, the son of Karl Pearson (Biography 9.1). Egon was educated at Cambridge University and closely followed in his father's footsteps. Early on, he joined his father's department at University College, London; in 1933, when his father resigned, he took over one of the two new positions created as replacements, the other one going to R. A. Fisher (Biography 8.1). In this position, and as editor of *Biometrika,* he contributed importantly to statistics (he himself published some 133 papers); above all, he is known, along with Jerzy Neyman (Biography 8.2), as the developer of the modern theory of hypothesis testing (as it is found in this chapter). The Neyman-Pearson approach differed considerably from Fisher's and this difference gave rise to a lifelong and bitter controversy. Unlike Fisher who viewed hypothesis testing as a procedure by which a *researcher* could *form an opinion* about some population parameter, Neyman-Pearson viewed it as a means by which a *decision maker* operating under uncertainty could *make a clear choice* between two alternatives, while at the same time controlling the chances for error (and minimizing costs associated therewith).

While Fisher hypothesized the value of only one parameter, Neyman-Pearson explicitly formulated *two* rival hypotheses, H_0 and H_A—a procedure suggested to them by William S. Gosset (Biography 7.1). This first step contrasts sharply with Fisher's total neglect of an alternative hypothesis. (If, under the Fisher approach, the parameter in question is judged unlikely to be true, what is the truth?) In addition, Neyman-Pearson introduced the formal acceptance/rejection rule employed throughout this chapter, as well as the notion of the two error types (with probabilities α and β), and they investigated the costs of committing these errors. (Fisher, who moved in the world of the research scientist, perhaps understandably had shown no concern about such costs. Indeed, there seems to be a qualitative difference between the cost to, say, a drug manufacturer of a wrong *decision* and the cost to, say, a genetic researcher of a wrong *opinion.*) Finally, Neyman-Pearson introduced the concept of the power of a hypothesis test and noted how the cost of making observations (which depends on sample size) can be traded against the costs of type I or type II errors.

If you would like to find additional information on Pearson, consult the *Encyclopedia of Statistical Sciences* (New York: Wiley-Interscience, 1982–86), vol. 6, pp. 650–53.

Summary

1. When people make decisions, they inevitably do so on the basis of beliefs they hold concerning the true state of the world. Every one of these beliefs originates as a *hypothesis,* a proposition tentatively advanced as possibly true. *Hypothesis testing* is a systematic approach to assessing beliefs about reality; it is confronting a belief with evidence and deciding whether that belief can be maintained as reasonable or must be discarded as untenable. Four major steps are involved.

2. Step 1 is the formulation of two opposing hypotheses, called the *null hypothesis* (H_0) and the *alternative hypothesis* (H_A). They are mutually exclusive and also collectively exhaustive of the possible states of reality. These hypotheses can be stated in various forms. While the null hypothesis may be stated either as an exact or inexact one, the alternative hypothesis is always stated as an inexact one (two-sided or one-sided).

3. Step 2 is the selection of a *test statistic*—the statistic to be computed from a simple random sample taken from the population of interest and to be used for establishing the probable truth or falsity of the null hypothesis.

4. In Step 3, a *decision rule* is derived that specifies in advance for all possible values of a test statistic that might be computed from a sample whether the null hypothesis should be accepted or whether it should be rejected in favor of the alternative hypothesis. Such a rule is needed because the test statistic is a random variable; simply as a result of sampling error, it can take on all kinds of values apparently contradicting the null hypothesis even when H_0 is true. One escapes the dilemma by accepting the null hypothesis for all sample results (even seemingly contradictory ones) that are highly likely and nonsurprising when H_0 is true and by rejecting it for results that are highly unu-

sual and surprising when H_0 is true. The cutoff point between values of the test statistic that are "surprising" (that have a "low" probability of occurring when H_0 is true) and those that are "not surprising" (that have a "high" probability of occurring when H_0 is true) must necessarily be an arbitrary one. "Surprising" results that lead to the rejection of the null hypothesis are also called *statistically significant* results. "Nonsurprising" results that lead to the acceptance of the null hypothesis are called *not statistically significant* results. Among all the sample results that are possible when the null hypothesis is true, the (arbitrary) maximum proportion, α, that is considered sufficiently unusual to reject the null hypothesis is called the significance level of the test.

When the null hypothesis is rejected for values of the test statistic located in either tail of that statistic's sampling distribution (because the alternative hypothesis is two-sided), the hypothesis test is called a *two-tailed test*. When, instead, the null hypothesis is rejected only for very low (or very high) values of the test statistic located entirely in one tail of that statistic's sampling distribution (because the alternative hypothesis is one sided), the hypothesis test is called a *one-tailed test*. Such a test can be *lower-tailed* or *upper-tailed*, depending on the location of the rejection region.

5. In Step 4, a simple random sample is selected, the actual value of the test statistic is computed, and it is confronted with the decision rule. The null hypothesis, accordingly, is accepted or rejected. We must note, however, that either outcome can be a correct or an erroneous interpretation of the true state of reality. The erroneous rejection of a null hypothesis that is in fact true is called a *type I error;* it occurs with a probability of α. The erroneous acceptance of a null hypothesis that is in fact false is called a *type II error;* it occurs with a probability of β. Given sample size, *n,* anything that reduces α automatically raises β. The two complementary probabilities, $1 - \alpha$ (with respect to α) and $1 - \beta$ (with respect to β) are, respectively, referred to as the *confidence level* and the *power* of the hypothesis test.

6. The development of the theory of hypothesis testing began at least eight centuries ago.

7. The modern theory can be applied to both large-sample and small-sample situations.

8. Modern hypothesis-testing procedures are still subject to considerable controversy. Critics are concerned about serious violations of assumptions, a rigid interpretation of the "acceptance" criterion, the nonpublication of nonsignificant results, and more.

Key Terms

α risk
acceptance number
acceptance region
alternative hypothesis
β risk
confidence level of the hypothesis test
critical value
cutoff point
decision rule
error of acceptance
error of rejection
exact hypothesis
fiducial interval
hypothesis
hypothesis testing
inexact hypothesis
lower-tailed test
not statistically significant result

null hypothesis
observed significance level
one-sided hypothesis
one-tailed test
pooled variance
power of the hypothesis test
p value
rejection region
significance level of a hypothesis test
size of a hypothesis test
statistically significant result
test statistic
two-sided hypothesis
two-tailed test
type I error
type II error
upper-tailed test

Questions and Problems

The computer program noted at the end of this chapter can be used to work many of the subsequent problems.

1. Reconsider Practice Problem 1 in this chapter on p. 274. What would have been the result of that test if a significance level of $\alpha = .001$ had been set? Of $\alpha = .25$?

2. Reconsider Practice Problem 2 on p. 275. What would have been the result of that test if the sample size had been 500 rods? Or 16 rods?

3. The buyer of shirts for a department store wants to test whether shirts with sleeve labels of "33 inches" really meet that specification. A random sample of $n = 100$ from 10,000 incoming shirts is to be taken; the desired significance level is $\alpha = .025$. The sample shows a mean length of $\overline{X} = 34$ inches, with a standard deviation of $s = 2$ inches. Make the test.

4. An economist wants to test whether the average salary of aircraft mechanics really is $600 per month as has been alleged. A random sample of $n = 100$ from the nation's 29,952 mechanics is taken; the desired significance level is $\alpha = .05$. The sample shows a mean salary of $\overline{X} = \$657$, with a standard deviation of $s = \$22$. Make the test.

5. In order to rule on a potential federal grant, a government official must confirm that the average family income in an Illinois county is $12,357, as alleged. A random sample of $n = 36$ families from the county is taken; the desired significance level is $\alpha = .001$. The sample shows a mean income of $13,950, with a standard deviation of $3,972. Make the test.

6. The manager of a firm wants to test the advertising claim of a competitor: "There are 50 chunks of beef in every can of Eric's Homemade Stew." A random sample of 50 cans of Eric's stew is acquired; a hypothesis test at the $\alpha = .025$ level of significance is to be performed. The sample shows a mean of 48 chunks of beef per can, with a standard deviation of 2 chunks. Make the test.

7. The manager of an airport has the distinct impression that the monthly maintenance cost of planes used in agricultural work does not equal the projected $500 per plane. A sample of 32 planes is taken; a hypothesis test at the $\alpha = .02$ level of significance is to be performed on the assumption that the population of cost figures is normally distributed. The sample shows costs of $592 per plane, with a standard deviation of $101. Make the test.

8. The mayor wants to know whether mean apartment rents in the town really are now higher than last year's $375 per month. A random sample of 50 of the town's 450 apartments is taken; the desired significance level is $\alpha = .01$. The sample shows current mean rents of $\overline{X} = \$399$ per month; it is assumed that the current population standard deviation equals last year's: $\sigma = \$25$. Make the test, assuming a normally distributed population.

9. The manager of a retail business wonders whether people other than customers are using the store's parking lot. She wants to test, at a significance level of $\alpha = .05$, whether cars in the lot are typically parked for over an hour. A random sample of 50 cars shows a mean parking time of 101 minutes, with a standard deviation of 21 minutes.

10. The Environmental Protection Agency is allowing a plant to dump its effluent into a river—as long as the effluent contains a maximum of 4 parts per million (ppm) of a certain toxic substance. During the course of one week, the EPA randomly samples the effluent and finds, in $n = 64$ samples, an average of $\overline{X} = 4.2$ ppm of toxic substance, with a standard deviation of $s = 1$ ppm. Is the plant in violation? Conduct an appropriate hypothesis test at $\alpha = .025$.

11. The Federal Aviation Administration believes that the mean number of takeoffs and landings at American airports last year was at most 50 per day. Make an appropriate hypothesis test of this belief at the $\alpha = .001$ level of significance, while using these sample data: $n = 100$, $\overline{X} = 71$, $s = 30$. (There are 13,000 airports in the United States.)

12. Once again consider problems 3–6. In each case, compute a *confidence interval* for the mean in question, always at the $100 (1 - \alpha)$ percent level of confidence.

13. Once again consider problems 7–10. In each case, compute a *confidence interval* for the mean in question, always at the $100 (1 - \alpha)$ percent level of confidence.

14. The buyer of tomatoes for a ketchup producer wants to test whether it is true that 80 percent of the tomatoes being received deserve the label "superior." A random sample of $n = 100$ from 5,000 incoming tomatoes is to be taken; the desired significance level is $\alpha = .05$. The sample shows 72 "superior" tomatoes. Make the test.

15. An advertiser wants to test a magazine publisher's claim that 25 percent of the magazine's readers are college students. A random sample of $n = 200$ from 2.1 million subscribers shows that 38 are college students. Perform an appropriate hypothesis test at the $\alpha = .01$ level of significance.

16. The manager of a firm wants to test the accountant's claim that the firm's cash-flow problem is due to the fact that at least 80 percent of accounts receivable are more than three months old. A random sample of $n = 50$ from 10,000 accounts is to be taken; the desired significance level is $\alpha = .001$. The sample shows 30 such delinquent accounts. Make the test.

17. A Bible salesman has been told that at least 50 percent of all families in Atlanta attend religious services regularly. He makes a random sample of 25 families and finds that 10 of them attend regularly.
  a. Does the finding support the claim at the $\alpha = .05$ level of significance? Explain.
  b. What would your answer be if the same result had been obtained from a random sample of $n = 100$? Explain.

18. A presidential candidate has decided to enter primaries in those states in which at least 20 percent of the voters support the candidate. Random samples of $n = 200$ in four states show the following numbers of supporters: New Hampshire 25, Florida 44, Illinois 31, California 17. Prepare hypothesis tests at the $\alpha = .05$ level of significance to determine where the candidate should enter.

19. An aircraft manufacturer finds that 12 percent of engine crankcases shrink too much when cooling down after being cast. A new process is proposed and tried: of 250 castings, some 27 are equally bad. At the $\alpha = .05$ level of significance, conduct a hypothesis test to find out whether the new process produces crankcases that are, at worst, as bad as, but possibly better than, the old process.

20. The manufacturer of Lycoming aircraft engines knows from long experience that 10 percent of these engines develop problems by the time they reach 2,000 flight hours. The manufacturer believes, however, that a new series of these engines will do better. Accordingly, a random sample of 225 of such engines is selected from the 9,000 owners of them. Of the sampled owners, 198 had used the engine for at least 2,000 hours and 22 of these had had problems. Perform an appropriate hypothesis test at the $\alpha = .05$ level of significance and determine whether the manufacturer's faith in the new engine series is justified.

21. Just prior to the election of President Kennedy, a political analyst believed that 53 percent of the voters would vote for Kennedy. Yet a Gallup poll of 1,500 voters showed only 765 (that is, 51 percent) so inclined. Conduct a hypothesis test of the analyst's belief at the $\alpha = .05$ level of significance. Then compare your result with that for problem 25 in Chapter 7.

22. Reconsider Chapter 7's Practice Problem 9 on p. 249. In it, a tire manufacturer wishes to estimate the difference between the mean lives of two types of tires. A 99-percent confidence interval of the difference in tire A and tire B lifetimes was computed as 4,900 miles $\pm$ 507.52 (as reaching, that is, from 4,392.48 miles to 5,407.52 miles). Perform a hypothesis test on the same example and discover the relationship between estimation and hypothesis testing.

23. An economist wants to test the claim that the wages of construction workers in New York (A) are different from those in Chicago (B). Random samples of workers, $n_A = 100$ and $n_B = 80$, are to be taken; the desired significance level is $\alpha = .025$. The samples show average weekly wages of $\overline{X}_A = \$500$ and $\overline{X}_B = \$410$, with sample standard deviations of $s_A = \$200$ and $s_B = \$100$. Make the test, assuming normally distributed populations with equal variances.

24. A firm's manager wants to know whether the delivery time of raw materials is truly longer with one shipping company (A) than another (B). Random samples of $n_A = 50$ and $n_B = 30$ shipments from among the thousands received in a month are to be taken; the desired significance level is .10. The samples show mean delivery times of $\overline{X}_A = 14$ days and $\overline{X}_B = 12$ days, with sample standard deviations of $s_A = 1$ day and $s_B = .4$ day. Make the test.

25. A government administrator wants to know whether there is any difference in the quality of training given to new air traffic controllers at two different training centers. Exams are administered to two random samples of recent trainees. At center A, the sample is $n_A = 36$, the mean score is $\overline{X}_A = 93$, the standard deviation is $s_A = 8$. At center B, the corresponding data are $n_B = 49$, $\overline{X}_B = 87$, $s_B = 3$. Conduct a hypothesis test at the $\alpha = .005$ level of significance, assuming normally distributed populations with equal variances.

26. A psychologist wants to test whether age influences IQ. A random sample of 50 middle-aged individuals whose IQ score at age 20 is available is to be taken; the desired significance level is .001. The sample shows a mean difference between the current and earlier score of 12 points, with a sample standard deviation of 8 points. Make the test.

27. The owner of a sawmill is about to purchase one of two tree lots, but wonders which lot has the larger trees. Two random samples of $n_A = n_B = 100$ are taken. The average diameter of trees in lot A is $\overline{X}_A = 48.25$ inches with a standard deviation of $s_A = 18.2$ inches. The corresponding data for the sampled trees in lot B are $\overline{X}_B = 39.95$ inches and $s_B = 3.2$ inches. Conduct a hypothesis test to find out whether lot A trees are larger than lot B trees, using a significance level of $\alpha = .05$.

28. The IRS commissioner wants to test whether the quality of work done in two regional offices is the same. Random samples of $n_A = 300$ and $n_B = 200$ tax returns are taken from each office and checked for accuracy; the desired significance level is .01. The samples show 39 errors in office A, 20 in office B. Conduct the test.

29. The study described in Analytical Example 8.1, "Antitrust Pork Barrel," also showed the following for the House Subcommittee on Monopolies and Commercial Law of the House Judiciary Committee: during the 1970–79 period, 18 of 25 formal complaints brought by the FTC were dismissed when a committee member's district was affected; 46 of 171 such complaints were dismissed when no committee member's district was involved. Test the null hypothesis of "no difference" in these proportions at the 10-percent level of significance.

30. The manager of a pizza parlor is testing the proposition that more of the phone orders received during the day (A) than during the night (B) are in fact being picked up. In a three-month test, 250 each of day and night phone orders are sampled; the desired significance level is $\alpha = .05$. It turns out that 240 of the day orders but only 220 of the night orders were picked up. Make the test.

31. An economist is testing the proposition that a larger percentage of firms in industry A than in industry B is making advertising expenditures in excess of 5 percent of sales revenue. A random sample of 200 firms from each industry reveals percentages of 18 and 19, respectively. Conduct an appropriate hypothesis test at a significance level of $\alpha = .01$.

32. An advertising agency wishes to test the claim that a larger percentage of adult women watches *Dallas* than *The Guiding Light* on TV. One random sample of 150 women shows 102 watching *Dallas*; another random sample of 250 women shows 182 watching *The Guiding Light*. Conduct an appropriate hypothesis test at a significance level of $\alpha = .10$.

33. A quality inspector wants to test the claim that the proportion of acceptable electronics components delivered by a foreign supplier (A) is at best equal to that coming from a domestic supplier (B). Random samples of $n_A = 100$ and $n_B = 150$ are to be taken from incoming shipments; the desired significance level is $\alpha = .005$. The samples show 90 good components coming from A and 105 from B. Make the test.

34. A manufacturer has been receiving complaints from customers about their orders arriving fully 12 days after being shipped. The manufacturer selects 20 of the following week's orders at random and ships them differently. A statistician is to test whether the new procedure is better or worse and to do so at a significance level of $\alpha = .05$. The mean delivery time in the sample turns out to be $\overline{X} = 9$ days, with a sample standard deviation of $s = 3$ days. Make the test.

35. A researcher wants to determine whether rats (and maybe even people) can live normal lives entirely with artificial blood. At a significance level of .001, a test is to be conducted on 16 rats that would normally live another 5 months. The mean remaining lifetime of the tested rats, whose natural blood was replaced with artificial blood, turns out to be $\overline{X} = 4.1$ months, with a sample standard deviation of $s = 1.6$ months. Make the test.

36. The process of producing steel is designed to use up 6 quarts of hydrochloric acid per ton of steel. Yet a random sample of 9 different tons of steel shows a mean usage of 9 quarts per ton, with a standard deviation of 1 quart. At a significance level of $\alpha = .05$, conduct an appropriate hypothesis test to determine whether the production process needs recalibration.

37. The mean operating temperature of an aircraft engine during level cruise is supposed to be 190° F. One particular engine was tested at randomly selected times. The mean temperature on these 25 tests was 193° F, with a standard deviation of 3° F. Should the pilot begin to worry about the engine running hot? Conduct an appropriate hypothesis test at a significance level of $\alpha = .05$.

38. A manager wishes to test the tensile strength of yarn that is to be used in the firm's new machines and that must be at least 25 lb. A random sample of 16 spools are taken from a number of incoming shipments; the average tensile strength is 23 lb., with a standard deviation of .2 lb. Conduct an appropriate hypothesis test at a significance level of α = .10 and tell the manager whether the yarn is suitable.

39. A manager wishes to test the weight of crankcases intended for aircraft engines that must be no heavier than 105 lb. A random sample of 20 crankcases is taken from the day's output at the foundry; their average weight is 107 lb., with a standard deviation of .1 lb. Conduct an appropriate hypothesis test at a significance level of α = .01 and tell the manager whether the crankcases are suitable.

40. A developer believes that the labor hours required to build a new home average 1,000 hours at most. A random sample of 11 new homes is taken, the mean is found to be $\overline{X}$ = 1,230 hours, with a standard deviation of s = 470 hours. Conduct an appropriate hypothesis test at a significance level of α = .025 to see whether the developer is correct.

41. The manager noted in problem 38 now wishes to test whether the tensile strength of yarn is alike for two suppliers. In addition to the sample noted earlier (n_A = 16, $\overline{X}_A$ = 23 lb., s_A = .2 lb.), a second sample of yarns coming from another firm is taken and the results are: n_B = 26, $\overline{X}_B$ = 24 lb., s_B = 2.3 lb. Conduct an appropriate hypothesis test at the α = .10 level of significance, while assuming that the two populations of tensile strengths are normally distributed and have equal variances.

42. A student wants to test, at the α = .02 level of significance, whether the final exam scores given by two different instructors, A and B, are alike. A random sample of each class is taken: n_A = 10, $\overline{X}_A$ = 88.2, s_A = 8; n_B = 15, $\overline{X}_B$ = 82.7, s_B = 2. Conduct the test, assuming the two populations of scores to be normally distributed with equal variances.

43. Determine the significance levels of the following hypothesis tests.
   a. Two-tailed tests: $z_{\alpha/2}$ = 1.96, $z_{\alpha/2}$ = .85, $z_{\alpha/2}$ = .15
   b. One-tailed tests: z_α = 1.96, z_α = 1.25, z_α = .15

44. As was noted in Biography 8.1 (R. A. Fisher), many statisticians like to report a *p-value* for each hypothesis test. On the assumption that the null hypothesis is true, this *p*-value measures the probability of getting a test statistic as extreme as or more extreme than the one actually computed from the sample data. For a believer in H_0, a large *p*-value signals "no surprise" (and acceptance of H_0), but the smaller the *p*-value is, the greater is the surprise (and the more do sample data point to the rejection of H_0 and the acceptance of H_A). Thus, in Figure 8.A, the crosshatched area represents the (arbitrarily chosen) maximum for probability α (the probability of making the Type I error of falsely rejecting a null hypothesis that is true), while the crosshatched plus dotted area represents the *p*-value corresponding to an actually observed z-value of $-.82$ in the this lower-tailed test.

Compute similar *p*-values for the results in Practice Problem 2 and end-of-chapter problems 30–32.

Figure 8.A

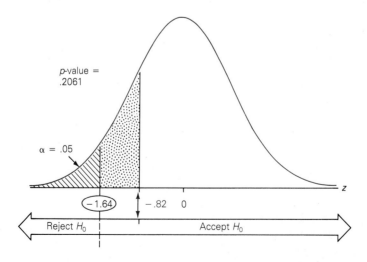

45. a. Similar to Figure 8.A, draw a graph depicting the difference between α value and p-value for a two-tailed hypothesis test.
   b. Then compute the p-values for Practice Problems 1 and and 4–7; for Analytical Example 8.1 (five parts); and for end-of-chapter problems 3– 7, 14, 15, 21–23, 25, 26, 28, and 29 (all of which are two-tailed tests).
   Hint: Part a *is* a difficult problem; if you can't do it, consult the answer at the back of the book, then do part b.

Selected Readings

Fisher, R. A. *Statistical Methods for Research Workers,* 14th ed. New York: Hafner Press, 1970. A classic work on statistical estimation, hypothesis testing, analysis of variance, correlation, and more.

Fisher, R. A. *The Design of Experiments,* 9th ed. New York: Hafner Press, 1971. A classic work on all aspects of experimentation, including randomized blocks, Latin squares, factorial designs, confounding, hypothesis testing, and more.

Fisher, R. A. *Statistical Methods and Scientific Inference,* 3rd ed. New York: Hafner, 1973.

Kohler, Heinz. *Statistics for Business and Economics,* 2nd ed., Glenview, Ill.: Scott, Foresman and Co., 1988. Chapter 9 discusses the relationship between error probabilities and sample size and introduces the operating characteristics curves of hypothesis tests.

Morrison, D., and R. E. Henkel. *The Significance Test Controversy.* Chicago: Aldine, 1970. Raises serious questions about the usefulness and sense of employing significance tests in social-science research.

Neyman, Jerzy, and E. S. Pearson. "On the Use and Interpretation of Certain Test Criteria for Purposes of Statistical Inference." *Biometrika,* 1928, pp. 175–240 and 263–94.

Neyman, Jerzy, and E. S. Pearson. "On the Problem of the Most Efficient Tests of Statistical Hypotheses." *Philosophical Transactions of the Royal Society,* 1933, pp. 289–337. This and the preceding entry are two crucial articles developing the modern theory of hypothesis testing.

Computer Program

The DAYSAL-2 personal-computer diskettes that accompany this book contain one program of interest to this chapter:

 9. Classical Hypothesis Tests lets you make large- and small-sample tests of the population mean or proportion or of the difference between two such means or proportions. You can select two-tailed, lower-tailed, or upper-tailed tests, and you can choose among numerous significance levels when you use this program to work problems.

CHAPTER NINE

Hypothesis Testing:
The Chi-Square Technique

Chapters 7 and 8 have amply demonstrated the importance of the normal probability distribution and of the t distribution for purposes of making estimates of unknown population parameters or testing hypotheses about them. There are occasions, however, when these distributions cannot be used; the *chi-square technique,* to be discussed in this chapter, is often employed in such circumstances. We will consider four major applications of this technique: (1) testing the alleged independence of two qualitative population variables, (2) making inferences about more than two population proportions, (3) making inferences about a population variance, and (4) conducting goodness-of-fit tests to assess the plausibility that sample data come from a population the elements of which conform to a specified type of probability distribution.

TESTING THE ALLEGED INDEPENDENCE OF TWO QUALITATIVE VARIABLES

Qualitative variables, as noted in Chapter 1, are characteristics (like type of business or job title) that are usually not expressed numerically because they differ in kind rather than in degree among the elementary units of a statistical population. There are many occasions when it is important to know whether two such variables are *statistically independent* of one another (when, therefore, the probability of the occurrence of one variable is unaffected by the occurrence of the other) or whether these variables, on the contrary, are *statistically dependent* (when, therefore, the probability of the occurrence of one *is* affected by the occurrence of the other). Consider how important it might be to know whether there exists a relationship of independence or dependence between heart attacks and lifestyle, between the occurrence of disease and having been vaccinated, between equipment failure and prior years of service, between the percentage of defective units produced and the time of day of their production, between the causes of airplane accidents and the extent of pilot qualification, between the educational level (or age or sex or geographic location) of consumers and their brand preferences, between the severity of auto accidents and the location of their occurrence, between people's major in college and their type of employment later on, between attitudes toward a piece of legislation and people's political affiliation, between sex and type of job held—the list is endless.

Statistical Independence Reviewed

Table 9.1 is designed to help us review the nature of statistical independence. We imagine that a census has revealed that 100 pilots are residing in a city and that they have been classified by sex (male, female) and by the type of license held (private, commercial, airline-transport). We note that the proportion of males in the entire pilot population (80/100) is the same as that among private pilots only (56/70) or among commercial pilots only (16/20) or among airline-transport pilots only (8/10). Similarly, the proportion of females in the entire pilot population (20/100) is identical with that in any of the three subcategories. Because sex and type of license held are, thus, statistically independent, the special multiplication law for independent events (Box 4.J on page 143) can be applied to any of the probabilities involved. For example, the probability of finding a male private pilot is simply the product of the probabilities of finding a male pilot and of finding a private pilot:

$$p(\text{male } and \text{ private pilot}) = p(\text{male pilot}) \cdot p(\text{private pilot})$$

$$\frac{56}{100} = \frac{80}{100} \cdot \frac{70}{100}.$$

Similarly, the probablity of finding a female airline-transport pilot is the product of the probabilities of finding a female pilot and of finding an airline-transport pilot:

$$p(\text{female } and \text{ airline-transport pilot}) = p(\text{female pilot}) \cdot p(\text{airline-transport pilot})$$

$$\frac{2}{100} = \frac{20}{100} \cdot \frac{10}{100}.$$

Table 9.1   Pilots Residing in City, Classified by Sex and by Type of License

This table exemplifies the statistical independence of two qualitative population variables, A and B (here sex and type of pilot's license held). When such independence exists, the proportion of units having one attribute (such as female sex) is the same in the total population as in any part of the population having the other attribute (such as commercial-pilot-license holder). Hence, the frequency for any particular pair of attributes can be found by multiplying the respective frequencies for the two individual attributes and dividing the product by the total number of units observed. Note how

$$\frac{20 \text{ female pilots} \times 20 \text{ commercial pilots}}{100 \text{ pilots}} = 4 \text{ female commercial pilots.}$$

(A) Sex	(B) Type of License			Total
	Private	Commercial	Airline-Transport	
Male	56	16	8	80
Female	14	4	2	20
Total	70	20	10	100

The foregoing discussion suggests an easy way of testing *census* data for statistical independence:

1. Set up a **contingency table,** such as Table 9.1, that classifies data according to two or more categories associated with each of two qualitative variables that may or may not depend on each other statistically. (Such a table, thus, shows all possible combinations or *contingencies,* which accounts for its name.)
2. For each cell of the table, check whether the joint probability (such as 56/100 for male *and* private pilot) equals the product of the corresponding unconditional probabilities (such as 80/100 for male pilot times 70/100 for private pilot). If the answer is yes, the characteristics in question are independent; if it is no, they are dependent.

Typically, however, population frequencies are unknown and only sample data are available. In that case, the type of test just suggested will not work because many a joint probability may then fail to equal the product of the relevant unconditional probabilities merely as a result of sampling error and not because of any statistical dependence between the two qualitative variables of interest in the population. It is in such circumstances that the chi-square technique finds its first application.

A Contingency-Table Test

Suppose the Federal Aviation Administration wanted to determine whether there existed any link between the official causes of aviation accidents and the type of license held by the involved pilots-in-command. The experience of the past five years, summarized in Table 9.2, might be viewed as a sample of all possible accidents. The observed frequencies given in the various cells of this table can subsequently be compared with the frequencies one would expect if the two variables in question (accident cause and type of license held) were statistically independent of one another. Such

Table 9.2 Five-Year Sample of Nonmilitary Aviation Accidents, Classified by Cause and by Type of Pilot's License: Observed Frequencies

This contingency table shows observed frequencies (typically symbolized by f_o) for various combinations of aviation-accident causes (A) and pilot's-license types (B). Setting up such a table is the first step in testing for independence between the two variables, A and B.

(A) Official Accident Cause	(B) License Held by Pilot-in-Command		Total
	Student or Private	Commercial or Airline-Transport	
(1) Thunderstorm	50	20	70
(2) Icing	40	10	50
(3) Equipment failure	25	6	31
(4) Controller error	25	5	30
(5) Pilot incapacitation	80	6	86
(6) Other	30	3	33
Total	250	50	300

Table 9.3   Five-Year Sample of Nonmilitary Aviation Accidents, Classified by Cause (A) and by Type of Pilot's License (B): Expected Frequencies, Assuming Independence Between A and B.

This contingency table shows expected frequencies (typically symbolized by f_e) for various combinations of aviation-accident causes (A) and pilot's license types (B) under the assumption of independence. Setting up such a table is the second step in testing for independence between the two variables, A and B.

(A) Official Accident Cause	(B) License Held by Pilot-in-Command		Total
	Student or Private	Commercial or Airline-Transport	
(1) Thunderstorms	[70(250)/300] = 58.33	[70(50)/300] = 11.67	70
(2) Icing	[50(250)/300] = 41.67	[50(50)/300] = 8.33	50
(3) Equipment failure	[31(250)/300] = 25.83	[31(50)/300] = 5.17	31
(4) Controller error	[30(250)/300] = 25.00	[30(50)/300] = 5.00	30
(5) Pilot incapacitation	[86(250)/300] = 71.67	[86(50)/300] = 14.33	86
(6) Other	[33(250)/300] = 27.50	[33(50)/300] = 5.50	33
Total	250	50	300

expected frequencies are shown in Table 9.3. On the assumption of independence between variables A and B, each of the expected frequencies in Table 9.3 has been calculated as

$$\text{expected frequency, } f_e = \frac{\text{row total} \times \text{column total}}{\text{sample size}}.$$

This procedure corresponds to that described in the caption of Table 9.1.

For purposes of conducting the test of independence, it is convenient to show both the observed and expected frequencies in the same table, which is accomplished in Table 9.4. This table, in turn, can be used to calculate the **chi-square statistic,** which was introduced in 1900 by Karl Pearson (see Biography 9.1 at the end of the chapter). The chi-square statistic is admirably suited to assess the issue of independence or dependence between qualitative variables. Contrary to the common practice of denoting population parameters by Greek letters and sample statistics by Roman letters, this statistic is denoted by the lowercase Greek letter χ, which is pronounced "ki" as in "kite." According to one commonly used formula (that is shown in Box 9.A), chi square (χ^2) equals the sum of all the ratios that can be constructed by taking the difference between each observed and expected frequency, squaring it, and dividing this squared deviation by the expected frequency.

9.A A Chi-Square Statistic

$$\chi^2 = \sum \frac{(f_o - f_e)^2}{f_e}.$$

where f_o is an observed frequency and f_e is the associated expected frequency.

Table 9.4   Five-Year Sample of Nonmilitary Aviation Accidents, Classified by Cause (A) and by Type of Pilot's License (B): Observed Frequencies vs. Expected Frequencies, Assuming Independence Between A and B.

This contingency table combines the information contained in Tables 9.2 and 9.3. Frequencies actually observed are shown in the upper-left portion of each cell; frequencies expected on the assumption of independence between variables A and B are shown in the lower-right portion.

(A) Official Accident Cause	(B) License Held by Pilot-in-Command		Total
	Student or Private (1)	Commercial or Airline-Transport (2)	
(1) Thunderstorm	50 / 58.33	20 / 11.67	70
(2) Icing	40 / 41.67	10 / 8.33	50
(3) Equipment failure	25 / 25.83	6 / 5.17	31
(4) Controller error	25 / 25.00	5 / 5.00	30
(5) Pilot incapacitation	80 / 71.67	6 / 14.33	86
(6) Other	30 / 27.50	3 / 5.50	33
Total	250	50	300

For our example, the value of χ^2 can be calculated as 14.88, which is shown in Table 9.5.

Note: The preceding χ^2 formula makes intuitive sense. We are interested in assessing the differences between the observed and expected; hence, it is reasonable to calculate the difference between f_o and f_e for each cell of the contingency table. Yet we are unable to reach an overall assessment by then summing these differences because such a sum necessarily equals zero, as Table 9.5 shows. For this reason, the deviations are squared. That procedure has the added advantage of magnifying large deviations (which are probably the result of some kind of dependence between the variables studied) relative to small deviations (which are more likely the result of sampling error). Consider, in Table 9.5, the deviation of 8.33 relative to that of .83 (a tenfold difference in magnitude). When squared, the numbers become 69.39 and .69 (a 100-

Table 9.5 The Calculation of Chi Square

This table shows the calculation of the chi-square statistic from the data embodied in Table 9.4. Note: because the χ^2 formula calls for the squaring of deviations between the observed and expected, χ^2 can never be negative but can take on values only of zero or above.

Row, Column	Observed Frequency, f_o	Expected Frequency (given independence), f_e	Deviation, $f_o - f_e$	Squared Deviation, $(f_o - f_e)^2$	Standardized Squared Deviation, $\dfrac{(f_o - f_e)^2}{f_e}$
(1, 1)	50	58.33	−8.33	69.39	1.19
(1, 2)	20	11.67	8.33	69.39	5.95
(2, 1)	40	41.67	−1.67	2.79	.07
(2, 2)	10	8.33	1.67	2.79	.33
(3, 1)	25	25.83	− .83	.69	.03
(3, 2)	6	5.17	.83	.69	.13
(4, 1)	25	25.00	0	0	0
(4, 2)	5	5.00	0	0	0
(5, 1)	80	71.67	8.33	69.39	.97
(5, 2)	6	14.33	−8.33	69.39	4.84
(6, 1)	30	27.50	2.50	6.25	.23
(6, 2)	3	5.50	−2.50	6.25	1.14
Total	300	300	0		$\chi^2 = 14.88$

fold difference in magnitude). Why, finally, does the formula divide each squared difference by the expected frequency? This division converts absolute squared deviations into relative ones and puts all the cells on an equal footing. Consider the first two rows of Table 9.5. The absolute value of the deviation (and, therefore, of its square) is the same in each case, but a deviation of 8.33 from an expected value of 58.33 is surely less significant than an identical deviation from an expected value of 11.67; this fact is evidenced by the different entries in the last column (1.19 versus 5.95).

Also note: If we had taken a different sample, all the observed frequencies would have been different, and a different value of χ^2 would have been calculated as well. Thus, it becomes crucial to know something about the sampling distribution of this statistic. That knowledge, in turn, will enable us to test the hypothesis of independence between any two qualitative variables that are of interest to us and to do so while specifying a maximum acceptable α risk (of rejecting a hypothesis of independence even though it is true) which, given sample size, implies a specific β risk (of accepting a hypothesis of independence even though it is false).

Sampling distributions of chi square. The nature of the sampling distribution of chi square depends on the number of degrees of freedom associated with the problem under investigation. This number equals the number of expected-frequency values one is free to set before the constraints of the problem dictate the remaining values. Consider Table 9.4. Given the column and row totals (which are the constraints of that problem), we are free to set only any 5 of the 12 expected-frequency values. Once we have calculated, say, the first five f_e values in column (1)—once we have calculated, that is, 58.33, 41.67, 25.83, 25.00, and 71.67—all the remaining f_e values are automatically determined by the constraints. Because the first column sums to 250, the last f_e value of column (1) must equal 27.50. Given the six column (1) f_e values and the row totals, the remaining column (2) f_e values, in turn, must necessarily be precisely what the table shows them to be. Thus, the problem involves 5 degrees of freedom. In a contingency-table test, this number can most easily be found as shown in Box 9.B. In our example, this number, of course, is

$$(6 - 1) \cdot (2 - 1) = 5.$$

9.B Number of Degrees of Freedom (d.f.) in a Contingency-Table Test

$$d.f. = (\text{number of rows} - 1) \cdot (\text{number of columns} - 1)$$

Because the χ^2 random variable is a sum of squares and if certain conditions are met, the sampling distribution of the χ^2 statistic for n degrees of freedom can be represented by the probability distribution of the sum of the squares of n independent standard normal random variables and is then referred to as a **chi-square distribution** (with n degrees of freedom). The conditions just referred to are (1) a sample sufficiently large so that each expected frequency equals 5 or more (in a contingency-table test, this result can often be assured by combining two or more rows or columns, as will be shown below) and (2) an underlying population that is normally distributed.

To understand what a chi-square distribution is, consider a population of values, x, that are normally distributed. The corresponding standard normal deviates, $z = [(x - \mu)/\sigma]$, would be normally distributed as well. Yet the squares of these deviates, z^2, would not be normally distributed. Their distribution would be continuous and unimodal, but strongly skewed to the right, with values ranging from zero to positive infinity. Under the conditions just cited, this distribution would equal the sampling distribution of χ^2 with 1 degree of freedom.

Now consider two independent standard normal random variables, z_a and z_b. If we squared each and added the squares, the distribution of this expression $(z_a^2 + z_b^2)$ would be similar to the skewed distribution just discussed and, again under the cited conditions, would equal the sampling distribution of χ^2 with 2 degrees of freedom. In the same way, we can imagine finding the probability distributions for the sum of 4, 10, 20, or any other number of squared z values; in each case, these distributions would correspond (under the cited conditions) to the sampling distribution of χ^2 with 4, 10, 20, or any other number of degrees of freedom. Figure 9.1 shows such chi-square distributions for 2, 4, 10, and 20 degrees of freedom. What is not shown in the graph is that the expected value or mean of every chi-square distribution equals the number of degrees of freedom, while the variance equals twice that number, as shown in Box 9.C on page 310.

Figure 9.1 Four Members of the Chi-Square Distribution Family

A different chi-square distribution exists for each possible number of degrees of freedom *(d.f.).* Because the χ^2 random variable is a sum of squares, negative values are impossible; all values, therefore, range between zero and positive infinity. Even though all chi-square distributions are, thus, skewed to the right, the extent of skewness decreases with increasing numbers of degrees of freedom. Chi-square distributions with 30 degrees of freedom or more very closely approach the normal distribution, a fact that is already evident by comparing panels (a) through (d).

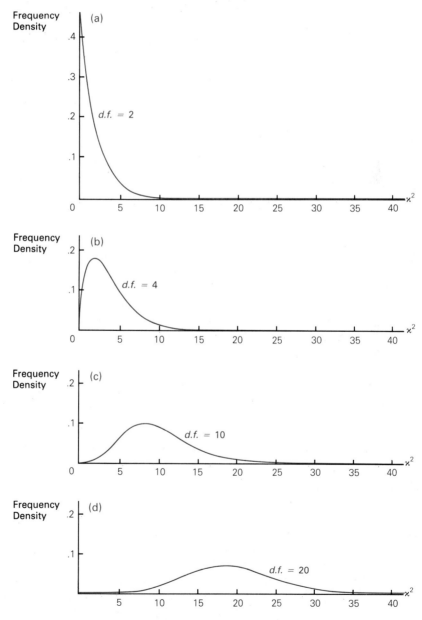

9.C Summary Measures of the Chi-Square Distribution

$$E_{\chi^2} = \mu_{\chi^2} = d.f.$$

$$\sigma^2_{\chi^2} = 2 \, d.f.$$

$$\sigma_{\chi^2} = \sqrt{2 \, d.f.}$$

where μ_{χ^2} is the mean, $\sigma^2_{\chi^2}$ is the variance, and σ_{χ^2} is the standard deviation of the chi-square distribution, while d.f. are the associated degrees of freedom.

The chi-square table. For the convenience of conducting χ^2 tests, critical values of χ^2 that correspond to specified upper-tail areas of the distribution and to specified numbers of degrees of freedom have been tabulated, as in Appendix Table H. Consider our example that involves 5 degrees of freedom. As the fifth row of Appendix Table H shows, .99 of the area under the appropriate chi-square distribution lies to the right of $\chi^2 = .554$; .98 of the area lies to the right of $\chi^2 = .752$, and so on, until .001 of the area lies to the right of $\chi^2 = 20.517$. If we cared to specify a significance level of $\alpha = .05$, the critical chi-square value would be $\chi^2_\alpha = 11.070$. This value can also be designated as $\chi^2_{.05,5}$ to remind us instantly of the significance level ($\alpha = .05$) and the degrees of freedom (5 *d.f.*). Our choice of $\alpha = .05$ implies that, due to sampling error, only 5 percent of the χ^2 values we would compute by taking repeated samples from a population for which the hypothesis of independence was true would exceed our critical value of 11.07. In our example, the computed value of χ^2 equals 14.88 (see Table 9.5). Thus, we are justified in suspecting that more than sampling error is involved and in *rejecting* the hypothesis of independence between accident cause and type of pilot's license held. But remember: our willingness to live with an α risk of .05 implies a willingness to make such a rejection erroneously at most 5 percent of the time when using this procedure over and over again.

Summary of procedure. We can summarize the procedure just introduced by showing how it relates to the hypothesis-testing steps introduced in Chapter 8.

Step 1: Formulating two opposing hypotheses.

H_0: The two variables (accident cause and type of pilot's license) are independent.
H_A: The two variables are dependent.

Step 2: Selecting a test statistic. We select

$$\chi^2 = \sum \frac{(f_o - f_e)^2}{f_e}.$$

Step 3: Deriving a decision rule. We set a significance level of $\alpha = .05$; hence (for 5 degrees of freedom), the critical value of χ^2_α is $\chi^2_{.05,5} = 11.070$ (Appendix Table H). Thus, the decision rule must be: "Accept H_0 if $\chi^2 \leq 11.070$." Note: χ^2 tests of independence are always upper-tailed because a perfect fit between f_o and f_e makes $\chi^2 = 0$. Because deviations between f_o and f_e are being squared, the greater the deviations are (positive or negative), the greater becomes the computed value of χ^2. The decision rule is illustrated in Figure A; the critical value is encircled.

Figure A

Step 4: Selecting a sample, computing the test statistic, and confronting it with the decision rule. After taking the sample, as noted above (Table 9.2), an actual χ^2 of 14.88 is calculated (Table 9.5). This value corresponds to the colored arrow in Figure A; it suggests that the null hypothesis should be *rejected.* At the 5-percent significance level, it appears that the accident cause and the type of pilot's license are statistically *dependent* variables. The following section considers another example of a χ^2 test for independence.

Practice Problem 1

Sex and Brand Preferences

An advertising agency wants to know whether the sex of consumers is independent of their preferences for 4 brands of coffee. (The answer will determine whether different ads must be created for men's and women's magazines.) A test of independence is to be conducted at the 5-percent significance level using a simple random sample of 100 persons.

Step 1: Formulating two opposing hypotheses.

H_0: Sex and brand preference are independent variables.
H_A: Sex and brand preference are dependent variables.

Step 2: Selecting a test statistic:

$$\chi^2 = \sum \frac{(f_o - f_e)^2}{f_e}.$$

Step 3: Deriving a decision rule. Given $\alpha = .05$ and 3 degrees of freedom (2 sexes and 4 brands make for a 2-row, 4-column contingency table and, hence, 1×3 degrees of freedom), the critical $\chi^2_{.05,3} = 7.815$ (Appendix Table H), making the decision rule: "Accept H_0 if $\chi^2 \leq 7.815$."

Step 4: Selecting a sample, computing the test statistic, and confronting it with the decision rule. The sample reveals, we assume, the frequencies given in Table A. Thus,

x^2 is calculated as shown in Table B. Note that some categories have been combined in order to avoid expected frequencies of less than 5 (in accordance with the rule noted earlier on page 308). The result suggests that the null hypothesis should be *rejected.* Sex and coffee brand preference are statistically dependent variables.

Table A

	Brand Preference				
Sex	A	B	C	D	Total
(1) Male	18	25	15	2	60
(2) Female	32	5	1	2	40
Total	50	30	16	4	100

Table B

Row, Column	Observed Frequency f_o	Expected Frequency (if H_0 is true), f_e	$f_o - f_e$	$(f_o - f_e)^2$	$\dfrac{(f_o - f_e)^2}{f_e}$
(1, A)	18	30.0	−12	144	4.80
(1, B)	25	18.0	7	49	2.72
(1, C)	15 ⎫ 17	9.6 ⎫ 12	5	25	2.08
(1, D)	2 ⎭	2.4 ⎭			
(2, A)	32	20.0	12	144	7.20
(2, B)	5	12.0	−7	49	4.08
(2, C)	1 ⎫ 3	6.4 ⎫ 8	−5	25	3.13
(2, D)	2 ⎭	1.6 ⎭			
Total	100	100	0		$x^2 = 24.01$

III

MAKING INFERENCES ABOUT MORE THAN TWO POPULATION PROPORTIONS

As the previous examples have shown, when expected frequencies are being calculated for the various cells of a contingency table, one effectively assumes that certain *proportions* (such as those of males among all pilots or of females among all consumers) are the same for all the categories of some variable (such as type of pilot's license held or preferred brand of product). It is not surprising, therefore, that the same technique that allows us to test for the independence of two qualitative population variables can also be used to test whether a number of population proportions are equal to each other or are equal to any predetermined set of values. The next two sections provide illustrations.

Case 1

Are Population Proportions Equal to Each Other?

Consider the process of assembling television sets. Management may be interested in testing the hypothesis that the proportion of defective units produced (which has been .05 in the past) would be the same for each of 6 possible assembly-line speeds. The company statistician may be asked to perform a test at the 1-percent significance level, taking 6 samples of 100 TV sets each, while different assembly-line speeds are being maintained. The statistical procedure may well be the following four steps.

Step 1: Formulating two opposing hypotheses.

H_0: The population proportion of defectives is the same for each of 6 assembly-line speeds; that is, $\pi_1 = \pi_2 = \pi_3 = \pi_4 = \pi_5 = \pi_6 = .05$.

H_A: The population proportion of defectives is not the same for each of 6 assembly-line speeds; that is, at least one of the equalities in H_0 does not hold.

Step 2: Selecting a test statistic:

$$\chi^2 = \sum \frac{(f_e - f_e)^2}{f_e}.$$

Step 3: Deriving a decision rule. Given $\alpha = .01$ and 5 degrees of freedom (2 possible product qualities and 6 possible assembly-line speeds make for 1×5 *d.f.*), the critical $\chi^2_{.01,5} = 15.086$ (Appendix Table H), making the decision rule: "Accept H_0 if $\chi^2 \leq 15.086$."

Step 4: Selecting the samples, computing the test statistic, and confronting it with the decision rule. The samples reveal, we assume, the observed frequencies given in the upper-left corners of the cells in combined contingency Table C, while the expected frequencies (assuming H_0 is true) are given in the lower-right corners. Accordingly, the value of the χ^2 statistic is calculated as shown in Table D. As a result, H_0 should be *accepted* at the 1-percent level of significance. The population proportion of defectives is the same for each assembly-line speed tested. |||

Table C

Product Quality	Assembly-Line Speed (units per hour)						
	A = 60	B = 70	C = 80	D = 90	E = 100	F = 110	Total
(1) Defective	6 / 5	4 / 5	5 / 5	5 / 5	6 / 5	4 / 5	30
(2) Acceptable	94 / 95	96 / 95	95 / 95	95 / 95	94 / 95	96 / 95	570
Total	100	100	100	100	100	100	600

Table D

Row, Column	Observed Frequency, f_o	Expected Frequency (if H_0 is true), f_e	$f_o - f_e$	$(f_o - f_e)^2$	$\dfrac{(f_o - f_e)^2}{f_e}$
(1, A)	6	5	1	1	.20
(1, B)	4	5	−1	1	.20
(1, C)	5	5	0	0	0
(1, D)	5	5	0	0	0
(1, E)	6	5	1	1	.20
(1, F)	4	5	−1	1	.20
(2, A)	94	95	−1	1	.01
(2, B)	96	95	1	1	.01
(2, C)	95	95	0	0	0
(2, D)	95	95	0	0	0
(2, E)	94	95	−1	1	.01
(2, F)	96	95	1	1	.01
Total	600	600	0		$\chi^2 = .84$

Case 2

Are Population Proportions Equal to a Given Set of (Unequal) Values?

Consider three firms, A, B, and C, that currently hold shares of 10, 40, and 50 percent of the market for a particular product. The smallest firm (A) wants to change the design of its product and wonders whether the subsequent market shares will be the same as before. A statistical test is to be conducted at the 5-percent significance level after asking a random sample of 300 consumers to indicate preferences among the newly designed product of A and the traditional products of B and C.

Step 1: Formulating two opposing hypotheses.

H_0: The population proportions (market shares) will be the same as in the past; that is, $\pi_A = .10$, $\pi_B = .40$, $\pi_C = .50$.

H_A: The population proportions will not be the same as in the past; that is, they will differ from those given in H_0.

Step 2: Selecting a test statistic:

$$\chi^2 = \sum \frac{(f_o - f_e)^2}{f_e}.$$

Step 3: Deriving a decision rule. Given $\alpha = .05$ and 2 degrees of freedom, the critical $\chi^2_{.05,2} = 5.991$ (Appendix Table H), making the decision rule: "Accept H_0 if $\chi^2 \leq 5.991$."

Step 4: Selecting the sample, computing the test statistic, and confronting it with the decision rule. The sample reveals, we assume, the observed frequencies shown in Table E; the expected frequencies are calculated on the assumption of the null hypothesis

Table E

	Number Preferring Product				$\dfrac{(f_o - f_e)^2}{f_e}$
Product	Observed Frequency, f_o	Expected Frequency (if H_0 is true), f_e	$f_o - f_e$	$(f_o - f_e)^2$	
new A	90	300(.10) = 30	60	3600	120.0
old B	90	300(.40) = 120	−30	900	7.5
old C	120	300(.50) = 150	−30	900	6.0
Total	300	300	0		$\chi^2 = 133.5$

being true. Accordingly, H_0 should be *rejected.* The market shares are unlikely to be the same as in the past.

    Note: Another example of this procedure is provided by Analytical Example 9.1, "Were Mendel's Data Fudged?" |||

MAKING INFERENCES ABOUT A POPULATION VARIANCE

We noted in Chapter 7 that the sample variance, s^2, is an unbiased estimator of the population variance, σ^2, provided that the selections of sample elements are statistically independent events. Prior to the selection of an actual sample, the sample variance, like every sample summary measure, is a random variable. As it turns out, probabilities concerning the sample variance can be established with the help of χ^2 distributions because the sample variance, s^2, which equals $(\Sigma X^2 - n\bar{X}^2)/n - 1$ (see Box 3.M on p. 98), can easily be converted into a chi-square random variable with $n - 1$ degrees of freedom, as shown in Box 9.D.

9.D Alternative Chi-Square Statistic

$$\chi^2 = \frac{s^2(n - 1)}{\sigma^2}.$$

where s^2 is the sample variance, n is sample size, and σ^2 is the population variance.

Probability Intervals for the Sample Variance

Given the Box 9.D expression of χ^2, one can quickly establish a probability interval for s^2 with the help of Appendix Table H. Consider, for example, a pharmaceutical firm that produces tablets the weights of which are normally distributed with a population standard deviation of $\sigma = 2$ milligrams (hence, the population variance is $\sigma^2 = 4$). If a simple random sample of 25 tablets is taken from this population, what is the probability of finding a sample variance between 3 and 5? For the lower limit of $s_L^2 = 3$, we get a lower χ^2 value of

$$\chi_L^2 = \frac{3(24)}{4} = 18.$$

For the upper limit of $s_U^2 = 5$, we get an upper χ^2 value of

$$\chi_U^2 = \frac{5(24)}{4} = 30.$$

The probability sought, accordingly, equals the probability of finding χ^2 values within the range of 18 to 30. An inspection of the appropriate row of Appendix Table H (for 24 degrees of freedom) reveals that an upper-tail area of slightly more than .8 lies to the right of $\chi^2 = 18$, while an upper-tail area of slightly less than .2 lies to the right of $\chi^2 = 30$. Thus, an area of about $.8 - .2 = .6$ lies between these values; this value of about .6 is also the probability of finding a sample variance between 3 and 5 in our example.

The same procedure can also be used to determine the limits below and above which lie specified percentages (say, 5 percent each) of all possible s^2 values. For our example (with 24 degrees of freedom), the lower χ^2 value would then be $\chi_{.95,24}^2 = 13.848$, and the upper χ^2 value would be $\chi_{.05,24}^2 = 36.415$. Accordingly, the lower s^2 limit would be

$$s_L^2 = \frac{\sigma^2 \cdot \chi_L^2}{n - 1} = \frac{4(13.848)}{24} = 2.31$$

while the upper s^2 limit would be

$$s_U^2 = \frac{\sigma^2 \cdot \chi_U^2}{n - 1} = \frac{4(36.415)}{24} = 6.07.$$

Thus, 5 percent of all s^2 values would lie below 2.31, another 5 percent would lie above 6.07, and 90 percent would lie between these values. The general formula for establishing a probability interval for the sample variance is shown in Box 9.E.

9.E Probability Interval for the Sample Variance, s^2

$$\frac{\sigma^2 \cdot \chi_L^2}{n - 1} \leq s^2 \leq \frac{\sigma^2 \cdot \chi_U^2}{n - 1}$$

where σ^2 is the population variance, n is sample size, and χ_L^2 and χ_U^2 are lower and upper chi-square values.

Confidence Intervals for the Population Variance

Unlike in the example just employed, the population variance, σ^2, is typically unknown and the problem is to make an estimate of it. In that case, the procedure used above can easily be adjusted to yield a confidence interval estimate of σ^2. By appropriately rewriting the expression found in Box 9.E, we get that in Box 9.F.

> ### 9.F Confidence Interval for the Population Variance, σ^2
>
> $$\frac{s^2(n - 1)}{\chi_U^2} \leq \sigma^2 \leq \frac{s^2(n - 1)}{\chi_L^2}$$
>
> where χ_U^2 = value of χ^2 variable with $n - 1$ degrees of freedom such that larger values have a probability of $\alpha/2$; χ_L^2 = value of χ^2 variable with $n - 1$ degrees of freedom such that smaller values have a probability of $\alpha/2$; and $\alpha = 1$ − confidence level.

Consider again our pharmaceutical firm, but this time imagine that it wants to *estimate* the variance in the population of tablet weights. A sample of 30 tablets might yield a sample variance, $s^2 = 3$. What would a 90-percent confidence interval of σ^2 be?

Given 29 degrees of freedom, and $\alpha = .10$, Appendix Table H yields $\chi_U^2 = \chi_{.05,29}^2 = 42.557$ and $\chi_L^2 = \chi_{.95,29}^2 = 17.708$. Accordingly, the interval sought is

$$\frac{3(29)}{42.557} \leq \sigma^2 \leq \frac{3(29)}{17.708}$$

or

$$2.04 \leq \sigma^2 \leq 4.91.$$

With 90-percent confidence, the population variance can be said to lie between 2.04 and 4.91 milligrams squared; hence, the population standard deviation is between 1.43 and 2.22 milligrams, the square roots of these values.

Testing Hypotheses About the Population Variance

It is easy to adapt the foregoing technique further in order to test various hypotheses about the population variance. These hypotheses can be one-sided or two-sided and, in the former case, lower-tailed or upper-tailed (just as noted in Chapter 8 with respect to other population parameters). Three examples will illustrate the possibilities.

Practice Problem 2

A Bank's Waiting-Line Policy (a lower-tailed test)
The manager of a bank is thinking of introducing a "single-line" policy (that directs all customers to enter a single waiting line in the order of their arrival and that, in turn, "feeds" them to different tellers as they become available). Although such a policy does not change the average time customers must wait, the manager prefers it because it decreases waiting-time variability. The manager's critics, however, claim that this variability will be at least as great as for a policy of multiple independent lines (which in the past had a standard deviation of $\sigma_0 = 8$ minutes per customer). A hypothesis test at the 2-percent significance level is to settle the issue. It is to be based on the experience of a random sample of 30 customers subjected to the new policy.

Step 1: Formulating two opposing hypotheses.

$H_0: \sigma^2 \geq 64$
$H_A: \sigma^2 < 64$

Note: Because σ_0 has equaled 8 in the past, the hypothesized value of the population variance, σ_0^2, becomes 64.

Step 2: Selecting a test statistic. The bank's statistician selects

$$\chi^2 = \frac{s^2(n-1)}{\sigma_0^2}.$$

Step 3: Deriving a decision rule. Given a desired significance level of $\alpha = .02$ and 29 degrees of freedom, Appendix Table H suggests a critical value of $\chi^2_{.98,29} = 15.574$ (this being a lower-tailed test). Thus, the decision rule must be: "Accept H_0 if $\chi^2 \geq 15.574$." The critical value is encircled in Figure B.

Figure B

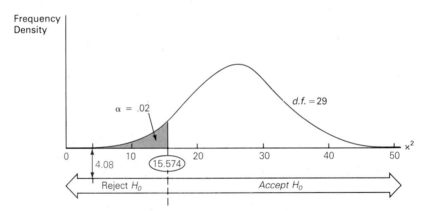

Step 4: Selecting a sample, computing the test statistic, and confronting it with the decision rule. After taking a sample of 30 customers, the statistician finds the sample single-line waiting times to have a standard deviation of $s = 3$ minutes per customer. Accordingly, the computed value of the test statistic equals:

$$\chi^2 = \frac{s^2(n-1)}{\sigma_0^2} = \frac{3^2(30-1)}{64} = 4.08.$$

This value corresponds to the colored arrow in Figure B; it suggests that the null hypothesis should be *rejected*. At the 2-percent significance level, the sample result is statistically significant. The observed divergence from the hypothesized value of $\sigma_0 = 8$ minutes is unlikely to be the result of chance factors operating during sampling; it is more likely to be the result of the manager being right: a single line feeding to many tellers does reduce waiting-time variability. III

Altimeter Readouts (an upper-tailed test)

A manufacturer of precision instruments is bringing out a new model of a radar altimeter (an instrument that measures aircraft height above the ever-changing ground rather than above mean sea level). The standard deviation of altitude readouts on the old model was $\sigma_0 = 5$ feet. A hypothesis test at the 5 percent significance level is to be conducted concerning management's claim that the readout variability on the new model is equal to or less than that on the old model. The test involves getting 20 readings from a known height above the ground.

Step 1: Formulating two opposing hypotheses.

$H_0: \sigma^2 \leq 25$
$H_A: \sigma^2 > 25$

Step 2: Selecting a test statistic. The company statistician selects

$$\chi^2 = \frac{s^2(n-1)}{\sigma_0^2}.$$

Step 3: Deriving a decision rule. Given a desired significance level of $\alpha = .05$ and 19 degrees of freedom, Appendix Table H suggests a critical value of $\chi^2_{.05,19} = 30.144$ (this being an upper-tailed test). Thus, the decision rule must be: "Accept H_0 if $\chi^2 \leq 30.144$." The critical value is encircled in Figure C.

Figure C

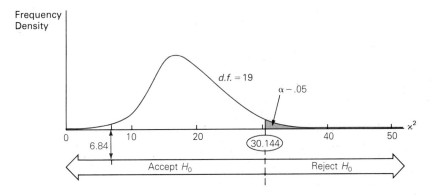

Step 4: Selecting a sample, computing the test statistic, and confronting it with the decision rule. After taking the 20 readings, an engineer finds them to have a standard deviation of $s = 3$ feet. Accordingly, the computed value of the test statistic equals:

$$\chi^2 = \frac{s^2(n-1)}{\sigma_0^2} = \frac{3^2(20-1)}{25} = 6.84.$$

This value corresponds to the colored arrow in Figure C; it suggests that the null hypothesis should be *accepted*. At the 5-percent significance level, the sample result is not statistically significant. The new model is equal to or better than the old one, as claimed. |||

||| Practice Problem 4

The Diameters of Fuel-Tank Lids (a two-tailed test)

An aircraft manufacturer is concerned about variability in the diameters of lids used to seal the fuel tanks that are located inside aircraft wings. Only a narrow range of diameters is acceptable. Lids that fit too tightly prevent air from entering the tanks as the fuel is being used, creating a vacuum and, ultimately, causing collapse of the wing structure. Lids that fit too loosely can allow fuel to be sucked out of the tank during flight, which is equally undesirable from the point of view of flight safety. A test at the 2-percent significance level is to be conducted with a random sample of 20 fuel-tank lids to see whether the population variance of lid diameters equals .0001 inches squared, as specified by engineers.

Step 1: Formulating two opposing hypotheses.

$$H_0: \sigma^2 = .0001$$
$$H_A: \sigma^2 \neq .0001$$

Step 2: Selecting a test statistic. The company statistician selects

$$\chi^2 = \frac{s^2(n - 1)}{\sigma_0^2}.$$

Step 3: Deriving a decision rule. Given a desired significance level of $\alpha = .02$ and a two-tailed test, a lower and upper critical χ^2 value must be established (so that .01 of the area under the χ^2 distribution lies below and above each of these values). According to Appendix Table H (and for 19 degrees of freedom) these values are $\chi^2_{.99,19} = 7.633$ and $\chi^2_{.01,19} = 36.191$. Thus, the decision rule must be: "Accept H_0 if $7.633 \leq \chi^2 \leq 36.191$." The critical values are encircled in Figure D.

Figure D

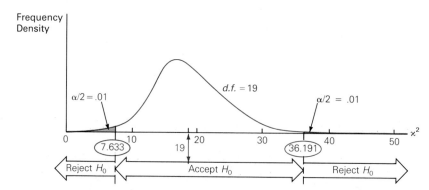

Step 4: Selecting a sample, computing the test statistic, and confronting it with the decision rule. After measuring the 20 diameters, an engineer finds them to have a standard deviation of $s = .01$ inches. Accordingly, the computed value of the test statistic equals:

$$\chi^2 = \frac{s^2(n - 1)}{\sigma_0^2} = \frac{(.01)^2(19)}{.0001} = 19.$$

This value corresponds to the colored arrow in Figure D; it suggests that the null hypothesis should be *accepted*. At the 2-percent significance level, the sample result is not statistically significant. The lids do meet specifications. ‖‖

CONDUCTING GOODNESS-OF-FIT TESTS

A **goodness-of-fit test** is a statistical test to determine the likelihood that sample data have been generated from a population that conforms to a specified type of probability distribution. For this purpose, the test compares the entire shapes of two (discrete or continuous) probability distributions: one describing known sample data and the other one describing hypothetical population data. The aim of the test might be limited to identifying only the family to which the underlying distribution belongs, or it might go further, seeking even to identify a particular member of that family. Accordingly, the null hypothesis to be tested might be rather general, such as "The sample data come from a population that is normally distributed." It can also be more specific: "The sample data come from a normally distributed population with a mean of 100 and a standard deviation of 10." In every case, the alternative hypothesis claims that "the sample data come from some other kind of population."

Note: In this type of test, a null hypothesis can clearly be false in many different ways. In our example, the sample data might be coming from all sorts of populations that are not normally distributed or even from one that is normally distributed but has population parameters that differ from those specified. As a result, it is difficult to calculate the β risk (of accepting a null hypothesis that is false) for a goodness-of-fit test, unless one first specifies in what particular way the null hypothesis is false. The best approach to protecting against a type II error, therefore, lies in taking a fairly large sample.

It is easy to see why one might want to conduct a goodness-of-fit test: a knowledge of the underlying population distribution is crucial whenever the statistical procedures used rely on sample data. Someone about to construct a table of random numbers, for example, would want to be sure that the population of numbers from which table entries were being selected was uniformly distributed. Someone who wanted to build a queuing model might want to be sure that the underlying population values were Poisson-distributed. Someone engaged in small-sample hypothesis testing with the help of the t distribution would grind out nonsense unless the underlying population values were normally distributed. In the following sections, we will consider a number of examples of how observed frequency distributions based on sample data can be used to make inferences about the distribution of the underlying population data. Given a single sample, a perfect match between the two is highly unlikely, of course; but on the average, when many samples are taken, we expect that the sample data will reveal to us the nature of the population distribution.

Goodness of Fit to the Binomial Distribution

The following example illustrates the procedure of testing whether sample data come from a binomially distributed population.

| | Practice Problem 5

Housing-Code Violations

A city inspector investigates the compliance of landlords with six conditions specified in the housing code. A simple random sample of 200 apartments reveals the data of Table F. The inspector wants to conduct a hypothesis test at the 5-percent significance level to determine whether the sample comes from a population in which the number of actual violations per apartment (out of six possible violations) is a binomially distributed random variable. Accordingly, the inspector proceeds with Steps 1 to 4.

Table F

Number of Possible Violations per Apartment	Observed Frequency, f_o
0	31
1	51
2	70
3	32
4	9
5	5
6	2
Total	200

Step 1: Formulating two opposing hypotheses.

H_0: The number of violations per apartment in the population of all city apartments is binomially distributed with a probability of success in any one trial of $\pi = .3$. [Note: the inspector estimates the parameter, π, by noting that the mean of a binomial random variable is $n \cdot \pi$ (see Box 5.C on p. 184) and that the mean number of violations per apartment in the sample equals

$$n \cdot \pi = \frac{0(31) + 1(51) + 2(70) + 3(32) + 4(9) + 5(5) + 6(2)}{200} = 1.8$$

Hence, π can be estimated as $1.8/6 = .3$]

H_A: The number of violations per apartment in the population of all city apartments is not correctly described by H_0.

Step 2: Selecting a test statistic. The inspector selects

$$\chi^2 = \sum \frac{(f_o - f_e)^2}{f_e}.$$

Step 3: Deriving a decision rule. The inspector first establishes in Table G the expected frequencies for the various numbers of possible violations per apartment, assuming H_0 is true. This process involves finding the binomial probabilities for $n = 6$ and $\pi = .3$ from Appendix Table C and multiplying each by the total number of observations (of 200). Subsequently, if expected frequencies in any class are less than 5 (which is true for 5 and 6 violations per apartment in this example), adjacent classes are combined as needed to yield minimum expected frequencies of 5 in every class (a requirement for using the χ^2 test that was first noted on page 308). The number of degrees of freedom must be established next.

Table G

Number of Possible Violations per Apartment	Binomial Probability, p, for $n = 6$ and $\pi = .3$	Expected Frequency (if H_0 is true), $f_e = 200p$
0	.1176	23.52
1	.3025	60.50
2	.3241	64.82
3	.1852	37.04
4	.0595	11.90 ⎫
5	.0102	2.04 ⎬14.04
6	.0007	.14 ⎭
Total		200*

*allowing for rounding error.

In goodness-of-fit tests, the number of degrees of freedom equals the number of classes, k, for which observed and expected frequencies are to be compared (possibly after some adjacent classes have been combined to assure that all expected frequencies are at least equal to 5) minus 1 minus the number of population parameters that are being estimated (see Box 9.G).

9.G Number of Degrees of Freedom *(d.f.)* in a Goodness-of-Fit Test

$d.f.$ = number of classes (adjusted) − 1 − number of estimated parameters

The negative 1 in the Box 9.G expression reflects the fact that once all but one of the expected frequencies have been calculated, the last one is predetermined by the fact that the sum of expected frequencies must equal sample size (that is, the sum of observed frequencies). The further reduction of the number of degrees of freedom by 1 for each estimated parameter provides more precise results, as more advanced texts show.

Practice Problem 5 (continued)

In our example, 5 classes remain after 3 of 7 classes are combined (note the bracket in Table G), while 1 parameter (π) must be estimated. Hence, there are $5 - 1 - 1 = 3$ degrees of freedom. Given a desired significance level of $\alpha = .05$ and 3 degrees of freedom, Appendix Table H suggests a critical value of $\chi^2_{.05,3} = 7.815$. Thus, the decision rule must be: "Accept H_0 if $\chi^2 \leq 7.815$."

Table H

Number of Possible Violations	Observed Frequency f_o	Expected Frequency (if H_0 is true), f_e	$f_o - f_e$	$(f_o - f_e)^2$	$\dfrac{(f_o - f_e)^2}{f_e}$
0	31	23.52	7.48	55.950	2.379
1	51	60.50	−9.50	90.250	1.492
2	70	64.82	5.18	26.832	.414
3	32	37.04	−5.04	25.402	.686
4–6	16	14.08	1.92	3.686	.262
Total	200	200	0		$\chi^2 = 5.233$

Step 4: Computing the test statistic and confronting it with the decision rule. The computation of the test statistic is given in Table H. In view of the value of χ^2 computed in Table H, the null hypothesis should be *accepted.* Any discrepancy between the observed set of frequencies and the set of frequencies expected from a binomially distributed population is most likely the result of sampling error. |||

Goodness of Fit to the Normal Distribution

The following example illustrates how one can test whether sample data come from a normally distributed population.

Practice Problem 6

Trading in the Futures Market

A financial analyst wishes to determine whether the daily volume of futures contracts traded at U.S. commodities exchanges is still normally distributed with a mean of 50 million contracts and a standard deviation of 10 million contracts, as indicated by a study conducted a decade ago. A test at the 2 percent level of significance is desired. The analyst collects the data of Table I during the next 90 business days. Subsequently, the analyst proceeds with Steps 1 to 4.

Table I

Number of Futures Contracts Traded per Day (millions)	Observed Frequency, f_o
under 10	5
10–under 20	9
20–under 30	15
30–under 40	23
40–under 50	20
50–under 60	8
60–under 70	6
70–under 80	3
80 and above	1
Total	90

Step 1: Formulating two opposing hypotheses.

H_0: The daily volume of futures contracts traded is normally distributed with a mean of 50 million contracts and a standard deviation of 10 million contracts.

H_A: The daily volume of futures contracts traded is not correctly described by H_0.

Step 2: Selecting a test statistic. The analyst selects

$$\chi^2 = \sum \frac{(f_o - f_e)^2}{f_e}.$$

Step 3: Deriving a decision rule. The analyst establishes in Table J the expected frequencies for the various classes of daily volumes, assuming H_0 is true. This process

Table J

Number of Futures Contracts Traded per Day (millions): upper class limit, x	Normal Deviate, $z = \dfrac{x - \mu}{\sigma} = \dfrac{x - 50}{10}$	Area Under Standard Normal Curve Left of x	Area of Class Interval, p	Expected Frequency (if H_0 is true), $f_e = 90p$
10	−4.0	.0000	.0000	0
20	−3.0	.0014	.0014	.126
30	−2.0	.0228	.0214	1.926 $\}$14.283
40	−1.0	.1587	.1359	12.231
50	0	.5000	.3413	30.717
60	1.0	.8413	.3413	30.717
70	2.0	.9772	.1359	12.231
80	3.0	.9986	.0214	1.926 $\}$14.283
∞	∞	1.0000	.0014	.126
Total			1.0000	90

involves finding areas under the normal curve for the various classes in Table I with the help of Appendix Table E and then multiplying each of these probabilities by the total number of observations (of 90). To bring each expected frequency up to a minimum of 5 (a requirement first noted on page 308), the analyst combines several classes, as shown, leaving $k = 4$ classes for the χ^2 test. Since no parameters were estimated, there are $4 - 1 - 0 = 3$ degrees of freedom, which implies (for $\alpha = .02$) a critical value of $\chi^2_{.02,3} = 9.837$ (Appendix Table H). Thus, the decision rule must be: "Accept H_0 if $\chi^2 \leq 9.837$."

Step 4: Computing the test statistic and confronting it with the decision rule. In view of the χ^2 value computed in Table K, H_0 should be *rejected.* This decision does not necessarily mean, of course, that the daily numbers of futures contracts traded are not normally distributed, but it does mean (at the 2-percent level of significance) that the distribution, if it is normal, does not conform to the historical $\mu = 50$ and $\sigma = 10$. [For a further discussion of this issue see problem 43 in the "Questions and Problems" section at the end of this chapter and its solution at the back of this book.]

Table K

Number of Futures Contracts Traded per Day (millions)	Observed Frequency, f_o	Expected Frequency (if H_0 is true), f_e	$f_o - f_e$	$\dfrac{(f_o - f_e)^2}{f_e}$
under 40	52	14.283	37.717	99.599
40–under 50	20	30.717	−10.717	3.739
50–under 60	8	30.717	−22.717	16.801
60 and above	10	14.283	−4.283	1.284
Total	90	90	0	$\chi^2 = 121.423$

|||

Goodness of Fit to Any Specified Distribution

Goodness-of-fit tests can be performed with respect to any specified distribution, not just with respect to members of some well-known distribution family. This fact is illustrated by the following final example on the television viewing habits of people in 1950 and 1980.

||| Practice Problem 7

Television Viewing Habits

A TV network polled 100 viewers in 1950 and 1980 and noted viewing habits as indicated in Table L. It now wants to test, at the 5-percent level of significance, whether 1980 viewing habits of the population at large differ from 1950. (The answer may affect its ability to sell advertising time to various sponsors.)

Table L

Hours of TV Watched per Week	Observed Frequency, f_o	
	1950	1980
0–under 3	3	9
3–under 5	4	3
5–under 7	10	9
7–under 15	20	21
15–under 25	45	44
25 or more	18	14
Total	100	100

Step 1: Formulating two opposing hypotheses.

H_0: 1980 viewing habits equal those of 1950.
H_A: 1980 viewing habits differ from those of 1950.

Step 2: Selecting a test statistic. The network manager selects

$$\chi^2 = \sum \frac{(f_o - f_e)^2}{f_e}.$$

Step 3: Deriving a decision rule. Under the null hypothesis, the manager expects the frequencies of 1950 to prevail also in 1980; thus, the data in the 1950 column become the expected frequencies for 1980, but the first two classes must be combined to yield a minimum value of 5. Thus, there are 5 remaining classes and $5 - 1 - 0 = 4$ degrees of freedom, making the critical $\chi^2_{.05,4} = 9.488$ (Appendix Table H). Thus, the decision rule is: "Accept H_0 if $\chi^2 \leq 9.488$."

Step 4: Computing the test statistic and confronting it with the decision rule. The computation of χ^2 is shown in Table M. In view of this χ^2 value, H_0 should be *accepted.* 1980 viewing habits equal those of 1950.

Table M

Hours of TV Watched per Week	1980 Observed Frequency, f_o (1980 values)	1980 Expected Frequency (if H_0 is true) f_e (1950 values)	$f_o - f_e$	$\frac{(f_o - f_e)^2}{f_e}$
0–under 5	12	7	5	3.571
5–under 7	9	10	−1	.100
7–under 15	21	20	1	.050
15–under 25	44	45	−1	.022
25 and more	14	18	−4	.889
Total	100	100	0	$\chi^2 = 4.632$

ANALYTICAL EXAMPLE

9.1 Were Mendel's Data Fudged?

In the 1860s, the Austrian monk Gregor Mendel wrote an important paper about inheritance. He postulated the existence of entities, now called *genes,* that determine how various characteristics are passed on from one generation to the next. For example, according to one of his theories (now known as Mendel's Second Law of Inheritance), the cross-fertilization of two pure strains of pea plants—one producing only round yellow seeds, the other one only wrinkled green seeds—would produce a first generation of hybrids with nothing but round yellow seeds. Yet a mating of these hybrids with one another would yield plants with round yellow, round green, wrinkled yellow, as well as wrinkled green seeds, and would produce these in definite proportions of $9:3:3:1$. Mendel performed numerous experiments to back up his theories. One experiment involved the just-mentioned law and yielded the data for the second-generation hybrids given in the accompanying Table A.

It is easy to perform a χ^2 test of the null hypothesis that this sample of 556 observations has come from a population obeying Mendel's Second Law. The theoretical proportions of $9:3:3:1$ translate into the expected frequencies given in the accompanying Table B, which also shows the calculation of χ^2. If one were to perform a χ^2 test at the 5-percent level of significance (there being 3 degrees of freedom), the critical value of $\chi^2_{.05,3}$ would be 7.815. Thus, the null hypothesis would be *accepted;* the experimental data would clearly vindicate Mendel's Second Law.

But note the tiny value of χ^2 calculated in Table B. The fit between expected and observed frequencies is almost perfect. The great statistician Ronald A. Fisher (Biography 8.1) thought this was too good to be true, especially after noting that such unusually close agreement between the observed and the expected could be found in *all* of Mendel's experiments! Fisher, therefore, made an experiment of his own. For each of Mendel's experiments, he computed the χ^2 value. Then he pooled the results by adding all the χ^2 values and also adding all the numbers of degrees of freedom. (This is a legitimate procedure as long as all the experiments are independent of one another. Thus, if one experiment yields $\chi^2 = .47$ and has 3 degrees of freedom, while another one yields $\chi^2 = .61$ and has 5 degrees of freedom, the two have a pooled χ^2 of $.47 + .61 = 1.08$ with $3 + 5 = 8$ degrees of freedom.) For all of Mendel's data, Fisher found a pooled χ^2 value of 41.6 with 84 degrees of freedom. More importantly, such a low pooled χ^2 value or a lower one

Table A

Pea Type	Observed Frequency, f_o
Round yellow	315
Round green	108
Wrinkled yellow	101
Wrinkled green	32
Total	556

Table B

Pea Type	Observed Frequency, f_o	Expected Frequency (if H_0 is true), f_e	$f_o - f_e$	$\dfrac{(f_o - f_e)^2}{f_e}$
Round yellow	315	312.75	2.25	.016
Round green	108	104.25	3.75	.135
Wrinkled yellow	101	104.25	−3.25	.101
Wrinkled green	32	34.75	−2.75	.218
Total	556	556	0	$\chi^2 = .47$

would occur, Fisher found, only 7 times in 100,000 experiments. That is, if 100,000 scientists each repeated Mendel's experiments, only 7 of them would manage to match Mendel's own low χ^2; 99,993 would get results with a larger χ^2. Thus, Mendel was either very, very lucky indeed or his data were fudged. It is hard to escape the latter conclusion. (Indeed, Fisher be-

lieved that Mendel's gardening assistant, who was overly eager to please his master, is to blame.)

Source: J. H. Bennett, ed., *Experiments in Plant Hybridisation* (Edinburgh: Oliver and Boyd, 1965), especially pp. 23 and 78. This book reprints Mendel's original article, along with commentary by Ronald A. Fisher.

BIOGRAPHY

9.1 Karl Pearson

Karl Pearson (1857–1936) was born in London, England, the son of a successful trial lawyer. While attending University College, London, and then the universities of Heidelberg, Berlin, and Cambridge (where he received a law degree), he exhibited a phenomenal range of interests, moving from mathematics and physics to philosophy and religion to history and law to German folklore, socialism, and Darwinism! Much of this study had little to do with those things for which Pearson is now remembered, but early in his life he was simply overwhelmed with all there is to know and noted that "not one subject in the universe is unworthy of study."

Pearson's interest in analytical statistics was kindled only in the late 1880s after he had become a professor of applied mathematics and mechanics at University College. (Later, he was named the first Galton professor of eugenics there.) His 1892 book, *The Grammar of Science,* illustrates his growing conviction that analytical statistics lies at the foundation of all knowledge. Going beyond Adolphe Quetelet (Biography 1.1) who believed that almost all phenomena can be described by the normal distribution (provided only that the number of cases examined is large enough), Pearson derived a system of generalized frequency curves that recognized the importance of asymmetrical ones. (In that connection, he noted that true variability among individuals was a concept very different from chance variation among errors made while estimating a single value. He introduced the term *standard deviation* and the symbol σ for the former and called the latter *probable error*.) Having established ways of fitting all kinds of curves to all kinds of observations, Pearson searched for a criterion that could measure the good-

ness of fit. Thus, in a famous paper published in 1900, he introduced chi square.

When Quetelet and his followers wanted to demonstrate the closeness of agreement between the frequencies in a distribution of observed data and the frequencies calculated on the assumption of a normal distribution, they merely printed the two series side by side, and that was that! Readers could look at these and reach their own conclusions. They had no measure of discrepancy between the observed and the expected. By introducing chi square, Pearson provided such a measure, and he worked out its distribution as well. As this chapter shows, chi square has turned out to be an enormously useful statistic, and it now occupies a major position in statistical theory.

Note: Unbeknownst to Pearson, a German—Friedrich Helmert—had discovered a chi-square distribution in 1875, when studying the sampling distribution of the sample variance, while sampling from a normal population. Pearson, of course, discovered it in a different context—namely, that of a goodness-of-fit problem—and he later extended its application to the analysis of frequencies in contingency tables. Pearson himself was unclear, however, on the proper number of degrees of freedom to use; he always used $k - 1$. As was later shown by Ronald A. Fisher (Biography 8.1), a more accurate result is achieved when this number is reduced by one for each estimated parameter.

Pearson also is responsible for developing considerably further the idea of correlation introduced by Francis Galton (Biography 11.1). He generalized Galton's conclusions and methods, derived the formula now called "Pearson's product moment," derived a simple routine for the computation of regression equations, and much more.

Most important, perhaps, is the fact that Pearson aroused the scientific world from a state of total indifference to statistical studies and convinced thousands in all fields of the necessity to gather and analyze data. He showed statistics to be a general method that was applicable to all sciences. Undoubtedly, there was no science for which Pearson himself demonstrated this fact more conclusively than biology. In 1900, Pearson became the cofounder (with Galton and Weldon) of *Biometrika,* a journal devoted to the statistical study of biological problems. Pearson edited the journal until his death and made it into the world's leading medium for the discussion of statistical theory and practice. The first issue of the journal carried a picture of a statue of Charles Darwin with the words, *"Ignoramus, in hoc signo laboremus"* (We are ignorant; so let us work). These words pretty much sum up Pearson's own philosophy of life. In the pursuit of that vast universe of knowledge that so overwhelmed him in his youth, he published hundreds of articles, and he facilitated the work of other statisticians by creating (and publishing in *Biometrika*) the types of tables found in the appendix of this book (the existence of which we all now take for granted). In his relentless pursuit of the truth, with the help of newly developing statistical techniques, Pearson also became embroiled in many controversies, often bitter and prolonged. As a young man, Pearson was something of a crusader, battling for such causes as socialism, the emancipation of women, and the ethics of free thought. This spirit was still evident later in his life when, as head of the Eugenics Laboratory, he battled the medical profession on such issues as the causes of tuberculosis or the effects of alcoholism on future generations. (Using statistical analysis, Pearson showed tuberculosis more related to hereditary factors than to environmental ones, which made the widespread use of sanatoria look foolish. He also showed that, contrary to common opinion, alcoholic parents were not producing mentally or physically deficient children.)

Sources: Dictionary of Scientific Biography, vol. 10 (New York: Scribner's, 1974), pp. 447–73; *International Encyclopedia of Statistics,* vol. 2 (New York: Free Press, 1978), pp. 691–98.

Summary

1. The chi-square technique can often be employed for purposes of estimation or hypothesis testing when use of the normal or *t* distributions is inadmissible. The first major application of the technique involves tests of the alleged statistical independence of two qualitative population variables. Typically, sample data are classified according to two or more categories associated with each of these variables, and the *chi-square statistic* is constructed on the basis of observed frequencies and frequencies expected on the assumption of statistical independence. Different sampling distributions of the chi-square statistic exist for different numbers of degrees of freedom; critical values of χ^2 that correspond to specified upper-tail areas of such distributions have been tabulated in the chi-square table (Appendix Table H). This table can be used in tests of statistical independence.

2. A second major application of the χ^2 technique is in making inferences about more than two population proportions. This technique can be used to test the equality or inequality of proportions.

3. Third, the χ^2 technique allows us to make inferences about a population variance. Confidence intervals for a population variance (and also probability intervals for a sample variance) can be established with the help of an appropriate χ^2 distribution. Hypotheses about a population variance can also be tested with the help of the χ^2 statistic.

4. Fourth, the χ^2 technique allows us to conduct goodness-of-fit tests to assess the plausibility that sample data come from a population the elements of which conform to a specified type of probability distribution. This technique can be used to test the goodness of fit of sample data to the binomial distribution, to the normal distribution, and more.

Key Terms

chi-square distribution
chi-square statistic

contingency table
goodness-of-fit test

Questions and Problems

The computer program noted at the end of this chapter can be used to work many of the subsequent problems.

1. Using the sample data of Table 9.A, conduct an appropriate hypothesis test of independence between the two stated variables (A and B) and do so at the 2-percent level of significance.
2. Using the sample data of Table 9.B, conduct an appropriate hypothesis test of independence between the two stated variables (A and B) and do so at the 10-percent level of significance.
3. Using the sample data of Table 9.C, conduct an appropriate hypothesis test of independence between the two stated variables and do so at the 1-percent level of significance.

Table 9.A Auto Accidents (Illinois sample, 1986)

| (A) Severity of Accident | (B) Location of Accident | | | Total |
	Freeway (1)	Rural Road (2)	City Road (3)	
(1) Property Damage	10	20	20	50
(2) Injury	10	10	5	25
(3) Fatality	10	10	5	25
Total	30	40	30	100

Table 9.B Heart-Attack Victims (New York sample, 1986)

| (A) Age | (B) Sex | | Total |
	Male	Female	
< 30	6	4	10
30–60	38	42	80
> 60	6	4	10
Total	50	50	100

Table 9.C Raw-Material Units (Peoria plant sample, 1986)

| Supplier | Type of Defect | | | | | | Total |
	A	B	C	D	E	F	
(1) Jones	5	0	3	12	50	30	100
(2) Smith	18	32	42	68	0	40	200
(3) Green	27	38	35	70	20	110	300
Total	50	70	80	150	70	180	600

Table 9.D U.S. Economic Performance (1887–1986)

Unemployment	Inflation			Total
	(A) Abated	(B) Unchanged	(C) Accelerated	
(1) Lower	5	5	10	20
(2) Unchanged	5	35	20	60
(3) Higher	20	0	0	20
Total	30	40	30	100

4. Using the sample data of Table 9.D, conduct an appropriate hypothesis test of independence between the two stated variables and do so at the 5-percent level of significance.

5. In each of the following situations, determine whether a null hypothesis of independence between the stated variables should be accepted on the basis of a χ^2 test.

   a. The variables are sex of viewers and their most preferred of 10 TV shows. Desired $\alpha = .001$; a sample yields $\chi^2 = 41.72$.

   b. The variables are marital status (single, married) and type of car owned (foreign, domestic). Desired $\alpha = .01$; a sample yields $\chi^2 = 2.39$.

   c. The variables are type of soap used (4 choices) and most preferred TV show (10 choices). Desired $\alpha = .02$; a sample yields $\chi^2 = 137.02$.

   d. The variables are college major (12 choices) and type of employment (3 choices). Desired $\alpha = .05$; a sample yields $\chi^2 = 55.19$.

6. In each of the following situations, determine whether a null hypothesis of independence between the stated variables should be accepted on the basis of a χ^2 test.

   a. The variables are income level (5 levels) and political affiliation (3 choices). Desired $\alpha = .10$; a sample yields $\chi^2 = 105.00$.

   b. The variables are social class (3 classes) and newspaper read (5 choices). Desired $\alpha = .01$; a sample yields $\chi^2 = 3.52$.

   c. The variables are sex and age at heart attack (3 classes). Desired $\alpha = .001$; a sample yields $\chi^2 = 1$ [as in problem 2, which students will find on page 331, where α was .10 and H_0 was accepted].

   d. The variables are women's preferred clothing stores (3 choices) and women's ages (3 classes). Desired $\alpha = .05$; a sample yields $\chi^2 = 7.78$.

7. Using the sample data of Table 9.E, conduct an appropriate hypothesis test of independence between the two stated variables and do so at the 2-percent level of significance.

8. Using the sample data of Table 9.F, conduct an appropriate hypothesis test of independence between the two stated variables and do so at the 2-percent level of significance.

Table 9.E American Stock Exchange (1986 sample)

Stock Price per Share	Dividend Yield		
	0 to under 3%	3 to under 10%	10% and more
(1) $0–25	27	52	12
(2) $25.01–$75	38	46	3
(3) $75.01 and more	19	3	40

Table 9.F Patients at General Hospital (Boston, 1986 sample)

Maternity Benefits Provided by Insurance	Days New Mothers Stayed in Maternity Ward					
	0–2	3–4	5–6	7	8	9 and more
(1) Poor	4	10	6	5	1	0
(2) Fair	2	27	38	17	6	5
(3) Excellent	0	38	59	20	15	7

9. Using the sample data of Table 9.G, conduct an appropriate hypothesis test of independence between the two stated variables and do so at the 5-percent level of significance.

Table 9.G Magazine Subscribers (U.S. sample, 1986)

Awareness of Scott Oxygen-Systems Ad	Magazine		
	The Pilot	Flying	Aviation Consumer
Don't Remember	100	125	132
Remember Vaguely	52	65	32
Remember Well	15	43	136

10. Using the sample data of Table 9.H, conduct an appropriate hypothesis test of independence between the stated variables and do so at the .1-percent level of significance.

Table 9.H Shoppers (Hartford suburbs, 1986 sample)

Distance of Residence	Preferred Shopping Center			
	A	B	C	D
0 to under 5 miles	40	55	27	18
5 to under 15 miles	128	72	12	3
15 miles and more	178	12	6	3

11. Using the sample data of Table 9.I, conduct an appropriate hypothesis test of independence between the stated variables and do so at the 10-percent level of significance.

Table 9.I Atlanta Residents (1986 sample)

Type of Chicken Consumed	Regularly Watch "Falconcrest"	
	Yes	No
Purdue Brand	255	127
Regular Unbranded	301	609
None	62	146

12. Using the sample data of Table 9.J, conduct an appropriate hypothesis test of independence between the stated variables and do so at the 5-percent level of significance.

Table 9.J Individuals With Auto Insurance (U.S. sample, 1986)

Action	Age			
	Under 25	25–35	36–55	Over 55
Filed Claim	86	92	121	67
No Claim	92	121	603	139

13. The Massachusetts Department of Motor Vehicles found that 163 of the last 550 accidents were serious and 49 of these involved drivers wearing seat belts. Yet 222 of the drivers in nonserious accidents wore belts. Conduct an appropriate hypothesis test of independence between the severity of accidents and the wearing of seat belts and do so at the 1-percent level of significance.

14. Using the sample data of Table 9.K, conduct an appropriate hypothesis test of independence between the stated variables and do so at the 10-percent level of significance.

Table 9.K   **Number of Sales (U.S. market, 1986)**

Type of Sales Letter Used	Response	
	Sale	No Sale
A	17	83
B	33	67
C	51	49
D	23	77

15. Using the sample data of Table 9.L, conduct an appropriate hypothesis test of independence between the stated variables and do so at the 1-percent level of significance.
16. For problems 7 through 11, calculate the expected value, variance, and standard deviation of the χ^2 random variable.

17. For problems 12 through 15, calculate the expected value, variance, and standard deviation of the χ^2 random variable.
18. An economist wants to test (at the 2-percent significance level) whether the proportion of firms planning to increase investment spending next year is the same in each of four industries as it is overall. A sample of 200 firms reveals the frequencies listed in Table 9.M. Make the test.
19. A manager wants to determine (at the 10-percent significance level) whether the proportion of defective pie crusts produced is the same for each of four oven temperatures as it is overall. A sample of 200 crusts reveals the frequencies listed in Table 9.N. Make the test.
20. The manager of a travel bureau wants to determine (at the 5-percent significance level) whether the proportion of potential customers who prefer the Massachusetts historical tour to the Bay State nature tour is the same for four ethnic groups as it is overall. A sample reveals the frequencies listed in Table 9.O. Make the test.
21. Someone flips a penny, a nickel, and a dime, each 50 times. The number of heads are 20, 25, and 30, respectively. Make a test on the fairness of the

Table 9.L   **Number of New Policies Sold (U.S. market, 1986)**

Life-Insurance Contract	Region			
	Northeast	Midwest	South	West
Whole Life	69	72	23	56
Level Term	23	29	38	77
Decreasing Term	105	93	105	19

Table 9.M   **Number of Firms (U.S. sample, 1986)**

Investment Plans	Industry				
	A	B	C	D	Total
(1) Increased Spending	10	20	30	40	100
(2) Unchanged or Decreased Spending	40	30	20	10	100
Total	50	50	50	50	200

Table 9.N Number of Pie Crusts (January 1986 sample)

Product Quality	Oven Temperature				Total
	A = 300°	B = 350°	C = 400°	D = 450°	
(1) Defective	10	0	20	40	70
(2) Perfect	40	50	30	10	130
Total	50	50	50	50	200

Table 9.O Potential Travelers (U.S. sample, 1986)

	Ethnic Group			
	A = French	B = English	C = Spanish	D = German
(1) Massachusetts Historical	57	33	29	76
(2) Bay State Nature	103	105	132	77

coins and do so at the 5-percent level of significance. (Hint: Test whether the sample data are consistent with an equality of proportions, of presumably .5, for the population of all possible coin flips.)

22. A sales manager wants to determine, at the 10-percent level of significance, whether the proportion of people who own the Polaroid SX-70 camera is the same in all parts of the country. A sample shows that the camera is owned by 10 of 30 Northerners, by 20 of 50 Easterners, by 30 of 60 Southerners, and by 30 of 90 Westerners. Make the test.

23. An engineer wants to determine, at the 1-percent level of significance, whether the proportion of defective pieces produced is the same for two machines. A sample shows that machine A produces 24 defectives out of 400, and machine B produces 42 out of 600. Make the test.

24. A car dealer wants to determine, at the 2-percent level of significance, whether the proportion of people under 30 who own cars is the same for 3 makes of cars. A sample shows that 27 of 120 car A owners are under 30; the corresponding numbers are 56 of 203 car B owners, and 13 of 31 car C owners. Make the test.

25. An insurance industry analyst wants to determine, at the .1-percent level of significance, whether the proportion of people who renew their policies only during the last week of the grace period is the same for 4 companies. A sample shows that 17 out of 52 customers of company A fall into this category; the corresponding numbers for company B are 12 out of 19, for company C they are 30 out of 51, and for company D they are 22 out of 25. Make the test.

26. A personnel manager wants to determine whether the taking of sick days by workers is explained by the workers' ages. The last 100 sick days taken are viewed as a random sample; they were distributed as in row 1 of Table 9.P. The percentage of the firm's labor force in the various age groups is given in row 2. A test at the 5-percent level of significance is desired. Make the test, showing your calculations.

27. A lawyer wants to know whether the educational level of jurors chosen at the county court reflects the makeup of the county population that is eligible for jury duty. The last 100 jurors chosen are viewed as a random sample; their makeup is given in row 1 of Table 9.Q. Row 2 shows countywide percentages. A test at the 10-percent level of signif-

Table 9.P

Distribution of	Age Groups				Total
	Under 25 (A)	25–35 (B)	36–50 (C)	51 and Over (D)	
(1) 100 Sick Days (number)	50	15	15	20	100
(2) Firm's Labor Force	20%	30%	30%	20%	100%

Table 9.Q

Distribution of	Educational Level				Total
	Elementary School (A)	High School (B)	Some College (C)	College Degree (D)	
(1) 100 Jurors (number)	12	64	12	12	100
(2) County Population Eligible for Jury Duty	20%	50%	10%	20%	100%

icance is desired. Make the test, showing your calculations. (What if a 1-percent level of significance had been chosen?)

28. A lawyer wants to know whether the available professional and managerial jobs are distributed among the races in accordance with the racial make-up of the population (which is 80-percent white, 15-percent black, 5-percent others). A random sample of 1,000 professional/managerial jobs shows whites to hold 792, blacks 98, others the rest. Make the test at the 5-percent level of significance.

29. A random sample of 500 deaths in Massachusetts shows that 43 percent died within 3 months after their birthday, 29 percent died in the next 3 months after that, but only 9 percent died in the 3 months preceding their birthday. Conduct an appropriate hypothesis test of the apparent link between time of death and birthday and do so at the 1-percent level of significance.

30. A pharmaceutical firm is bringing out a new sleeping pill; it is tested on a random sample of 20 people. The sample standard deviation of the time to sleep turns out to be 11 minutes. An 80-percent confidence interval is desired for the population variance of the time to sleep. Construct it.

31. The manager of a chain of muffler-repair shops takes a random sample of 30 days from last year's sales records. The sample shows mean sales of 51.7 mufflers per day, with a standard deviation of 11.3 mufflers. Construct a 96-percent confidence interval for the population variance of daily muffler sales.

32. A product's diameter must have a maximum variance of .006 squared inches. The production manager takes a random sample of 25 units from the week's output. The sample shows a standard deviation of diameters of .07 inches. Construct a 90-percent confidence interval for the population variance of diameters and, thus, determine whether the production process turns out a product meeting specifications.

33. An aircraft altimeter must have readouts with a maximum variance of 25 squared feet. An inspector takes a random sample of 10 new altimeters, the sample variance is 26 squared feet. Construct a

98-percent confidence interval for the population variance of readouts and, thus, determine whether the altimeters meet specifications.

34. A machine that fills cereal boxes is supposed to do so with a maximum standard deviation of .01 ounces. An inspector takes a random sample of 15 boxes that were filled during the day; the sample standard deviation is .013 ounces. Construct a 96-percent confidence interval for the population standard deviation and, thus, determine whether the machine meets specifications.

35. In each of the following cases, indicate whether the hypothesis concerning the population variance is lower-tailed, upper-tailed, or two-sided; then determine the critical and actual χ^2 values and show whether the null hypothesis should be accepted or rejected.

  a. H_0: $\sigma^2 \geq 64$; $n = 26$, $\alpha = .01$, $s^2 = 10$.
  b. H_0: $\sigma^2 \geq .64$; $n = 15$, $\alpha = .02$, $s^2 = .1$.
  c. H_0: $\sigma^2 \leq 100$; $n = 5$, $\alpha = .001$, $s^2 = 110$.

36. For many years, the Federal Aviation Administration has been giving a pilot's exam with a mean score of 85 and a variance of 64. It is introducing a completely new exam and wants to know whether the population of exam scores for the new exam has the same variance. A hypothesis test at the 2-percent significance level is desired. When a sample of 15 persons takes the exam, the sample variance equals 36. Make the test.

37. The manager of a firm has found the variance in the firm's daily net cash flow to be very important for proper cash management. In the past, this figure was $\sigma^2 = 100,000$ dollars squared. The manager has reason to suspect, however, that this figure might have changed drastically. A hypothesis test at the 4-percent level of significance is to settle this issue. Table 9.R shows the net-cash-flow figures over a recent two-week period that are to be used in the test. Make the test.

38. A manufacturer claims that the light emitted by a cathode ray tube used in computer terminals is so uniform that the light intensity variance is .0001 squared units per pixel. A user takes a random sample of 14 pixels (dots on the screen) and measures the light intensity, finding a sample standard deviation of .0094 units per pixel. Can the manufacturer's claim be accepted? Make a test at the 4-percent level of significance.

39. A builder claims that the compressive strength of his concrete has a variance of at least 80 squared pounds per square inch. A skeptic takes a random

Table 9.R

Day	Net Cash Flow (thousands of dollars)
1	−17
2	+22
3	+ 5
4	0
5	+ 8
6	− 3
7	+13
8	+20
9	+25
10	−41
11	0
12	− 4
13	+ 8
14	+13

sample of 25 batches of concrete; the sample standard deviation is 8 psi. Make a hypothesis test at the 5-percent level of significance.

40. A quality inspector wishes to test the null hypothesis that the number of defective items found in a box of 3 is a binomially distributed random variable with a probability of success in any one trial of $\pi = .2$. A test at the 10-percent level of significance is desired; a sample of 200 boxes yields the results given in Table 9.S. Make the test.

Table 9.S

Number of Defectives Found in a Box of 3	Observed Frequency, f_o
0	46
1	60
2	58
3	36
Total	200

41. One hundred high-school seniors have each applied to three colleges. Of these students, 24 received no acceptances, 36 received one acceptance, 26 received two acceptances, and the rest received three acceptances. A guidance counselor wants to test, at the 5-percent level of significance, whether the number of acceptances is a binomially distributed random variable with a probability of success in any one trial of $\pi = .25$. Make the test.

42. An economist wants to determine (at the 5-percent level of significance) whether the annual incomes of the nation's lawyers can be described by a normal curve with $\mu = \$45,000$ and $\sigma = \$10,000$. A simple random sample of 100 lawyers yields the results given in Table 9.T. Make the test.

Table 9.T

Annual Incomes of Lawyers (thousands of dollars)	Observed Frequency, f_o
under 20	2
20—under 30	5
30—under 40	10
40—under 50	60
50—under 60	10
60—under 70	5
70—under 80	3
80—under 90	0
90 and over	5
Total	100

43. Reconsider Practice Problem 6 about trading in the futures market (on page 324). How would you determine whether the data fit a normal distribution, if the population parameters $\mu = 50$ and $\sigma = 10$

were *not* given to you? Make the test and show your computations in detail.

44. A medical doctor checks the record of 200 patients who have taken a certain drug and makes the observations given in Table 9.U. A test at the .1-percent level of significance is to be conducted to see whether these data could possibly come from a larger population that does conform with the manufacturer's claim, in which 50 percent of users improve, 30 percent show no change, 15 percent show minor deterioration, and 5 percent show major deterioration. Make the test.

Table 9.U

Effect of Drug on Patients	Observed Frequency, f_o
Improvement	90
No change	40
Minor negative effect	30
Major negative effect	40
Total	200

45. A polling organization recorded the opinions given in Table 9.V at the two indicated times. It wishes to test, at the 2-percent significance level, whether public opinion at the more recent date equals that at the earlier date. Make the test.

Table 9.V

Opinion Expressed About the President's Performance	Observed Frequency, f_o	
	1 Year After Inauguration	3 Years After Inauguration
Does superb job	50	68
Does good job	25	33
Does fair job	15	22
Does lousy job	10	27
Total	100	150

Selected Readings

Kohler, Heinz. *Statistics for Business and Economics,* 2nd ed., Glenview, Ill.: Scott, Foresman and Co., 1988. Chapter 10 discusses goodness-of-fit tests to other distributions, such as the Poisson or uniform distributions. Chapter 11 discusses other nonparametric hypothesis tests, such as the Wilcoxon rank-sum test, the sign test, the signed-rank test, the number-of-runs test, the Kruskal-Wallis test, and more.

Pearson, Karl. "On the Criterion That a Given System of Deviations From the Probable in the Case of a Correlated System of Variables is Such That It Can Be Reasonably Supposed to Have Arisen From Random Sampling." *Philosophical Magazine,* 5th series (1900), pp. 157–75. Presents the chi-square test of goodness of fit.

Computer Program

The DAYSAL-2 personal-computer diskettes that accompany this book contain one program of interest to this chapter:

10. Chi-Square Tests lets you make (a) contingency-table tests to check on the alleged independence of 2 qualitative variables, (b) tests for more than 2 population proportions, (c) tests about the population variance, and (d) goodness-of-fit tests to check whether an observed distribution fits a specified theoretical distribution.

PART FOUR

Analytical Statistics: Advanced Inference

10 Analysis of Variance
11 Simple Regression and Correlation
12 Multiple Regression and Correlation

CHAPTER TEN

Analysis of Variance

We learned in Chapter 8 that independent random samples taken from two appropriate quantitative populations can help us answer questions such as these: Are the mean lifetimes of two types of aircraft radios the same? Is the mean yield of fruit trees sprayed with gypsy-moth parasites larger than that of trees sprayed with traditional pesticides? Is the mean number of cavities associated with the use of toothpaste 1 smaller than that associated with the use of toothpaste 2? We also learned in that chapter that the normal probability distribution (in the case of large samples) and Student's t distribution (in the case of small samples) are ideally suited to help us perform any desired hypothesis test about the comparative magnitudes of two population means. In this chapter, we will take this type of analysis one step further. We will imagine ourselves wanting to compare the means not merely of two, but of more than two quantitative populations, and we will learn that such extended comparisons are not performed very well by the earlier procedures.

The desire to compare the means of more than two statistical populations is common indeed. Consider the farmer who wants to know whether there is any difference in the average crop yield associated with the use of five types of fertilizer (or with five alternative quantities of a given type of fertilizer that might be applied, or even with five alternative times at which a given quantity of a given type might be applied). Consider the industrial manager who wants to know whether the average lifetime (or delivery time) of components is the same regardless of which one of seven outside contractors supplies them or the manager who wants to test whether the average number of units assembled in a day differs among seven possible production methods (or workers or machines) that may be employed. Consider the seller who wonders whether average sales vary among six alternative window displays, TV ads, types of packaging, or price levels. Think of the teacher who ponders whether the average scores of students would differ among four teaching methods or even whether a dozen alternative styles or colors of textbook print might be associated with different degrees of comprehension. Finally, consider the government agency that might wish to know whether average housing prices or average pollution levels or average price levels in general differ among the 50 states. In all these cases and a million more, sample evidence can be employed to make inferences about the population means of interest. Yet tempting as it may be, stringing together the results of several two-sample tests is not wise.

To see why it is not wise to combine the results of several two-sample tests for a multiple-sample test, imagine that we wanted to compare the mean numbers of cavities associated with five types of toothpaste, 1 through 5, and that we hypothesized an equality of these means. If our samples were small, we could perform a series of t tests, comparing the mean associated with the use of 1 vs. 2, then 1 vs. 3, 1 vs. 4, and 1 vs. 5, then 2 vs. 3, 2 vs. 4, 2 vs. 5, and, finally, 3 vs. 4, 3 vs. 5, and 4 vs. 5. We would be performing a total of 10 separate tests of the type introduced in Chapter 8. Yet if each of these tests were performed at the $\alpha = .05$ level of significance, the maximum probability of making a type I error in the entire series of tests would not be 5 percent, but considerably larger (more than 40 percent). If the null hypothesis of equal means were in fact true, the chances of falsely rejecting equality in at least one of the 10 pairwise comparisons outlined above can be found in Appendix Table C for binomial probabilities. For $n = 10$ and $\pi = .05$, for probability of $x \geq 1$ equals $1 - .5987 = .4013$. This probability figure is hardly acceptable. This chapter is devoted to an alternative procedure that avoids compounding the probability of a type I error when making comparisons of many means.

The **analysis of variance,** frequently referred to by the contraction **ANOVA,** is a statistical technique specially designed to test whether the means of more than two quantitative populations are equal. (It can, however, also be used to test the equality of only two means and then yields the same result as the normal-distribution or t-distribution tests of Chapter 8.) The technique involves taking an independent simple random sample from each of several populations of interest (such as the users of toothpastes 1, 2, 3, and so on) and then analyzing the data. Like the t test, the ANOVA test assumes that the sampled populations are normally distributed and have identical variances. (The latter assumption is often referred to as that of **homoscedasticity,** an awkward word created by the Greek *homo* = "same" and *skedastikos* = "scatter.") It turns out that the analysis-of-variance test is quite robust with respect to the normality assumption (even moderate departures from this assumption do not change the results much), but any violation of the equal-variances assumption seriously affects the validity of the test. This fact is not surprising because the basis of the entire test is the development, from the sample data, of two independent estimates of what is assumed to be the common variance, σ^2, of the populations of interest. A first estimate of σ^2 is based on the variation *among the sample means.* This estimate is denoted by s_A^2; it is an unbiased estimate of σ^2 only if the population means are in fact equal. A second estimate of σ^2, based on the variation *of individual sample observations within each sample,* is denoted by s_W^2 and is a weighted average of the individual sample variances—which (as we noted in Chapter 7) always provide an unbiased estimate of σ^2. It can easily be seen that the ratio, s_A^2/s_W^2, will be close to 1 if, and only if, the population means are equal to each other. On the other hand, the more the value of this ratio diverges from 1 (and, in principle, this ratio of squares can take on any value between zero and positive infinity), the greater is the probability that the population means are not equal to each other. Thus, an analysis of *variances* helps us test hypotheses about the equality of *means* (which accounts for the seemingly inappropriate name of the test). Figure 10.1 presents these ideas in another way.

Figure 10.1   Basic ANOVA Concepts

In each panel, the density functions shown refer to three populations, consisting, say, of the numbers of cavities observed, X, among all the users of toothpastes 1, 2, and 3, respectively. The populations are normally distributed and have identical variances, σ^2. The positions of the density functions, however, differ in the three panels. In panel a, the population means (μ_1, μ_2, and μ_3) are wide apart. In panel b, they are closer together. In panel c, they are identical; hence, the three density functions have merged into one. If a sample of five were taken from each of these populations, the individual sample observations might have values corresponding to the positions of the small circles underneath each graph. The corresponding sample means are given by the colored squares. It is immediately clear that the common population variance ($\sigma^2 = \sigma_1^2 = \sigma_2^2 = \sigma_3^2$) would be estimated well by the dispersion among the five individual observations (small circles) within each sample as well as by the average of these sample variances, s_W^2. It is also clear that this common population variance would not be estimated well (but would be vastly exaggerated) by the dispersion among the three sample means (colored squares) in panel a and, to only a slightly lesser extent, in panel b. Only in panel c, wherein the colored squares are not any more dispersed than any one set of small circles, could one derive, from the dispersions among the sample means, an unbiased estimate, s_A^2, of σ^2. *Conclusion:* Whenever the estimate, s_A^2, of the common population variance that is derived from the dispersion *among* sample means is considerably larger than such an estimate, s_W^2, that is based on the dispersion *within* samples, the populations have unequal means, as panels a and b show. Whenever the two estimates are about the same, as in c, the populations have equal means.

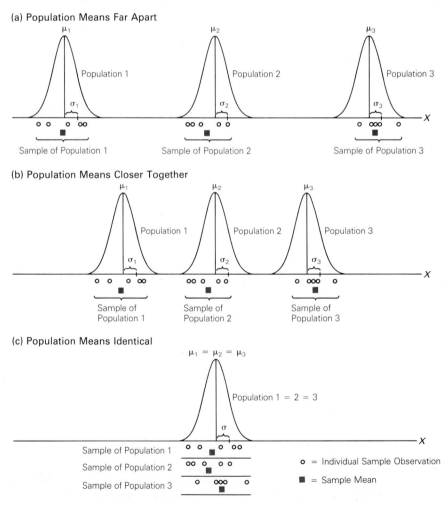

ALTERNATIVE VERSIONS OF ANOVA

Several major versions of the analysis of variance exist. These versions of ANOVA can most easily be distinguished by noting that all experiments, in one way or another, involve the assignment of different "treatments" (such as different types of toothpaste) to different experimental units (such as people) in order to see whether effects differ. (All the situations noted above that involve comparisons of means are experiments in this sense.) Yet is is all too easy to mix up or confound the effects of the treatments (such as types of toothpaste) with the effects of other extraneous factors that are also operating (such as the users' ethnic backgrounds and, hence, dietary habits). That is why good experimental designs carefully control extraneous factors, either by randomization or by blocking.

Extraneous factors are controlled by **randomization** when they are allowed to operate but when the experimenter assures, by virtue of the random assignment of experimental units to treatments, that each treatment has an equal chance of being enhanced or handicapped by these factors. (Randomization is achieved, for example, when each toothpaste is assigned to a mixed group of people whose composition with respect to ethnic background is roughly identical.) Extraneous factors are controlled by **blocking** when their effect is eliminated because the experimenter forms blocks of experimental units within each of which all units are as alike as possible with respect to these factors. (Blocking occurs, for example, when each toothpaste is tested with persons of Italian background but also with persons of German background, and so on.)

From these considerations, a number of experimental designs have emerged. The **completely randomized design** uses randomization as a control device. It creates one treatment group for each treatment (such as toothpastes 1, 2, and 3) and assigns each experimental unit to one of these groups by a random process (such that, for example, a randomly chosen third of all participants ends up using toothpaste 1, another such third uses toothpaste 2, and the last third uses toothpaste 3). In the end, only one factor (the treatment) is considered as possibly affecting the variable of interest (such as the number of cavities observed). The analysis of this type of problem, taken up in the next section, is referred to as **one-factor ANOVA,** or **one-way ANOVA.**

Another favorite experimental plan, the **randomized block design,** uses blocking as an additional control device. It divides the available experimental units into distinctly different, but internally homogeneous blocks (each of which contains as many units as there are treatments or some multiple thereof) and then randomly matches each treatment with one or more units within any given block. Within any one block (among persons of Italian background, say) only the treatment (such as the type of toothpaste used) is believed to have an effect (as on the number of cavities observed), but it is now also possible to observe potential differences for any given treatment (such as toothpaste 1) among blocks. (For example, different numbers of cavities may be observed when toothpaste 1 is used by spice-loving Italians than when it is used by sweets-loving Germans.) The analysis of this type of problem is referred to as **two-factor ANOVA,** or **two-way ANOVA** (the two factors being the treatments and the blocks), but will not be considered in this book.[1]

---

[1]For a discussion of more advanced forms of ANOVA, consult the *Selected Readings* at the end of this chapter.

ONE-WAY ANOVA

It is easiest to introduce the technique of one-way ANOVA with the help of an example. Consider a 10-year study in which a sample of 15 people has been observed while using toothpaste 1, 2, or 3, respectively. Let us assume that five of the participants have been randomly assigned to each of the treatments and that the study has provided the data given in Table 10.1. (This study is the subject of Practice Problem 1 at the end of this section.)

A statistician about to analyze these data would formulate opposing hypotheses such as:

H_0: The mean number of cavities for all users of toothpaste 1 is the same as that for all users of toothpaste 2 or 3; that is, $\mu_1 = \mu_2 = \mu_3$.

H_A: At least one of the population means is different from the others. (At least one of the equalities doesn't hold.)

The analyst would proceed by studying the variation in the sample data listed in the $r = 5$ rows and $c = 3$ columns of our table. This variation has two components.

Variation Among Columns: Explained by Treatments

The variation among the $c = 3$ sample means ($\overline{X}_1$, $\overline{X}_2$, and $\overline{X}_3$), which summarize the data associated with each of the treatments, is referred to as **explained variation,** or **treatments variation,** because it is attributable not to chance, but to inherent differences among the treatment populations. As noted above, the measurement of this variation constitutes a first estimate, s_A^2, of the population variance, σ^2; this estimate is based on the following considerations.

Given the assumed normality of the sampled populations, the sampling distribution of each sample mean, $\overline{X}_j$ (where j represents treatments and can equal 1, 2, or 3 in our case) will be normally distributed and will (as a quick review of Box 7.A on page 230 can confirm) itself have a variance of

$$\sigma_{\overline{X}_j}^2 = (\sigma_j^2/n_j). \tag{1}$$

Given the assumed equality of population variances ($\sigma_1^2 = \sigma_2^2 = \sigma_3^2$) and equal sample sizes ($n_1 = n_2 = n_3$), the value of $\sigma_{\overline{X}_j}^2$ is the same for each j; hence, we can drop the j-subscripts in equation 1. Furthermore, n, the number of observations in each sample (which equals 5 in our case) also equals r, the number of rows in Table 10.1; hence, the common population variance sought equals

$$\sigma^2 = n\sigma_{\overline{X}}^2 = r\sigma_{\overline{X}}^2. \tag{2}$$

The variance of the sampling distribution of $\overline{X}$, or $\sigma_{\overline{X}}^2$, can itself be estimated by the variance of the individual sample means, $\overline{X}_j$, about their mean, which is called the **grand mean, $\overline{\overline{X}}$** (pronounced "X double bar" and calculated, as Table 10.1 shows, as the arithmetic mean of the sample means). In accordance with the sample variance

Table 10.1   The Completely Randomized Design: Numbers of Cavities Observed During 10-Year Period

Observation, i	Treatment, j (type of toothpaste used)		
	1	2	3
1	19	20	18
2	15	25	12
3	22	22	16
4	17	19	17
5	19	23	15
Totals:	92	109	78
Sample Means	$\bar{X}_1 = \dfrac{92}{5} = 18.4$	$\bar{X}_2 = \dfrac{109}{5} = 21.8$	$\bar{X}_3 = \dfrac{78}{5} = 15.6$
Grand Mean	$\bar{\bar{X}} = \dfrac{18.4 + 21.8 + 15.6}{3} = 18.6$		

formula (in Box 3.I on page 95), this estimate of the variance of the sampling distribution of $\bar{X}$ can be calculated as

$$\sigma_{\bar{X}}^2 \cong s_{\bar{X}}^2 = \frac{\Sigma(\bar{X}_j - \bar{\bar{X}})^2}{c - 1}. \tag{3}$$

(Note: The individual sample observations, called X in Box 3.I, are in equation 3 now themselves sample means, $\bar{X}_j$, while the earlier sample mean, $\bar{X}$, is in equation 3 the mean of these means, $\bar{\bar{X}}$. In addition, the number of individual observations, called n in Box 3.I, is now equal to c, the number of columns, each of which provides us with one sample mean.)

From equations 2 and 3, it follows that

$$\sigma^2 = r\sigma_{\bar{X}}^2 \cong \frac{r\Sigma(\bar{X}_j - \bar{\bar{X}})^2}{c - 1}. \tag{4}$$

The numerator of the ratio in equation 4—the sum of the squared deviations between each sample mean and the grand mean, multiplied by the number of observations made for each treatment—is referred to as the **treatments sum of squares, or TSS.** In our example, TSS can be calculated as

$$\begin{aligned} TSS &= r\Sigma(\bar{X}_j - \bar{\bar{X}})^2 \\ &= 5[(18.4 - 18.6)^2 + (21.8 - 18.6)^2 + (15.6 - 18.6)^2] \\ &= 96.4. \end{aligned}$$

The denominator of the ratio in equation 4 equals the degrees of freedom associated with the estimation of σ^2 with the help of the variation among sample means. Thus,

$$d.f. = c - 1 = 3 - 1 = 2$$

in our case. (This number of degrees of freedom indicates that two of the three sample means are free to vary, given the grand mean.)

The equation 4 ratio as a whole, finally, is also called the **treatments mean square (TMS),** or **explained variance,** and is identical to the population variance estimate symbolized by s_A^2 above (where the subscript A was used to remind us that this estimate of σ^2 is based on the observed variation *among* samples).

$$\sigma^2 \cong \frac{r\Sigma(\overline{X}_j - \overline{\overline{X}})^2}{c - 1} = TMS = s_A^2 \tag{5}$$

In our case, $TMS = s_A^2 = (96.4/2) = 48.2$.

Variation Within Columns: Due to Error

The variation of the sample data within each of the $c = 3$ columns (or samples) about the respective sample means is referred to as **unexplained variation,** or **residual variation,** or simply as (experimental or sampling) **error;** it is attributed to chance. As noted above, the measurement of this variation constitutes a second estimate, s_W^2, of the population variance, σ^2; this estimate is based on the following considerations.

From each of the j samples in Table 10.1, a (probably different) sample variance can be derived by using the familiar Box 3.I formula (page 95) such that the variance of sample j equals

$$s_j^2 = \frac{\Sigma(X_{ij} - \overline{X}_j)^2}{n_j - 1}, \tag{6}$$

where X_{ij} is the sample observation in row i and column j, $\overline{X}_j$ is the mean of sample j, and n_j is the number of observations in sample j (which also equals r, the number of rows in Table 10.1).

To get a single estimate of σ^2, one can, in turn, take the weighted average of the j sample variances; in our case, wherein all the samples are of equal size, taking this weighted average requires only the summing of the $c = 3$ sample variances and division by c. Thus, the second estimate of the population variance can be calculated as

$$\sigma^2 \cong \frac{\Sigma\Sigma(X_{ij} - \overline{X}_j)^2}{(r - 1) \cdot c}. \tag{7}$$

The numerator of the ratio in equation 7—the sum of each sample's sum of squared deviations of individual observations from that sample's mean—is referred to as the **error sum of squares,** or **ESS.** Its calculation for our example, as given in Table 10.2, is

$$ESS = \Sigma\Sigma(X_{ij} - \overline{X}_j)^2 = 27.2 + 22.8 + 21.2 = 71.2.$$

Table 10.2   Calculation of the Error Sum of Squares

Observation, i	Sample 1 $(X_{i1} - \bar{X}_1)^2$	Sample 2 $(X_{i2} - \bar{X}_2)^2$	Sample 3 $(X_{i3} - \bar{X}_3)^2$
1	$(19 - 18.4)^2 = .36$	$(20 - 21.8)^2 = 3.24$	$(18 - 15.6)^2 = 5.76$
2	$(15 - 18.4)^2 = 11.56$	$(25 - 21.8)^2 = 10.24$	$(12 - 15.6)^2 = 12.96$
3	$(22 - 18.4)^2 = 12.96$	$(22 - 21.8)^2 = .04$	$(16 - 15.6)^2 = .16$
4	$(17 - 18.4)^2 = 1.96$	$(19 - 21.8)^2 = 7.84$	$(17 - 15.6)^2 = 1.96$
5	$(19 - 18.4)^2 = .36$	$(23 - 21.8)^2 = 1.44$	$(15 - 15.6)^2 = .36$
Total	27.20	22.80	21.20

$$ESS = 27.2 + 22.8 + 21.2 = 71.2$$

The denominator of the ratio in equation 7 equals the degrees of freedom associated with the estimation of σ^2 with the help of the variation within the samples. Thus,

$$d.f. = (r - 1) \cdot c = (5 - 1) \cdot 3 = 12$$

in our case. [This number of degrees of freedom indicates that, given a sample mean, $\bar{X}_j$, only $r - 1 = 4$ of the 5 observations in the sample are free to vary; there being $c = 3$ samples, however, the total number of values free to vary is $4(3) = 12$.]

The ratio in equation 7 as a whole, finally, is called the **error mean square (EMS),** or **unexplained variance,** and is identical to the population variance estimate symbolized by s_W^2 above (where the subscript W was used to remind us that this estimate of σ^2 is based on the observed variation *within* samples).

$$\sigma^2 \cong \frac{\Sigma\Sigma(X_{ij} - \bar{X}_j)^2}{(r - 1)c} = EMS = s_W^2 \tag{8}$$

In our case, $EMS = s_W^2 = (71.2/12) = 5.93$.

The ANOVA Table

Our computations so far are conveniently summarized in Table 10.3. This **ANOVA table** shows, for each source of variation, the sum of squares, the degrees of freedom, and the *ratio* of the sum of squares to the degrees of freedom. This ratio is called the mean square and is the desired estimate of the population variance, as equations **5** and **8** have just shown. The last row of the table also introduces the concept of the **total sum of squares,** or *Total SS,* which can be calculated independently by summing the squared deviations of each individual sample observation (regardless of the sample to which it belongs) from the mean of all observations, $\bar{\bar{X}}$. Thus, the data of Table 10.1 can be used to calculate

$$Total\ SS = (19 - 18.6)^2 + (15 - 18.6)^2 + (22 - 18.6)^2 + \ldots + (15 - 18.6)^2$$
$$= 167.6.$$

Table 10.3   The One-Way ANOVA Table

Source of Variation	Sum of Squares (1)	Degrees of Freedom (2)	Mean Square (3) = (1) ÷ (2)	Test Statistic (4)
Treatments	$TSS = 96.4$	$c - 1 = 2$	$TMS = \dfrac{96.4}{2} = 48.2$	$F = \dfrac{TMS}{EMS} = \dfrac{48.2}{5.93} = 8.13$
Error	$ESS = 71.2$	$(r - 1)c = 12$	$EMS = \dfrac{71.2}{12} = 5.93$	
Total	$Total\ SS = 167.6$	$rc - 1 = 14$		

In one-way ANOVA, this *Total SS* value is, however, also equal to the sum of the treatments sum of squares and the error sum of squares:

$$Total\ SS = TSS + ESS = 96.4 + 71.2 = 167.6.$$

Table 10.3 also shows that the total degrees of freedom equal the sum of those associated with the two separate sources of variation:

$$\text{total degrees of freedom} = (c - 1) + (r - 1)c = c - 1 + rc - c = rc - 1.$$

This sum is $5(3) - 1 = 14$ in our case and equals the total number of observations (15) minus 1. (This *d.f.* reflects the fact that, given the grand mean, all but one of the $rc = 15$ values in Table 10.1 are free to vary.)

The next section, finally, discusses the meaning of the last column of our ANOVA table and reveals the result of our hypothesis test.

The *F* Distribution

When discussing basic ANOVA concepts earlier in this chapter, we noted that the ratio of the two independent estimates of the common population variance—then denoted by s_A^2/s_W^2 and now found to equal *TMS/EMS*—would be close to 1 whenever the null hypothesis of equal population means was true. Indeed, this ratio is used as the ANOVA test statistic and is denoted by *F,* in honor of Ronald A. Fisher (Biography 8.1) who developed its probability distribution in the 1920s.[2] Thus, the *F* statistic is given in Box 10.A.

The probability distribution of *F* helps us decide whether any given divergence of *F* from 1 (note that $F = 8.13$ in Table 10.3) is significant enough to warrant the rejection of the null hypothesis of equal population means. Appendix Table I provides crit-

---

[2]Strictly speaking, he developed a Z distribution that can be transformed into the F distribution by the fact that $F = e^{2Z}$, where $e = 2.71828$. One of the earliest tabulations of F distributions appeared in George W. Snedecor, *Calculation and Interpretation of Analysis of Variance and Covariance* (Ames, Iowa: Collegiate Press, 1934), which also explains why the statistic is sometimes referred to as *Snedecor's F.* Biography 12.1 tells more of Snedecor's contributions.

10.A The *F* Statistic

$$F = \frac{s_A^2}{s_W^2} = \frac{TMS}{EMS} = \frac{\text{explained variance}}{\text{unexplained variance}}$$

where s_A^2 is the variation *among* sample means, s_W^2 is the variation of individual sample observations *within* samples, *TMS* is treatments mean square, and *EMS* is error mean square.

ical values of F_α for significance levels of $\alpha = .05, .025,$ and $.01$. Note: in contrast to the *t* and χ^2 statistics, the **F statistic** (being a ratio of two variances) is associated not with a single number of degrees of freedom, but with a pair of them. One of these (equal to $c - 1 = 2$ in our example) goes with the numerator of the ratio; the other one (equal to $(r - 1)c = 12$ in our example) goes with the denominator. As Appendix Table I shows, with 2 numerator degrees of freedom and 12 denominator degrees of freedom, the critical value of $F_{.05}$ is 3.89, of $F_{.025}$ is 5.10, and of $F_{.01}$ is 6.93. This means there is only a 5-percent, 2.5-percent, or 1-percent chance, respectively, that a computed value of F will exceed the indicated magnitudes when the null hypothesis is true.

Thus, regardless of which of the three significance levels our analyst had selected, the actual value of $F = 8.13$ would exceed the critical value, and H_0 would be *rejected*.

Figure 10.2 pictures a number of **F distributions** for selected pairs of degrees of freedom. In each case, the first *d.f.* number that appears within the parentheses refers

Figure 10.2 *F* Distributions

A different *F*-distribution curve can be drawn for each possible pair of numerator and denominator degrees of freedom that is associated with the two variance estimates making up the *F* ratio. All the members of this continuous-probability-distribution family are positively skewed between values of zero and positive infinity, but *F* distributions tend towards normality as both numbers of degrees of freedom become large. Indeed, the normal, *t*, and χ^2 distributions are special cases of the *F* distribution family. (Note: The numerator degrees of freedom and the denominator degrees of freedom happen to be equal to one another in each of the three examples given here. There is no logical necessity for this, as Table 10.3 and numerous other examples in this chapter illustrate.)

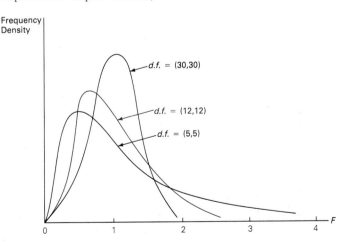

to the numerator; the second one refers to the denominator of the F ratio. Extreme caution is advised: an F distribution with 2 numerator degrees of freedom and 12 denominator degrees of freedom must not be confused with one that has 12 numerator degrees of freedom and 2 denominator degrees of freedom. The shapes of these distributions are entirely different. This fact becomes evident when inspecting Appendix Table I. Note, for example, how the critical value of $F_{.05}$ equals 3.89 for a numerator/denominator degrees of freedom pair of 2, 12 but equals 19.41 for a pair of 12, 2.

||| Practice Problem 1

The Effect of Three Toothpastes on Cavity Development

Figure A pictures the F distribution applicable to our experiment (which seeks to determine whether 3 toothpastes are equally effective by randomly assigning 5 different participants to use each toothpaste). The critical value of $F_\alpha = 3.89$ for a test at the 5-percent significance level is encircled. Thus, if an experiment of this type were to be repeated over and over again, while the null hypothesis of equal population means was true, the computed value of F, shown by the colored arrow for our example, would exceed 3.89 only 5 percent of the time. This result leads us to reject the null hypothesis and to conclude the three toothpastes do have different effects.

Figure A

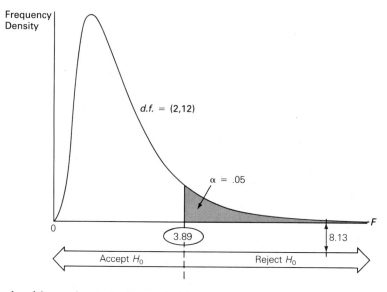

What should one do, given the probable truth of the alternative hypothesis that the three toothpastes are *not* equally effective? One possible reaction is recommending use of toothpaste 3 because sample 3 yielded the lowest mean number of cavities (Table 10.1). Yet the sample results are still clouded by sampling error. It is conceivable that toothpaste 1 is better than 3 but that an unfortunate sample was drawn from population 1 and a lucky one from population 3. Thus, further analysis is in order, and we will examine the possibilities in the next section of this chapter. |||

DISCRIMINATING AMONG DIFFERENT POPULATION MEANS

Whenever the analysis of variance leads to the rejection of the null hypothesis of equal population means and, thus, suggests that there are differences among them, the analyst inevitably asks *which* of the means differ. Consider the *one-way* ANOVA problem above concerning the effectiveness of three different brands of toothpaste. It concluded that they differed, but one would surely want to know whether $\mu_1 = \mu_2$ while μ_3 is different, or whether $\mu_1 = \mu_3$ while μ_2 is different, or whether, perhaps, $\mu_2 = \mu_3$ while μ_1 is different. Yet, given sampling error, one would be hesitant to make that determination solely on the basis of the known sample means. In this final section of the chapter, we discuss three different approaches to discriminating among three or more population means, some of which are known to be different from the others.

Establishing Confidence Intervals for Individual Population Means

The first approach makes use of the technique, introduced in Chapter 7, of estimating an individual mean. As noted in Box 7.D on page 247, the limits of a confidence interval for a population mean can be estimated, from a small sample using the t distribution, as

$$\mu = \overline{X} \pm \left(t\frac{s}{\sqrt{n}} \right)$$

In this expression, $s/\sqrt{n}$ serves as an estimate of the unknown population variance, σ^2, but in the ANOVA situation (in which several samples are taken from different populations that have identical variances, σ^2) a better estimate of σ^2 can be obtained by pooling the sample results and using the unexplained variance (or error mean square, *EMS*) in place of the single sample variance, s^2. If we let n equal sample size (the number of observations made about any treatment, j, and, thus, used to calculate any one sample mean), we get the formula given in Box 10.B.

10.B Limits of a Confidence Interval of a Population Mean for *j*th Treatment

$$\mu_j = \overline{X}_j \pm \left(t\sqrt{\frac{EMS}{n}} \right)$$

where $\overline{X}_j$ is the sample mean for *j*th treatment, *EMS* is the error mean square, and n is sample size, while t is the t-statistic (Appendix Table G) associated with a chosen confidence level and the degrees of freedom of *EMS*.

In this formula, μ_j and $\overline{X}_j$ refer to the population and sample mean, respectively, for the *j*th treatment. To find the value of t for any given confidence level—say, 95 percent—we must employ the degrees of freedom associated with *EMS*—for example, $(r - 1)c$ in the case of one-way ANOVA.

In our earlier example (Table 10.3), $EMS = 5.93$; $(r - 1)c = 12$; accordingly, the value of $t_{.025,12}$ can be found in Appendix Table G as 2.179. Thus, the following confidence intervals can be calculated from the sample means of Table 10.1.

For the population mean of toothpaste 1, the 95-percent confidence-interval limits are the following:

$$\mu_1 = 18.4 \pm \left(2.179\sqrt{\frac{5.93}{5}}\right)$$
$$\mu_1 = 18.4 \pm 2.37$$

Therefore, the confidence interval is: $16.03 \leq \mu_1 \leq 20.77$.

For the population mean of toothpaste 2, the 95-percent confidence-interval limits are

$$\mu_2 = 21.8 \pm 2.37.$$

Therefore, the confidence interval is: $19.43 \leq \mu_2 \leq 24.17$.

For the population mean of toothpaste 3, the 95-percent confidence-interval limits are

$$\mu_3 = 15.6 \pm 2.37.$$

Therefore, the confidence interval is: $13.23 \leq \mu_3 \leq 17.97$.

This analysis suggests that μ_1 and μ_2 may well be the same (the confidence intervals overlap) and that μ_1 and μ_3 are the same, but μ_2 and μ_3 clearly differ (the confidence intervals do not overlap).

Establishing Confidence Intervals for the Difference Between Two Population Means

A second approach also uses Chapter 7 material but focuses not on one mean at a time, but on the difference between two means. For this purpose, we refer to Box 7.E on page 249, according to which the limits of a confidence interval for the difference between two population means can be estimated from small and independent samples as

$$\mu_1 - \mu_2 \cong (\overline{X}_1 - \overline{X}_2) \pm \left(t\sqrt{\frac{s_1^2}{n_1} + \frac{s_2^2}{n_2}}\right).$$

The square root can be replaced by $\sqrt{2EMS/n}$ because the common population variance is estimated by the error mean square and because variability is additive for the two samples. The result is the formula in Box 10.C.

10.C Limits of a Confidence Interval of the Difference Between Two Population Means

$$\mu_1 - \mu_2 \cong (\overline{X}_1 - \overline{X}_2) \pm \left(t\sqrt{\frac{2EMS}{n}}\right)$$

where the $\overline{X}$s are the sample means, EMS is the error mean square, n is sample size, while t is the t-statistic (Appendix Table G) associated with a chosen confidence level and the degrees of freedom of EMS.

The value of t is again found as in the previous section; hence, the 95-percent confidence intervals can be calculated as follows.

For the difference between the population means of toothpastes 1 and 2, the 95-percent confidence-interval limits are

$$\mu_1 - \mu_2 \cong (18.4 - 21.8) \pm \left(2.179 \sqrt{\frac{2(5.93)}{5}} \right) = -3.4 \pm 3.36.$$

Therefore, the confidence interval is: $-6.76 \le (\mu_1 - \mu_2) \le -.04$. The fact that the interval does not overlap zero and lies entirely below it suggests that a null hypothesis, $H_0: \mu_1 = \mu_2$, should be rejected at the 5-percent significance level (that toothpaste 1 produces *fewer* cavities than toothpaste 2).

For the difference between the population means of toothpastes 1 and 3, the 95-percent confidence-interval limits are

$$\mu_1 - \mu_3 \cong (18.4 - 15.6) \pm 3.36 = 2.8 \pm 3.36.$$

Therefore, the confidence interval is: $-.56 \le (\mu_1 - \mu_3) \le 6.16$. Because this interval does overlap zero, the null hypothesis, $H_0: \mu_1 = \mu_3$, can be accepted at the 5-percent significance level.

For the difference between the population means of toothpastes 2 and 3, the 95-percent confidence-interval limits are

$$\mu_2 - \mu_3 \cong (21.8 - 15.6) \pm 3.36 = 6.2 \pm 3.36.$$

Therefore, the confidence interval is: $2.84 \le (\mu_2 - \mu_3) \le 9.56$. The fact that this interval does not overlap zero and lies entirely above it suggests that the null hypothesis, $H_0: \mu_2 = \mu_3$, should be rejected at the 5-percent significance level (that toothpaste 2 produces *more* cavities than toothpaste 3). Caution is advised: The stated confidence level of 95 percent applies to each test individually, but not to the series of all three. We cannot state with a 95-percent degree of confidence that toothpaste 1 is better than toothpaste 2, which is worse than toothpaste 3, which is as good as toothpaste 1. The chance of making at least one erroneous rejection in the *series* of null hypotheses exceeds 5 percent (for reasons indicated in the third paragraph of this chapter on page 343).

Tukey's *HSD* Test

Another approach yet is **Tukey's *HSD* test** that seeks out "honestly significant differences" among paired sample means. Imagine that an analysis of variance has rejected equality among all the population means of interest, as in our one-way example (Table 10.3). We call k the number of such means to be compared in pairwise fashion, such as the $k = 3$ treatment means in our earlier example. Thus, the number of comparisons equals

$$\frac{k(k - 1)}{2}$$

which, as we have already noted, also equals 3 in our case. According to Tukey, the difference between any two population means is statistically significant at level α if the absolute difference between the corresponding sample means equals or exceeds his honestly significant difference, *HSD*, which is calculated according to the formula given in Box 10.D.

10.D Tukey's *HSD*

$$HSD = q_\alpha \sqrt{\frac{EMS}{n}}$$

where *EMS* is the error mean square, n is sample size, and q_α is found in Appendix Table J.

A key element in the formula is a number denoted by q_α that varies with the number of means to be compared, k, with the error degrees of freedom, and with the desired significance level, α. Appendix Table J provides values of q_α for different values of k, for various error degrees of freedom, and for α levels of .05 and .01. The ANOVA test itself, of course, provides the value of the error mean square, *EMS*, that also appears in the formula, and n equals the number of observations used to calculate the sample mean (which corresponds to r, the number of rows, in Table 10.1).

For our example, with $k = 3$ means, error degrees of freedom $= 12$, and $\alpha = .05$, Appendix Table J yields $q_\alpha = 3.77$. Given $r = 5$ observations per sample and *EMS* $= 5.93$,

$$HSD = 3.77\sqrt{\frac{5.93}{5}} = 4.11.$$

Now consider the (absolute) differences between our sample means from Table 10.1:

$$\overline{X}_1 \text{ vs. } \overline{X}_2: 3.4$$
$$\overline{X}_1 \text{ vs. } \overline{X}_3: 2.8$$
$$\overline{X}_2 \text{ vs. } \overline{X}_3: 6.2$$

Only the last number equals or exceeds *HSD* $= 4.11$; the test, therefore, concludes that population means μ_2 and μ_3 are different.

Summary

1. The *analysis of variance*, frequently referred to by the contraction *ANOVA*, is a statistical technique specially designed to test whether the means of more than two quantitative populations are equal. An independent simple random sample is taken from each of several populations (that are assumed to be normally distributed and to have identical variances). From the sample data, in turn, two independent estimates are derived of σ^2, the assumed common variance of the populations of interest. A first estimate, s_A^2, is based on the variation *among* the sample means; it is an unbiased estimate of σ^2 only if the population means happen to be equal. A second estimate, s_W^2, is based on the variation of individual sample observations *within* each sample; this weighted average of individual sample variances is always an unbiased estimate of σ^2. Therefore, the ratio s_A^2/s_W^2 is close to 1 if the population means are equal to each other; the more it diverges from 1, the greater is the probability that the population means are not equal to each other. Thus, an analysis of *variance* helps us test hypotheses about *means*.

2. Alternative versions of ANOVA reflect alternative designs of experiments. *One-way ANOVA* is performed with data derived from experiments based on the completely randomized design that controls extraneous factors by creating one treatment group for each treatment and assigning each experimental unit to one of these groups by a random process; thus, only one factor (the treatment) is considered as possibly affecting the variable of interest. The one-way ANOVA table summarizes the one-way analysis of variance and contains the ANOVA test statistic, F. Because F equals the ratio of s_A^2/s_W^2, the probability distribution of F (first developed by Ronald A. Fisher) helps us decide whether any given divergence of F from 1 is significant enough to warrant rejection of a null hypothesis of equal population means.

3. Whenever the analysis of variance leads to the rejection of the null hypothesis of equal population means, further analysis can be conducted to determine how these means differ. This further analysis can involve the establishment of confidence intervals for individual population means (or for the difference between them); it can also involve the application of Tukey's *HSD* test.

Key Terms

analysis of variance (ANOVA)
ANOVA table
blocking
completely randomized design
error
error mean square *(EMS)*
error sum of squares *(ESS)*
explained variance
explained variation
F distributions
F statistic
grand mean
homoscedasticity
one-factor ANOVA

one-way ANOVA
randomization
randomized block design
residual variation
total sum of squares *(Total SS)*
treatments mean square *(TMS)*
treatments sum of squares *(TSS)*
treatments variation
Tukey's *HSD* test
two-factor ANOVA
two-way ANOVA
unexplained variance
unexplained variation

Questions and Problems

The computer program noted at the end of the chapter can be used to work most of the subsequent problems.

1. A manager wants to test, at the 2.5-percent level of significance, whether the mean delivery time of components supplied by five outside contractors is the same. A random sample of past records reveals the delivery times (in days after orders were placed) given in Table 10.A. Perform the test desired by the manager, showing your computations in detail, along with an ANOVA table.

2. A government agency wants to test, at the 1-percent level of significance, whether the average cost of a given market basket of goods is the same in 3

Table 10.A   Delivery Times of 5 Contractors (days)

Observation	Supplier				
	1	2	3	4	5
1	3	4	2	7	2
2	3	1	4	1	3
3	2	3	3	1	4
4	5	5	1	2	5
5	3	2	2	3	4

cities. A random sample of 8 stores in each of the cities reveals the data given in Table 10.B. Perform the test desired by the agency, showing your computations in detail, along with an ANOVA table.

Table 10.B The Average Cost of a Given Market Basket

Store	Boston (1)	Chicago (2)	Los Angeles (3)
1	$73.64	$67.05	$72.20
2	69.27	77.50	53.20
3	63.88	75.30	52.22
4	77.50	68.15	55.55
5	74.11	73.11	56.80
6	67.82	69.72	54.72
7	68.28	76.00	68.39
8	70.03	71.85	51.20

3. An economist wants to test, at the 2.5-percent level of significance, whether mean housing prices are the same regardless of which of 3 air-pollution levels typically prevails. A random sample of house purchases in 3 areas yields the price data (in thousands of dollars) given in Table 10.C. Perform the test desired by the economist, showing your computations in detail, along with an ANOVA table.

Table 10.C Mean Housing Prices (thousands of dollars)

Observation	Air Pollution Level		
	Low	Moderate	High
1	120	61	40
2	68	59	55
3	40	110	73
4	95	75	45
5	83	80	64

4. An advertising manager wants to test whether average sales are the same regardless of which of four available displays is used to advertise a product. Independent random samples of stores using displays A–D, respectively, show the following monthly sales (in dollars).
A: 102, 113, 98, 123, 141, 152, 173, 129
B: 53, 66, 88, 79, 41, 52, 71, 50
C: 172, 189, 193, 205, 252, 98, 87, 79
D: 130, 131, 133, 141, 129, 118, 106, 100
Perform the desired test at the 5-percent level of significance, showing your ANOVA table. What assumptions are you making?

5. An office manager wants to test whether output per worker is the same regardless of which of three degrees of office-crowding prevails. Independent random samples from three offices show the following data (in typed pages per day).
Office A (no crowding): 56, 63, 71, 46, 61, 66
Office B (some crowding): 32, 41, 43, 46, 38, 49
Office C (severe crowding): 28, 32, 30, 26, 25, 31
Perform the desired test at the 2.5-percent level of significance, showing your ANOVA table. What conditions must hold in order for your results to be valid?

6. A corporate executive wants to test whether the average total cost of producing lawnmowers is the same in four parts of the country. Independent random samples of seven firms in each of the four regions show the following data (in dollars per mower).
Region A: 88, 92, 85, 80, 78, 81, 80
Region B: 92, 96, 98, 91, 88, 97, 106
Region C: 57, 88, 102, 99, 62, 49, 59
Region D: 76, 79, 80, 81, 88, 59, 93
Perform the desired test at the 1-percent level of significance, showing your ANOVA table. What conditions must hold in order for your results to be valid?

7. An educator wants to test whether the average reading score of children is the same regardless of the typeface used in a text. The identical book is printed in three different ways and used in three different schools. Subsequent independent random samples in the three schools show the following data (reading scores).
Typeface A: 88, 92, 66, 55, 73, 82, 86
Typeface B: 75, 56, 88, 93, 89, 76, 66
Typeface C: 56, 63, 78, 82, 82, 88, 96
Perform the desired test at the 5-percent level of significance, showing your ANOVA table. What conditions must hold in order for your results to be valid?

8. An economist wants to test whether the average total cost of producing lawnmowers is, on average, the same for small, medium-sized, and large firms. Independent random samples of 10 firms in each category show the following data (in dollars per mower).
A. Small Firms: 101, 99, 89, 98, 105, 110, 82, 107, 105, 98
B. Medium-sized Firms: 73, 82, 88, 86, 110, 57, 69, 83, 66, 77
C. Large Firms: 49, 52, 66, 51, 53, 57, 69, 76, 81, 60
Perform the desired test at the 2.5-percent level of significance, showing your ANOVA table. What

conditions must hold in order for your results to be valid?

9. The manufacturer of dog food wants to test whether dogs, on average, like three new foods equally well. Each of the food is fed to six different dogs at their regular mealtime; their consumption is recorded as follows (in ounces).
Food A: 16, 18, 22, 10, 19, 23
Food B: 3, 6, 18, 29, 36, 48
Food C: 12, 22, 32, 28, 10, 6
Perform the desired test at the 1-percent level of significance, showing your ANOVA table. What conditions must hold in order for your results to be valid?

10. An airline manager wants to test whether airplanes, on average, get the same minutes of flight time per gallon (mpg) with three different brands of gas. An appropriate experiment yields the following data (in mpg).
A. 100 octane, low-lead: 50, 59, 51, 48, 43, 63
B. 130 octane, low-lead: 59, 63, 66, 71, 65, 73
C. 180 octane, no-lead: 63, 76, 81, 79, 80, 85
Perform the desired test at the 5-percent level of significance, showing your ANOVA table. What conditions must hold in order for your results to be valid?

11. A police chief wants to test whether, on average, the incidence of crime is the same in four city districts. The number of police calls received in the last ten days is used as a sample:
District A: 11, 17, 9, 3, 20, 18, 15, 13, 8, 15
District B: 8, 15, 7, 0, 12, 16, 10, 11, 6, 10
District C: 6, 10, 6, 1, 6, 8, 8, 8, 3, 7
District D: 29, 33, 12, 13, 33, 22, 17, 15, 8, 20
Perform the desired test at the 2.5-percent level of significance, showing your ANOVA table. What conditions must hold in order for your results to be valid?

12. A doctor wants to test whether dieting men, on average, lose the same weight during a week regardless of which of three diets are involved. The following sample data are available (in pounds of weight loss per week).
Diet A: 5.1, 6.7, 7.0, 5.3, 9.1, 4.3
Diet B: 6.1, 7.3, 8.1, 9.1, 10.3, 12.1
Diet C: 7.3, 8.9, 9.4, 12.0, 15.7, 9.5
Perform the desired test at the 1-percent level of significance, showing your ANOVA table. What conditions must hold in order for your results to be valid?

13. An economist wants to test whether the debt-to-equity ratio is, on average, the same for firms in four industries. Independent random samples of

six firms in each industry show the following results:
Industry A: .1, .2, .3, 0, .1, .2
Industry B: .3, .2, .4, .5, .3, .1
Industry C: .1, .2, .3, .1, 0, 0
Industry D: .9, .8, .1, 1.2, 1.3, .4
Perform the desired test at the 5-percent level of significance, showing your ANOVA table. What conditions must hold in order for your results to be valid?

14. An analyst wants to test whether the price per share, on average, differs among the New York Stock Exchange, the American Stock Exchange, and the over-the-counter market. Independent random samples of eight stocks from each market yield the following (in dollars per share).
A. NYSE: 45, 56, 82, 49, 53, 61, 48, 51
B. ASE: 17, 19, 27, 22, 31, 41, 15, 16
C. OTC: 30, 19, 82, 49, 31, 19, 16, 51
Perform the desired test at the 2.5-percent level of significance, showing your ANOVA table. What conditions must hold in order for your results to be valid?

15. An agricultural researcher wants to test whether the mean yield associated with 3 types of soil is the same. Independent random samples yield the following (in pints of strawberries per season per unit of land):
Soil A: 15, 13, 8, 17, 9
Soil B: 10, 12, 15, 15, 11
Soil C: 14, 9, 6, 14, 12
Perform the desired test at the 2.5-percent level of significance, showing your ANOVA table. What conditions must hold in order for your results to be valid?

16. A manager wants to test whether the mean amount of delinquent debt repaid is the same for 3 categories of amount overdue. Independent random samples yield the following (in dollars paid, following a nasty letter):
A. overdue under $500: $10; $30; $10; $200
B. overdue $500 to under $1,000: $100; $300; $50; $500
C. overdue $1,000 and more: $500; $1,000; $100; $1,200
Perform the desired test at the 5-percent level of significance, showing your ANOVA table. What conditions must hold in order for your results to be valid?

17. A manager wants to test whether the average quality of a paint job is the same for 5 types of surface to which it is applied. Independent random samples yield the following scores:

A (Pine): 80, 86, 89, 93, 100
B (Oak): 90, 94, 95, 97, 100
C (Cedar): 73, 69, 88, 92, 96
D (Plastic): 60, 73, 72, 79, 86
E (Metal): 100, 85, 73, 84, 61
Perform the desired test at the 2.5-percent level of significance, showing your ANOVA table. What conditions must hold in order for your results to be valid?

18. The manager of a fast-food chain wants to test whether average monthly sales are the same in 6 months of the year. Independent random samples yield the following (in thousand of dollars):
A. July: 15.1, 8.9, 22.3, 5.9, 9.2
B. August: 20.7, 23.6, 56.6, 25.0, 19.2
C. September: 13.2, 13.2, 27.8, 25.0, 19.2
D. October: 12.8, 10.9, 25.8, 16.7, 10.3
E. November: 8.7, 9.9, 25.9, 10.2, 9.3
F. December: 5.2, 9.6, 13.0, 2.1, 5.0
Perform the desired test at the 1-percent level of significance, showing your ANOVA table. What conditions must hold in order for your results to be valid?

19. An engineer wants to test whether the tensile strength of steel rods is, on average, the same for 3 production pressures. Independent random samples yield the following (in pounds per square inch):
A. low pressure: 600; 700; 800; 900
B. medium pressure: 700; 800; 900; 1,000
C. high pressure: 800; 900; 1,000; 500
Perform the desired test at the 5-percent level of significance, showing your ANOVA table. What conditions must hold in order for your results to be valid?

20. An economist wants to test whether the amount of taxes figured by different tax preparers is, on average, the same for 5 types of return. Independent random samples yield the following (in dollars per return):
Return A: 250.97 − 147.31 − 0
Return B: 1,900.00 − 1,782.57 − 1,300.28
Return C: 2,878.00 − 2,878.00 − 1,300.99
Return D: 9,822.87 − 9,822.87 − 7,929.25
Return E: 22,357.00 − 39,852.00 − 18,999.53
Perform the desired test at the 2.5-percent level of significance, showing your ANOVA table. What conditions must hold in order for your results to be valid?

21. An economist wants to test whether the price of a 1-minute advertisement is, on average, the same for 3 sizes of radio stations. Independent random samples yield the following (in dollars per minute):

A. 1,000 watts station: 100, 150, 90, 150
B. 5,000 watts station: 200, 250, 120, 290
C. 10,000 watts station: 500, 600, 220, 850
Perform the desired test at the 1-percent level of significance, showing your ANOVA table. What conditions must hold in order for your results to be valid?

22. A mayor wants to test whether property-tax assessments are, on average, the same for 5 types of property. Independent random samples yield the following (in thousands of dollars of assessed value):
A. 20.5, 19.8, 21.0, 22.5
B. 30.7, 39.8, 43.0, 29.9
C. 89.0, 92.0, 87.0, 73.5
D. 100.7, 110.8, 120.0, 105.6
E. 23.2, 37.9, 120.0, 67.3
Perform the desired test at the 5-percent level of significance, showing your ANOVA table. What conditions must hold in order for your results to be valid?

23. The Environmental Protection Agency wants to test whether, on average, sulfur dioxide levels in the air over Massachusetts are found to be the same by 4 laboratories that analyze air samples. Independent random samples yield the following (in parts per million reported):
A. 3.1, 5.6, 11.7
B. 3.1, 6.5, 7.1
C. 5.9, 11.6, 23.0
D. 1.0, 6.7, 2.3
Perform the desired test at the 2.5-percent level of significance, showing your ANOVA table. What conditions must hold in order for your results to be valid?

24. An economist wants to test whether annual price increases are, on average, the same for 3 types of firms. Independent random samples yield the following (in percent per year):
A. Small Firms: 12, 10, 9, 9, 5, 6, 9, 7
B. Medium-Sized Firms: 8, 8, 9, 10, 6, 9, 11, 8
C. Large Firms: 3, 5, 6, 4, 8, 10, 12, 11
Perform the desired test at the 1-percent level of significance, showing your ANOVA table. What conditions must hold in order for your results to be valid?

25. An engineer wants to test whether the average tensile strength of plastic sheets is the same for 3 pressures applied during production. Independent random samples yield the following (in pounds per square inch):
A. Low Pressure: 27, 29, 35, 39, 40, 39
B. Medium Pressure: 18, 19, 20, 15, 14, 17
C. High Pressure: 9, 7, 8, 6, 5, 4

Perform the desired test at the 5-percent level of significance, showing your ANOVA table. What conditions must hold in order for your results to be valid?

26. An engineer wants to test whether the average strength of plywood boards is the same for 4 types of wood. Independent random samples yield the following (in pounds per square inch):
A. Birch: 50, 75, 56, 63
B. Fir: 71, 82, 79, 87
C. Larch: 22, 36, 29, 31
D. Pine: 22, 36, 31, 27
Perform the desired test at the 2.5-percent level of significance, showing your ANOVA table. What conditions must hold in order for your results to be valid?

27. The captain of a fishing fleet wants to test whether the average weight of fishes caught is the same in 3 fishing areas. Independent random samples yield the followiing (in pounds per fish):
A. Casco Bay: 5.4, 6.6, 4.3, 3.3
B. Georges Bank: 3.3, 6.9, 9.0, 10.3
C. Long Island Sound: 3.0, 2.1, 2.6, .9
Perform the desired test at the 1-percent level of significance, showing your ANOVA table. What conditions must hold in order for your results to be valid?

28. The manager of a new radio station wants to test whether the average approval score for a station theme song is the same for 3 alternative songs. Independent random samples yield the following scores:
A. 50, 60, 63, 99, 100, 100
B. 75, 75, 81, 75, 75, 81
C. 99, 100, 100, 50, 50, 0
Perform the desired test at the 5-percent level of significance, showing your ANOVA table. What conditions must hold in order for your results to be valid?

29. A psychologist wants to test whether the average job-satisfaction score is the same for 3 types of work environment. Independent random samples yield the following scores:
A. Unstructured: 50, 50, 70, 80, 89, 92
B. Some Structure: 60, 70, 80, 92, 88, 70
C. Highly Structured: 70, 60, 80, 80, 100, 92
Perform the desired test at the 2.5-percent level of significance, showing your ANOVA table. What conditions must hold in order for your results to be valid?

30. An economist wants to test whether the average price per pound of soap is the same for 3 brands. Independent random samples yield the following (in dollars per pound):

A. 5.60, 7.20, 5.90, 8.33, 6.71
B. 10.25, 11.50, 7.90, 12.00, 14.00
C. 2.20, 1.70, 2.90, 3.10, 2.90
Perform the desired test at the 1-percent level of significance, showing your ANOVA table. What conditions must hold in order for your results to be valid?

31. A supervisor wants to test whether the average productivity score is the same for 3 types of work. Independent random samples yield the following:
A. Relaxed: 88, 92, 61, 90, 99, 100
B. Some Stress: 66, 76, 85, 85, 76, 60
C. Highly Stressful: 66, 76, 88, 75, 89, 93
Perform the desired test at the 5-percent level of significance, showing your ANOVA table. What conditions must hold in order for your results to be valid?

32. A farm manager wants to test whether the average yield per acre of wheat is the same for 3 types of fertilizer. Independent random samples yield the following (in bushels per acre):
A. 24, 29, 31, 36, 39, 43
B. 66, 70, 71, 29, 23, 24
C. 87, 85, 81, 15, 12, 10
Perform the desired test at the 1-percent level of significance, showing your ANOVA table. What conditions must hold in order for your results to be valid?

33. An economist wants to test whether the average number of strikes is the same in 4 industries. Independent random samples yield the following (numbers per firm per year):
A. 0, 2, 3, 0, 1, 0
B. 1, 2, 0, 0, 0, 1
C. 4, 3, 6, 1, 0, 0
D. 3, 5, 2, 10, 3, 6
Perform the desired test at the 5-percent level of significance, showing your ANOVA table. What conditions must hold in order for your results to be valid?

34. An economist wants to test whether the average number of strikes was the same in 4 years. Independent random samples yield the following (numbers per firm per year):
A. 1983: 1, 2, 0, 4, 3, 6
B. 1984: 0, 0, 1, 1, 0, 0
C. 1985: 2, 1, 3, 3, 5, 2
D. 1986: 5, 2, 2, 10, 3, 6
Perform the desired test at the 5-percent level of significance, showing your ANOVA table. What conditions must hold in order for your results to be valid?

35. An orchardist wants to test whether the average yield of apple trees is the same for 3 types of trees.

Independent random samples yield the following (pounds per tree):

A. 189, 291, 192, 210, 287, 245
B. 210, 223, 231, 320, 338, 341
C. 272, 287, 260, 240, 327, 192

Perform the desired test at the 2.5-percent level of significance, showing your ANOVA table. What conditions must hold in order for your results to be valid?

36. An orchardist wants to test whether the average yield of apple trees is the same for 4 types of weather, measured by the time of last winter frost. Independent random samples yield the following (pounds per tree):

A. January: 189, 192, 210, 231
B. February: 291, 287, 245, 236
C. March: 348, 352, 370, 341
D. April: 260, 240, 327, 333

Perform the desired test at the 2.5-percent level of significance, showing your ANOVA table. What conditions must hold in order for your results to be valid?

37. A shopper wants to test whether the average price of electric dryers is the same for 3 brands. Independent random samples yield the following (in dollars per dryer):

A. 129, 150, 135
B. 300, 310, 315
C. 250, 250, 250

Perform the desired test at the 1-percent level of significance, showing your ANOVA table. What conditions must hold in order for your results to be valid?

38. A shopper wants to test whether the average price of electric dryers is the same for 3 cities. Independent random samples yield the following (in dollars per dryer):

A. 129, 220, 175, 150, 200, 250
B. 220, 160, 175, 250, 300, 310
C. 150, 150, 150, 100, 200, 179

Perform the desired test at the 1-percent level of significance, showing your ANOVA table. What conditions must hold in order for your results to be valid?

39. An economist wants to test whether the average price of a divorce is the same in 3 states. Independent random samples yield the following (lawyer's fees in dollars):

A. 333, 401, 510, 792, 853, 980
B. 792; 890; 642; 1,560; 1,710; 1,900
C. 200, 200, 210, 510, 600, 850

Perform the desired test at the 5-percent level of significance, showing your ANOVA table. What conditions must hold in order for your results to be valid?

40. Reconsider problem 2; calculate a 99-percent confidence interval for each of the three treatment population means.

41. Reconsider problem 7; calculate a 95-percent confidence interval for each of the three treatment population means.

42. Reconsider problem 4; calculate a 95-percent confidence interval for each of the four treatment population means.

43. Reconsider problem 9; calculate a 99-percent confidence interval for each of the three treatment population means.

44. Reconsider problem 6; calculate a 99-percent confidence interval for each of the four treatment population means.

45. Reconsider problem 13; calculate a 95-percent confidence interval for each of the four treatment population means.

Selected Readings

Anderson, V. L., and R. A. McLean. *Design of Experiments: A Realistic Approach.* New York: Marcel Dekker, 1974.

Cochran, W. G., and G. M. Cox. *Experimental Designs,* 2nd ed. New York: Wiley, 1957.

Davies, Owen L., ed. *The Design and Analysis of Industrial Experiments,* 2nd ed. New York: Hafner, 1956.

Fisher, Ronald A. *Statistical Methods for Research Workers,* 14th ed. New York: Hafner, 1970.

Fisher, Ronald A. *The Design of Experiments,* 9th ed. New York: Hafner, 1971. A classic, first published in 1935.

Kohler, Heinz. *Statistics for Business and Economics,* 2nd ed. Glenview, Ill.: Scott Foresman and Co., 1988. Chapter 12 discusses two-way ANOVA in detail; Chapter 12 of the accompanying *Student Workbook,* three-way ANOVA.

Computer Program

The DAYSAL-2 personal-computer diskettes that accompany this book contain one program of interest to this chapter:

12. Analysis of Variance lets you perform one-way ANOVA, two-way ANOVA without and with interaction, as well as three-way ANOVA, which is introduced in the *Student Workbook*. You can also develop confidence intervals for all the means—and the differences between them—that are inevitably computed during ANOVA. Numerous confidence levels are available.

CHAPTER ELEVEN

Simple Regression and Correlation

In many situations that are of interest to business administrators or to economists or to the practitioners of almost any other field of human endeavor, the value of one variable is associated with the value of another variable in some systematic way. Consider how quantity demanded (or supplied) varies with price, how output varies with input, cost varies with output, saving varies with income, income varies with education, interest rate varies with money supply, investment varies with interest rate, unemployment varies with inflation. Consider the link between the frequency of repairs and the age of equipment, between advertising expenditures and sales, R&D spending and corporate profits, corporate profits and stock prices, entry barriers and economic concentration, grade-point averages of graduates and their starting salaries, pre-employment test scores and subsequent job performance, assembly-line speeds and defective units produced, the unemployment rate at election time and the percentage of votes captured by an incumbent president. The list goes on, but one thing is instantly clear: an awareness of such linkages between variables can be of great usefulness to researchers and decision makers because it enables them to predict, from a knowledge of the value of one variable, the value of the other variable. Regression and correlation analysis aid such an endeavor by establishing (a) what types of linkages between variables exist and (b) how strong these linkages are. A word of caution must, however, be sounded at the outset: the techniques to be described in this chapter and the next are designed to determine the existence and strength of associations between variables, *but they are unable to prove anything about possible cause-and-effect relationships.* If high R&D spending, for example, were shown to be related to high corporate profit, this relationship would not prove that high R&D spending was *causing* high corporate profit. In fact, the opposite might be true (high profit might give rise to high R&D spending), or neither variable might be causally related to the other, despite their association. Nevertheless, a knowledge of such an association would be very helpful in that it would allow one to estimate, say, an unknown R&D spending figure from a known value of corporate profit. The knowledge of such an association would also be useful as a first step in a broader investigation of cause-and-effect relations. Indeed, the discovery of a systematic association between two variables frequently provides the initial impetus to further studies about causation. Close-Up 11.1, "Smoking and Health," provides a case in point.

The first major section of this chapter introduces **regression analysis,** a statistical method the central focus of which is the establishment of an *equation* that allows the unknown value of one variable to be estimated from the known value of one or more other variables. When a single variable is used to estimate the value of an unknown variable (a procedure that might be employed with respect to all of the two-variable examples given above), the method is referred to as **simple regression analysis;** thus, the term *simple* is not to be interpreted as a synonym for *easy.* When, on the other hand, several variables are used to estimate the value of an unknown variable (a procedure that will be introduced in the next chapter), the method is referred to as **multiple regression analysis.** The origin of the term *regression* itself is described in Biography 11.1 (at the end of the chapter), which introduces Sir Francis Galton who first developed the quantitative expressions that show how two or more variables are related and that are now referred to as *regression equations.*

The second major section of this chapter introduces **simple correlation analysis,** a statistical method the central focus of which is the establishment of an *index* that provides, in a single number, an instant measure of the strength of association between two variables. Depending on the size of this quantitative measure, one can tell how closely two variables move together and, therefore, how reliably one can estimate one with the help of the other.

SIMPLE REGRESSION ANALYSIS

It is easiest to introduce the nature of regression analysis with the help of the extended example introduced in Practice Problem 1. Before we turn to it, however, it is necessary to understand a number of basic concepts.

Basic Concepts

To begin with, we must become familiar with the names typically given to the variables under study and with the nature of the relationship between them.

Independent versus dependent variable. In regression analysis, the variable the value of which is assumed known (and that is being used to explain or predict the value of the other variable) is symbolized by X and is referred to as the **independent, explanatory,** or **predictor variable** and, on occasion, even as the **regressor.** In contrast, the variable the value of which is assumed unknown (and that is being explained or predicted on the basis of its relationship with the other variable) is symbolized by Y and is referred to as the **dependent, explained,** or **predicted variable** and, on occasion, even as the **regressand.**

Deterministic versus stochastic relationships. The relationship between any two variables, X and Y, can be one of two types. It can be a precise, exact, or **deterministic relationship,** such that the value of Y is uniquely determined whenever the value of X is specified. It can, in contrast, be an imprecise, inexact, or **stochastic relationship,** such that many possible values of Y can be associated with any one value of X. The former type of relationship abounds in the physical sciences; the latter one (which alone will concern us in this chapter and the next) is common in the social sciences.

Panel (a) of Figure 11.1 illustrates a deterministic relationship that is described by the equation $Y = a + bX$. In this particular example, Y equals degrees Fahrenheit and X equals degrees Celsius, while a and b are constants that represent, respectively, the vertical intercept and the slope of the straight line. Note that for any given value of X only one possible value of Y exists. Thus, 0° Celsius corresponds to 32° Fahrenheit, 50° C corresponds to 122° F, and 100° C corresponds to 212° F. Just like the three fat dots shown in the graph, every other conceivable pair of X and Y values lies precisely on the straight line that represents the equation $Y = 32 + 1.8X$.

Panel (b) of Figure 11.1, in contrast, illustrates a stochastic relationship that is described by the equation $\hat{Y}_X = a + bX$. In this particular example, $\hat{Y}_X$ (pronounced "Y hat sub X") equals the value of Y (sales) that is predicted from a knowledge of X (advertising expenditures), while a and b once again represent, respectively, the vertical intercept and the slope of the straight line. This time, however, it is possible to observe different values of Y for any given X. Points A and B, for example, represent those two pairs of observations, among the 25 pairs plotted in the graph, that involve advertising expenditures of $X = \$1,000$ per month. Yet sales were not the same in these two instances: sales equaled 130 units in one case (point A) and 200 units in the other (point B). Thus, anyone who would use the equation of the straight line in panel (b) to estimate the volume of sales associated with $\$1,000$ of advertising expenditures would get an imprecise answer of $\hat{Y}_X = 100 + .05(1,000) = 150$ (which equals the height of the straight line at $X = 1,000$). Neither one of the actual observations of Y at $X = 1,000$, however, equals $\hat{Y}_X$. Instead, both can be found as

$$Y = \hat{Y}_X + e = a + bX + e,$$

where e is an error term that can take on negative, zero, or positive values and that reflects the net effect on Y of numerous other variables besides X that are excluded from the analysis. (Observation A, for example, might refer to the experience of one firm, while observation B refers to that of another firm that finds itself in very different circumstances even though it is also spending $\$1,000$ on advertising.) For observation A, the value of $e = -20$; for observation B, $e = +50$. As will be shown below, regression analysis seeks to establish a line such as the color **regression line** in panel (b) that summarizes the stochastic relationship between X and Y while also minimizing the errors made when the equation of that line is employed to estimate Y from X.

Direct versus inverse relationships. The relationship between any two variables, X and Y, can be classified in still another way. If the values of the dependent variable, Y, *increase* with larger values of the independent variable, X, theirs is said to be a **direct relationship**—in which case, b, the slope of the regression line, is positive. If instead the values of the dependent variable, Y, *decrease* with larger values of the independent variable, X, theirs is an **inverse relationship**—in which case, the value of b is negative. Panels (a) and (b) of Figure 11.2 graphically illustrate the difference between a

Figure 11.1   Deterministic versus Stochastic Relationship

In a deterministic relationship, shown in panel (a), the value of Y is uniquely determined whenever the value of X is specified. The equation of the straight line, $Y = 32 + 1.8X$, predicts Y with precision for any given X. In a stochastic relationship, shown in panel (b), many possible values of Y can be associated with any one value of X. The equation of the straight line, $\hat{Y}_X = 100 + .05X$, therefore, predicts Y only roughly for any given X. This line in panel (b) is a *regression line*.

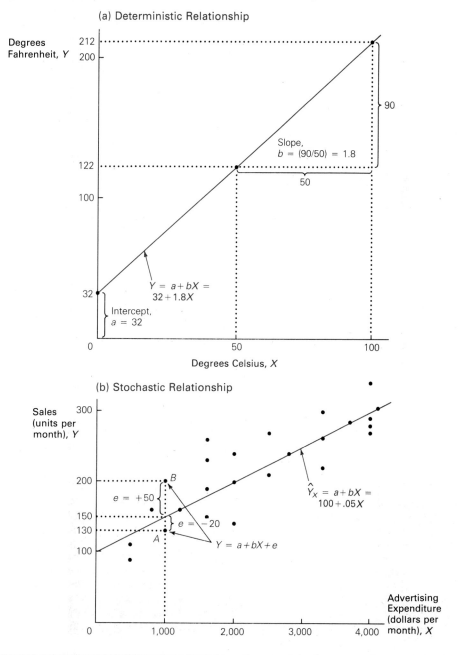

(a) Deterministic Relationship

(b) Stochastic Relationship

Figure 11.2 Alternative Relationships Between *X* and *Y*

Each dot in these graphs represents a hypothetical pair of observations about an independent variable, *X*, and a dependent variable, *Y*. The color lines summarize the nature of their relationship. It can be linear, as in panels (a) and (b), or curvilinear, as in panels (c) through (h). In addition, the relationship can be direct, as in panels (a), (e), and (g); it can be inverse, as in panels (b) and (f); or it can be a combination of both, as in panels (c), (d), and (h). Naturally, the values of the constants, such as *a, b,* and *c,* differ from one equation to the next.

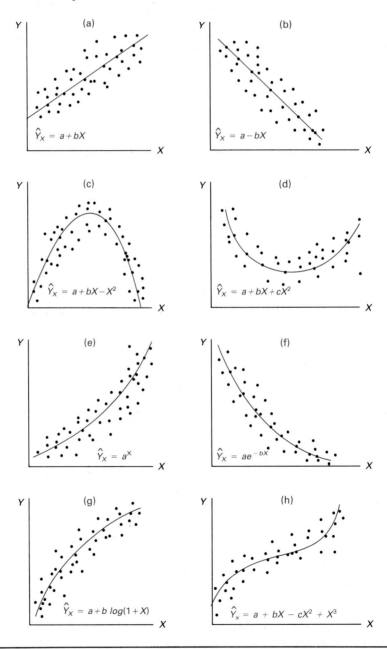

direct and an inverse relationship. The straight lines of panels (a) and (b) might summarize, respectively, observations about the familiar macroeconomic consumption function or the equally familiar microeconomic demand line. Yet relationships between our two variables need not necessarily be *linear*.

As the remaining panels of Figure 11.2 indicate, the association between two variables is often better described as *curvilinear*. Yet, because it is easier to explain, and also because such an analysis can easily be extended to the curvilinear case (as the final section of this chapter will show), the bulk of this chapter will focus on the analysis of linear relationships.

Practice Problem 1

A Preliminary Graphical Test of the Association Between Education and Income

Suppose someone desired to investigate the possible association between the extent of formal education attained by the residents of a country and their annual incomes. Because it would be too cumbersome to gather relevant data for every adult member of society, a simple random sample might be taken. To minimize our burden of calculation, we assume that only 20 individuals are included in the sample and that the result is that given in Table 11.1. Typically, however, it is difficult to tell much about the nature of the relationship between the variables of interest from such tabular data. This

Table 11.1 Sample Data on Education and Income

Individual	Education (years), X	Income (dollars per year), Y
A	2	5,012
B	4	9,680
C	8	28,432
D	8	8,774
E	8	21,008
F	10	26,565
G	12	25,428
H	12	23,113
I	12	22,500
J	12	19,456
K	12	21,690
L	13	24,750
M	14	30,100
N	14	24,798
O	15	28,532
P	15	26,000
Q	16	38,908
R	16	22,050
S	17	33,060
T	21	48,276

Figure 11.3   The Scatter Diagram

The 20 fat dots found in this scatter diagram represent the data pairs of Table 11.1 and help us visualize the relationship of interest to us. The diagram suggests the existence of a direct relationship between X and Y: more education seems to be associated with higher income. This fact may well be summarized best by a positively sloping straight line.

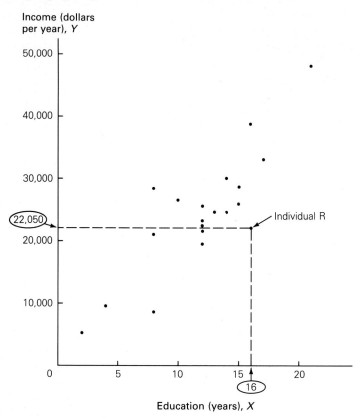

difficulty is all the more present when the sample data are more extensive than in our example. The need to make at least a preliminary judgment about what type of regression line might be an appropriate summary of the data (and the panels of Figure 11.2 indicate just some of the many possibilities), it is customary to plot the available data pairs as dots in a **scatter diagram** in such a way that values of the independent variable are measured along the horizontal axis and associated values of the dependent variable along the vertical axis. The data pairs of Table 11.1, for example, appear as the 20 fat dots of Figure 11.3. Note how the information provided by individual R (16 years of education and an annual income of $22,050) has been highlighted in the graph. The scatter of dots provides an important clue: A direct relationship seems to exist between education (X) and income (Y); this association may well be summarized best by a (positively sloping) straight line of the type found in panel (a) of Figure 11.2. (The situation just described in this Practice Problem will be referred to throughout this chapter and in many of the following Practice Problems.) ‖

Finding the Regression Line: The Use of Eyesight

One conceivable way of establishing a regression line (with the help of which one might predict the value of Y for any given value of X) involves nothing more complicated than using one's eyesight, a ruler, and a pen to draw a line in the scatter diagram that seems to "fit" the scattered dots. The procedure has been used in Figure 11.4. Having drawn the line, one can, of course, easily determine its equation from the graph. In this case, the vertical intercept of the line is seen to equal a = $10,000, while its slope can be determined by placing a triangle underneath the line and calculating

Figure 11.4   A Regression Line Drawn by Sight

The regression line shown here has been drawn by sight with a view toward making it somehow "fit" the scatter of dots that were first seen in Figure 11.3. Given this line, one would estimate a person's annual income as $10,000 plus an additional $1,333.33 for each year of education.

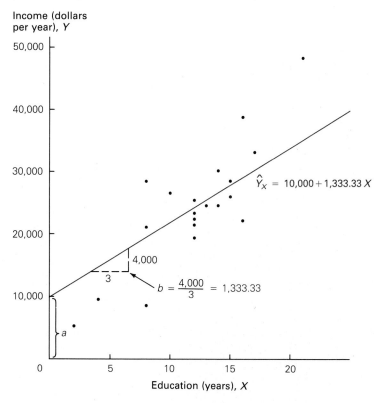

the ratio of its vertical side to its horizontal side (which comes to $b = \$1,333.33$ in this case).

This regression line, therefore, suggests that a person's annual income can be estimated as \$10,000 plus an additional \$1,333.33 for each year of education. Naturally, this is only true on average, but not for any particular individual. Thus, the income of individual R (which we know to be \$22,050) would be overestimated from a knowledge of this person's 16-year education and this regression line as $\hat{Y}_{16} = 10,000 + 1,333.33(16) = 31,333.33$. Similarly, the income of individual T (which Table 11.1 shows to be \$48,276) would be underestimated from a knowledge of this person's 21-year education as $\hat{Y}_{21} = 10,000 + 1,333.33(21) = 38,000$.

Given the nature of the problem (our desire to summarize a set of scattered dots by a straight line), such overestimation or underestimation is, of course, inevitable. Yet there is another problem with the above procedure, and this problem is the reason why statisticians reject regression lines drawn with the help of eyesight alone as unacceptably crude. The problem is that such a procedure cannot be duplicated, except by sheer accident. Different people, faced with an identical scatter of dots (such as that in Figure 11.3), would undoubtedly use different judgment and draw regression lines with different intercepts and slopes. Which of these different lines could we trust to be the best summary of the data plots? On what basis, furthermore, could we establish confidence intervals for the estimates we would make with the help of such a regression line? Because satisfactory answers to these questions do not exist, statisticians establish regression lines (and their associated equations) in a different way.

Finding the Regression Line: The Method of Least Squares

The commonly used approach to fitting a line to the sample data displayed in the scatter diagram is called the **method of least squares.** The line is derived in such a way that the sum of the squares of the vertical deviations between the line and the individual data plots is minimized. As we noted above, these vertical deviations represent the errors associated with using the regression line to predict Y with the help of X. Figure 11.5 illustrates the nature of the squares the sum of which is being minimized.[1]

Why do statisticians use this particular method of establishing a regression line? Their desire to minimize errors is understood easily enough, but why minimize the *squares* of the errors? One reason is that a summation of unsquared deviations can easily yield a number close to zero (and, thus, give the impression of low error) even when errors are common and huge—simply because positive errors (arising when observed data points lie above the line) and negative errors (arising when observed data points lie below the line) tend to cancel each other. Another reason will be noted later in this section (see footnote 2).

The regression line calculated from sample data by the method of least squares is called the **sample regression line,** or **estimated regression line;** the values of a and b found in its equation, $\hat{Y}_X = a + bX$, are referred to as the **estimated regression coefficients.** By minimizing $\Sigma(Y - \hat{Y}_X)^2 = \Sigma(Y - a - bX)^2$, these coefficients can be calculated as shown in Box 11.A.

---

[1]A supplemental section in Chapter 11 of the *Student Workbook* that accompanies this text shows the mathematics of the procedure and derives the formulas given in Box 11.A.

Figure 11.5 The Method of Least Squares

This scatter diagram displays eight data plots, only two of which lie precisely on the regression line drawn. The six remaining plots diverge from the line by varying distances, as shown by the solid vertical line segments. The method of least squares establishes the regression line, among all the conceivable ones, that minimizes the sum of the squares of such vertical deviations or errors. These squares are represented by the boxes in the graph; it is their combined area that the line minimizes. For any given data set, there exists only one line that minimizes this error sum of squares, $ESS = \Sigma(Y - \hat{Y}_X)^2$, where Y is a point such as A (an individual value of the dependent variable, Y, that is actually observed for a given value of the independent variable, X), while $\hat{Y}_X$ is a point such as B (the value of Y that is estimated for the given value of X with the help of the regression line).

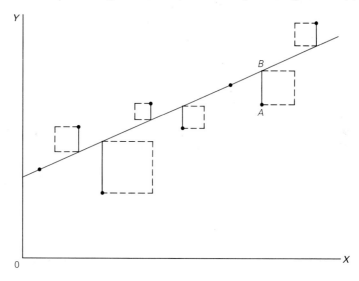

11.A Estimated Regression Coefficients, Using Method of Least Squares

$$b = \frac{\Sigma XY - n\overline{X}\,\overline{Y}}{\Sigma X^2 - n\overline{X}^2} \qquad\qquad a = \overline{Y} - b\overline{X}$$

where b is the slope and a is the intercept of the estimated regression line, while X's are observed values of the independent variable ($\overline{X}$ being their mean) and Y's are the associated observed values of the dependent variable ($\overline{Y}$ being *their* mean), and n is sample size.

||| Practice Problem 2

Using the Least-Squares Method to Determine the Estimated Regression Line for Education and Income

The calculation of the sample regression coefficients, using the formulas in Box 11.A, is performed in Table 11.2 with the help of our sample data given in Table 11.1. Note that the last three columns of Table 11.2 are not needed for the calculation of our regression coefficients (these columns will be used later) and that coefficient b must

Table 11.2   Calculating the Estimated Regression Line by the Method of Least Squares

Individual	X	Y	XY	X^2	$\hat{Y}_X$	$Y - \hat{Y}_X$	Y^2
A	2	5,012	10,024	4	5,839.23	− 827.23	25,120,144
B	4	9,680	38,720	16	9,534.23	145.77	93,702,400
C	8	28,432	227,456	64	16,924.23	11,507.77	808,378,624
D	8	8,774	70,192	64	16,924.23	−8,150.23	76,983,076
E	8	21,008	168,064	64	16,924.23	4,083.77	441,336,064
F	10	26,565	265,650	100	20,619.23	5,945.77	705,699,225
G	12	25,428	305,136	144	24,314.23	1,113.77	646,583,184
H	12	23,113	277,356	144	24,314.23	−1,201.23	534,210,769
I	12	22,500	270,000	144	24,314.23	−1,814.23	506,250,000
J	12	19,456	233,472	144	24,314.23	−4,858.23	378,535,936
K	12	21,690	260,280	144	24,314.23	−2,624.23	470,456,100
L	13	24,750	321,750	169	26,161.73	−1,411.73	612,562,500
M	14	30,100	421,400	196	28,009.23	2,090.77	906,010,000
N	14	24,798	347,172	196	28,009.23	−3,211.23	614,940,804
O	15	28,532	427,980	225	29,856.73	−1,324.73	814,075,024
P	15	26,000	390,000	225	29,856.73	−3,856.73	676,000,000
Q	16	38,908	622,528	256	31,704.23	7,203.77	1,513,832,464
R	16	22,050	352,800	256	31,704.23	−9,654.23	486,202,500
S	17	33,060	562,020	289	33,551.73	−491.73	1,092,963,600
T	21	48,276	1,013,796	441	40,941.73	7,334.27	2,330,572,176
	$\Sigma X =$ 241	$\Sigma Y -$ 488,132	$\Sigma XY =$ 6,585,796	$\Sigma X^2 =$ 3,285		$\Sigma(Y - \hat{Y}_X) = 0$	$\Sigma Y^2 =$ 13,734,414,590

$$\overline{X} = \frac{\Sigma X}{n} = \frac{241}{20} = 12.05 \qquad \overline{Y} = \frac{\Sigma Y}{n} = \frac{488,132}{20} = 24,406.60$$

$$b = \frac{\Sigma XY - n\overline{X}\,\overline{Y}}{\Sigma X^2 - n\overline{X}^2} = \frac{6,585,796 - 20(12.05)(24,406.60)}{3,285 - 20(145.2025)} = \frac{703,805.40}{380.95} = 1,847.50$$

$$a = \overline{Y} - b\overline{X} = 24,406.60 - 1,847.50(12.05) = 2,144.23$$

be calculated before a because b appears in the formula for a. Thus, our estimated regression line comes to $\hat{Y}_X = 2,144.23 + 1,847.50\,X$, which is quite different from the one drawn freehand in Figure 11.4. This least-squares line is graphed in Figure 11.6, along with the original scatter of dots. Note how one can estimate, with the help of this line, the annual income of someone with 5 years or 20 years of education (and no such persons were included in the original sample) as

$$\hat{Y}_5 = 2,144.23 + 1,847.50(5) = 11,381.73, \text{ and}$$
$$\hat{Y}_{20} = 2,144.23 + 1,847.50(20) = 39,094.23.$$

These numbers represent the height of the colored line above the X values of 5 and 20, respectively. |||

Figure 11.6 A Regression Line Derived by the Method of Least Squares

The regression line shown here has been fitted by the method of least squares to the scatter of dots first seen in Figure 11.3. Given this line, one would estimate a person's annual income as $2,144.23 plus an additional $1,847.50 for each year of education. Two features of this and every least-squares regression line should be noted. First, the line goes through point $(\overline{X}, \overline{Y})$—the respective means of the X and Y observations in the sample, here equal to 12.05 and 24,406.60. Second, the sum of the vertical deviations of Ys from the regression line is zero: $\Sigma(Y - \hat{Y}_X) = 0$ (a fact that is illustrated in the right-hand side of Table 11.2 and serves as an excellent check on the accuracy of calculations). In these two senses, the line can be said to "go through the center" of the scatter diagram.

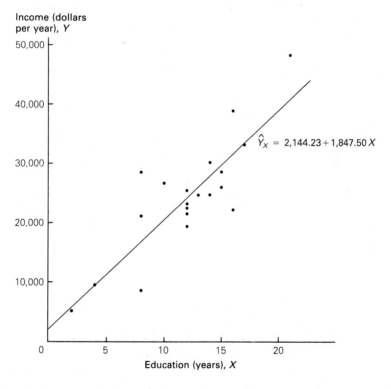

Caution: Most people recognize that it would be unwise to generalize the relationship summarized in our regression line to completely different times and places. What is true for 1985 may not be true for 1995; what is true for the United States may not be true for France. Fewer people realize, however, that it is equally unwise to use a regression line to make estimates of Y for values of X that lie beyond the range of the data (here below 2 or above 21 years of education, respectively) from which the estimated regression line has been derived. Such extrapolation can easily lead to absurd results (such as the association of certain low incomes with negative years of education or of certain high levels of education with near-infinite incomes). Such absurd results could be obtained because the apparently linear relationship that we see in our scatter diagram may well fail to hold for smaller or larger values of X (and even if it does hold, additional data would probably lead us to estimate a much flatter or steeper regression line with quite a different intercept). In addition, we must become aware of another matter yet that concerns the quality, even within the range of observed X values, of the least-squares regression line itself. To that matter we now turn.

The Estimated versus the True Regression Line

Let us return briefly to our above example (Practice Problems 1 and 2) and recall that we set out to discover the nature of a possible association between the extent of formal education attained by the residents of a country and their annual incomes. Does the estimated regression line of Figure 11.6 provide an answer to our original quest? It certainly appears to do just that. It establishes the existence of a direct relationship between the two variables and also specifies, by means of the equation of the regression line, the mathematical details of that relationship. But now consider this sobering thought: the regression line that appears in Figure 11.6 was fitted to data (first given in Table 11.1) that were obtained not by taking a full-fledged census of the population of interest (of all the adult residents of the country), but by taking a tiny simple random sample of them. What if we were to take another such sample from the same population? Surely, we would acquire a different set of data, would end up with a different scatter diagram, and would derive a different estimated regression line! We could repeat the process; for every new sample, we would estimate yet another regression line. Which one of these could be trusted to reveal the true relationship (if there is a relationship) that exists between X and Y in the population as a whole and that we would find if we took a census of the entire population of adult individuals, recorded data on education and income for every one of them, and derived a least-squares regression line on that basis?

Imagine a scatter diagram for the entire adult population of our country. For each and every value of X (and not just for some of them), it would record not merely a few values of Y (as in Figure 11.6), but a great multitude of them. The least-squares regression line that we might derive from such census data is called the **population regression line,** or **true regression line** (in contrast to the *sample* or *estimated regression line* that we have met above). Such a line might be described by the equation $E(Y) = \alpha + \beta X$, where $E(Y)$ is the expected value of Y for any given X. In this expression, α and β are the **true regression coefficients,** the vertical intercept and the slope, respectively, of the true regression line. These population parameters are the values that are really of interest to us rather than the *sample* or *estimated regression coefficients, a* and *b,* which, as a result of sampling error, may seriously distort the true underlying relationship between X and Y. (Conceivably, the value of b might be 1,847.50, as in Figure 11.6, while β is zero, indicating that, unlike in this particular sample, there exists no relationship whatsoever between X and Y *in the population as a whole.*)

As it turns out, statisticians have devised ways of making inferences concerning the true regression line from sample data, but the validity of these inferences rests on some or all of the following assumptions:

Assumption 1. It is assumed that the values of X are known without error and that the different sample observations about Y that are associated with any given X are statistically independent of each other. (The fact that an observed value of Y is low, for example, is no reason for another Y observation to be low as well.)

Figure 11.7 Three Conditional Probability Distributions of Y

The solid color lines shown here represent the three conditional probability distributions of Y that are associated with X values of 5, 10, and 15, respectively. Each of these frequency curves rises above the Y vs. X plane into the third dimension; each is also a normal curve clustering around a central value, $\mu_{Y \cdot X}$; hence, the probability of finding Y values declines progressively and equally the farther one moves along the dashed lines that parallel the Y axis either above or below any one of the three means. The dashed color line that connects all the conditional means in the Y vs. X plane is the true regression line.

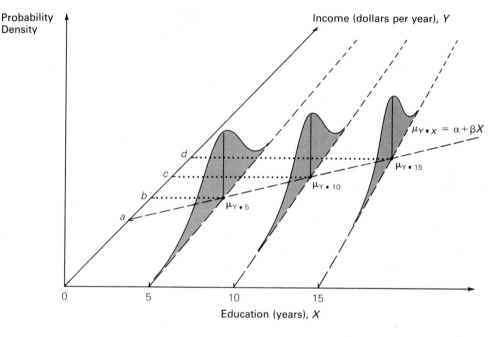

Assumption 2. It is assumed that every population of Y values (a different one of which is associated with every possible value of X) is normally distributed. Because each of these normal curves of Y values is associated with a specific value of X, each curve is referred to as a **conditional probability distribution of Y** and is said to have a **conditional mean of Y ($\mu_{Y \cdot X}$)** and a **conditional standard deviation of Y ($\sigma_{Y \cdot X}$).** This assumption is most easily understood with the help of Figure 11.7. The graph is three-dimensional, containing the now familiar Y vs. X plane but also measuring, vertically above this plane the probability density of Y values.

Imagine plotting, for specified values of $X = 5$, $X = 10$, and $X = 15$, all the associated values of Y found in the entire population. Given $X = 5$, the conditional mean of Y values might equal distance 0b; it is designated as $\mu_{Y \cdot 5}$. Given $X = 10$, the conditional mean of Y values might equal distance 0c; it is designated as $\mu_{Y \cdot 10}$. Finally, given $X = 15$, the conditional mean of Y values might equal distance 0d; it is designated as $\mu_{Y \cdot 15}$. In each case, many individual Y values would be smaller or larger than this mean, of course, but the second assumption of linear regression analysis asserts that the frequency curve of Y values for any given X is a normal curve centered on $\mu_{Y \cdot X}$, such as each of the three solid color lines shown in Figure 11.7.

Assumption 3. It is assumed that all conditional probability distributions of Y (a different one of which is associated with every possible value of X) have the same conditional standard deviation of Y. This value, $\sigma_{Y \cdot X}$, is also referred to as the **population standard error of the estimate** (of Y, given X). This third assumption, thus, is the familiar one of homoscedasticity; we have met it in previous chapters, but here it postulates that the scatter of observed Y values above and below each conditional mean is the same. (Note how the normal curves shown in Figure 11.7 look alike even though they are centered on different means.)

Assumption 4. It is assumed that all the conditional means, $\mu_{Y \cdot X}$, lie on a straight line that is the true regression line and is described by the equation

$$E(Y) = \mu_{Y \cdot X} = \alpha + \beta X.$$

The dashed color line in the Y vs. X plane of Figure 11.7 is such a line. Its Y-intercept equals α = distance $0a$, and its slope equals β.

In the following sections, we will show how a number of important inferences can be made about population data from sample data when all these assumptions are fulfilled and sometimes even when only some of them are fulfilled.

Estimating the Average Value of Y, Given X

Practice Problem 3

Estimating the Average Value of Income, Given Education

Suppose we wanted to estimate the value of a conditional mean, $\mu_{Y \cdot X}$, such as $\mu_{Y \cdot 10}$, the mean income of all those people in the population from which the Table 11.1 sample data were drawn who had 10 years of education. Just as the sample mean, $\overline{X}$, has been used in an earlier chapter to estimate the population mean, μ, we can now use $\hat{Y}_{10}$, a point on the sample regression line, as an estimate of $\mu_{Y \cdot 10}$, a point on the population regression line. Our least-squares sample regression line, depicted in Figure 11.6, is described by the equation,

$$\hat{Y}_X = a + bX = 2{,}144.23 + 1{,}847.50X.$$

Hence, $\hat{Y}_{10} = 2{,}144.23 + 1{,}847.50(10) = 20{,}619.23.$ |||

This number turns out to be the best *point estimate* of $\mu_{Y \cdot 10}$ that is available.[2] Yet we may wish for more than that; we may wish to acknowledge explicitly the existence

---

[2]The reason that $\hat{Y}_X$ is the best point estimate of $\mu_{Y \cdot X}$ is: The method of least squares produces estimated regression coefficients, a and b, that are unbiased, efficient, and consistent estimators of the corresponding true regression coefficients, α and β. The concepts of unbiasedness, efficiency, and consistency were first introduced in Chapter 7 and may be reviewed there. In the present context, if all possible random samples of a given size were taken from the population of interest and if a least-squares regression equation were calculated for each of these, the average value of the estimated coefficient a would equal the true coefficient α, while the average value of b would equal β (unbiasedness). In addition, the coefficients estimated by the method of least squares have the smallest variance for a given sample size among all available unbiased estimators (efficiency), and they get progressively closer to α and β as sample size is increased (consistency).

of sampling error. If we had drawn a different sample, we would have estimated a different regression equation (perhaps one stating that $\hat{Y}_X = 1{,}800.50 + 2{,}001.73X$) and, thus, would have made a different point estimate of $\mu_{Y\cdot 10}$ (in this case equal to $\hat{Y}_{10} = 21{,}817.80$). This degree of uncertainty about the value of $\mu_{Y\cdot X}$ can be made explicit by presenting not a point estimate, but an interval estimate. In the context of regression analysis, such a confidence interval within which a population param-eter that is being estimated presumably lies is typically referred to as a **prediction interval.** We will discuss its determination separately for the small-sample case ($n < 30$) and the large-sample case ($n \geq 30$).

Establishing a prediction interval for $\mu_{Y\cdot X}$ *from a small sample* (n < 30). We noted in Chapter 7 that the limits of a confidence interval for the population mean can be estab-lished in the small-sample case, using the t distribution, as

$$\mu = \overline{X} \pm (t\sigma_{\overline{X}})$$

and that $\sigma_{\overline{X}}$ can be estimated by $s/\sqrt{n}$ (Box 7.D on page 247). In regression analysis, the variable being estimated is Y rather than X; hence, an expression analogous to the previous one is

$$\mu_{Y\cdot X} = \hat{Y}_X \pm t\sigma_{\hat{Y}_X}.$$

The term, $\sigma_{\hat{Y}_X}$, is the standard error of $\hat{Y}_X$; it measures the variability (caused by the selection of different samples) of all the possible $\hat{Y}_X$ values (of which the above values of 20,169.23 and 21,817.80 are just two examples) at the X value for which the prediction interval of the conditional mean is to be established. This variability has two components. One of these, the variability in the mean of observed Y values, depends on the conditional standard deviation, $\sigma_{Y\cdot X}$, of the relevant population of Ys, as well as on sample size, n. (In terms of Figure 11.7, this variability in the mean of observed Y values depends on the shape of the frequency curve at the relevant X and on the number of sample observations made.) A second source of variability in $\hat{Y}_X$, however, is the distance of the specified X from $\overline{X}$, as illustrated in Figure 11.8. The graph fea-tures, in color, the (unknown) true regression line and, in black, estimated regression lines fitted to four different samples drawn from the same population. Note how these lines diverge increasingly the farther the specified value of X lies from $\overline{X}$. As a result, the estimated values of $\hat{Y}_X$ become more varied, too.

As a matter of fact, the value of $\sigma_{\hat{Y}_X}$ cannot be calculated directly because all pos-sible values of $\hat{Y}_X$ are never determined. In practice the standard error of $\hat{Y}_X$ is esti-mated as

$$\sigma_{\hat{Y}_X} \cong s_{Y\cdot X}\sqrt{\frac{1}{n} + \frac{(X - \overline{X})^2}{\Sigma X^2 - n\overline{X}^2}}$$

and this complicated expression reflects both of the components of $\hat{Y}_X$ variability that have just been noted. In this expression $s_{Y\cdot X}$ is the **sample standard error of the**

Figure 11.8   One Source of Variability in $\hat{Y}_x$

This graph features a true regression line (in color) with the equation $\mu_{Y \cdot X} = \alpha + \beta X$. It also shows estimated regression lines (in black) that might have been fitted to four different samples and are described by the equation $\hat{Y}_X = a + bX$ (where a and b, of course, differ from one line to the next). Note how the computed values of $\hat{Y}_{10}$ (the four fat dots above $X = 10$) are closer together than the computed values of $\hat{Y}_{21}$ (the four fat dots above $X = 21$) even though they are derived from identical samples. This illustrates one source in the variability of $\hat{Y}_X$: the distance of the specified X from $\overline{X}$.

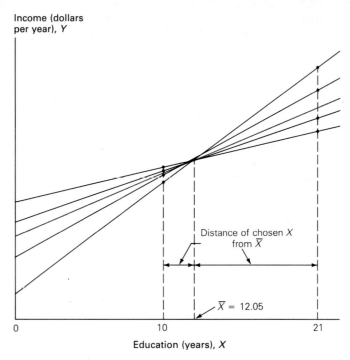

estimate about the regression line that (as the subscript reminds us) gives estimates of Y for any specified X. This standard error equals

$$s_{Y \cdot X} = \sqrt{\frac{\Sigma(Y - \hat{Y})^2}{n - 2}}$$

which is mathematically equivalent to

$$s_{Y \cdot X} = \sqrt{\frac{\Sigma Y^2 - a\Sigma Y - b\Sigma XY}{n - 2}}$$

and measures the dispersion of the values of Y observed in a sample around the least-squares regression line estimated from this sample.

Figure 11.9 The Sample Standard Error of the Estimate About the Regression Line

This graph copies the scatter diagram and the estimated regression line of Figure 11.6. It also illustrates (on the right-hand side) the kind of dispersion that is being measured by the sample standard error of the estimate about the regression line. Note how the dashed arrows (that parallel the color estimated regression line) project points *A* and *C* (the two extreme deviations of observed *Y* values above and below the regression line) onto the vertical line at points *B* and *D*. All other observed *Y* values clearly diverge from the regression line to a lesser extent (and would be projected to various points on the vertical line lying between *B* and *D*). The sample standard error of the estimate about the regression line, $s_{Y \cdot X}$, measures this type of dispersion of observed *Y* values around the estimated regression line. But caution is advised: this sample standard error of the estimate about the regression line should not be confused with the sample standard deviation of observed *Y* values, s_Y, that measures the dispersion of these values around their sample mean (as shown on the left-hand side of this graph).

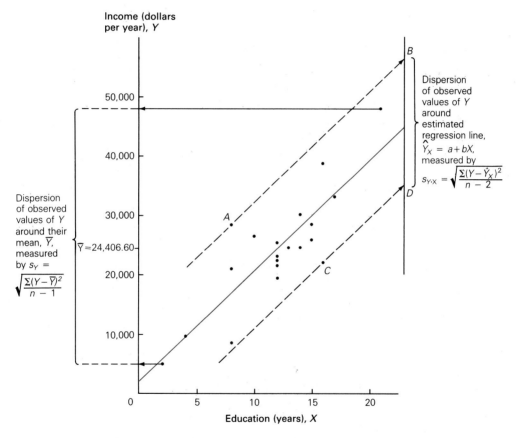

Note: The divisor, $n - 2$, is used in the formula because this makes $s_{Y \cdot X}^2$ an unbiased estimator of the variance of *Y* values around the true regression line. The reduction of n by 2 indicates the loss of 2 degrees of freedom because the coefficients *a* and *b* (that allow the calculation of $\hat{Y}_X$) are being calculated from the same data as $s_{Y \cdot X}$. The nature of this measure of dispersion is illustrated with the help of Figure 11.9.

The formula given in Box 11.B, finally, puts together the various strands of our discussion and becomes the basis for calculating a prediction interval for the conditional mean in our example.

11.B The Limits of the Prediction Interval for the Average Value of Y, Given X (small-sample case)

$$\mu_{Y \cdot X} = \hat{Y}_X \pm \left(t_{\alpha/2} \cdot s_{Y \cdot X} \sqrt{\frac{1}{n} + \frac{(X - \overline{X})^2}{\Sigma X^2 - n\overline{X}^2}} \right)$$

where Xs are observed values of the independent variable ($\overline{X}$ being their mean), while $\hat{Y}_X$ is the estimated value of dependent variable Y, $s_{Y \cdot X}$ is the sample standard error of the estimate of Y, n is sample size, and the t-statistic is found in Appendix Table K for $n - 2$ degrees of freedom.

Practice Problem 4

Establishing a Prediction Interval for the Average Value of Income, Given Education

As we return to our example from Practice Problems 1–3, we must recall having earlier determined a point estimate of $\mu_{Y \cdot 10}$ as $\hat{Y}_X = 20{,}619.23$. To calculate, say, a 95-percent prediction interval for this conditional mean, we must now find the appropriate value of $t_{.025}$ from Appendix Table G. Table 11.1 shows $n = 20$; hence, we have $n - 2 = 20 - 2 = 18$ degrees of freedom; $t_{.025,18} = 2.101$.

Next we can calculate, with the help of Table 11.2,

$$s_{Y \cdot X} = \sqrt{\frac{\Sigma Y^2 - a\Sigma Y - b\Sigma XY}{n - 2}}$$

$$= \sqrt{\frac{13{,}734{,}414{,}590 - 2{,}144.23(488{,}132) - 1{,}847.50(6{,}585{,}796)}{18}}$$

$$= \sqrt{\frac{520{,}489{,}201.64}{18}}$$

$$= \sqrt{28{,}916{,}066.76} = 5{,}377.37.$$

Next, we can calculate:

$$\sqrt{\frac{1}{n} + \frac{(X - \overline{X})^2}{\Sigma X^2 - n\overline{X}^2}} = \sqrt{\frac{1}{20} + \frac{(10 - 12.05)^2}{3{,}285 - 20(12.05)^2}}$$

$$= \sqrt{.05 + .011} = .247.$$

Thus, the limits of our prediction interval are

$$\mu_{Y \cdot 10} = 20{,}619.23 \pm [2.101(5{,}377.37).247]$$
$$= 20{,}619.23 \pm 2{,}790.57,$$

and the 95-percent prediction interval is $17{,}828.66 \leq \mu_{Y \cdot 10} \leq 23{,}409.80$. We can conclude with 95-percent confidence that the average income of all the people within our population who have 10 years of education lies between $17,828.66 and $23,409.80. But note: the true conditional mean could lie outside the confidence limits just computed. If we follow the above procedure of sampling and computation again and again, however, it will yield ever new, but similar intervals, and they will contain the true conditional mean 95 percent of the time. ▯▯▯

Figure 11.10 shows graphically the prediction interval we have just computed. It also shows that if one calculated similar 95-percent prediction intervals for all other values of X and connected the limits of these intervals graphically, one could envelope the entire estimated regression line in a 95-percent **confidence band** (the width of which would be narrowest for $X = \overline{X}$ and would become ever broader the farther the X values were from their mean).

Figure 11.10   A 95-Percent Confidence Band for the Conditional Mean

The color line in this graph is the least-squares estimated regression line of $\hat{Y}_X = 2{,}144.23 + 1{,}847.50X$ that was fitted to the sample data of Table 11.1 and first shown in Figure 11.6. Segment AC represents the 95-percent prediction interval for $\mu_{Y \cdot 10}$ that reaches from $17,828.66 (point A) past the point estimate of $20,619.23 (point B) to $23,409.80 (point C). Similar intervals can easily be calculated for other values of X. For example, they reach from $6,582.21 to $16,181.25 for $X = 5$, from $21,880.33 to $26,932.89 for $\overline{X} = 12.05$, and from $33,844.59 to $44,343.87 for $X = 20$. Thus, it is possible to envelope the entire estimated regression line in a 95-percent (nonlinear) confidence band and be that confident that any given conditional mean will lie within its limits.

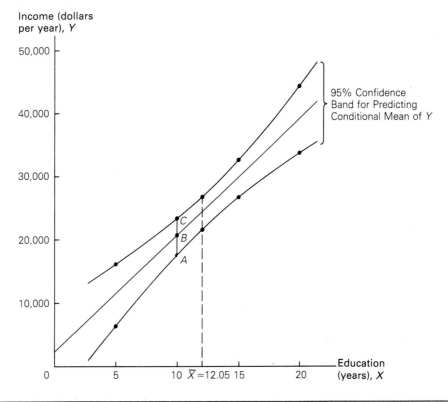

Establishing a prediction interval for $\mu_{Y \cdot X}$ *from a large sample* (n ≥ 30). We also noted in Chapter 7 (in Box 7.C on page 241) that the limits of a confidence interval for the population mean can be established in the large-sample case using the normal distribution as

$$\mu = \overline{X} \pm (z\sigma_{\overline{X}})$$

and that $\sigma_{\overline{X}}$ can be estimated by $s/\sqrt{n}$. In regression analysis, the analogous expression is

$$\mu_{Y \cdot X} = \hat{Y}_X \pm (z\sigma_{\hat{Y}_X})$$

and $s_{Y \cdot X}/\sqrt{n}$ can be used to estimate $\sigma_{\hat{Y}_X}$ in the large-sample case. Thus, the result is the formula in Box 11.C,

---

11.C The Limits of the Prediction Interval for the Average Value of Y, Given X (large-sample case)

$$\mu_{Y \cdot X} = \hat{Y}_X \pm \left(z_{\alpha/2} \cdot \frac{s_{Y \cdot X}}{\sqrt{n}} \right)$$

where $\hat{Y}_X$ is the estimated value of Y, the dependent variable, $s_{Y \cdot X}$ is the sample standard error of the estimate of Y, n is sample size, and z is the standard normal deviate from Appendix Table F.

Practice Problem 5

Establishing a Prediction Interval for the Average Wheat Output, Given Fertilizer Input

Suppose that a sample of 100 acres has revealed the association between pounds of fertilizer input per acre *(X)* and bushels of wheat output per acre *(Y)* that is summarized by the estimated regression line: $\hat{Y}_X = 100 + .1X$. Let $s_{Y \cdot X} = 10$ bushels and assume that a 95-percent prediction interval is desired for the conditional mean of Y, given $X = 100$ pounds. Clearly, the point estimate is $\mu_{Y \cdot 100} = 100 + .1(100) = 110$ bushels per acre. The appropriate value of z can be found in Appendix Table F as 1.96. Thus, the interval limits are:

$$\mu_{Y \cdot 100} = 110 \pm \left[1.96 \left(\frac{10}{\sqrt{100}} \right) \right]$$
$$= 110 \pm 1.96 \text{ bushels,}$$

and the interval is $108.04 \leq \mu_{Y \cdot 100} \leq 111.96$ bushels of wheat. |||

Predicting an Individual Value of Y, Given X

Suppose now that we wished to predict, at a given X, not the average value of Y, or $\mu_{Y \cdot X}$, but an individual value, $I_{Y \cdot X}$, such as the next one we are likely to encounter in the process of sampling. In our example from Practice Problem 1, we may wish to predict the income of the next member of our population who has an education of, say, 10 years.

The point estimate for an individual Y is the same as for the average Y; in our case it equals once again $I_{Y \cdot X} = \hat{Y}_{10} = 20{,}619.23$. The prediction interval is once again determined differently for small and large samples.

Establishing a prediction interval for $I_{Y \cdot X}$ *from a small sample* (n < 30). The formula given in Box 11.B has to be amended only slightly, as shown in Box 11.D. The only difference is the "+ 1" appearing in the square root; its presence acknowledges the existence of a third component yet in the variability of $\hat{Y}_X$: the dispersion, illustrated in Figure 11.9, of individual Y values around the estimated regression line.

11.D The Limits of the Prediction Interval for an Individual Value of Y, Given X (small-sample case)

$$I_{Y \cdot X} = \hat{Y}_X \pm \left(t_{\alpha/2} \cdot s_{Y \cdot X} \sqrt{\frac{1}{n} + \frac{(X - \overline{X})^2}{\Sigma X^2 - n\overline{X}^2} + 1} \right)$$

where Xs are observed values of the independent variable ($\overline{X}$ being their mean), while $\hat{Y}_X$ is the estimated value of dependent variable Y, $s_{Y \cdot X}$ is the sample standard error of the estimate of Y, n is sample size, and the t-statistic is found in Appendix Table G for $n - 2$ degrees of freedom.

||| Practice Problem 6

Establishing a Prediction Interval for an Individual Value of Income, Given Education
Using the Box 11.D formula and the values previously calculated, we can quickly establish the limits of a 95-percent prediction interval for an individual value of Y, given an X of 10 years, as

$$I_{Y \cdot 10} = 20{,}619.23 \pm [2.101(5{,}377.37)\sqrt{.05 + .011 + 1}]$$
$$= 20{,}619.23 \pm 11{,}636.79.$$

Thus, the interval is: $\$8{,}982.44 \leq I_{Y \cdot 10} \leq \$32{,}256.02$. In other words, we can be 95-percent confident that the next person encountered in sampling who has 10 years of education will have an income that lies within this range. |||

This interval, it should be noted, is considerably wider than that previously calculated for $\mu_{Y \cdot 10}$. Indeed, as Figure 11.11 shows, the entire confidence band for individual values of Y, given X, lies outside that for average values of Y.

Figure 11.11   A 95-Percent Confidence Band for the Individual Y, Given X

The color line in this graph is the least-squares estimated regression line of $\hat{Y}_X = 2{,}144.23 + 1{,}847.50X$ that was fitted to the sample data of Table 11.1 and first shown in Figure 11.6. Segment AC represents the 95-percent prediction interval for $I_{Y·10}$ that reaches from \$8,982.44 (point A) past the point estimate of \$20,619.23 (point B) to \$32,256.02 (point C). Similar intervals for individual Y values can easily be calculated given other specified values of X; in all cases, they are wider than those for $\mu_{Y·X}$. Note how the solid confidence band for predicting individual values of Y lies beyond the dashed confidence band (copied from Figure 11.10) for average values of Y.

Establishing a prediction interval for $I_{Y·X}$ *from a large sample* ($n \geq 30$). In this case, the formula given in Box 11.C has to be amended only slightly, as shown in Box 11.E.

11.E The Limits of the Prediction Interval for an Individual Value of Y, Given X (large-sample case)

$$I_{Y·X} = \hat{Y}_X \pm (z_{\alpha/2} \cdot s_{Y·X})$$

where $\hat{Y}_X$ is the estimated value of Y, the dependent variable, $s_{Y·X}$ is the sample standard error of the estimate of Y, and z is the standard normal deviate from Appendix Table F.

Practice Problem 7

Establishing a Prediction Interval for an Individual Value of Wheat Output, Given Fertilizer Input

Consider applying the Box 11.E formula to the earlier fertilizer-and-wheat example in Practice Problem 5, the goal now being to predict with 95-percent confidence the output of the next acre (rather than the average output of many acres) on which $X = 100$ pounds of fertilizer was applied. In that case, the interval limits are:

$$I_{Y \cdot 100} = 110 \pm [1.96(10)]$$
$$= 110 \pm 19.6 \text{ bushels,}$$

and the interval is $90.4 \leq I_{Y \cdot 100} \leq 129.6$ bushels of wheat. Once again the interval is considerably wider than that for $\mu_{Y \cdot 100}$. III

Making Inferences About the True Regression Coefficients

Often the making of inferences about α and β, the coefficients of the true regression line, is even more important to researchers and decision makers than the prediction, for any specified value of the independent variable, X, of an associated average or individual value of the dependent variable, Y. As was noted above, when the method of least squares is employed, the estimated regression coefficients, a and b, are un-biased, efficient, and consistent estimators of α and β; yet, here, too, it is often desirable to establish confidence intervals. The establishment of confidence intervals is particularly common with respect to β, the slope of the true regression line, because there is always the danger that a positive or negative b is a fluke resulting from the selection of a highly improbable sample and that β is in fact zero (meaning there is no association whatsoever between X and Y in the population as a whole).

Establishing a confidence interval for β. In our small-sample case, the limits of a confidence interval for β can be established using the t distribution as

$$\beta = b \pm (t\sigma_b)$$

where σ_b is the standard error of slope b and

$$\sigma_b = \frac{\sigma_{Y \cdot X}}{\sqrt{\Sigma X^2 - n\overline{X}^2}}$$

but where σ_b can be estimated by s_b and

$$s_b = \frac{s_{Y \cdot X}}{\sqrt{\Sigma X^2 - n\overline{X}^2}}.$$

Box 11.F gives the formula for the limits of the confidence interval for β.

11.F The Limits of the Confidence Interval for β (small-sample case)

$$\beta = b \pm \left(t_{\alpha/2} \cdot \frac{s_{Y \cdot X}}{\sqrt{\Sigma X^2 - n\overline{X}^2}} \right)$$

where b is the estimated (slope) regression coefficient, Xs are observed values of the independent variable ($\overline{X}$ being their mean), while $s_{Y \cdot X}$ is the sample standard error of the estimate of dependent variable Y, n is sample size, and the t-statistic is found in Appendix Table G for $n - 2$ degrees of freedom.
Note: For large samples, the normal deviate $z_{\alpha/2}$ replaces $t_{\alpha/2}$.

Practice Problem 8

Establishing a Confidence Interval for β in the Case of Education and Income
In Table 11.2, we calculated $b = 1,847.50$. As noted in Practice Problem 4, the value of t for a 95-percent confidence interval and $n - 2 = 18$ degrees of freedom is $t_{.025,18} = 2.101$. Also in Practice Problem 4, we found $s_{Y \cdot X} = 5,377.37$. Thus, our interval limits are

$$\beta = 1,847.50 \pm \left(2.101 \frac{5,377.37}{\sqrt{3,285 - 20(12.05)^2}} \right)$$

$$= 1,847.50 \pm [2.101(275.51)] = 1,847.50 \pm 578.84.$$

Thus, the interval is $1,268.66 \le \beta \le 2,426.34$, and we can be 95-percent confident that the true value of β lies within this range. In other words, for the country's population as a whole, each additional year of education yields between $1,269 and $2,426 of additional annual income.

Note again: β may lie outside this range, but if we follow the above procedure of sampling and computation a large number of times, it will yield ever new but similar intervals, and they will contain the true β 95 percent of the time.

Note also: The confidence interval just calculated is distressingly imprecise, but nothing can increase the degree of precision except an increase in sample size. In that case, of course, the t distribution approaches the normal one, and the normal deviate, z, can replace t in our Box 11.F formula. |||

Establishing a confidence interval for α. In our small-sample case, the limits of a confidence interval for α can be established in a similar fashion as that for β.

$$\alpha = a \pm (t\sigma_a)$$

but σ_a—the standard error of intercept a—can be estimated by s_a and

$$s_a = s_b \sqrt{\frac{\Sigma X^2}{n}}.$$

Thus, we can calculate the confidence interval for α using the formula in Box 11.G.

11.G The Limits of the Confidence Interval for α (small-sample case)

$$\alpha = a \pm \left(t_{\alpha/2} \frac{s_{Y \cdot X}}{\sqrt{\Sigma X^2 - n\overline{X}^2}} \sqrt{\frac{\Sigma X^2}{n}} \right)$$

where a is the estimated (intercept) regression coefficient, Xs are observed values of the independent variable ($\overline{X}$ being their mean), while $s_{Y \cdot X}$ is the sample standard error of the estimate of dependent variable Y, n is sample size, and the t-statistic is found in Appendix Table G for $n - 2$ degrees of freedom.

Note: For large samples, the normal deviate $z_{\alpha/2}$ replaces $t_{\alpha/2}$.

Practice Problem 9

Establishing a Confidence Interval for α in the Case of Education and Income
Using once again the calculations of Table 11.2 and those of Practice Problem 8, we get the following 95-percent confidence-interval limits for our example about education and income:

$$\alpha = 2,144.23 \pm \left(2.101(275.51) \sqrt{\frac{3,285}{20}} \right)$$
$$= 2,144.23 \pm 7,418.50$$

Thus, the interval is $-5,274.27 \leq \alpha \leq 9,562.73$. In other words, we can be 95-percent confident that the vertical intercept of the true regression line between education and income lies within this income range. |||

Testing hypotheses about β. A commonly used alternative to establishing a confidence interval for β is testing a hypothesis about it. Sometimes the key question is not the precise value of β, but merely whether the value is zero. If it is zero—no matter what the value of X—the value of $\mu_{Y \cdot X}$ equals α, the true regression line is parallel to the X axis, and regression analysis is of no use in making predictions of Y from X. The usual four steps of hypothesis testing are involved.

Practice Problem 10

A Two-Tailed Test of β in the Case of Education and Income
Step 1: The set of opposing hypotheses typically is that for a two-tailed test:

H_0: $\beta = 0$ (because Y is not associated with X in the population)
H_A: $\beta \neq 0$ (because Y is associated with X in the population).

Step 2: For any small-sample case, such as our limited data set from Practice Problem 1, the Student *t*-statistic is used as the test statistic. Given Box 11.F above, and setting β equal to zero,

$$t = \frac{b}{\dfrac{s_{Y \cdot X}}{\sqrt{\Sigma x^2 - n\overline{X}^2}}}$$

Step 3: Given once again a desired confidence level of 95 percent (that is, a significance level of α = .05, where α clearly does *not* equal the vertical intercept of the true regression line), Appendix Table G implies (for $n - 2 = 18$ degrees of freedom) critical values of $\pm t_{\alpha/2} = \pm t_{.025,18} = \pm 2.101$. Thus, the decision rule must be: "Accept H_0 if $-2.101 \le t \le +2.101$."

Step 4: In our example, the actual value of *t* is computed as

$$t = \frac{1{,}847.50}{\dfrac{5{,}377.37}{\sqrt{3{,}285 - 20(12.05)^2}}} = \frac{1{,}847.50}{275.51} = 6.71.$$

Accordingly, the null hypothesis should be *rejected.* At the 5-percent level of significance, the sample result (a positive slope *b*) is statistically significant. The observed divergence of *b* from the zero value of β hypothesized in the null hypothesis is unlikely to be caused merely by chance factors operating during sampling. There is good reason to embrace the alternative hypothesis: more education does mean more income. Note: This conclusion is also evident by the nature of the 95-percent confidence interval for β calculated in Practice Problem 8; that interval does not contain β = 0 as it would have to if H_0: β = 0 were true. |||

||| Practice Problem 11

An Upper-Tailed Test of β in the Case of Education and Income
When one is fairly certain that *Y* varies directly with *X,* an upper-tailed test can be used instead of a two-tailed test.

Step 1: The set of opposing hypotheses becomes:

H_0: β ≤ 0
H_A: β > 0

Step 2: The test statistic remains the same.

Step 3: If we again use α = .05 as a significance level, the critical value is $t_{\alpha} = t_{.05,18} = 1.734$ (again from Appendix Table G and for 18 degrees of freedom), making the decision rule: "Accept H_0 if $t \le 1.734$."

Step 4: Given the computed value of $t = 6.71$, H_0 should be *rejected.* Thus, this test also indicates that the slope of the true regression line is greater than zero. In the population as a whole, more education does mean more income. Note: When *Y* is inversely related to *X,* an analogous *lower-tailed test* can be employed, with opposing claims of H_0: β ≥ 0 and H_A: β < 0.

Analytical Examples 11.1, "Evaluating a New Food Product," and 11.2, "Of Bulls, Bears, and Beauty" provide typical illustrations of the use of simple regression analysis in business and economics. ▐▐▐

SIMPLE CORRELATION ANALYSIS

While simple regression analysis establishes a precise equation that links two variables, simple correlation analysis seeks to establish a general index about their strength of association. Consider the three panels of Figure 11.12. When all the points in a scatter diagram lie precisely on the estimated regression line, as in panel (a), the index to be discussed in this section will show the variables to be *perfectly correlated* (and in that case, Y can always be predicted with precision from a knowledge of the regression line and any value of X). When, on the other hand, the points in a scatter diagram are so widely scattered as to make X completely worthless as a predictor of Y, as in panel (b), the index will show the variables to be *completely uncorrelated.* When, finally, many scattered points diverge from the regression line, but not without suggesting some definite association between the variables, such as the direct association in panel (c), the index will show the variables to be *imperfectly correlated.*

Several different indexes of association between quantitative variables are used by statisticians, and we will discuss them in order of importance.

The Coefficient of Determination

The most important measure of how well the estimated regression line fits the sample data on which it is based is the **sample coefficient of determination**—which equals the proportion of the total variation in the values of the dependent variable, Y, that can be explained by means of the association of Y with X as measured by the estimated

Figure 11.12 **Different Degrees of Correlation**

The three panels of this graph illustrate different degrees of correlation. Depending on whether the scatter of data points around a regression line is nonexistent, considerable, or moderate, correlation between the variables is said to be (a) perfect, (b) zero, or (c) weak.

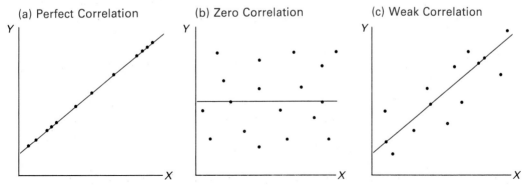

Figure 11.13   Total, Explained, and Unexplained Deviation in Y

The total deviation (here AC) of any observed value of Y from the mean of all such observations ($\bar{Y}$) always equals the sum of the explained deviation (here BC) and the unexplained deviation (here AB).

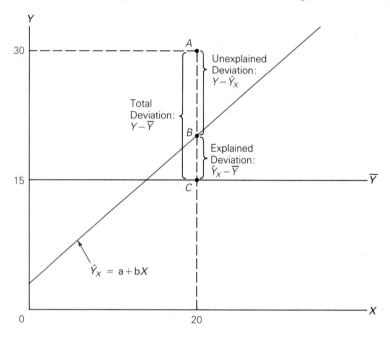

regression line. Imagine that Figure 11.13 is a scatter diagram with all the dots removed except point A (which corresponds to the data pair of $X = 20$, $Y = 30$). Assume that the color line represents the estimated regression line calculated from the complete data set and that the average value of Y is $\bar{Y} = 15$, as illustrated by the horizontal line. The difference between any particular observation of the dependent variable, such as $Y = 30$ when $X = 20$ (point A) and the mean of the dependent variable, such as $\bar{Y} = 15$ (point C), is the **total deviation of Y** (distance AC).

There are, of course, many such deviations in a scatter diagram, one between each individual observation of Y (most of which are not shown here) and the common mean of all these Ys. Now imagine squaring all these total deviations and summing them. The resulting sum of squares of total deviations, $\Sigma(Y - \bar{Y})^2$, is called the **total variation of Y,** or the **total sum of squares *(Total SS).***

Next focus on the difference between the regression estimate of our observed Y, or $\hat{Y}_{20}$ (point B) and the average Y (point C); this difference is the **explained deviation of Y** (distance BC). We can see where this name originates: the positive deviation, BC, of $\hat{Y}_X$ from $\bar{Y}$ is explained by the fact that Y rises with X and the value of $X = 20$ is above average. (If we had chosen $X = 5$ for our example, we would explain a similar negative deviation in an analogous fashion—namely, by $X = 5$ being below average.) Now imagine squaring all these explained deviations in the scatter diagram and summing them. The resulting sum of squares of explained deviations, $\Sigma(\hat{Y}_X - \bar{Y})^2$, is the **explained variation of Y,** or the **regression sum of squares *(RSS).***

Finally, focus on the difference between our particular observation of Y when $X = 20$ (point A) and the corresponding regression estimate, $\hat{Y}_{20}$ (point B); this difference is the **unexplained deviation of Y** (distance AB). Again, we can see the reason for this name: the fact that Y rises with X as shown by the regression line explains why our observed Y (point A) lies above $\overline{Y}$ to the extent of BC, but it does not explain the additional deviation of AB. Now imagine squaring all these unexplained deviations in the scatter diagram and summing them. The resulting sum of squares of unexplained deviations, $\Sigma(Y - \hat{Y}_X)^2$, is the **unexplained variation of Y,** or the **error sum of squares (ESS).** (The method of least squares, of course, guarantees that this error component is minimized.)

Just as the total deviation is the sum of explained and unexplained deviations, so the relationships given in Box 11.H hold as well.

11.H Components of Total Variation

$$\begin{array}{ccc} \text{Total} & = & \text{Explained} & + & \text{Unexplained} \\ \text{Variation} & & \text{Variation} & & \text{Variation} \end{array}$$

or

$$\begin{array}{ccc} \text{Total Sum} & = & \text{Regression Sum} & + & \text{Error Sum} \\ \text{of Squares} & & \text{of Squares} & & \text{of Squares} \\ \textit{(Total SS)} & & \textit{(RSS)} & & \textit{(ESS)} \end{array}$$

or

$$\Sigma(Y - \overline{Y})^2 = \Sigma(\hat{Y}_X - \overline{Y})^2 + \Sigma(Y - \hat{Y}_X)^2$$

where Ys are observed values of the dependent variable ($\overline{Y}$ being their mean) and $\hat{Y}_X$ is the estimated value of Y, given a value of independent variable X.

Given the relationships in Box 11.H, the meaning of the sample coefficient of determination, usually symbolized by r^2, is quite clear—as shown in Box 11.I.

11.I Sample Coefficient of Determination

$$r^2 = \frac{\text{explained variation}}{\text{total variation}} = \frac{RSS}{Total\ SS} = \frac{\Sigma(\hat{Y}_X - \overline{Y})^2}{\Sigma(Y - \overline{Y})^2}$$

Note: Because *Total SS* − *ESS* = *RSS*, we can also write

$$r^2 = \frac{Total\ SS - ESS}{Total\ SS} = 1 - \frac{ESS}{Total\ SS} = 1 - \frac{\Sigma(Y - \hat{Y}_X)^2}{\Sigma(Y - \overline{Y})^2}.$$

where Ys are observed values of the dependent variable ($\overline{Y}$ being their mean) and $\hat{Y}_X$ is the estimated value of Y, given a value of independent variable X.

When X and Y are perfectly correlated, as in panel (a) of Figure 11.12, every $Y = \hat{Y}_X$; hence, $\Sigma(Y - \hat{Y}_X)^2 = 0$, and $r^2 = 1 - 0 = 1$. When instead X and Y are completely uncorrelated, the regression line has a slope of zero, as in panel (b) of Figure 11.12. Then $\hat{Y}_X = \overline{Y}$, the last fraction in Box 11.I equals one, and $r^2 = 1 - 1 = 0$. Thus, the value of r^2 always lies between 1 (perfect correlation) and 0 (no correlation). The formula in Box 11.J is one of several mathematical equivalents to the formulas in Box 11.I and is easier and less cumbersome to employ.

11.J Sample Coefficient of Determination: Alternative Formulation

$$r^2 = \frac{a\Sigma Y + b\Sigma XY - n\overline{Y}^2}{\Sigma Y^2 - n\overline{Y}^2}$$

where Xs are observed values of the independent variable, Ys are observed associated values of the dependent variable ($\overline{Y}$ being their mean), while n is sample size, and a and b are the estimated (intercept and slope) regression coefficients, respectively.

||| Practice Problem 12

Calculating the Sample Coefficient of Determination (r^2) for Education and Income
To calculate r^2 for our example, we could apply the Box 11.I formula to the data of Table 11.2, but it would be less cumbersome to apply the formula in Box 11.J, using values already derived during regression analysis. Thus, the sample coefficient of determination can be calculated as

$$r^2 = \frac{2,144.23(488,132) + 1,847.5(6,585,796) - 20(24,406.6)^2}{13,734,414,590 - 20(24,406.6)^2} = .71.$$

In other words, 71 percent of the total variation in Y (income) observed in Figure 11.6 can be explained by the association of Y (income) with X (education) as estimated by the regression line shown there. Naturally, statisticians are always pleased with a high r^2 because it indicates that they have explained a large proportion of the variation in the variable of interest.

   Note: The sample coefficient of determination varies from sample to sample; it is only an estimate of the underlying **population coefficient of determination (ρ^2).** The symbol ρ^2 is the lowercase Greek rho, squared.

   Analytical Example 11.3, "The Inflation Rate Versus Its Predictability," illustrates how research results are routinely assessed with the help of the coefficient of determination. |||

The Coefficient of Correlation

The square root of the sample coefficient of determination, $\sqrt{r^2}$, is a common alternative index of the degree of association between two quantitative variables. This measure is called the **sample coefficient of correlation (r)** and is a point estimator of the **population coefficient of correlation (ρ).** Just like r^2, the coefficient r takes on absolute values between 0 and 1, but it is common to take advantage of the fact that the square root of any number can be positive or negative and to place a plus sign or a minus sign in front of r to denote, respectively, a direct relationship or an inverse relationship between X and Y. Thus, $r = 0$ denotes no correlation, while $r = +1$ or $r = -1$ denotes, respectively, perfect correlation between directly related variables or perfect correlation between inversely related variables. In our example, therefore, the $r^2 = .71$ translates into $r = \sqrt{.71} = +.84$ because the relationship between education and income is a direct one. Note: An alternative formula for calculating r without prior regression analysis is given in Box 11.K.

11.K Pearson's Sample Coefficient of Correlation

$$r = \frac{\Sigma XY - n\overline{X}\,\overline{Y}}{\sqrt{(\Sigma X^2 - n\overline{X}^2)(\Sigma Y^2 - n\overline{Y}^2)}}$$

where Xs are observed values of the independent variable ($\overline{X}$ being their mean), Ys are observed associated values of the dependent variable ($\overline{Y}$ being *their* mean), and n is sample size.

||| Practice Problem 13

Calculating Pearson's Sample Coefficient of Correlation (r) for Education and Income
When applied to our Table 11.2 data, the formula in Box 11.K yields

$$r = \frac{6{,}585{,}796 - 20(12.05)(24{,}406.6)}{\sqrt{[3{,}285 - 20(12.05)^2][13{,}734{,}414{,}590 - 20(24{,}406.6)^2]}}$$

$$= \frac{703{,}805.4}{\sqrt{(380.95)(1{,}820{,}772{,}118.8)}} = \frac{703{,}805.4}{836{,}434.7} = .84,$$

and this value confirms the earlier calculation. Income is strongly correlated with education. **|||**

Finally, three notes of caution must be sounded. First, the correlation coefficient discussed here is capable of detecting only linear associations between variables. It is quite possible to calculate $r = 0$ from a set of data such as that given in panel (c) of Figure 11.2—thus, falsely suggesting no association between X and Y. In fact, there is no *linear* association, but quite a pronounced curvilinear one.

Second, statisticians prefer to use r^2 rather than r as a measure of association because fairly high absolute values of r (such as .70) can give the false impression that there exists a strong association between Y and X (but in this case, $r^2 = .49$; thus, the association of Y with X explains less than half of the variation of Y). Only at the extremes, when $r = \pm 1$ or when $r = 0$ (and, therefore, $r = r^2$) does the value of r directly convey what proportion of the variation of Y is explained by X. Analytical Example 11.4, "Predicting Changes in the GNP," illustrates the use of these coefficients.

Third, we must reemphasize that the correlation coefficient measures only the strength of a statistical relationship between two variables, but the discovery of a strong statistical relationship between X and Y need not imply a causal one. Perhaps X does cause Y; perhaps Y is causing X; perhaps each is part cause, part effect of the other. Perhaps the two variables move together by pure chance. Indeed, the existence of a high positive or negative correlation between two variables that have no logical connection with one another is common and is called **nonsense correlation** or **spurious correlation.** By pure chance one might discover, for example, a correlation coefficient of $r = +.98$ between the numbers of ministers and illegitimate pregnancies in the 50 states or a correlation coefficient of $r = -.85$ between the numbers of whales caught in the world's oceans and Congressional expenditures on stationery. Sometimes, of course, such statistical associations reflect the operation of some third factor (such as population differences) that affects both of the variables (producing simultaneously more ministers and also more illegitimate pregnancies in states where population is larger). Analytical Example 11.5 "Snowfall and Unemployment," provides a vivid illustration of the problem.

AN IMPORTANT EXTENSION

We must, finally, consider an important extension of the foregoing analysis: a new application of the previous chapter's analysis of variance.

Testing β With Analysis of Variance

In Practice Problem 10, we tested the hypothesis that $\beta = 0$. The analysis of variance, introduced in Chapter 10, provides an alternative way of testing this hypothesis. The ANOVA procedure yields the same result as the t test employed above, but it has one major advantage: it can be applied to multiple regression problems as well.

The ANOVA test makes use of the three sums of squares discussed previously:

1. the regression sum of squares, $RSS = \Sigma(\hat{Y}_X - \overline{Y})^2$,
2. the error sum of squares, $ESS = \Sigma(Y - \hat{Y}_X)^2$, and
3. the total sum of squares, $Total\ SS = \Sigma(Y - \overline{Y})^2$.

‖‖‖ Practice Problem 14

An ANOVA Test of β in the Case of Education and Income

The sums of squares can be calculated from the data of Table 11.2 and the formulas already noted, but they can also be derived from mathematically equivalent expressions that yield the following for our example:

$$
\begin{aligned}
RSS &= \Sigma(\hat{Y}_X - \overline{Y})^2 = a\Sigma Y + b\Sigma XY - n\overline{Y}^2 \\
&= 2{,}144.23(488{,}132) + 1{,}847.5(6{,}585{,}796) - 20(24{,}406.6)^2 \\
&= 1{,}046{,}667{,}278.36 + 12{,}167{,}258{,}110 - 11{,}913{,}642{,}471.2 \\
&= 1{,}300{,}282{,}917.16. \\
ESS &= \Sigma(Y - \hat{Y}_X)^2 = \Sigma Y^2 - a\Sigma Y - b\Sigma XY \\
&= 520{,}489{,}201.64
\end{aligned}
$$

(as already noted in the calculation of $s_{Y \cdot X}$ in Practice Problem 4).

Table 11.3 The ANOVA Table

Source of Variation	Sum of Squares (1)	Degrees of Freedom (2)	Mean Square (3) = (1) ÷ (2)	Test Statistic (4)
Regression (variation explained by X)	$RSS =$ 1,300,282,917.16	$m = 1$	$RMS =$ 1,300,282,917.16	$F = \dfrac{RMS}{EMS} = 44.97$
Error (unexplained variation)	$ESS =$ 520,489,201.64	$n - 2 = 18$	$EMS =$ 28,916,066.76	
Total	$Total\ SS =$ 1,820,772,118.80	$n - 1 = 19$		

$$Total\ SS = \Sigma(Y - \overline{Y})^2 = \Sigma Y^2 - n \cdot \overline{Y}^2$$
$$= 1,820,772,118.80$$

(as already noted in the calculation of r using the Box 11.K formula in Practice Problem 13). These sums of squares have also been entered in column 1 of Table 11.3. The appropriate degrees of freedom, which appear in column 2, equal the number of independent variables, m, for the regression sum of squares; they equal sample size, n, minus 2 for the error sum of squares (there are n errors and 2 constraints—namely, the estimated regression coefficients, a and b); they equal $n - 1$ for the total sum of squares [there are n deviations subject to the constraint that $\Sigma(Y - \overline{Y}) = 0$].

From this information, the regression mean square and error mean square of column 3 can be calculated; the error mean square has, in fact, been derived in Practice Problem 4 in connection with the calculation of $s_{Y \cdot X}$. These mean squares are sample variances that can be used to calculate F, as shown in column 4.

Suppose we want to test H_0: $\beta = 0$ at the 5-percent level of significance. We can find the critical value of F in Appendix Table I for 1 numerator $d.f.$ and 18 denominator $d.f.$ as $F_{.05} = 4.41$. Thus, there is only a 5-percent chance that the computed value of F exceeds 4.41 if H_0 is true. Accordingly, as in our t test in Practice Problem 10, we should *reject* H_0. We conclude that $\beta \neq 0$; X (education) does help explain the variation in Y (income) in the population as a whole. |||

||| ANALYTICAL EXAMPLES

11.1 Evaluating a New Food Product

Americans are said to be eager for on-the-run meals, provided only that they are also palatable and nutritious. The General Foods Corporation decided to bring out such a product, called H. In order to test the palatability and nutritive quality of H, it arranged for a 28-day rat-feeding experiment (because rats have metabolic processes similar to people), and it used regression techniques to evaluate the experimental data to draw conclusions about marketing such a product.

In order to provide randomization, 30 rats were divided into 10 blocks of 3 rats each such that the rats within any one block had about the same initial

(a)

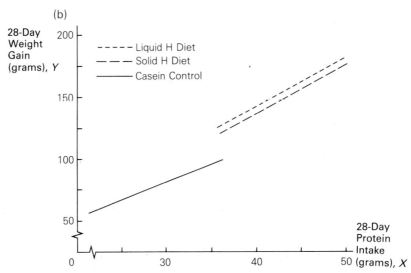

(b)

weight. One member of each trio was then assigned, at random, to a diet of either liquid H or solid H or to a control diet of casein. Each of the three diets contained about 9 percent protein, and the rats could eat all they wanted. Ultimately, the rats' final and initial weights were compared.

Panel (a) of the accompanying figure shows a scatter diagram of the results; panel (b) shows the regression line for the 3 dietary groups, relating 28-day weight gain, Y, to protein intake, X.

The equations were: for liquid H, $\hat{Y}_X = -4.124 + 3.72\ X$; for solid H, $\hat{Y}_X = -8.478 + 3.66\ X$; for casein control, $\hat{Y}_X = -3.644 + 2.91\ X$.

The company statisticians concluded (a) from the position of the H lines to the right of the casein line that rats ate more H than casein because H was more palatable (at least to rats) and (b) from the higher slope of the H lines that the protein efficiency of H was superior to that of the control diet (protein contained in H resulted in greater weight gain). The difference between the slopes of the H lines (3.72 vs. 3.66) was not statistically significant; the difference between the H-line slopes and the slope of the casein line (2.91) was statistically significant.

Source: Elisabeth Street and Mavis B. Carroll, "Preliminary Evaluation of a New Food Product," in Judith M. Tanur et al., eds., *Statistics: A Guide to the Unknown* (San Francisco: Holden-Day, 1972), pp. 220–28.

11.2 Of Bulls, Bears, and Beauty

Stock market analysts sometimes characterize the riskiness of a particular stock (such as TWA or IBM or Procter & Gamble) with the help of a sample regression line that relates the stock's rate of return to the average rate of return on all stocks, as in Figure A.

Depending on whether the slope, β, of the true regression line is believed to exceed, equal, or fall short of unity, the stock is referred to as an *aggressive, marketlike,* or *defensive* stock. In a bull market (when the average rate of return on all stocks is rising), an aggressive stock gains more than the average; a defensive stock gains less. In a bear market (when the average rate of return on all stocks is falling), an aggres-

sive stock loses more than the average; a defensive stock loses less. Clearly, an investor is well advised to own aggressive stocks in a bull market, but defensive stocks in a bear market. Thus, stock market analysts are always engaging in "beta analysis."

Recently, an economist derived 1971–1984 rates-of-return on particular types of *paintings* from components of the well-known art-market price index produced by Sotheby's London auction house and compared these returns on particular assets with the average rate-of-return on a market portfolio of bonds, real estate, and stocks. As Figure B shows, differences in riskiness similar to these connected with particular stocks emerged.

Figure A

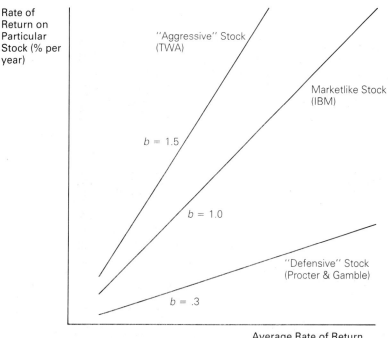

Rate of Return on Particular Stock (% per year)

"Aggressive" Stock (TWA)

Marketlike Stock (IBM)

$b = 1.5$

$b = 1.0$

"Defensive" Stock (Procter & Gamble)

$b = .3$

Average Rate of Return on All Stocks (% per year)

Figure B

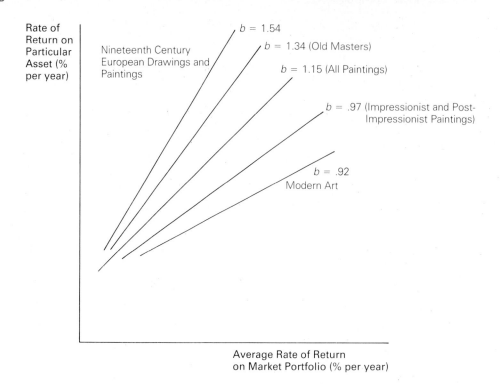

Sources: Charles L. Olson and Mario J. Picconi, *Statistics for Business Decision Making* (Glenview, IL: Scott, Foresman and Company, 1983), p. 496, and Michael F. Bryan, "Beauty and the Bulls: The Investment Characteristics of Paintings," *Federal Reserve Bank of Cleveland Economic Review,* I, 1985, pp. 2–10.

11.3 The Inflation Rate Versus Its Predictability

Two researchers suspected the existence of a direct relationship between the level of inflation and its variability (and, thus, an inverse relation between inflation and its predictability). Using a sample of 40 economies during the 1970s, they recorded each country's average rate of inflation, $X = \bar{I}$ (which ranged from 4.89 percent in Germany to 59.25 percent in Uruguay) as well as each country's standard deviation of this inflation rate, $Y = SDI$. The following regression equation was estimated from these data:

$$SDI = a + b\bar{I} = -.119 + .507\,\bar{I}.$$

The coefficient of determination was $r^2 = .84$, and a similar coefficient for ρ^2 was confirmed. Thus, it was confirmed that higher levels of inflation are strongly and positively related to inflation variability (and, thus, reduce inflation predictability).

Source: R. W. Hafer and Gail Heyne-Hafer, "The Relationship Between Inflation and Its Variability: International Evidence from the 1970s," *Journal of Macroeconomics,* Fall 1981, pp. 571–77.

11.4 Predicting Changes in the GNP

There are two major types of policies that governments can pursue in order to affect the Gross National Product (GNP). Monetary policy works through changes in the money supply, ΔM; and fiscal policy works through changes in government expenditures (ΔG) or in tax collections (ΔT). One economist wanted to test whether a knowledge of ΔM or ΔG is better for predicting an associated change in GNP, ΔY.

Using U.S. data about quarterly changes between 1953 and 1969, the accompanying scatter diagrams were produced. Coefficients of determination and correlation were calculated from the data; they are shown along with the two graphs. It was concluded that ΔY can be predicted more accurately from ΔM than from ΔG—but note: even so only 44 percent of the variation in ΔY is explained by ΔM. Caution: This is shown by $r^2 = .44$ in panel (a), not by $r = .44$ in panel (b).

Source: Leonall C. Andersen, "Statistics for Public Financial Policy," in Judith M. Tanur et al., eds., *Statistics: A Guide to the Unknown* (San Francisco: Holden-Day, 1972), pp. 321–25.

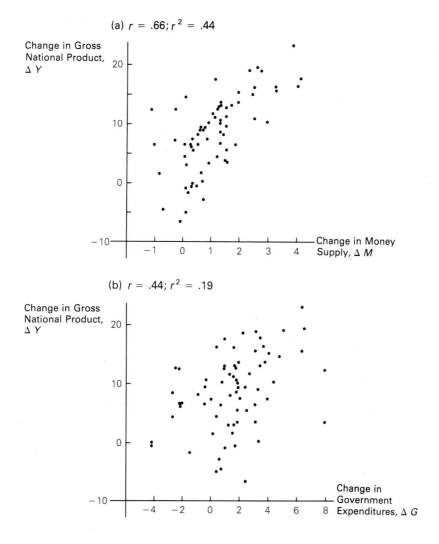

(a) $r = .66$; $r^2 = .44$

(b) $r = .44$; $r^2 = .19$

11.5 Snowfall and Unemployment

A splendid example of spurious correlation is provided by the data in the accompanying table. A scatter diagram reveals a strong positive association between the two variables; their relationship can be summarized by the least-squares regression line, $\hat{Y}_X = .3764 + .0955X$, which is also shown in the accompanying graph. The coefficient of determination equals $r^2 = .967$, implying that 96.7 percent of the variation in U.S. unemployment between 1973 and 1982 can be explained by variations in snowfall at Amherst, Massachusetts. Yet, unlike the association between smoking and health (Close-Up 11.1), this association is totally unreasonable. Nobody would be prepared to argue that the strong association between U.S. unemployment and the extent of snowfall in a small New England town is anything but pure coincidence.

Sources: Economic Report of the President (Washington, D.C.: U.S. Government Printing Office, 1983), p. 199, and data collected by the author.

Year	Snowfall at Amherst, Massachusetts (inches)	U.S. Unemployment Rate (percent)
1973	45	4.9
1974	59	5.6
1975	82	8.5
1976	80	7.7
1977	71	7.1
1978	60	6.1
1979	55	5.8
1980	69	7.1
1981	79	7.6
1982	95	9.7

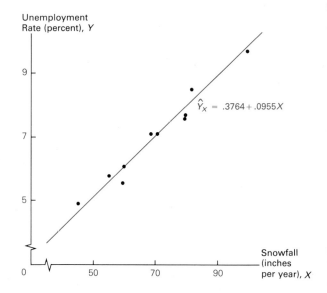

CLOSE-UP

11.1 Smoking and Health

The long-standing controversy on smoking and health provides an excellent example of how regression and correlation analysis has been employed to discover possible *causal* connections between variables yet has been unable to provide definite proof. After hundreds of years of tobacco use, governments throughout the world now condemn the practice as a serious health hazard, claiming that smoking tobacco causes cancer, heart attacks, and many other deleterious effects. (In the early 1980s, some 340,000 deaths per year were blamed on tobacco smoking in the United States alone.) Such conclusions have been based on scatter diagrams, showing a strong positive association between different countries' per capita cigarette consumption and lung-cancer deaths. They have been based on life expectancy data such as those in the accompanying graph, showing that only 45 percent of heavy smokers, but 65 percent of nonsmokers survive to age 60 (and similarly, for all other ages). They have been based on prospective studies that select

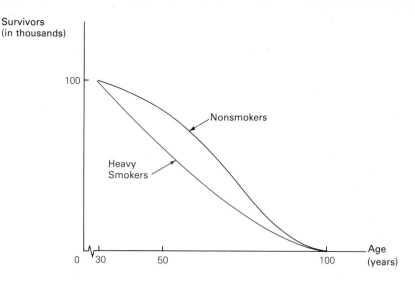

large numbers of people who either smoke or don't smoke and whose health is then monitored until death. (One such study has shown heavy smokers to die from lung cancer at a rate of 1.66 per 1,000 per year, while the corresponding rate for nonsmokers was .07. The corresponding numbers for deaths from heart attacks were 5.99 and 4.22.)

Yet despite the mounting evidence against smoking, there are many who remain unconvinced, citing such problems as selection bias in the studies noted above.

(Sir Ronald Fisher—Biography 8.1—was among these dissenters.)

Sources: R. Doll, "Etiology of Lung Cancer," *Advances in Cancer Research* 3(1955):1–50; R. A. Fisher, *Smoking: The Cancer Controversy* (Edinburgh: Oliver and Boyd, 1959); Report of the U.S. Surgeon General, *Smoking and Health,* 1964; B. W. Brown, Jr., "Statistics, Scientific Method, and Smoking," in Judith M. Tanur et al., eds., *Statistics: A Guide to the Unknown* (San Francisco: Holden-Day, 1972), pp. 40–51.

BIOGRAPHY

11.1 Francis Galton

Francis Galton (1822–1911) was born in Birmingham, England. Intellectually precocious, he turned to studying mathematics, and then medicine, at an unusually early age. Yet he never finished his studies. When his father died and left him a fortune at age 22, he took to travelling: down the Danube to the Black Sea, to Egypt, and on to then-unexplored parts of southwest Africa. Upon his return, he settled in London for the rest of his life (and promptly won the Royal Geographic Society's gold medal for his earlier exploits). He never held any academic or professional posts, but he was driven by an endless curiosity about humanity and nature, and he never ceased to roam every nook and cranny of the world of thought. A rich flow of original ideas yielded 16 books and more than 200 articles and, eventually, brought him knighthood shortly before his death.

Consider just these few examples of his accomplishments: He was a meteorologist who systematically charted weather patterns and in the process discovered and named the anticyclone. He was a psychologist who measured sensory acuity and character traits and sowed the seeds of mental testing. Like his cousin, Charles Darwin, he was a biologist and as such established fingerprinting as an infallible means of human identification. (His taxonomy of prints is used to this day.) Using data on notable families, he also studied the inheritance of talent (artistic, athletic, scholarly) and concluded, in such works as *Hereditary Genius* (1869) and *Natural Inheritance* (1889), that inheritance played a significant role in transmitting talent, even after making due allowance for environmental factors. Indeed, he became convinced that a "eugenics" program (he coined the word) "to foster talent and healthiness and suppress stupidity and sickliness" was crucial for the promotion of a high-quality society. He, therefore, endowed the Galton chair of eugenics at University College, London; the first occupant of that chair, Karl Pearson (Biography 9.1), extended and refined much of Galton's work—particularly Galton's work in statistics.

Galton had a passion for collecting numerical data and exhibited an astonishing cleverness in analyzing them. Thus, he became the originator of the modern regression and correlation techniques discussed in this chapter (that were foreshadowed by Carl Friedrich Gauss—Biography 6.1). In 1875, Galton experimented with sweet peas to determine the law of inheritance of size. He selected 7 groups of different sizes, persuaded friends in different parts of England to plant 10 seeds from each group, and eventually compared the sizes of parental seeds with those of their offspring. When he plotted the results, as in the accompanying figure, he discovered that instead of each offspring being just like its parent (which would put his pairwise data plots

precisely on the 45° line in our graph), dwarf peas produced larger offspring, while giant peas produced smaller offspring (note how the actual data plots diverge from the 45° line).[1] Galton called this phenomenon "reversion" (toward some average ancestral type) and later "regression toward mediocrity." Thus, Galton argued, for purposes of predicting the transmission of natural characteristics from one generation to the next, the 45° line in the accompanying figure was inferior to another or *regression* line going through the scatter of dots. Nowadays, this tendency of those members of any population who, with respect to a given characteristic, are in an extreme position (below or above the population mean) at one time to be in a less extreme position at a later time (either personally or by means of their offspring) is called the **regression effect.** It is found to apply widely and not only to the sizes of sweet peas from one generation to the next. Thus, as Galton also discovered, on average, extremely short fathers tend to have somewhat taller sons; extremely tall fathers tend to have somewhat shorter sons. The subset of all students with extremely poor grades on exam 1 tends to have better grades (closer to the average of all students) on exam 2, while the subset of all students with extremely good grades on exam 1 tends to have worse grades (again closer to the average of all students) on exam 2. And firms with the worst profit picture in year 1, on average, are not

[1]Francis Galton, "Typical Laws of Heredity," *Proceedings of the Royal Institute* 8(1877):282–301.

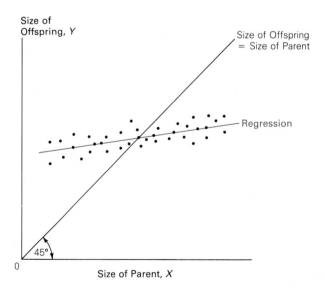

among the worst in year 2, while those with the highest profit in year 1, on average, are not in the top group in year 2.

But caution is advised: it is one thing merely to observe this regression effect; it is quite another to succumb to the **regression fallacy** and incorrectly attribute the regression effect (as Galton did) to the operation of some important unseen factor (such as his "tendency toward mediocrity"). The real explanation is that extreme values (whether of size, grades, or profit) often occur by pure chance; the odds are that they do not occur twice in a row.

Sources: Dictionary of Scientific Biography, vol. 5 (New York: Scribner's, 1972), pp. 265–67; *International Encyclopedia of Statistics,* vol. 1 (New York: Free Press, 1978), pp. 359–64.

Summary

1. In many situations that are of interest to researchers or decision makers, the value of one variable is associated with the value of another variable in some systematic way. Regression and correlation analysis is concerned with such linkages. The central focus of *regression analysis* is the establishment of an equation that allows the unknown value of one variable to be estimated from the known value of one or more other variables. (A single variable is used to do the estimating in *simple* regression analysis; several variables are used in *multiple* regression analysis.) The central focus of *simple correlation analysis* is the establishment of an index that provides, in a single number, an instant picture of the strength of association between two variables.

2. An understanding of regression analysis requires an understanding of the basic concepts of independent vs. dependent variables, deterministic vs. stochastic relationships, and direct vs. inverse relationships. The technique of simple regression analysis for two variables that are related in linear fashion begins with the plotting of sample data in a scatter diagram. In principle, regression lines can be fitted to the scatter diagram (a) by eyesight or (b) by the method of least squares, but (b) is preferred to (a) because the result of (b) alone is reliably reproducible and usable for making a variety of inferences. An *estimated regression line* shows the association between two variables present in a sample, while the *true regression line* shows the association present in the population. Certain basic assumptions must hold before one can make valid inferences about the true regression line with the help of the estimated regression line. Inferences can be made concerning the average value of Y, given X (for both the small-sample and large-sample case), concerning an individual value of Y, given X (for both the small-sample and large-sample case), and concerning the true regression coefficients (using confidence intervals or hypothesis tests).

3. Simple correlation analysis provides a variety of general indices measuring the strength of linear association between two variables. These indexes include (a) the sample coefficient of determination, r^2 (which equals the proportion of the total variation in the values of Y that can be explained by means of the association of Y with X as measured by the estimated regression line) and (b) the sample coefficient of correlation, r.

4. The analysis of variance can play a role in regression analysis (such as in testing hypotheses concerning β, the slope of the true regression line).

Key Terms

conditional mean of Y ($\mu_{Y \cdot X}$)
conditional probability distribution of Y
conditional standard deviation of Y ($\sigma_{Y \cdot X}$)
confidence band
dependent variable
deterministic relationship
direct relationship
error sum of squares *(ESS)*

estimated regression coefficients
estimated regression line
explained deviation of Y
explained variable
explained variation of Y
explanatory variable
independent variable
inverse relationship

method of least squares	sample coefficient of correlation (r)
multiple regression analysis	sample coefficient of determination (r^2)
nonsense correlation	sample regression line
population coefficient of correlation (ρ)	sample standard error of the estimate $(s_{Y \cdot X})$
population coefficient of determination (ρ^2)	scatter diagram
population regression line	simple correlation analysis
population standard error of the estimate	simple regression analysis
predicted variable	spurious correlation
prediction interval	stochastic relationship
predictor variable	total deviation of Y
regressand	total sum of squares *(Total SS)*
regression analysis	total variation of Y
regression effect	true regression coefficients
regression fallacy	true regression line
regression line	unexplained deviation of Y
regression sum of squares *(RSS)*	unexplained variation of Y
regressor	

Questions and Problems

The computer program noted at the end of the chapter can be used to work most of the subsequent problems.

1. The following has been hypothesized about U.S. presidential elections: the percentage of the popular vote, Y, that goes to the candidate put up by the party of the incumbent president depends on the change in the unemployment rate during the election year, X, in such a way that any increase in unemployment detracts from, and any decrease in unemployment adds to, the vote for the incumbent-party candidate. On the basis of the data provided in Table 11.A, test this hypothesis by (a) pre-

paring a scatter diagram with axes like those in Figure A, (b) estimating a regression line with the equation, $\hat{Y}_X = a + bX$, and (c) drawing this line in your diagram. Do your results make sense?

2. An avionics manufacturer wants to establish the relationship between the monthly sales of long-range navigation units, Y, and the number of ads placed in monthly magazines, X. Given the data of Table 11.B, determine an estimated regression equation.

Figure A

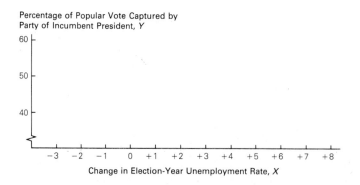

Percentage of Popular Vote Captured by Party of Incumbent President, Y

Change in Election-Year Unemployment Rate, X

Table 11.A

Election Year	Party of Incumbent President[a]	Change in Election-Year Unemployment Rate, X	Percentage of Popular Vote Captured by Party of Incumbent President, Y
1892	1	−2.4	48.3
1896	0	+0.7	47.8
1900	1	−1.5	53.2
1904	1	+1.5	60.0
1908	1	+5.2	54.5
1912	1	−2.1	54.7
1916	0	−3.4	51.7
1920	0	+3.8	36.1
1924	1	+2.6	54.3
1928	1	+0.9	58.8
1932	1	+7.7	40.9
1936	0	−3.2	62.5
1940	0	−2.6	55.0
1944	0	−0.7	53.8
1948	0	−0.1	52.4
1952	0	−0.3	44.6
1956	1	−0.3	57.8
1960	1	0.0	49.9
1964	0	−0.5	61.3
1968	0	−0.2	49.6
1972	1	−0.3	61.8
1976	1	−0.8	49.4
1980	0	+1.3	48.4
1984	1	−2.1	59.0

[a]Democrats = 0; Republicans = 1.

Table 11.B

Loran Units Sold, Y		Number of Ads, X		Loran Units Sold, Y		Number of Ads, X	
7	9	20	126	18	11	187	172
4	13	36	195	9	12	70	160
8	17	36	158	11	6	81	50
7	7	50	77	12	9	111	96
18	7	200	100	16	12	200	122
19	16	195	155	16	15	180	130
9	10	140	98	14	11	167	40

3. The manager of the city water department wants to establish the relationship between monthly household water consumption, Y, and household size, X. Given the data of Table 11.C, determine an estimated regression equation.

Table 11.C

Gallons of Water Used, Y	Household Size, X
650	2
1,200	7
1,300	9
430	4
1,400	12
900	6
1,800	9
640	3
793	3
925	2

4. A realtor wants to establish the relationship between the number of weeks homes are on the market prior to sale, Y, and the asking price, X. Given the data of Table 11.D, determine an estimated regression equation.

Table 11.D

Weeks to Sale, Y	Asking Price (thousands), X	Weeks to Sale, Y	Asking Price (thousands), X
6.5	20	8.6	99
6.8	80	10.6	99
7.0	100	15.0	125
8.6	99	15.0	130
12.1	125	19.0	180
9.0	140	12.5	120
9.5	110	27.0	200

5. A marketing manager wants to establish the relationship between the number of cereal boxes sold, Y, and the shelf space devoted to them, X. Given the data of Table 11.E, determine an estimated regression equation.

Table 11.E

Number of Boxes Sold, Y	Feet of Shelf Space, X	Number of Boxes Sold, Y	Feet of Shelf Space, X
145	3	125	7
151	6	190	5
235	9	210	10
120	5	118	8
272	13	100	7
300	15	390	12
110	2	210	6

6. The manager of an insurance company wants to establish the relationship between people's life insurance, Y, and their salaries, X. Given the data of Table 11.F, determine an estimated regression equation.

Table 11.F

Life Insurance in Force, Y (thousands of dollars)	Annual Salary, X (thousands of dollars)
50	10
80	29
100	30
130	31
150	36
150	40
200	40
300	29
300	50
350	127

7. The manager of a railroad wants to establish the relationship between fuel costs per ton mile, Y, and the number of cars on a freight train, X. Given the data of Table 11.G, determine an estimated regression equation.

8. The operator of a small airport wants to establish the relationship between annual maintenance costs of airplanes, Y, and the airplanes' age, X. Given the data of Table 11.H, determine an estimated regression equation.

Table 11.G

Fuel Costs per Ton-Mile, Y (cents)	Cars on Train, X (number)
16.3	20
15.0	30
14.3	36
13.0	40
15.1	49
12.0	51
12.0	60
12.1	63
14.7	70
15.2	70
12.1	81
11.0	97
9.2	100
10.1	120
6.2	152
4.8	156

Table 11.H

Annual Maintenance Cost, Y (dollars per plane)	Age of Plane, X (years)
562	1
622	1
1,922	1
3,502	2
222	3
2,050	5
751	5
1,200	6
512	8
723	10
1,980	12
2,050	12
751	15
3,171	20
622	21

9. An economist wants to establish the relationship between the unemployment rate, Y, and the Treasury bill rate, X. Given the data of Table 11.I, determine an estimated regression equation.

Table 11.I

Unemployment Rate, Y (percent of labor force)	Treasury Bill Rate, X (percent per year)
6.7	9.7
7.3	9.8
8.9	7.6
9.1	6.1
7.2	10.2
5.2	12.7
6.9	14.3
6.9	7.9
7.1	8.9

10. A farmer wants to establish the relationship between the per-acre yield of corn, Y, and the average July temperature in Iowa, X. Given the data of Table 11.J, determine an estimated regression equation.

Table 11.J

Corn Yield per Acre, Y (bushels)	Average July Temperature, X (degrees Fahrenheit)
115	91
119	95
126	101
91	88
107	92
89	89
111	95
90	87

11. An economist wants to establish the relationship between the price per gallon of gasoline, Y, and the price per barrel of crude oil, X. Given the data of Table 11.K, determine an estimated regression equation.

12. A marketing manager wants to establish the relationship between the second-year sales of sales representatives, Y, and their first-year sales, X. Given the data of Table 11.L, determine an estimated regression equation.

Table 11.K

Price per Gallon of Gasoline, Y (dollars)	Price per Barrel of Crude Oil, X (dollars)
.39	3.50
.42	4.29
.56	5.66
.79	18.21
.82	19.33
.97	29.10
1.11	36.12
1.22	36.05
1.36	42.12
1.22	30.50

Table 11.M

Sales, Y (units)	Competitor's Price, X (dollars per unit)
520	13
550	13
600	15
610	15
620	16
724	21
680	21
300	14
962	40
270	12

Table 11.L

Second-Year Sales, Y (units)	First-Year Sales, X (units)
69	170
75	133
86	86
111	161
129	112
133	133
134	136
136	82
140	60
152	152
161	83
170	97

Table 11.N

Fire Damage, Y (thousands of dollars)	Distance to Fire Station, X (miles)
12.6	1.1
18.9	.9
33.0	2.0
4.1	.7
150.0	1.2
77.0	.8
6.1	3.1
9.2	3.2
10.2	7.0
300.0	1.3

13. A marketing manager wants to establish the relationship between sales, Y, and the price of a similar product produced by a competitor, X. Given the data of Table 11.M, determine an estimated regression equation.

14. An insurance company executive wants to establish the relationship between fire damage, Y, and distance of the victims' property from the fire station, X. Given the data of Table 11.N, determine an estimated regression equation.

15. An advertising executive wants to establish the relationship between the firm's market share, Y, and its annual spending on television ads, X. Given the data of Table 11.O, determine an estimated regression equation.

Table 11.O

Market Share, Y (percent)	Annual Spending on TV ads, X (thousands of dollars)
8.8	23
12.7	33
13.8	36
15.0	39
17.2	42
28.6	69
30.8	88
41.2	127
53.0	130

16. An economist wants to establish the relationship between annual sales of new cars, Y, and the stocks of old cars, X. Given the data of Table 11.P, determine an estimated regression equation.

Table 11.P

New Car Sales, Y (millions per year)	Old Car Stock, X (millions at mid-year)
9.3	68.9
8.4	80.4
8.6	95.2
10.7	104.7
9.0	104.6
8.5	105.8
8.0	106.9

17. An economist wants to establish the relationship between the demand for electricity, Y, and the price of natural gas, X. Given the data of Table 11.Q, determine an estimated regression equation.

Table 11.Q

Electricity Demand, Y (million megawatt-hours)	Natural-Gas Price, X (dollars per unit)
173	33
161	36
142	22
193	41
211	38
219	42
260	50
250	51
181	30
129	25

18. An economist wants to establish the relationship between the number of airplanes sold per year, Y, and their price, X. Given the data of Table 11.R, determine an estimated regression equation.

19. An economist wants to establish the relationship between U.S. personal consumption expenditures, Y, and gross national product, X. Given the data of Table 11.S, determine an estimated regression equation.

Table 11.R

Airplanes Sold, Y (number per year)	Price per Plane, X (thousands of dollars)
451	49
437	63
411	89
389	111
359	141
350	150
345	155
331	169
267	233
207	293
87	413
85	415

Table 11.S

Personal Consumption Expenditures, Y (billions of dollars)	Gross National Product, X (billions of dollars)
452.00	737.20
461.40	756.60
482.00	800.30
500.50	832.50
528.00	876.40
557.50	929.30
585.70	984.80
602.70	1011.40
634.40	1058.10
657.90	1087.60
672.10	1085.60
696.80	1122.40
737.10	1185.90
767.90	1254.30
762.80	1246.30
779.40	1231.60
823.10	1298.20
864.30	1369.70
903.20	1438.60
927.60	1479.40
931.80	1475.00
956.80	1513.80
970.20	1485.40
1011.40	1534.80

20. An economist wants to establish the relationship between U.S. gross private domestic investment, Y, and Moody's Aaa corporate bond yields, X. Given the data of Table 11.T, determine an estimated regression equation.

21. Reconsider problem 1. Make a *point estimate* for the voting percentage captured by the incumbent-party candidate if election-year unemployment goes from 2 to 7 percent (hence, the change is $+5$ points); then figure a *98-percent prediction interval* for the voting percentage received, on the average, by candidates who find themselves in this predicament. Would your answer differ for a particular incumbent-party candidate you might be advising? What would it be?

22. Reconsider problem 2. Make a *point estimate* for monthly sales if there are 150 ads. Then compute *95-percent prediction intervals* for $\mu_{Y \cdot 150}$ and $I_{Y \cdot 150}$.

23. Reconsider problem 3. Make a *point estimate* of household water consumption if household size is 5 persons. Then compute *90-percent prediction intervals* for $\mu_{Y \cdot 5}$ and $I_{Y \cdot 5}$.

24. Reconsider problem 4. Make a *point estimate* of weeks to sale if the asking price is $138,000. Then compute *95-percent prediction intervals* for $\mu_{Y \cdot 138}$ and $I_{Y \cdot 138}$.

25. Reconsider problem 5. Make a *point estimate* of boxes sold if 11 feet of shelf space are available. Then compute *98-percent prediction intervals* for $\mu_{Y \cdot 11}$ and $I_{Y \cdot 11}$.

26. Reconsider problem 6. Make a *point estimate* of life insurance in force if annual salary is $80,000. Then compute *99-percent prediction intervals* for $\mu_{Y \cdot 80}$ and $I_{Y \cdot 80}$.

27. Reconsider problem 7. Make a *point estimate* of fuel costs per ton-mile if there are 140 cars on the train. Then compute *99.5-percent prediction intervals* for $\mu_{Y \cdot 140}$ and $I_{Y \cdot 140}$.

28. Reconsider problem 8. Make a *point estimate* of annual maintenance cost if the plane is 7 years old. Then compute *99.8-percent prediction intervals* for $\mu_{Y \cdot 7}$ and $I_{Y \cdot 7}$.

29. Reconsider problem 9. Make a *point estimate* of the unemployment rate when the Treasury Bill rate is 14 percent. Then compute *99.9-percent prediction intervals* for $\mu_{Y \cdot 14}$ and $I_{Y \cdot 14}$. Also ask yourself this: Does the observed relationship between Y and X make sense?

30. Reconsider problem 10. Make a *point estimate* of corn yield if the average July temperature is an even 90°F. Then compute *90-percent prediction intervals* for $\mu_{Y \cdot 90}$ and $I_{Y \cdot 90}$.

31. Reconsider problem 11. Make a *point estimate* of the gasoline price if the crude oil price is $25.

Table 11.T

Gross Private Domestic Investment, Y (billions of dollars)	Moody's Aaa Corporate Bond Yields, X (percent per annum)
75.90	4.41
74.80	4.35
85.40	4.33
90.90	4.26
97.40	4.40
113.50	4.49
125.70	5.13
122.80	5.51
133.30	6.18
149.30	7.03
144.20	8.04
166.40	7.39
195.00	7.21
229.80	7.44
228.70	8.57
206.10	8.83
257.90	8.43
324.10	8.02
386.60	8.73
423.00	9.63
401.90	11.94
474.90	14.17
414.50	13.79
471.30	12.04

Then compute *95-percent prediction intervals* for $\mu_{Y \cdot 25}$ and $I_{Y \cdot 25}$.

32. Reconsider problem 12. Make a *point estimate* of the second-year sales if first-year sales are 100 units. Then compute *98-percent prediction intervals* for $\mu_{Y \cdot 100}$ and $I_{Y \cdot 100}$.

33. Reconsider problem 1. Determine *95-percent confidence intervals* for β and α. Show your computations.

34. Determine *95-percent confidence intervals* for β and α in problems 2, 4, and 6.

35. Determine *95-percent confidence intervals* for β and α in problems 3, 5, and 7.

36. Determine *95-percent confidence intervals* for β and α in problems 8, 10, and 12.

37. Determine *95-percent confidence intervals* for β and α in problems 9, 11, and 13.

38. Determine *95-percent confidence intervals* for β and α in problems 14, 16, and 18.

39. Reconsider problem 1. Make a *two-tailed hypothesis test* (H_0: $\beta = 0$), using a confidence level of 95 percent. Show your computations.

40. Using a confidence level of 95 percent, make *two-tailed hypothesis tests* of H_0: $\beta = 0$ for problems 2, 4, 6, 8, and 10.

41. Calculate the sample coefficients of determination and of correlation for problems 1, 3, 5, 7, and 9. For problem 1, explain verbally what r^2 means.

42. Calculate the sample coefficients of determination and of correlation for problems 12, 14, 16, 18, and 20.

43. At the 5-percent level of significance, make an ANOVA test of H_0: $\beta = 0$ for
  a. problem 1.
  b. problem 3.

44. At the 5-percent level of significance, make an ANOVA test of H_0: $\beta = 0$ for
  a. problem 6.
  b. problem 8.

45. A personnel manager wants to establish the relationship between the weekly number of grievances filed by workers, Y, and the average weekly humidity index, X. Given the data of Table 11.U, determine an estimated linear regression equation. Also determine the values of $s_{Y \cdot X}$, r^2, and F (in an ANOVA test of H_0: $\beta = 0$).

Table 11.U

Grievances Filed, Y (number per week)	Average Weekly Humidity Index, X
11	.7
9	1.4
11	2.1
13	2.7
22	3.7
18	4.2
22	5.0
41	5.6
62	6.9
50	7.3
88	7.7
68	7.8
109	8.1
125	8.4
142	8.5
96	8.6
112	8.8
132	9.0

Selected Readings

Kohler, Heinz. *Statistics for Business and Economics,* 2nd ed. Glenview, Ill.: Scott, Foresman and Co., 1988. Chapter 13 takes regression analysis beyond the discussion here to include *curvilinear* regression.

Neter, J., W. Wasserman, and M. H. Kutner. *Applied Linear Regression Models.* Homewood, Il.: Richard D. Irwin, 1983. A good text.

Walker, Helen M. *Studies in the History of Statistical Method.* Baltimore: Williams and Wilkins, 1929. An excellent history of regression and correlation theory in chap. 5.

Wonnacott, Thomas H. and Ronald J. Wonnacott. *Regression: A Second Course in Statistics.* New York: Wiley, 1981. A superb text devoted to regression and correlation analysis at a more advanced level.

Computer Program

The DAYSAL-2 personal-computer diskettes that accompany this book contain one program of interest to this chapter:

13. Regression and Correlation Analysis lets you enter any 2 sets of data and view a scatter diagram of their association, along with all summary measures of central tendency and of dispersion. You can perform simple regression analysis that provides not only regression coefficients, standard errors, and *t*-values, but also a correlation matrix, an ANOVA table, confidence intervals for α and β coefficients, hypothesis tests about them, and prediction intervals, at numerous confidence levels, for average and individual Y-values associated with each of the X-values originally entered. You can also take advantage of a "live regression" feature: You specify any X-value . . . and the computer provides an instant point estimate of Y, along with limits of prediction intervals for the average and individual Y at your chosen confidence level. You can use data files A–P that contain numerous data on the U.S. economy.

CHAPTER TWELVE

Multiple Regression and Correlation

Simple regression analysis, we learned in Chapter 11, helps us predict the value of one variable from a knowledge of that of another variable, while simple correlation analysis helps us measure the overall strength of association between two such variables. **Multiple regression analysis,** in contrast, is a technique that employs several independent variables to predict the value of a dependent variable; hence, each of these predictor variables explains part of the total variation of the dependent variable. **Multiple correlation analysis,** accordingly, measures the overall strength of association among all these variables.

There are plenty of good reasons for utilizing the more complex techniques to be discussed in this chapter; foremost among them is the fact that the world is a complicated network of interdependencies, which is rarely captured very well by models involving two variables only. Typically, the value of any one variable is influenced not only by one other variable, but by two, three, or even a multitude of other variables. People's income, for example, may indeed, depend on their education (as noted in Chapter 11), but it can also depend on their job experience, perhaps on their age, and, in the presence of discriminatory practices, even on their sex or race. The demand for a good may well depend on the good's price but, in addition, may vary with the income and taste of consumers, the prices of substitutes or complements, and a host of other factors. A firm's sales may, indeed, vary with advertising expenditures but also with the number of competitors it has, with the size of the community in which it is located, and even with the local unemployment rate. A farmer's output per acre may be a function of fertilizer input but will surely also vary with the quantity of labor used, with the amounts of pesticide used, and even with the extent of rainfall.

Examples such as these could be multiplied without end, but their message is already clear: simple regression and correlation analysis is rarely adequate for exploring relationships, even between two variables of interest. Such simple analysis always leaves open the possibility of finding a fuller explanation of the dependent variable by considering more than one independent variable. (A substantial amount of the variation in Y = income that is not explained by X_1 = education may, for example, be explained by X_2 = job experience, X_3 = age, or even X_4 = sex.) In addition, the precise effect of a particular independent variable (such as education) on a dependent variable of interest (such as income) may be seriously distorted by the results of simple regression and correlation analysis because the influence of other independent variables (such as job experience, age, or sex) has been ignored. The sample results noted in Table 11.1 on page 369, for example, are ambiguous if high or low income can also be caused by high or low job experience and not only by high or low education. If the sampled high-income individuals just happened to have above-average job experience, while the low-income individuals had below-average job experience, a simple regression that ignored job experience but used these data would exaggerate both the income-raising effect of high education and the income-depressing effect of low education. In other words, if only a single independent variable is used in the regression equation, but several such variables are in fact operating in the real-life situation that this equation attempts to describe, the influence of all the ignored independent variables is absorbed into the coefficient attached to that single variable—which makes the coefficient misleading. The problem noted here is once again that of controlling for extraneous factors.

Especially in the fields of business and economics, the influence of X on Y can rarely be established by *experimental control,* the literal control of all aspects of the environment in a laboratory wherein the effect of X on Y might be observed, other things being equal. As we shall see, however, multiple regression analysis, unlike simple regression analysis, allows us to exercise *statistical control* and to determine the influence of any X on Y for specified constant values of other variables that might affect Y as well. (In business and economics texts, an "other things being equal" or *ceteris paribus* clause is frequently inserted in sentences to warn the reader that a given relationship between two variables was found while exercising statistical control and might not be evident when other variables fail to remain constant.)

LINEAR MULTIPLE REGRESSION: TWO EXPLANATORY VARIABLES

The techniques of multiple regression are straightforward extensions of those of simple regression. Let us consider the case in which one dependent variable, Y, is related, in linear fashion, to two independent variables, X_1 and X_2. (Y might be income, X_1 might be education, and X_2 might be job experience.) A first goal of the analysis of such a case is the establishment of an *estimated multiple-regression equation,* such as

$$\hat{Y} = a + b_1 X_1 + b_2 X_2.$$

This equation gives us the estimated value, $\hat{Y}$, of the dependent variable for any specified pair of values of the independent variables. In contrast to our experience in Chapter 11, there now exist three estimated regression coefficients, a, b_1, and b_2. Their meaning is easy to comprehend: a is the estimated value of Y when $X_1 = X_2 = 0$; b_1 gives us the change in Y (also referred to as the *partial* change or *net* change in Y) associated with a unit change in X_1 when X_2 is held constant; while b_2, similarly, equals the change in Y associated with a unit change in X_2 when X_1 is held constant. The values of b_1 and b_2 are called the **estimated partial-regression coefficients;** they are, in fact, the partial derivatives of Y with respect to either X_1 or X_2. [Thus, $b_1 = (\delta Y/\delta X_1)$ and $b_2 = (\delta Y/\delta X_2)$ where δ, the lowercase Greek *delta,* stands for "the change in."] But note: Unlike the last chapter's two-variable simple regression equation (that related Y to X), our three-variable multiple regression equation (that relates Y to X_1 and X_2) does not correspond to a line in a two-dimensional scatter diagram (such as Figure 11.6 on page 375); it corresponds to a plane in three-dimensional space.

The Regression Plane

In our three-variable case, three observations are made for each sample unit: one for the value of Y, one for X_1, and one for X_2. These observations, too, can be depicted in a scatter diagram, but the scatter diagram has to be three-dimensional, as Figure 12.1 indicates. In this particular graph, the value of Y associated with any sample unit is measured vertically from the origin, 0; the value of X_1 is measured toward the right, and the value of X_2 is measured toward the left. Thus, a sample observation of $Y =$ distance 0*A*, $X_1 =$ distance 0*B*, and $X_2 =$ distance 0*C* appears as the color sample point, labeled P.

Now imagine that there were lots of sample observations, such as all the colored dots suspended in the three-dimensional space of Figure 12.2. The three-variable multiple regression technique estimates the above-noted multiple regression equation in such a way that all the estimates derived from it fall on a surface, such as shaded area *ABCD* in our graph, that is called the **regression plane** and that is positioned among the sample points in such a way as to minimize the sum of the squared vertical devia-

Figure 12.1   **Sample Point in Three-Dimensional Scatter Diagram**

When three observations are associated with each sample unit (such as $Y =$ distance 0*A*, $X_1 =$ distance 0*B*, and $X_2 =$ distance 0*C*), any set of these observations can be depicted like sample point *P,* as a dot suspended in three-dimensional space.

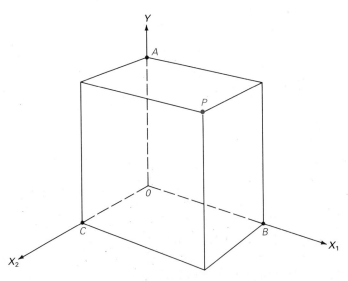

Figure 12.2 The Regression Plane

Three-variable linear multiple-regression analysis estimates an equation ($\hat{Y} = a + b_1X_1 + b_2X_2$) in such a way that all the estimates of Y made with its help (and depicted by crosses in this graph) fall on a surface, such as *ABCD,* that is called the *regression plane* and that is positioned among the sample points (the color dots) in such a way as to minimize the sum of the squared vertical deviations between these sample points and their associated estimates. Note how point P from Figure 12.1 reappears here at a position above the regression plane and is projected by a vertical line to the cross lying on the plane directly below it. Similarly, a point such as Q is suspended below the plane and is projected by a vertical line to the cross lying on the plane directly above it.

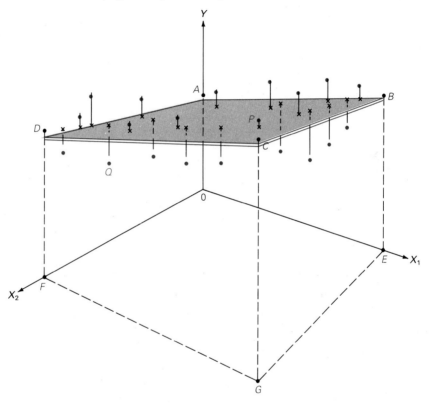

tions between these sample points and their associated estimates. In Figure 12.2, the observed sample points (or the actual values of Y for any given combination of X_1 and X_2) are represented by the colored dots; the associated estimates, $\hat{Y}$, are represented by the black crosses lying on the shaded plane. These crosses, of course, are positioned immediately below or above the colored dots, depending on whether the latter are suspended above or below the regression plane.

We can also note the values of the regression coefficients in Figure 12.2. The value of a estimates Y for $X_1 = X_2 = 0$; hence, it is still the Y intercept, equal to distance $0A$ in our graph. The value of b_1 is the slope of the regression plane when holding X_2 constant (at any desired level). Imagine cutting the regression plane parallel to the X_1 axis at $X_2 = 0$. The cut would trace line AB in the YX_1 plane and show the value of Y rising from $0A$ at $X_1 = 0$ to EB at $X_1 = 0E$. The slope of line AB, relative to $0E$, equals

b_1. Imagine instead cutting the regression plane parallel to the X_1 axis at $X_2 = 0F$. The cut would trace line DC and show the value of Y rising from FD at $X_1 = 0$ to GC at $X_1 = 0E$. The slope of line DC, relative to FG, also equals b_1. Thus, b_1 always indicates how Y changes with X_1, while *not* changing X_2.

Finally, b_2 is the slope of the regression plane when holding X_1 constant (again at any desired level). Imagine cutting the regression plane parallel to the X_2 axis at $X_1 = 0$. That cut would trace line AD in the YX_2 plane and show the value of Y rising (even though slightly) from $0A$ at $X_2 = 0$ to FD at $X_2 = 0F$. The slope of line AD, relative to $0F$, equals b_2. Imagine instead cutting the regression plane parallel to the X_2 axis at $X_1 = 0E$. The cut would trace line BC and show the value of Y rising from EB at $X_2 = 0$ to GC at $X_2 = 0F$. The slope of line BC, relative to EG, also equals b_2. Thus, b_2 always indicates how Y changes with X_2, while not changing X_1.

Least Squares Revisited

As we have just noted, the regression plane is positioned so as to minimize the sum of squared errors, or $\Sigma(Y - \hat{Y})^2$. In Figure 12.2, the heights of the color sample points above the X_1X_2 plane represent the values of Y; the heights of the crosses on the regression plane immediately below or above the color dots represent the associated values of $\hat{Y}$ that are estimated from the equation given above. The vertical differences, $Y - \hat{Y}$, are the deviations or errors, of course; the sum of their squares is to be minimized. Ronald A. Fisher (Biography 8.1), George Snedecor (see Biography 12.1 at the end of this chapter) and others have shown that when the following three so-called *normal equations* are satisfied, the sum of the squares of the deviations is minimized.[1]

$$\Sigma Y = na + b_1\Sigma X_1 + b_2\Sigma X_2$$
$$\Sigma X_1 Y = a\Sigma X_1 + b_1\Sigma X_1^2 + b_2\Sigma X_1 X_2$$
$$\Sigma X_2 Y = a\Sigma X_2 + b_1\Sigma X_1 X_2 + b_2\Sigma X_2^2.$$

From these equations one can, in turn, derive the expressions given in Box 12.A.

12.A Estimated Regression Coefficients (two independent variables)

$$b_1 = \frac{[\Sigma X_2^2 - n\overline{X}_2^2][\Sigma X_1 Y - n\overline{X}_1\overline{Y}] - [(\Sigma X_1 X_2 - n\overline{X}_1\overline{X}_2)(\Sigma X_2 Y - n\overline{X}_2\overline{Y})]}{[\Sigma X_1^2 - n\overline{X}_1^2][\Sigma X_2^2 - n\overline{X}_2^2] - [\Sigma X_1 X_2 - n\overline{X}_1\overline{X}_2]^2}$$

$$b_2 = \frac{[\Sigma X_1^2 - n\overline{X}_1^2][\Sigma X_2 Y - n\overline{X}_2\overline{Y}] - [(\Sigma X_1 X_2 - n\overline{X}_1\overline{X}_2)(\Sigma X_1 Y - n\overline{X}_1\overline{Y})]}{[\Sigma X_1^2 - n\overline{X}_1^2][\Sigma X_2^2 - n\overline{X}_2^2] - [\Sigma X_1 X_2 - n\overline{X}_1\overline{X}_2]^2}$$

$$a = \overline{Y} - b_1\overline{X}_1 - b_2\overline{X}_2$$

where X_1 and X_2 are observed values of the two independent variables ($\overline{X}_1$ and $\overline{X}_2$ being their means), Ys are associated observed values of the independent variable ($\overline{Y}$ being their mean), and n is sample size.

---

[1]A mathematical proof is given in the "Supplementary Topics" section of Chapter 12 of the *Student Workbook* that accompanies this text.

Table 12.1 Sample Data on Education, Job Experience, and Income

Individual	Education (years), X_1	Job Experience (years), X_2	Income (thousands of dollars per year), Y
A	2	9	5.0
B	4	18	9.7
C	8	21	28.4
D	8	12	8.8
E	8	14	21.0
F	10	16	26.6
G	12	16	25.4
H	12	9	23.1
I	12	18	22.5
J	12	5	19.5
K	12	7	21.7
L	13	9	24.8
M	14	12	30.1
N	14	17	24.8
O	15	19	28.5
P	15	6	26.0
Q	16	17	38.9
R	16	1	22.1
S	17	10	33.1
T	21	17	48.3

Practice Problem 1

Regressing Income on Education and Job Experience

Consider the hypothetical study of the link between income and education, first introduced in Chapter 11's Practice Problem 1, but now suppose that income is believed to be linked to years of job experience as well. Conceivable sample data, based on Table 11.1 on page 369, are presented in Table 12.1 (but Y is now measured in *thousands* of dollars per year, instead of *just plain dollars,* as in Table 11.1).

Given the formulas of Box 12.A, the Table 12.1 data must be subjected to a number of intermediate calculations before the multiple regression equation can be determined. These calculations are shown in Table 12.2, along with a variety of others that will be utilized later in this chapter. From these calculations, we can determine

$$b_1 = \frac{[(566.55)(704.285)] - [(-49.65)(271.905)]}{[(380.95)(566.55)] - (-49.65)^2} = 1.933;$$

$$b_2 = \frac{[(380.95)(271.905)] - [(-49.65)(704.285)]}{[(380.95)(566.55)] - (-49.65)^2} = .649;$$

$$a = 24.415 - 1.933(12.05) - .649(12.65) = -7.088.$$

When inserted into the normal equations given above, these values will be found to satisfy the equations (except for rounding errors). Consequently, the estimated regression equation is

$$\hat{Y} = -7.088 + 1.933X_1 + .649X_2.$$

The values of the estimated partial regression coefficients can be interpreted easily enough: $b_1 = 1.933$ implies that an estimated increase of $1,933 of income is associated with each additional year in education, if job experience is held constant at any level, while $b_2 = .649$ implies that an estimated increase of $649 of income is associated with each additional year of job experience, if education is similarly held constant. Not much should be made of $a = -7.088$ (and the apparent implication that a person's income would be *negative* $7,088 with zero education and zero job experience). As noted in Chapter 11, such extrapolation beyond the range of sample data is not appropriate. Yet the equation can be used to predict the income of someone whose education and job experience does fall within the observed ranges of these variables. Thus, one could

Table 12.2   Intermediate Calculations for Multiple Regression

Individual	X_1	X_2	Y	X_1Y	X_2Y	X_1X_2	X_1^2	X_2^2	Y^2
A	2	9	5.0	10.0	45.0	18	4	81	25.00
B	4	18	9.7	38.8	174.6	72	16	324	94.09
C	8	21	28.4	227.2	596.4	168	64	441	806.56
D	8	12	8.8	70.4	105.6	96	64	144	77.44
E	8	14	21.0	168.0	294.0	112	64	196	441.00
F	10	16	26.6	266.0	425.6	160	100	256	707.56
G	12	16	25.4	304.8	406.4	192	144	256	645.16
H	12	9	23.1	277.2	207.9	108	144	81	533.61
I	12	18	22.5	270.0	405.0	216	144	324	506.25
J	12	5	19.5	234.0	97.5	60	144	25	380.25
K	12	7	21.7	260.4	151.9	84	144	49	470.89
L	13	9	24.8	322.4	223.2	117	169	81	615.04
M	14	12	30.1	421.4	361.2	168	196	144	906.01
N	14	17	24.8	347.2	421.6	238	196	289	615.04
O	15	19	28.5	427.5	541.5	285	225	361	812.25
P	15	6	26.0	390.0	156.0	90	225	36	676.00
Q	16	17	38.9	622.4	661.3	272	256	289	1,513.21
R	16	1	22.1	353.6	22.1	16	256	1	488.41
S	17	10	33.1	562.7	331.0	170	289	100	1,095.61
T	21	17	48.3	1,014.3	821.1	357	441	289	2,332.89
	241 $= \Sigma X_1$	253 $= \Sigma X_2$	488.3 $= \Sigma Y$	6,588.3 $= \Sigma X_1Y$	6,448.9 $= \Sigma X_2Y$	2,999 $= \Sigma X_1X_2$	3,285 $= \Sigma X_1^2$	3,767 $= \Sigma X_2^2$	13,742.27 $= \Sigma Y^2$

$$\bar{X}_1 = \frac{\Sigma X_1}{n} = \frac{241}{20} = 12.05 \qquad \bar{X}_2 = \frac{\Sigma X_2}{n} = \frac{253}{20} = 12.65 \qquad \bar{Y} = \frac{\Sigma Y}{n} = \frac{488.3}{20} = 24.415$$

predict the income of someone with 11 years of education and 20 years of job experience as \$27,155 because

$$\hat{Y} = -7.088 + 1.933(11) + .649(20) = 27.155.$$

Similarly, one can predict the income of someone with 8 years of education and 2 years of job experience as \$9,674 because then

$$\hat{Y} = -7.088 + 1.933(8) + .649(2) = 9.674.$$

In fact, such a person is included in our sample (individual D) and is earning about \$8,800, as Table 12.1 confirms.

Note: The simple regression equation of Figure 11.6 on page 375 predicts D's income as $\hat{Y}_8 = 2,144.23 + 1,847.5(8) = \$16,924.23$—that is, with a considerably larger error, which brings us to the next topic. ∭

The Sample Standard Error of the Estimate of Y

We can once again evaluate the quality of our estimated regression equation by computing the *sample standard error of the estimate of Y* which in this case measures the dispersion of observed Y values about the regression *plane*. Because two independent variables are being used to estimate Y, the standard error is designated not as $s_{Y \cdot X}$ (as in Chapter 11), but as $s_{Y \cdot X_1 X_2}$, or, more simply, as $s_{Y \cdot 12}$. In addition, because 3 degrees of freedom are being lost when estimating the 3 regression coefficients, the divisor in the formula is adjusted accordingly from $n - 2$ (noted in Chapter 11) to $n - 3$ (which makes $s_{Y \cdot 12}^2$ an unbiased estimator of the variance of Y about the regression plane). Thus,

$$s_{Y \cdot 12} = \sqrt{\frac{\Sigma(Y - \hat{Y})^2}{n - 3}},$$

which is mathematically equivalent to

$$s_{Y \cdot 12} = \sqrt{\frac{\Sigma Y^2 - a\Sigma Y - b_1 \Sigma X_1 Y - b_2 \Sigma X_2 Y}{n - 3}}.$$

∭ **Practice Problem 2**

Calculating the Sample Standard Error of the Estimate of Income
In our example, the sample standard error of the estimate of Y can be calculated, using the previous formula, as

$$s_{Y \cdot 12} = \sqrt{\frac{13,742.27 - (-7.088)(488.3) - 1.933(6,588.3) - .649(6,448.9)}{17}}$$

$$= \sqrt{\frac{282.82}{17}} = \sqrt{16.64} = 4.079.$$

Compare this standard error of $4,079 with that of $5,377, calculated in Chapter 11's Practice Problem 4 for the simple regression relating income to education alone. It is obvious that our multiple regression provides a better explanation of the data. Much less of the variation in Y is left unexplained here. Therefore, predictions of Y with the help of our multiple regression equation are better than those made with the help of our simple regression equation. ‖

Making Inferences

Provided that certain conditions can be assumed to hold, we can once again use our regression equation to make a variety of inferences. The necessary assumptions are analogous to the four assumptions described in Chapter 11 (pages 376 to 378); we review them here, along with one addition:

Assumption 1: It is assumed that the values of X_1 and X_2 are known without error and that the different sample observations about Y that are associated with any X_1X_2 combination are statistically independent of each other.

Assumption 2: It is assumed that every population of Y values (a different one of which is associated with every possible X_1X_2 combination) is normally distributed, with a conditional mean of $\mu_{Y \cdot X_1X_2}$ (also denoted as $\mu_{Y \cdot 12}$) and a conditional standard deviation of $\sigma_{Y \cdot X_1X_2}$ (also denoted as $\sigma_{Y \cdot 12}$).

Assumption 3: It is assumed that all conditional probability distributions of Y (a different one of which is associated with every possible X_1X_2 combination) have the same conditional standard deviation, $\sigma_{Y \cdot 12}$ (homoscedasticity exists).

Assumption 4: It is assumed that all the conditional means, $\mu_{Y \cdot 12}$, lie on a surface (similar to *ABCD* in Figure 12.2) that is the *true regression plane* and is described by the equation $E(Y) = \mu_{Y \cdot 12} = \alpha + \beta_1X_1 + \beta_2X_2$. (The estimated regression coefficients a, b_1, and b_2 are, of course, estimators of α, β_1, and β_2). As in the simple regression case, this assumption is still referred to as the assumption of linearity (even though the line has become a plane).

Assumption 5: It is assumed that there exists no exact linear relationship between different independent variables. (If different Xs are perfectly linearly correlated, one cannot calculate the parameter estimates because two or more of the normal equations are then not independent. If different Xs are imperfectly but highly correlated, one can calculate the regression coefficients, but one cannot isolate the individual effects on Y of each of these highly linearly correlated variables.) As the scatter diagram of Figure 12.3 indicates, there certainly exists no such perfect linear correlation between the independent variables in our example.

Estimating the average value of Y, *given* X$_1$ *and* X$_2$. Suppose we wanted to estimate the value of a conditional mean, $\mu_{Y \cdot X_1X_2}$, such as $\mu_{Y \cdot 11,20}$ (the average income of those members of our population with 11 years' education and 20 years' job experience). The best *point estimate* equals 27.155 (that is, $27,155), as noted above. For a small

Figure 12.3   The Relation of X_1 to X_2

This scatter diagram illustrates the fulfillment of Assumption 5 noted in the text: There exists no perfect linear correlation between the values of the two independent variables listed in Table 12.1. (In fact, the simple correlation coefficient between X_1 and X_2 is $r_{1 \cdot 2} = -.11$.)

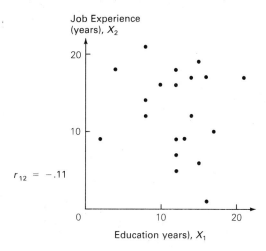

sample, such as ours, a prediction interval is computed by using the formula in Box 12.B, which involves matrix notation.

Matrix algebra, however, is too complex for the intended level of this text; therefore, we cannot use Box 12.B directly. It has, however, been built into the DAYSAL-2 computer program that accompanies the text, and Practice Problem 3 can be solved by computer.[2]

12.B The Limits of a Prediction Interval for the Average Value of Y, Given X_1 and X_2 (small-sample case, $n < 30$)

$$\mu_{Y \cdot 12} = \hat{Y} \pm (t_{\alpha/2} \cdot s_{Y \cdot 12} \cdot \sqrt{M})$$

where $\hat{Y}$ is the estimated value of dependent variable Y, $s_{Y \cdot 12}$ is the sample standard error of the estimate of Y, and M is the product of three matrices.

Note: Given sample size n and m independent variables, the value of $t_{\alpha/2}$ is found in Appendix Table G for $n - m - 1$ degrees of freedom. In the large-sample case ($n \geq 30$), the normal deviate, z, replaces t.

---

[2]Without the use of matrix algebra or computers, the prediction interval for the average Y can still be *approximated* by substituting, for the parenthetical expression in Box 12.B, the value $\left(t_{\alpha/2} \cdot \dfrac{s_{Y \cdot 12}}{\sqrt{n}} \right)$.

Practice Problem 3

Estimating Average Income, Given Education and Job Experience
Assuming that a 95-percent prediction interval is desired, the interval limits in our example can be calculated for 11 years' education and 20 years' job experience, using the DAYSAL-2 computer program, as:

$$23.8753 \le \mu_{Y \cdot 11,20} \le 30.44026.$$

We can be 95-percent confident that the height of the regression plane at $X_1 = 11$ and $X_2 = 20$ lies within these limits (that, on the average, people with 11 years of education and 20 years of work experience have annual incomes between \$23,875.30 and \$30,440.26). |||

Predicting an individual value of Y, given X_1 and X_2. What if we had wanted to predict $I_{Y \cdot X_1 X_2}$ such as $I_{Y \cdot 11,20}$ (the particular income of the next member of our population encountered in sampling who falls into the 11 years' education and 20 years' job-experience category)? The best point estimate still equals 27.155, as noted in the previous section. A small-sample prediction interval is computed by using the formula in Box 12.C, which once again employs matrix algebra.[3]

12.C The Limits of a Prediction Interval for the Individual Value of Y, Given X_1 and X_2 (small-sample case, $n < 30$)

$$I_{Y \cdot 12} = \hat{Y} \pm (t_{\alpha/2} \cdot s_{Y \cdot 12} \sqrt{1 + M})$$

where $\hat{Y}$ is the estimated value of dependent variable Y, $s_{Y \cdot 12}$ is the sample standard error of the estimate of Y, and M is the product of three matrices.

Note: Given sample size n and m independent variables, the value of $t_{\alpha/2}$ is found in Appendix Table G for $n - m - 1$ degrees of freedom. In the large-sample case ($n \ge 30$), the normal deviate, z, replaces t.

Practice Problem 4

Estimating Individual Income, Given Education and Job Experience
Assuming that a 95-percent prediction interval is desired, the interval limits in our example can be calculated for 11 years' education and 20 years' job experience, using the DAYSAL-2 computer program, as:

$$17.95563 \le I_{Y \cdot 11,20} \le 36.35993.$$

---

[3]Without the use of matrix algebra or computers, the prediction interval for the individual Y can still be *approximated* by substituting, for the parenthetical expression in Box 12.C, the value $\left(t_{\alpha/2} \cdot s_{Y \cdot 12} \sqrt{\dfrac{n + 1}{n}} \right)$.

We can be 95-percent confident that the next person with the characteristics of $X_1 = 11$ and $X_2 = 20$ whom we encounter in sampling has an annual income between $17,955.63 and $36,359.93. |||

Establishing confidence intervals for β_1 *and* β_2. In multiple regression, no less than in simple regression, it is crucial to determine whether a nonzero value of any sample coefficient, b (which estimates the corresponding population coefficient, β), is statistically significant or is instead the result of a sampling error that is perfectly consistent with a β value of zero. Limits of confidence intervals for β values can be established using the t distribution, in a fashion analogous to that noted in Chapter 11, as $\beta_1 = b_1 \pm t\sigma_{b_1}$ and $\beta_2 = b_2 \pm t\sigma_{b_2}$. The two standard errors, σ_{b_1} and σ_{b_2}, must again be estimated by s_{b_1} and s_{b_2}, which yields the formulas of Box 12.D.

12.D The Limits of Confidence Intervals for β_1 and β_2 (small-sample case, $n < 30$)

$$\beta_1 \cong b_1 \pm (t_{\alpha/2} \cdot s_{b_1})$$

$$= b_1 \pm \left(t_{\alpha/2} \cdot s_{Y\cdot12} \sqrt{\frac{\Sigma X_2^2 - n\overline{X}_2^2}{[\Sigma X_1^2 - n\overline{X}_1^2][\Sigma X_2^2 - n\overline{X}_2^2] - [\Sigma X_1 X_2 - n\overline{X}_1\overline{X}_2]^2}} \right)$$

$$\beta_2 \cong b_2 \pm (t_{\alpha/2} \cdot s_{b_2})$$

$$= b_2 \pm \left(t_{\alpha/2} \cdot s_{Y\cdot12} \sqrt{\frac{\Sigma X_1^2 - n\overline{X}_1^2}{[\Sigma X_1^2 - n\overline{X}_1^2][\Sigma X_2^2 - n\overline{X}_2^2] - [\Sigma X_1 X_2 - n\overline{X}_1\overline{X}_2]^2}} \right)$$

where b_1 and b_2 are estimated partial regression coefficients (and s_{b_1} and s_{b_2} are their standard errors), Xs are observed values of independent variables ($\overline{X}$s being their means), while $s_{Y\cdot12}$ is the sample standard error of the estimate of Y.

Note: Given sample size n and m independent variables, the value of $t_{\alpha/2}$ is found in Appendix Table G for $n - m - 1$ degrees of freedom.

||| Practice Problem 5

Establishing Confidence Intervals for β_1 and β_2 in the Regression of Income on Education and Job Experience

Given the data of Table 12.2 and the above-noted values of $b_1 = 1.933$, $b_2 = .649$, $s_{Y\cdot12} = 4.079$, and $t_{\alpha/2} = 2.11$ (for a 95-percent confidence level), we can use the Box 12.D formulas to calculate the following confidence-interval limits:

$$\beta_1 = 1.933 \pm \left[2.11(4.079) \sqrt{\frac{3,767 - 20(12.65)^2}{[3,285 - 20(12.05)^2][3,767 - 20(12.65)^2] - [2,999 - 20(12.05)(12.65)]^2}} \right]$$

$$= 1.933 \pm \left[2.11(4.079) \sqrt{\frac{566.55}{(380.95)(566.55) - (-49.65)^2}} \right]$$

$$= 1.933 \pm [2.11(.210)] = 1.933 \pm .443$$

$$\beta_2 = .649 \pm \left[2.11(4.079) \sqrt{\frac{3,285 - 20(12.05)^2}{[3,285 - 20(12.05)^2][3,767 - 20(12.65)^2] - [2,999 - 20(12.05)(12.65)]^2}} \right]$$

$$= .649 \pm \left[2.11(4.079) \sqrt{\frac{380.95}{(380.95)(566.55) - (-49.65)^2}} \right] = .649 \pm [2.11(.172)] = .649 \pm .363$$

Thus, we can be 95-percent confident that the values of β_1 and β_2 are not zero but lie within these intervals:

$$1.49 \leq \beta_1 \leq 2.376$$
$$.286 \leq \beta_2 \leq 1.012 \quad |||$$

t *values*. Frequently, the information derived in the previous section is provided by a **t value,** or **t ratio,** which is nothing else but the ratio of an estimated partial regression coefficient to its standard error.

||| Practice Problem 6

Using t Values for Practice Problem 5
For b_1:

$$t_{b_1} = \frac{b_1}{s_{b_1}} = \frac{1.933}{.210} = 9.205$$

and for b_2:

$$t_{b_2} = \frac{b_2}{s_{b_2}} = \frac{.649}{.172} = 3.773.$$

Because these t values exceed the critical t value of 2.11 established for a 95-percent confidence level and $n - 3 = 17$ degrees of freedom, one concludes that both b_1 and b_2 are statistically significant at that level. Typically, the computed t values appear in parentheses underneath the coefficients of the regression equation, as shown:

$$\hat{Y} = -7.088 + 1.933 X_1 + .649 X_2$$
$$(9.205) \qquad (3.773) \quad |||$$

Note: For almost any sample size, the critical t value for a 95-percent confidence level approximates 2. As the appropriate column of Appendix Table G shows, the critical t equals 2.228 for $n = 10$ and 1.96 for $n = \infty$. This convenient fact leads to the following *rule of thumb:*

If the absolute value of the computed t value is greater than 2 (we ignore a possible minus sign), the parameter estimate, b, is significant at the 95-percent level of confidence.

An examination of the parenthetical values underneath the above equation, with this rule of thumb in mind, instantly reveals the significance of the coefficients.

p *values.* There exists a third alternative yet for noting the degree of significance of an estimated regression coefficient. As noted in Biography 8.1, one can calculate a **p value** that gives the probability that a test statistic (such as a *t* value) as extreme as or more extreme than the one actually observed could have occurred by chance (merely as a result of sampling error), assuming some hypothesis about a parameter is true (assuming, for example, that the true regression coefficient, β, equals zero).

||| Practice Problem 7

Using *p* Values for Practice Problem 5

In our example, what is the probability of finding *t* values at least as large as those given in the previous section if the corresponding β values are zero? The 17 degrees-of-freedom row in Appendix Table G is of little help with respect to t_{b_1} since the largest *t* value listed there equals 3.965. A computer calculation, however, shows the tiny probabilities involved:

$$p = .000 \text{ for } t_{b_1} \geq 9.205 \text{ if } \beta_1 = 0 \text{ and}$$

$$p = .001 \text{ for } t_{b_2} \geq 3.773 \text{ if } \beta_2 = 0. \quad |||$$

Testing the overall significance of the regression. A test of the overall significance of the regression, rather than a test of the significance of individual coefficients, amounts to testing the hypothesis that *all* of the true regression coefficients are zero and that, therefore, *none* of the independent variables helps explain the variation of the dependent one.

||| Practice Problem 8

An Overall Significance Test of the Regression of Income on Education and Job Experience

In our example containing two independent variables (education and job experience) the opposing hypotheses are the following:

$H_0: \beta_1 = \beta_2 = 0$
$H_A:$ At least one $\beta \neq 0$.

Once again, the test can be performed most easily by constructing an ANOVA table, such as Table 12.3. Column 1 contains the regression sum of squares, the error sum of squares, and the total sum of squares:

$$RSS = \Sigma(\hat{Y} - \overline{Y})^2 = a\Sigma Y + b_1\Sigma X_1 Y + b_2\Sigma X_2 Y - n\overline{Y}^2$$
$$= -7.088(488.3) + 1.933(6,588.3) + .649(6,448.9) - 20(24.415)^2$$
$$= 1,537.61.$$

$$ESS = \Sigma(Y - \hat{Y})^2 = \Sigma Y^2 - a\Sigma Y - b_1\Sigma X_1 Y - b_2\Sigma X_2 Y$$
$$= 13,742.27 - (-7.088)(488.3) - 1.933(6,588.3) - .649(6,448.9) = 282.82.$$

$$Total\ SS = \Sigma(Y - \overline{Y})^2 = \Sigma Y^2 - n\overline{Y}^2 = 13,742.27 - 20(24.415)^2 = 1,820.43.$$

Table 12.3   The ANOVA Table

Source of Variation	Sum of Squares (1)	Degrees of Freedom (2)	Mean Square (3) = (1) ÷ (2)	Test Statistic (4)
Variation Explained by Regression	$RSS = 1,537.61$	$m = 2$	$RMS = 768.81$	$F = \dfrac{RMS}{EMS} = 46.20$
Unexplained Variation	$ESS = 282.82$	$n - m - 1 =$ $n - 3 = 17$	$EMS = 16.64$	
Total	$Total\ SS = 1,820.43$	$n - 1 = 19$		

Column 2 lists the appropriate degrees of freedom: equal to the number of independent variables, m, for the regression sum of squares; equal to sample size, n, minus 3 for the error sum of squares (there are n errors and 3 constraints—namely, the estimated regression coefficients, a, b_1, and b_2); and equal to $n - 1$ for the total sum of squares [there are n deviations subject to the single constraint that $\Sigma(Y - \bar{Y}) = 0$].

From this information, the regression and error mean squares of column 3 can be calculated (the latter has, in fact, been derived earlier in connection with the calculation in Practice Problem 2 of $s_{Y\cdot 12}$, which equals $\sqrt{EMS}$). Column 4, finally, contains the F ratio.

If we wish to test our hypothesis at the 5-percent level of significance, Appendix Table I provides a critical value of $F_{.05}$ (for 2 numerator and 17 denominator degrees of freedom) of 3.59. Thus, we should *reject* H_0 resoundingly. Far from being of no help, our regression is highly significant, indeed; a knowledge of X_1 and X_2 (years of education and job experience) does explain almost all of the variation of Y (income) in the population from which we drew our sample. ▮

MULTIPLE CORRELATION ANALYSIS

The concept of correlation can also be extended easily from an analysis dealing with one independent variable to one dealing with several of them. The aim is once again to calculate a single value which, in this case, is to describe the overall strength of association between a dependent variable and two or more independent ones. To distinguish between the coefficients of simple and multiple correlation, capital letters are used to designate the coefficients of multiple correlation.

The Coefficient of Multiple Determination

The **sample coefficient of multiple determination** is denoted by R^2 and equals the proportion of the total variation in the values of the dependent variable, Y, that is explained by the multiple regression of Y on X_1, X_2, and possibly additional independent variables (X_3, X_4, and so on). This coefficient describes, therefore, how well the regression *plane* fits the data; its definition is given in Box 12.E.

12.E Sample Coefficient of Multiple Determination

$$R^2 = \frac{\text{explained variation}}{\text{total variation}} = \frac{RSS}{Total\ SS} = \frac{\Sigma(\hat{Y} - \bar{Y})^2}{\Sigma(Y - \bar{Y})^2}$$

where Ys are observed values of the dependent variable ($\bar{Y}$ being their mean), while $\hat{Y}$s are estimated values of Ys.

As in the case of simple regression, R^2 can take on values between 0 and 1. The closer R^2 is to 0, the worse is the fit of the regression plane to the data; the closer it is to 1, the better is the fit.

The value of R^2 can, however, also be calculated directly from a knowledge of the regression coefficients, as shown in Box 12.F.

12.F Sample Coefficient of Multiple Determination: Alternative Formulation

$$R^2 = \frac{a\Sigma Y + b_1\Sigma X_1 Y + b_2\Sigma X_2 Y - n\bar{Y}^2}{\Sigma Y^2 - n\bar{Y}^2}$$

where a and b's are estimated partial regression coefficients, Xs are observed values of the independent variables, Ys are associated observed values of the dependent variable ($\bar{Y}$ being their mean), and n is sample size.

||| Practice Problem 9

Calculating the Sample Coefficient of Multiple Determination for the Regression of Income on Education and Job Experience

Using the values in Table 12.3 and the Box 12.E formula, we can immediately calculate

$$R^2_{Y\cdot12} = \frac{1,537.61}{1,820.43} = .84$$

for our example (which is an improvement over the value of $r^2_{Y\cdot1} = .71$ calculated in Practice Problem 12 of Chapter 11 that related income only to education).

Using the Box 12.F formula, R^2 can be calculated alternatively for our example as

$$R^2_{Y\cdot12} = \frac{-7.088(488.3) + 1.933(6,588.3) + .649(6,448.9) - 20(24.415)^2}{13,742.27 - 20(24.415)^2} = .84,$$

which confirms the calculation.

Note: We need not test the statistical significance of our sample R^2 for fear that the corresponding population coefficient, ρ^2 (lowercase Greek rho squared), might be zero. Such a test is not needed because of the large F ratio noted in Table 12.3. Our analysis of variance, by rejecting the null hypothesis of H_0: $\beta_1 = \beta_2 = 0$, implicitly rejected a similar hypothesis of H_0: $\rho^2 = 0$. |||

The Coefficient of Multiple Correlation

The **sample coefficient of multiple correlation** is denoted by R and equals the square root of R^2. Its sign is always considered positive. In our example, $R_{Y\cdot12} = \sqrt{.84} = .92$. This value, too, has risen substantially from that of $r_{Y\cdot1} = .84$, found in Chapter 11's Practice Problem 13.

The Adjusted Coefficient of Multiple Determination

As a look at Box 12.E confirms, the calculation of the sample coefficient of multiple determination does not involve any adjustment for degrees of freedom. As a result, there is a tendency for R^2 to be "too large," or to overestimate the true or population coefficient. This bias can be removed by calculating an **adjusted sample coefficient of multiple determination, $\overline{R}^2$,** using the formula in Box 12.G, wherein n equals sample size and m is the number of independent variables in the regression.[4]

12.G Adjusted Sample Coefficient of Multiple Determination

$$\overline{R}^2 = 1 - \left[\frac{ESS}{Total\ SS} \left(\frac{n-1}{n-m-1} \right) \right]$$

An alternative, but mathematically equivalent formula for $\overline{R}^2$ is given in Box 12.H, wherein s_Y^2 is the sample variance of the dependent variable, Y, around its mean, and $s_{Y\cdot12\ldots m}^2$ is the variance of Y around the estimated regression line or plane.

12.H Adjusted Sample Coefficient of Multiple Determination: Alternative Formulation

$$\overline{R}^2 = \frac{s_Y^2 - s_{Y\cdot12\ldots m}^2}{s_Y^2}$$

||| Practice Problem 10

Adjusting the Sample Coefficient of Multiple Determination for the Regression of Income on Education and Job Experience
Using the Table 12.3 data and Box 12.G, the adjusted coefficient in our example can be calculated as:

$$\overline{R}^2_{Y\cdot12} = 1 - \left[\frac{282.82}{1,820.43} \left(\frac{19}{17} \right) \right] = 1 - .17 = .83.$$

---

[4]The formula in Box 12.G can easily be seen as an adjustment of that in Box 12.E by noting that $RSS/Total\ SS$ is the same as $1 - (ESS/Total\ SS)$.

Alternatively, however, in order to use the Box 12.H formula, we can use the Table 12.2 data to calculate

$$s_Y^2 = \frac{\Sigma Y^2 - n\bar{Y}^2}{n - 1} = \frac{13,742.27 - 20(24.415)^2}{19} = \frac{1,820.43}{19} = 95.81.$$

We have already calculated $s_{Y\cdot12} = \sqrt{\dfrac{282.82}{17}}$ in Practice Problem 2 of this chapter; hence,

$$s_{Y\cdot12}^2 = \frac{282.82}{17} = 16.64$$

and

$$\bar{R}_{Y\cdot12}^2 = \frac{95.81 - 16.64}{95.81} = .83. \;\; ||||$$

Two things should be noted:

1. While an unadjusted coefficient of determination always increases when a relevant and previously ignored independent variable is added to a regression model (note the increase in our example from $r_{Y\cdot1}^2 = .71$ to $R_{Y\cdot12}^2 = .84$), $\bar{R}^2$ can actually decrease and will do so when the addition of a new independent variable does not reduce the error sum of squares *(ESS)* enough to offset the loss of one degree of freedom that is occasioned by the increase in m. (In that case, in an ANOVA table such as Table 12.3, a lower *ESS* can nevertheless raise *EMS* and lower *F.*) If that happens, the addition of the new independent variable was not meaningful.
2. The formulas given in Boxes 12.G and 12.H can also be used to calculate adjusted coefficients of simple determination. For example, the coefficient of $r_{Y\cdot1}^2 = .71$ calculated in Practice Problem 12 of Chapter 11 can be adjusted to

$$\bar{r}_{Y\cdot1}^2 = 1 - \left[\frac{520,489,201.64}{1,820,772,118.80}\left(\frac{19}{18}\right) \right] = .70$$

by using the data of Table 11.3 and the Box 12.G formula.

Partial Correlation

Our regression of income *(Y)* on education (X_1) in Chapter 11 produced, as we just recalled, an unadjusted simple coefficient of determination of $r_{Y\cdot1}^2 = .71$. Earlier in the current chapter, the addition to our model of a second explanatory variable (job experience, X_2) yielded a regression equation with considerably greater predictive power and produced an unadjusted multiple coefficient of determination of $R_{Y\cdot12}^2 = .84$. It is common practice to quantify the improvement achieved by such higher dimensional regression analysis. The result is an index of association between the dependent variable and just *one* of the independent variables, given a prior accounting of the effects of one or more other independent variables. The index is calculated as the proportional decrease in the previously unexplained variation (or increase in the previously explained variation) resulting from the higher dimensional regression. Box 12.I defines the new measure, called the **sample coefficient of partial determination,** for the case in which a regression with one independent variable is replaced by one with two such variables.[5]

12.I Sample Coefficient of Partial Determination

$$r^2_{Y2 \cdot 1} = \frac{\text{reduction in previously unexplained variation in } Y}{\text{previously unexplained variation in } Y} = \frac{R^2_{Y \cdot 12} - r^2_{Y \cdot 1}}{1 - r^2_{Y \cdot 1}}$$

Note the subscripts in the Box 12.I formula: The coefficient of simple determination (calculated for a regression of Y on X_1 only) is denoted by $r^2_{Y \cdot 1}$; the subscript tells us that the effect of variables other than X_1 (such as X_2, X_3, etc.) is totally being ignored. The coefficient of multiple determination (calculated for a regression of Y on X_1 and X_2) is denoted by $R^2_{Y \cdot 12}$; here the subscript tells us that the mutual effect of X_1 and X_2 is being considered. Finally, the measure of partial correlation that shows the correlation of Y on X_2, when X_1 is present but constant, is denoted by $r^2_{Y2 \cdot 1}$; the subscript reminds us that the (additional) effect of X_2 on Y is being considered, given that the effect of X_1 on Y has already been accounted for. Such coefficients of partial determination, like the simple ones, range from -1 to $+1$ in value; they take on a sign corresponding to that of the associated parameter estimate (b_2 in this case).

Practice Problem 11

Calculating the Sample Coefficient of Partial Determination to Show the Correlation of Income With Job Experience, When the Effect of Education is Present but Constant

For our example, we can use the Box 12.I formula to calculate:

$$r^2_{Y2 \cdot 1} = \frac{.84 - .71}{1 - .71} = .45.$$

This result implies that 45 percent of the variation in Y (income) that was left unexplained by the simple regression of Y on X_1 (education) in Chapter 11 has been explained by the addition here in Chapter 12 of X_2 (job experience) as an explanatory variable. |||

---

[5]The coefficient is defined analogously for additional moves to ever higher dimensional regressions. Thus, going from two to three independent variables produces

$$r^2_{Y3 \cdot 12} = \frac{R^2_{Y \cdot 123} - R^2_{Y \cdot 12}}{1 - R^2_{Y \cdot 12}}.$$

Indeed, we can confirm this result independently by comparing the error sum of squares found in the ANOVA tables. The value of *ESS* changed from 520.49 (million) in Table 11.3 (on p. 397) to 282.82 in Table 12.3, which is a 45-percent reduction (except for rounding error). It is easy to see why the sample coefficient of partial determination, or its square root, the **sample coefficient of partial correlation,** is often used to evaluate the merit of potentially adding an explanatory variable to a regression analysis. In our case, the addition of X_2 is worthwhile because X_2 is quite effective in sharpening the analysis.

LINEAR MULTIPLE REGRESSION AND CORRELATION WITH THREE OR MORE EXPLANATORY VARIABLES (COMPUTER APPLICATIONS)

When simple regression analysis turns into multiple regression analysis, hand calculations become quite burdensome. We have seen how cumbersome the calculations are even for the relatively simple case of two independent variables; they are all the more cumbersome when there are three or more independent variables. The principles involved remain the same as the number of variables rises, but the calculations, luckily, can nowadays be performed by computer.

When three independent variables are involved, the computer finds an estimated least-squares regression equation of the form,

$$\hat{Y} = a + b_1X_1 + b_2X_2 + b_3X_3,$$

and this procedure can be extended without end when additional predictor variables are included in the analysis (by adding b_4X_4, b_5X_5, and so on to the equation).

The moment the analysis involves four variables (one dependent and three independent ones), the data can no longer be pictured in a scatter diagram. The human mind can envision two dimensions (note Figure 11.3 on p. 370) and even three dimensions (note Figure 12.2), but it balks at four dimensions and more. As mathematicians put it, the estimation of a least-squares regression equation with three or more independent variables is analogous to the fitting of a **regression hyperplane** to the sample data. Such a plane exists in four or higher-dimensional space, but no one has ever seen it!

Higher-dimensional regression analysis still includes the calculation of the standard error of the estimate, but it is now symbolized by $s_{Y\cdot123}$, by $s_{Y\cdot1234}$, and so on, depending on how many independent variables are being considered. The calculation of these standard errors is again a straightforward extension of that noted above. While

$$s_{Y\cdot12} = \sqrt{\frac{\Sigma Y^2 - a\Sigma Y - b_1\Sigma X_1Y - b_2\Sigma X_2Y}{n - 3}},$$

the calculation of $s_{Y\cdot123}$ adds $-b_3\Sigma X_3Y$ to the numerator and changes the denominator to $n - 4$. And so it goes.

Prediction intervals, too, are calculated as noted above (in Box 12.B or 12.C, for example), except that the degrees of freedom for $t_{\alpha/2}$ change as they do for the standard error, from $n - 3$ to $n - 4$, and so on (always being equal to $n - m - 1$, where n

is sample size and m is the number of independent variables). The use of the computer saves an enormous amount of time, especially when sample size and the number of independent variables are large, but it does more than that. It also produces a greater level of accuracy by (a) preventing a cascade of errors once a single one has been made and (b) being able to handle a large number of significant figures.

Consider the data of Table 12.4. They were fed into a computer, which promptly produced the accompanying printout.

"Canned" computer programs are widely available; all of them, when fed sample data, quickly determine and print out sample regression coefficients (a, b_1, b_2, etc.) along with their standard errors (s_a, s_{b_1}, s_{b_2}, etc.) and t values (which, of course are the ratios of each estimated regression coefficient to its standard error, or a/s_a, b_1/s_{b_1}, b_2/s_{b_2}, etc.) Some programs print p values instead of t ratios. Most of them also provide the standard error of the estimate of Y, the ANOVA table, the sample coefficient of determination, and more.[6]

The following sections describe how to interpret and use the computer results.

The Estimated Multiple Regression Equation

We can quickly formulate our desired multiple regression equation as

$$\hat{Y} = -2.983 \quad + 2.099\ X_1 + \quad 1.197 X_2 - \quad .3107\ X_3$$
$$(-1.265) \quad (15.817) \quad (\ 8.151) \quad (-5.381)$$

The interpretation of this equation is straightforward:

1. A b_1 of 2.099 means that each extra year of education (X_1) raises estimated annual income by \$2,099, if the values of the other two variables (X_2 and X_3) are held constant.
2. A b_2 of 1.197 means that each extra year of work experience (X_2) raises estimated annual income by \$1,197, if the values of the other two variables (X_1 and X_3) are held constant.
3. A b_3 of $-.3107$ means that each extra year of age (X_3) *lowers* estimated annual income by \$310.70, if the values of the other two variables (X_1 and X_2) are held constant.

The corresponding t values (in parentheses below the coefficients) all pass the rule-of-thumb test of 95-percent significance noted earlier in the chapter. As noted previously (and for the reasons noted), we need not concern ourselves with the intercept of -2.983, nor with its t value that does not pass the 95-percent significance test.

A note of caution for those about to study the results of other people's regression analyses: Instead of t values, some publications print standard errors or even p values underneath the estimated coefficients. It is mandatory to read the fine print, lest utter confusion prevail.

---

[6]See, for example, the DAYSAL-2 floppy disks that accompany this text and are designed to be used with personal computers.

Table 12.4   Sample Data on Income, Using Education, Job Experience, and Age as Independent Variables

Individual	Income (thousands of dollars per year), Y	Education (years), X_1	Job Experience (years), X_2	Age (years), X_3
A	5.0	2	9	29
B	9.7	4	18	50
C	28.4	8	21	41
D	8.8	8	12	55
E	21.0	8	14	34
F	26.6	10	16	36
G	25.4	12	16	61
H	23.1	12	9	29
I	22.5	12	18	64
J	19.5	12	5	30
K	21.7	12	7	28
L	24.8	13	9	29
M	30.1	14	12	35
N	24.8	14	17	59
O	28.5	15	19	65
P	26.0	15	6	30
Q	38.9	16	17	40
R	22.1	16	1	23
S	33.1	17	10	58
T	48.3	21	17	44

```
Problem title : Textbook Table 12.4
The Number of Paired Observations, n .... = 20
Dependent Variable ...................... = Y
Independent Variables  .................. = X1 X2 X3

Regression Results

Variable        Coefficient    Std. Error    t-Value

Intercept        -2.98295       2.35732      -1.26540
X1 (Education)    2.09943       0.13274      15.81660
X2 (Job Exp.)     1.19746       0.14690       8.15145
X3 (Age)         -0.31067       0.05773      -5.38144

The ANOVA Table

Regression sum of squares (RSS) ......... = 1720
Regression degrees of freedom (m) ....... = 3
Regression mean square (RMS) ............ = 573.3333

Error sum of squares (ESS) .............. = 100.4275
Error degrees of freedom (n-m-1) ....... = 16
Error mean square (EMS) ................. = 6.27672

Total sum of squares (SS) ............... = 1820.428
Total degrees of freedom (n-1) .......... = 19

Sample standard error of estimate of Y... = 2.505338
F statistic (RMS/EMS) ................... = 91.34282

Sample coeff. of multiple determination,
  R² = (RSS/SS) ......................... = .9448331
Adjusted sample coefficient of multiple
  determination, R̄² ..................... = .9344892
Sample coefficient of nondetermination
  (ESS/SS) .............................. = .055167
```

The Standard Error of the Estimate of Y

Note that the standard error of the estimate of Y equals $s_{Y \cdot 123} = 2.505$. Compare this standard error of the estimate of Y with $s_{Y \cdot 12} = 4.079$ from this chapter's Practice Problem 2 or with $s_{Y \cdot X} = 5.377$ from Chapter 11's Practice Problem 4. Clearly, there has been further improvement in our predictive ability.

Predicting the Average or an Individual Value of Y

Using a computer program like DAYSAL-2, we can quickly produce prediction intervals for any average or individual Y. Suppose that we wanted such predictions for $X_1 = 11$ years of education, $X_2 = 20$ years of job experience, and $X_3 = 45$ years of age. In both cases, the point estimate comes to:

$$\hat{Y} = -2.983 + 2.099(11) + 1.197(20) - .3107(45) = 30.0645.$$

We predict, that is, that a person with 11 years of education, 20 years of job experience, and an age of 45 will have an annual income of $30,064.50. Yet the limits of a 95-percent prediction interval for the *average* person with these characteristics equal:

$$27.74799 \leq \mu_{Y \cdot 123} \leq 32.41188.$$

We can be 95-percent confident that people with the noted characteristics will, on the average, have annual incomes between $27,747.99 and $32,411.88. Similarly, the limits of a 95-percent prediction interval for the *next* person encountered with these characteristics equal:

$$24.27924 \leq I_{Y \cdot 123} \leq 35.88063.$$

We can be 95-percent confident that such a person's annual income will lie somewhere between $24,279.24 and $35,880.63.

Confidence Intervals for the True Regression Coefficients

Using the Box 12.D formula, we can also produce confidence intervals for any one of the true regression coefficients from our computer printout. Choosing, once again, 95-percent intervals, the limits are:

$$\alpha = a \pm (t \, s_a) = -2.983 \pm [2.12(2.357)] = -2.983 \pm 4.997$$
$$\beta_1 = b_1 \pm (t \, s_{b_1}) = 2.099 \pm [2.12(.133)] = 2.099 \pm .282$$
$$\beta_2 = b_2 \pm (t \, s_{b_2}) = 1.197 \pm [2.12(.147)] = 1.197 \pm .312$$
$$\beta_3 = b_3 \pm (t \, s_{b_3}) = -.3107 \pm [2.12(.0577)] = -.3107 \pm .1223$$

Thus, we conclude:

$$-7.98 \leq \alpha \leq 2.014$$
$$1.817 \leq \beta_1 \leq 2.381$$
$$.885 \leq \beta_2 \leq 1.509$$
$$-.433 \leq \beta_3 \leq -.1884$$

Only the true intercept, α (which concerns us little), might conceivably be zero.

Hypothesis Tests About True Regression Coefficients

We can use an alternative approach as well to test whether any one of our parameters might equal zero; we can look at the associated t value. If the parameter is zero, the t value follows a t distribution with $n - m - 1$ degrees of freedom (16 in our case); thus, our critical t value is 2.12 (as noted before). All of the printed t values (except that for the intercept) have absolute value in excess of 2.12; hence, we can be (at least) 95-percent confident that the associated parameters β_1, β_2, and β_3 (but not the intercept, α) are not zero.

Testing the Overall Significance of the Regression

Our computer printout provides all the information needed to create an ANOVA table like Table 12.3. It also gives the F value directly. The likelihood of having gotten such a large F by pure chance is tiny, indeed.

The Sample Coefficient of Multiple Determination

The value of $R^2_{Y \cdot 123}$ is given directly as .9448. We can also check it with the help of the Box 12.E formula: the printout indicates that $RSS = 1{,}720$ and $ESS = 100.4275$. Hence, $Total\ SS = 1{,}820.4275$ and

$$R^2_{Y \cdot 123} = \frac{1{,}720}{1{,}820.4275} = .9448.$$

The *adjusted* sample coefficient of multiple determination, of course, must equal (according to Box 12.G):

$$\overline{R}^2_{Y \cdot 123} = 1 - \left[\frac{100.4275}{1{,}820.4275} \left(\frac{19}{16} \right) \right] = .9345.$$

Many computer programs print this adjusted R-squared as well.

Examples

In this section we consider just three examples that illustrate how the linear-multiple-regression technique is used to answer innumerable questions every day.

Application 1: Housing Prices and Proposition 13*

On June 7, 1978, California voters approved a statewide property-tax-limitation initiative, known as Proposition 13. This initiative led to a substantial *and differential* reduction in property taxes among localities, and one would expect corresponding impacts on housing prices.

Source: Kenneth T. Rosen, "The Impact of Proposition 13 on House Prices in Northern California: A Test of the Interjurisdictional Capitalization Hypothesis," *Journal of Political Economy* (February 1982), pp. 191–200.

A multiple regression equation was estimated from data on the San Francisco Bay Area, relating the change in post-Proposition 13 mean house prices, Y, to a number of independent variables (t values appear in parentheses):

$$\hat{Y} = -.171 + 7.275\,X_1 + .5468\,X_2 + .00073\,X_3 + .0638\ X_4$$
$$\quad (3.21) \quad (2.97) \qquad (2.32) \qquad (1.34) \qquad (3.26)$$
$$\qquad\qquad - .0043\,X_5 + .857\,X_6$$
$$\qquad\qquad (2.24) \qquad (1.80)$$
$$\overline{R}^2 = .89.$$

The *independent variables* in the above equation were defined as follows:

X_1 = post-Proposition 13 decrease in property-tax bill on mean house;
X_2 = mean square footage of house;
X_3 = median income;
X_4 = mean age of house;
X_5 = transportation time to San Francisco;
X_6 = housing-quality index.

Partial interpretation. As the coefficient of X_1 indicates, each $1 decrease in relative property taxes increased relative property values by about $7. Thus, "the results of this regression provide strong confirmation that the differential interjurisdictional tax reductions of Proposition 13 were partially capitalized in the year following the effective date of the statewide initiative. The capitalization rate implied by this equation is about 7, which is precisely the magnitude that one would expect given an interest rate of about 12–15 percent."

[Anyone who expects an eternal annual return of $1 (for example, as a result of a $1 property-tax reduction) will find that a one-time receipt of $7 is precisely equivalent to this future flow of returns, provided the applicable interest rate is (1/7)100, or 14.29 percent. This is so because $7, when invested at this rate, will generate a $1 return annually, forever. For a more complete discussion of capitalization, the reader may wish to consult Heinz Kohler, *Intermediate Microeconomics: Theory and Applications*, 2nd ed. (Glenview, Ill.: Scott, Foresman and Co., 1986), chapter 15.]

Application 2: The Price of Heroin and the Incidence of Crime*

It has often been claimed that there is an important relationship between drug abuse and crime. This theory was tested recently with the help of monthly data from New York City in the early 1970s. A number of multiple regression equations were estimated, relating the incidence of various crimes (Y) to various independent variables, such as the per-gram retail price of heroin (X_1), the average temperature (X_2), and a time trend (X_3). The accompanying table summarizes selected results (t values appear in parentheses).

Source: George F. Brown, Jr. and Lester P. Silverman, "The Retail Price of Heroin: Estimation and Applications," *Journal of the American Statistical Association* (September 1974), pp. 595–606.

Crime	a = Constant	b_1 = Coefficient of X_1	b_2 = Coefficient of X_2	b_3 = Coefficient of X_3	R^2
Murder	51.66	1.45 (2.89)	.04 (.22)	.05 (.07)	.523
Rape	113.26	.07 (.10)	.81 (3.03)	3.84 (4.11)	.670
Robbery	6,351.3	58.67 (2.24)	−14.18 (−1.40)	−87.84 (−2.48)	.314
Aggravated Assault	1,188.3	9.88 (1.73)	19.74 (8.94)	9.52 (1.23)	.813
Burglary	13,251.0	61.19 (1.68)	25.11 (1.79)	−213.05 (−4.33)	.605
Larceny over $50	9,220.7	15.13 (.51)	49.83 (4.33)	−202.09 (−5.02)	.806
Auto Theft	6,235.3	46.07 (2.38)	26.58 (3.55)	−157.92 (−6.04)	.771
Total[a]	36,412.0	192.48 (1.83)	107.93 (2.66)	−647.50 (−4.56)	.655

[a]Total includes above plus manslaughter, simple assault, and larceny under $50.

Partial interpretation. The data in each row can be used to formulate a regression equation, such as, for murder,

$$\hat{Y} = 51.66 + 1.45\,X_1 + .04\,X_2 + .05\,X_3.$$

Focusing on the b_1 column, we invariably find higher heroin prices being associated with higher numbers of crimes. All else being equal, a $1-per-gram increase in the retail price of heroin, for instance, was associated with and possibly even caused one additional murder, fifty-nine additional robberies, and forty-six additional auto thefts per month.

Application 3: Revenue-Sharing and the Flypaper Effect*

It has often been claimed that a disproportionate amount of intergovernmental aid dispensed by higher levels of government is subsequently used by the recipient lower levels of government to increase spending rather than reduce taxes. Revenue-sharing money seems to stick to the public sector like flies stick to flypaper! A ready explanation exists: Bureaucrats have an insatiable appetite for ever-increasing budgets, because

Source: Paul Gary Wyckoff, "A Bureaucratic Theory of Flypaper Effects," Federal Reserve Bank of Cleveland *Working Paper 8501,* June 1985.

bigger budgets (unlike lower taxes) enhance the personal prestige, income, and power of the bureaucrats. When they get additional funds, they will always find some new project on which to spend them. Recently, the flypaper effect was tested with the help of data pertaining to 105 small city governments in Michigan. A multiple regression ($\bar{R}^2$ = .633) was estimated relating governmental capital expenditures (on construction, equipment, and land) to 9 independent variables, one of which was the proportion of the total income of the median voter (private plus revenue-sharing funds) that was provided by revenue sharing. The estimated coefficient was positive (more revenue-sharing funds, more government spending), it was significant (with a t-value of 2.087), and its size suggested that a large flypaper effect was occurring.

BIOGRAPHY

12.1 George Snedecor

George Waddell Snedecor (1882–1974) was born in Memphis, Tennessee. He studied mathematics and physics at the Universities of Alabama and Michigan, then became a professor of mathematics at Iowa State University. While there, he offered the first statistics course and began a famous collaboration with Henry A. Wallace, then editor of *Wallace's Farmer* in Des Moines, but later vice-president of the United States. They were interested in agricultural research, jointly organized a seminar to study multiple regression, and did pioneering work in the use of punch cards and punchcard machines. Jointly, they published *Correlation and Machine Calculation* in 1925, and they established a Mathematical Statistical Service in 1927 and a now famous Statistical Laboratory in 1933.

In 1931, Snedecor had invited Ronald A. Fisher (Biography 8.1) to Ames and their meeting set in motion developments that produced many of the techniques that are discussed in this chapter. Among the visible signs of this development was the founding, at Iowa State, of the first Department of Statistics in the United States, Snedecor's presidency of the American Statistical Association, and the publication by Snedecor of two famous works: *Calculation and Interpretation of Analysis of Variance and Covariance* (1934) and *Statistical Methods* (1937). The latter work, ultimately co-authored with William G. Cochran, went through seven editions before the death of both authors and sold more than 125,000 copies.

Source: O. Kempthorne, "George W. Snedecor," *International Statistical Review* 42(1974):319–21.

Summary

1. *Multiple-regression analysis* is a technique that uses several independent variables (rather than a single one) to estimate the value of a dependent variable; *multiple-correlation analysis* measures the strength of association among all these variables. Unlike simple regression analysis, multiple-regression analysis allows us to exercise statistical control over extraneous factors and to determine the influence of any X on Y for specified constant values of other variables that might affect Y.

2. The techniques of multiple regression are straightforward extensions of those of simple regression. In the presence of two explanatory variables, they yield an estimated multiple-regression equation of the form, $\hat{Y} = a + b_1 X_1 + b_2 X_2$. The bs are the *estimated partial-regression coefficients;* they give the partial change in Y that is associated with a unit change in one of the independent variables when the other one is held constant. The above three-variable multiple-regression equation (involving Y,

X_1, and X_2) can be pictured as a *regression plane* in a three-dimensional scatter diagram. In a manner analogous to that for simple regression, the corresponding equation can be estimated, the standard error of the estimate of Y can be calculated, and various inferences (that are valid only if five crucial assumptions are fulfilled) can be made. Such inferences include the estimation of an average (or of an individual) value of Y for a given X_1 and X_2 and the establishment of confidence intervals for the βs. In making inferences, t values and p values can be used, and the overall significance of a multiple regression can also be tested by means of an analysis of variance. (This analysis of variance is the testing of the hypothesis that *all* the true regression coefficients are zero and, therefore, that *none* of the independent variables helps explain the variation of the dependent one.)

3. *The sample coefficient of multiple determination,* denoted by R^2, is the most important measure of how well an estimated regression plane (or hyperplane) fits the sample data on which it is based. It equals the proportion of the total variation in the values of the dependent variable, Y, that is explained by the multiple regression of Y on X_1, on X_2, and

possibly on additional independent variables (X_3, X_4, and so on). The square root of R^2 is sometimes used as an alternative; it is called the *sample coefficient of multiple correlation.* Because R^2 overestimates ρ^2, the corresponding population coefficient, an *adjusted sample coefficient of multiple determination,* $\overline{R}^2$—which is not so biased—is often calculated. Another important measure employed in multiple correlation analysis, paradoxical as it may sound, is the *sample coefficient of partial determination,* which is an index of association between the dependent variable and just *one* of the independent variables, given a prior accounting of the effects of one or more other independent variables. The index measures the proportional decrease in the previously unexplained variation in Y resulting from the addition of another independent variable to the regression model; thus, it can be used to evaluate the potential merit of adding an explanatory variable to the analysis.

4. While the principles remain the same, the calculations involved in multiple-regression analysis become extremely burdensome once the analysis encompasses more than two explanatory variables. Luckily, the modern computer comes to the rescue.

Key Terms

adjusted sample coefficient of multiple determination ($\overline{R}^2$)
estimated partial-regression coefficient
multiple-correlation analysis
multiple-regression analysis
p value
regression hyperplane
regression plane
sample coefficient of multiple correlation (R)
sample coefficient of multiple determination (R^2)
sample coefficient of partial correlation
sample coefficient of partial determination
t ratio
t value

Questions and Problems

The computer program noted at the end of this chapter can be used to work many of the subsequent problems.

1. Reconsider Chapter 11's problem 2. Using the data of Table 11.B and the data on the interest rate charged credit customers, X_2, given below, determine a multiple-regression equation, including t-values, R^2, $\overline{R}^2$, the ANOVA table, and the critical $F_{.05}$. X_2 (percent per year): 17, 18, 17, 17, 14, 14, 17, 17, 16, 16, 19, 19, 14, 15, 13, 16, 16, 16, 14, 14, 14, 15, 15, 18, 18, 17, 16, 17 (for successive columns).

2. Reconsider Chapter 11's problem 3. Using the data of Table 11.C and the additional data on household income, X_2, given below, determine a multiple regression equation, including t-values, R^2, $\overline{R}^2$, the ANOVA table, and the critical $F_{.05}$. X_2 (annual household income, thousands of dollars): 19.2, 22.7, 33.0, 17.2, 9.6, 8.2, 11.9, 27.3, 33.0, 14.9.

3. Reconsider Chapter 11's problem 4. Using the data of Table 11.D and the additional data on the selling realtor's sex, D_1, given below, determine a multiple-regression equation, including t-values, R^2, $\bar{R}^2$, the ANOVA table, and the critical $F_{.05}$.
D_1 ($= 0$ if male; $= 1$ if female): 1, 1, 1, 1, 0, 1, 1, 1, 0, 0, 0, 0, 0, 0 (for successive columns).

4. Reconsider Chapter 11's problem 5. Using the data of Table 11.E and the additional data on the price of cereal, X_2, given below, determine a multiple-regression equation, including t-values, R^2, $\bar{R}^2$, the ANOVA table, and the critical $F_{.05}$.
X_2 (dollars per box): 2.20, 2.22, 1.50, 2.20, 1.60, 1.40, 3.00, 3.00, 1.70, 1.70, 2.90, 2.90, 1.70, 2.30 (for successive columns).

5. Reconsider Chapter 11's problem 6. Using the data of Table 11.F and the additional data on age, X_2, given below, determine a multiple regression-equation, including t-values, R^2, $\bar{R}^2$, the ANOVA table, and the critical $F_{.05}$.
X_2 (age in years): 60, 33, 49, 42, 43, 39, 48, 29, 27, 30.

6. Reconsider Chapter 11's problem 7. Using the data of Table 11.G and the additional data on terrain, D_1, given below, determine a multiple regression equation, including t-values, R^2, $\bar{R}^2$, the ANOVA table, and the critical $F_{.05}$.
D_1 ($= 0$ if flat; $= 1$ if mountainous): 1, 1, 1, 1, 1, 0, 0, 0, 1, 1, 0, 0, 0, 0, 0, 0.

7. Reconsider Chapter 11's problem 8. Using the data of Table 11.H and the additional data on hours flown, X_2, given below, determine a multiple regression equation, including t-values, R^2, $\bar{R}^2$, the ANOVA table, and the critical $F_{.05}$.
X_2 (annual hours flown): 56, 62, 190, 350, 22, 200, 75, 120, 53, 72, 198, 205, 118, 299, 21.

8. Reconsider Chapter 11's problem 9. Using the data of Table 11.I and the additional data on the tax rate, X_2, given below, determine a multiple regression equation, including t-values, R^2, $\bar{R}^2$, the ANOVA table, and the critical $F_{.05}$.
X_2 (average federal tax rate, percent): 21, 22, 23, 25, 18, 19, 18, 19, 20.

9. Reconsider Chapter 11's problem 10. Using the data of Table 11.J and the additional data on fertilizer input, X_2, given below, determine a multiple regression equation, including t-values, R^2, $\bar{R}^2$, the ANOVA table, and the critical $F_{.05}$.
X_2 (pounds of fertilizer per acre): 200, 200, 220, 180, 190, 180, 190, 180.

10. Reconsider Chapter 11's problem 11. Using the data of Table 11.K and the additional data on real per capita income, X_2 given below, determine a multiple regression equation, including t-values, R^2, $\bar{R}^2$, the ANOVA table, and the critical $F_{.05}$.
X_2 (per capita income in thousands of 1972 dollars): 3.39, 3.62, 3.86, 4.32, 4.67, 5.08, 5.48, 5.97, 6.62, 7.33.

11. Reconsider Chapter 11's problem 12. Using the data of Table 11.L and the additional data on sales territory, D_1, given below, determine a multiple regression equation, including t-values, R^2, $\bar{R}^2$, the ANOVA table, and the critical $F_{.05}$.
D_1 (sales territory, $= 0$ if North; $= 1$ if South): 0, 0, 0, 0, 1, 0, 0, 1, 1, 0, 1, 1.

12. Reconsider Chapter 11's problem 13. Using the data of Table 11.M and the additional data on the price of a complementary good, X_2, given below, determine a multiple regression equation, including t-values, R^2, $\bar{R}^2$, the ANOVA table, and the critical $F_{.05}$.
X_2 (price of complementary good, dollars per unit): 50, 45, 40, 39, 38, 25, 31, 70, 20, 70 (for successive columns).

13. Reconsider Chapter 11's problem 14. Using the data of Table 11.N and the additional data on property type, D_1, given below, determine a multiple regression equation, including t-values, R^2, $\bar{R}^2$, the ANOVA table, and the critical $F_{.05}$.
D_1 (property type, $= 0$ if commercial, $= 1$ if residential): 1, 1, 0, 1, 0, 0, 1, 1, 1, 0.

14. Reconsider Chapter 11's problem 15. Using the data of Table 11.O and the additional data on the number of competitors, X_2, given below, determine a multiple regression equation, including t-values, R^2, $\bar{R}^2$, the ANOVA table, and the critical $F_{.05}$.
X_2 (number of competitors): 61, 58, 57, 60, 49, 33, 30, 25, 18.

15. Reconsider Chapter 11's problem 16. Using the data of Table 11.P and the additional data on import quotas, D_1, given below, determine a multiple regression equation, including t-values, R^2, $\bar{R}^2$, the ANOVA table, and the critical $F_{.05}$.
D_1 (import quotas, $0 =$ not in effect; $1 =$ in effect): 0, 0, 0, 1, 1, 1, 1.

16. Reconsider Chapter 11's problem 17. Using the data of Table 11.Q and the additional data on electricity price, X_2, given below, determine a multiple regression equation, including t-values, R^2, $\bar{R}^2$, the ANOVA table, and the critical $F_{.05}$.

X_2 (electricity price, cents per kilowatt hour): 2.1, 2.3, 2.5, 2.6, 2.8, 2.9, 2.9, 4.6, 5.3, 7.9.

17. Reconsider Chapter 11's problem 18. Using the data of Table 11.R and the additional data on aircraft type, D_1, given below, determine a multiple regression equation, including t-values, R^2, $\bar{R}^2$, the ANOVA table, and the critical $F_{.05}$.

D_1 (aircraft type, 0 = single-engine, 1 = multiengine): 0, 0, 0, 0, 1, 1, 1, 1, 1, 1, 1, 1.

18. Reconsider Chapter 11's problem 45. Using the data of Table 11.U and the additional data on type of plant, D_1, determine a multiple regression equation, including t-values, R^2, $\bar{R}^2$, the ANOVA table, and the critical $F_{.05}$.

D_1 (type of plant, 0 = old; 1 = new): 0, 0, 0, 0, 0, 0, 0, 0, 0, 0, 1, 1, 1, 1, 1, 1, 1, 1.

19. The manager of an art gallery wants to determine the relationship between the auction price of paintings, Y, the number of bidders, X_1, and the type of painting, D_1. Given the data of Table 12.A, determine a multiple regression equation, including t-values, R^2, $\bar{R}^2$, the ANOVA table, and the critical $F_{.025}$.

20. A publisher wants to determine the relationship between the number of book copies sold, Y, the number of pages in the book, X_1, the age of the author, X_2, the advertising expenditure made, X_3, and the number of similar books on the market, X_4. Given the data of Table 12.B, determine a multiple regression equation, including t-values, R^2, $\bar{R}^2$, the ANOVA table, and the critical $F_{.025}$.

Table 12.A

Auction Price, Y (thousands of dollars)	Number of Bidders, X_1	Type of Painting, D_1 (0 = Modern Art; 1 = Old Masters)
150	12	0
147	7	0
299	8	1
500	3	0
840	7	1
77	12	0
22	9	1
179	15	1
200	33	0
20	3	0

21. A textbook publisher wants to determine the relationship between the number of book copies sold, Y, the number of pages in the book, X_1, the number of analytical examples, X_2, the number of end-of-chapter problems, X_3, the ratio of favorable to unfavorable prepublication reviews, X_4, and the nature of the support package, D_1. Given the data of Table 12.C, determine a multiple-regression equation, including t-values, R^2, $\bar{R}^2$, the ANOVA table, and the critical $F_{.025}$.

Table 12.B

Copies Sold, Y (thousands)	Pages, X_1 (number)	Author's Age, X_2 (years)	Advertising Expense, X_3 (thousands of dollars)	Competing Books, X_4 (number)
5	352	22	5	7
13	490	29	9	3
23	250	38	5	20
36	670	67	7	39
57	805	70	8	61
72	390	52	6	10
98	510	50	12	7
102	711	31	22	2
122	905	29	39	5
190	472	75	30	0

Table 12.C

Copies Sold, Y (thousands)	Pages, X_1 (number)	Analytical Examples, X_2 (number)	Problems, X_3 (number)	Favorable to Unfavorable Reviews, X_4 (ratio)	Support Package, D_1 (Study Guide, Intstructor's Manual, Computer Disk, 0 = unavailable, 1 = available)
7	390	0	0	5.1	0
11	472	0	0	4.6	0
14	512	0	0	6.0	0
17	408	15	100	2.0	0
29	610	15	100	3.1	1
39	830	30	900	.2	1
52	310	15	500	.1	1
6	910	92	900	10.3	0
3	211	0	10	5.0	0
2	172	0	0	12.0	0

22. An engineer wants to determine whether there is a relationship between the tensile strength of plastic sheets, Y, and the respective quantities of ingredient 1, X_1, and 2, X_2. Given the data of Table 12.D, determine a multiple regression equation, including t-values, R^2, $\overline{R}^2$, the ANOVA table, and the critical $F_{.01}$.

Table 12.D

Tensile Strength, Y (psi)	Ingredient 1, X_1 (units)	Ingredient 2, X_2 (units)
31.7	12	88
52.0	13	144
45.6	14	126
29.1	15	81
49.3	16	136
37.0	17	102
48.0	18	133

23. A marketing manager wants to determine the relationship between average monthly purchases with the company's credit card, Y, the cardholder's family income, X_1, and the cardholder's family size, X_2. Given the data of Table 12.E, determine a multiple regression equation, including t-values, R^2, $\overline{R}^2$, the ANOVA table, and the critical $F_{.01}$.

Table 12.E

Average Purchases, Y (dollars per month)	Family Income, X_1 (thousands of dollars per month)	Family Size, X_2 (number of persons)
22.50	6.9	1
33.77	3.1	2
52.41	2.6	3
60.00	3.7	4
68.72	2.4	1
75.53	3.9	2
82.99	4.7	3
91.07	.8	4
98.99	.6	2
107.23	1.7	1

24. A politician wants to determine the relationship between per capita government expenditure, Y, in a state's counties, the county per capita income, X_1, and the county population density, X_2. Given the data of Table 12.F, determine a multiple regression equation, including t-values, R^2, $\overline{R}^2$, the ANOVA table, and the critical $F_{.01}$.

25. Compute the *sample standard error of the estimate of Y* for problems 1 through 8.

Table 12.F

Per Capita Government Expenditure in County, Y (dollars per year)	Per Capita Income in County, X_1 (thousands of dollars per year)	Population Density in County, X_2 (persons per square mile)
92.70	3.2	52
95.60	4.3	39
110.00	6.2	48
115.29	8.7	27
120.37	9.1	93
130.40	15.6	117
140.82	20.3	985
190.67	25.6	1,033
233.12	29.7	10,611
293.15	30.1	22

26. Compute the *sample standard error of the estimate of Y* for problems 9 through 16.

27. Compute the *sample standard error of the estimate of Y* for problems 17 through 24.

28. Reconsider problem 24. Given $X_1 = 15.6$ and $X_2 = 117$, compute a *95-percent prediction interval* for
  a. the average value of Y.
  b. the individual value of Y.

29. Reconsider problem 23. Given $X_1 = 3.9$ and $X_2 = 2$, compute a *95-percent prediction interval* for
  a. the average value of Y.
  b. the individual value of Y.

30. Reconsider problem 22. Given $X_1 = 17$ and $X_2 = 102$, compute a *95-percent prediction interval* for
  a. the average value of Y.
  b. the individual value of Y.

31. Reconsider problem 21. Given $X_1 = 830, X_2 = 30, X_3 = 900, X_4 = .2$, and $D_1 = 1$, compute a *95-percent prediction interval* for
  a. the average value of Y.
  b. the individual value of Y.

32. Reconsider problem 20. Given $X_1 = 390, X_2 = 52, X_3 = 6$, and $X_4 = 10$, compute a *95-percent prediction interval* for
  a. the average value of Y.
  b. the individual value of Y.

33. Reconsider problem 19. Given $X_1 = 12$ and $D_1 = 0$, compute a *95-percent prediction interval* for
  a. the average value of Y.
  b. the individual value of Y.

34. Reconsider your answer to problem 18. Use it to compute *95-percent confidence intervals for* β_1 *and* β_2.

35. Reconsider your answer to problem 17. Use it to compute *95-percent confidence intervals for* β_1 *and* β_2.

36. Reconsider your answer to problem 16. Use it to compute *95-percent confidence intervals for* β_1 *and* β_2.

37. Reconsider your answer to problem 15. Use it to compute *95-percent confidence intervals for* β_1 *and* β_2.

38. Given $r_{Y\cdot1}^2 = .89, r_{Y\cdot2}^2 = .10, R_{Y\cdot12}^2 = .95$, what is $r_{Y1\cdot2}^2$ and $r_{Y2\cdot1}^2$?

39. Given $r_{Y\cdot1}^2 = .715, r_{Y\cdot2}^2 = .072, r_{Y\cdot3}^2 = .011$, and $R_{Y\cdot12}^2 = .845$, what is $r_{Y1\cdot2}^2$ and $r_{Y2\cdot1}^2$?

40. In each case listed, interpret the meaning of the estimated regression coefficients; then assess the overall significance of the regression.

  a. One economist ranked U.S. utilities by assets, selected the 25 largest companies, and regressed their 1983 net income, Y (in millions of dollars) against assets, X_1 (in billions of dollars) and employees, X_2 (numbers). The result (*t* values in parentheses):

$$\hat{Y} = 371.01330 + 8.88493\,X_1 - .00137\,X_2$$
$$\quad (3.57713) \quad (.57669) \quad (-.60151)$$
$$r_{X_1X_2} = .99497 \qquad R_{Y\cdot12}^2 = .01826$$
$$RSS = 18{,}821.51 \qquad ESS = 1{,}011{,}779$$

  b. One economist selected the top 29 companies on the *Fortune 500* list and regressed the 1984 stock trading activity, Y (the percent of the company's outstanding shares traded during the average week) against the company's net profit margin, X_1 (percent) and debt/equity ratio, X_2 (percent). The result (*t* values in parentheses):

$$\hat{Y} = 1.92869 - .07573\,X_1 - .00312\,X_2$$
$$\quad (7.25846)(-2.41312) \qquad (-1.15834)$$
$$r_{X_1X_2} = .00362 \qquad R_{Y\cdot12}^2 = .21652$$
$$RSS = 2.442634 \qquad ESS = 8.838742$$

41. A local tax assessor wants to develop a regression model that can explain the value of houses, Y, measured in thousands of dollars, by the age of houses (X_1), by their square footage (X_2), by the number of bathrooms (X_3), by the absence (0) or presence (1) of an attached garage (D_1), and by the absence (0) or presence (1) of a view (D_2). A random sample of 15 houses is used to gather observations, with these results (*standard errors* in parentheses):

$$\hat{Y} = 64.42 - .5415\, X_1 + .00845\, X_2$$
$$(40.54)\quad (.5060)\qquad (.02166)$$

$$+ .76\, X_3 - 5.22\, D_1 + 12.16\, D_2$$
$$(12.01)\quad (18.77)\qquad (18.15)$$

The error sum of squares is 8,337; the total sum of squares is 9,773.

a. Comment on the significance of the regression coefficients.

b. Comment on the overall significance of this regression.

42. Consider the auto-theft multiple-regression equation given in Application 2, "The Price of Heroin and the Incidence of Crime": $\hat{Y} = 6{,}235.3 + 46.07\, X_1 + 26.58\, X_2 - 157.92\, X_3$.

a. Interpret the coefficients of X_1 and X_2.

b. For $X_1 = \$30$, $X_2 = 100°F$, and $X_3 = 50$, make a point estimate of the number of auto thefts.

43. Use the U.S. data (in billions of current dollars) in Table 12.G to work the following problems.

Table 12.G

Year	Personal Consumption Expenditures, C	Disposable Personal Income, DPI
1939	67.0	70.0
1940	71.0	75.3
1941	80.8	92.2
1942	88.6	116.6
1943	99.4	133.0
1944	108.2	145.6
1945	119.5	149.1
1946	143.8	158.9
1947	161.7	168.7
1948	174.7	188.0
1949	178.1	187.9
1950	192.0	206.6

Source: Economic Report of the President (Washington, D.C.: U.S. Government Printing Office, 1982), p. 261.

a. Estimate a simple regression equation of the form $\hat{C} = a + b\,DPI$.

b. Interpret the coefficient of DPI.

c. Graph this aggregate consumption function in a scatter diagram, showing war years (1942–45) as crosses, peace years as dots.

d. Because people's consumption behavior during the war years was obviously different from that during peace years (probably as a result of rationing, moral suasion, dislocation, and the unavailability of goods), estimate two separate consumption functions, for war years only and for peace years only. Graph them as well and interpret your results.

e. Compute the standard error of the estimate of C for the three equations.

f. Given the following information from a computer printout, test the overall significance of your three regressions by computing F; also compute r^2. For *all years*, $ESS = 1{,}496.547$ and *Total SS* $= 21{,}986.24$; for *war years*, $ESS = 40.02113$ and *Total SS* $= 516.1975$; for *peace years*, $ESS = 76.09928$ and *Total SS* $= 19{,}115.82$.

44. Reread the discussion of paintings in Analytical Example 11.2, "Of Bulls, Bears, and Beauty" (on page 399). The values of b given in Figure B were derived from a series of regression equations, one of which (on modern art) was (with t-values in parentheses):

$$R_P = .061 + .92\, R_M + 2.70\, I - .37\, D + e$$
$$(3.10)\quad (2.64)\qquad (3.11)\ (-4.87)$$
$$R^2 = .80$$
$$F = 13.02$$

In this equation, R_P is the rate of return on paintings in excess of a risk-free rate of return; R_M is the rate of return on the market portfolio in excess of a risk-free rate of return; I is the rate of inflation in excess of the expected rate, and D is a dummy variable measuring the special (art-price depressing) effects of a 1975 British tax on art and U.S. legalization of gold holdings.

a. Assess the significance of the regression coefficients and of the regression as a whole.

b. Assess the meaning of the coefficient of I.

45. The study noted in Application 3, "Revenue Sharing and the Flypaper Effect," provides the following estimated regression coefficients (and t-values): $a = -12.325$ (-1.439), $b_1 = -.704$ (-1.434), $b_2 = -1.354$ (-2.719), $b_3 = 1.972$ (1.873), $b_4 = 1.079$ (2.087), $b_5 = .627$ (3.464), $b_6 = -.0000835$ $(-.747)$, $b_7 = -.008704$ $(-.665)$, $b_8 = -.00949$ $(-.678)$, and $b_9 = -.05642$ (-1.424). Compute the coefficients' standard errors.

Selected Readings

Allen, D. M., and F. B. Cady. *Analyzing Experimental Data by Regression.* Belmont, Calif.: Wadsworth, 1982.

Chatterjee, S., and B. Price. *Regression Analysis by Example.* New York: Wiley, 1977.

Daniel, C., F. S. Wood, and J. W. Gorman. *Fitting Equations to Data: Computer Analysis of Multifactor Data,* 2nd ed., New York: Wiley, 1980.

Draper, N., and H. Smith. *Applied Regression Analysis,* 2nd ed., New York: Wiley, 1981.

Kohler, Heinz. *Statistics for Business and Economics,* 2nd ed., Glenview, Ill.: Scott, Foresman and Co., 1988.

Chapter 14 contains more advanced treatment of multiple regression analysis, including stepwise multiple regression, dummy variables, analysis of residuals, and more.

Mendenhall, William, and James T. McClave. *A Second Course in Business Statistics: Regression Analysis.* San Francisco: Dellen, 1981. Chapter 5 discusses matrix algebra and its application to multiple regression (as, for example, in the formulas of Box 12.B and 12.C).

Computer Program

The DAYSAL-2 personal-computer diskettes that accompany this book contain one program of interest to this chapter:

13. Regression and Correlation Analysis can perform multiple-regression analysis that provides not only regression coefficients, standard errors, and *t*-values, but also a correlation matrix, an ANOVA table, confidence intervals for α and β coefficients, hypothesis tests about them, and prediction intervals, at numerous confidence levels, for average and individual *Y*-values associated with each set of *X*-values originally entered.

You can view a scatter diagram of the association between any two variables, along with all the summary measures of central tendency and of dispersion. You can take advantage of a "live regression" feature, specifying any set of *X*-values, while the computer provides an instant point estimate of *Y*, along with limits of prediction intervals for the average and individual *Y* at your chosen confidence level. You can select followup programs of backwards elimination, make Durbin-Watson tests, or multicollinearity tests. You can use data files *Q–T* that contain data on the U.S. economy.

PART FIVE

Special Topics for Business and Economics

13   Time Series and Forecasting
14   Decision Theory

CHAPTER THIRTEEN

Time Series and Forecasting

Every organization must plan for the future; businesses and governments are no exception. Implicitly or explicitly, all such planning involves the making of predictions, and such predictions are inevitably linked to experiences gained in the past. An electric power producer, for example, may study how demand has grown in the past decades, project this growth into the future, and on that basis plan the construction of new generating capacity. A clothing manufacturer, likewise, may use past sales data as a guide to what can be expected and place new orders for raw materials and equipment accordingly. A government agency may study the past behavior of national income to estimate its future course and, thus, the tax revenues likely to be available.

The business and economic data about the past that are available to decision makers usually fall into one of two categories; most of them are either cross-section or time-series data. **Cross-section data** are numerical values pertaining to units of different populations that have been observed simultaneously at the same point in time (in the case of a *stock variable*) or during the same period of time (in the case of a *flow variable*). Examples are listings of the closing prices of different securites traded at the New York Stock Exchange on June 15, 1987 or listings of the profits of the different U.S.-based multinationals during the (Jan. 1–Dec. 31) 1981 period (as in Table 2.1 on page 39). Such cross-section data will *not* concern us in this chapter. We will focus instead on **time-series data,** or numerical values pertaining to units of a given population that have been observed repeatedly at different points in time (in the case of a stock variable) or during different periods of time (in the case of a flow variable). Examples are listings of the closing prices of a given security at the end of each of 30 successive business days or listings of the profits of a given multinational company during each of the last 15 years. It is precisely the latter type of chronological sequence of numerical data (each one of which is associated with a different moment or period of time) that becomes the raw material for **time-series analysis,** a statistical procedure that employs time-series data, usually for the purpose of explaining past or forecasting future events.

THE COMPOSITION OF A TIME SERIES

Several models seek to describe the typical time series. The most popular one, no doubt, is the **classical time-series model;** it attempts to explain the pattern observed in an actual time series by the presence of four components: trend, cyclical, seasonal, and irregular components.[1]

The Trend

The **trend component** of a time series is denoted by T and is evidenced as a relatively smooth, progressively upward or downward movement of the variable of interest, Y, over an extended period of time. The trend is viewed as the consequence of long-range gradual change in such factors as population size or composition, technology, or consumer preferences, and is typically computed from data that cover a minimum of 20 years.

As a result of this trend factor, a firm's sales of avionics equipment, for example, may be rising steadily at a rate of 50 units per year. (*Avionics equipment* is electronic instruments used by aircraft, such as communication and navigation units, distance-measuring units, radar transponders, and more.) Figure 13.1 illustrates this trend component in the firm's sales (with trend sales, T_t, measured vertically against time, t, measured horizontally). Note how anyone in 1981 with a knowledge of initial sales (point a) would, on the basis of this trend line, have predicted quarterly sales of 300 units four years hence—that is, in the winter of 1985 (point b). Yet this may not be what actually happened. We also have to take into account the cyclical component in the firm's sales.

Cyclical Fluctuations

The **cyclical component** of a time series is denoted by C and is evidenced by wide up-and-down swings of the variable of interest around the trend, with the swings lasting from one to several years each and typically differing in length and amplitude from one cycle to the next. Such irregular, but recurring, swings are seen to reflect the endless ebb and flow of economic activity in general, the eternal *business cycle* of boom and bust, that is bound to affect any particular variable in the economy.

Returning to our example, consider the influence on avionics sales of alternating sequences of boom and bust in the economy in general. If such business cycles occurred, avionics sales might lie above the trend in good years and below it in bad

---

[1]As will be shown below, all four components are likely to be present in data that pertain to periods shorter than a year, such as weekly, monthly, or quarterly data. When data pertaining to periods of a year or longer are being used, however, the seasonal component disappears from the model.

Figure 13.1   Trend Component in Avionics Sales

The line in this graph illustrates the trend component in a firm's sales that has been determined, we assume, on the basis of data pertaining to at least 20 years, but only some of which are shown here. Taking into account the trend only, sales of 100 units in the winter of 1981 (point *a*) rise by 50 units per year—for example, to 450 units in the winter of 1988 (point *c*). Note: *W, S, S,* and *F* refer to the winter, spring, summer, and fall quarters of the various years.

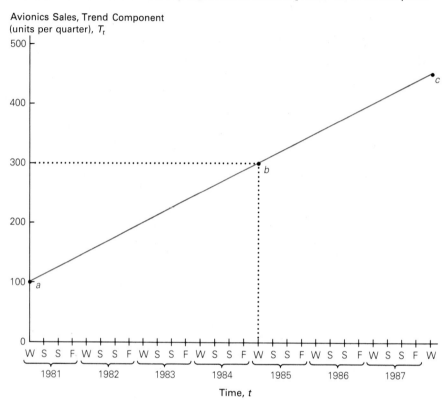

years, as depicted in panel (a) of Figure 13.2. (The cyclical component is measured as a proportion of the trend, which is now depicted as the horizontal line.) Thus, "booms" raise avionics sales to 1.3 times the expected in the winter of 1981 (point *a*), and again to well above the trend in late 1983 (point *c*) and in the winter of 1988 (point *f*). Yet, by the same token, economic "busts" lower sales below their trend, to 82 and 61 percent of the trend, respectively, in mid-1982 and mid-1985 (points *b* and *e*). By combining the information in panel (a) with that given in Figure 13.1, we can produce the trend/cyclical time series of sales given in panel (b). Note how the trend sales for the winter of 1985 (300 units according to point *b* of either Figure 13.1 or Figure 13.2) do not materialize. Panel (a) of Figure 13.2 shows how the (assumed) business cycle trough at that time depresses sales to 64 percent of their trend level (point *d*); thus, trend and cycle together produce sales of 300(.64) = 192 units only, as represented by point *c* in panel (b) of Figure 13.2. Thus, the combined trend/cyclical component of

Figure 13.2 Recognizing a Cyclical Component in Avionics Sales

The influence of the business cycle on a particular time series, shown in panel (a), will make the movement of the series over time diverge from the trend. Compare the color line in panel (b) with the black line representing the trend.

(a) The Cyclical Component in Sales

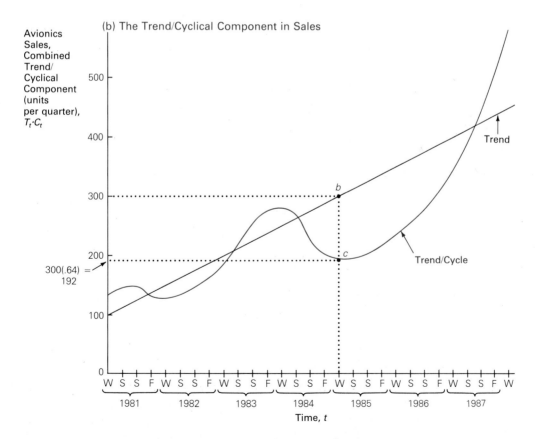

(b) The Trend/Cyclical Component in Sales

our time series can always be found by multiplying T_t (expressed in physical units sold) by C_t (expressed as a proportion of T_t). Yet trend and cycle do not tell the whole story; seasonal fluctuations play a role as well.

Seasonal Fluctuations

The **seasonal component** of a time series is denoted by S and is evidenced by narrow up-and-down swings of the variable of interest around the trend/cyclical components, with the swings predictably repeating each other within periods of one year or less. Such regularly recurring intrayear variations often reflect the influence of the weather and the calendar on economic activity. Consider how the sales of coal, gas, and oil (or of heavy clothing and snow-removal equipment) rise predictably each winter but fall during other seasons. Consider how the sales of air conditioners, gardening supplies, and swimming pools rise during the spring and summer and stagnate in the fall and winter. Consider the booming activity of tax-preparation firms just before April 15, of resorts after schools let out, of caterers of June weddings, or of sellers of Christmas-tree ornaments before December 24.

Once again we can depict the story graphically, as in Figure 13.3. Panel (a) shows (assumed) seasonal fluctuations in avionics sales, this time around the longer term trend/cyclical component of the time series. Note how winter sales always are 80 percent of average quarterly sales, while spring and summer sales rise to 100 and 120 percent, respectively, with fall sales returning to the average. The color line in panel (b) shows the actual time series, taking into account trend, cycle, and seasonal factors; the trend-only sales and the trend/cycle components are shown in black for comparison. Note how neither the 300-unit trend sales for the winter of 1985 (point b) nor the 192-unit trend/cycle sales (point c) materialize. The assumed seasonal low at that time, represented by point a in panel (a) of Figure 13.3, depresses sales to 80 percent of their trend/cyclical level; thus, trend, cycle, and season together produce sales of $300(.64)(.8) = 154$ units, rounded, which is represented by point d in panel (b) of Figure 13.3. Thus, the combined trend/cycle/seasonal component of our time series can always be found by multiplying T_t (expressed in physical units sold) by C_t (expressed as a proportion of T_t) and by S_t (expressed as a proportion of $T_t \cdot C_t$). Still, there is more to the story than trends, cycles, and seasons, as we will see in the next section.

Irregular Variations

The **irregular component** of a time series is denoted by I and is evidenced by random movements of the variable of interest around the trend/cyclical/seasonal components. Such movements are viewed as arising from completely unpredictable and probably nonrecurring chance events, such as fads in fashion, strikes, natural disasters, or wars.

The effects of such unsystematic influences on our avionics sales might be summarized by panel (a) of Figure 13.4 (it being assumed that these factors average to zero in the long run). Note how the winter 1985 sales are only 95 percent of what would be expected as a result of trend, cyclical, and seasonal factors, perhaps because of a strike in the industry (point a). The color line in panel (b) shows the actual time series, taking into account all the components considered in the classical model. This line

Figure 13.3 Recognizing a Seasonal Component in Avionics Sales

The influence of seasonal factors on a particular time series, shown in panel (a), will make the movement of the series over time diverge from the path that trend alone or trend plus cyclical factors alone would suggest. Compare the color line in panel (b) with the solid and dashed black lines that represent, respectively, trend sales alone or trend/cyclical sales alone.

(a) The Seasonal Component in Sales

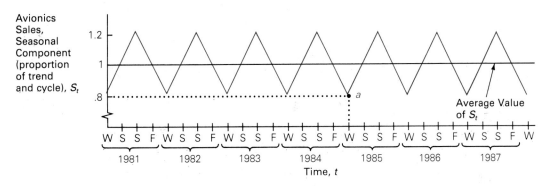

(b) The Trend/Cyclical/Seasonal Component in Sales

Figure 13.4   Recognizing an Irregular Component in Avionics Sales

The influence of irregular factors on a particular time series, shown in panel (a), will make the movement of the series over time diverge from the path that trend, cyclical, and seasonal factors alone would suggest [and that is depicted by the color line in panel (b) of Figure 13.3]. The color line shown here shows the actual avionics sales the company statistician would record.

(a) The Irregular Component of Sales

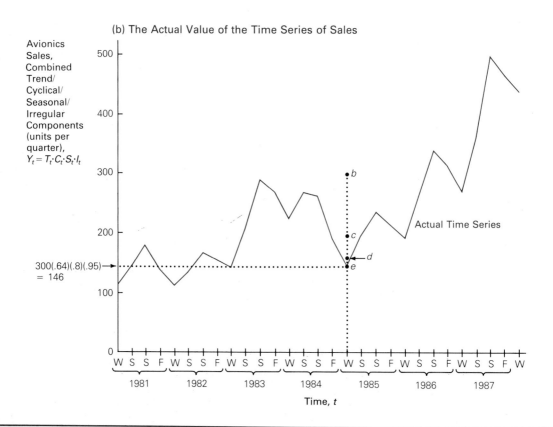

(b) The Actual Value of the Time Series of Sales

differs from the color line in panel (b) of Figure 13.3 by the inclusion of the irregular factors. Note how, for the winter of 1985, neither the 300-unit trend sales materialize (point *b*), nor the 192-unit trend/cycle sales (point *c*), nor even the 154-unit trend/cycle/seasonal sales (point *d*). Actual sales are 146 units (point *e*). The winter of 1985 is a bad time for avionics sales, indeed: The recession depresses sales to 64 percent below the trend (from *b* to *c*), the winter season, as usual, depresses sales to 80 percent of that (from *c* to *d*), and the strike reduces them further to 95 percent of that (from *d* to *e*). Thus, the classical time-series model asserts, the actual value of a time series, Y_t, at time *t*, can always be found by multiplying its trend value, T_t (expressed in physical units), by C_t (expressed as a proportion of T_t), by S_t (expressed as a proportion of $T_t \cdot C_t$), and by I_t (expressed as a proportion of $T_t \cdot C_t \cdot S_t$). Indeed, for this reason, the classical time-series model discussed here is generally referred to as a **multiplicative time-series model.**

13.A The Classical Multiplicative Time-Series Model

$$Y_t = T_t \cdot C_t \cdot S_t \cdot I_t$$

Note: The multiplicative model is well suited for a great variety of business and economic data and, therefore, has been widely accepted as the standard for analyzing time series. Yet, on occasion, time series are better represented by alternative models that allow for the possibility of components interacting in additive fashion or in a mixture of additive and multiplicative ways. In the **additive time-series model,**

$$Y_t = T_t + C_t + S_t + I_t,$$

and all the components are expressed in the same physical units. Examples of **mixed time-series models** are

$$Y_t = T_t \cdot C_t + S_t \cdot I_t$$

and

$$Y_t = T_t \cdot C_t \cdot I_t + S_t.$$

Analytical Example 13.1, "The Role of Time Series in Social Experiments," provides one example of the usefulness of time-series analysis.

FORECASTING: EXTRAPOLATING TRENDS

Business and economic **forecasting** is making statements about one or more unknown, uncertain, and, typically, future events. As this section and those that follow will show, many such statements are derived with the help of time-series data. Indeed, the trend component of a time series is often considered the most valuable forecasting tool, especially for long-term projections. If one has available a trend line, similar to that given in Figure 13.1, that summarizes the historical movement of a time series over an extended period of time, it is tempting to predict the future by *extrapolation,* by extending that line beyond the range of past and current data. In the beginning of his or her analysis, however, such a trend line is not available to the forecaster.

The forecaster has available some actual time series, such as the quarterly data embodied in the color line of panel (b) of Figure 13.4, which clearly reflect a mixture of trend, cyclical, seasonal, and irregular influences. To the extent that forecasts are to be made by means of a trend line, however, *annual* data are typically used: they automatically exclude all the seasonal fluctuations (that, by definition, arise within years only) and many irregular fluctuations as well. The first two columns of Table 13.1 might represent such an annual time series available to a would-be forecaster. (In order to simplify calculations, we restrict the example to 20 years; in real-life cases, trend calculations are preferably performed with much longer time series.) To find a trend line for these data, the forecaster has a variety of methods available.

Moving Averages

One way of portraying a trend is the construction of a **moving-averages series,** which is a series of numbers obtained by successively averaging overlapping groups of two or more consecutive values in a time series and replacing the central value in each group by the group's average. The average is "moving" because an ever-new average is calculated by adding a more recent time-series value to the group and dropping the oldest one.

||| Practice Problem 1

Constructing a 3-Year Moving-Averages Series for the Avionics Sales
of Butler Aviation
Consider constructing a 3-year moving average for the time series given in columns 1 and 2 of Table 13.1. The first three numbers (330, 241, and 200) are averaged first, and their average (257) is used to replace the central value of the group (241). This explains the first numerical entry in column 3. Then a new average is calculated, this time after adding to the group of three the 1971 time series value (499), while dropping the 1968 value (330). The (rounded) average of the new group (241, 200, and 499) equals 313, and it replaces the central value of the new group (200), which accounts for the second numerical entry in column 3. And so it goes. |||

Now let us picture what has been accomplished. We graph the original time series, along with the moving-averages series, in Figure 13.5 on page 460. The moving-average line displays considerably fewer fluctuations, which explains why any statistical procedure that dampens (or averages out) fluctuations in a time series is referred to as a **smoothing technique.** Having used the technique, one is free to imagine that the line of moving averages, such as the color line in Figure 13.5, reflects the systematic movement of the underlying series over time, such as its trend. But caution is advised: Our original series of annual data, by definition, is free of seasonal fluctuations because such fluctuations are intrayear changes. Our series, perhaps, is free of most irregular fluctuations as well because such fluctuations are often short-lived. Given these facts, if our annual data series contained a *regular* 3-year cycle, the averaging out of these

Table 13.1 The Avionics Sales of Butler Aviation and the Moving-Average Method

Year, t (1)	Units Sold, Y_t (2)	3-Year Moving Average, (3)	8-Year Moving Average Unadjusted (4)	8-Year Moving Average Adjusted (5)
1968	330	—		—
			—	
1969	241	257		—
			—	
1970	200	313		—
			—	
1971	499	340		—
			387	
1972	322	440		392
			396	
1973	500	474		437
			478	
1974	601	501		503
			528	
1975	401	468		546
			564	
1976	401	567		586
			607	
1977	899	633		611
			615	
1978	598	761		613
			610	
1979	787	684		643
			675	
1980	666	674		709
			743	
1981	569	599		736
			729	
1982	561	682		762
			794	
1983	915	808		844
			894	
1984	947	884		—
			—	
1985	791	951		—
			—	
1986	1,114	1,165		—
			—	
1987	1,591	—		—

recurring cycles by our construction of a 3-year moving-averages series would make sense. Yet cycles are not that regular; their periodicity could vary from 3 years in the 1950s and 1960s to 8 years in the 1970s or 1980s, and in that case we could not conclude that the Figure 13.5 color line has accurately captured the underlying trend. Nor would the result be any more certain if we constructed an 8-year moving-averages series instead, but when we do, we learn two additional lessons.

Figure 13.5   A 20-Year Time Series of Avionics Sales and 3-Year Moving Averages

The original 20-year time series of avionics sales—given in columns 1 and 2 of Table 13.1 and plotted here as the black line—shows considerably more fluctuations than a 3-year moving-averages series constructed with its help (color line). For this reason, the construction of a moving-averages series is referred to as a *smoothing technique*. The color line may be interpreted as a trend with the help of which values pertaining to future years (shaded) might be forecast. (Note the projection of the color line by the forecaster to some point such as *a* or *b*.)

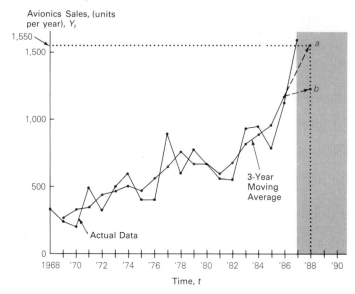

As we construct moving averages for even-numbered groups of time-series data (rather than odd-numbered ones), the procedure gets more complicated (which is the first lesson). As we construct moving averages for larger groups of data, the resulting series becomes even smoother (which is the second lesson). We can illustrate both of these effects by constructing an 8-year moving-averages series from our data, as shown in the last columns of Table 13.1.

||| Practice Problem 2

Constructing an 8-Year Moving-Averages Series for the Avionics Sales of Butler Aviation

As we average the first group of 8 data—the 1968–75 values of column 2—we derive an average of 387, but placing it in the center of the group of 1968–75 values puts it right in the middle between 1971 and 1972. To which year does the average belong? The assignment of time-series values to proper times is crucial indeed.[2] In our present case, the calculated average really belongs to a year that includes the second half of 1971 and the first half of 1972; therefore, the entry of 387 in column 4 is printed

---

[2]This fact is vividly illustrated by the Close-Up, "The Importance of Proper Timing" that is contained in the "Supplementary Topics" section of Chapter 13 of the *Student Workbook* that accompanies this text.

between the other rows of our table. The same is true of all the other, similarly derived entries in that column. The entry of 396, for example (which is the average of the 1969–76 values), belongs to the second half of 1972 and to the first half of 1973. To find the moving average associated with a particular (whole) calendar year, one averages adjacent averages. Thus, the averaging of 387 (which half belongs to the first half of 1972) and of 396 (which half belongs to the second half of 1972) gives us 392, and that number can be regarded as the 8-year moving average properly belonging to 1972. The remaining entries in column 5 of Table 13.1 have been similarly derived. ⦀

Figure 13.6 is a graph of the original time series and the 8-year moving-averages series. When compared with Figure 13.5, it clearly shows how the longer-term moving-averages series produces an even smoother picture of long-term change.

Disadvantages of moving averages. Although it is certainly possible to use a moving-averages series as an approximation of the underlying trend component of a time series, there are disadvantages. For one thing, many a moving-averages series still is quite irregular in appearance; the 3-year series of Figure 13.5 exemplifies this fact. Worse yet, it is possible for a moving-averages series to exhibit strong cyclical fluctuations even though there are none whatsoever in the original time-series data. Analytical Example 13.2 describes this so-called *Slutsky-Yule effect*. Indeed, a moving-averages series can produce all kinds of false impressions about the underlying data. To name a few

Figure 13.6 A 20-Year Time Series of Avionics Sales and 8-Year Moving Averages

The original 20-year time series of avionics sales (given in Table 13.1 and plotted here as the black line) shows considerably more fluctuations than an 8-year moving-averages series constructed with its help (color line). Note that the 8-year series, in turn, is even smoother than the 3-year series depicted in Figure 13.5. Unfortunately, it is also shorter; as a result, it is all the more risky to "cast forward," by means of the dashed lines, the past movement of the (color) "trend" into the (shaded) future.

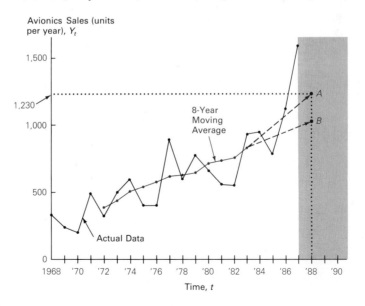

additional ones, the series can lie consistently above or below the original data (namely, when they are growing or declining exponentially), it will anticipate or prolong changes in the original data (and thus show a different timing of turning points), it will be extremely sensitive to unusually large or small values in the times series (as any average is bound to be). Worst of all, the very procedure of calculating moving averages makes it impossible to construct this "trend line" for the earliest and latest years of one's time series (because the average centered on any one year must be calculated with the help of a number of preceding and following values). Note how Table 13.1 contains no entry in the first and last rows of column 3, nor in the first and last *four* rows of column 5. Accordingly, the moving-averages lines in Figures 13.5 and 13.6 are considerably shorter than the 20-year period for which the original time-series data are available. This fact leads directly, of course, to the most serious disadvantage of all, and to that we now turn.

Forecasting with moving averages. Suppose we wanted to forecast, in 1987, avionics sales for 1988 or later years. How would we do so with the help of our 3-year moving-averages series (that ends in 1986) or with the help of our 8-year moving-averages series (that ends in 1983)? Would we dare project our color "trend line" straight into the (shaded) future region depicted in our graphs? Note how such a projection of the 3-year moving-averages series brings us to point *a* in Figure 13.5, leading us to forecast 1988 trend sales of 1,550 units, if we project the last segment of the color line by means of the upper dashed arrow. (This procedure is nothing more complicated than extending the last segment of the color line, while leaving its slope unchanged.) Yet we project sales, perhaps, to point *b,* if we take into account the generally lower slope of the entire color line. On the other hand, a projection of the last segment of the 8-year moving-averages series brings us to point *A* in Figure 13.6, leading us to forecast 1988 trend sales of 1,230 units, while a projection of the general slope of that line leads us, perhaps, to point *B*. Even assuming we had good reason to choose one of these "trend lines" over the other, how reliable would such a forecast be? It would be excellent if one could assume that the behavior of the "trend" observed in the past will not change in the near future, but such is never true with certainty. Close-Up 13.1, "The Dangers of Extrapolation," issues an important warning on the subject.

Exponential Smoothing

Much of the uncertainty of making a forecast that is connected with the moving-averages method is eliminated by another and much more popular forecasting procedure, called **exponential smoothing.** It operates like a thermostat and produces self-correcting forecasts by means of a built-in adjustment mechanism that corrects for earlier forecasting errors. We will consider two versions of the procedure.

Single-parameter exponential smoothing. The simplest approach to exponential smoothing produces the next period's forecast value, F_{t+1}, directly from the current actual value of the time series, Y_t, and the current forecast value, F_t, by using a **smoothing constant,** α, as shown in Box 13.B.

13.B Single-Parameter Exponential Smoothing

$$F_{t+1} = \alpha Y_t + (1 - \alpha)F_t.$$

The parameter, α, is a value betwen 0 and 1 that is chosen by forecasters to indicate the weight they want to attach to the most recent value of the time series. As long as only one such parameter is being employed, the procedure is called **single-parameter exponential smoothing.**

||| Practice Problem 3

Predicting Future Avionics Sales Using Single-Parameter Exponential Smoothing
Consider yourself in 1968 ($t = 1$) knowing only that current avionics sales equal $Y_1 = 330$ units, that current sales had been forecast as $F_1 = 320$ units, and that the current sales for 1968, thus, exceed the forecast by 10 units, as in row 1 of Table 13.2. If you choose $\alpha = .20$ as your smoothing constant, you can use the Box 13.B equation to forecast the 1969 value as

$$F_2 = \alpha\, Y_1 + (1 - \alpha)F_1 = .20(330) + (1 - .20)320 = 322.$$

Table 13.2 Forecasting Avionics Sales by Single-Parameter Exponential Smoothing

Year, t (1)	Actual Sales, Y_t (2)	$\alpha = .20$ Forecast Sales, F_t (3)	$\alpha = .20$ Forecasting Error, $Y_t - F_t$ (4) = (2) − (3)	$\alpha = .50$ Forecast Sales, F_t (5)	$\alpha = .50$ Forecasting Error, $Y_t - F_t$ (6) = (2) − (5)
1968	330	320	10	320	10
1969	241	322	−81	325	−84
1970	200	306	−106	283	−83
1971	499	285	214	242	257
1972	322	328	−6	371	−49
1973	500	327	173	347	153
1974	601	362	239	424	177
1975	401	410	−9	513	−112
1976	401	408	−7	457	−56
1977	899	407	492	429	470
1978	598	505	93	664	−66
1979	787	524	263	631	156
1980	666	577	89	709	−43
1981	569	595	−26	688	−119
1982	561	590	−29	629	−68
1983	915	584	331	595	320
1984	947	650	297	755	192
1985	791	709	82	851	−60
1986	1,114	725	389	821	293
1987	1,591	803	788	968	623
1988	—	961	—	1,280	—

This value is the second entry in column 3. Note that the previous error of $+10$ has raised the subsequent forecast by α times the error, or by $+2$. (Thus, the next forecast is being adjusted by a fraction of and in the direction of the previous forecast error). As in Table 13.1, you might find actual 1969 ($t = 2$) sales of 241, however. Accordingly, you would record an error of -81 and make a new forecast for 1970 of

$$F_3 = \alpha\, Y_2 + (1 - \alpha)F_2 = .20(241) + (1 - .20)322 = 305.8.$$

Once again, you would have adjusted the previous forecast (of 322) by α times the previous error (of -81), or by -16.2 in this case. The remainder of the table can be filled out in similar fashion; all entries have been rounded. Note: What if no forecast had existed for 1968? Then the actual 1968 value of 330 would have been used as the 1969 forecast, and the equation would have been employed thereafter. |||

Fine-tuning the smoothing constant. As an inspection of column 4 of Table 13.2 indicates, many of our forecasts were fairly wide off the mark; that is, they differed significantly from actual sales. Whenever this happens, forecasters adjust the value of α, increasing it when they want to give more weight to current experience (and less to the past) or decreasing it when they want to do the opposite. By trial and error, they can in this way select a smoothing constant that minimizes the forecasting error for the particular time series with which they are working. (Typically, a very stable time series calls for the use of a large α; an extremely volatile one calls for a smaller α.) Columns 5 and 6 of Table 13.2 indicate the forecasting record that would have been achieved with $\alpha = .5$, and Figure 13.7 shows graphically how actual and forecast values would have diverged under the two alternative procedures.

Advantages of exponential smoothing. The popularity of the exponential-smoothing method arises from two facts: (1) the calculations are simple, and (2) next to no data storage is required. To appreciate the latter, consider that only one number (namely F_{t+1}, the most recent forecast made) has to be remembered, and only until a new forecast has been made, because any given forecast embodies the entire time series in it. As we have noted,

$$F_{t+1} = \alpha Y_t + (1 - \alpha)F_t$$

but, in turn,

$$F_t = \alpha(Y_{t-1}) + (1 - \alpha)F_{t-1}.$$

Thus,

$$
\begin{aligned}
F_{t+1} &= \alpha Y_t + (1 - \alpha)[\alpha(Y_{t-1}) + (1 - \alpha)F_{t-1}] \\
&= \alpha Y_t + \alpha(1 - \alpha)Y_{t-1} + (1 - \alpha)^2 F_{t-1}.
\end{aligned}
$$

If we continued this breakdown for ever earlier periods, we would discover all the preceding Y values (Y_t, Y_{t-1}, Y_{t-2}, Y_{t-3}, and so on) hidden in the most recent forecast, and they would appear with successive weights of α, $\alpha(1 - \alpha)$, $\alpha(1 - \alpha)^2$, $\alpha(1 - \alpha)^3$, . . . , and so on. These weights decrease *exponentially* (the more distant the value of the time series, the less its forecasting weight), which explains the name of this technique.

To sum up, the exponential-smoothing technique produces a weighted average of all past time-series values (with weights decreasing exponentially as one goes back in

Figure 13.7 Avionics Sales, Actual and Single-Parameter Exponentially Smoothed

This graph indicates how an actual time series might be smoothed exponentially—in this case, by employing a smoothing constant of $\alpha = .2$ or $\alpha = .5$, respectively. The method can also be employed to make forecasts—for example, of A or B in 1988.

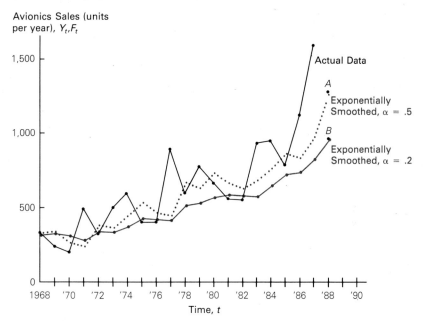

time), and the average so constructed serves as a forecast for the next period. The method is cheap and fast and is often employed when multiple forecasts for a large set of items are required—as, perhaps, in a firm's inventory-control system, which might require demand forecasts for tens of thousands of items.

Two-parameter exponential smoothing. Whenever there is a strong upward trend in a time series, single-parameter exponential smoothing can be relied upon to produce generally too-low forecasts, as is evident from an inspection of Figure 13.7. **Two-parameter exponential smoothing** is a technique that might eliminate this problem by explicitly taking into account the influence of the trend. The technique employs three equations: (1) to produce a smoothed time-series value, Y_t^*, for the current time period, (2) to produce a smoothed trend-component value, T_t^*, for the current time period, and (3) to produce a forecast, F_{t+1}, for the next time period. These equations are given in Box 13.C.

13.C Two-Parameter Exponential Smoothing

$$Y_t^* = \alpha Y_t + (1 - \alpha)(Y_{t-1}^* + T_{t-1}^*) \tag{1}$$
$$T_t^* = \gamma(Y_t^* - Y_{t-1}^*) + (1 - \gamma)T_{t-1}^* \tag{2}$$
$$F_{t+1} = Y_t^* + T_t^* \tag{3}$$

In the equations in Box 13.C, the parameter γ (the lowercase Greek *gamma*) is the **trend smoothing constant,** also chosen to lie between 0 and 1, while the current trend is defined as the difference between the current and preceding smoothed value of the series, $Y_t^* - Y_{t-1}^*$.

Practice Problem 4

Predicting Future Avionics Sales Using Two-Parameter Exponential Smoothing

We can apply the technique to our familiar example, now using Table 13.3. Columns 1 and 2 are copies of Table 13.2; the column 3 entries have been calculated using equation (1) of Box 13.C, arbitrarily using a smoothing constant of $\alpha = .20$ and a value of $Y_2^* = Y_1 = 330$. The column (4) entries have been calculated using equation (2) of Box 13.C, arbitrarily using a trend smoothing constant of $\gamma = .40$ and an initial smoothed trend value of $T_2^* = Y_2 - Y_1 = -89$. All of the above implies (assumed)

Table 13.3   Forecasting Avionics Sales by Two-Parameter Exponential Smoothing ($\alpha = .20$, $\gamma = .40$)

Year, t (1)	Actual Sales, Y_t (2)	Equation (1): Smoothed Sales, Y_t^* (3)	Equation (2): Smoothed Trend, T_t^* (4)	Equation (3): Forecast Sales, F_t (5)	Forecasting Error, $Y_t - F_t$ (6) = (2) − (5)
1968	330	—	—	—	—
1969	241	330	−89	—	—
1970	200	233	−92	241	−41
1971	499	213	−63	141	358
1972	322	184	−49	150	172
1973	500	208	−20	135	365
1974	601	271	13	188	413
1975	401	307	22	284	117
1976	401	343	28	329	72
1977	899	477	70	371	528
1978	598	557	74	547	51
1979	787	662	86	631	156
1980	666	732	80	748	−82
1981	569	763	60	812	−243
1982	561	771	39	823	−262
1983	915	831	47	810	105
1984	947	892	53	878	69
1985	791	914	41	945	−154
1986	1,114	987	54	955	159
1987	1,591	1,151	98	1,041	550
1988	—	—	—	1,249	—

Figure 13.8   Avionics Sales, Actual and Two-Parameter Exponentially Smoothed

This graph indicates how an actual time series might be smoothed exponentially—in this case, by employing two parameters ($\alpha = .2$, $\gamma = .4$). The method can also be employed to make forecasts—as of 1,249 units (point *A*) in 1988.

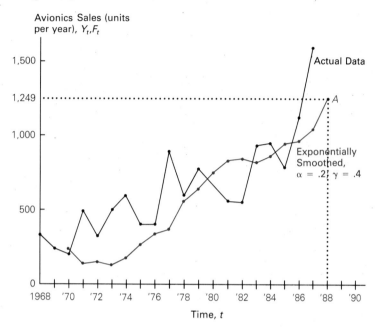

initial values of $F_3 = 241$ and $Y_3 - F_3 = -41$. Now we employ our equations to forecast 1971 sales. Thus, in 1970 ($t = 3$),

$$Y_3^* = .20(200) + (.80)(330 - 89) = 232.8,$$

and

$$T_3^* = .40(233 - 330) + (.60)(-89) = -92.2.$$

Thus, our 1971 forecast is

$$F_4 = 232.8 - 92.2 = 140.6.$$

The remaining entries in Table 13.3 have been similarly calculated (but always rounded to whole numbers). We can also picture our forecasting record, as in Figure 13.8. As the graph shows, and as column (6) of Table 13.3 confirms, that record is not too good and certainly makes us wonder whether we should trust the 1988 forecast of 1,249 units (point *A* in the graph). The use of different smoothing constants might improve that record, however, and this outcome could be determined by trial and error. ▮

Other exponential-smoothing procedures. A great variety of other exponential-smoothing procedures exist. These include **three-parameter exponential smoothing** (a method that adds seasonal smoothing to trend smoothing), the adjustment of the smoothing constant, α, from one period to the next, and even the consideration of nonlinear relationships between values.

Figure 13.9   Alternative Trend Lines Suggested by Scatter Diagrams

The black dots in these four panels depict alternative scatters of some variable, Y_t, over time, t. The color lines represent trends that seem to fit the data best. Panel (a) might represent the steady *absolute* growth in a firm's sales (at a rate of 50 units per year). Panel (b) might depict a steady *percentage* growth of sales (at a rate of 2 percent per year), and panel (c) might show a steady percentage decline in sales (at a rate of 3 percent a year). Panel (d) shows a mixture, with growth at an increasing rate followed by growth at a decreasing rate.

Panel (b) depicts well the trend in such time series as the U.S. population, the U.S. GNP, the number of passengers carried by U.S. airlines, the U.S. per capita consumption of synthetic fibers, or that of frozen vegetables. Panel (c) depicts well the trend in such time series as the number of passengers carried by U.S. railroads, the U.S. per capita consumption of cotton, or that of fresh vegetables. Panel (d) might be a so-called Gompertz or logistic (Pearl-Reed) growth curve; it depicts the production and sales behavior of many a new product that experiences spectacular growth, followed by market saturation.

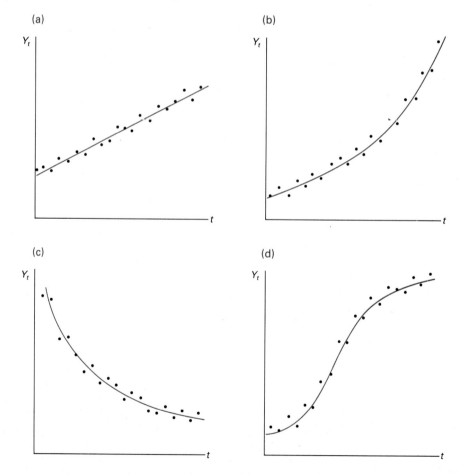

Figure 13.10   Scatter Diagram of Avionics Sales and Fitted Regression Line

This scatter diagram is based on the data in columns 1 and 2 of Table 13.4; the solid color line was fitted by the method of least squares. With its help, one would forecast 1990 trend sales of 1,234 units (point *a*).

Fitting Regression Lines

It is also possible to estimate a trend, and then to make forecasts with its help, by fitting a least-squares regression line to the available time-series data. This approach is dangerous business, however, because the crucial assumptions underlying that procedure (and noted in Chapters 11 and 12) may well not be applicable. In that case, unbiased regression coefficients can, perhaps, still be estimated, but the making of all kinds of valid inferences ceases to be possible. Yet such inferences would be particularly desirable in a forecasting situation: Precarious as it is, the projection of the dependent variable (for example, of avionics sales), by means of a regression line, to time periods beyond the range of data from which that line was calculated is in this case the whole point of estimating the regression line in the first place. Thus, one would want to have reliable confidence intervals for any estimated future values at the very time that such intervals are unlikely to be available. Nevertheless, because the procedure is frequently employed, let us consider how regression analysis might be used in the service of forecasting.

   In all cases, good judgment suggests first drawing a scatter diagram of the variable of interest against time. Such a diagram will suggest a basic type of trend line that is likely to fit the data best; Figure 13.9 shows several possibilities. The black dots in Figure 13.10, in turn, constitute a scatter diagram for our avionics-sales data; we will illustrate the fitting of a linear regression line to these data presently.

Practice Problem 5

Letting a Linear Regression Line Approximate the Trend in Avionics Sales
The values pertaining to our example are given in columns 1 and 2 of Table 13.4.
Typically, the time values (that serve as the independent variable) are transformed or
coded as shown in column 3, and a regression line such as the following is then
estimated:

$$\hat{Y}_X = a + bX$$

Table 13.4   Fitting a Trend Line by Linear Regression

Year, t (1)	Units Sold, Y_t (2)	Time, X (3)	XY (4)	X^2 (5)	$\hat{Y}_X$ (6)	$Y - \hat{Y}_X$ (7)
1968	330	0	0	0	200	130
1969	241	1	241	1	247	−6
1970	200	2	400	4	294	−94
1971	499	3	1,497	9	341	158
1972	322	4	1,288	16	388	−66
1973	500	5	2,500	25	435	65
1974	601	6	3,606	36	482	119
1975	401	7	2,807	49	529	−128
1976	401	8	3,208	64	576	−175
1977	899	9	8,091	81	623	276
1978	598	10	5,980	100	670	−72
1979	787	11	8,657	121	717	70
1980	666	12	7,992	144	764	−98
1981	569	13	7,397	169	811	−242
1982	561	14	7,854	196	858	−297
1983	915	15	13,725	225	905	10
1984	947	16	15,152	256	952	−5
1985	791	17	13,447	289	999	−208
1986	1,114	18	20,052	324	1,046	68
1987	1,591	19	30,229	361	1,093	498
	12,933 = ΣY	190 = ΣX	154,123 = ΣXY	2,470 = ΣX^2		$\Sigma(Y - \hat{Y}_X) = 0$ (except for rounding)

$$\bar{Y} = \frac{\Sigma Y}{n} = \frac{12,933}{20} = 646.65$$

$$\bar{X} = \frac{\Sigma X}{n} = \frac{190}{20} = 9.5$$

In this equation, $\hat{Y}_X$ is not equal to the actual time-series value (in our case, actual sales, Y_t) but is the computed or trend value. The calculations proceed according to the Box 11.A formulas given on p. 373; the preliminaries are indicated in columns 4–7 of Table 13.4. Hence,

$$b = \frac{\Sigma XY - n\bar{X}\bar{Y}}{\Sigma X^2 - n\bar{X}^2} = \frac{154{,}123 - 20(9.5)646.65}{2{,}470 - 20(9.5)^2}$$

$$= 47 \text{ (rounded)};$$

$$a = \bar{Y} - b\bar{X} = 646.65 - 47(9.5) = 200 \text{ (rounded)}.$$

Therefore, our estimated trend line can be described as:

$$\hat{Y}_X = 200 + 47X,$$

where X is measured in years and $X = 0$ is located at mid-1968. This line is the solid color line in Figure 13.10; with its help, one might estimate 1990 trend sales as 1,234 units (point *a*). |||

From annual to quarterly data. Most trend lines are fitted to annual data, but it may be desirable to have quarterly or even monthly trend values. These can be easily determined by modifying our above regression equation. Suppose our manager required quarterly information on sales (at annual rates). In that case, we would need X values measured in quarters from a zero base, and each of them would be centered in the middle of a quarter. If $X = 0$ were placed in the middle of the first or winter quarter of 1968 (that is, at mid-February, 1968), the base would shift backward in time by 1.5 quarters (the original base was mid-1968, at the end of the second or spring quarter). Our earlier calculation showed sales rising annually by $b = 47$ units; hence, they are rising by $b/4$ or 11.75 units per quarter, which would be the slope of our new regression line for quarterly data. Hence, the new line's intercept would have to equal 200 (or mid-1968 trend sales) minus 1.5(11.75), or 182.375. Thus, the transformed regression line (measuring sales quarterly at annual rates) would be

$$\hat{Y}_X = 182.375 + 11.75X,$$

where X is measured in quarters and $X = 0$ is located at the middle of the first or winter quarter of 1968. Note how this equation accurately predicts the 200-unit annual trend sales of 1968 that constitutes the intercept of the annual-data regression line in Figure 13.10:

$$\hat{Y}_{1.5} = 182.375 + 11.75(1.5) = 200.$$

ANALYTICAL EXAMPLES

13.1 The Role of Time Series in Social Experiments

On many occasions, throughout every year, some kind of social experiment is being initiated. It may involve a tax cut designed to put an end to a recession or a cut in the money supply to stop inflation; it may involve a price cut or a change in product design to increase sales; it may involve the assignment of more police to a city district to deter crime or to the freeways to put an end to drunken driving. Sooner or later, one question naturally arises: how effective has the program been?

Unfortunately, these types of social experiments can rarely be conducted like laboratory experiments. Typically, one cannot assign human subjects randomly to a control group or to an experimental one, supply a "treatment" (tax cut, tight money, lower price, new product design, more police patrols) to the latter group only, and ultimately attribute observed differences between the groups to the treatment. In the realm of social policy, such controlled experiments happen but rarely. More likely than not, for example, *all* citizens get a tax cut at the same time, and later on it is difficult to know whether any subsequent behavior change was the result of the tax cut or would have happened anyway for all kinds of other reasons. Nevertheless, a comparison of time-series data from before and after the introduction of a new social policy is usually employed to evaluate the effectiveness of that policy. Clearly, this is tricky business.

Consider the case of Connecticut's crackdown on speeding that began on December 23, 1955. Traffic deaths in 1955 had been 324; in 1956, they dropped to 284, as shown in panel (a). The governor attributed the saving of 40 lives to the crackdown. Yet he implicitly assumed that there would have been no change in traffic deaths in the absence of his program, and he was surely wrong on that, as panel (b) indicates. Throughout the 1950s, Connecticut traffic deaths had been zigzagging up and down. The 1955 figure had been an all-time high; thus, chances were high that the number would drop in the next year even without any crackdown (once again, the regression effect!).

Thus, an evaluation of social quasi-experiments by means of time-series analysis must carefully consider the observed effect—panel (a)—*in the context of the general trend*—panel (b). Any single rise or fall in a variable that is part of a zigzagging time series can hardly be attributed to the policy with which that rise or fall happens to have been associated. The same is true for any rise embedded in a generally upward trend or any decline that is part of a generally downward trend. On the other hand, a rise, at the time of a policy change, in a long-standing downward trend is quite another matter; so is a decline at such a time in a long-term upward trend. Thus, time-series analysis is far from useless when it comes to evaluating social experiments. It just needs to be applied with caution.

Source: Donald T. Campbell, "Measuring the Effects of Social Innovations by Means of Time Series," in Judith M. Tanur et al., eds., *Statistics: A Guide to the Unknown* (San Francisco: Holden-Day, 1972), pp. 120–29.

(a)

(b)

13.2 The Slutsky-Yule Effect

While the method of moving averages will produce a fairly reliable estimate of a time-series value at any particular time, it also produces a misleading *pattern* of points over an extended period of time. It *creates* periodicities where there are none in the original data. This **Slutsky-Yule effect** can be illustrated most effectively by noting how a time series that consists of nothing but random numbers (and, therefore, contains no oscillations) can give birth to an oscillatory moving-averages series that shows clearly visible, yet artificially created and, therefore, misleading waves.

Consider the hypothetical time series plotted in panel (a). It is nothing else but a plot of the 50 digits found in row 1 of Appendix Table A, "Random Numbers," or, what time-series analysts refer to as pattern-less "white noise." Now imagine taking 10-period moving averages of this series and plotting these, as in panel (b). The oscillations are obvious, yet it would be wrong to look for any causal link here (such as the influence of strikes or holidays on our time series).

In fact, the reason for the waves is fairly simple. Because the original numbers vary randomly, the sum of any ten of these (and the average of that group) is bound to fluctuate. When that sum is unusually high, there are lots of large numbers in the group; chances are lower that a larger number will be added next than that a smaller number will be added next. But a large number will be dropped, so the sum (and average) probably decline after a peak. The reverse is true when we start with an unusually small sum (and average); thus, the two probably rise after a trough. If all this reminds you of the *regression effect,* discussed in Biography 11.1 of Francis Galton, it should.

Source: Based on Eugen E. Slutsky, "The Summation of Random Causes as the Source of Cyclic Processes," *Econometrica* 5(1937):105–46; G. Udny Yule, "On a Method of Investigating Periodicities in Disturbed Series, with Special Reference to Wolfer's Sunspot Numbers," Royal Society of London, *Philosophical Transactions* (1927), pp. 267–98.

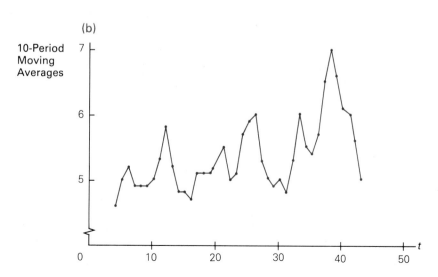

13.1 The Dangers of Extrapolation

Even when the past data of a time series exhibit a clearly visible trend, forecasting the future on that basis (or, for that matter, making inferences about a more distant and unknown past) is dangerous business. Thus, President Lincoln, upon observing the 1790–1860 trend in U.S. population, predicted a U.S. population of 252 million by 1930. In fact, the actual 1930 population was 123 million. On the other hand, a 1938 presidential commission doubted that the U.S. population would ever reach 140 million, yet that figure was reached and surpassed by 1945.

No one, perhaps, has ever put the matter more clearly than Mark Twain when he wrote these lines in 1874 in *Life on the Mississippi:*

In the space of one hundred and seventy-six years the Lower Mississippi has shortened itself two hundred and forty-two miles. That is an average of a trifle over one mile and a third per year.

Therefore, any calm person, who is not blind or idiotic, can see that in the Old Oölitic Silurian Period, just a million years ago next November, the Lower Mississippi River was upward of one million three hundred thousand miles long, and stuck out over the Gulf of Mexico like a fishing-rod. And by the same token any person can see that seven hundred and forty-two years from now the Lower Mississippi will be only a mile and three-quarters long, and Cairo and New Orleans will have joined their streets together, and be plodding comfortably along under a single mayor and a mutual board of aldermen. There is something fascinating about science. One gets such wholesale returns of conjecture out of such a trifling investment of fact.

Source: Darrell Huff, *How to Lie with Statistics* (New York: Norton, 1954), p. 142.

Summary

1. All economic organizations must plan for the future; inevitably, this planning requires the making of predictions that, in turn, are linked to experiences gained in the past. Most business and economic data are either *cross-section data* (which come from different populations observed at the same time) or *time-series data* (which come from the same population observed at different times). *Time-series analysis* employs time-series data for the purpose of explaining past or forecasting future events.

2. The *classical time-series model* attempts to explain the pattern observed in an actual time series by the presence of (1) trend, (2) cyclical, (3) seasonal, and (4) irregular components. The trend component, T, is a relatively smooth, progressively upward or downward movement of the variable of interest over an extended period of time. The cyclical component, C, is evidenced by wide up-and-down swings of the variable of interest around the trend, with the swings lasting from one to several years each and typically differing in length and amplitude from one cycle to the next. The seasonal component, S, is evidenced by narrow up-and-down swings of the variable of interest around the trend/cyclical components, with the swings predictably repeating each other within periods of one year or less. The irregular component, I, finally, adds random movements to all of the above movement. Because the classical model expresses the actual time-series value, Y, as the product of the four components just noted (making $Y = T \cdot C \cdot S \cdot I$), this model is also referred to as a *multiplicative time-series model.*

3. Business and economic forecasting is making statements about unknown, uncertain, and, typically, future events. Many such statements are derived with the help of time-series data. The trend component is often considered the most valuable forecasting tool, especially for long-term projections. The trend can be estimated in a variety of ways, including the construction of a *moving-averages series* (by successively averaging overlapping groups of two or more consecutive values in a time series and replacing the

central value in each group by the group's average), the employment of single or multiple-parameter *exponential-smoothing techniques* (that produce self-correcting forecasts in the form of weighted averages of all past time-series values, with weights de-

creasing exponentially as one goes back in time), and the fitting of least-squares (linear or curvilinear) regression lines (which, however, can be problematic because of the possible violation of assumptions that underlie regression analysis).

Key Terms

additive time-series model
classical time-series model
cross-section data
cyclical component
exponential smoothing
forecasting
irregular component
mixed time-series models
moving-averages series
multiplicative time-series model
seasonal component

single-parameter exponential smoothing
Slutsky-Yule effect
smoothing constant
smoothing technique
three-parameter exponential smoothing
time-series analysis
time-series data
trend component
trend smoothing constant
two-parameter exponential smoothing

Questions and Problems

The computer programs noted at the end of this chapter can be used to work many of the subsequent problems.

1. An analyst has determined the values given in Table 13.A for the sales of a company. Assuming that the classical multiplicative time-series model applies, find the missing values. Explain how you found them.
2. The following time series represents the U.S. gross federal debt at the end of 1939–86 (in billions of dollars): 48.2 — 50.7 — 57.5 — 79.2 — 142.6 — 204.1 — 260.1 — 271.0 — 257.1 — 252.0 — 252.6 — 256.9 — 255.3 — 259.1 — 266.0 — 270.8 — 274.4 — 272.8 — 272.4 — 279.7 — 287.8 — 290.9

— 292.9 — 303.3 — 310.8 — 316.8 — 323.2 — 329.5 — 341.3 — 369.8 — 367.1 — 382.6 — 409.5 — 437.3 — 468.4 — 486.2 — 544.1 — 631.9 — 709.1 — 780.4 — 833.8 — 914.3 — 1,003.9 — 1,147.0 — 1,381.9 — 1,576.7 — 1,841.1 — 2,074.2.

a. Estimate the trend in the form of a 5-year moving-averages series.
b. On the basis of the slope of the trend line between the last two numbers estimated in (a), forecast the U.S. gross federal debt at the end of 1989.

Table 13.A

Quarter, t	Actual Sales, Y_t	Trend Component, T_t	Cyclical Component, C_t	Seasonal Component, S_t	Irregular Component, I_t
Winter	$93,060	$50,000	.99	2.00	?
Spring	?	55,000	.95	.50	1.02
Summer	22,344	60,000	.95	?	.98
Fall	69,998.50	65,000	?	1.10	1.10

3. Using the data of problem 2,
   a. Estimate the trend in the form of a 10-year moving-averages series.
   b. On the basis of the slope of the trend line between the last two numbers estimated in (a), forecast the U.S. gross federal debt at the end of 1989.

4. The following time series represents new U.S. housing units started in 1959–84 (in thousands of units): 1,554 — 1,296 — 1,365 — 1,493 — 1,635 — 1,561 — 1,510 — 1,196 — 1,322 — 1,545 — 1,500 — 1,469 — 2,085 — 2,379 — 2,058 — 1,353 — 1,171 — 1,548 — 2,002 — 2,036 — 1,760 — 1,313 — 1,100 — 1,072 — 1,713 — 1,751.
   a. Estimate the trend in the form of a 3-year moving-averages series.
   b. On the basis of the slope of the trend line between the last two numbers estimated in (a), forecast new U.S. housing starts in 1989.

5. Using the data of problem 4,
   a. Estimate the trend in the form of a 6-year moving-averages series.
   b. On the basis of the slope of the trend line between the last two numbers estimated in (a), forecast new U.S. housing starts in 1989.

6. The following time series represents the service industry portion of the U.S. gross national product from 1947–83 (in billions of 1972 dollars): 55.9 — 57.5 — 57.6 — 59.7 — 60.8 — 61.6 — 62.7 — 62.9 — 67.6 — 70.9 — 74.1 — 76.2 — 80.8 — 83.5 — 86.6 — 90.3 — 94.0 — 98.8 — 103.1 — 109.0 — 115.0 — 118.8 — 124.0 — 126.7 — 128.4 — 136.5 — 144.8 — 147.9 — 148.5 — 154.7 — 164.3 — 174.2 — 183.0 — 189.1 — 197.6 — 200.2 — 206.8.
   a. Estimate the trend in the form of a 7-year moving-averages series.
   b. On the basis of the slope of the trend line between the last two numbers estimated in (a), forecast the service industry portion of U.S. real GNP in 1989.

7. Using the data of problem 6,
   a. Estimate the trend in the form of a 14-year moving-averages series.
   b. On the basis of the slope of the trend line between the last two numbers estimated in (a), forecast the service industry portion of U.S. real GNP in 1989.

8. The following time series represents the mining industry portion of the U.S. gross national product from 1947–83 (in billions of 1972 dollars): 6.8 — 9.4 — 8.1 — 9.3 — 10.2 — 10.1 — 10.6 — 10.9 —

12.4 — 13.4 — 13.5 — 12.4 — 12.3 — 12.6 — 12.7 — 12.8 — 13.1 — 13.4 — 13.5 — 14.2 — 14.6 — 15.3 — 16.1 — 17.6 — 17.4 — 19.0 — 21.7 — 32.2 — 38.8 — 43.0 — 47.4 — 52.0 — 66.8 — 96.0 — 132.3 — 125.1 — 112.4.
   a. Estimate the trend in the form of a 9-year moving-averages series.
   b. On the basis of the slope of the trend line between the last two numbers estimated in (a), forecast the mining industry portion of U.S. real GNP in 1989.

9. Using the data of problem 8,
   a. Estimate the trend in the form of a 4-year moving-averages series.
   b. On the basis of the slope of the trend line between the last two numbers estimated in (a), forecast the mining industry portion of U.S. real GNP in 1989.

10. The following time series represents U.S. federal government transfer payments from fiscal year 1960 to fiscal year 1986 (in billions of dollars): 20.6 — 23.6 — 25.1 — 26.5 — 27.4 — 28.4 — 31.8 — 37.2 — 42.7 — 48.7 — 55.0 — 67.7 — 76.1 — 87.2 — 101.8 — 131.4 — 153.8 — 166.6 — 178.7 — 197.8 — 234.6 — 273.7 — 304.5 — 338.3 — 340.7 — 361.0 — 377.6.
   a. Estimate the trend in the form of a 3-year moving-averages series.
   b. On the basis of the slope of the trend line between the last two numbers estimated in (a), forecast the U.S. federal government transfers in fiscal 1989.

11. Using the data of problem 10,
   a. Estimate the trend in the form of a 4-year moving-averages series.
   b. On the basis of the slope of the trend line between the last two numbers estimated in (a), forecast the U.S. federal government transfers in fiscal 1989.

12. Reconsider the data of problem 2. After equating the first actual value (48.2) with the 1939 forecast value, employ the method of single-parameter exponential smoothing with $\alpha = .2$ to develop a 1987 forecast. Show your computations.

13. Reconsider the data of problem 2. After equating the first actual value (48.2) with the 1939 forecast value, employ the method of single-parameter exponential smoothing with $\alpha = .8$ to develop a 1987 forecast. Which is better to use, $\alpha = .2$ (in problem 12) or $\alpha = .8$ (in problem 13)? Explain.

14. Reconsider the data of problem 4. After equating the first actual value (1,554) with the 1959 forecast value, employ the method of single-parameter exponential smoothing with α = .1 to develop a 1985 forecast. Show your computations.

15. Reconsider the data of problem 4. After equating the first actual value (1,554) with the 1959 forecast value, employ the method of single-parameter exponential smoothing with α = .9 to develop a 1985 forecast. Which is better to use, α = .1 (as in problem 14) or α = .9 (as in problem 15)? Explain.

16. Reconsider the data of problem 6. After equating the first actual value (55.9) with the 1947 forecast value, employ the method of single-parameter exponential smoothing with α = .3 to develop a 1984 forecast. Show your computations.

17. Reconsider the data of problem 6. After equating the first actual value (55.9) with the 1947 forecast value, employ the method of single-parameter exponential smoothing with α = .7 to develop a 1984 forecast. Which is better, α = .3 (as in problem 16) or α = .7 (as in problem 17)? Explain.

18. Reconsider the data of problem 8. After equating the first actual value (6.8) with the 1947 forecast value, employ the method of single-parameter exponential smoothing with α = .4 to develop a 1984 forecast. Show your computations.

19. Reconsider the data of problem 8. After equating the first actual value (6.8) with the 1947 forecast value, employ the method of single-parameter exponential smoothing with α = .5 to develop a 1984 forecast. Which is better, α = .4 (as in problem 18) or α = .5 (as in problem 19)? Explain.

20. Reconsider the data of problem 10. After equating the first actual value (20.6) with the 1960 forecast value, employ the method of single-parameter exponential smoothing with α = .15 to develop a 1987 forecast. Show your computations.

21. Reconsider the data of problem 10. After equating the first actual value (20.6) with the 1960 forecast value, employ the method of single-parameter exponential smoothing with α = .8 to develop a 1987 forecast. Which is better, α = .15 (as in problem 20) or α = .8 (as in problem 21)? Explain.

22. Reconsider the data of problem 2. Using two-parameter exponential smoothing (α = .2, γ = .3), prepare a 1987 forecast. Show the entire series.

23. Reconsider the data of problem 4. Using two-parameter exponential smoothing (α = .1, γ = .4), prepare a 1985 forecast. Show the entire series.

24. Reconsider the data of problem 6. Using two-parameter exponential smoothing (α = .3, γ = .6), prepare a 1984 forecast. Show the entire series.

25. Reconsider the data of problem 2. Estimate a linear regression line; then make a forecast for 1989.

26. Given the regression determined in problem 25,
   a. test the overall significance with the help of an ANOVA table (using a critical $F_{.05}$).
   b. determine the value of r^2.

27. Given the regression determined in problem 25,
   a. compute the sample standard error of the estimate of Y.
   b. determine 95-percent confidence intervals for α and β.
   c. make 2-tailed tests of the hypotheses H_0: α = 0 and H_0: β = 0.

28. Reconsider the data of problem 6. Estimate a linear regression line; then make a forecast for 1989, along with a 95-percent prediction interval for the individual Y value.

29. Figure 13.10 gives an annual trend equation. Convert it to a monthly equation (that gives month-to-month trend sales at annual rates).

30. Consider the annual trend equation determined in problem 28. Convert it to a monthly equation.

Selected Readings

Armstrong, J. Scott. *Long-Range Forecasting: From Crystal Ball to Computer.* New York: Wiley, 1978. An excellent, wide-ranging discussion with lots of examples.

Brown, Robert G. *Smoothing, Forecasting, and Prediction of Discrete Time Series.* Englewood Cliffs, N.J.: Prentice-Hall, 1963. Shows, among other things, certain desirable theoretical properties of exponential smoothing.

Kohler, Heinz. *Statistics for Business and Economics,* 2nd. ed. Glenview, Ill.: Scott, Foresman and Co., 1988. Chapter 15 contains additional material on forecasting, including the use of curvilinear regression, seasonal indices, econometric models, and more. Chapter 16 focuses on index numbers.

Morgenstern, Oskar. *On the Accuracy of Economic Observations,* 2nd ed. Princeton, N.J.: Princeton University Press, 1963. A must reading for anyone working with business and economic statistics.

Computer Programs

The DAYSAL-2 personal computer diskettes that accompany this book contain three programs of interest to this chapter:

13. Regression and Correlation Analysis is described in detail at the end of Chapters 11 and 12.

14. Time Series and Forecasting lets you create moving-averages series, prepare forecasts by single-param-eter or two-parameter exponential smoothing. The latter two programs also allow you to test a wide range of alpha values or of alpha-gamma combinations for the best results. Data files U-Z can be used for all of these options.

19. General Graphics, lets you draw time-series line graphs on graphs of your own design.

CHAPTER FOURTEEN

Decision Theory

Only rarely do the actions of business and economic decision makers lead to unique consequences that are known with certainty in advance. In countless situations, such decision makers face a serious problem: They are called upon to choose among two or more alternative courses of action at a time when the associated consequences cannot be foreseen with certainty because they depend on the nature of some future event over which the decision maker has no control. Under such circumstances, it is difficult to make a rational decision. Consider the farmer who must choose now among plantings of alternative crops but knows that some of these crops will flourish if the season turns out to be cool and dry, while others will do so only if it is hot and wet. Consider the manufacturer who must decide now whether to introduce a new production process or to continue using the old one but who also knows that the profit consequences of either decision will differ greatly, depending on whether future oil prices (or future wages) are low or high. Consider the marketing manager whose advice about introducing a new product (or modifying the style, packaging, or labeling of an old product) is desired now but who knows the consequences of any of these actions to be dependent on as-yet-unknown levels of future demand. Examples such as these can be multiplied without end because the ultimate outcome of any choice is likely to be affected by chance. Think of the real-estate developer who must make a commitment now to a small- or medium-sized or large-scale project; the consulting firm that must decide now to install a small or a large computer; the oil company that must drill now or sell its rights; the record producer who must initiate a major sales campaign now or abort the production of a record altogether; the independent TV producer who can submit the pilot of a new program to a network now or sell the rights to someone else; the job applicant who must decide now to work for a fixed salary or on a commission basis. In all these cases and many more, the wisdom or foolishness of any present decision will ultimately be determined by (uncontrollable) future events: by whether the demand for housing (or consulting services) turns out to be high or low, by whether oil is found at the drilling site or the test hole is dry, by whether the new record (or TV show) becomes a hit or is a failure, by whether the new employee is successful at making sales or not. In this final chapter of the book, we will consider a variety of methods, collectively referred to as **decision theory,** that can be employed in the systematic analysis and solution of decision-making problems that arise because uncertainty exists about future events over which the decision maker has no control but which are bound to influence the ultimate outcome of a decision.

Note: Some writers, such as the economist Frank Knight (1885–1972), have introduced a fine distinction between decision making under *uncertainty* and decision making under *risk.* In either case, the ultimate outcome of a present decision is affected by future random events. In the case of decision making under uncertainty, however, the probabilities associated with all of these events, or even the nature of these events, are unknown at the time of decision making; in the case of decision making under risk, both the nature and the probabilities of these events are known in advance. Like most writers, we make no such distinction here and refer to all situations in which the ultimate outcome of a decision maker's choice depends on chance as **decision making under uncertainty.**

BASIC CONCEPTS

Whenever decisions are being made in the context of uncertainty, certain identical elements are present; it is helpful to focus on these elements systematically when searching for a solution to a decision-making problem.

Actions, Events, Payoffs

First, each decision maker has specific decision alternatives available. In the language of decision theory, the decision maker must make a choice among various mutually exclusive **actions,** a complete list of which is symbolized by $A_1, A_2, \ldots, A_n$. A business executive, for example, might be called upon to manufacture and market a new product by taking one of two actions: A_1 = constructing a small plant; A_2 = constructing a large plant.

Second, future occurrences, not under the control of the decision maker, will affect the outcome of any present action taken by the decision maker. These occurrences are commonly referred to as **states of nature** or, more simply, as **events;** a collectively exhaustive list of mutually exclusive events is symbolized by $E_1, E_2, \ldots, E_n$. The production and marketing of a new product, for example, might be linked with one of two events: E_1 = an environment of weak demand; E_2 = an environment of strong demand. As we shall see, probability values about such events may or may not be employed by a decision maker at the time when a choice among actions is taken.

Third, positive or negative net benefits are associated with each possible action/event combination. These net benefits are the joint outcome of choice and chance and are commonly referred to as **payoffs.** Payoffs can be measured in any kind of unit appropriate to the problem at hand: in dollars, in time, or even in utility, as we will see. The payoffs from constructing a small plant, for example, might be annual profits of $8 million if demand is weak or $5 million if demand is strong (the smaller payoff resulting from the need, perhaps, to then run the plant above its designed capacity if demand is strong, which leads to extremely high unit costs of production). The corresponding payoffs from a large plant might be $-\$2$ million or $+\$12$ million instead (the smaller payoff again, perhaps, occurring because the running of a large plant *below* its capacity also involves unusually high unit costs of production).

The Payoff Table

A decision-making situation in the context of uncertainty can be summarized in a variety of ways; one popular summary is the **payoff table,** a tabular listing of the payoffs associated with all possible combinations of actions and events. Table 14.1 is such a tabular summary for our example. Each row corresponds to one of the two possible actions; each column corresponds to one of the two possible events. The four cells of the table show the associated annual profits (in millions of dollars).

Note: In this payoff table neither action is unambiguously superior to the other in the sense that it produces payoffs that are as good as or better than those of the alternative action no matter which event occurs. If such an unambiguously superior action exists, it is called a **dominant action,** and its existence turns the alternative into an **inadmissible action** that need not be considered further because of its obvious inferiority. In our example, A_1 would be dominant if, all else being equal, the A_1/E_2 cell contained an entry of 12 instead of 5; if A_1 were dominant, A_2 would be inadmissible. Under those circumstances, the decision maker could choose A_1 without hesitation and would reap a profit of \$8 million if E_1 occurred or of \$12 million if E_2 occurred. The choice of A_2 would lead to worse or at best equally good results, as the A_2 row of Table 14.1 indicates.

The Decision Tree

Another popular summary of a decision-making situation is the **decision tree.** This graphical device is similar to the tree diagram encountered in Chapter 4; it illustrates a decision problem by showing, in chronological order from left to right, every poten-

Table 14.1 A Payoff Table
This table shows the annual profits (in millions of dollars) associated with four possible action/event combinations.

Actions	Events	
	E_1 = weak demand for new product	E_2 = strong demand for new product
A_1 = constructing a small plant	8	5
A_2 = constructing a large plant	-2	12

tial action, event, and payoff. The decision tree diagram is particularly useful when a decision problem involves a sequence of many decisions that extend over a long stretch of time.

It is customary in decision-tree diagrams to denote any point of *choice* (at which the decision maker is in control) by a square symbol from which branches representing the possible actions of the decision maker (and, therefore, called **action branches** or **decision branches**) emanate toward the right. Such an **action point** (also variously referred to as a **decision point, decision node,** or **decision fork**) is illustrated in panel (a) of Figure 14.1, along with the relevant branches. In contrast, any point of *chance* (at which the decision maker exercises no control, but "nature" is in charge) is denoted symbolically by a circle from which branches representing the possible events confronting the decision maker (and, therefore, called **event branches** or **state-of-nature branches**) emanate toward the right. Such an **event point** (also variously referred to as a **state-of-nature point, state-of-nature node,** or **state-of-nature fork**) is illustrated in panel (b) of Figure 14.1, again with the associated branches. Panel (c), finally, shows the entire decision tree that summarizes our example.

Summarizing a decision problem, either with the help of a payoff table or a decision tree, is helpful, but it is only the first step toward a solution. A best plan of action still remains to be chosen; the following sections consider some of the criteria that might be employed to find it.

DECISION MAKING WITHOUT PROBABILITIES

Consider a situation, such as that summarized by Table 14.1 or Figure 14.1, in which a decision must be made in the face of uncertainty, but nothing is known about the likelihood of occurrence of those alternative future events that are certain to affect the eventual outcome of the present decision. Imagine that the decision maker does not even care to guess what the event probabilities might be. In such a case, it is common practice to employ one of three decision criteria: *maximin* (or *minimax*), *maximax* (or *minimin*), or *minimax regret.*

Maximin (or Minimax)

The first of the nonprobabilistic criteria embodies a decidedly conservative approach to decision making. It guarantees that the decision maker does no worse than achieve the best among the poorest possible outcomes. This criterion takes one of two forms, depending on whether the decision maker aims to maximize benefit or minimize cost. If maximization of benefit is the objective (as in our example above), the criterion is called **maximin** because the decision maker is to find the lowest possible (or minimum) benefit associated with each possible action, identify the highest (or maximum) benefit among these minima, and then choose the action associated with this maximum of minima (which explains the name given to this criterion). Table 14.2 illustrates how the maximin criterion would be applied to our example.

Figure 14.1 A Decision Tree

As this illustration shows, the possible payoffs associated with a decision made under uncertainty result from a mixture of choice (square symbol) and chance (circular symbols). Note that the information contained in this diagram is precisely the same as that found in the payoff table (Table 14.1).

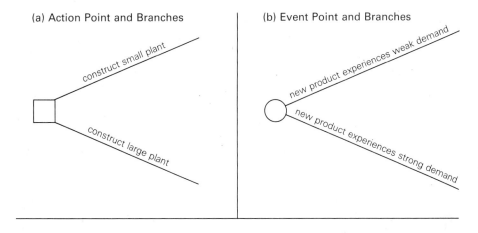

(a) Action Point and Branches
construct small plant
construct large plant

(b) Event Point and Branches
new product experiences weak demand
new product experiences strong demand

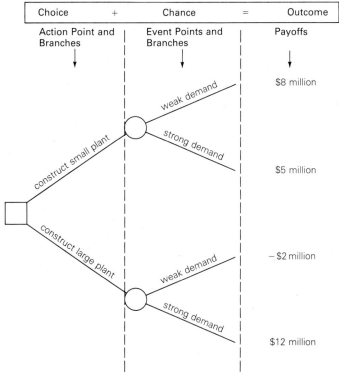

(c) Actions, Events, and Payoffs

Choice	+	Chance	=	Outcome
Action Point and Branches		Event Points and Branches		Payoffs

construct small plant

weak demand — $8 million

strong demand — $5 million

construct large plant

weak demand — −$2 million

strong demand — $12 million

Table 14.2   Maximizing Benefit: The Maximin Criterion

This table shows the annual profits (in millions of dollars) associated with four possible action/event combinations. The lowest of the possible profits associated with each action is shown in the last column; the maximin criterion suggests taking the best of these (encircled), or action A_1. Under the circumstances, the firm can do no worse than earn a profit of $5 million. It might even earn $8 million, if E_1 rather than E_2 occurs.

| Actions | Events | | Row Minimum |
	E_1 = weak demand for new product	E_2 = strong demand for new product	
A_1 = constructing a small plant	8	5	(5)←Maximin
A_2 = constructing a large plant	−2	12	−2

If minimization of cost is the objective (as when a firm must choose, perhaps, between the installation of alternative antipollution devices), the analogous criterion is often referred to as **minimax** because the decision maker is to find the highest possible (or maximum) cost associated with each possible action, identify the lowest (or minimum) cost among these maxima, and then choose the action associated with this minimum of maxima (which explains the name given to the criterion in this instance). Table 14.3 illustrates how the minimax criterion would be applied.

Table 14.3   Minimizing Cost: The Minimax Criterion

This table shows the annual costs (in millions of dollars) associated with four possible action/event combinations. The highest of the possible costs associated with each action is shown in the last column; the minimax criterion suggests taking the best of these (encircled), or action A_1. Under the circumstances, the firm can do no worse than incur a cost of $8 million. It might even pay as little as $5 million, if E_1 rather than E_2 occurs.

| Actions | Events | | Row Maximum |
	E_1 = low level of demand and production	E_2 = high level of demand and production	
A_1 = install small antipollution system	5	8	(8)←Minimax
A_2 = install large antipollution system	10	4	10

Table 14.4   Maximizing Benefit: The Maximax Criterion

This table shows the annual profits (in millions of dollars) associated with four possible action/event combinations. The highest of the possible profits associated with each action is shown in the last column; the maximax criterion suggests taking the best of these (encircled), or action A_2. Under the circumstances, the firm might earn a profit of \$12 million, but it might also have a loss of \$2 million, if E_1 rather than E_2 occurs.

Actions	Events		Row Maximum
	E_1 = weak demand for new product	E_2 = strong demand for new product	
A_1 = constructing a small plant	8	5	8
A_2 = constructing a large plant	-2	12	(12) ←Maximax

Note that the maximin (or minimax) criterion is ideally suited to the born pessimist. If you always expect the worst, if you always suspect that "nature" or chance will work against you, then you cannot do better than employ this criterion. It will give you nothing worse than the best among all the worst things that can happen: the highest of the lowest-possible profits in Table 14.2 and the lowest of the highest-possible costs in Table 14.3.

Maximax (or Minimin)

The second among the nonprobabilistic criteria goes to the other extreme and seeks to achieve the best of the best possible outcomes. This criterion, too, comes in two forms. If maximization of benefit is the objective, the criterion is called **maximax** because the decision maker is to find the highest possible (or maximum) benefit associated with each possible action, identify the highest (or maximum) benefit among these maxima, and then choose the action associated with this maximum of maxima (as the name suggests). Table 14.4 illustrates how the maximax criterion would be applied to our first example.

If minimization of cost is the objective (as in our second example), the analogous criterion is often referred to as **minimin** because the decision maker is to find the lowest possible (or minimum) cost associated with each possible action, identify the lowest (or minimum) cost among these minima, and then choose the action associated with this minimum of minima (as the name suggests). Table 14.5 illustrates how the minimin criterion would be applied to our second example.

Note that the maximax (or minimin) criterion is ideally suited to the born optimist. If you always expect the best, if you are always convinced that "nature" or chance will be on your side, then you might go after the best of the best in the fashion just shown: the highest of the highest-possible profits in Table 14.4 and the lowest of the lowest-possible costs in Table 14.5.

Table 14.5 Minimizing Cost: The Minimin Criterion

This table shows the annual costs (in millions of dollars) associated with four possible action/event combinations. The lowest of the possible costs associated with each action is shown in the last column; the minimin criterion suggests taking the best of these (encircled), or action A_2. Under the circumstances, the firm might incur costs as low as $4 million, but it might also have costs of $10 million, if E_1 rather than E_2 occurs.

| | Events | | |
Actions	E_1 = low level of demand and production	E_2 = high level of demand and production	Row Minimum
A_1 = install small antipollution system	5	8	5
A_2 = install large antipollution system	10	4	④ ←Minimin

Minimax Regret

The third one among the nonprobabilistic criteria is just a bit more complicated. To understand it, we must first define the concept of **opportunity loss (OL)** or **regret.** When the decision maker aims to maximize benefit, the opportunity loss equals the difference between (a) the optimal payoff for a given event (the highest benefit in a given event column of our tables) and (b) the actual payoff achieved as a result of taking a specified action and the subsequent occurrence of that event. Consider, for example, the E_1 column of Table 14.4. If E_1 occurs, the optimal payoff is $8 million. If the decision maker has previously chosen A_1, the actual payoff will equal $8 million as well; hence, the opportunity loss will equal $8 million − $8 million = 0. The decision maker has no reason to regret anything because given the occurrence of E_1, the best possible action had, in fact, been taken. If, on the other hand, the decision maker has previously chosen A_2, the actual payoff will equal − $2 million; hence, the opportunity loss will equal $8 million − (− $2 million) = $10 million. The decision maker has plenty of reason for regret because given the occurrence of E_1, if only A_1 had been chosen instead of A_2, the decision maker would be better off by $10 million (having a gain of $8 million instead of a loss of $2 million). Similar opportunity-loss values can be calculated for the E_2 column, of course, as is shown in Table 14.6 (which is also referred to as a **regret table**). The application of the criterion of **minimax regret** is also illustrated in that table. According to this criterion, the decision maker finds the highest possible (or maximum) regret value associated with each possible action, identifies the lowest (or minimum) value among these maxima, and then chooses the action associated with this minimum of maximum regret values. (Note the encircled figure in Table 14.6.)

The minimax regret criterion is applied in the same way to cost-minimization problems, but the regret table itself is computed in a slightly different way. When the deci-

Table 14.6   Maximizing Benefit: The Minimax-Regret Criterion

This table, based on Table 14.4, shows the opportunity loss or regret values (in millions of dollars) associated with four possible action/event combinations. The highest of the possible regret values associated with each action is shown in the last column; the minimax-regret criterion suggests taking the least painful of these (encircled), or action A_1. Under the circumstances, the decision maker might come to regret the lost opportunity of earning $7 million in extra profit, but the regret might also be zero (if E_1 rather than E_2 occurs) and it could not be worse than $7 million (as it would be if A_2 were chosen and E_1 were to occur).

	Events		
Actions	E_1 = weak demand for new product	E_2 = strong demand for new product	Row Maximum
A_1 = constructing a small plant	$8 - 8 = 0$	$12 - 5 = 7$	⑦←Minimax Regret
A_2 = constructing a large plant	$8 - (-2) = 10$	$12 - 12 = 0$	10

sion maker aims to minimize cost, the opportunity loss equals the difference between (a) the actual cost incurred as a result of taking a specified action and the subsequent occurrence of an event and (b) the minimum cost achievable for that event (the lowest cost in a given event column of our tables). Table 14.7 gives the solution for our example.

Table 14.7   Minimizing Cost: The Minimax-Regret Criterion

This table, based on Table 14.5, shows the opportunity loss or regret values (in millions of dollars) associated with four possible action/event combinations. The highest of the possible regret values associated with each action is shown in the last column; the minimax-regret criterion suggests taking the least painful of these (encircled), or action A_1. Under the circumstances, the decision maker might come to regret the lost opportunity of reducing costs by $4 million more, but the regret might also be zero (if E_1 rather than E_2 occurs) and it could not be worse than $4 million (as it would be if A_2 were chosen and E_1 were to occur).

	Events		
Actions	E_1 = low level of demand and production	E_2 = high level of demand and production	Row Maximum
A_1 = install small antipollution system	$5 - 5 = 0$	$8 - 4 = 4$	④←Minimax Regret
A_2 = install large antipollution system	$10 - 5 = 5$	$4 - 4 = 0$	5

Criticisms

When we survey the above solutions to our decision-making problems, we discover that the results differ depending on the criterion that is applied. In our benefit-maximizing example, the maximin criterion suggests A_1 (the construction of a small plant), the maximax criterion suggests A_2 (the construction of a large plant), and the minimax regret criterion suggests A_1 as well. Similarly, different results were obtained in the cost-minimizing case. Such differences, however, should not surprise us; they merely reflect the underlying differences in decision-making philosophies that stress, respectively, being careful, being daring, or minimizing future regret.

There exist, however, certain other problems with the above criteria that might make us think twice about applying them.

Undue sensitivity to irrelevant factors. The results obtained by the use of these criteria are easily affected by irrelevant factors. Consider Table 14.2 and imagine a government offering a fixed subsidy of $8 million, *regardless of the firm's action,* if demand should be weak. One would think that this incentive should not affect the firm's action, but under the maximin criterion it does: The E_1 column entries change from 8 and -2 to 16 and 6, the row minima change from 5 and -2 to 5 and 6, and the maximin action changes from A_1 to A_2.

Consider the same subsidy in the context of Table 14.3. The E_1 column entries (being costs) change from 5 and 10 to -3 and 2, the row maxima change from 8 and 10 to 8 and 4, and the minimax action changes from A_1 to A_2.

Although a fixed subsidy would not change the minimax-regret action, some other irrelevant factor easily might. Imagine that Table 14.2 contained a third possible action, A_3 = buying an established plant, with payoffs (in millions of dollars) of -5 and $+20$, respectively, for E_1 and E_2. The existence of this third alternative would change the Table 14.6 entries in the E_1 column from 0 and 10 to 0, 10, and 13, and it would change the E_2 column from 7 and 0 to 15, 8, and 0. Hence, the row maxima would become 15, 10, and 13, and the minimax-regret action would change from A_1 to A_2. But why should the mere existence of another (and rejected) alternative change the choice between the remaining alternatives?

Undue reliance on extreme values. Another factor that has worried decision makers is the fact that the above criteria ignore all payoff values except certain extreme ones found in a given row. Consider again Table 14.2 and imagine that the 12 changed to 12,000. Despite the drastic change (and the sudden possibility of making a profit of $12 *billion* by constructing a large plant) the maximin criterion would continue to counsel that action A_1 be taken (that a small plant be built in order to avoid a possible $2 million loss).

Lack of consideration for the probabilities of events. All of the criteria ignore the probabilities of events, but, critics argue, even guesses could often be extremely helpful. If, in the example just cited (Table 14.2 with the 12 changed to 12,000), E_1 had a low probability and E_2 had a high one, it would surely be foolish to apply the maximin criterion and forgo an almost certain $12 billion gain in order to avoid a highly unlikely $2 million loss. Indeed, many decision makers prefer the criteria discussed in the next sections because they do take probabilities into account.

DECISION MAKING WITH PROBABILITIES: PRIOR ANALYSIS

Many times a decision maker can develop fairly good estimates for the probabilities of alternative future events. Decision criteria that make use of such estimates include the *maximum-likelihood criterion,* the *maximization* (or minimization) *of expected monetary value* (depending on whether the objective is the maximization of some benefit or the minimization of some cost), the *minimization of expected opportunity loss* or regret value, and the *maximization of expected utility.* In this section, we will discuss the use of these criteria in **prior analysis,** when decision making is based solely on whatever information on event probabilities is available prior to the gathering of *new* experimental or sample evidence about them. As noted in the Chapter 4 section, "Revising Probabilities: Bayes's Theorem," such probabilities are called *prior probabilities.* They are usually purely subjective (and, thus, representative of nothing else but a personal degree of belief in the likelihood of some event), but they can also be objective (and, thus, indicative of the relative frequency with which the event in question is bound to occur if logic or past experience can be used as a guide). Another Chapter 4 section, "The Meaning of Probability," can be used to review the concepts of subjective and objective probability.

Maximum Likelihood

A decision maker using the **maximum-likelihood criterion** simply ignores all the events that might occur except the most likely one and selects the action that produces the optimal result (maximum benefit or minimum cost) associated with this most likely event. Table 14.8, which is an adaptation of Table 14.1, illustrates the procedure.

Critics argue that the use of this criterion amounts to "playing ostrich" because so much that might happen is being ignored. In this example, the best of all possible outcomes (12) is never even considered; in other cases, the preoccupation with the

Table 14.8 Maximizing Benefit: The Maximum-Likelihood Criterion

This table shows the annual profits (in millions of dollars) associated with four possible action/event combinations. Given the assumed prior probabilities of the events [$p(E_1)$ = .7 and $p(E_2)$ = .3], the maximum-likelihood criterion leads a decision maker to focus entirely on the shaded E_1 column (because .7 is greater than .3) and to select the optimal result in that column (encircled). Thus, A_1 is the action chosen by those who employ the maximum-likelihood criterion.

	Events	
Actions	E_1 = weak demand for new product $p(E_1)$ = .7	E_2 = strong demand for new product $p(E_2)$ = .3
A_1 = constructing a small plant	⑧	5
A_2 = constructing a large plant	−2	12

event of the highest probability might lead the decision maker to focus on an event with a probability of .1, while ignoring other events with a combined probability of .9 (but individual probabilities of less than .1). This result is considered most unsatisfactory by critics of this criterion.

Proponents of this criterion, on the other hand, argue that people often do behave in precisely this fashion. They look at nothing but the most likely event, as evidenced, for example, in the famous *cobweb cycles* in agricultural and labor markets. Thus, farmers assume that the *most likely* future price of hogs is the one equal to the present price; hence, a low price in period 1 (because it is used as an indicator of a low price in period 2) discourages the raising of hogs in period 1 and leads to a shortage and an unexpectedly *high* price in period 2. This high price in period 2, in turn, encourages the raising of hogs in period 2 and leads to a surplus and an unexpectedly low price in period 3. And so it goes, low and high prices following one another eternally.[1]

Expected Monetary Value

A decision maker using the **expected-monetary-value criterion** determines an expected monetary value for each possible action and then selects the action with the optimal expected monetary value (the largest, if the objective is to maximize some benefit; the smallest if the objective is to minimize some cost). The **expected monetary value *(EMV)*** of an action equals the sum of the weighted payoffs associated with that action, the weights being the probabilities of the alternative events that produce the various possible payoffs. Tables 14.9 and 14.10, which are based on our earlier examples, illustrate the procedure.

Note how a decision maker who, when confronted with the type of situation given in Table 14.9, consistently chooses A_1 will, on average, gain $7.1 million per year because $8 million will be gained 70 percent of the time and $5 million 30 percent of the time. In contrast, a consistent choice of nonoptimal A_2 would, in the long run, yield a gain of only $2.2 million per year because $2 million would be lost 70 percent of the time and $12 million would be gained 30 percent of the time in this type of situation. Similarly, a decision maker who, when confronted with the type of situation given in Table 14.10, consistently chooses A_1 will, on average, incur costs of only $5.9 million per year, while the nonoptimal choice of A_2 would, in the long run, produce costs of $8.2 million per year.

Note also that the probabilities used in these examples are assumed to be the best guesses of the decision maker. It may well happen, however, that a decision maker has no idea about event probabilities whatsoever. In that case, many decision theorists recommend that equal probabilities be assigned to all events. Given a list of collectively exhaustive and mutually exclusive events, the probability assigned to each then equals 1 divided by the number of such events. This procedure is variously referred to as the **equal-likelihood criterion,** the **criterion of insufficient reason, Bayes's postulate,** or (because he popularized it, even though it was first suggested by Thomas Bayes) the **Laplace criterion.**

---

[1]For a more detailed discussion of cobweb cycles, see Heinz Kohler, *Intermediate Microeconomics: Theory and Applications,* 2nd. ed. (Glenview, Ill.: Scott, Foresman and Company, 1986), pp. 196–201 and 302–3.

Table 14.9 Maximizing Benefit: The Expected-Monetary-Value Criterion

This table shows the annual profits (in millions of dollars) associated with four possible action/event combinations; it also shows the expected monetary value of each possible action, based on the assumed prior probabilities of the events $[p(E_1) = .7$ and $p(E_2) = .3]$. Because the objective is to maximize benefit, the *largest EMV* is optimal; thus, A_1 is the action chosen by those who employ the expected-monetary-value criterion.

Actions	Events		Expected Monetary Value, *EMV*
	E_1 = weak demand for new product $p(E_1) = .7$	E_2 = strong demand for new product $p(E_2) = .3$	
A_1 = constructing a small plant	8	5	$8(.7) + 5(.3) =$ (7.1)←optimum
A_2 = constructing a large plant	−2	12	$-2(.7) + 12(.3) = 2.2$

The use of the expected-monetary-value criterion can also be depicted graphically. Consider the decision tree of Figure 14.2 on page 492. It is based on Figure 14.1 and, as the caption indicates, the decision-tree analysis leads to the same decision as Table 14.9. Because the graphical process of finding the optimal action involves starting at the terminal (ultimate-payoff) points of the tree branches (at the "top" or right side of the tree) and then working backwards along the branches to find expected monetary values for each fork, this type of graphical solution of a decison problem is called **backward induction.**

Table 14.10 Minimizing Cost: The Expected-Monetary-Value Criterion

This table shows the annual costs (in millions of dollars) associated with four possible action/event combinations; it also shows the expected monetary value of each possible action, based on the assumed prior probabilities of the events $[p(E_1) = .7$ and $p(E_2) = .3]$. Because the objective is to minimize cost, the *smallest EMV* is optimal; thus, A_1 is the action chosen by those who employ the *EMV* criterion.

Actions	Events		Expected Monetary Value, *EMV*
	E_1 = low level of demand and production $p(E_1) = .7$	E_2 = high level of demand and production $p(E_2) = .3$	
A_1 = install small antipollution system	5	8	$5(.7) + 8(.3) =$ (5.9)←optimum
A_2 = install large antipollution system	10	4	$10(.7) + 4(.3) = 8.2$

Figure 14.2   A Decision Tree and the Expected-Monetary-Value Criterion

The application of the expected-monetary-value criterion is often depicted by means of a decision tree. Event probabilities are attached to each event branch and, by a process termed *backward induction,* the ultimate payoff values appearing at the tips of the branches are translated, using these probabilities, into expected monetary values at each fork of the tree (here, $7.1 million at event point *b,* $2.2 million at event point *c,* and, ultimately, $7.1 million at action point *a*). When maximizing benefit, a decision maker employing the *EMV* criterion will take the action leading to the highest *EMV* (in this case, the decision maker will follow the branch from action point *a* to event point *b*). By refusing to follow the alternative action branch (from *a* to *c*) that leads to the lower *EMV,* the benefit-maximizing decision maker is said to "cut off" or "prune" the nonoptimal path through the tree. (Note the red line suggesting a cut by the pruning shears.) This pruning leaves intact only the optimal path from *a* to *b* and beyond and makes the expected monetary value of the decision maker's action equal to $7.1 million as well. Caution: Branch pruning can take place only at action points, never at event points, because only the former are controlled by the decision maker.

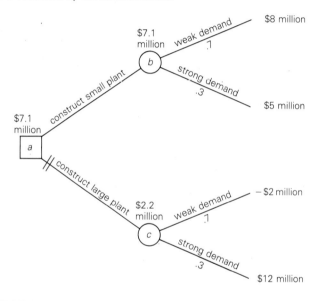

Expected Opportunity Loss

A decision maker using the **expected-opportunity-loss criterion** determines an expected opportunity loss (or expected regret value) for each possible action and then selects the action with the smallest of these values. This procedure *always* yields the same result as the expected-monetary-value criterion. The **expected opportunity loss (EOL),** or **expected regret value,** of an action equals the sum of the weighted opportunity-loss values associated with that action—the weights being the probabilities of the alternative events that produce the various possible opportunity losses. Tables 14.11 and 14.12, which are based on our familiar examples, illustrate the procedure.

Note how a decision maker who, when confronted with the type of situation given in Table 14.11, consistently chooses A_1 will, on average, incur an opportunity loss of

Table 14.11 Maximizing Benefit: The Expected-Opportunity-Loss Criterion

This table, based on Table 14.4, shows the opportunity-loss or regret values (in millions of dollars) associated with four possible action/event combinations; it also shows the expected opportunity loss for each possible action, based on the assumed prior probabilities of the events [$p(E_1) = .7$ and $p(E_2) = .3$]. The smallest *EOL* is optimal; thus, A_1 is the action chosen by those who employ the expected-opportunity-loss criterion. It is no accident that this result is identical with that given by the expected-monetary-value criterion (Table 14.9).

	Events		
Actions	E_1 = weak demand for new product $p(E_1) = .7$	E_2 = strong demand for new product $p(E_2) = .3$	Expected Opportunity Loss, *EOL*
A_1 = constructing a small plant	$8 - 8 = 0$	$12 - 5 = 7$	$0(.7) + 7(.3) =$ ②.1 ←optimum
A_2 = constructing a large plant	$8 - (-2) = 10$	$12 - 12 = 0$	$10(.7) + 0(.3) = 7$

$2.1 million per year because that loss will be zero 70 percent of the time and equal to $7 million 30 percent of the time. In contrast, a consistent choice of nonoptimal A_2 would, in the long run, produce an opportunity loss of $7 million per year because such a loss would equal $10 million 70 percent of the time and be zero 30 percent of the time. A similar analysis can be made with respect to Table 14.12.

Table 14.12 Minimizing Cost: The Expected-Opportunity-Loss Criterion

This table, based on Table 14.5, shows the opportunity loss or regret values (in millions of dollars) associated with four possible action/event combinations; it also shows the expected opportunity loss for each possible action, based on the assumed prior probabilities of the events [$p(E_1) = .7$ and $p(E_2) = .3$]. The smallest *EOL* is optimal; thus, A_1 is the action chosen by those who employ the expected-opportunity-loss criterion. It is no accident that this result is identical with that given by the expected-monetary-value criterion (Table 14.10).

	Events		
Actions	E_1 = low level of demand and production $p(E_1) = .7$	E_2 = high level of demand and production $p(E_2) = .3$	Expected Opportunity Loss, *EOL*
A_1 = install small antipollution system	$5 - 5 = 0$	$8 - 4 = 4$	$0(.7) + 4(.3) =$ ①.2 ←optimum
A_2 = install large antipollution system	$10 - 5 = 5$	$4 - 4 = 0$	$5(.7) + 0(.3) = 3.5$

Expected Utility

All of the decision-making criteria discussed so far have employed *monetary* outcome measures. Yet critics have argued that using money to measure outcomes is a mistake, that people will invariably take those actions that maximize their welfare or *utility*, and that actions that maximize monetary benefit or minimize monetary cost may well not coincide with those that maximize utility. The kind of situation depicted in Table 14.13 can be used to illustrate what these critics have in mind.

Would the typical decision maker really be indifferent between actions A_1 (investing $100 million in vineyards) and A_2 (investing $100 million in the auto industry) just because their expected monetary values are equal? Critics argue that the decision maker would not be indifferent because A_2, unlike A_1, involves *risk*. Choosing A_2 over A_1 is equivalent to taking a gamble, to giving up a sure thing (a $15 million return from action A_1 regardless of which event occurs) for an uncertain thing (*either* a $25 million return if E_1 occurs *or* a $5 million return if E_2 occurs). Although such a gamble would be a **fair gamble** (in which the expected monetary value of what is given up— that is, the $15 million certain to be received from action A_1—is precisely equal to the expected monetary value of what is received—that is, the $15 million received, on the average, from action A_2), most people would not take the gamble. If critics are right, most people would be even less inclined to take an **unfair gamble** in which the expected monetary value of what is given up exceeds the expected monetary value of what is received. (By changing all the entries of 15 in the A_1 row of Table 14.13 to 16, for example, action A_2 would be turned into an unfair gamble.) Indeed, many people, critics assert, would even reject a **more-than-fair gamble** in which the expected monetary value of what is given up is less than the expected monetary value of what is received. (By changing all the entries of 15 in the A_1 row of Table 14.13 to 14, for example, action A_2 would be turned into a more-than-fair gamble.)

Table 14.13 Doubts About the Expected-Monetary-Value Criterion

This table shows the annual profits (in millions of dollars) associated with four possible action/event combinations; it also shows the expected monetary value of each possible action, based on the assumed prior probabilities of the events [$p(E_1)$ = $p(E_2)$ = .5]. According to the *EMV* criterion, a decision maker would be indifferent between the two actions, but critics argue that most people would decisively opt for A_1.

Actions	Events		Expected Monetary Value, *EMV*
	E_1 = oil price rises moderately $p(E_1)$ = .5	E_2 = oil price rises sharply $p(E_2)$ = .5	
A_1 = investing $100 million in vineyards	15	15	15(.5) + 15(.5) = 15 ←optimum?
A_2 = investing $100 million in the auto industry	25	5	25(.5) + 5(.5) = 15

Table 14.14   The St. Petersburg Game

The St. Petersburg game has an expected monetary value of gain of infinity (equal to the sum of the last column of this table).

Number of Toss (1)	Payoff if Heads First Appears at Given Toss (2)	Probability of Heads First Appearing at Given Toss (3)	Expected Monetary Value = Payoff × Probability (4) = (2) × (3)
1	$1	.5	$.50
2	2	$(.5)^2$	.50
3	4	$(.5)^3$	.50
4	8	$(.5)^4$	.50
5	16	$(.5)^5$	.50
.	.	.	.
.	.	.	.
.	.	.	.

The St. Petersburg paradox. The apparent paradox of people refusing to take fair gambles was first solved some 250 years ago. A Swiss mathematician, Daniel Bernoulli (1700–1782) studied gamblers at the casinos of St. Petersburg, and he considered the game described below between two persons, A and B:

> A fair coin is tossed until heads appears; if heads appears on the first toss, A pays B $1; if heads appears for the first time on the second toss, A pays B $2; if heads appears first on the third toss, A pays B $4; and so on, with A always paying 2^{n-1} at the nth toss if heads appears.

If playing this game is to be a fair gamble, what fee, Bernoulli asked, should B be willing to pay A for the privilege of playing this game? Because the player of a fair game is never asked to pay more than the expected monetary value of gain, this value can be calculated easily. Consider Table 14.14. Given a probability of .5 for heads to appear on the first toss, the expected monetary value of gain is $1(.5) = $.50 if the game ends after the first toss. Given a probability of $(.5)^2$ for heads to appear first on the second toss, the expected monetary value of gain is $2(.5)^2 = $.50 if the game ends after the second toss. And so it goes. Thus, the expected monetary value of gain during the entire game is the sum of the expected monetary values of all possible outcomes, or the infinite series $.50 + $.50 + . . . = ∞. Yet people are clearly not willing to pay such an infinite sum of money for the privilege of playing this game (even though such a payment would make the gamble precisely fair).

One probably could solve the paradox by postulating that gamblers cannot possibly be convinced to take this game seriously. How could they be so gullible as to believe that payoff would actually be made should they be lucky enough to win a large sum? (If heads did not appear until the 42nd toss, for example, the required payoff would equal more than the entire U.S. gross national product!) Bernoulli, however, had a different idea. He argued that people making decisions under uncertainty were not attempting to maximize expected *monetary* values, but maximized expected *utilities*

Figure 14.3   The Utility of Money

This graph illustrates a person's utility function that is characterized by declining marginal utility of money. The total utility of money, U, here is related to the amount of money, $, by the equation $U = \sqrt{\$}$. Other relationships are, of course, also possible.

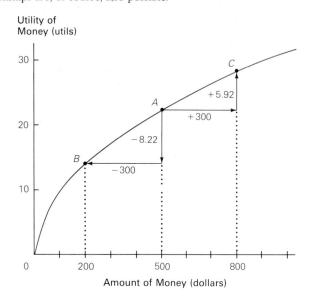

instead. He thought that the **total utility of money** (the overall welfare people derived from the possession of money) was rising the more money people had, while the **marginal utility of money** (the increase in total utility that was associated with a one dollar increase in the quantity of money, for example) was declining the more money people had. Any person starting with $500, for example, would, therefore, place a smaller subjective value on gaining an extra sum than on losing an equal amount. Any game with an equal probability of gaining and losing a given amount was, therefore, fair in monetary terms, but unfair in utility terms. The game's expected utility was negative. No wonder people refused to play it!

Figure 14.3 contains a hypothetical person's **utility-of-money function**—the relationship, that is, between alternative amounts of money the person might possess and the different utility totals associated with these amounts. It is assumed, in accordance with Bernoulli's postulate, that the total utility of money rises with greater amounts of it, while its marginal utility (shown by the slope of the total utility curve) declines.

In Figure 14.3, the total utility of money, U, is related to the amount of money, $, by the equation $U = \sqrt{\$}$. Thus, a person with $500 is assumed to receive $\sqrt{500}$, or 22.36 *utils* (units of utility) from it (point A). The same person would, however, receive only 5.92 extra utils from an added $300 (when moving from A to C) but would lose 8.22 utils by a loss of $300 (when moving from A to B). Given this type of utility function, the St. Petersburg paradox is easily solved. While the expected monetary value of the game equals infinity, its expected utility equals

$$EU = \sqrt{1}\,(.5) + \sqrt{2}\,(.5)^2 + \sqrt{4}\,(.5)^3 + \sqrt{8}\,(.5)^4 + \sqrt{16}\,(.5)^5 + \dots$$
$$= 1(1/2) + \sqrt{2}\,(1/4) + 2(1/8) + 2\sqrt{2}\,(1/16) + 4(1/32) + \dots$$
$$= 1/2 + \sqrt{2}\,(1/4) + 1/4 + \sqrt{2}\,(1/8) + 1/8 + \dots$$
$$= 1 + \sqrt{2}\,(1/2) = 1.707 \text{ utils.}$$

The equation $U = \sqrt{\$}$ implies $U^2 = \$$; hence an expected utility of 1.707 translates into a dollar equivalent of $(1.707)^2 = \$2.91$. This $2.91, then, is the far-less-than-infinite amount that a person who maximized expected utility, and possessed the $U = \sqrt{\$}$ utility function, would pay for the privilege of playing the St. Petersburg game.

Note: Someone else who also maximized expected utility, but possessed a different utility function, would act quite differently. This difference brings us to an important conclusion: Even if people, as is likely, maximize the expected utility rather than the expected monetary value of an action when facing uncertainty, we cannot predict their behavior without a knowledge of their utility functions. Indeed, the shape of these functions reveals important information about people's attitudes concerning the spread of possible outcomes of their action around the action's expected value. The extent of such spread (as from 15 down to 5 and up to 25 for action A_2 in Table 14.13) in fact measures the *risk* of an action, and people can view risk in one of three ways. They can be averse to it; they can be neutral toward it; they can seek it out.

Risk aversion. Imagine for a moment that the investor whose choices were pictured in Table 14.13 possesses the type of utility function postulated by Bernoulli. This type of utility function is illustrated in panel (a) of Figure 14.4. Action A_1, as we noted earlier, would bring the investor a profit of $15,000 a year, regardless of what happens to the price of oil. This amount of money is associated, according to point B on the utility function, with a total utility of $0b$. Action A_2, on the other hand, would bring the investor, with equal probability, a profit of $5,000 a year *or* of $25,000 a year. As we can see from points A and C, respectively, the associated utilities equal $0a$ and $0c$. The *expected* utility from an equally weighted $5,000 or $25,000 a year, however, equals the sum of half of $0a$ plus half of $0c$, and this sum is shown in the graph by utility $0d$. It corresponds to point D, located on the dashed line connecting A and C and above the $15,000 expected monetary value of the $5,000 or $25,000 gamble. (Note: The expected utility of receiving $5,000 with a probability of .2 or $25,000 with a probability of .8 could similarly be read off of dashed line AC, but at a point above the $5,000(.2) + $25,000(.8) = \$21,000$ expected monetary value of this different gamble.)

Whenever a person in this way considers the utility (as at point B) of a certain prospect of money to be higher than the expected utility (as at point D) of an uncertain prospect of equal expected monetary value, the person is said to hold an attitude of **risk aversion.** This attitude is always present when a person's marginal utility of money (shown here by the slope of utility function $0ABC$) declines with larger amounts of money.

Risk aversion is, indeed, quite common. Consider how people do all kinds of things, small and large, to *escape* gambles: they place person-to-person calls instead of station-to-station calls; at airports, banks, and post offices, they prefer single lines feeding to many clerks to the chance of getting into a slow or fast line; they diversify their assets and do not "place all eggs in one basket"; they buy plenty of insurance; they

Figure 14.4   Attitudes Toward Risk

This set of graphs illustrates three basic attitudes toward risk: risk aversion in panel (a), risk neutrality in panel (b), and risk seeking in panel (c). These attitudes correspond, respectively, to declining, constant, and increasing marginal utilities of money.

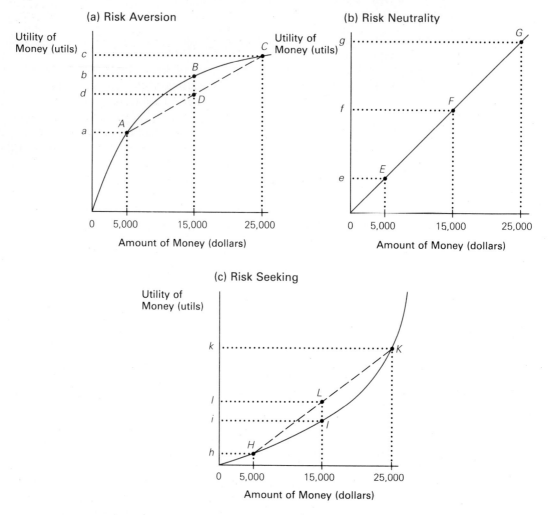

reject fair gambles, as our risk-averse investor would by taking action A_1 (to gain utility 0*b*) rather than action A_2 (that has an equal expected monetary value but a smaller expected utility 0*d*).

Risk neutrality. Let us imagine instead that our investor's utility function is illustrated by panel (b) of Figure 14.4. In this case, the utility of action A_1, with its certain payoff of $15,000 a year, would equal 0*f*, corresponding to point *F*. Action A_2 would bring, according to points *E* and *G*, utility of 0*e* or 0*g*, and with equal probability. The expected utility of this gamble, however, would also equal 0*f*. (A straight line between

E and *G* leads to point *F* above the $15,000 expected monetary value of the $5,000 or $25,000 gamble.)

Whenever a person in this way considers the utility (as at point *F*) of a certain prospect of money to be equal to the expected utility (as at point *F*) of an uncertain prospect of equal expected monetary value, the person is said to hold an attitude of **risk neutrality.** This attitude is always present when a person's marginal utility of money (shown here by the slope of utility function 0*EFG*) remains constant with larger amounts of money.

Risk neutrality is not very common, but it is precisely the implicit assumption made by those who advocate the use of the expected-monetary-value criterion. If all people were risk neutral, our investor would be truly indifferent between actions A_1 and A_2 because, as we saw, their expected monetary values are equal. By the same token, a risk-neutral person would be willing to pay $100 for the privilege of taking each of the following fair gambles, for the expected monetary value of each of these is also $100:

    a. a 99-percent chance of getting $101.01 and a 1-percent chance of getting nothing;
    b. a 50-percent chance of getting $101 and a 50-percent chance of getting $99;
    c. a 1-percent chance of getting $10,000 and a 99-percent chance of getting nothing;
    d. a 1-percent chance of getting $1 million, a 1-percent chance of losing $990,000, and a 98-percent chance of getting nothing.

After thinking about it, you might pay the price for (a) and (b) or something close to it, but do you have the sweepstakes mentality to go after (c)? Are you ready for the Russian roulette of (d)?

Risk seeking. Now imagine that our investor's utility function is illustrated by panel (c) of Figure 14.4. In this case, the utility of action A_1, with its certain payoff of $15,000 a year, would equal 0*i,*corresponding to point *I.* Action A_2 would bring, according to points *H* and *K*, utility of 0*h* or 0*k,* and with equal probability. By the now familiar procedure, we can establish the expected utility of this gamble as 0*l,* corresponding to point *L* on dashed line *HK.*

Whenever a person in this way considers the utility (as at point *I*) of a certain prospect of money to be lower than the expected utility (as at point *L*) of an uncertain prospect of equal expected monetary value, the person is said to hold an attitude of **risk seeking.** This attitude is always present when a person's marginal utility of money (shown here by the slope of utility function 0*HIK*) rises with larger amounts of money.

Risk seeking, like risk neutrality, is not very common. If all people were risk seekers, our investor would, of course, prefer action A_2 to A_1, and all of us would constantly seek out and accept huge riches-or-ruin gambles. In fact, of course, most of us have the opposite inclination and seek to buy insurance.

Conclusion. An evaluation of decision outcomes in terms of utility rather than money may well be a superior approach. If the investor depicted in Table 14.13, for example, were risk-averse and had Bernoulli's utility function, the decision problem could be laid out as in Table 14.15. This approach would explain why the investor would prefer A_1 to A_2, even though A_1 and A_2 have equal expected monetary values.

A decision maker using the **expected-utility criterion,** thus, determines the expected utility for each possible action and then selects the action that maximizes expected utility. The **expected utility *(EU)*** of an action equals the sum of the weighted

Table 14.15   Maximizing Expected Utility

This table, based on Table 14.13, shows the utility associated with four possible action/event combinations for a risk-averse person with a utility function of $U = \sqrt{\$}$. The table also shows the expected utility of each possible action, based on the assumed prior probabilities of the events $[p(E_1) = p(E_2) = .5]$. Because the objective is to maximize expected utility, the largest EU is optimal; thus, A_1 is the action chosen by those who employ the expected-utility criterion.

	Events		
Actions	$E_1 =$ oil price rises moderately $p(E_1) = .5$	$E_2 =$ oil price rises sharply $p(E_2) = .5$	Expected Utility, *EU*
$A_1 =$ investing \$100 million in vineyards	$\sqrt{15} = 3.87$	$\sqrt{15} = 3.87$	$3.87(.5) + 3.87(.5) = \boxed{3.87}$ ←optimum
$A_2 =$ investing \$100 million in the auto industry	$\sqrt{25} = 5$	$\sqrt{5} = 2.24$	$5(.5) + 2.24(.5) = 3.62$

utilities associated with that action—the weights being the probabilities of the alternative events that produce the various possible utility payoffs.

Note: Modern theorists have advanced utility theory far beyond Bernoulli and have devised a variety of ingenious methods of identifying a decision maker's utility function. Readers interested in pursuing the matter may begin by consulting the classic work of von Neumann and Morgenstern listed in the "Selected Readings" section at the end of this chapter.

Note also: Real-world examples of the use of decision trees to maximize benefit or minimize cost (and of making a decision on the basis of expected monetary values alone or by considering expected utilities instead) are given in Analytical Examples 14.1, "Decision Analysis Comes of Age," and 14.2, "The Decision to Seed Hurricanes— A Second Look."

DECISION MAKING WITH PROBABILITIES: POSTERIOR ANALYSIS

In the previous section, we discussed various forms of decision making that made use of prior probabilities concerning events over which a decision maker has no control. **Posterior analysis** is a form of decision making under uncertainty that starts out with prior probabilities, proceeds to gather additional experimental or sample evidence about event probabilities, and then uses this new evidence to transform the set of prior probabilities, by means of Bayes's theorem, into a revised set of posterior probabilities with the help of which a final decision is reached. The procedure of revising probabilities, first discussed in the Chapter 4 section, "Revising Probabilities: Bayes's Theorem," can be sketched as follows:

Prior Probability		New Information		Application of Bayes's Theorem		Posterior Probability
	+		+		→	

Gathering New Information

Consider our above example concerning a decision on constructing either a small or a large plant to manufacture and sell a new product (Table 14.9). The decision maker's prior probabilities for the two possible events (E_1 = weak demand and E_2 = strong demand) equaled $p(E_1) = .7$ and $p(E_2) = .3$; they might have been based on nothing but a hunch. Instead of proceeding with the analysis on that basis (as illustrated in Tables 14.9 and 14.11, or in Figure 14.2), the decision maker might decide to gather new information about the event probabilities. The required research can take many different forms. In this particular case, a consumer survey might be conducted. In other instances, such as those noted at the beginning of this chapter, other approaches might be more appropriate: obtaining a long-range weather forecast, perhaps, or a forecast of future prices and wages, sampling inputs bought or output produced for quality, building a pilot plant to test a new production process, conducting a seismic test (that can indicate the likely presence of oil deposits), taking an aptitude test (that can indicate future job performance), or even paying a bribe (that can provide inside information on the likely reaction of a TV network to a new pilot that might be offered to it). Unfortunately, none of this new information will be perfect. Even the best of aptitude tests will flunk some job applicants who later turn out to be superb, while passing others who later turn out to be incompetent. Even the best of seismic tests can deny the presence of oil in a field that is already producing oil, while confirming the presence of oil at a site that is later proven to be dry. Thus, the decision maker who wants to know about weak or strong demand for the new product will still have to contend with uncertainty, even if a consumer survey is conducted and a fancy market-research report is received. Indeed, the market-research firm's past record may be that indicated by Table 14.16. As the caption explains, although the firm's record is not perfect, its "batting average" is quite impressive; one can place a fairly high degree of confidence in the research firm's reports. When it predicts weak demand, it is correct 90 percent of the time; when it predicts strong demand, it is correct 80 percent of the time.

Table 14.16 The Past Record of a Market-Research Firm

This table shows the probabilities that a firm's market research will discover the true state of affairs about the demand for a new product. Based on past performance, the conditional probability is .9 that the firm's research result indicates weak demand when demand is in fact weak, and it is .1 that the result indicates strong demand when demand is in fact weak. Similarly, the conditional probability is .2 that the firm's research result indicates weak demand when demand is in fact strong, and it is .8 that the result indicates strong demand when demand is in fact strong.

Market-Research Result	True State of Affairs	
	E_1 = weak demand	E_2 = strong demand
R_1 = demand will be weak	$p(R_1|E_1) = .9$	$p(R_1|E_2) = .2$
R_2 = demand will be strong	$p(R_2|E_1) = .1$	$p(R_2|E_2) = .8$

Applying Bayes's Theorem

Let us now suppose that our decision maker does consult the research firm and is given a report concerning the likely intensity of future demand. What happens next depends on the content of the report.

A report of weak demand. If the report indicates weak demand (the research result is R_1), the data contained in the first row of Table 14.16 come into play as *Bayes's theorem* (see Box 4.K on page 151 or Box 4.L on page 151) is applied. According to the Box 4.K formula, the posterior probability of an event is computed as

$$p(E|R) = \frac{p(E)\, p(R|E)}{p(E)\, p(R|E) + p(\overline{E})\, p(R|\overline{E})}$$

where E denotes the event, $\overline{E}$ its complement, and R the research result.

In our case, therefore (using the above-noted prior probabilities and the data of Table 14.16),

$$p(E_1|R_1) = \frac{p(E_1)\, p(R_1|E_1)}{p(E_1)\, p(R_1|E_1) + p(E_2)\, p(R_1|E_2)} = \frac{.7(.9)}{.7(.9) + .3(.2)} = .91,$$

and

$$p(E_2|R_1) = \frac{p(E_2)\, p(R_1|E_2)}{p(E_2)\, p(R_1|E_2) + p(E_1)\, p(R_1|E_1)} = \frac{.3(.2)}{.3(.2) + .7(.9)} = .09.$$

An alternative calculation is provided by the Box 4.L formula as

$$p(E|R) = \frac{p(E \text{ and } R)}{p(R)}.$$

In our case, according to Box 4.I on page 141, the joint probability of E_1 *and* R_1 equals $p(E_1)p(R_1|E_1)$, or $.7(.9) = .63$, while the joint probability of E_2 *and* R_1 equals $p(E_2)p(R_1|E_2)$, or $.3(.2) = .06$. These being mutually exclusive events, we can calculate the unconditional probability of our result, $p(R_1)$ as $.63 + .06 = .69$. Therefore,

$$p(E_1|R_1) = \frac{p(E_1 \text{ and } R_1)}{p(R_1)} = \frac{.63}{.69} = .91,$$

and

$$p(E_2|R_1) = \frac{p(E_2 \text{ and } R_1)}{p(R_1)} = \frac{.06}{.69} = .09.$$

These results confirm those already derived.

A report of strong demand. What if the research firm's report had predicted a strong demand for the new product (the research result had been R_2)? In that case, of course, an analogous set of calculations would have been performed:

$$p(E_1|R_2) = \frac{p(E_1)\,p(R_2|E_1)}{p(E_1)\,p(R_2|E_1) + p(E_2)\,p(R_2|E_2)} = \frac{.7(.1)}{.7(.1) + .3(.8)} = .23;$$

$$p(E_2|R_2) = \frac{p(E_2)\,p(R_2|E_2)}{p(E_2)\,p(R_2|E_2) + p(E_1)\,p(R_2|E_1)} = \frac{.3(.8)}{.3(.8) + .7(.1)} = .77.$$

Finding the Optimal Strategy

If our decision maker were risk-neutral and cared to maximize expected *monetary* value (which we will assume), an optimal strategy could now be mapped out with the help of these posterior probabilities. An **optimal strategy** is a complete plan specifying the actions to be taken at each available action point, if the expected (monetary or utility) payoff is to be the best one available. The procedure is illustrated in Figure 14.5 on page 504. It corresponds to the example discussed so far in which a decision maker, at action point *a,* has decided to make a consumer survey in order to gauge the likely intensity of future demand. This decision, inevitably, takes us to event point *b.* At event point *b,* the survey will predict either weak or strong future demand, taking us to either action point *c* or action point *d.* Once located at either *c* or *d,* the decision maker must then act and construct either a small plant or a large one. As a result, one of the event points labeled *f* through *i* is reached, and chance will decide whether demand is in fact weak or strong. The right-hand column shows the payoffs, which are equal to those given throughout this chapter (starting with Table 14.1) but are reduced by an assumed $100,000 cost of taking the survey.

We can find the optimal strategy by using the process of backward induction. Consider event point *f.* Given the conditional probabilities of $p(E_1|R_1) = .91$ and $p(E_2|R_1) = .09$ that were calculated above, the decision maker can calculate the expected monetary value at point *f* as $7.9(.91) + $4.9(.09) = 7.63 (million). This number is indicated in color above event point *f.*

Now consider event point *g.* Given the same conditional probabilities as at *f,* the decision maker can calculate the expected monetary value at point *g* as $-2.1(.91) + $11.9(.09) = -$.84$ (million). This number is indicated in color above event point *g.* Surely, a decision maker at action point *c* would want to move toward *f* rather than *g;* thus, we can prune the branch going from *c* to *g* (note the color cut) and carry the color *EMV* figure at *f* to *c.*

We now turn to event point *h.* The probabilities are now different as we assume having received a survey report indicating strong demand (R_2). Given the conditional probabilities calculated above, the decision maker can calculate the expected monetary value at point *h* as $7.9(.23) + $4.9(.77) = 5.59 (million). This number is indicated in color above point *h.*

Now consider event point *i.* Given the same conditional probabilities as at point *h,* the decision maker can calculate the expected monetary value at point *i* as $-2.1(.23) + $11.9(.77) = 8.68 (million). This number is indicated in color above event point *i.* Surely, a decision maker at action point *d* would want to move towards *i* rather than *h;*

Figure 14.5   Posterior Decision-Tree Analysis

This decision tree summarizes the development of an optimal strategy, while making use of posterior probabilities. (Note: All dollar amounts are in millions.)

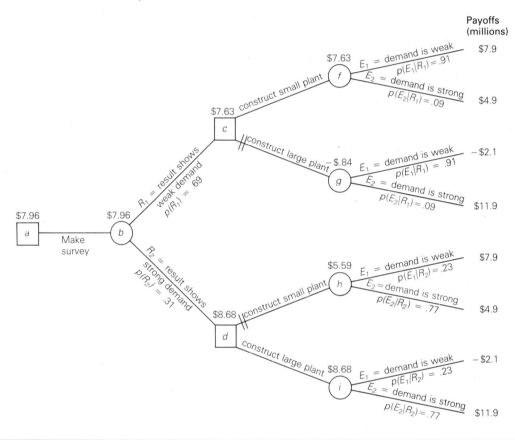

Payoffs
(millions)

thus, we can prune the branch going from *d* to *b* (note the color cut) and carry the color *EMV* figure at *i* to *d*.

We must, finally, consider event point *b*. As we noted earlier, the unconditional probability of getting survey result R_1 equals .69. There being only two possible survey results, the unconditional probability of getting survey result R_2 can, thus, be found quickly as $p(R_2) = 1 - p(R_1) = 1 - .69 = .31$. This conclusion can, however, also be checked by adding the joint probability of E_1 *and* R_2 [equal to $p(E_1) \cdot p(R_2|E_1) = .7(.1) = .07$] to the joint probability of E_2 *and* R_2 [equal to $p(E_2) \cdot p(R_2|E_2) = .3(.8) = .24$]. As we can see, the same result is obtained: $p(R_2) = .07 + .24 = .31$. Thus, we can calculate the expected monetary value at event point *b* as $\$7.63(.69) + \$8.68(.31) = \$7.96$ (million), a figure that can be placed above *b* as well as *a*.

The grand conclusion is at hand. The optimal strategy (that carries an expected monetary value of $7.96 million) is the following:

Given that a survey is made,

    a. if the result shows weak demand, to construct a small plant (at *c*) and let chance (at *f*) determine the reward: of $7.9 million if the survey result was correct, of $4.9 million if it was incorrect.

    b. if the result shows strong demand, to construct a large plant (at *d*) and let chance (at *i*) determine the reward: of −$2.1 million if the survey result was incorrect, of $11.9 million if it was correct.

Note: This course would be the best strategy, given the information available at the time. Obviously, this information can change over time and, if it is not too late, the decision can be changed accordingly. Information can change as a result of some deliberate action taken by the decision maker (such as the making of a second, independent survey) or as a result of some unpredicted event (such as a scandal that seriously hurts the firm's image). In the former case, new posterior probabilities can be calculated; in the latter case, the decision maker may wish to revise the prior probabilities as well. To illustrate the former contingency, imagine that a second (independent) survey were to confirm (rather than contradict) a report of weak demand obtained by a first one. The conditional probability of E_1, given the first R_1 (which was shown to be .91) would then change to become the conditional probability of E_1, given the first R_1 *and* a second R_1^*, and would be calculated as

$$p(E_1|R_1 \text{ and } R_1^*) = \frac{p(E_1)\,p(R_1|E_1)\,p(R_1^*|E_1)}{p(E_1)\,p(R_1|E_1)\,p(R_1^*|E_1) + p(E_2)\,p(R_1|E_2)\,p(R_1^*|E_2)}$$

$$= \frac{.7(.9)(.9)}{.7(.9)(.9) + .3(.2)(.2)} = .98.$$

As more and more information comes in, therefore, the posterior probability depends less and less on the prior one. In this case, the prior .7 changed to .91 after the first survey, and we now have seen it change to .98 after the second. Caution: Evidence can, of course, go either way; hence, a posterior probability can decline below the prior one as well as rise above it.

    Because the crucial insights of the Rev. Thomas Bayes have become such an integral part of modern decision theory, he can rightly be called the "father of decision theory" (see Biography 14.1 at the end of this chapter).

THE VALUE OF INFORMATION

In the previous section, we made an implicit assumption: that it was worthwhile to the decision maker to spend $100,000 on a survey of consumers. In fact, this assumption has been borne out, as we can see when comparing the expected monetary values of the optimal strategies that have emerged, respectively, from prior analysis ($7.10 million in Figure 14.2) and posterior analysis ($7.96 million in Figure 14.5). Yet this result was a lucky circumstance; it is time to consider the issue systematically.

The Expected Value of Perfect Information

We must note, first of all, that there exists a maximum amount, called the **expected value of perfect information,** or **EVPI,** that any decision maker can be expected to pay for obtaining perfect information about future events and, thus, for eliminating uncertainty completely. When the objective is the maximization of some benefit, this maximum equals the difference between the expected benefit when the optimal decision is made with perfect information and the expected benefit when that decision is made under uncertainty. (When the objective is the minimization of some cost, this maximum equals the difference between the expected cost when the optimal decision is made under uncertainty and the expected cost when that decision is made with perfect information.) Depending on whether the expected value of the optimal action under uncertainty is based on prior or posterior probabilities, one can also distinguish a **prior EVPI** and a **posterior EVPI.**

Prior EVPI. Consider the case illustrated in Table 14.9 and Figure 14.2. The expected value of that decision under uncertainty equals $7.1 million. If the decision maker, however, were offered perfect information by someone about the guaranteed occurrence of either E_1 or E_2, things would be different. If told that E_1 will occur with certainty, the decision maker would instantly take action A_1 and gain $8 million. If told that E_2 will occur with certainty, the decision maker would instantly take action A_2 and gain $12 million. There is a problem, however. The purveyor of the perfect information may not reveal it until *after* being paid; otherwise, the decision maker would have the information and could use it while paying nothing for it. To determine the expected value of a decision with perfect information, the decision maker must guess, before possessing the information, about the likelihood of receiving a report that E_1 will occur, $p(R_1)$, and of receiving a report that E_2 will occur, $p(R_2)$. Suppose that such prior probabilities, like $p(E_1)$ and $p(E_2)$ earlier, are also set to equal .7 and .3, respectively. Then the expected value of the decision with perfect information is

$$\$8(.7) \; + \; \$12(.3) \; = \; \$9.2 \text{ (million)}.$$

Given an expected value of the decision under uncertainty of $7.1 million, the

$$\text{prior } EVPI \; = \; \$9.2 \; - \; \$7.1 \; = \; \$2.1 \text{ million}.$$

This amount is the maximum the decision maker would pay to the holder of perfect information for revealing it. Note that this figure also equals the expected opportunity loss of the optimal decision under uncertainty (Table 14.11). This equality of prior *EVPI* and *EOL* is always the case and explains why some decision makers prefer the expected-opportunity-loss criterion (Table 14.11) over the expected-monetary-value criterion (Table 14.9); the former automatically indicates the *EVPI.* (Nevertheless, because the computations are less complicated, most decision makers use the *EMV* criterion that, of course, always points to the same optimal action.)

Posterior EVPI. Now consider the case illustrated in Figure 14.5. The expected value of that decision under uncertainty equals $7.96 million because this decision maker has the advantage of additional information (the survey result). Accordingly, we can determine a

$$\text{posterior } EVPI \; = \; \$9.20 \; - \; \$7.96 \; = \; \$1.24 \text{ million}.$$

After having already paid $100,000 for a survey of consumers, this decision maker would pay a purveyor of perfect information only this smaller maximum amount. Analytical Example 14.3," The Value of Perfect Information: The Case of the U.S. Cattle Industry," provides an interesting real-world application.

The Expected Value of Sample Information

In fact, no one will ever be able to provide *perfect* information about future events. Decision makers, therefore, are much more interested in another concept, the **expected value of sample information,** or **EVSI,** which equals the maximum amount any decision maker can be expected to pay for obtaining admittedly imperfect information about future events and, thus, for reducing (rather than eliminating) uncertainty. When the objective is the maximization of some benefit, this maximum equals the difference between the expected benefit when an optimal decision is made with the sample information (and before paying for it) and the expected benefit when that decision is made without it. (When the objective is the minimization of some cost, this maximum equals the difference between the expected cost when the optimal decision is made without the sample information and the expected cost when that decision is made with it, but again before paying for it.)

Consider our example. The optimal decision with the survey information (Figure 14.5) is expected to yield $7.96 million only *after* paying an assumed $100,000 for conducting the survey. Hence, the expected gross benefit with survey information equals $8.06 million. The optimal decision without the survey information (Figure 14.2) is expected to yield $7.10 million. Hence, the

$$EVSI = \$8.06 - \$7.10 = \$.96 \text{ million.}$$

Our decision maker would pay a maximum of $960,000 for the survey. Since $100,000 was paid in fact, a shrewd survey firm could have extracted an additional $860,000 for its (imperfect) information.

The Efficiency of Sample Information

We just noted how our decision maker, before obtaining any sample information, would place a maximum value of $2.1 million on perfect information. At the same time, the expected value of (inevitably imperfect) sample information was shown to equal $.96 million. The ratio of the latter to the former, multiplied by 100, is called the **efficiency of sample information,** or **ESI.** Thus,

$$ESI = \frac{EVSI}{(\text{prior}) \ EVPI} \cdot 100.$$

In our example, the *ESI* can be calculated as:

$$ESI = \frac{\$.96 \text{ million}}{\$2.10 \text{ million}} \cdot 100 = 45.7.$$

This *ESI* value indicates that the consumer survey provided only 45.7 percent of what perfect information would have been worth. Some decision makers employ this type of efficiency rating to decide whether they should look for some other type of information. Presumably, they seek other information when the rating is low, but decide that the information is as good as one can hope to obtain when the rating is high. Naturally, what is "low" and what is "high" is a debatable point.

PREPOSTERIOR ANALYSIS

When decision makers are deciding the worth of obtaining additional (experimental or sample) information before proceeding to prior or posterior analysis, they are said to be engaged in **preposterior analysis.** This analysis can be simple or complicated.

When the new (and inevitably imperfect) information about future events is being offered to them at a price exceeding the (prior) expected value of perfect information, a decision can instantly be made in the negative; it is then better to stick with prior analysis. In the above example, for instance, it would clearly not be worthwhile to acquire (assuredly imperfect) information at a price of $2.1 million or more because that sum equals the prior *EVPI.*

When the new (and inevitably imperfect) information about future events is being offered at a price below the prior *EVPI,* however, it becomes necessary to determine how that price compares with the expected value, *EVSI,* of the sample information that is being offered. On that basis, one can decide whether to buy the information or proceed without it. In our above example, the information was offered, we assumed, for $100,000, while being worth, we calculated, $960,000; hence, the decision to make the survey (a decision that was treated as a foregone conclusion in Figure 14.5) was validated in retrospect. It is customary, however, to make the decision about possibly acquiring additional information in a more systematic way, either with the help of a decision tree (that now includes the choice between posterior and prior analysis) or in tabular form. The decision-tree approach is referred to as **extensive-form analysis;** the tabular approach is **normal-form analysis.**

Extensive-Form Analysis

Figure 14.6 provides an example of the extensive form of preposterior analysis; it identifies the optimal strategy by means of backward induction on a decision tree. The lower branches of this particular tree—those emanating from action point *e*—correspond precisely to Figure 14.2 and illustrate prior analysis. The upper branches—those emanating from action point *a*—correspond to Figure 14.5 and illustrate posterior analysis. The tree as a whole now also depicts the initial choice (at *a*) of making or not making a survey. Because the former route leads to *b* (and an expected monetary value, *after* paying the $100,000 survey cost, of $7.96 million), while the latter route leads to *e* (and a lower expected monetary value of $7.10 million), the decision maker would prune the branch leading from *a* to *e* and beyond, as indicated here.

Thus, the optimal strategy is the following:

a. Do make a survey.
b. If the result shows weak demand, construct a small plant (and let chance at *f* determine the reward).
c. If the result shows strong demand, construct a large plant (and let chance at *i* determine the reward).

Figure 14.6   Preposterior Decision-Tree Analysis

This decision tree summarizes the development of an optimal strategy that includes the initial choice between engaging in prior analysis only or engaging in posterior analysis. Thus, the tree combines Figures 14.2 and 14.5. (Note: All dollar amounts are in millions.)

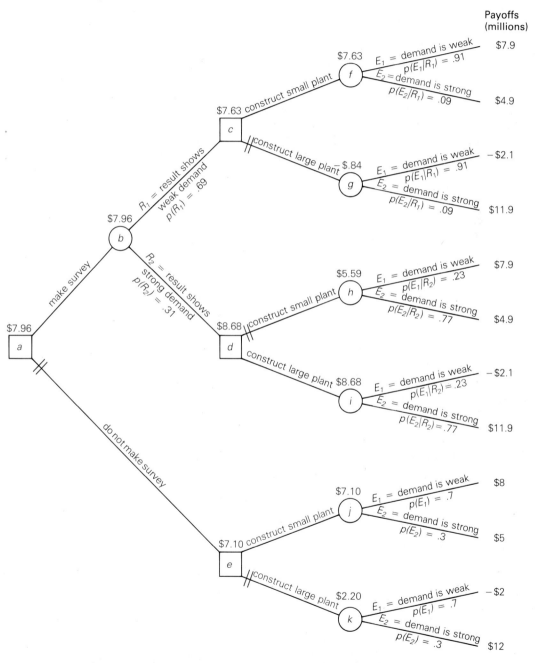

Note: The difference between the expected monetary values at *b* and *e,* or $860,000, represents the *additional* amount (on top of the actual $100,000 survey cost) that the decision maker could pay for this information before "making a survey" and "not making a survey" would become a matter of indifference. Thus, the *EVSI* equals $960,000, as noted above. Therefore, one could show that the "make survey" branch, rather than the "do not make survey" branch, would be pruned from this decision tree if the survey cost exceeded this sum and equaled, say, $1 million, rather than $100,000.

Normal-Form Analysis

The normal form of preposterior analysis always leads to the same conclusion as the extensive form, but uses a different route; it is the systematic computation of an expected payoff value for every conceivable strategy and the subsequent selection of the strategy with the best of these values as the optimal one. As we can see by inspecting Figure 14.6, our example includes 6 potential strategies, which we can label S_1 through S_6. They are the following:

S_1: making a survey; given a report of weak demand, constructing a small plant and, given a report of strong demand, constructing a small plant as well.
S_2: making a survey; given a report of weak demand, constructing a small plant and, given a report of strong demand, constructing a large plant.
S_3: making a survey; given a report of weak demand, constructing a large plant and, given a report of strong demand, constructing a small plant.
S_4: making a survey; given a report of weak demand, constructing a large plant and, given a report of strong demand, constructing a large plant as well.
S_5: making no survey; constructing a small plant.
S_6: making no survey; constructing a large plant.

Normal-form analysis requires that for each possible strategy an expected payoff value be calculated, as is shown in Tables 14.17 and 14.18 for strategies S_1 and S_6, respectively.

Observe the logic of Table 14.17. The first two rows indicate that the probability of a report of weak demand is .69 and that the subsequent construction of a small plant (which is what S_1 requires in that case) can have two consequences: (a) that demand

Table 14.17 Calculation of Expected Payoff for Strategy S_1

First-Stage Event Probability (1)	Second-Stage Event Probability (2)	Joint Probability (3) = (1) × (2)	Payoff (millions of dollars) (4)	Joint Probability × Payoff (millions of dollars) (5) = (3) × (4)
$p(R_1) = .69$	$p(E_1\|R_1) = .91$	.69(.91) = .63	$7.9	.63(7.9) = $4.98
$p(R_1) = .69$	$p(E_2\|R_1) = .09$	.69(.09) = .06	4.9	.06(4.9) = .29
$p(R_2) = .31$	$p(E_1\|R_2) = .23$	.31(.23) = .07	7.9	.07(7.9) = .55
$p(R_2) = .31$	$p(E_2\|R_2) = .77$	.31(.77) = .24	4.9	.24(4.9) = 1.18
		1.00		Expected Payoff → $7.00

Table 14.18   Calculation of Expected Payoff for Strategy S_6

Event Probability (1)	Payoff (millions of dollars) (2)	Probability × Payoff (millions of dollars) (3) = (1) × (2)
$p(E_1) = .7$	− $2	$.7(-\$2) = -\$1.4$
$p(E_2) = .3$	$12	$.3(\$12) = \quad 3.6$

Expected Payoff → $2.2

Table 14.19   Normal-Form Analysis: Expected Payoffs (in millions of dollars) of Six Strategies

S_1: $7.00	S_4: $2.10
S_2: 7.98	S_5: 7.10
S_3: 1.12	S_6: 2.20

is in fact weak (which carries a probability of .91) and that a payoff of $7.9 million is received or (b) that demand is in fact strong (which carries a probability of .09) and that a payoff of $4.9 million is received. The last two rows indicate, similarly, that the probability of a report of strong demand is .31 and that the subsequent construction of a small plant (which is what S_1 requires in that case as well) can also have two consequences: (a) that demand is in fact weak (which carries a probability of .23) and that a payoff of $7.9 million is received or (b) that demand is in fact strong (which carries a probability of .77) and that a payoff of $4.9 million is received. The rest of the table is self-explanatory. So is the much simpler Table 14.18, which gives the normal-form analysis of strategy S_6.

Table 14.19 shows our results for S_1 and S_6 and those of similar calculations for the other strategies as well. Given that S_2 has the highest expected value (encircled), S_2 is the optimal strategy. This result corresponds precisely to the result of our extensive-form analysis (that is, to the portion of our Figure 14.6 decision tree that has not been pruned away). Note: Except for rounding error, the expected value of the optimal strategy has been found to be the same in the two forms of analysis.

Pros and Cons

It is easy to see why most people prefer the extensive form of analysis: it is so much easier computationally. The pruning of the decision tree leaves only the optimal strategy behind; the expected values of all the nonoptimal strategies need not be computed at all.[2]

---

[2]These values *could* be computed easily enough, however, even from a decision tree. Consider Figure 14.6. Find the weighted average of the *EMV*s at f and h, the weights being $p(R_1) = .69$ and $p(R_2) = .31$, and you have the *EMV* of strategy S_1. The similarly weighted average of the *EMV*s at f and i equals the *EMV* of S_2; of the *EMV*s at g and h equals the *EMV* of S_3, and of the *EMV*s at g and i equals the *EMV* of S_4. The *EMV*s of S_5 and S_6 are, of course, given directly at points j and k.

There are occasions, however, when the manifold and detailed calculations produced by the normal-form analysis come in handy. The most important of these occasions arises when a decision maker wants to determine how sensitive the selection of the optimal strategy is to the initial choice of prior probabilities [$p(E_1) = .7$ and $p(E_2) = .3$ in our example]. Many times, a decision maker may feel uncomfortable with probabilities based on subjective feelings and may, therefore, want to test whether slight or moderate changes in these values would change the ultimate result. In our example, for instance, there are three leading contenders for optimal strategy: S_1, S_2, and S_5 (as Table 14.19 shows). What would have happened, a decision maker might wonder, if the prior probabilities had been $p(E_1) = p(E_2) = .5$ instead of $p(E_1) = .7$ and $p(E_2) = .3$? Would S_1 have become the optimal strategy? Or S_5? Or even S_4? To answer such questions, the systematic and admittedly tedious calculations of the normal-form analysis might be repeated with the new probability values.

SEQUENTIAL SAMPLING

A logical extension of preposterior analysis has emerged from the pioneering work of Abraham Wald (see Biography 14.2 at the end of this chapter). Wald recognized that if a decision maker (as in Figure 14.6) can make the choice of either gathering information or doing entirely without it before deciding upon a course of action, that decision maker surely also has the option, after getting information, to seek still more information, and still more after that, and so on, before a final decision is made. In principle, the number of information-gathering stages could be infinite; the need to make a decision, of course, dictates an end to that process long before that. From this thought emerged the technique of **sequential sampling,** in which the sample size is not fixed in advance, but a decision is made, after each bit of information is received, either to continue sampling or to put a stop to it.

Wald argued that much of traditional sampling was wasteful in that people were forever scrutinizing complete samples of fixed size even when an examination of the first few units in the sample already indicated the decision to be made. Wald showed how the sequential-sampling technique could replace traditional hypothesis testing. After each observation, a test statistic could be computed, and once it reached predetermined limits, the null hypothesis could be accepted or rejected. Depending on what the observations turned out to be, the statistician might make the decision to accept or to reject after only a few observations or after many, but the crucial advantage of the procedure was that, on average, it would require a smaller number of observations than a fixed-size sample required, while attaining the same probability of making a Type I or Type II error. Clearly, this advantage is a crucial one, especially in the costly case of destructive sampling.

Tables are now available for carrying out various sequential-sampling plans. In the realm of industrial quality control, for example, such a table might state the set of alternative hypotheses (such as "H_0: the proportion of defectives is .01 or less" and "H_A: the proportion of defectives is greater than .01") and the probability of error (such as "the Type I error probability of rejecting H_0 although it is true $= .05$" and "the Type II error probability of accepting H_0 although it is false $= .10$") and then give appropriate instructions depending on the number of items sampled so far (such as "accept H_0 if the number of defectives is 5 or below," "reject H_0 if the number of defectives is 20 or more," and "inspect another unit").

THE GREAT CONTROVERSY: CLASSICAL VS. BAYESIAN STATISTICS

A great controversy erupted among statisticians in the early 1960s, and it continues unabated to this day. As a result of the development of decision theory by Wald and others, more and more statisticians began to replace traditional and so-called "classical" methods of analysis (that relied on *objective probabilities* exclusively and attempted to *make inferences*) by Bayesian methods (that incorporated personal judgment in the form of *subjective probabilities* and attempted to *make decisions* in the face of uncertainty). The quarrel between the two camps has turned out to be a heated one and is reminiscent of (and not entirely unrelated to) the Neyman-Pearson vs. Fisher controversy noted in Biographies 8.1–8.3. The essence of the controversy can, perhaps, be seen most easily by an example.

An Example

Suppose there are two urns, I and II, each containing 6 balls. The contents of the urns are as follows:

Urn I: 5 white balls and 1 black ball.
Urn II: 1 white ball and 5 black balls.

You are told all of the above, you are given one of the urns (but not told which), and you are asked to draw a ball from it at random. It turns out to be white. You are asked to determine the probability that the urn from which you drew was urn I. (Note: The wider applications of this general example should be obvious. The urns might represent two shipments of raw materials or the outputs of two plants; the white and black balls might represent satisfactory and defective items, respectively.)

It is easy enough to state the answer in a general way. The probability that you drew from urn I, given that a white ball was drawn, or $p(\text{I}|\text{W})$, equals the ratio of the joint probability of encountering urn I and a white ball, or $p(\text{I } and \text{ W})$, to the unconditional probability of drawing a white ball, $p(\text{W})$. Thus,

$$p(\text{I}|\text{W}) = \frac{p(\text{I } and \text{ W})}{p(\text{W})}. \tag{1}$$

According to the general multiplication law (Box 4.I on page 141, $p(\text{I } and \text{ W}) = p(\text{I}) \cdot p(\text{W}|\text{I})$; hence,

$$p(\text{I}|\text{W}) = \frac{p(\text{I})\, p(\text{W}|\text{I})}{p(\text{W})}. \tag{2}$$

The probability of drawing a white ball can occur in two mutually exclusive ways, by means of urn I or urn II:

$$p(\mathrm{W}) = p(\mathrm{I} \; and \; \mathrm{W}) + p(\mathrm{II} \; and \; \mathrm{W}), \tag{3}$$

and this probability, in turn, equals

$$p(\mathrm{W}) = p(\mathrm{I}) \, p(\mathrm{W}|\mathrm{I}) + p(\mathrm{II}) \, p(\mathrm{W}|\mathrm{II}). \tag{4}$$

Combining equations 2 and 4,

$$p(\mathrm{I}|\mathrm{W}) = \frac{p(\mathrm{I}) \, p(\mathrm{W}|\mathrm{I})}{p(\mathrm{I}) \, p(\mathrm{W}|\mathrm{I}) + p(\mathrm{II}) \, p(\mathrm{W}|\mathrm{II})} \tag{5}$$

Equation 5 is our answer stated in general terms. Do we know enough to be specific? We know that $p(\mathrm{W}|\mathrm{I}) = (5/6)$ because urn I contains 6 balls, 5 of which are white. We also know that $p(\mathrm{W}|\mathrm{II}) = (1/6)$ because urn II contains 6 balls, 1 of which is white. But we know nothing else. The best we can say is:

$$p(\mathrm{I}|\mathrm{W}) - \frac{p(\mathrm{I})(5/6)}{p(\mathrm{I})(5/6) + p(\mathrm{II})(1/6)}. \tag{6}$$

Classical statisticians, who like to rely on objective probabilities exclusively, would stop right here and despair of finding an answer unless someone supplies values for $p(\mathrm{I})$ and $p(\mathrm{II})$—the unconditional probabilities of being given urn I or urn II.

Bayesian statisticians would not give up as easily. Under most circumstances, they would argue, people hold some degree of belief concerning all kinds of probabilities, and if they are totally ignorant on the matter (as, perhaps, in this case), they can always invoke Bayes's postulate and assume that $p(\mathrm{I}) = p(\mathrm{II}) = (1/2)$. In that case, the answer is

$$p(\mathrm{I}|\mathrm{W}) = \frac{(1/2)(5/6)}{(1/2)(5/6) + (1/2)(1/6)} = \frac{5}{6}. \tag{7}$$

Note that the answer has risen, as a result of picking a white ball, from a prior 1/2 to a posterior 5/6. Note also that a similar increase in probability would have occurred if the original subjective probabilities had been different, such as $p(\mathrm{I}) = 1/3$ and $p(\mathrm{II}) = 2/3$. Then the answer would have been

$$p(\mathrm{I}|\mathrm{W}) = \frac{(1/3)(5/6)}{(1/3)(5/6) + (2/3)(1/6)} = \frac{5}{7}. \tag{8}$$

The Unresolved Issue

Classical statisticians are apt to reject all techniques that make use of subjective probabilities as nonsensical. They would, therefore, relegate to the ash heap most of the decision theory presented in this chapter. (The legitimacy of subjective probability values, however, is not the only issue. The cardinal measurement of utilities is seen as an equally weak link in the Bayesian analysis, for example.)

The Bayesians are far from speechless, however. They argue that real-life people act precisely as modern decision theory suggests. They hold beliefs about various probabilities; they obtain new information (if it is not too costly); they revise their beliefs; they finally act. Indeed, say the Bayesians, if one rigidly stuck to the use of objective probabilities only, many problems (especially in business and economics) could not be solved at all. Moreover, even classical methods are riddled with *subjective* judgment; the claim of exclusive use of objective data is patently false. Consider, the Bayesians say, how the statistician's subjective judgment determines how a hypothesis is formulated, what kind of probability distribution is employed, what the maximum error probability is to be (almost without thought, the significance level is set at $\alpha = .05$ or $.01$), and what kinds of data are to be collected.

More likely than not, the controversy will be with us for some time to come.

ANALYTICAL EXAMPLES

14.1 Decision Analysis Comes of Age

As recently as the early 1970s, decision analysis was still an experimental management technique. The idea that a choice facing a decision maker can be expressed as a mathematical function of probability and utility numbers (that measured, respectively, the decision maker's uncertainties and value judgments) and that the best action was the one with the highest expected utility had just begun to move out of business schools and into practical application in the business world. By the early 1980s, however, decision making with the help of quantitative models that incorporate personal judgment had gained acceptance in many large corporations and government departments.

For example, in 1974, the AIL Division of Cutler-Hammer, Inc. was offered the opportunity to acquire the defense-market rights to a new flight-safety-system patent. The inventor claimed he had a strong patent position as well as technical superiority, but the market for the product depended on legislative action and was very uncertain. Management had to decide fast whether to purchase the rights to the new patent or to reject the offer; it used standard decision-tree techniques. Its initial analysis is illustrated in the accompanying diagram.

The immediate choice (at action point a) was the purchase of a six-month option on the patent rights or rejection of the offer. Six months hence, it was judged, there was (at event point b) a 71-percent chance of

exercising the option and a 29-percent chance of not doing so. Three years down the road (at event point c), there was a 15-percent chance of getting, and an 85-percent chance of not getting, a first defense contract. Five years down the road (at event point d), finally, there was a 25-percent chance of getting a second defense contract and a 75-percent chance of being able to sublicense. Thus, there were five possible outcomes. Using the technique of backward induction, the expected monetary values given in color were computed, and the tree was pruned as shown. Purchasing the option was shown to be worth $100,425; rejecting the offer was worth $0.

Upon further analysis, however, management discovered a third possible action at a—rejecting the offer and seeking a sublicense later on. That approach was shown to yield an expected value of $49,800, about half of the $100,425 expected monetary value of the option-purchase route. Nevertheless, being risk-averse, management chose the newly discovered third alternative: its analysis showed the *EMV* of $49,800 to be composed of a 94-percent chance of zero gain or loss and a 6-percent chance of gaining $830,000. On the other hand, as we can figure from the diagram, the *EMV* of $100,425 is provided by a riskier route composed of a 29-percent chance of losing $125,000, a 60-percent chance of losing $700,000, and an 11-percent chance of a positive return with an expectation of

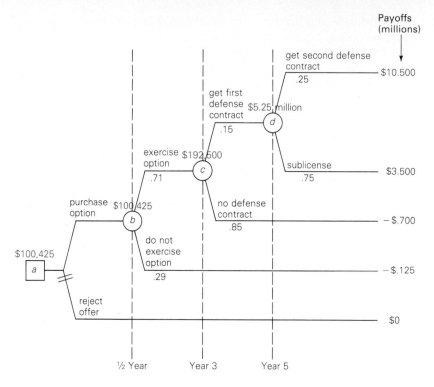

Payoffs (millions)

get second defense contract .25 — $10.500

get first defense contract $5.25 million
$5.25 million
.15
d

sublicense .75 — $3.500

exercise option $192,500
.71
c

purchase option $100,425
b

no defense contract
.85 — −$.700

do not exercise option
.29

$100,425
a

reject offer — $0

— −$.125

½ Year   Year 3   Year 5

$5.25 million. Thus, management effectively maximized expected utility rather than monetary value.

Many other companies have also used the decision-tree technique. Ford has determined in this way whether to produce its own tires and whether to stop producing convertibles; Honeywell used it to decide whether to pursue certain new weapons programs; Pillsbury used the technique to determine whether to switch from a box to a bag for a certain grocery product (and whether it was worthwhile to make a market test of the issue); Southern Railway employed it to choose whether to electrify part of its system; Gulf Oil used it to decide whether to explore certain sites; ITT used decision trees in deciding whether to make certain capital investments.

Source: Jacob W. Ulvila and Rex V. Brown, "Decision Analysis Comes of Age," *Harvard Business Review,* September-October 1982, pp. 130–41. For a fascinating discussion of the use of decision trees in legal disputes, see also Samuel E. Bodily, "When should you go to court?" *Harvard Business Review,* May-June 1981, pp. 103–13.

14.2 The Decision to Seed Hurricanes—A Second Look

Analytical Example 6.2 on page 211 analyzed the decision to seed hurricanes with the help of the normal curve. Decision-tree analysis, however, can be employed with equal success and was, in fact, used in that decision. The accompanying diagram is a copy of the government's analysis.

The diagram is almost self-explanatory. The government, at action point *a,* can decide to seed or not to seed. Subsequently, at event points *b* or *c,* changes in wind speed will occur, with varying probabilities. The ultimate "payoff" is here measured in millions of dollars of property damage. Naturally, the government wants to minimize this cost; the branch from *a* to *c* is pruned, and the decision is to go ahead and seed.

Can you figure out how the expected monetary values at *b* and *c* were calculated?

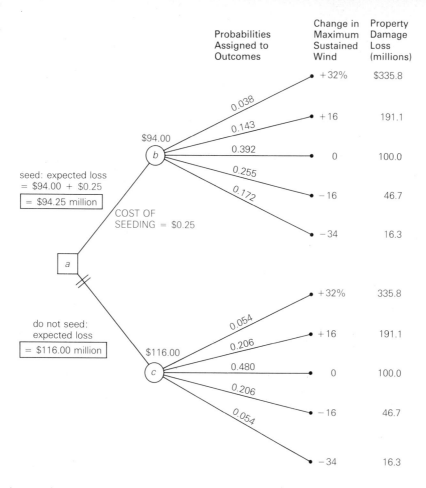

Source: R. A. Howard, J. E. Matheson, D. W. North, "The Decision to Seed Hurricanes," *Science,* June 16, 1972, pp. 1191–1202. (This article is well worth reading, especially for its expanded discussion of the value of gathering additional information.)

14.3 The Value of Perfect Information: The Case of the U.S. Cattle Industry

Two economists recently studied the willingness of risk-averse decision makers to pay for price information under uncertainty. They focussed on the U.S. cattle industry; the accompanying table on page 518 shows some of their results. To an individual producer, the value of knowing the realized cattle price (column 1) beforehand ranged, in constant dollars, from 8 cents per hundredweight (cwt) to a $4.24 per cwt (column 2), with a mean value of 46 cents per cwt for the period as a whole.

Multiplying the column 2 values by mean bimonthly production provides, in turn, a maximum estimate of the value of perfect price information to the entire industry: $296 million per period. This figure is a *maximum* for this reason: While the output response of an individual producer to the receipt of perfect information can be expected to have a negligible impact on market price, the output response by the entire industry (if everyone got the information) would change price markedly and depress the information value per unit of output.

Period	Realized Price ($ per cwt) (1)	Value of Perfect Information ($ per cwt) (2)
1978 Jan./Feb.	20.82	.59
Mar./Apr.	23.92	1.38
May/June	26.15	1.57
July/Aug.	25.20	1.63
Sept./Oct.	26.34	.80
Nov./Dec.	26.30	.42
1979 Jan./Feb.	30.33	1.13
Mar./Apr.	34.23	4.24
May/June	32.21	2.55
July/Aug.	29.03	.83
Sept./Oct.	29.28	.29
Nov./Dec.	28.05	.11
1980 Jan./Feb.	27.85	.08
Mar./Apr.	25.92	.23
May/June	24.73	.80
July/Aug.	25.65	.36
Sept./Oct.	24.55	.12

Source: Terry Roe and Frances Antonovitz, "A Producer's Willingness to Pay for Information Under Price Uncertainty: Theory and Applications," *Southern Economic Journal,* October 1985, pp. 382–91. See also the more advanced discussion by the same authors in "A Theoretical and Empirical Approach to the Value of Information in Risky Markets," *The Review of Economics and Statistics,* February 1986, pp. 105–14.

BIOGRAPHIES

14.1 Thomas Bayes

Thomas Bayes (1702–61) was born in London, England, as the eldest son of the Rev. Joshua Bayes, one of the first nonconformist ministers to be publicly ordained in England. The younger Bayes was privately educated (as was the usual practice among nonconformists at the time) in languages, literature, and science. Quite possibly, he was tutored in mathematics by Abraham de Moivre (Biography 4.1). Like his father, Thomas Bayes also became a Presbyterian minister and spent 30 years in that capacity. In that period, he published (nonmathematical) treatises under the pseudonym of John Noon, and they got him elected a Fellow of the Royal Society. Bayes's mathematical work is contained in two papers published posthumously due to the efforts of a friend, Richard Price. One of the papers, "An Essay Towards Solving a Problem in the Doctrine of Chances" reversed de Moivre's focus of reasoning from the population to the sample and dealt with inferences from the sample to the population. The paper put forth a number of mathematical propositions; as "Proposition 9," it presented what is now known as *Bayes's theorem* (first discussed in Chapter 4 of this book). The essay is, perhaps, one of the least understood but most famous and controversial contributions ever made in the history of science. Many call it a masterpiece of mathematical elegance. In the words of Ronald Fisher (Biography 8.1), Bayes's "mathematical contributions . . . show him to have been in the first rank of independent thinkers."[1] Certainly, Bayes has become, through his famous theorem, the father of modern decision theory. For this reason,

[1] Ronald Fisher, *Statistical Methods and Scientific Inference,* 2nd ed., rev. (New York: Hafner, 1959), p. 8.

the optimal strategy selected in expected-payoff-optimizing decision making problems is universally referred to as **Bayes's strategy.**

Sources: Biometrika, December 1958, pp. 293–315 (includes a reproduction of the famous essay); *International Encyclopedia of Statistics,* vol. 1 (New York: Free Press, 1978), pp. 7–9.

14.2 Abraham Wald

Abraham Wald (1902–50) was born in Klausenburg, Hungary (which later became Cluj, Rumania). He was privately educated and self-taught (a consequence of complications arising from his family's Jewish orthodoxy). In the 1930s, he moved to Vienna, worked at the Austrian Institute for Business Cycle Research, and, at the time of the Nazi annexation, accepted an invitation by the Cowles Commission to work in the United States. That move surely saved him from death in a gas chamber—a fate that befell all but one member of his numerous family. Eventually, Wald taught statistics at Columbia University (his students remember him as a superb teacher of unrivalled clarity and precision), but his career was cut short by an airplane crash during a lecture tour of India.

Above all else, Wald's name will forever be linked with statistical decision theory and sequential analysis. In one of his earliest and finest papers, he introduced many fundamental concepts of decision theory, along with a mathematical structure for single-sample decision making that was sufficiently general to embrace estimation, hypothesis testing, and even the design of experiments. As he states in the Preface to his monumental *Statistical Decision Functions* (1950): "A major advance beyond previous results is the treatment of the design of experimentation as a part of the general decision problem." This tendency of Wald's (and of later decision theorists) to view all of statistics as a science of decision making under uncertainty (and the user of statistical data as someone in a decision-making

situation who seeks to make the one of several possible decisions that has the best expected payoff) invoked the ire of Ronald Fisher (Biography 8.1). Fisher thought that decision theory was appropriate, perhaps, for industry but certainly not for scientific work (in which inferences about the truth were to be made and not where payoffs were to be achieved through optimal decisions). Yet Wald's approach fit in nicely with the Neyman-Pearson theory of hypothesis testing (see Biographies 8.2 and 8.3).

Although he had forerunners, Wald was the first to formulate mathematically and solve generally the problem of sequential tests of statistical hypotheses. If evidence in sequentially unfolding data is sharply one-sided, Wald asked, why not stop an inquiry early? Why insist on observing and analyzing all the data contained in a single sample of predetermined fixed size? This line of work ultimately gave rise to a classic work, *Sequential Analysis* (1947).

Nor is this all. During his short life, Wald made major contributions in other fields as well. (He published more than 90 books and papers!) These include topology, measure and set theory, lattice theory, econometrics (the seasonal correction of time series, formulas for approximating economic index numbers), mathematical economics (indifference surfaces, proofs of the existence and uniqueness of solutions to the Walrasian system of equations), and much more.

Sources: Oskar Morgenstern, "Abraham Wald, 1902–1950," *Econometrica,* October 1951, pp. 361–67; *International Encyclopedia of Statistics,* vol. 2 (New York: Free Press, 1978), pp. 1235–38.

Summary

1. *Decision theory* consists of a variety of methods that can be employed in the systematic analysis and solution of decision-making problems that arise because uncertainty exists about future events over which the decision maker has no con-

trol, but which are bound to influence the ultimate outcome of a decision.
2. Certain identical elements are present in every decision problem involving uncertainty. They include the mutually exclusive decision alternatives, or *ac-*

tions, available to the decision maker, the mutually exclusive future occurrences, or *events,* that will affect the outcome of any present action taken but that are not under the control of a decision maker, and the positive or negative net benefits, or *payoffs,* that are associated with each possible action/event combination and that are, thus, the joint outcome of choice and chance.

A *payoff table* summarizes a decision-making situation in the context of uncertainty; it consists of a tabular listing of the payoffs associated with all possible combinations of actions and events. A *decision tree* provides an alternative summary; it shows graphically, and in chronological order from left to right, every potential action, event, and payoff. In such a tree diagram, any point of choice (at which the decision maker is in control) is denoted by a square symbol and is called an *action point* (from which *action branches* representing the possible actions emanate). In contrast, any point of chance (at which the decision maker exercises no control but at which "nature" is in charge) is denoted by a circle and is called an *event point* (from which *event branches* representing the possible events emanate).

3. When nothing is known about the likelihood of those alternative future events that are certain to affect the eventual outcome of a present decision, and if the decision maker does not even care to guess what the event probabilities might be, one of three decision criteria is commonly employed: maximin (or minimax), maximax (or minimin), or minimax regret.

  a. According to the *maximin criterion,* a decision maker who seeks to maximize some benefit finds the minimum benefit associated with each possible action, identifies the maximum among these minima, and chooses the action associated with this maximum of minima. (The same criterion is called the *minimax criterion* when a decision maker seeks to minimize some cost; such a decision maker finds the maximum cost associated with each possible action, identifies the minimum among these maxima, and chooses the action associated with this minimum of maxima.)

  b. According to the *maximax criterion,* a decision maker who seeks to maximize some benefit finds the maximum benefit associated with each possible action, identifies the maximum among these maxima, and chooses the action associated with this maximum of maxima. (The same criterion is called the *minimin criterion* when a

decision maker seeks to minimize some cost; such a decision maker finds the minimum cost associated with each possible action, identifies the minimum among these minima, and chooses the action associated with this minimum of minima.)

  c. According to the *minimax-regret criterion,* a decision maker finds the maximum regret value associated with each possible action, identifies the minimum among these maxima, and chooses the action associated with this minimum of maximum regret values.

The above criteria can be criticized for a number of reasons, including undue sensitivity of the results to irrelevant factors, undue reliance on extreme values, and, of course, a total lack of consideration for the probabilities of events.

4. A form of decision making under uncertainty that employs probabilities, but only those available prior to the gathering of new experimental or sample evidence about the likelihood of alternative future events, is called *prior analysis.* Once again, a number of alternative criteria can be employed.

  a. According to the *maximum-likelihood criterion,* a decision maker simply ignores all the events that might occur except the most likely one and selects the action that produces the optimal result (maximum benefit or minimum cost) associated with this most likely event.

  b. A decision maker using the *expected-monetary-value criterion* determines an expected monetary value for each possible action and selects the action with the optimal expected monetary value (the optimal *EMV* is the largest one if the objective is to maximize some benefit; the optimal *EMV* is the smallest one if the objective is to minimize some cost). This criterion is often employed with the help of a decision tree; expected monetary values are then computed (from ultimate-payoff values and event-branch probabilities) for each fork of the tree by a process termed *backward induction.*

  c. A decision maker using the *expected-opportunity-loss criterion* determines an expected opportunity loss for each possible action and selects the action with the smallest of these values. This criterion necessarily always yields the same result as the expected-monetary-value criterion.

Observation shows that people making decisions under uncertainty often fail to follow the *EMV* criterion, a fact that can be explained by studying people's attitudes toward risk. For a risk-averse person, the marginal utility of money declines with larger amounts of money; for a risk-neutral person, it is constant; for a risk-seeking person, it rises. Except for the risk-neutral person, the optimization of expected monetary value does not produce maximum welfare. For the risk averter and the risk seeker, therefore, a different decision criterion is superior—namely that of *maximizing expected utility*.

5. A form of decision making under uncertainty that starts out with prior probabilities, proceeds to gather additional or sample evidence about event probabilities, and then uses this new evidence to transform the set of prior probabilities, by means of Bayes's theorem, into a revised set of posterior probabilities (with the help of which a final decision is reached) is called *posterior analysis*.

6. The maximum amount a decision maker can be expected to pay for obtaining perfect information about future events—and, thus, for eliminating uncertainty completely—is called the *expected value of perfect information,* EVPI. It always equals the expected opportunity loss of the optimal decision under uncertainty. The maximum amount a decision maker can be expected to pay for obtaining admittedly imperfect information about future events—and, thus, for reducing (rather than eliminating) uncertainty—is called the *expected value*

of sample information, EVSI. The ratio of the *EVSI* to the (prior) *EVPI,* multiplied by 100, measures the efficiency of sample information, or *ESI*.

7. A form of decision making under uncertainty that decides the worth of obtaining additional (experimental or sample) information before proceeding to prior and posterior analysis is called *preposterior analysis*. This analysis can be performed by using either one of two alternative approaches that always yield the same result: (a) the decision-tree approach *(extensive-form analysis)* and (b) the tabular approach *(normal-form analysis)*.

8. A logical extension of preposterior analysis is *sequential sampling,* in which the sample size is not fixed in advance but in which a decision is made, after each bit of information is received, either to continue sampling or to put a stop to it.

9. With the development of decision theory (and its widespread use of subjective probabilities), a great controversy among statisticians has emerged. Classicists wish to rely exclusively on objective probabilities and view statistics as a means to the making of inferences; Bayesians freely incorporate personal judgment by means of subjective probabilities and are apt to view statistics as a collection of techniques for the making of decisions in the face of uncertainty.

Key Terms

action branches
action point
actions
backward induction
Bayes's postulate
Bayes's strategy
criterion of insufficient reason
decision branches
decision fork
decision making under uncertainty
decision node
decision point
decision theory
decision tree
dominant action
efficiency of sample information *(ESI)*
equal-likelihood criterion
event branches
event point
events
expected monetary value *(EMV)*

expected-monetary-value criterion
expected opportunity loss *(EOL)*
expected-opportunity-loss criterion
expected regret value
expected utility *(EU)*
expected-utility criterion
expected value of perfect information *(EVPI)*
expected value of sample information *(EVSI)*
extensive-form analysis
fair gamble
inadmissible action
Laplace criterion
marginal utility of money
maximax
maximin
maximum-likelihood criterion
minimax
minimax regret
minimin
more-than-fair gamble
normal-form analysis

opportunity loss *(OL)*
optimal strategy
payoffs
payoff table
posterior analysis
posterior expected value of perfect information
    (posterior *EVPI*)
preposterior analysis
prior analysis
prior expected value of perfect information
    (prior *EVPI*)
regret
regret table

risk aversion
risk neutrality
risk seeking
sequential sampling
state-of-nature branches
state-of-nature fork
state-of-nature node
state-of-nature point
states of nature
total utility of money
unfair gamble
utility-of-money function

Questions and Problems

The computer program noted at the end of the chapter can be used to work some of the subsequent problems.

1. Depending on whether future demand is low, moderate, or high, a real-estate developer foresees profits (in millions of dollars) of 4, 5, or 6 from a small project; of 1, 6, or 10 from a medium-sized project; and of -5, 0, or 30 from a large project. Determine the best action under the criterion of
a. maximin.
b. maximax.

2. Depending on whether future demand is low, moderate, or high, a producer of aircraft ELTs (emergency locator transmitters) foresees profits (in thousands of dollars) of 15, 20, or 25 from a small production run; of 6, 20, or 35 from a medium-sized run; and of -10, 0, or 69 from a large run. Determine the best action under the criterion of
a. maximin.
b. maximax.

3. Table 14.A shows profits (in million dollars) that are associated with 5 actions and 6 events. Determine the best action under the criterion of
a. maximin.
b. maximax.

4. Table 14.B shows profits (in million dollars) that are associated with 4 actions and 5 events. Determine the best action under the criterion of
a. maximin.
b. maximax.

5. Table 14.C shows benefits that are associated with 5 actions and 4 events. Determine the best action under the criterion of
a. maximin.
b. maximax.

6. The Zoning Board of Appeals has granted permission to a builder to erect between 1 and 5 homes on a certain lot, but whatever the chosen number

Table 14.A

Actions \ Events	E_1	E_2	E_3	E_4	E_5	E_6
A_1	0	-15	-35	-50	23	67
A_2	-15	-35	-50	23	67	102
A_3	-35	-50	23	67	102	139
A_4	-49	23	67	102	139	150
A_5	23	67	102	139	145	-100

Table 14.B

Actions \ Events	E_1	E_2	E_3	E_4	E_5
A_1	93	75	60	−10	0
A_2	0	17	65	107	2
A_3	5	6	7	79	−5
A_4	−53	89	0	−12	112

Table 14.C

Actions \ Events	E_1	E_2	E_3	E_4
A_1	22	39	56	−17
A_2	33	34	36	36
A_3	12	−50	−1	79
A_4	−60	12	19	70
A_5	0	0	12	62

is, the homes must all be built now. The builder's cost is $100,000 per home; up to the number demanded, homes built can be sold at 25 percent above cost; any surplus homes can be sold to a real-estate firm at 60 percent of cost. How many homes should the profit-seeking builder erect, given that 1 home is the minimum and the criterion is
a. maximin.
b. maximax.

7. Rework problem 6, but let the builder's cost be $60,000 per home, while regular sales are at 50 percent above cost and surplus sales at 80 percent of cost.

8. Rework problem 6, but let the builder's cost be $120,000 per home, while regular sales are at 15 percent above cost and surplus sales at 50 percent of cost.

9. An advertising manager believes that a given volume of sales can be achieved at different costs, depending on whether the ads are disseminated by means of direct mail, newspapers, or television.

However, it is also believed that the state of the economy (recession, inflation, stagflation) will influence the cost figures such that the cost (in thousands of dollars) of a direct-mail campaign will be 75, 20, or 40; of a newspaper campaign will be 50, 60, or 30; and of a TV campaign will be 55, 40, or 45. Determine the best action under the criterion of
a. minimax.
b. minimin.

10. A given volume of output can be produced by one of three different methods, but the associated cost will vary with the state of the economy (recession, inflation, stagflation) because prices of different productive inputs will be affected differently. The projected costs of method 1 are (in thousands of dollars) 15, 19, 29; the corresponding figures for method 2 are 29, 19, 14 and for method 3 are 25, 17, 25. Determine the best action under the criterion of
a. minimax.
b. minimin.

11. Consider Table 14.A and assume the entries are costs. Determine the best action under the criterion of
   a. minimax.
   b. minimin.
12. Consider Table 14.B and assume the entries are costs. Determine the best action under the criterion of
   a. minimax.
   b. minimin.
13. Consider Table 14.C and assume the entries are costs. Determine the best action under the criterion of
   a. minimax.
   b. minimin.
14. Reconsider problem 1, but this time determine the best action under the criterion of minimax regret.
15. Reconsider problem 2, but this time determine the best action under the criterion of minimax regret.
16. Reconsider problem 3, but determine the best action under the maximum-likelihood criterion, assuming that $p(E_1) = .1$, $p(E_2) = .2$, $p(E_3) = .3$, $p(E_4) = .1$, $p(E_5) = .05$, and $p(E_6) = .25$.
17. Reconsider problem 4, but determine the best action under the maximum-likelihood criterion, assuming that $p(E_1) = .4$, $p(E_2) = .2$, $p(E_3) = .05$, $p(E_4) = .05$, and $p(E_5) = .3$.
18. Reconsider problem 2. Determine the best action under the expected-monetary-value criterion, assuming $p(E_1) = .1$, $p(E_2) = .7$, and $p(E_3) = .2$.
19. A business executive has determined that keeping the business in its present location is a disaster and that it must be moved to one of two new locations, East or West. Moving East will produce future net earnings with a present value of either $50 million or $10 million; moving the business West will produce future net earnings with a present value of either $100 million or zero—in each case, depending on whether the attempt to break into the new market will be a success or a failure.[3] Assuming that the chances for success or failure at the new locations are judged to be equal, determine the best action under the criterion of expected monetary value.
20. Analytical Example 14.2 ends with a question. What is the answer?
21. An independent producer of TV films wants to market a new soap opera. The rights can be sold to a distributor for $125 million now, or the program can be offered to a TV network for review with these possible results (in the judgment of the film producer): a 60-percent chance of rejection (which ruins all further chances of a sale to anyone and spells a $30 million loss) or a 40-percent chance of getting a contract (which means a $300 million profit). Use a decision tree to determine the optimal action, given the desire to maximize expected monetary value.
22. Reconsider problem 19. Determine the best action under the criterion of expected opportunity loss.
23. Reconsider problem 19. Determine the best action under the criterion of expected utility, assuming the executive has a utility function of $U = \sqrt{\$}$.
24. Conduct an extensive-form (preposterior) analysis for an oil company, given the following information. It can drill for oil on a given site now (and the prior probabilities of $E_1 =$ finding oil or $E_2 =$ not finding oil are considered equal) or sell its rights to the site for $10 million. If it drills and finds oil, a net payoff of $100 million is expected; if no oil is found, that payoff equals $-$15 million. It can instead contract for a seismic test. If the test predicts oil, it can drill (and get a net payoff of $90 million if oil is in fact found and of $-$25 million if none is found) or sell the rights (for $20 million). If the test predicts no oil, it can drill (and get a net payoff of $90 million if oil is in fact found and of $-$25 million if none is found) or sell the rights for $1 million. The "track record" of the seismic-test company is given in Table 14.D.
25. Consider the answer to problem 24. If no test is made, what is the (prior) expected value of perfect information, assuming $p(R_1) = p(E_1)$ and also that $p(R_2) = p(E_2)$?
26. Consider the answer to problem 24. What is the expected value of the sample information?
27. Reconsider problem 21. Suppose that a consulting firm is willing to offer advice on the network's likely reaction for a $1 million fee. The consulting firm's "track record" is given in Table 14.E. Use a decision tree to determine the optimal strategy if the film producer definitely decides to buy the advice and to then take the action (sell the rights or offer the film to the network) that maximizes expected monetary value.
28. Table 14.19 gives expected payoffs for six different strategies, but the computation is shown only for strategies S_1 and S_6 (in Tables 14.17 and 14.18). Make a similar computation for S_3.
29. Rework problem 28, but for S_4.
30. Rework problem 28, but for S_5.

---

[3]In order to make comparable dollars of revenue and cost that pertain to different periods of time, it is customary to calculate payoffs in terms of "present value." For a detailed discussion of this concept, see Heinz Kohler, *Intermediate Microeconomics: Theory and Applications,* 2nd ed. (Glenview, Ill.: Scott, Foresman and Company, 1986), chapter 15.

Table 14.D

Report Issued	Subsequent Events	
	E_1 = oil is found	E_2 = oil is not found
R_1 = oil present R_2 = no oil present	$p(R_1|E_1) = .6$ $p(R_2|E_1) = .4$	$p(R_1|E_2) = .2$ $p(R_2|E_2) = .8$

Table 14.E

Advice Given	Subsequent Events	
	E_1 = Rejection	E_2 = Contract Offer
R_1 = network will reject R_2 = network will offer contract	$p(R_1|E_1) = .8$ $p(R_2|E_1) = .2$	$p(R_1|E_2) = .3$ $p(R_2|E_2) = .7$

Selected Readings

Aitchison, John. *Choice Against Chance: An Introduction to Statistical Decision Theory.* Reading, Mass.: Addison-Wesley, 1970.

Bayes, Thomas. "An Essay Towards Solving a Problem in the Doctrine of Chances." *Philosophical Transactions of the Royal Society* 53(1763):370–418. The posthumous publication containing the Bayes's theorem. Now most easily available in: Thomas Bayes. *Facsimiles of Two Papers by Bayes.* New York: Hafner, 1963.

Bell, D. E., R. L. Keeney, H. Raiffa. *Conflicting Objectives in Decisions.* New York: Wiley, 1977.

Berkson, Joseph. "My Encounter With Neo-Bayesianism." *International Statistical Review* 45(1977):1–8. Presents arguments against the Bayesian approach.

Box, George E. P. and George C. Tiao. *Bayesian Inference in Statistical Analysis.* Reading, Mass.: Addison-Wesley, 1973.

Brown, Rex V., Andrew S. Kahr, C. Peterson. *Decision Analysis for the Manager.* New York: Holt, Rinehart and Winston, 1974.

Efron, B. "Why Isn't Everyone a Bayesian?" *The American Statistician,* February 1986, pp. 1–11. A discussion, with numerous comments by critics, of the Fisher versus Neyman-Pearson-Wald controversy.

Fisher, Ronald A. *Statistical Methods and Scientific Inference,* 2nd ed., rev. New York: Hafner, 1959. In this, his last book, Fisher strongly rejects the Bayesian approach to statistics as well as the view of statistics as a decision-making science traceable to Wald.

Harvard Business Review. *Statistical Decision Series,* Parts I–IV. Boston: 1951–70.

Kempthorne, Oscar, and Leroy Folks. *Probability, Statistics, and Data Analysis.* Ames: Iowa State University Press, 1971.

Kohler, Heinz. *Intermediate Microeconomics: Theory and Applications,* 2nd. ed. (Glenview, Ill.: Scott, Foresman and Company, 1986). Includes presentations of basic utility theory (Chapter 3), the economics of uncertainty (Chapter 16), and (in the accompanying Student Workbook) game theory.

Lindley, Dennis V. *Bayesian Statistics: A Review.* Philadelphia: Society for Industrial and Applied Mathematics, 1972.

Luce, Robert D., and Howard Raiffa. *Games and Decisions.* New York: Wiley, 1958.

Menges, G. "Inference and Decision." *Selecta Statistica Canadiana,* vol. I. New York: Wiley, 1973. On the Fisher-Wald controversy concerning the nature of statistical science.

Morris, W. T. *Management Science: A Bayesian Introduction.* Englewood Cliffs, N.J.: Prentice-Hall, 1968.

Neumann, John von, and Oskar Morgenstern. *The Theory of Games and Economic Behavior,* rev. ed. Princeton: Princeton University Press, 1953. The classic work on *game theory* that studies decision making under uncertainty that is complicated by a conscious conflict of wills so that the payoff to an action depends not only on the decision maker's choice and "nature," but also on the conscious choices made by other people.

Raiffa, Howard. *Decision Analysis: Introductory Lectures on Choices Under Uncertainty.* Reading, Mass.: Addison-Wesley, 1968.

Savage, Leonard J. "The Theory of Statistical Decision." *Journal of the American Statistical Association* (March 1951), pp. 55–67. A classic article, including a stimulating exposition of Wald's work.

Schlaifer, Robert. *Probability and Statistics for Business Decisions: An Introduction to Managerial Economics Under Uncertainty.* New York: McGraw-Hill, 1959. On Bayesian inference, with emphasis on decision theory.

Schlaifer, Robert. *Analysis of Decisions Under Uncertainty.* New York: McGraw-Hill, 1969.

Stigler, Stephen M. "Who Discovered Bayes's Theorem?" *The American Statistician* (November 1983), pp. 290–96. A fascinating article that suggests that Bayes, perhaps, was not the first to discover the theorem bearing his name.

Tsokos, Chris P., and I. N. Shimi, eds. *The Theory and Applications of Reliability With Emphasis on Bayesian and Nonparametric Methods.* New York: Academic Press, 1977.

Wald, Abraham. *Sequential Analysis.* New York: Wiley, 1947, and *Statistical Decision Functions.* New York: Wiley, 1950. These books are crucial landmarks in the development of statistics as a decision making science.

Winkler, Robert L. *Introduction to Bayesian Inference and Decision.* New York: Holt, Rinehart and Winston, 1972.

Zellner, Arnold. *An Introduction to Bayesian Inference in Econometrics.* New York: Wiley, 1971.

Computer Program

The DAYSAL-2 personal-computer diskettes contain one program of interest to this chapter:

16. Expected Monetary Values helps you calculate expected monetary values for decision-tree analysis.

APPENDIX TABLES

A   Random Numbers

B   The Greek Alphabet

C   Binomial Probabilities for Individual Values of x

D   Binomial Probabilities for Cumulative Values of x

E   Standard Normal Curve Areas

F   Critical Normal Deviate Values

G   Student t Distributions

H   Chi-Square Distributions

I F Distributions

J   Values of q_α in Tukey's HSD Test

Appendix Table A
Random Numbers

71274	84346	75444	85690	35384	87841	97411	78698	46796	33552
64017	01373	14665	31891	80997	14321	47741	59980	87739	38174
43747	17686	11045	15549	52779	65135	00275	95434	36337	24041
59688	48689	41591	47042	83615	93034	25077	64835	67798	30547
95016	73467	11447	59500	94921	15166	69217	26267	11316	22651
65207	30591	65947	58339	00952	32111	45459	14986	57395	34492
34510	78657	08883	49489	85619	52912	01662	49854	78354	30631
56299	60624	91572	31734	18159	18927	31314	59682	41320	88602
02113	12579	86172	03819	69968	02616	72687	42699	04792	16510
00884	87979	45184	61572	20086	14498	29640	94263	90964	29278
89367	53577	97412	19603	57234	63055	49059	35761	72007	22751
99781	56740	42659	46617	21828	99831	45987	63450	66919	78252
92024	12100	76013	12587	86340	74880	79979	35906	38122	64917
82861	09215	87342	72789	76132	24468	93065	78968	03321	48081
11286	13011	67982	74101	44961	25468	14247	95934	50711	24492
68674	24686	14460	61242	92310	86810	87702	69811	53996	99517
24882	20749	94139	28785	74402	18561	79069	56838	30020	99707
21740	51134	39298	92203	66230	30636	58169	78982	50057	39908
46901	34825	28673	10404	97777	30782	04680	15319	84125	40937
92686	81702	74149	76326	01101	96278	90855	55145	89705	51199
98743	59366	94797	96803	69876	87533	19675	59246	65348	06606
11463	25619	38107	91053	58416	02720	86563	27443	99598	04074
76975	18636	54975	67422	57101	68857	35389	35641	34505	71552
67359	50379	81053	97357	00717	59504	34480	77127	23243	48682
34084	48031	27227	48912	10797	88917	93126	52945	79457	66528
94553	44441	83166	40056	73935	52103	13972	56781	31900	95037
20545	43211	34500	92233	53497	39401	78535	82360	57410	32060
05646	01152	13235	39168	69214	83852	00144	08105	94247	37189
54022	58295	96122	87620	74774	06884	81689	68392	25776	08748
32291	30700	32561	83579	94582	77930	06826	96855	97751	42664
66212	75061	65891	96896	15107	12985	97616	64241	08592	81036
98059	44951	23078	92793	80756	52799	20340	62969	81775	31065
23157	52179	24394	39833	89427	58771	26992	00649	25475	50888
28911	81929	91368	49372	43335	44465	43257	66893	34761	60423
85959	90369	02100	28727	83001	84166	20473	35305	38088	54795
34459	31400	58760	17157	73816	55527	54133	24605	56153	35354
04073	57781	36894	93000	57834	29343	98195	58425	97275	71392
22126	91330	95667	75737	36869	55209	41663	04943	59401	17039
10288	61685	25302	84097	13088	86840	04020	43046	01043	43157
75431	32853	72907	99432	65482	23011	70466	87386	67471	77629
90800	17425	28042	53770	98924	31863	84115	82488	23239	82185
19083	89475	05207	41284	83405	55825	31117	59821	96455	63796
10405	67911	77238	46262	42766	07215	02391	47316	78724	41170
34711	77325	99768	63455	44335	91028	27740	86163	81474	08159
73334	61941	16883	05012	63191	35763	60157	09617	25501	44989
79452	68381	71937	23274	60273	47091	82876	24641	03825	50894
13864	28746	32434	88325	99996	96130	39471	74020	56077	22133
73082	50271	83240	80065	09328	02940	41686	32758	89467	73553
43060	88221	35010	79829	71520	80453	95049	66352	77495	83256
15172	42061	33264	63832	48528	23258	13520	83222	45659	39074

Source: The Rand Corporation, *A Million Random Digits with 100,000 Normal Deviates* (Glencoe, Illinois: The Free Press, 1955), p. 377.

Appendix Table B
The Greek Alphabet

A	α	alpha	H	η	eta	N	ν	nu	T	τ	tau
B	β	beta	Θ	θ	theta	Ξ	ξ	xi	Y	υ	upsilon
Γ	γ	gamma	I	ι	iota	O	o	omicron	Φ	φ	phi
Δ	δ	delta	K	κ	kappa	Π	π	pi	X	χ	chi
E	ε	epsilon	Λ	λ	lambda	P	ρ	rho	Ψ	ψ	psi
Z	ζ	zeta	M	μ	mu	Σ	σ	sigma	Ω	ω	omega

Appendix Table C
Binomial Probabilities for Individual Values of x

Entries in the table give the probability of x successes in n trials of a binomial experiment, where π is the probability of a success in one trial. For example, with two trials and $\pi = .40$, the probability of two successes is .1600.

						π						
n	x	.05	.10	.15	.20	.25	.30	.35	.40	.45	.50
1	0	.9500	.9000	.8500	.8000	.7500	.7000	.6500	.6000	.5500	.5000
	1	.0500	.1000	.1500	.2000	.2500	.3000	.3500	.4000	.4500	.5000
2	0	.9025	.8100	.7225	.6400	.5625	.4900	.4225	.3600	.3025	.2500
	1	.0950	.1800	.2550	.3200	.3750	.4200	.4550	.4800	.4950	.5000
	2	.0025	.0100	.0225	.0400	.0625	.0900	.1225	.1600	.2025	.2500
3	0	.8574	.7290	.6141	.5120	.4219	.3430	.2746	.2160	.1664	.1250
	1	.1354	.2430	.3251	.3840	.4219	.4410	.4436	.4320	.4084	.3750
	2	.0071	.0270	.0574	.0960	.1406	.1890	.2389	.2880	.3341	.3750
	3	.0001	.0010	.0034	.0080	.0156	.0270	.0429	.0640	.0911	.1250
4	0	.8145	.6561	.5220	.4096	.3164	.2401	.1785	.1296	.0915	.0625
	1	.1715	.2916	.3685	.4096	.4219	.4116	.3845	.3456	.2995	.2500
	2	.0135	.0486	.0975	.1536	.2109	.2646	.3105	.3456	.3675	.3750
	3	.0005	.0036	.0115	.0256	.0469	.0756	.1115	.1536	.2005	.2500
	4	.0000	.0001	.0005	.0016	.0039	.0081	.0150	.0256	.0410	.0625
5	0	.7738	.5905	.4437	.3277	.2373	.1681	.1160	.0778	.0503	.0312
	1	.2036	.3280	.3915	.4096	.3955	.3602	.3124	.2592	.2059	.1562
	2	.0214	.0729	.1382	.2048	.2637	.3087	.3364	.3456	.3369	.3125
	3	.0011	.0081	.0244	.0512	.0879	.1323	.1811	.2304	.2757	.3125
	4	.0000	.0004	.0022	.0064	.0146	.0284	.0488	.0768	.1128	.1562
	5	.0000	.0000	.0001	.0003	.0010	.0024	.0053	.0102	.0185	.0312
6	0	.7351	.5314	.3771	.2621	.1780	.1176	.0754	.0467	.0277	.0156
	1	.2321	.3543	.3993	.3932	.3560	.3025	.2437	.1866	.1359	.0938
	2	.0305	.0984	.1762	.2458	.2966	.3241	.3280	.3110	.2780	.2344
	3	.0021	.0146	.0415	.0819	.1318	.1852	.2355	.2765	.3032	.3125
	4	.0001	.0012	.0055	.0154	.0330	.0595	.0951	.1382	.1861	.2344
	5	.0000	.0001	.0004	.0015	.0044	.0102	.0205	.0369	.0609	.0938
	6	.0000	.0000	.0000	.0001	.0002	.0007	.0018	.0041	.0083	.0156
7	0	.6983	.4783	.3206	.2097	.1335	.0824	.0490	.0280	.0152	.0078
	1	.2573	.3720	.3960	.3670	.3115	.2471	.1848	.1306	.0872	.0547
	2	.0406	.1240	.2097	.2753	.3115	.3177	.2985	.2613	.2140	.1641
	3	.0036	.0230	.0617	.1147	.1730	.2269	.2679	.2903	.2918	.2734
	4	.0002	.0026	.0109	.0287	.0577	.0972	.1442	.1935	.2388	.2734
	5	.0000	.0002	.0012	.0043	.0115	.0250	.0466	.0774	.1172	.1641
	6	.0000	.0000	.0001	.0004	.0013	.0036	.0084	.0172	.0320	.0547
	7	.0000	.0000	.0000	.0000	.0001	.0002	.0006	.0016	.0037	.0078

Source: The National Bureau of Standards, *Tables of the Binomial Probability Distribution,* Applied Mathematics Series, no. 6 (Washington, D.C.: U.S. Government Printing Office, 1949). The original contains probabilities for values of n up to 49, and for values of π in increments of .01. All entries were calculated by the formula in Box 5.B.

Appendix Table C (continued)

n	x	.05	.10	.15	.20	.25	π .30	.35	.40	.45	.50
8	0	.6634	.4305	.2725	.1678	.1001	.0576	.0319	.0168	.0084	.0039
	1	.2793	.3826	.3847	.3355	.2670	.1977	.1373	.0896	.0548	.0312
	2	.0515	.1488	.2376	.2936	.3115	.2965	.2587	.2090	.1569	.1094
	3	.0054	.0331	.0839	.1468	.2076	.2541	.2786	.2787	.2568	.2188
	4	.0004	.0046	.0185	.0459	.0865	.1361	.1875	.2322	.2627	.2734
	5	.0000	.0004	.0026	.0092	.0231	.0467	.0808	.1239	.1719	.2188
	6	.0000	.0000	.0002	.0011	.0038	.0100	.0217	.0413	.0703	.1094
	7	.0000	.0000	.0000	.0001	.0004	.0012	.0033	.0079	.0164	.0312
	8	.0000	.0000	.0000	.0000	.0000	.0001	.0002	.0007	.0017	.0039
9	0	.6302	.3874	.2316	.1342	.0751	.0404	.0207	.0101	.0046	.0020
	1	.2985	.3874	.3679	.3020	.2253	.1556	.1004	.0605	.0339	.0176
	2	.0629	.1722	.2597	.3020	.3003	.2668	.2162	.1612	.1110	.0703
	3	.0077	.0446	.1069	.1762	.2336	.2668	.2716	.2508	.2119	.1641
	4	.0006	.0074	.0283	.0661	.1168	.1715	.2194	.2508	.2600	.2461
	5	.0000	.0008	.0050	.0165	.0389	.0735	.1181	.1672	.2128	.2461
	6	.0000	.0001	.0006	.0028	.0087	.0210	.0424	.0743	.1160	.1641
	7	.0000	.0000	.0000	.0003	.0012	.0039	.0098	.0212	.0407	.0703
	8	.0000	.0000	.0000	.0000	.0001	.0004	.0013	.0035	.0083	.0176
	9	.0000	.0000	.0000	.0000	.0000	.0000	.0001	.0003	.0008	.0020
10	0	.5987	.3487	.1969	.1074	.0563	.0282	.0135	.0060	.0025	.0010
	1	.3151	.3874	.3474	.2684	.1877	.1211	.0725	.0403	.0207	.0098
	2	.0746	.1937	.2759	.3020	.2816	.2335	.1757	.1209	.0763	.0439
	3	.0105	.0574	.1298	.2013	.2503	.2668	.2522	.2150	.1665	.1172
	4	.0010	.0112	.0401	.0881	.1460	.2001	.2377	.2508	.2384	.2051
	5	.0001	.0015	.0085	.0264	.0584	.1029	.1536	.2007	.2340	.2461
	6	.0000	.0001	.0012	.0055	.0162	.0368	.0689	.1115	.1596	.2051
	7	.0000	.0000	.0001	.0008	.0031	.0090	.0212	.0425	.0746	.1172
	8	.0000	.0000	.0000	.0001	.0004	.0014	.0043	.0106	.0229	.0439
	9	.0000	.0000	.0000	.0000	.0000	.0001	.0005	.0016	.0042	.0098
	10	.0000	.0000	.0000	.0000	.0000	.0000	.0000	.0001	.0003	.0010
11	0	.5688	.3138	.1673	.0859	.0422	.0198	.0088	.0036	.0014	.0005
	1	.3293	.3835	.3248	.2362	.1549	.0932	.0518	.0266	.0125	.0054
	2	.0867	.2131	.2866	.2953	.2581	.1998	.1395	.0887	.0513	.0269
	3	.0137	.0710	.1517	.2215	.2581	.2568	.2254	.1774	.1259	.0806
	4	.0014	.0158	.0536	.1107	.1721	.2201	.2428	.2365	.2060	.1611
	5	.0001	.0025	.0132	.0388	.0803	.1321	.1830	.2207	.2360	.2256
	6	.0000	.0003	.0023	.0097	.0268	.0566	.0985	.1471	.1931	.2256
	7	.0000	.0000	.0003	.0017	.0064	.0173	.0379	.0701	.1128	.1611
	8	.0000	.0000	.0000	.0002	.0011	.0037	.0102	.0234	.0462	.0806
	9	.0000	.0000	.0000	.0000	.0001	.0005	.0018	.0052	.0126	.0269
	10	.0000	.0000	.0000	.0000	.0000	.0000	.0002	.0007	.0021	.0054
	11	.0000	.0000	.0000	.0000	.0000	.0000	.0000	.0000	.0002	.0005

Appendix Table C (continued)

						π					
n	x	.05	.10	.15	.20	.25	.30	.35	.40	.45	.50
12	0	.5404	.2824	.1422	.0687	.0317	.0138	.0057	.0022	.0008	.0002
	1	.3413	.3766	.3012	.2062	.1267	.0712	.0368	.0174	.0075	.0029
	2	.0988	.2301	.2924	.2835	.2323	.1678	.1088	.0639	.0339	.0161
	3	.0173	.0853	.1720	.2362	.2581	.2397	.1954	.1419	.0923	.0537
	4	.0021	.0213	.0683	.1329	.1936	.2311	.2367	.2128	.1700	.1208
	5	.0002	.0038	.0193	.0532	.1032	.1585	.2039	.2270	.2225	.1934
	6	.0000	.0005	.0040	.0155	.0401	.0792	.1281	.1766	.2124	.2256
	7	.0000	.0000	.0006	.0033	.0115	.0291	.0591	.1009	.1489	.1934
	8	.0000	.0000	.0001	.0005	.0024	.0078	.0199	.0420	.0762	.1208
	9	.0000	.0000	.0000	.0001	.0004	.0015	.0048	.0125	.0277	.0537
	10	.0000	.0000	.0000	.0000	.0000	.0002	.0008	.0025	.0068	.0161
	11	.0000	.0000	.0000	.0000	.0000	.0000	.0001	.0003	.0010	.0029
	12	.0000	.0000	.0000	.0000	.0000	.0000	.0000	.0000	.0001	.0002
13	0	.5133	.2542	.1209	.0550	.0238	.0097	.0037	.0013	.0004	.0001
	1	.3512	.3672	.2774	.1787	.1029	.0540	.0259	.0113	.0045	.0016
	2	.1109	.2448	.2937	.2680	.2059	.1388	.0836	.0453	.0220	.0095
	3	.0214	.0997	.1900	.2457	.2517	.2181	.1651	.1107	.0660	.0349
	4	.0028	.0277	.0838	.1535	.2097	.2337	.2222	.1845	.1350	.0873
	5	.0003	.0055	.0266	.0691	.1258	.1803	.2154	.2214	.1989	.1571
	6	.0000	.0008	.0063	.0230	.0559	.1030	.1546	.1968	.2169	.2095
	7	.0000	.0001	.0011	.0058	.0186	.0442	.0833	.1312	.1775	.2095
	8	.0000	.0000	.0001	.0011	.0047	.0142	.0336	.0656	.1089	.1571
	9	.0000	.0000	.0000	.0001	.0009	.0034	.0101	.0243	.0495	.0873
	10	.0000	.0000	.0000	.0000	.0001	.0006	.0022	.0065	.0162	.0349
	11	.0000	.0000	.0000	.0000	.0000	.0001	.0003	.0012	.0036	.0095
	12	.0000	.0000	.0000	.0000	.0000	.0000	.0000	.0001	.0005	.0016
	13	.0000	.0000	.0000	.0000	.0000	.0000	.0000	.0000	.0000	.0001
14	0	.4877	.2288	.1028	.0440	.0178	.0068	.0024	.0008	.0002	.0001
	1	.3593	.3559	.2539	.1539	.0832	.0407	.0181	.0073	.0027	.0009
	2	.1229	.2570	.2912	.2501	.1802	.1134	.0634	.0317	.0141	.0056
	3	.0259	.1142	.2056	.2501	.2402	.1943	.1366	.0845	.0462	.0222
	4	.0037	.0349	.0998	.1720	.2202	.2290	.2022	.1549	.1040	.0611
	5	.0004	.0078	.0352	.0860	.1468	.1963	.2178	.2066	.1701	.1222
	6	.0000	.0013	.0093	.0322	.0734	.1262	.1759	.2066	.2088	.1833
	7	.0000	.0002	.0019	.0092	.0280	.0618	.1082	.1574	.1952	.2095
	8	.0000	.0000	.0003	.0020	.0082	.0232	.0510	.0918	.1398	.1833
	9	.0000	.0000	.0000	.0003	.0018	.0066	.0183	.0408	.0762	.1222
	10	.0000	.0000	.0000	.0000	.0003	.0014	.0049	.0136	.0312	.0611
	11	.0000	.0000	.0000	.0000	.0000	.0002	.0010	.0033	.0093	.0222
	12	.0000	.0000	.0000	.0000	.0000	.0000	.0001	.0005	.0019	.0056
	13	.0000	.0000	.0000	.0000	.0000	.0000	.0000	.0001	.0002	.0009
	14	.0000	.0000	.0000	.0000	.0000	.0000	.0000	.0000	.0000	.0001

Appendix Table C (continued)

n	x	.05	.10	.15	.20	.25	.30	.35	.40	.45	.50
15	0	.4633	.2059	.0874	.0352	.0134	.0047	.0016	.0005	.0001	.0000
	1	.3658	.3432	.2312	.1319	.0668	.0305	.0126	.0047	.0016	.0005
	2	.1348	.2669	.2856	.2309	.1559	.0916	.0476	.0219	.0090	.0032
	3	.0307	.1285	.2184	.2501	.2252	.1700	.1110	.0634	.0318	.0139
	4	.0049	.0428	.1156	.1876	.2252	.2186	.1792	.1268	.0780	.0417
	5	.0006	.0105	.0449	.1032	.1651	.2061	.2123	.1859	.1404	.0916
	6	.0000	.0019	.0132	.0430	.0917	.1472	.1906	.2066	.1914	.1527
	7	.0000	.0003	.0030	.0138	.0393	.0811	.1319	.1771	.2013	.1964
	8	.0000	.0000	.0005	.0035	.0131	.0348	.0710	.1181	.1647	.1964
	9	.0000	.0000	.0001	.0007	.0034	.0116	.0298	.0612	.1048	.1527
	10	.0000	.0000	.0000	.0001	.0007	.0030	.0096	.0245	.0515	.0916
	11	.0000	.0000	.0000	.0000	.0001	.0006	.0024	.0074	.0191	.0417
	12	.0000	.0000	.0000	.0000	.0000	.0001	.0004	.0016	.0052	.0139
	13	.0000	.0000	.0000	.0000	.0000	.0000	.0001	.0003	.0010	.0032
	14	.0000	.0000	.0000	.0000	.0000	.0000	.0000	.0000	.0001	.0005
	15	.0000	.0000	.0000	.0000	.0000	.0000	.0000	.0000	.0000	.0000
16	0	.4401	.1853	.0743	.0281	.0100	.0033	.0010	.0003	.0001	.0000
	1	.3706	.3294	.2097	.1126	.0535	.0228	.0087	.0030	.0009	.0002
	2	.1463	.2745	.2775	.2111	.1336	.0732	.0353	.0150	.0056	.0018
	3	.0359	.1423	.2285	.2463	.2079	.1465	.0888	.0468	.0215	.0085
	4	.0061	.0514	.1311	.2001	.2252	.2040	.1553	.1014	.0572	.0278
	5	.0008	.0137	.0555	.1201	.1802	.2099	.2008	.1623	.1123	.0667
	6	.0001	.0028	.0180	.0550	.1101	.1649	.1982	.1983	.1684	.1222
	7	.0000	.0004	.0045	.0197	.0524	.1010	.1524	.1889	.1969	.1746
	8	.0000	.0001	.0009	.0055	.0197	.0487	.0923	.1417	.1812	.1964
	9	.0000	.0000	.0001	.0012	.0058	.0185	.0442	.0840	.1318	.1746
	10	.0000	.0000	.0000	.0002	.0014	.0056	.0167	.0392	.0755	.1222
	11	.0000	.0000	.0000	.0000	.0002	.0013	.0049	.0142	.0337	.0667
	12	.0000	.0000	.0000	.0000	.0000	.0002	.0011	.0040	.0115	.0278
	13	.0000	.0000	.0000	.0000	.0000	.0000	.0002	.0008	.0029	.0085
	14	.0000	.0000	.0000	.0000	.0000	.0000	.0000	.0001	.0005	.0018
	15	.0000	.0000	.0000	.0000	.0000	.0000	.0000	.0000	.0001	.0002
	16	.0000	.0000	.0000	.0000	.0000	.0000	.0000	.0000	.0000	.0000
17	0	.4181	.1668	.0631	.0225	.0075	.0023	.0007	.0002	.0000	.0000
	1	.3741	.3150	.1893	.0957	.0426	.0169	.0060	.0019	.0005	.0001
	2	.1575	.2800	.2673	.1914	.1136	.0581	.0260	.0102	.0035	.0010
	3	.0415	.1556	.2359	.2393	.1893	.1245	.0701	.0341	.0144	.0052
	4	.0076	.0605	.1457	.2093	.2209	.1868	.1320	.0796	.0411	.0182
	5	.0010	.0175	.0668	.1361	.1914	.2081	.1849	.1379	.0875	.0472
	6	.0001	.0039	.0236	.0680	.1276	.1784	.1991	.1839	.1432	.0944
	7	.0000	.0007	.0065	.0267	.0668	.1201	.1685	.1927	.1841	.1484
	8	.0000	.0001	.0014	.0084	.0279	.0644	.1134	.1606	.1883	.1855
	9	.0000	.0000	.0003	.0021	.0093	.0276	.0611	.1070	.1540	.1855
	10	.0000	.0000	.0000	.0004	.0025	.0095	.0263	.0571	.1008	.1484
	11	.0000	.0000	.0000	.0001	.0005	.0026	.0090	.0242	.0525	.0944
	12	.0000	.0000	.0000	.0000	.0001	.0006	.0024	.0081	.0215	.0472
	13	.0000	.0000	.0000	.0000	.0000	.0001	.0005	.0021	.0068	.0182
	14	.0000	.0000	.0000	.0000	.0000	.0000	.0001	.0004	.0016	.0052
	15	.0000	.0000	.0000	.0000	.0000	.0000	.0000	.0001	.0003	.0010
	16	.0000	.0000	.0000	.0000	.0000	.0000	.0000	.0000	.0000	.0001
	17	.0000	.0000	.0000	.0000	.0000	.0000	.0000	.0000	.0000	.0000

Appendix Table C (continued)

n	x	.05	.10	.15	.20	.25	π .30	.35	.40	.45	.50
18	0	.3972	.1501	.0536	.0180	.0056	.0016	.0004	.0001	.0000	.0000
	1	.3763	.3002	.1704	.0811	.0338	.0126	.0042	.0012	.0003	.0001
	2	.1683	.2835	.2556	.1723	.0958	.0458	.0190	.0069	.0022	.0006
	3	.0473	.1680	.2406	.2297	.1704	.1046	.0547	.0246	.0095	.0031
	4	.0093	.0700	.1592	.2153	.2130	.1681	.1104	.0614	.0291	.0117
	5	.0014	.0218	.0787	.1507	.1988	.2017	.1664	.1146	.0666	.0327
	6	.0002	.0052	.0301	.0816	.1436	.1873	.1941	.1655	.1181	.0708
	7	.0000	.0010	.0091	.0350	.0820	.1376	.1792	.1892	.1657	.1214
	8	.0000	.0002	.0022	.0120	.0376	.0811	.1327	.1734	.1864	.1669
	9	.0000	.0000	.0004	.0033	.0139	.0386	.0794	.1284	.1694	.1855
	10	.0000	.0000	.0001	.0008	.0042	.0149	.0385	.0771	.1248	.1669
	11	.0000	.0000	.0000	.0001	.0010	.0046	.0151	.0374	.0742	.1214
	12	.0000	.0000	.0000	.0000	.0002	.0012	.0047	.0145	.0354	.0708
	13	.0000	.0000	.0000	.0000	.0000	.0002	.0012	.0045	.0134	.0327
	14	.0000	.0000	.0000	.0000	.0000	.0000	.0002	.0011	.0039	.0117
	15	.0000	.0000	.0000	.0000	.0000	.0000	.0000	.0002	.0009	.0031
	16	.0000	.0000	.0000	.0000	.0000	.0000	.0000	.0000	.0001	.0006
	17	.0000	.0000	.0000	.0000	.0000	.0000	.0000	.0000	.0000	.0001
	18	.0000	.0000	.0000	.0000	.0000	.0000	.0000	.0000	.0000	.0000
19	0	.3774	.1351	.0456	.0144	.0042	.0011	.0003	.0001	.0000	.0000
	1	.3774	.2852	.1529	.0685	.0268	.0093	.0029	.0008	.0002	.0000
	2	.1787	.2852	.2428	.1540	.0803	.0358	.0138	.0046	.0013	.0003
	3	.0533	.1796	.2428	.2182	.1517	.0869	.0422	.0175	.0062	.0018
	4	.0112	.0798	.1714	.2182	.2023	.1491	.0909	.0467	.0203	.0074
	5	.0018	.0266	.0907	.1636	.2023	.1916	.1468	.0933	.0497	.0222
	6	.0002	.0069	.0374	.0955	.1574	.1916	.1844	.1451	.0949	.0518
	7	.0000	.0014	.0122	.0443	.0974	.1525	.1844	.1797	.1443	.0961
	8	.0000	.0002	.0032	.0166	.0487	.0981	.1489	.1797	.1771	.1442
	9	.0000	.0000	.0007	.0051	.0198	.0514	.0980	.1464	.1771	.1762
	10	.0000	.0000	.0001	.0013	.0066	.0220	.0528	.0976	.1449	.1762
	11	.0000	.0000	.0000	.0003	.0018	.0077	.0233	.0532	.0970	.1442
	12	.0000	.0000	.0000	.0000	.0004	.0022	.0083	.0237	.0529	.0961
	13	.0000	.0000	.0000	.0000	.0001	.0005	.0024	.0085	.0233	.0518
	14	.0000	.0000	.0000	.0000	.0000	.0001	.0006	.0024	.0082	.0222
	15	.0000	.0000	.0000	.0000	.0000	.0000	.0001	.0005	.0022	.0074
	16	.0000	.0000	.0000	.0000	.0000	.0000	.0000	.0001	.0005	.0018
	17	.0000	.0000	.0000	.0000	.0000	.0000	.0000	.0000	.0001	.0003
	18	.0000	.0000	.0000	.0000	.0000	.0000	.0000	.0000	.0000	.0000
	19	.0000	.0000	.0000	.0000	.0000	.0000	.0000	.0000	.0000	.0000
20	0	.3585	.1216	.0388	.0115	.0032	.0008	.0002	.0000	.0000	.0000
	1	.3774	.2702	.1368	.0576	.0211	.0068	.0020	.0005	.0001	.0000
	2	.1887	.2852	.2293	.1369	.0669	.0278	.0100	.0031	.0008	.0002
	3	.0596	.1901	.2428	.2054	.1339	.0716	.0323	.0123	.0040	.0011
	4	.0133	.0898	.1821	.2182	.1897	.1304	.0738	.0350	.0139	.0046
	5	.0022	.0319	.1028	.1746	.2023	.1789	.1272	.0746	.0365	.0148
	6	.0003	.0089	.0454	.1091	.1686	.1916	.1712	.1244	.0746	.0370
	7	.0000	.0020	.0160	.0545	.1124	.1643	.1844	.1659	.1221	.0739
	8	.0000	.0004	.0046	.0222	.0609	.1144	.1614	.1797	.1623	.1201
	9	.0000	.0001	.0011	.0074	.0271	.0654	.1158	.1597	.1771	.1602
	10	.0000	.0000	.0002	.0020	.0099	.0308	.0686	.1171	.1593	.1762
	11	.0000	.0000	.0000	.0005	.0030	.0120	.0336	.0710	.1185	.1602
	12	.0000	.0000	.0000	.0001	.0008	.0039	.0136	.0355	.0727	.1201
	13	.0000	.0000	.0000	.0000	.0002	.0010	.0045	.0146	.0366	.0739
	14	.0000	.0000	.0000	.0000	.0000	.0002	.0012	.0049	.0150	.0370
	15	.0000	.0000	.0000	.0000	.0000	.0000	.0003	.0013	.0049	.0148
	16	.0000	.0000	.0000	.0000	.0000	.0000	.0000	.0003	.0013	.0046
	17	.0000	.0000	.0000	.0000	.0000	.0000	.0000	.0000	.0002	.0011
	18	.0000	.0000	.0000	.0000	.0000	.0000	.0000	.0000	.0000	.0002
	19	.0000	.0000	.0000	.0000	.0000	.0000	.0000	.0000	.0000	.0000
	20	.0000	.0000	.0000	.0000	.0000	.0000	.0000	.0000	.0000	.0000

Appendix Table D
Binomial Probabilities for Cumulative Values of x

Entries in this table give the probability of *x* or fewer successes in *n* trials of a binomial experiment, where π is the probability of a success in one trial. For example, with *n* = 5 trials and π = .20, the probability of two or fewer successes is .9421. (Probability values in blank spaces equal 1.0000.)

n	*x*	.05	.10	.20	.30	.40	.50
1	0	0.9500	0.9000	0.8000	0.7000	0.6000	0.5000
	1	1.0000	1.0000	1.0000	1.0000	1.0000	1.0000
2	0	0.9025	0.8100	0.6400	0.4900	0.3600	0.2500
	1	0.9975	0.9900	0.9600	0.9100	0.8400	0.7500
	2	1.0000	1.0000	1.0000	1.0000	1.0000	1.0000
3	0	0.8574	0.7290	0.5120	0.3430	0.2160	0.1250
	1	0.9927	0.9720	0.8960	0.7840	0.6480	0.5000
	2	0.9999	0.9990	0.9920	0.9730	0.9360	0.8750
	3	1.0000	1.0000	1.0000	1.0000	1.0000	1.0000
4	0	0.8145	0.6561	0.4096	0.2401	0.1296	0.0625
	1	0.9860	0.9477	0.8192	0.6517	0.4752	0.3125
	2	0.9995	0.9963	0.9728	0.9163	0.8208	0.6875
	3	1.0000	0.9999	0.9984	0.9919	0.9744	0.9375
	4		1.0000	1.0000	1.0000	1.0000	1.0000
5	0	0.7738	0.5905	0.3277	0.1681	0.0778	0.0313
	1	0.9774	0.9185	0.7373	0.5282	0.3370	0.1875
	2	0.9988	0.9914	0.9421	0.8369	0.6826	0.5000
	3	1.0000	0.9995	0.9933	0.9692	0.9130	0.8125
	4		1.0000	0.9997	0.9976	0.9898	0.9688
	5			1.0000	1.0000	1.0000	1.0000
6	0	0.7351	0.5314	0.2621	0.1176	0.0467	0.0156
	1	0.9672	0.8857	0.6554	0.4202	0.2333	0.1094
	2	0.9978	0.9841	0.9011	0.7443	0.5443	0.3438
	3	0.9999	0.9987	0.9830	0.9295	0.8208	0.6563
	4	1.0000	0.9999	0.9984	0.9891	0.9590	0.8906
	5		1.0000	0.9999	0.9993	0.9959	0.9844
	6			1.0000	1.0000	1.0000	1.0000
7	0	0.6983	0.4783	0.2097	0.0824	0.0280	0.0078
	1	0.9556	0.8503	0.5767	0.3294	0.1586	0.0625
	2	0.9962	0.9743	0.8520	0.6471	0.4199	0.2266
	3	0.9998	0.9973	0.9667	0.8740	0.7102	0.5000
	4	1.0000	0.9998	0.9953	0.9712	0.9037	0.7734
	5		1.0000	0.9996	0.9962	0.9812	0.9375
	6			1.0000	0.9998	0.9984	0.9922
	7				1.0000	1.0000	1.0000
8	0	0.6634	0.4305	0.1678	0.0576	0.0168	0.0039
	1	0.9428	0.8131	0.5033	0.2553	0.1064	0.0352
	2	0.9942	0.9619	0.7969	0.5518	0.3154	0.1445
	3	0.9996	0.9950	0.9437	0.8059	0.5941	0.3633
	4	1.0000	0.9996	0.9896	0.9420	0.8263	0.6367
	5		1.0000	0.9988	0.9887	0.9502	0.8555
	6			0.9999	0.9987	0.9915	0.9648
	7			1.0000	0.9999	0.9993	0.9961
	8				1.0000	1.0000	1.0000

Source: For values of *n* up to 20, the National Bureau of Standards, *Tables of the Binomial Probability Distribution*, Applied Mathematics Series, No. 6 (Washington D.C.: U.S. Government Printing Office, 1949); for *n* = 50 and *n* = 100, Lawrence L. Lapin, *Statistics for Modern Business Decisions*, 3rd ed. (New York: Harcourt, Brace, Jovanovich, 1982), pp. 830–32.

Appendix Table D (continued)

n	x	.05	.10	.20	π .30	.40	.50
9	0	0.6302	0.3874	0.1342	0.0404	0.0101	0.0020
	1	0.9288	0.7748	0.4362	0.1960	0.0705	0.0195
	2	0.9916	0.9470	0.7382	0.4628	0.2318	0.0898
	3	0.9994	0.9917	0.9144	0.7297	0.4826	0.2539
	4	1.0000	0.9991	0.9804	0.9012	0.7334	0.5000
	5		0.9999	0.9969	0.9747	0.9006	0.7461
	6		1.0000	0.9997	0.9957	0.9750	0.9102
	7			1.0000	0.9996	0.9962	0.9805
	8				1.0000	0.9997	0.9980
	9					1.0000	1.0000
10	0	0.5987	0.3487	0.1074	0.0282	0.0060	0.0010
	1	0.9139	0.7361	0.3758	0.1493	0.0464	0.0107
	2	0.9885	0.9298	0.6778	0.3828	0.1673	0.0547
	3	0.9990	0.9872	0.8791	0.6496	0.3823	0.1719
	4	0.9999	0.9984	0.9672	0.8497	0.6331	0.3770
	5	1.0000	0.9999	0.9936	0.9526	0.8338	0.6230
	6		1.0000	0.9991	0.9894	0.9452	0.8281
	7			0.9999	0.9999	0.9877	0.9453
	8			1.0000	1.0000	0.9983	0.9893
	9					0.9999	0.9990
	10					1.0000	1.0000
11	0	0.5688	0.3138	0.0859	0.0198	0.0036	0.0005
	1	0.8981	0.6974	0.3221	0.1130	0.0302	0.0059
	2	0.9848	0.9104	0.6174	0.3127	0.1189	0.0327
	3	0.9984	0.9815	0.8369	0.5696	0.2963	0.1133
	4	0.9999	0.9972	0.9496	0.7897	0.5328	0.2744
	5	1.0000	0.9997	0.9883	0.9218	0.7535	0.5000
	6		1.0000	0.9980	0.9784	0.9006	0.7256
	7			0.9998	0.9957	0.9707	0.8867
	8			1.0000	0.9994	0.9941	0.9673
	9				1.0000	0.9993	0.9941
	10					1.0000	0.9995
	11						1.0000
12	0	0.5404	0.2824	0.0687	0.0138	0.0022	0.0002
	1	0.8816	0.6590	0.2749	0.0850	0.0196	0.0032
	2	0.9804	0.8891	0.5583	0.2528	0.0834	0.0193
	3	0.9978	0.9744	0.7946	0.4925	0.2253	0.0730
	4	0.9998	0.9957	0.9274	0.7237	0.4382	0.1938
	5	1.0000	0.9995	0.9806	0.8821	0.6652	0.3872
	6		0.9999	0.9961	0.9614	0.8418	0.6128
	7		1.0000	0.9994	0.9905	0.9427	0.8062
	8			0.9999	0.9983	0.9847	0.9270
	9			1.0000	0.9998	0.9972	0.9807
	10				1.0000	0.9997	0.9968
	11					1.0000	0.9998
	12						1.0000
13	0	0.5133	0.2542	0.0550	0.0097	0.0013	0.0001
	1	0.8646	0.6213	0.2336	0.0637	0.0126	0.0017
	2	0.9755	0.8661	0.5017	0.2025	0.0579	0.0112
	3	0.9969	0.9658	0.7473	0.4206	0.1686	0.0461
	4	0.9997	0.9935	0.9009	0.6543	0.3530	0.1334
	5	1.0000	0.9991	0.9700	0.8346	0.5744	0.2905
	6		0.9999	0.9930	0.9376	0.7712	0.5000
	7		1.0000	0.9988	0.9818	0.9023	0.7095
	8			0.9998	0.9960	0.9679	0.8666
	9			1.0000	0.9993	0.9922	0.9539
	10				0.9999	0.9987	0.9888
	11				1.0000	0.9999	0.9983
	12					1.0000	0.9999
	13						1.0000

Appendix Table D (continued)

n	x	.05	.10	.20	.30	.40	.50
14	0	0.4877	0.2288	0.0440	0.0068	0.0008	0.0001
	1	0.8470	0.5846	0.1979	0.0475	0.0081	0.0009
	2	0.9699	0.8416	0.4481	0.1608	0.0398	0.0065
	3	0.9958	0.9559	0.6982	0.3552	0.1243	0.0287
	4	0.9996	0.9908	0.8702	0.5842	0.2793	0.0898
	5	1.0000	0.9985	0.9561	0.7805	0.4859	0.2120
	6		0.9998	0.9884	0.9067	0.6925	0.3953
	7		1.0000	0.9976	0.9685	0.8499	0.6047
	8			0.9996	0.9917	0.9417	0.7880
	9			1.0000	0.9983	0.9825	0.9102
	10				0.9998	0.9961	0.9713
	11				1.0000	0.9994	0.9935
	12					0.9999	0.9991
	13					1.0000	0.9999
	14						1.0000
15	0	0.4633	0.2059	0.0352	0.0047	0.0005	0.0000
	1	0.8290	0.5490	0.1671	0.0353	0.0052	0.0005
	2	0.9638	0.8159	0.3980	0.1268	0.0271	0.0037
	3	0.9945	0.9444	0.6482	0.2969	0.0905	0.0176
	4	0.9994	0.9873	0.8358	0.5155	0.2173	0.0592
	5	0.9999	0.9978	0.9389	0.7216	0.4032	0.1509
	6	1.0000	0.9997	0.9819	0.8689	0.6098	0.3036
	7		1.0000	0.9958	0.9500	0.7869	0.5000
	8			0.9992	0.9848	0.9050	0.6964
	9			0.9999	0.9963	0.9662	0.8491
	10			1.0000	0.9993	0.9907	0.9408
	11				0.9999	0.9981	0.9824
	12				1.0000	0.9997	0.9963
	13					1.0000	0.9995
	14						1.0000
16	0	0.4401	0.1853	0.0281	0.0033	0.0003	0.0000
	1	0.8108	0.5147	0.1407	0.0261	0.0033	0.0003
	2	0.9571	0.7892	0.3518	0.0994	0.0183	0.0021
	3	0.9930	0.9316	0.5981	0.2459	0.0651	0.0106
	4	0.9991	0.9830	0.7982	0.4499	0.1666	0.0384
	5	0.9999	0.9967	0.9183	0.6598	0.3288	0.1051
	6	1.0000	0.9995	0.9733	0.8247	0.5272	0.2272
	7		0.9999	0.9930	0.9256	0.7161	0.4018
	8		1.0000	0.9985	0.9743	0.8577	0.5982
	9			0.9998	0.9929	0.9417	0.7728
	10			1.0000	0.9984	0.9809	0.8949
	11				0.9997	0.9951	0.9616
	12				1.0000	0.9991	0.9894
	13					0.9999	0.9979
	14					1.0000	0.9997
	15						1.0000
17	0	0.4181	0.1668	0.0225	0.0023	0.0002	0.0000
	1	0.7922	0.4818	0.1182	0.0193	0.0021	0.0001
	2	0.9497	0.7618	0.3096	0.0774	0.0123	0.0012
	3	0.9912	0.9174	0.5489	0.2019	0.0464	0.0064
	4	0.9988	0.9779	0.7582	0.3887	0.1260	0.0245
	5	0.9999	0.9953	0.8943	0.5968	0.2639	0.0717
	6	1.0000	0.9992	0.9623	0.7752	0.4478	0.1662
	7		0.9999	0.9891	0.8954	0.6405	0.3145
	8		1.0000	0.9974	0.9597	0.8011	0.5000
	9			0.9995	0.9873	0.9081	0.6855
	10			0.9999	0.9968	0.9652	0.8338
	11			1.0000	0.9993	0.9894	0.9283
	12				0.9999	0.9975	0.9755
	13				1.0000	0.9995	0.9936
	14					0.9999	0.9988
	15					1.0000	0.9999
	16						1.0000

Appendix Table D (continued)

n	x	.05	.10	.20	π .30	.40	.50
18	0	0.3972	0.1501	0.0180	0.0016	0.0001	0.0000
	1	0.7735	0.4503	0.0991	0.0142	0.0013	0.0001
	2	0.9419	0.7338	0.2713	0.0600	0.0082	0.0007
	3	0.9891	0.9018	0.5010	0.1646	0.0328	0.0038
	4	0.9985	0.9718	0.7164	0.3327	0.0942	0.0154
	5	0.9998	0.9936	0.8671	0.5344	0.2088	0.0481
	6	1.0000	0.9988	0.9487	0.7217	0.3743	0.1189
	7		0.9998	0.9837	0.8593	0.5634	0.2403
	8		1.0000	0.9957	0.9404	0.7368	0.4073
	9			0.9991	0.9790	0.8653	0.5927
	10			0.9998	0.9939	0.9424	0.7597
	11			1.0000	0.9986	0.9797	0.8811
	12				0.9997	0.9942	0.9519
	13				1.0000	0.9987	0.9846
	14					0.9998	0.9962
	15					1.0000	0.9993
	16						0.9999
	17						1.0000
19	0	0.3774	0.1351	0.0144	0.0011	0.0001	0.0000
	1	0.7547	0.4203	0.0829	0.0104	0.0008	0.0000
	2	0.9335	0.7054	0.2369	0.0462	0.0055	0.0004
	3	0.9868	0.8850	0.4551	0.1332	0.0230	0.0022
	4	0.9980	0.9648	0.6733	0.2822	0.0696	0.0096
	5	0.9998	0.9914	0.8369	0.4739	0.1629	0.0318
	6	1.0000	0.9983	0.9324	0.6655	0.3081	0.0835
	7		0.9997	0.9767	0.8180	0.4878	0.1796
	8		1.0000	0.9933	0.9161	0.6675	0.3238
	9			0.9984	0.9674	0.8139	0.5000
	10			0.9997	0.9895	0.9115	0.6762
	11			0.9999	0.9972	0.9648	0.8204
	12			1.0000	0.9994	0.9884	0.9165
	13				0.9999	0.9969	0.9682
	14				1.0000	0.9994	0.9904
	15					0.9999	0.9978
	16					1.0000	0.9996
	17						1.0000
20	0	0.3585	0.1216	0.0115	0.0008	0.0000	0.0000
	1	0.7358	0.3917	0.0692	0.0076	0.0005	0.0000
	2	0.9245	0.6769	0.2061	0.0355	0.0036	0.0002
	3	0.9841	0.8670	0.4114	0.1071	0.0160	0.0013
	4	0.9974	0.9568	0.6296	0.2375	0.0510	0.0059
	5	0.9997	0.9887	0.8042	0.4164	0.1256	0.0207
	6	1.0000	0.9976	0.9133	0.6080	0.2500	0.0577
	7		0.9996	0.9679	0.7723	0.4159	0.1316
	8		0.9999	0.9900	0.8867	0.5956	0.2517
	9		1.0000	0.9974	0.9520	0.7553	0.4119
	10			0.9994	0.9829	0.8725	0.5881
	11			0.9999	0.9949	0.9435	0.7483
	12			1.0000	0.9987	0.9790	0.8684
	13				0.9997	0.9935	0.9423
	14				1.0000	0.9984	0.9793
	15					0.9997	0.9941
	16					1.0000	0.9987
	17						0.9998
	18						1.0000

Appendix Table D (continued)

n	x	.05	.10	.20	π .30	.40	.50
50	0	0.0769	0.0052	0.0000	0.0000	0.0000	0.0000
	1	0.2794	0.0338	0.0002	0.0000	0.0000	0.0000
	2	0.5405	0.1117	0.0013	0.0000	0.0000	0.0000
	3	0.7604	0.2503	0.0057	0.0000	0.0000	0.0000
	4	0.8964	0.4312	0.0185	0.0002	0.0000	0.0000
	5	0.9622	0.6161	0.0480	0.0007	0.0000	0.0000
	6	0.9882	0.7702	0.1034	0.0025	0.0000	0.0000
	7	0.9968	0.8779	0.1904	0.0073	0.0001	0.0000
	8	0.9992	0.9421	0.3073	0.0183	0.0002	0.0000
	9	0.9998	0.9755	0.4437	0.0402	0.0008	0.0000
	10	1.0000	0.9906	0.5836	0.0789	0.0022	0.0000
	11		0.9968	0.7107	0.1390	0.0057	0.0000
	12		0.9990	0.8139	0.2229	0.0133	0.0002
	13		0.9997	0.8894	0.3279	0.0280	0.0005
	14		0.9999	0.9393	0.4468	0.0540	0.0013
	15		1.0000	0.9692	0.5692	0.0955	0.0033
	16			0.9856	0.6839	0.1561	0.0077
	17			0.9937	0.7822	0.2369	0.0164
	18			0.9975	0.8594	0.3356	0.0325
	19			0.9991	0.9152	0.4465	0.0595
	20			0.9997	0.9522	0.5610	0.1013
	21			0.9999	0.9749	0.6701	0.1611
	22			1.0000	0.9877	0.7660	0.2399
	23				0.9944	0.8438	0.3359
	24				0.9976	0.9022	0.4439
	25				0.9991	0.9427	0.5561
	26				0.9997	0.9686	0.6641
	27				0.9999	0.9840	0.7601
	28				1.0000	0.9924	0.8389
	29					0.9966	0.8987
	30					0.9986	0.9405
	31					0.9995	0.9675
	32					0.9998	0.9836
	33					0.9999	0.9923
	34					1.0000	0.9967
	35						0.9987
	36						0.9995
	37						0.9998
	38						1.0000

Appendix Table D (continued)

n	x	.05	.10	.20	π .30	.40	.50	n	x	.20	.30	π .40	.50
100	0	0.0059	0.0000	0.0000	0.0000	0.0000	0.0000	100	36	0.9999	0.9201	0.2386	0.0033
	1	0.0371	0.0003	0.0000	0.0000	0.0000	0.0000		37	1.0000	0.9470	0.3068	0.0060
	2	0.1183	0.0019	0.0000	0.0000	0.0000	0.0000		38		0.9660	0.3822	0.0105
	3	0.2578	0.0078	0.0000	0.0000	0.0000	0.0000		39		0.9790	0.4621	0.0176
	4	0.4360	0.0237	0.0000	0.0000	0.0000	0.0000		40		0.9875	0.5433	0.0284
	5	0.6160	0.0576	0.0000	0.0000	0.0000	0.0000						
	6	0.7660	0.1172	0.0001	0.0000	0.0000	0.0000		41		0.9928	0.6225	0.0443
	7	0.8720	0.2061	0.0003	0.0000	0.0000	0.0000		42		0.9960	0.6967	0.0666
	8	0.9369	0.3209	0.0009	0.0000	0.0000	0.0000		43		0.9979	0.7635	0.0967
	9	0.9718	0.4513	0.0023	0.0000	0.0000	0.0000		44		0.9989	0.8211	0.1356
	10	0.9885	0.5832	0.0057	0.0000	0.0000	0.0000		45		0.9995	0.8689	0.1841
	11	0.9957	0.7030	0.0126	0.0000	0.0000	0.0000		46		0.9997	0.9070	0.2421
	12	0.9985	0.8018	0.0253	0.0000	0.0000	0.0000		47		0.9999	0.9362	0.3086
	13	0.9995	0.8761	0.0469	0.0001	0.0000	0.0000		48		0.9999	0.9577	0.3822
	14	0.9999	0.9274	0.0804	0.0002	0.0000	0.0000		49		1.0000	0.9729	0.4602
	15	1.0000	0.9601	0.1285	0.0004	0.0000	0.0000		50			0.9832	0.5398
	16		0.9794	0.1923	0.0010	0.0000	0.0000		51			0.9900	0.6178
	17		0.9900	0.2712	0.0022	0.0000	0.0000		52			0.9942	0.6914
	18		0.9954	0.3621	0.0045	0.0000	0.0000		53			0.9968	0.7579
	19		0.9980	0.4602	0.0089	0.0000	0.0000		54			0.9983	0.8159
	20		0.9992	0.5595	0.0165	0.0000	0.0000		55			0.9991	0.8644
	21		0.9997	0.6540	0.0288	0.0000	0.0000		56			0.9996	0.9033
	22		0.9999	0.7389	0.0479	0.0001	0.0000		57			0.9998	0.9334
	23		1.0000	0.8109	0.0755	0.0003	0.0000		58			0.9999	0.9557
	24			0.8686	0.1136	0.0006	0.0000		59			1.0000	0.9716
	25			0.9125	0.1631	0.0012	0.0000		60				0.9824
	26			0.9442	0.2244	0.0024	0.0000		61				0.9895
	27			0.9658	0.2964	0.0046	0.0000		62				0.9940
	28			0.9800	0.3768	0.0084	0.0000		63				0.9967
	29			0.9888	0.4623	0.0148	0.0000		64				0.9982
	30			0.9939	0.5491	0.0248	0.0000		65				0.9991
	31			0.9969	0.6331	0.0398	0.0001		66				0.9996
	32			0.9984	0.7107	0.0615	0.0002		67				0.9998
	33			0.9993	0.7793	0.0913	0.0004		68				0.9999
	34			0.9997	0.8371	0.1303	0.0009		69				1.0000
	35			0.9999	0.8839	0.1795	0.0018						

Appendix Table E
Standard Normal Curve Areas

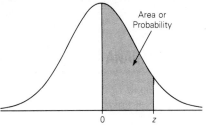

Area or
Probability

Entries in this table give the area under the curve between the mean and z standard deviations above the mean. For example, for $z = 2.25$, the area under the curve between the mean and z is .4878

z	.00	.01	.02	.03	.04	.05	.06	.07	.08	.09
.0	.0000	.0040	.0080	.0120	.0160	.0199	.0239	.0279	.0319	.0359
.1	.0398	.0438	.0478	.0517	.0557	.0596	.0636	.0675	.0714	.0753
.2	.0793	.0832	.0871	.0910	.0948	.0987	.1026	.1064	.1103	.1141
.3	.1179	.1217	.1255	.1293	.1331	.1368	.1406	.1443	.1480	.1517
.4	.1554	.1591	.1628	.1664	.1700	.1736	.1772	.1808	.1844	.1879
.5	.1915	.1950	.1985	.2019	.2054	.2088	.2123	.2157	.2190	.2224
.6	.2257	.2291	.2324	.2357	.2389	.2422	.2454	.2486	.2518	.2549
.7	.2580	.2612	.2642	.2673	.2704	.2734	.2764	.2794	.2823	.2852
.8	.2881	.2910	.2939	.2967	.2995	.3023	.3051	.3078	.3106	.3133
.9	.3159	.3186	.3212	.3238	.3264	.3289	.3315	.3340	.3365	.3389
1.0	.3413	.3438	.3461	.3485	.3508	.3531	.3554	.3577	.3599	.3621
1.1	.3643	.3665	.3686	.3708	.3729	.3749	.3770	.3790	.3810	.3830
1.2	.3849	.3869	.3888	.3907	.3925	.3944	.3962	.3980	.3997	.4015
1.3	.4032	.4049	.4066	.4082	.4099	.4115	.4131	.4147	.4162	.4177
1.4	.4192	.4207	.4222	.4236	.4251	.4265	.4279	.4292	.4306	.4319
1.5	.4332	.4345	.4357	.4370	.4382	.4394	.4406	.4418	.4429	.4441
1.6	.4452	.4463	.4474	.4484	.4495	.4505	.4515	.4525	.4535	.4545
1.7	.4554	.4564	.4573	.4582	.4591	.4599	.4608	.4616	.4625	.4633
1.8	.4641	.4649	.4656	.4664	.4671	.4678	.4686	.4693	.4699	.4706
1.9	.4713	.4719	.4726	.4732	.4738	.4744	.4750	.4756	.4761	.4767
2.0	.4772	.4778	.4783	.4788	.4793	.4798	.4803	.4808	.4812	.4817
2.1	.4821	.4826	.4830	.4834	.4838	.4842	.4846	.4850	.4854	.4857
2.2	.4861	.4864	.4868	.4871	.4875	.4878	.4881	.4884	.4887	.4890
2.3	.4893	.4896	.4898	.4901	.4904	.4906	.4909	.4911	.4913	.4916
2.4	.4918	.4920	.4922	.4925	.4927	.4929	.4931	.4932	.4934	.4936
2.5	.4938	.4940	.4941	.4943	.4945	.4946	.4948	.4949	.4951	.4952
2.6	.4953	.4955	.4956	.4957	.4959	.4960	.4961	.4962	.4963	.4964
2.7	.4965	.4966	.4967	.4968	.4969	.4970	.4971	.4972	.4973	.4974
2.8	.4974	.4975	.4976	.4977	.4977	.4978	.4979	.4979	.4980	.4981
2.9	.4981	.4982	.4982	.4983	.4984	.4984	.4985	.4985	.4986	.4986
3.0	.4986	.4987	.4987	.4988	.4988	.4989	.4989	.4989	.4990	.4990
3.1	.4990	.4991	.4991	.4991	.4992	.4992	.4992	.4992	.4993	.4993
3.2	.4993	.4993	.4994	.4994	.4994	.4994	.4994	.4995	.4995	.4995
3.3	.4995	.4995	.4995	.4996	.4996	.4996	.4996	.4996	.4996	.4997
3.4	.4997	.4997	.4997	.4997	.4997	.4997	.4997	.4997	.4997	.4998
3.6	.4998	.4998	.4999	.4999	.4999	.4999	.4999	.4999	.4999	.4999
3.9	.5000									

Source: The National Bureau of Standards, *Tables of Normal Probability Functions,* Applied Mathematics Series, no. 23 (Washington, D.C.: U.S. Government Printing Office, 1953). The original contains probabilities for values of z from 0 to 8.285, mostly in increments of .0001, and for areas from $\mu - z$ to $\mu + z$.

Appendix Table F
Critical Normal Deviate Values

| (A) For Statistical Estimation | | (B) For Hypothesis Testing | | | |
| | | Two-Tailed Tests | | One-Tailed Tests | |
Confidence Level, C	Normal Deviate, z	Significance Level, α	Normal Deviate, $z_{\alpha/2}$	Significance Level, α	Normal Deviate, z_α
.80	1.2817	.10	1.645	.10	1.2817
.90	1.645	.05	1.96	.05	1.645
.95	1.96	.025	2.24	.025	1.96
.98	2.3267	.01	2.575	.01	2.3267
.99	2.575	.005	2.81	.005	2.575
.998	3.08	.001	3.27	.001	3.08
.999	3.27				

Source: The National Bureau of Standards, *Tables of Normal Probability Functions,* Applied Mathematics Series, No. 23 (Washington, D.C.: U.S. Government Printing Office, 1953).

Appendix Table G
Student *t* Distributions

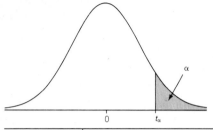

The following table provides the values of t_α that correspond to a given upper-tail area α and a specified number of degrees of freedom. For example, for an upper-tail area of .05 and 9 degrees of freedom, the critical value of $t_\alpha = 1.833$.

Degrees of Freedom	Critical tail areas (= α for one-tailed tests, = $\alpha/2$ for two-tailed tests)									
	.4	.25	.1	.05	.025	.01	.005	.0025	.001	.0005
1	0.325	1.000	3.078	6.314	12.706	31.821	63.657	127.32	318.31	636.62
2	.289	.816	1.886	2.920	4.303	6.965	9.925	14.089	22.327	31.598
3	.277	.765	1.638	2.353	3.182	4.541	5.841	7.453	10.214	12.924
4	.271	.741	1.533	2.132	2.776	3.747	4.604	5.598	7.173	8.610
5	0.267	0.727	1.476	2.015	2.571	3.365	4.032	4.773	5.893	6.869
6	.265	.718	1.440	1.943	2.447	3.143	3.707	4.317	5.208	5.959
7	.263	.711	1.415	1.895	2.365	2.998	3.499	4.029	4.785	5.408
8	.262	.706	1.397	1.860	2.306	2.896	3.355	3.833	4.501	5.041
9	.261	.703	1.383	1.833	2.262	2.821	3.250	3.690	4.297	4.781
10	0.260	0.700	1.372	1.812	2.228	2.764	3.169	3.581	4.144	4.587
11	.260	.697	1.363	1.796	2.201	2.718	3.106	3.497	4.025	4.437
12	.259	.695	1.356	1.782	2.179	2.681	3.055	3.428	3.930	4.318
13	.259	.694	1.350	1.771	2.160	2.650	3.012	3.372	3.852	4.221
14	.258	.692	1.345	1.761	2.145	2.624	2.977	3.326	3.787	4.140
15	0.258	0.691	1.341	1.753	2.131	2.602	2.947	3.286	3.733	4.073
16	.258	.690	1.337	1.746	2.120	2.583	2.921	3.252	3.686	4.015
17	.257	.689	1.333	1.740	2.110	2.567	2.898	3.222	3.646	3.965
18	.257	.688	1.330	1.734	2.101	2.552	2.878	3.197	3.610	3.922
19	.257	.688	1.328	1.729	2.093	2.539	2.861	3.174	3.579	3.883
20	0.257	0.687	1.325	1.725	2.086	2.528	2.845	3.153	3.552	3.850
21	.257	.686	1.323	1.721	2.080	2.518	2.831	3.135	3.527	3.819
22	.256	.686	1.321	1.717	2.074	2.508	2.819	3.119	3.505	3.792
23	.256	.685	1.319	1.714	2.069	2.500	2.807	3.104	3.485	3.767
24	.256	.685	1.318	1.711	2.064	2.492	2.797	3.091	3.467	3.745
25	0.256	0.684	1.316	1.708	2.060	2.485	2.787	3.078	3.450	3.725
26	.256	.684	1.315	1.706	2.056	2.479	2.779	3.067	3.435	3.707
27	.256	.684	1.314	1.703	2.052	2.473	2.771	3.057	3.421	3.690
28	.256	.683	1.313	1.701	2.048	2.467	2.763	3.047	3.408	3.674
29	.256	.683	1.311	1.699	2.045	2.462	2.756	3.038	3.396	3.659
30	0.256	0.683	1.310	1.697	2.042	2.457	2.750	3.030	3.385	3.646
40	.255	.681	1.303	1.684	2.021	2.423	2.704	2.971	3.307	3.551
60	.254	.679	1.296	1.671	2.000	2.390	2.660	2.915	3.232	3.460
120	.254	.677	1.289	1.658	1.980	2.358	2.617	2.860	3.160	3.373
∞	.253	.674	1.282	1.645	1.960	2.326	2.576	2.807	3.090	3.291

Confidence Level C = 1 − α										
one-tailed test	.60	.75	.90	.95	.975	.99	.995	.9975	.999	.9995
two-tailed test	.20	.50	.80	.90	.95	.98	.99	.995	.998	.999

Note: A significance level α is associated with, for example, an upper-tail area α in an upper-tailed test, but with an upper- *as well as* lower-tail area $\alpha/2$ in a two-tailed test. The confidence level always equals C = 1 − α. Therefore, a significance level of, say, α = .05, goes with an upper-tail area of .05 in an upper-tailed test (C = .95), but with an upper-tail area of .025 in a two-tailed test (C = .95).

Source: E. S. Pearson and H. O. Hartley, *Biometrika Tables for Statisticians*, vol. I. (Cambridge: Cambridge University Press, 1966), p. 146.

Appendix Table H
Chi-Square Distributions

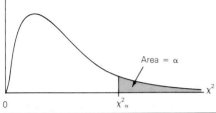

This table provides values of χ_α^2 that correspond to a given upper-tail area, α, and a specified number of degrees of freedom. For example, for an upper-tail area of .05 and 4 degrees of freedom, the critical value of $\chi_{.05}^2$ equals 9.488.

Degrees of Freedom	Critical Values for Upper-Tail Area, α													
	.99	.98	.95	.90	.80	.70	.50	.30	.20	.10	.05	.02	.01	.001
1	$.0^3157$	$.0^3628$	.00393	.0158	.0642	.148	.455	1.074	1.642	2.706	3.841	5.412	6.635	10.827
2	.0201	.0404	.103	.211	.446	.713	1.386	2.408	3.219	4.605	5.991	7.824	9.210	13.815
3	.115	.185	.352	.584	1.005	1.424	2.366	3.665	4.642	6.251	7.815	9.837	11.345	16.266
4	.297	.429	.711	1.064	1.649	2.195	3.357	4.878	5.989	7.779	9.488	11.668	13.277	18.467
5	.554	.752	1.145	1.610	2.343	3.000	4.351	6.064	7.289	9.236	11.070	13.388	15.086	20.515
6	.872	1.134	1.635	2.204	3.070	3.828	5.348	7.231	8.558	10.645	12.592	15.033	16.812	22.457
7	1.239	1.564	2.167	2.833	3.822	4.671	6.346	8.383	9.803	12.017	14.067	16.622	18.475	24.322
8	1.646	2.032	2.733	3.490	4.594	5.527	7.344	9.524	11.030	13.362	15.507	18.168	20.090	26.125
9	2.088	2.532	3.325	4.168	5.380	6.393	8.343	10.656	12.242	14.684	16.919	19.679	21.666	27.877
10	2.558	3.059	3.940	4.865	6.179	7.267	9.342	11.781	13.442	15.987	18.307	21.161	23.209	29.588
11	3.053	3.609	4.575	5.578	6.989	8.148	10.341	12.899	14.631	17.275	19.675	22.618	24.725	31.264
12	3.571	4.178	5.226	6.304	7.807	9.034	11.340	14.011	15.812	18.549	21.026	24.054	26.217	32.909
13	4.107	4.765	5.892	7.042	8.634	9.926	12.340	15.119	16.985	19.812	22.362	25.472	27.688	34.528
14	4.660	5.368	6.571	7.790	9.467	10.821	13.339	16.222	18.151	21.064	23.685	26.873	29.141	36.123
15	5.229	5.985	7.261	8.547	10.307	11.721	14.339	17.322	19.311	22.307	24.996	28.259	30.578	37.697
16	5.812	6.614	7.962	9.312	11.152	12.624	15.338	18.418	20.465	23.542	26.296	29.633	32.000	39.252
17	6.408	7.255	8.672	10.085	12.002	13.531	16.338	19.511	21.615	24.769	27.587	30.995	33.409	40.790
18	7.015	7.906	9.390	10.865	12.857	14.440	17.338	20.601	22.760	25.989	28.869	32.346	34.805	42.312
19	7.633	8.567	10.117	11.651	13.716	15.352	18.338	21.689	23.900	27.204	30.144	33.687	36.191	43.820
20	8.260	9.237	10.851	12.443	14.578	16.266	19.337	22.775	25.038	28.412	31.410	35.020	37.566	45.315
21	8.897	9.915	11.591	13.240	15.445	17.182	20.337	23.858	26.171	29.615	32.671	36.343	38.932	46.797
22	9.542	10.600	12.338	14.041	16.314	18.101	21.337	24.939	27.301	30.813	33.924	37.659	40.289	48.268
23	10.196	11.293	13.091	14.848	17.187	19.021	22.337	26.018	28.429	32.007	35.172	38.968	41.638	49.728
24	10.856	11.992	13.848	15.659	18.062	19.943	23.337	27.096	29.553	33.196	36.415	40.270	42.980	51.179
25	11.524	12.697	14.611	16.473	18.940	20.867	24.337	28.172	30.675	34.382	37.652	41.566	44.314	52.620
26	12.198	13.409	15.379	17.292	19.820	21.792	25.336	29.246	31.795	35.563	38.885	42.856	45.642	54.052
27	12.879	14.125	16.151	18.114	20.703	22.719	26.336	30.319	32.912	36.741	40.113	44.140	46.963	55.476
28	13.565	14.847	16.928	18.939	21.588	23.647	27.336	31.391	34.027	37.916	41.337	45.419	48.278	56.893
29	14.256	15.574	17.708	19.768	22.475	24.577	28.336	32.461	35.139	39.087	42.557	46.693	49.588	58.302
30	14.953	16.306	18.493	20.599	23.364	25.508	29.336	33.530	36.250	40.256	43.773	47.962	50.892	59.703

Source: Ronald A. Fisher and Frank Yates, *Statistical Tables for Biological, Agricultural and Medical Research,* 6th ed. (New York, Hafner, 1963), p. 47.

Note: When the number of degrees of freedom exceeds 30, the χ^2 distribution can be approximated by the normal distribution, but the original Fisher/Yates table provides values up to 70 degrees of freedom.

Appendix Table I
F Distribution

Area or Probability = α

Entries in the table give F_{α} values, where α is the area or probability in the upper tail of the F distribution. For example, with 12 numerator degrees of freedom, 15 denominator degrees of freedom, and a .05 area in the upper tail, $F_{.05} = 2.48$.

$F_{.05}$ Values

Denominator Degrees of Freedom	Numerator Degrees of Freedom																		
	1	2	3	4	5	6	7	8	9	10	12	15	20	24	30	40	60	120	∞
1	161.4	199.5	215.7	224.6	230.2	234.0	236.8	238.9	240.5	241.9	243.9	245.9	248.0	249.1	250.1	251.1	252.2	253.3	254.3
2	18.51	19.00	19.16	19.25	19.30	19.33	19.35	19.37	19.38	19.40	19.41	19.43	19.45	19.45	19.46	19.47	19.48	19.49	19.50
3	10.13	9.55	9.28	9.12	9.01	8.94	8.89	8.85	8.81	8.79	8.74	8.70	8.66	8.64	8.62	8.59	8.57	8.55	8.53
4	7.71	6.94	6.59	6.39	6.26	6.16	6.09	6.04	6.00	5.96	5.91	5.86	5.80	5.77	5.75	5.72	5.69	5.66	5.63
5	6.61	5.79	5.41	5.19	5.05	4.95	4.88	4.82	4.77	4.74	4.68	4.62	4.56	4.53	4.50	4.46	4.43	4.40	4.36
6	5.99	5.14	4.76	4.53	4.39	4.28	4.21	4.15	4.10	4.06	4.00	3.94	3.87	3.84	3.81	3.77	3.74	3.70	3.67
7	5.59	4.74	4.35	4.12	3.97	3.87	3.79	3.73	3.68	3.64	3.57	3.51	3.44	3.41	3.38	3.34	3.30	3.27	3.23
8	5.32	4.46	4.07	3.84	3.69	3.58	3.50	3.44	3.39	3.35	3.28	3.22	3.15	3.12	3.08	3.04	3.01	2.97	2.93
9	5.12	4.26	3.86	3.63	3.48	3.37	3.29	3.23	3.18	3.14	3.07	3.01	2.94	2.90	2.86	2.83	2.79	2.75	2.71
10	4.96	4.10	3.71	3.48	3.33	3.22	3.14	3.07	3.02	2.98	2.91	2.85	2.77	2.74	2.70	2.66	2.62	2.58	2.54
11	4.84	3.98	3.59	3.36	3.20	3.09	3.01	2.95	2.90	2.85	2.79	2.72	2.65	2.61	2.57	2.53	2.49	2.45	2.40
12	4.75	3.89	3.49	3.26	3.11	3.00	2.91	2.85	2.80	2.75	2.69	2.62	2.54	2.51	2.47	2.43	2.38	2.34	2.30
13	4.67	3.81	3.41	3.18	3.03	2.92	2.83	2.77	2.71	2.67	2.60	2.53	2.46	2.42	2.38	2.34	2.30	2.25	2.21
14	4.60	3.74	3.34	3.11	2.96	2.85	2.76	2.70	2.65	2.60	2.53	2.46	2.39	2.35	2.31	2.27	2.22	2.18	2.13
15	4.54	3.68	3.29	3.06	2.90	2.79	2.71	2.64	2.59	2.54	2.48	2.40	2.33	2.29	2.25	2.20	2.16	2.11	2.07
16	4.49	3.63	3.24	3.01	2.85	2.74	2.66	2.59	2.54	2.49	2.42	2.35	2.28	2.24	2.19	2.15	2.11	2.06	2.01
17	4.45	3.59	3.20	2.96	2.81	2.70	2.61	2.55	2.49	2.45	2.38	2.31	2.23	2.19	2.15	2.10	2.06	2.01	1.96
18	4.41	3.55	3.16	2.93	2.77	2.66	2.58	2.51	2.46	2.41	2.34	2.27	2.19	2.15	2.11	2.06	2.02	1.97	1.92
19	4.38	3.52	3.13	2.90	2.74	2.63	2.54	2.48	2.42	2.38	2.31	2.23	2.16	2.11	2.07	2.03	1.98	1.93	1.88
20	4.35	3.49	3.10	2.87	2.71	2.60	2.51	2.45	2.39	2.35	2.28	2.20	2.12	2.08	2.04	1.99	1.95	1.90	1.84
21	4.32	3.47	3.07	2.84	2.68	2.57	2.49	2.42	2.37	2.32	2.25	2.18	2.10	2.05	2.01	1.96	1.92	1.87	1.81
22	4.30	3.44	3.05	2.82	2.66	2.55	2.46	2.40	2.34	2.30	2.23	2.15	2.07	2.03	1.98	1.94	1.89	1.84	1.78
23	4.28	3.42	3.03	2.80	2.64	2.53	2.44	2.37	2.32	2.27	2.20	2.13	2.05	2.01	1.96	1.91	1.86	1.81	1.76
24	4.26	3.40	3.01	2.78	2.62	2.51	2.42	2.36	2.30	2.25	2.18	2.11	2.03	1.98	1.94	1.89	1.84	1.79	1.73
25	4.24	3.39	2.99	2.76	2.60	2.49	2.40	2.34	2.28	2.24	2.16	2.09	2.01	1.96	1.92	1.87	1.82	1.77	1.71
26	4.23	3.37	2.98	2.74	2.59	2.47	2.39	2.32	2.27	2.22	2.15	2.07	1.99	1.95	1.90	1.85	1.80	1.75	1.69
27	4.21	3.35	2.96	2.73	2.57	2.46	2.37	2.31	2.25	2.20	2.13	2.06	1.97	1.93	1.88	1.84	1.79	1.73	1.67
28	4.20	3.34	2.95	2.71	2.56	2.45	2.36	2.29	2.24	2.19	2.12	2.04	1.96	1.91	1.87	1.82	1.77	1.71	1.65
29	4.18	3.33	2.93	2.70	2.55	2.43	2.35	2.28	2.22	2.18	2.10	2.03	1.94	1.90	1.85	1.81	1.75	1.70	1.64
30	4.17	3.32	2.92	2.69	2.53	2.42	2.33	2.27	2.21	2.16	2.09	2.01	1.93	1.89	1.84	1.79	1.74	1.68	1.62
40	4.08	3.23	2.84	2.61	2.45	2.34	2.25	2.18	2.12	2.08	2.00	1.92	1.84	1.79	1.74	1.69	1.64	1.58	1.51
60	4.00	3.15	2.76	2.53	2.37	2.25	2.17	2.10	2.04	1.99	1.92	1.84	1.75	1.70	1.65	1.59	1.53	1.47	1.39
120	3.92	3.07	2.68	2.45	2.29	2.17	2.09	2.02	1.96	1.91	1.83	1.75	1.66	1.61	1.55	1.50	1.43	1.35	1.25
∞	3.84	3.00	2.60	2.37	2.21	2.10	2.01	1.94	1.88	1.83	1.75	1.67	1.57	1.52	1.46	1.39	1.32	1.22	1.00

Appendix Table I *(continued)*

$F_{.025}$ Values

Denominator Degrees of Freedom	Numerator Degrees of Freedom																		
	1	2	3	4	5	6	7	8	9	10	12	15	20	24	30	40	60	120	∞
1	647.8	799.5	864.2	899.6	921.8	937.1	948.2	956.7	963.3	968.6	976.7	984.9	993.1	997.2	1001	1006	1010	1014	1018
2	38.51	39.00	39.17	39.25	39.30	39.33	39.36	39.37	39.39	39.40	39.41	39.43	39.45	39.46	39.46	39.47	39.48	39.49	39.50
3	17.44	16.04	15.44	15.10	14.88	14.73	14.62	14.54	14.47	14.42	14.34	14.25	14.17	14.12	14.08	14.04	13.99	13.95	13.90
4	12.22	10.65	9.98	9.60	9.36	9.20	9.07	8.98	8.90	8.84	8.75	8.66	8.56	8.51	8.46	8.41	8.36	8.31	8.26
5	10.01	8.43	7.76	7.39	7.15	6.98	6.85	6.76	6.68	6.62	6.52	6.43	6.33	6.28	6.23	6.18	6.12	6.07	6.02
6	8.81	7.26	6.60	6.23	5.99	5.82	5.70	5.60	5.52	5.46	5.37	5.27	5.17	5.12	5.07	5.01	4.96	4.90	4.85
7	8.07	6.54	5.89	5.52	5.29	5.12	4.99	4.90	4.82	4.76	4.67	4.57	4.47	4.42	4.36	4.31	4.25	4.20	4.14
8	7.57	6.06	5.42	5.05	4.82	4.65	4.53	4.43	4.36	4.30	4.20	4.10	4.00	3.95	3.89	3.84	3.78	3.73	3.67
9	7.21	5.71	5.08	4.72	4.48	4.32	4.20	4.10	4.03	3.96	3.87	3.77	3.67	3.61	3.56	3.51	3.45	3.39	3.33
10	6.94	5.46	4.83	4.47	4.24	4.07	3.95	3.85	3.78	3.72	3.62	3.52	3.42	3.37	3.31	3.26	3.20	3.14	3.08
11	6.72	5.26	4.63	4.28	4.04	3.88	3.76	3.66	3.59	3.53	3.43	3.33	3.23	3.17	3.12	3.06	3.00	2.94	2.88
12	6.55	5.10	4.47	4.12	3.89	3.73	3.61	3.51	3.44	3.37	3.28	3.18	3.07	3.02	2.96	2.91	2.85	2.79	2.72
13	6.41	4.97	4.35	4.00	3.77	3.60	3.48	3.39	3.31	3.25	3.15	3.05	2.95	2.89	2.84	2.78	2.72	2.66	2.60
14	6.30	4.86	4.24	3.89	3.66	3.50	3.38	3.29	3.21	3.15	3.05	2.95	2.84	2.79	2.73	2.67	2.61	2.55	2.49
15	6.20	4.77	4.15	3.80	3.58	3.41	3.29	3.20	3.12	3.06	2.96	2.86	2.76	2.70	2.64	2.59	2.52	2.46	2.40
16	6.12	4.69	4.08	3.73	3.50	3.34	3.22	3.12	3.05	2.99	2.89	2.79	2.68	2.63	2.57	2.51	2.45	2.38	2.32
17	6.04	4.62	4.01	3.66	3.44	3.28	3.16	3.06	2.98	2.92	2.82	2.72	2.62	2.56	2.50	2.44	2.38	2.32	2.25
18	5.98	4.56	3.95	3.61	3.38	3.22	3.10	3.01	2.93	2.87	2.77	2.67	2.56	2.50	2.44	2.38	2.32	2.26	2.19
19	5.92	4.51	3.90	3.56	3.33	3.17	3.05	2.96	2.88	2.82	2.72	2.62	2.51	2.45	2.39	2.33	2.27	2.20	2.13
20	5.87	4.46	3.86	3.51	3.29	3.13	3.01	2.91	2.84	2.77	2.68	2.57	2.46	2.41	2.35	2.29	2.22	2.16	2.09
21	5.83	4.42	3.82	3.48	3.25	3.09	2.97	2.87	2.80	2.73	2.64	2.53	2.42	2.37	2.31	2.25	2.18	2.11	2.04
22	5.79	4.38	3.78	3.44	3.22	3.05	2.93	2.84	2.76	2.70	2.60	2.50	2.39	2.33	2.27	2.21	2.14	2.08	2.00
23	5.75	4.35	3.75	3.41	3.18	3.02	2.90	2.81	2.73	2.67	2.57	2.47	2.36	2.30	2.24	2.18	2.11	2.04	1.97
24	5.72	4.32	3.72	3.38	3.15	2.99	2.87	2.78	2.70	2.64	2.54	2.44	2.33	2.27	2.21	2.15	2.08	2.01	1.94
25	5.69	4.29	3.69	3.35	3.13	2.97	2.85	2.75	2.68	2.61	2.51	2.41	2.30	2.24	2.18	2.12	2.05	1.98	1.91
26	5.66	4.27	3.67	3.33	3.10	2.94	2.82	2.73	2.65	2.59	2.49	2.39	2.28	2.22	2.16	2.09	2.03	1.95	1.88
27	5.63	4.24	3.65	3.31	3.08	2.92	2.80	2.71	2.63	2.57	2.47	2.36	2.25	2.19	2.13	2.07	2.00	1.93	1.85
28	5.61	4.22	3.63	3.29	3.06	2.90	2.78	2.69	2.61	2.55	2.45	2.34	2.23	2.17	2.11	2.05	1.98	1.91	1.83
29	5.59	4.20	3.61	3.27	3.04	2.88	2.76	2.67	2.59	2.53	2.43	2.32	2.21	2.15	2.09	2.03	1.96	1.89	1.81
30	5.57	4.18	3.59	3.25	3.03	2.87	2.75	2.65	2.57	2.51	2.41	2.31	2.20	2.14	2.07	2.01	1.94	1.87	1.79
40	5.42	4.05	3.46	3.13	2.90	2.74	2.62	2.53	2.45	2.39	2.29	2.18	2.07	2.01	1.94	1.88	1.80	1.72	1.64
60	5.29	3.93	3.34	3.01	2.79	2.63	2.51	2.41	2.33	2.27	2.17	2.06	1.94	1.88	1.82	1.74	1.67	1.58	1.48
120	5.15	3.80	3.23	2.89	2.67	2.52	2.39	2.30	2.22	2.16	2.05	1.94	1.82	1.76	1.69	1.61	1.53	1.43	1.31
∞	5.02	3.69	3.12	2.79	2.57	2.41	2.29	2.19	2.11	2.05	1.94	1.83	1.71	1.64	1.57	1.48	1.39	1.27	1.00

Appendix Table I (continued)

$F_{.01}$ Values

Denominator Degrees of Freedom	\multicolumn{19}{c}{Numerator Degrees of Freedom}
	1	2	3	4	5	6	7	8	9	10	12	15	20	24	30	40	60	120	∞
1	4052	4999.5	5403	5625	5764	5859	5928	5981	6022	6056	6106	6157	6209	6235	6261	6287	6313	6339	6366
2	98.50	99.00	99.17	99.25	99.30	99.33	99.36	99.37	99.39	99.40	99.42	99.43	99.45	99.46	99.47	99.47	99.48	99.49	99.50
3	34.12	30.82	29.46	28.71	28.24	27.91	27.67	27.49	27.35	27.23	27.05	26.87	26.69	26.60	26.50	26.41	26.32	26.22	26.13
4	21.20	18.00	16.69	15.98	15.52	15.21	14.98	14.80	14.66	14.55	14.37	14.20	14.02	13.93	13.84	13.75	13.65	13.56	13.46
5	16.26	13.27	12.06	11.39	10.97	10.67	10.46	10.29	10.16	10.05	9.89	9.72	9.55	9.47	9.38	9.29	9.20	9.11	9.02
6	13.75	10.92	9.78	9.15	8.75	8.47	8.26	8.10	7.98	7.87	7.72	7.56	7.40	7.31	7.23	7.14	7.06	6.97	6.88
7	12.25	9.55	8.45	7.85	7.46	7.19	6.99	6.84	6.72	6.62	6.47	6.31	6.16	6.07	5.99	5.91	5.82	5.74	5.65
8	11.26	8.65	7.59	7.01	6.63	6.37	6.18	6.03	5.91	5.81	5.67	5.52	5.36	5.28	5.20	5.12	5.03	4.95	4.86
9	10.56	8.02	6.99	6.42	6.06	5.80	5.61	5.47	5.35	5.26	5.11	4.96	4.81	4.73	4.65	4.57	4.48	4.40	4.31
10	10.04	7.56	6.55	5.99	5.64	5.39	5.20	5.06	4.94	4.85	4.71	4.56	4.41	4.33	4.25	4.17	4.08	4.00	3.91
11	9.65	7.21	6.22	5.67	5.32	5.07	4.89	4.74	4.63	4.54	4.40	4.25	4.10	4.02	3.94	3.86	3.78	3.69	3.60
12	9.33	6.93	5.95	5.41	5.06	4.82	4.64	4.50	4.39	4.30	4.16	4.01	3.86	3.78	3.70	3.62	3.54	3.45	3.36
13	9.07	6.70	5.74	5.21	4.86	4.62	4.44	4.30	4.19	4.10	3.96	3.82	3.66	3.59	3.51	3.43	3.34	3.25	3.17
14	8.86	6.51	5.56	5.04	4.69	4.46	4.28	4.14	4.03	3.94	3.80	3.66	3.51	3.43	3.35	3.27	3.18	3.09	3.00
15	8.68	6.36	5.42	4.89	4.56	4.32	4.14	4.00	3.89	3.80	3.67	3.52	3.37	3.29	3.21	3.13	3.05	2.96	2.87
16	8.53	6.23	5.29	4.77	4.44	4.20	4.03	3.89	3.78	3.69	3.55	3.41	3.26	3.18	3.10	3.02	2.93	2.84	2.75
17	8.40	6.11	5.18	4.67	4.34	4.10	3.93	3.79	3.68	3.59	3.46	3.31	3.16	3.08	3.00	2.92	2.83	2.75	2.65
18	8.29	6.01	5.09	4.58	4.25	4.01	3.84	3.71	3.60	3.51	3.37	3.23	3.08	3.00	2.92	2.84	2.75	2.66	2.57
19	8.18	5.93	5.01	4.50	4.17	3.94	3.77	3.63	3.52	3.43	3.30	3.15	3.00	2.92	2.84	2.76	2.67	2.58	2.49
20	8.10	5.85	4.94	4.43	4.10	3.87	3.70	3.56	3.46	3.37	3.23	3.09	2.94	2.86	2.78	2.69	2.61	2.52	2.42
21	8.02	5.78	4.87	4.37	4.04	3.81	3.64	3.51	3.40	3.31	3.17	3.03	2.88	2.80	2.72	2.64	2.55	2.46	2.36
22	7.95	5.72	4.82	4.31	3.99	3.76	3.59	3.45	3.35	3.26	3.12	2.98	2.83	2.75	2.67	2.58	2.50	2.40	2.31
23	7.88	5.66	4.76	4.26	3.94	3.71	3.54	3.41	3.30	3.21	3.07	2.93	2.78	2.70	2.62	2.54	2.45	2.35	2.26
24	7.82	5.61	4.72	4.22	3.90	3.67	3.50	3.36	3.26	3.17	3.03	2.89	2.74	2.66	2.58	2.49	2.40	2.31	2.21
25	7.77	5.57	4.68	4.18	3.85	3.63	3.46	3.32	3.22	3.13	2.99	2.85	2.70	2.62	2.54	2.45	2.36	2.27	2.17
26	7.72	5.53	4.64	4.14	3.82	3.59	3.42	3.29	3.18	3.09	2.96	2.81	2.66	2.58	2.50	2.42	2.33	2.23	2.13
27	7.68	5.49	4.60	4.11	3.78	3.56	3.39	3.26	3.15	3.06	2.93	2.78	2.63	2.55	2.47	2.38	2.29	2.20	2.10
28	7.64	5.45	4.57	4.07	3.75	3.53	3.36	3.23	3.12	3.03	2.90	2.75	2.60	2.52	2.44	2.35	2.26	2.17	2.06
29	7.60	5.42	4.54	4.04	3.73	3.50	3.33	3.20	3.09	3.00	2.87	2.73	2.57	2.49	2.41	2.33	2.23	2.14	2.03
30	7.56	5.39	4.51	4.02	3.70	3.47	3.30	3.17	3.07	2.98	2.84	2.70	2.55	2.47	2.39	2.30	2.21	2.11	2.01
40	7.31	5.18	4.31	3.83	3.51	3.29	3.12	2.99	2.89	2.80	2.66	2.52	2.37	2.29	2.20	2.11	2.02	1.92	1.80
60	7.08	4.98	4.13	3.65	3.34	3.12	2.95	2.82	2.72	2.63	2.50	2.35	2.20	2.12	2.03	1.94	1.84	1.73	1.60
120	6.85	4.79	3.95	3.48	3.17	2.96	2.79	2.66	2.56	2.47	2.34	2.19	2.03	1.95	1.86	1.76	1.66	1.53	1.38
∞	6.63	4.61	3.78	3.32	3.02	2.80	2.64	2.51	2.41	2.32	2.18	2.04	1.88	1.79	1.70	1.59	1.47	1.32	1.00

Source: E. S. Pearson and H. O. Hartley, *Biometrika Tables for Statisticians*, vol. 1 (Cambridge: Cambridge University Press, 1966), pp. 171–73.

Appendix Table J
Values of q_α in Tukey's *HSD* Test

Error d.f.	α	k = Number of means or number of steps between ordered means										
		2	3	4	5	6	7	8	9	10	11	12
5	.05	3.64	4.60	5.22	5.67	6.03	6.33	6.58	6.80	6.99	7.17	7.32
	.01	5.70	6.98	7.80	8.42	8.91	9.32	9.67	9.97	10.24	10.48	10.70
6	.05	3.46	4.34	4.90	5.30	5.63	5.90	6.12	6.32	6.49	6.65	6.79
	.01	5.24	6.33	7.03	7.56	7.97	8.32	8.61	8.87	9.10	9.30	9.48
7	.05	3.34	4.16	4.68	5.06	5.36	5.61	5.82	6.00	6.16	6.30	6.43
	.01	4.95	5.92	6.54	7.01	7.37	7.68	7.94	8.17	8.37	8.55	8.71
8	.05	3.26	4.04	4.53	4.89	5.17	5.40	5.60	5.77	5.92	6.05	6.18
	.01	4.75	5.64	6.20	6.62	6.96	7.24	7.47	7.68	7.86	8.03	8.18
9	.05	3.20	3.95	4.41	4.76	5.02	5.24	5.43	5.59	5.74	5.87	5.98
	.01	4.60	5.43	5.96	6.35	6.66	6.91	7.13	7.33	7.49	7.65	7.78
10	.05	3.15	3.88	4.33	4.65	4.91	5.12	5.30	5.46	5.60	5.72	5.83
	.01	4.48	5.27	5.77	6.14	6.43	6.67	6.87	7.05	7.21	7.36	7.49
11	.05	3.11	3.82	4.26	4.57	4.82	5.03	5.20	5.35	5.49	5.61	5.71
	.01	4.39	5.15	5.62	5.97	6.25	6.48	6.67	6.84	6.99	7.13	7.25
12	.05	3.08	3.77	4.20	4.51	4.75	4.95	5.12	5.27	5.39	5.51	5.61
	.01	4.32	5.05	5.50	5.84	6.10	6.32	6.51	6.67	6.81	6.94	7.06
13	.05	3.06	3.73	4.15	4.45	4.69	4.88	5.05	5.19	5.32	5.43	5.53
	.01	4.26	4.96	5.40	5.73	5.98	6.19	6.37	6.53	6.67	6.79	6.90
14	.05	3.03	3.70	4.11	4.41	4.64	4.83	4.99	5.13	5.25	5.36	5.46
	.01	4.21	4.89	5.32	5.63	5.88	6.08	6.26	6.41	6.54	6.66	6.77
15	.05	3.01	3.67	4.08	4.37	4.59	4.78	4.94	5.08	5.20	5.31	5.40
	.01	4.17	4.84	5.25	5.56	5.80	5.99	6.16	6.31	6.44	6.55	6.66
16	.05	3.00	3.65	4.05	4.33	4.56	4.74	4.90	5.03	5.15	5.26	5.35
	.01	4.13	4.79	5.19	5.49	5.72	5.92	6.08	6.22	6.35	6.46	6.56
17	.05	2.98	3.63	4.02	4.30	4.52	4.70	4.86	4.99	5.11	5.21	5.31
	.01	4.10	4.74	5.14	5.43	5.66	5.85	6.01	6.15	6.27	6.38	6.48
18	.05	2.97	3.61	4.00	4.28	4.49	4.67	4.82	4.96	5.07	5.17	5.27
	.01	4.07	4.70	5.09	5.38	5.60	5.79	5.94	6.08	6.20	6.31	6.41
19	.05	2.96	3.59	3.98	4.25	4.47	4.65	4.79	4.92	5.04	5.14	5.23
	.01	4.05	4.67	5.05	5.33	5.55	5.73	5.89	6.02	6.14	6.25	6.34
20	.05	2.95	3.58	3.96	4.23	4.45	4.62	4.77	4.90	5.01	5.11	5.20
	.01	4.02	4.64	5.02	5.29	5.51	5.69	5.84	5.97	6.09	6.19	6.28
24	.05	2.92	3.53	3.90	4.17	4.37	4.54	4.68	4.81	4.92	5.01	5.10
	.01	3.96	4.55	4.91	5.17	5.37	5.54	5.69	5.81	5.92	6.02	6.11
30	.05	2.89	3.49	3.85	4.10	4.30	4.46	4.60	4.72	4.82	4.92	5.00
	.01	3.89	4.45	4.80	5.05	5.24	5.40	5.54	5.65	5.76	5.85	5.93
40	.05	2.86	3.44	3.79	4.04	4.23	4.39	4.52	4.63	4.73	4.82	4.90
	.01	3.82	4.37	4.70	4.93	5.11	5.26	5.39	5.50	5.60	5.69	5.76
60	.05	2.83	3.40	3.74	3.98	4.16	4.31	4.44	4.55	4.65	4.73	4.81
	.01	3.76	4.28	4.59	4.82	4.99	5.13	5.25	5.36	5.45	5.53	5.60
120	.05	2.80	3.36	3.68	3.92	4.10	4.24	4.36	4.47	4.56	4.64	4.71
	.01	3.70	4.20	4.50	4.71	4.87	5.01	5.12	5.21	5.30	5.37	5.44
∞	.05	2.77	3.31	3.63	3.86	4.03	4.17	4.29	4.39	4.47	4.55	4.62
	.01	3.64	4.12	4.40	4.60	4.76	4.88	4.99	5.08	5.16	5.23	5.29

Source: E. S. Pearson and H. O. Hartley, *Biometrika Tables for Statisticians*, vol. 1, (Cambridge: Cambridge University Press, 1966), pp. 192–93.

GLOSSARY

α risk in hypothesis testing, the probability of making a type I error **(8)**

absolute class frequency the absolute number of observations that fall into a given class **(2)**

absolute frequency distribution a tabular summary of a data set showing the absolute class frequencies in each of several collectively exhaustive and mutually exclusive classes **(2)**

acceptance number see **critical value**

acceptance region in a hypothesis test, that range of possible values of the test statistic which signals the acceptance of the null hypothesis **(8)**

action branches branches emanating from an action point in a decision-tree diagram and representing the possible actions of the decision maker **(14)**

action point in a decision-tree diagram, a point of choice (at which the decision maker is in control), symbolized by a square **(14)**

actions in decison theory, the mutually exclusive decision alternatives open to a decision maker **(14)**

addition law a law of probability theory that is used to compute the probability for the occurrence of a *union* of two or more events; according to the general law, for any two events $p(A \text{ } or \text{ } B) = p(A) + p(B) - p(A \text{ } and \text{ } B)$; according to the special law for two mutually exclusive events, $p(A \text{ } or \text{ } B) = p(A) + p(B)$ **(4)**

additive time-series model a time-series model that expresses the actual value of a time series as the sum of its components—for example, $Y = T + C + S + I$ **(13)**

adjusted sample coefficient of multiple determination a measure, denoted by $\overline{R}^2$ (and equal to R^2 adjusted for degrees of freedom), that is an unbiased estimator of the corresponding population coefficient **(12)**

alternative hypothesis this hypothesis is the second of two opposing hypotheses in a hypothesis test, symbolized by H_A, and often (but not necessarily) representing that proposition about an unknown population parameter that is tentatively assumed to be false **(8)**

analysis of variance (ANOVA) a statistical technique specially designed to test whether the means of more than two quantitative populations are equal **(10)**

analytical statistics a branch of the discipline that is concerned with developing and utilizing techniques for properly analyzing (or drawing inferences from) numerical information **(1)**

ANOVA table in the analysis of variance, a table that shows, for each source of variation, the sum of squares, the degrees of freedom and their ratio, the mean square (which is the desired estimate of the population variance), and more **(10)**

arithmetic mean (μ or $\overline{X}$) a measure of central tendency obtained from ungrouped data by adding the values observed and dividing their sum by the number of observations **(3)**

array see **ordered array**

β risk in hypothesis testing, the probability of making a type II error **(8)**

backward induction a process by which a decision problem is solved with the help of a decision-tree diagram and that involves the computation, with the help of ultimate payoff values and event-branch probability values, of expected payoff values at each fork of the tree **(14)**

bar chart a graphical portrayal of qualitative or discrete quantitative variables by a series of noncontiguous horizontal bars the lengths of which are proportional to the values that are to be depicted **(2)**

basic outcome any one of a random experiment's possible outcomes the occurrence of which rules out the occurrence of all the alternative outcomes **(4)**

Bayes's postulate see **equal-likelihood criterion**

Bayes's strategy the optimal strategy selected in decision-making problems that are solved by optimizing the expected payoff **(14)**

Bayes's theorem a formula for revising an initial subjective (prior) probability value on the basis of results obtained by an empirical investigation and for, thus, obtaining a new (posterior) probability value **(4)**

Bernoulli process a sequence of n identical trials of a random experiment such that each trial (a) produces one of two possible complementary outcomes that are conventionally called *success* and *failure* and (b) is independent of any other trial so that the probability of success or failure is constant from trial to trial **(5)**

bias see **systematic error**

bimodal frequency distribution a frequency distribution in which two different values occur with the highest frequency, or almost that **(3)**

binomial coefficient (c_x^n) a part of the binomial formula that gives the number of permutations of x successes and $n - x$ failures that can be achieved in n trials that satisfy the conditions of a Bernoulli process **(5)**

binomial formula a formula for calculating the probability of x successes in n trials of a random experiment that satisfies the conditions of a Bernoulli process **(5)**

binomial probability distribution a probability distribution that shows the probabilities associated with possible values of a discrete random variable that are generated by a Bernoulli process **(5)**

binomial probability tables tables that list binomial probabilities (probabilities of x successes in a Bernoulli process) for various combinations of possible values of n (number of trials) and π (probability of success in any one trial) **(5)**

binomial random variable the number of successes achieved in a Bernoulli process **(5)**

bivariate data set a data set containing information on two variables **(1)**

blocking a procedure that eliminates the effects of extraneous factors during an experiment by forming blocks of experimental units within each of which all units are as alike as possible with respect to those factors **(10)**

census a complete survey, in which observations are made about one or more characteristics of interest for every elementary unit that exists **(1)**

central limit theorem the theorem that states: if $\overline{X}$ is the mean of (or P is the proportion in) a random sample taken from a population and if the population values are *not* normally distributed, the sampling distribution of $\overline{X}$ (or P) nevertheless approaches a normal distribution as sample size increases—provided sample size, n, remains small relative to population size, N; this approximation is near-perfect for samples of $n \geq 30$ but $n < .05N$ (or when $n \cdot \pi$ as well as $n(1 - \pi) \geq 5$) **(7)**

chance error see **random error**

chance variable see **random variable**

Chebyshev's theorem the theorem that states: regardless of the shape of a population's frequency distribution, the proportion of observations falling within k standard deviations of the mean is at least $1 - (1/k^2)$, given that k is 1 or more **(3)**

chi-square distribution the sampling distribution of the χ^2 statistic **(9)**

chi-square statistic the sum of all the ratios that can be constructed by taking the difference between each observed and expected frequency (in, for example, a contingency table), squaring the difference, and dividing this squared deviation by the expected frequency **(9)**

classes groupings of data **(2)**

class frequency see **absolute**, **relative**, and **cumulative class frequency**

classical time-series model a model that attempts to explain the pattern observed in an actual time series by the presence of four components: trend, cyclical, seasonal, and irregular components **(13)**

class mark the midpoint of a class width **(2)**

class width the difference between the numerical lower and upper limit of a class of quantitative data **(2)**

clustered random sample a subset of a frame, or of an associated population, chosen by taking separate censuses in a randomly chosen subset of geographically distinct clusters into which the frame or population is naturally subdivided **(1)**

coefficient of variation (V or v) a measure of relative dispersion; equal to the ratio of the standard deviation to the arithmetic mean **(3)**

collectively exhaustive events different random events that jointly contain all the basic outcomes in the sample space **(4)**

column chart a graphical portrayal of qualitative or discrete quantitative variables by a series of noncontiguous vertical columns the lengths of which are proportional to the values that are to be depicted **(2)**

combinations different selections of items such that possible alternative sequences among the components of any one such selection are considered immaterial **(4)**

compatible events different random events that have at least some basic outcomes in common **(4)**

complementary events two random events such that precisely all those basic outcomes that are not contained in one are contained in the other **(4)**

completely randomized design see **randomized group design**

composite event any combination of two or more basic outcomes **(4)**

conditional mean of Y in regression analysis, the mean, $\mu_{Y \cdot X}$, of a conditional probability distribution of Y **(11)**

conditional probability a measure of the likelihood that a particular event occurs, given the fact that another event has already occurred or is certain to occur **(4)**

conditional probability distribution of Y in regression analysis, the normal curve for a given population of Y values that is associated with a specified value of X **(11)**

conditional standard deviation of Y in regression analysis, the standard deviation, $\sigma_{Y \cdot X}$, of a conditional probability distribution of Y **(11)**

confidence band in regression analysis, a band of values lying above and below an estimated regression line and indicating the limits of the prediction interval for each $\mu_{Y \cdot X}$ **(11)**

confidence intervals intervals within which population parameters presumably lie **(7)**

confidence level the percentage of interval estimates (obtained from repeated samples, each of size n, taken from a given population) that can be expected to contain the actual value of the parameter being estimated **(7)**

confidence level of the hypothesis test the probability (equal to $1 - \alpha$) of avoiding a type I error when the null hypothesis is true **(8)**

confidence limits the two limits that define the interval within which a population parameter that is being estimated presumably lies **(7)**

consistent estimator a sample statistic such that the probability of its being close to the parameter being estimated gets ever larger (and, therefore, approaches unity) as the sample size increases **(7)**

contingency table a table that classifies data according to two or more categories associated with each of two qualitative variables that may or may not be statistically independent **(9)**

continuous quantitative variable a quantitative variable observations about which can assume values at all points on a scale of values, with no breaks between possible values **(1)**

control charts graphical devices that highlight the average performance of data series and the dispersion around this average so that average and dispersion of the past become standards for controlling performance in the present **(3)**

convenience sample a subset of a frame, or of an associated population, resulting when expediency is the primary consideration in selecting elementary units for observation and simply the most easily accessible units are chosen **(1)**

correlation analysis a statistical method the central focus of which is the establishment of an *index* that provides, in a single number, an instant picture of the strength of association between two variables **(11)**

criterion of insufficient reason see **equal-likelihood criterion**

critical value in hypothesis testing, the value of a test statistic that divides all possible values into acceptance and rejection regions **(8)**

cross-section data numerical values pertaining to units of different populations that have been observed simultaneously at the same point in time (in the case of a stock variable) or during the same period of time (in the case of a flow variable) **(13)**

cumulative class frequency the sum of the (absolute or relative) class frequencies for all classes up to and including the class in question, beginning at either end of the frequency distribution **(2)**

cumulative frequency distribution a tabular summary of a data set showing for each of several collectively exhaustive and mutually exclusive classes the absolute number or proportion of observations that are less than or equal to the upper limits of the classes in question (LE type) or that are more than or equal to their lower limits (ME type) **(2)**

cumulative probability distribution a probability distribution that shows the probabilities of a random variable being less than or equal to any given possible value **(5)**

cutoff point see **critical value**

cyclical component in time-series analysis, wide up-and-down swings of the variable of interest around the trend, with the swings lasting from one to several years each and typically differing in length and amplitude from one cycle to the next **(13)**

data set a collection of observations about one or more characteristics of interest, for one or more elementary units **(1)**

datum a single observation about a specified characteristic of interest possessed by an elementary unit **(1)**

deciles values in a data set that divide the total of observations into ten parts, each of which contains .10 (or 10 percent) of the observed values **(3)**

decision branches see **action branches**

decision fork see **action point**

decision making under uncertainty any situation in which the ultimate outcome of a decision maker's choice depends on chance **(14)**

decision node see **action point**

decision point see **action point**

decision rule a hypothesis-testing rule that specifies in advance, for all possible values of a test statistic that might be observed in a sample, when the null hypothesis should be accepted and when it should be rejected in favor of the alternative hypothesis **(8)**

decision theory a variety of methods that can be employed in the systematic analysis and solution of decision-making problems that arise because uncertainty exists about future events over which the decision maker has no control, but which are bound to influence the ultimate outcome of a decision **(14)**

decision tree a summary of a decision-making situation in the context of uncertainty, showing graphically and in chronological order from left to right every potential action, event, and payoff **(14)**

deductive reasoning the drawing of inferences about an unknown part from a known whole **(1)**

degrees of freedom (d.f.) the number of values that can be freely chosen when calculating a statistic, equal to sample size minus 1 in the case of a single sample **(7)**

dependent events two random events such that the probability of one event is affected by the occurrence of the other event; hence, $p(A) \neq p(A|B)$ **(4)**

dependent variable in regression analysis, the variable (denoted by Y) the value of which is assumed unknown and that is being explained or predicted on the basis of its relationship with one or more other variables **(11)**

descriptive statistics a branch of the discipline that is concerned with developing and utilizing techniques for the careful collection and effective presentation of numerical information **(1)**

deterministic relationship a relationship between any two variables, X and Y, such that the value of Y is uniquely determined whenever the value of X is specified **(11)**

dichotomous qualitative variable a qualitative variable about which observations can be made in only two categories **(1)**

direct relationship a relationship between any two variables, X and Y, such that the values of Y increase with larger values of X **(11)**

discrete quantitative variable a quantitative variable such that observations about it can assume values only at specific points on a scale of values, with gaps between **(1)**

disjoint events see **mutually exclusive events**

dominant action in decision theory, an action that is unambiguously superior to an alternative action because it produces payoffs that are as good as or better than those of the alternative action no matter which event occurs **(14)**

efficiency of sample information (ESI) the ratio of the expected value of sample information to the (prior) expected value of perfect information, multiplied by 100 **(14)**

efficient estimator that sample statistic, among all the available unbiased estimators, which has the smallest variance for a given sample size **(7)**

elementary event see **simple event**

elementary unit a person or object possessing a characteristic in which a statistician is interested **(1)**

equal-likelihood criterion in decision theory, the assignment of equal prior probabilities to all possible events when absolutely nothing is known about the likelihood of occurrence of these events **(14)**

error see **random error,** and **unexplained variation**

error mean square (EMS) in the analysis of variance, the population variance estimate, s_W^2, based on the observed variation within samples; the denominator of the F statistic **(10)**

error of acceptance see **type II error**

error of rejection see **type I error**

error sum of squares (ESS) in the analysis of variance, the sum of all samples' sums of squared deviations of individual observations from their sample mean; the total sum of squares minus explained variation **(10,11)**

estimated partial-regression coefficient the coefficient of an independent variable in an estimated multiple-regression equation; it gives the net change in Y for a unit change in that independent variable, while holding other independent variables constant **(12)**

estimated regression coefficients   the values of *a* and *b* found in the equation of the estimated regression line, $\hat{Y}_X = a + bX$ **(11)**

estimated regression line   the regression line calculated from sample data by the method of least squares   **(11)**

estimation   the process of inferring the values of unknown population parameters from known sample statistics   **(7)**

estimator   the type of sample statistic that serves the purpose of making inferences about a given type of parameter   **(7)**

event branches   branches emanating from an event point in a decision-tree diagram and representing the possible events confronting the decision maker   **(14)**

event point   in a decision-tree diagram, a point of chance (at which the decision maker exercises no control but at which "nature" is in charge), symbolized by a circle   **(14)**

events   in decision theory, the mutually exclusive future occurrences that will affect the outcome of any present action taken but that are not under the control of a decision maker   **(14)**

exact hypothesis   a hypothesis that specifies a single value for an unknown parameter   **(8)**

expected monetary value *(EMV)*   the sum of the weighted payoffs associated with an action, the weights being the probabilities of the alternative events that produce the various possible payoffs   **(14)**

expected-monetary-value criterion   one of several probabilistic criteria for making decisions under uncertainty according to which a decision maker determines an expected monetary value for each possible action and selects the action with the optimal expected monetary value (the largest *EMV* is optimal if the objective is to maximize some benefit; the smallest *EMV* is optimal if the objective is to minimize some cost)   **(14)**

expected opportunity loss *(EOL)*   the sum of the weighted opportunity-loss values associated with an action, the weights being the probabilities of the alternative events that produce the various possible opportunity losses   **(14)**

expected-opportunity-loss criterion   one of several probabilistic criteria for making decisions under uncertainty according to which a decision maker determines an expected opportunity loss for each possible action and selects the action with the smallest of these values   **(14)**

expected regret value   see **expected opportunity loss**

expected utility *(EU)*   the sum of the weighted utilities associated with an action, the weights being the probabilities of the alternative events that produce the various possible utility payoffs   **(14)**

expected-utility criterion   one of several probabilistic criteria for making decisions under uncertainty according to which a decision maker determines the expected utility for each possible action and selects the action that maximizes it   **(14)**

expected value *(E_R)*   the weighted arithmetic mean of a random variable; that value of a random variable which one can expect to find on the average by numerous repetitions of the random experiment that generates the variable's actual value   **(5)**

expected value of perfect information *(EVPI)*   the maximum amount a decision maker can be expected to pay for obtaining perfect information about future events and, thus, for eliminating uncertainty completely (given the objective of maximizing some benefit, it equals the difference between the expected benefit when the optimal decision is made with perfect information and the expected benefit when that decision is made under uncertainty; given the objective of minimizing some cost, it equals the difference between the expected cost when the optimal decision is made under uncertainty and the expected cost when that decision is made with perfect information)   **(14)**

expected value of sample information *(EVSI)*   the maximum amount a decision maker can be expected to pay for obtaining admittedly imperfect information about future events and, thus, for reducing, rather than eliminating, uncertainty (given the objective of maximizing some benefit, it equals the difference between the expected benefit when an optimal decision is made with the sample information and before paying for it and the expected benefit when that decision is made without it; given the objective of minimizing some cost, it equals the difference between the expected cost when the optimal decision is made without the sample information and the expected cost when that decision is made with it, but before paying for it)   **(14)**

experiment   the collection of data from elementary units, while exercising control over some or all factors that may make these units different from one another and that may, therefore, affect the characteristic of interest being observed   **(1)**

explained deviation of *Y*   in regression analysis, the difference between the regression estimate of an individual *Y*, or $\hat{Y}_X$, and the mean, $\bar{Y}$, of all the observations of the dependent variable   **(11)**

explained variable   see **dependent variable**

explained variance   in the analysis of variance, the numerator of the F statistic   **(10)**

explained variation   in the analysis of variance, the variation among sample means; it is attributed to inherent differences among the sampled populations and equals the treatments sum of squares (1-way ANOVA)   **(10)**

explained variation of Y   in regression analysis, the sum of the squares of all the explained deviations of Y, or $\Sigma(\hat{Y}_X - \bar{Y})^2$, also called the **regression sum of squares**   **(11)**

explanatory variable   see **independent variable**

exponential smoothing   a forecasting procedure that produces self-correcting forecasts by means of a built-in adjustment mechanism that corrects for earlier forecasting errors   **(13)**

extensive-form analysis   a form of preposterior analysis that uses backward induction on a decision tree to identify the optimal strategy   **(14)**

factorial   the product of a series of positive whole numbers that descends from a given number, n, down to 1   **(4)**

factorial product   see **factorial**

fair gamble   a gamble in which the expected monetary value of what is given up is precisely equal to the expected monetary value of what is received   **(14)**

F distributions   the different members of the continuous probability distribution family of the F statistic that are distinguished by different numerator and denominator degrees of freedom associated with the two variance estimates making up this statistic   **(10)**

fiducial interval   in the theory of estimation, a range of values that can be trusted, with a specified probability, to contain the value of some parameter   **(8)**

first quartile   the .25 fractile or 25th percentile in a data set below which a quarter of all observations lie   **(3)**

forecasting   making statements about unknown, uncertain, and, typically, future events   **(13)**

fractile   a value in a data set below which a specified proportion of all values is found   **(3)**

frame   a complete listing of all elementary units relevant to a statistical investigation   **(1)**

frequency curve   a graphical portrayal, by a smooth curve, of an absolute or relative frequency distribution of a continuous quantitative variable   **(2)**

frequency density   the ratio of class frequency to class width   **(2)**

frequency distribution   see **absolute, relative,** and **cumulative frequency distribution**

frequency histogram   a graphical portrayal, as a series of contiguous rectangles, of an absolute or relative frequency distribution of a continuous quantitative variable in such a way that the areas of the rectangles correspond to the frequencies being depicted   **(2)**

frequency polygon   a graphical portrayal (as a many-sided figure) of an absolute or relative frequency distribution of a continuous quantitative variable   **(2)**

F statistic   in the analysis of variance, the ratio of the explained to the unexplained variance   **(10)**

goodness-of-fit test   a statistial test to determine the likelihood that sample data have been generated from a population that conforms to a specified type of probability distribution   **(9)**

grand mean ($\bar{\bar{X}}$)   the arithmetic mean of two or more means   **(10)**

histogram   see **frequency histogram**

homoscedasticity   the characteristic of different statistical populations having equal variances   **(10)**

hypothesis   a proposition tentatively advanced as possibly true   **(8)**

hypothesis testing   a systematic approach to assessing beliefs about reality; confronting a belief with evidence and deciding whether that belief can be maintained as reasonable or must be discarded as untenable   **(8)**

inadmissible action   in decision theory, an action that is unambiguously inferior to an alternative action because it produces payoffs that are at best as good as and often worse than those of the alternative action no matter which event occurs   **(14)**

incompatible events   see **mutually exclusive events**

independent events   two random events such that the probability of one event is not affected by the occurrence of the other event; hence, $p(A) = p(A|B)$ **(4)**

independent variable   in regression analysis, the variable (denoted by X) the value of which is assumed known and that is being used to explain or predict the value of the other variable   **(11)**

inductive reasoning   the drawing of inferences about an unknown whole from a known part   **(1)**

inexact hypothesis   a hypothesis that specifies a range of values for an unknown parameter   **(8)**

inferential statistics see **analytical statistics**

interfractile ranges distance measures of dispersion; differences between two values, called fractiles, in a data set **(3)**

interquartile range a distance measure of dispersion; the difference between the third and first quartile in a data set, containing the middle 50 percent of all values observed **(3)**

intersection of two events all the basic outcomes simultaneously contained in both one random event *and* another **(4)**

interval data numbers that by their size rank observations in order of importance and between which *intervals* or distances are comparable, while their ratios are meaningless **(1)**

interval estimate the estimate of a population parameter, expressed as a range of values within which the unknown parameter presumably lies **(7)**

inverse relationship a relationship between any two variables, X and Y, such that the values of Y decrease with larger values of X **(11)**

irregular component in time-series analysis, random movements of the variable of interest around the trend/cyclical/seasonal components **(13)**

joint probability a measure of the likelihood of the simultaneous occurrence of two or more events **(4)**

judgment sample a subset of a frame, or of an associated population, resulting when "expert" judgment plays a major role in selecting elementary units for observation in such a way that the chosen ones are "representative" of the whole **(1)**

Laplace criterion see **equal-likelihood criterion**

law of large numbers the law that states that: the probability of a significant deviation of an empirically determined probability value from a theoretically determined one is smaller the larger is the number of repetitions of the random experiment in question **(4)**

lower-tailed test a one-tailed hypothesis test in which the entire rejection region is found in the lower tail **(8)**

marginal probability see **unconditional probability**

marginal utility of money the change in the total utility of money that is associated with a unit change in the quantity of money **(14)**

matched-pairs sample a sample taken after each experimental unit in population A has been matched with a "twin" from population B so that any sample observation about a unit in population A automatically yields an associated observation about a unit in population B **(7)**

maximax one of several nonprobabilistic criteria for decision making under uncertainty according to which a decision maker who seeks to maximize some benefit finds the maximum benefit associated with each possible action, identifies the maximum among these maxima, and chooses the action associated with this maximum of maxima **(14)**

maximin one of several nonprobabilistic criteria for decision making under uncertainty according to which a decision maker who seeks to maximize some benefit finds the minimum benefit associated with each possible action, identifies the maximum among these minima, and chooses the action associated with this maximum of minima **(14)**

maximum-likelihood criterion one of several probabilistic criteria for making decisions under uncertainty according to which a decision maker simply ignores all the events that might occur except the most likely one and selects the action that produces the optimal result (maximum benefit or minimum cost) associated with this most likely event **(14)**

mean see **arithmetic mean**

mean absolute deviation (*MAD*) an average deviation measure of dispersion, obtained by averaging the absolute values of deviations of individual observations from the mean **(3)**

mean of squared deviations see **variance**

mean squared error the sum of an estimator's squared bias plus its variance **(7)**

measurement the assignment of numbers to characteristics that are being observed **(1)**

measures of central tendency (or location) values around which observations tend to cluster and that describe the location of the "center" of a data set **(3)**

measures of dispersion (or variability) numbers that indicate the spread or scatter of observations; they show the extent to which individual values in a data set differ from one another and, hence, differ from their central location **(3)**

measures of shape numbers that indicate either the degree of asymmetry or that of peakedness in a frequency distribution **(3)**

median (*M* or *m*) a measure of central tendency; that value in an ordered array of data above and below which an equal number of observations can be found **(3)**

median class   the class among grouped data that contains the median   **(3)**

method of least squares   in regression analysis, the method of fitting a line to data displayed in a scatter diagram in such a way that the sum of the squares of the vertical deviations between the line and the individual data plots is minimized   **(11)**

minimax   one of several nonprobabilistic criteria for making decisions under uncertainty according to which a decision maker who seeks to minimize some cost finds the maximum cost associated with each possible action, identifies the minimum among these maxima, and chooses the action associated with this minimum of maxima   **(14)**

minimax regret   one of several nonprobabilistic criteria for making decisions under uncertainty according to which a decision maker finds the maximum regret value associated with each possible action, identifies the minimum among these maxima, and chooses the action associated with this minimum of maximum regret values   **(14)**

minimin   one of several nonprobabilistic criteria for making decisions under uncertainty according to which a decision maker who seeks to minimize some cost finds the minimum cost associated with each possible action, identifies the minimum among these minima, and chooses the action associated with this minimum of minima   **(14)**

mixed time-series model   a time-series model that combines features of the multiplicative model and additive model—for example, $Y = T \cdot C \cdot I + S$ **(13)**

modal class   the class among grouped data that contains the mode   **(3)**

mode (*Mo* or *mo*)   a measure of central tendency; the most frequently occurring value in a set of data **(3)**

more-than-fair gamble   a gamble in which the expected monetary value of what is given up is less than the expected monetary value of what is received   **(14)**

moving-averages series   a series of numbers obtained by successively averaging overlapping groups of two or more consecutive values in a time series and replacing the central value in each group by the group's average   **(13)**

multimodal frequency distribution   a frequency distribution in which two or more different values occur with the highest frequency, or almost that **(3)**

multinomial qualitative variable   a qualitative variable observations about which can be made in more than two categories   **(1)**

multiple correlation analysis   correlation analysis that measures the strength of association among more than two variables   **(12)**

multiple regression analysis   regression analysis in which several (independent) variables are used to estimate the value of an unknown (dependent) variable   **(11,12)**

multiplication law   a law of probability theory that is used to compute the probability for an *intersection* of two or more events; according to the general law for any two events, $p(A \ and \ B) = p(A) \cdot p(B|A)$ and also $p(A \ and \ B) = p(B) \cdot p(A|B)$; according to the special law for independent events, $p(A \ and \ B) = p(A) \cdot p(B)$ **(4)**

multiplicative time-series model   a time-series model, such as the classical one, that expresses the actual value of a time series as the product of its components—for example, $Y = T \cdot C \cdot S \cdot I$ **(13)**

multivariate data set   a data set containing information on more than two variables   **(1)**

mutually exclusive events   different random events that have no basic outcomes in common **(4)**

nominal data   numbers that merely *name* or label differences in kind   **(1)**

nonresponse bias   a systematic tendency for selected elementary units with particular characteristics not to contribute data in a survey, while other such units, with other characteristics, do   **(1)**

nonsampling error   see **systematic error**

nonsense correlation   see **spurious correlation**

normal-form analysis   a tabular form of preposterior analysis that systematically computes an expected payoff value for every conceivable strategy and selects the strategy with the best of these values as the optimal one   **(14)**

normal probability distribution   a probability density function that is (a) single-peaked above the random variable's mean, median, and mode, (b) perfectly symmetric about this central value, and (c) characterized by tails extending indefinitely in both directions from the center, approaching but never touching the horizontal axis   **(6)**

not statistically significant result   in hypothesis testing, any sample result that leads to the continued acceptance of the null hypothesis because it has a high probability of occurring when that hypothesis is true   **(8)**

null hypothesis the first of two opposing hypotheses in a hypothesis test, symbolized by H_0, and often (but not necessarily) representing that proposition about an unknown population parameter that is tentatively assumed to be true **(8)**

objective probability a numerical measure of chance that estimates the likelihood of a specific occurrence (event A) in a repeatable random experiment (according to the classical approach, it is equal to the number of equally likely basic outcomes favorable to the occurrence of event A divided by the number of equally likely basic outcomes possible; according to the empirical approach, it is equal to the number of times event A did occur in the past during a large number of experiments divided by the maximum number of times event A could have occurred during these experiments) **(4)**

observational study see **survey**

observed significance level see *p* **value**

ogive a graphical portrayal of a cumulative frequency distribution **(2)**

one-factor ANOVA see **one-way ANOVA**

one-sided hypothesis an alternative hypothesis that holds for deviations from the null hypothesis in one direction only **(8)**

one-tailed test a hypothesis test in which the alternative hypothesis is one-sided and in which the null hypothesis, therefore, is rejected only for very low (or very high) values of the test statistic located entirely in one tail of that statistic's sampling distribution **(8)**

one-way ANOVA one of several versions of the analysis of variance (it is performed with data derived from experiments based on the completely randomized design that controls extraneous factors by creating one treatment group for each treatment and assigning each experimental unit to one of these groups by a random process; thus, only one factor (the treatment) is considered as possibly affecting the variable of interest) **(10)**

open-ended classes classes that have only one stated end point, the upper or lower limit **(2)**

opportunity loss *(OL)* when a decision maker maximizes benefit, the difference between (a) the optimal payoff for a given event and (b) the actual payoff achieved as a result of taking a specified action and the subsequent occurrence of that event; when a decision maker minimizes cost, the difference between (a) the actual cost incurred as a result of taking a specified action and the subsequent

occurrence of an event and (b) the minimum cost achievable for that event **(14)**

optimal strategy in decision theory, a complete plan specifying the actions to be taken at each conceivable action point, if the expected (monetary or utility) payoff is to be the best one available **(14)**

ordered array a listing of numerical data in order of ascending or descending magnitude **(2)**

ordinal data numbers that by their size *order* or rank observations on the basis of importance, while intervals between the numbers, or ratios of such numbers, are meaningless **(1)**

outcome space see **sample space**

outliers wild values or maverick data that are not believable because they differ greatly from the majority of observed values **(1)**

parameter a summary measure based on population data **(3)**

payoffs in decision theory, the positive or negative net benefits that are associated with each possible action/event combination and that are, thus, the joint outcome of choice and chance **(14)**

payoff table a summary of a decision-making situation in the context of uncertainty, consisting of a tabular listing of the payoffs associated with all possible combinations of actions and events **(14)**

percentile a value in a data set below which a specified percentage of all values is found **(3)**

permutations distinguishable ordered arrangements of items all of which have been drawn from a given group of items **(4)**

pictogram a set of symbols used to depict data **(2)**

pie chart a portrayal of divisions of some aggregate by a segmented circle in such a way that the sector areas are proportional to the sizes of the divisions in question **(2)**

point estimate the estimate of a population parameter, expressed as a single numerical value **(7)**

polygon see **frequency polygon**

pooled variance the weighted average of two sample variances (the weights being the degrees of freedom associated with each variance), used to estimate the variance (known to be identical) in each of two populations from which the samples were taken **(8)**

population the set of all possible observations about a specified characteristic of interest **(1)**

population coefficient of correlation an index of the degree of association between two variables, denoted by ρ and equal to the square root of the population coefficient of determination **(11)**

population coefficient of determination the most important measure of how well a true regression line fits the population data on which it is based; this coefficient is denoted by ρ^2 **(11)**

population regression line see **true regression line**

population standard error of the estimate in regression analysis, the standard deviation, $\sigma_{Y \cdot X}$, of a conditional probability distribution of Y **(11)**

posterior analysis a form of decision making under uncertainty that starts out with prior probabilities, proceeds to gather additional experimental or sample evidence about event probabilities, and then uses this new evidence to transform the set of prior probabilities, by means of Bayes's theorem, into a revised set of posterior probabilities with the help of which a final decision is reached **(14)**

posterior expected value of perfect information (posterior *EVPI*) the *EVPI* that is computed when the expected value of the optimal action under uncertainty is based on posterior probabilities **(14)**

posterior probability a prior probability modified on the basis of new information **(4)**

power of the hypothesis test the probability (equal to $1 - \beta$) of avoiding a type II error when the null hypothesis is false **(8)**

predicted variable see **dependent variable**

prediction interval in regression analysis, a confidence interval within which a population parameter that is being estimated presumably lies **(11)**

predictor variable see **independent variable**

preposterior analysis a form of decision making under uncertainty that decides the worth of obtaining additional (experimental or sample) information before proceeding to prior or posterior analysis **(14)**

prior analysis decision making under uncertainty that employs probabilities but only those available prior to the gathering of new experimental or sample evidence about the likelihood of alternative future events **(14)**

prior expected value of perfect information (prior *EVPI*) the *EVPI* that is computed when the expected value of the optimal action under uncertainty is based on prior probabilities **(14)**

prior probability a person's initial subjective estimate of the likelihood of an event, prior to any empirical investigation **(4)**

probability density function a smooth frequency curve that describes the probability distribution of a continuous random variable **(6)** ·

probability distribution a systematic listing of each possible value of a random variable, together with the associated likelihood of its occurrence **(5)**

probability sample see **random sample**

probability space see **sample space**

proportion (π or P) a number that describes the frequency of observations in a particular category as a fraction of all observations made **(3)**

***p* value** in hypothesis testing, the probability that a value of the test statistic as extreme as or more extreme than the one actually observed could have occurred by chance, assuming that a hypothesized parameter value is true **(8)**; in regression analysis, the probability that a t value as extreme as or more extreme than the one actually observed could have occurred by chance, assuming the true regression coefficient, β, equals zero **(12)**

qualitative variable a variable that is normally not expressed numerically because it differs in kind rather than degree among elementary units **(1)**

quantitative variable a variable that is normally expressed numerically because it differs in degree rather than kind among elementary units **(1)**

quartiles values in a data set that divide the total of observation into four quarters, each of which contains .25 (or 25 percent) of the observed values **(3)**

random error in surveys, the difference between the value of a variable obtained by taking a single random sample and the value obtained by taking a census (or by averaging the results of all possible random samples of like size) **(1,10)**

random event any subset of the sample space **(4)**

random experiment any activity that will result in one and only one of several well-defined outcomes but that does not allow us to tell in advance which of these will prevail in any particular instance **(4)**

randomization a procedure that lets extraneous factors operate during an experiment but that assures, by virtue of the random assignment of experimental units to experimental and control groups, that each treatment has an equal chance to be enhanced or handicapped by those factors **(10)**

randomized block design an experimental plan that divides the available experimental units into blocks of fairly homogeneous units (each block containing as many units as there are treatments or some multiple of that number) and then matches each treatment with one or more units within each block by a random process **(10)**

randomized group design an experimental plan that creates one treatment group for each treatment and assigns each experimental unit to one of these groups by a random process **(10)**

random-numbers table a listing of numbers generated by a random process in such a way that each possible digit is equally likely to precede or follow any other one **(1)**

random sample a subset of a frame, or of an associated population, chosen by a random process that gives each unit of the frame or associated population a known positive (but not necessarily equal) chance to be selected **(1)**

random variable any quantitative variable the numerical value of which is determined by a random experiment and, thus, by chance **(5)**

range a distance measure of dispersion; the difference between the largest and smallest observation in a set of ungrouped data or between the upper limit of the largest class and the lower limit of the smallest class for grouped data **(3)**

ratio data numbers that by their size rank observations in order of importance and between which intervals as well as *ratios* are meaningful **(1)**

regressand see **dependent variable**

regression analysis a statistical method the central focus of which is the establishment of an *equation* that allows the unknown value of one variable to be estimated from the known value of one or more other variables **(11)**

regression effect the tendency of those members of any population who, with respect to a given characteristic, are in an extreme position (below or above the population mean) at one time to be in a less extreme position at a later time (either personally or by means of their offspring) **(11)**

regression fallacy the incorrect attribution of the regression effect to the operation of some important unseen factor **(11)**

regression hyperplane the four- and higher-dimensional equivalent of a regression line **(12)**

regression line in regression analysis, a line that summarizes the stochastic relationship between an independent variable, X, and a dependent variable, Y, while also minimizing the errors made when the equation of that line is employed to estimate Y from X **(11)**

regression plane the three-dimensional equivalent of a regression line that minimizes the sum of the squared vertical deviations between the sample points suspended in Y vs. X_1 vs. X_2 space and their associated multiple regression equation estimates—all of which lie on this plane **(12)**

regression sum of squares *(RSS)* see **explained variation of Y**

regressor see **independent variable**

regret see **opportunity loss**

regret table in decision theory, a table showing the opportunity-loss values associated with each possible action/event combination **(14)**

rejection region in a hypothesis test, that range of possible values of the test statistic which signals the rejection of the null hypothesis **(8)**

relative class frequency the proportion of all observations that fall into a given class **(2)**

relative frequency distribution a tabular summary of a data set showing the relative class frequencies in each of several collectively exhaustive and mutually exclusive classes **(2)**

residual variation see **unexplained variation**

response bias a tendency for answers to survey questions to be wrong in some systematic way **(1)**

risk aversion an attitude according to which a person considers the utility of a certain prospect of money to be higher than the expected utility of an uncertain prospect of equal expected monetary value **(14)**

risk neutrality an attitude according to which a person considers the utility of a certain prospect of money to be equal to the expected utility of an uncertain prospect of equal expected monetary value **(14)**

risk seeking an attitude according to which a person considers the utility of a certain prospect of money to be lower than the expected utility of an uncertain prospect of equal expected monetary value **(14)**

sample a subset of a population or of the frame from which it is derived **(1)**

sample coefficient of correlation an index of the degree of linear association between two variables, denoted by r and equal to the square root of the sample coefficient of determination **(11)**

sample coefficient of determination the most important measure of how well an estimated regression line fits the sample data on which it is based, denoted by r^2 and equal to the proportion of the total variation in the values of Y that can be explained by the association of Y with X as measured by the estimated regression line **(11)**

sample coefficient of multiple correlation an index of the degree of linear association between more than two variables, denoted by R and equal to the square root of the sample coefficient of multiple determination **(12)**

sample coefficient of multiple determination the most important measure of how well an estimated regression plane (or hyperplane) fits the sample data on which it is based, denoted by R^2 and equal to the proportion of the total variation in the values of the dependent variable Y that is explained by the multiple regression of Y on X_1, X_2, and possibly additional independent variables $(X_3, X_4,$ and so on) **(12)**

sample coefficient of partial correlation the square root of the sample coefficient of partial determination **(12)**

sample coefficient of partial determination in multiple regression analysis, an index of association between the dependent variable and just *one* of the independent variables, given a prior accounting of the effects of one or more other independent variables **(12)**

sample point see **simple event**

sample regression line see **estimated regression line**

sample space a listing of all the basic outcomes of a random experiment **(4)**

sample standard error of the estimate in regression analysis, the standard deviation, such as $s_{Y \cdot X}$ or $s_{Y \cdot X_1 X_2}$, of the values of Y observed in a sample around the least-squares regression line (or plane) estimated from this sample **(11)**

sample survey a partial survey in which observations are made about one or more characteristics of interest for only a subset of all existing elementary units **(1)**

sampling distribution a probability distribution that shows the likelihood of occurrence associated with all the possible values of a sample statistic, which values would be obtained when drawing all possible simple random samples of a given size from a population **(7)**

sampling error see **random error**

scatter diagram a graphical display used in regression analysis, consisting of a scatter of dots, each dot of which represents one value of the independent variable, which is measured along the horizontal axis, and an associated value of the dependent variable, which is measured along the vertical axis **(11)**

seasonal component in time-series analysis, narrow up-and-down swings of the variable of interest around the trend/cyclical components, with the swings predictably repeating each other within periods of one year or less **(13)**

second quartile the .50 fractile or 50th percentile in a data set below which half of all observations lie; the median **(3)**

selection bias a systematic tendency to favor the inclusion in a survey of selected elementary units with particular characteristics, while excluding other such units with other characteristics **(1)**

sequential sampling a form of sampling in which the sample size is not fixed in advance but in which a decision is made, after each bit of information is received, either to continue sampling or to put a stop to it **(14)**

significance level of the hypothesis test among all the sample results that are possible when the null hypothesis is true, the (arbitrary) maximum proportion, α, that is considered sufficiently unusual to reject the null hypothesis **(8)**

simple correlation analysis correlation analysis that measures the strength of association between two variables only **(11)**

simple event any single one of a random experiment's basic outcomes **(4)**

simple random sample a subset of a frame, or of an associated population, chosen in such a fashion that every possible subset of like size has an equal chance of being selected **(1)**

simple regression analysis regression analysis in which a single variable is used to estimate the value of an unknown variable **(11)**

single-parameter exponential smoothing one of several exponential-smoothing techniques **(13)**

size of the hypothesis test see **significance level of the hypothesis test**

Slutsky-Yule effect the appearance of artificially created and, therefore, misleading waves in a moving-averages series even when there are no waves in the original time series **(13)**

smoothing constant a parameter employed in the exponential-smoothing formula **(13)**

smoothing technique a statistical procedure, such as the construction of a moving-averages series, that dampens (or averages out) fluctuations in a time series **(13)**

spurious correlation high positive or negative correlation between two variables that have no logical connection with one another **(11)**

standard deviation (σ or s) an average deviation measure of dispersion, equal to the positive square root of the variance **(3)**

standard error the standard deviation of any statistic **(7)**

standard error of the proportion the standard deviation calculated from a sampling distribution of the proportion **(7)**

standard error of the sample mean the standard deviation, $\sigma_{\overline{X}}$, of the sampling distribution of the sample mean, $\overline{X}$ **(7)**

standard error of the sample proportion the standard deviation, σ_P, of the sampling distribution of the sample proportion, P **(7)**

standard normal curve a normal probability density function with a mean of zero and a standard deviation of one **(6)**

standard normal deviate the deviation from the mean, μ_R, of any given value, x, of a normally distributed random variable, measured in units of standard deviations, σ_R; symbolized by z **(6)**

standard scores the number of standard deviations that particular observations lie below or above the mean of all observations in a data set **(3)**

state-of-nature branches see **event branches**

state-of-nature fork see **event point**

state-of-nature node see **event point**

state-of-nature point see **event point**

states of nature see **events**

statistic a summary measure based on sample data **(3)**

statistically significant result in hypothesis testing, any sample result that leads to the rejection of the null hypothesis because it has a low probability of occurring when that hypothesis is true **(8)**

statistical map a portrayal of data for areal units by such devices as differentiated cross-hatching or shading of these units on a geographic map **(2)**

statistics a branch of mathematics that is concerned with facilitating wise decision making in the face of uncertainty and that, therefore, develops and utilizes techniques for the careful collection, effective presentation, and proper analysis of numerical information **(1)**

stochastic relationship a relationship between any two variables, X and Y, such that many possible values of Y can be associated with any one value of X **(11)**

stratified random sample a subset of a frame, or of an associated population, chosen by dividing the latter into mutually exclusive and clearly distinguishable strata (which are internally homogeneous but differ greatly from one another with respect to some characteristic of interest) and then taking separate (simple or systematic) random samples from every stratum, often in such a way that the sizes of the separate samples vary with the importance of the different strata **(1)**

Student's *t* distribution a sampling distribution for a random variable, t, derived from a normally distributed population and defined as $(\overline{X} - \mu)/(s/\sqrt{n})$ **(7)**

subjective probability a numerical measure of chance that expresses a purely personal degree of belief in the likelihood of a specific occurrence in a unique random experiment **(4)**

survey the collection of data from elementary units without exercising any particular control over factors that may make these units different from one another and that may, therefore, affect the characteristic of interest being observed **(1)**

systematic error in surveys, the difference between the value of a variable obtained by taking a census (or by averaging the results of all random samples of a given size) and the true value **(1)**

systematic random sample a subset of a frame, or of an associated population, chosen by randomly selecting one of the first k elements and then including every kth element thereafter **(1)**

test statistic the type of statistic to be computed from a simple random sample taken from the population of interest in a hypothesis test and to be used for establishing the truth or falsity of the null hypothesis **(8)**

theory of probability the calculus of the likelihood of specific occurrences **(4)**

third quartile the .75 fractile of 75th percentile in a data set below which three quarters of all observations lie **(3)**

three-parameter exponential smoothing one of several exponential-smoothing techniques that includes seasonal as well as trend smoothing **(13)**

time-series analysis a statistical procedure that employs time-series data, usually for the purpose of explaining past or forecasting future events **(13)**

time-series data numerical values pertaining to units of a given population that have been observed repeatedly at different points in time (in the case of a stock variable) or during different periods of time (in the case of a flow variable) **(13)**

time-series line graph a graphical portrayal, by a continuous line, of data that are linked with time **(2)**

total deviation of *Y* in regression analysis, the difference between any particular observation of Y and the mean, $\overline{Y}$, of all such observations **(11)**

total sum of squares *(Total SS)* in the analysis of variance, the sum of the squared deviations of each individual sample observation from the mean of all observations **(10,11)**

total utility of money the overall welfare a person derives from the possession of a given quantity of money **(14)**

total variation of Y in regression analysis, the sum of the squares of all the total deviations of Y, or $\Sigma(Y - \bar{Y})^2$, also called the **total sum of squares** **(11)**

treatments mean square *(TMS)* in the analysis of variance, the population variance estimate, s_A^2, based on the observed variation among the treatments samples **(10)**

treatments sum of squares *(TSS)* in the analysis of variance, the sum of the squared deviations between each treatment sample mean and the grand mean, multiplied by the number of observations made for each treatment, multiplied by the number of observations per cell (which may well equal 1) **(10)**

treatments variation in the analysis of variance, variation among treatment sample means that is attributable to inherent differences among treatment populations **(10)**

tree diagram a device for defining the sample space of multiple-step random experiments **(4), see also decision tree (14)**

trend component in time-series analysis, a relatively smooth, progressively upward or downward movement of the variable of interest over an extended period of time **(13)**

trend smoothing constant a parameter employed in the multiple-parameter exponential smoothing formula **(13)**

true regression coefficients the values of α and β found in the equation of the true regression line, $E(Y) = \mu_{Y \cdot X} = \alpha + \beta X$ **(11)**

true regression line the regression line calculated from census data by the method of least squares **(11)**

Tukey's HSD test a test, employed as a follow-up to ANOVA, that seeks out "honestly significant differences" among paired sample means **(10)**

t **value** the ratio of an estimated partial regression coefficient to its standard error **(12)**

two-factor ANOVA see **two-way ANOVA**

two-parameter exponential smoothing one of several exponential-smoothing techniques that includes trend smoothing **(13)**

two-sided hypothesis an alternative hypothesis that holds for deviations from the null hypothesis in either direction **(8)**

two-tailed test a hypothesis test in which the alternative hypothesis is two-sided and in which the null hypothesis, therefore, is rejected for values of the test statistic located in either tail of that statistic's sampling distribution **(8)**

two-way ANOVA one of several versions of the analysis of variance [performed with data derived from experiments based on the randomized block design that controls extraneous factors by dividing the available experimental units into distinctly different, but internally homogeneous, blocks and then randomly matching each treatment with one or more units within any given block; thus, two factors (treatments and blocks) are considered as possibly affecting the variable of interest] **(10)**

type I error in hypothesis testing, the erroneous rejection of a null hypothesis that is in fact true **(8)**

type II error in hypothesis testing, the erroneous acceptance of a null hypothesis that is in fact false **(8)**

unbiased estimator a sample statistic that, on the average, across many samples, takes on a value equal to the population parameter that is being estimated with its help **(7)**

unconditional probability a measure of the likelihood that a particular event occurs, regardless of whether another event occurs **(4)**

unexplained deviation of Y in regression analysis, the difference between any particular observation of Y and its regression estimate, $\hat{Y}_X$ **(11)**

unexplained variance in the analysis of variance, the error mean square; the denominator of the F statistic **(10)**

unexplained variation in the analysis of variance, the variation of the sample data *within* each of the samples about the respective sample means; it is attributed to chance and equals the total sum of squares minus explained variation **(10)**

unexplained variation of Y in regression analysis, the sum of the squares of all the unexplained deviations of Y, or $\Sigma(Y - \hat{Y}_X)^2$, also called the **error sum of squares** **(11)**

unfair gamble a gamble in which the expected monetary value of what is given up exceeds the expected monetary value of what is received **(14)**

union of two events all the basic outcomes contained in either one random event *or* another, or, conceivably, in both **(4)**

univariate data set a data set containing information on one variable only   **(1)**

universe see **population**

upper-tailed test a one-tailed hypothesis test in which the entire rejection region is found in the upper tail   **(8)**

utility-of-money function the relationship between alternative amounts of money a person might possess and the different utility totals associated with these amounts   **(14)**

variables characteristics possessed by elementary units   **(1)**

variance (σ^2 or s^2) an average deviation measure of dispersion, obtained by averaging the squares of deviations of individual observations from the mean   **(3)**

Venn diagrams graphical devices that depict sample spaces and random events symbolically   **(4)**

weighted mean (μ_w or $\overline{X}_w$) a measure of central tendency; a form of arithmetic mean that gives different observations unequal weights in accordance with their unequal relative importance   **(3)**

ANSWERS TO ODD-NUMBERED QUESTIONS AND PROBLEMS

Chapter 1

1. a. Seven (as given in the headings of columns 2–8).

b. Qualitative: the headings of columns 2, 3, 4, and 7.

Quantitative: the headings of columns 5, 6, and 8.

c. Qualitative: all multinomial

Quantitative: all continuous

3. qualitative and dichotomous: *c, e*

qualitative and multinomial: *f*

quantitative and discrete: *a, b*

quantitative and continuous: *b, d*

Note: Item *b* is listed twice because dollar figures that measure the value of coins and paper bills in our pockets might be considered *discrete* because numbers are necessarily separated from each other by distances of one penny. One could find numbers such as $17.23 and $17.24, but never a $17.238. Dollar figures in other contexts, however, might be considered *continuous*. A tax rate might be expressed as 66.75 mills, which equals 6.675 cents on the dollar; an exchange rate might be stated as $8.9776 per unit of foreign currency; and the average cost of production might equal $17.5396 per unit of output.

5. Qualitative: *c, e, f*

Quantitative: *a, b, d*

7. multinomial: *c, e, f*

continuous: *a, b, d*

9. Nominal: *a, d, f;* ordinal: *c;* ratio: *b, e*

11. Nominal: *b, d*

13. Census: *a* (if small firm); sample: *a* (if large firm), *b* (the population is infinite), *c* for all of the reasons for sampling given in the text).

15. Answers can vary, but this is how the A. C. Nielsen Co. does it: Electronic devices, called *audimeters,* are attached to all TV sets owned by a sample of 1,700 homes and connected to a Nielsen computer via special telephone lines. The devices record when the TV sets are on or off, as well as the channels tuned. A Nielson rating of 25 means that 25 percent of sample households were tuned to the rated show for at least 6 minutes.

17. Convenience sample: *a;* judgment sample: *b;* probability sample: *c.*

19. Problem 16: none; problem 17c: clustered random sample; problem 18a: clustered random sample; problem 18b: systematic random sample.

21. See footnote 1 of this chapter. For $n = 5$ (and $N = 100$), the answer is 75,287,520. For $n = 50$, it is 100,891,344,545,564,200,000,000,000,000, a remarkable number calculated by computer. (The footnote formula here reappears in Chapter 4 as Box 4.F.)

23. See footnote 1 of this chapter. For $n = 3$ (and $N = 9$), the answer is 84. For $n = 1$, it is 9.

25. 13 Burroughs, 41 Getty Oil, 39 General Motors, 27 Digital Equipment, 02 American Cyanamid, 31 Exxon, 04 American Home Products, 20 Colgate-Palmolive, 43 Goodyear, 26 Deere.

27. 06 Evergreen Fund, 02 Amcap Fund, 15 Nicholas Fund, 04 American Capital Venture Fund, 05 Claremont Capital Corp., 08 Fidelity Destiny Fund.

29. 13 Mass Cap Development Fund, 20 Sigma Venture Shares, 22 Twentieth Century Select, 16 Over-the-Counter Securities Fund, 04 American Capital Venture Fund, 03 American Capital Pace Fund.

31. 25 Dart and Kraft, 41 Getty Oil, 30 Eastman Kodak, 93 Union Carbide, 57 Kmart, 31 Exxon.

33. Answers will vary, but students should have followed a procedure like the one discussed in the text in connection with the selection of 10 multinationals.

35. The value of k is $(100/4) = 25$; the sample is 15 Chase Manhattan, 40 General Telephone and Electric, 65 J. P. Morgan, 90 Texas Instruments.

37. By taking every 50th bill, starting with one chosen at random from serial numbers 00–49. If that random number were 17, the sample would consist of bills numbered 17, 67, 117, and so on.

39. A sample of one.

41. *Cluster 17:* 85 Standard Oil California, 86 Standard Oil Indiana, 87 Sun Co., 88 Tenneco, 89 Texaco. *Cluster 05:* 25 Dart and Kraft, 26 Deere, 27 Digital Equipment, 28 Dow Chemical, 29 DuPont de Nemours.

43. Stratified random sample; random error.

45. Selection bias: *d;* response bias: *a, e;* nonresponse bias: *b, c.*

Chapter 2

1. The answer can most easily be produced with the help of a so-called *stem-and-leaf* diagram as in Table A.1. The *ordered array* is given in Table A.2.

3. The *ordered array* is given in Table A.3.

5. For $n = 80$: $k = 1 + 3.3 \log 80 = 1 + 3.3(1.9031) = 7.28$. Seven classes. For $n = 400$: $k = 1 + 3.3 \log 400 = 1 + 3.3(2.6021) = 9.59$. Ten classes.

7. Arrangement 1 fails to make the classes *collectively exhaustive;* it does not cover the whole range of possible data; some of them, therefore, could not be classified (for example, 15.5 or 15.59). Arrangement 2 fails to make the classes *mutually exclusive;* classes overlap; some data, therefore, have two places to go (for example, 15.5).

Table A.1

Stems	Leafs														
15.3	3	9													
15.4	4														
15.5	1	3	7												
15.6	2	9													
15.7	1	2	3	4	5	5	7	7	9						
15.8	1	2	3	4	4	5	9								
15.9	0	0	0	1	1	1	2	2	2	3	3	4	5	6	8 8 8 9
16.0	0	0	0	1	1	1	4	5	7	8	9				
16.1	0	1	2	3	7	9									
16.2	0	1	4	5	7	9									
16.3	1	7	8												
16.4	0	2	3	7	9										
16.5	3														
16.6	0	3	5	8											
16.7															
16.8	1	8													

Table A.2

15.33	15.71	15.79	15.90	15.92	15.98	16.04	16.13	16.29	16.49
15.39	15.72	15.81	15.90	15.93	15.99	16.05	16.17	16.31	16.53
15.44	15.73	15.82	15.90	15.93	16.00	16.07	16.19	16.37	16.60
15.51	15.74	15.83	15.91	15.94	16.00	16.08	16.20	16.38	16.63
15.53	15.75	15.84	15.91	15.95	16.00	16.09	16.21	16.40	16.65
15.57	15.75	15.84	15.91	15.96	16.01	16.10	16.24	16.42	16.68
15.62	15.77	15.85	15.92	15.98	16.01	16.11	16.25	16.43	16.81
15.69	15.77	15.89	15.92	15.98	16.01	16.12	16.27	16.47	16.88

Table A.3

15,971	19,900	21,486	23,380	25,829
17,356	20,153	21,536	24,110	25,861
18,329	20,209	21,705	24,378	25,920
18,577	20,452	22,064	24,435	26,300
18,954	20,607	22,141	24,456	26,709
19,014	20,691	22,520	24,520	27,384
19,180	20,934	22,600	24,577	28,401
19,563	21,057	22,737	24,628	28,621
19,690	21,208	22,924	25,125	29,000
19,700	21,307	23,089	25,610	39,751
19,800				

9. a. See Table A.4.

Table A.4

Class (net weight in ounces) (1)	Absolute Class Frequency (number of cans in class) (2)
15.3 to under 15.4	2
15.4 to under 15.5	1
15.5 to under 15.6	3
15.6 to under 15.7	2
15.7 to under 15.8	9
15.8 to under 15.9	7
15.9 to under 16.0	18
16.0 to under 16.1	11
16.1 to under 16.2	6
16.2 to under 16.3	6
16.3 to under 16.4	3
16.4 to under 16.5	5
16.5 to under 16.6	1
16.6 to under 16.7	4
16.7 to under 16.8	0
16.8 to under 16.9	2
Total	80

b. See Table A.5.

Table A.5

Class (net weight in ounces) (1)	Absolute Class Frequency (number of cans in class) (2)
15.3 to under 16.1	53
16.1 to under 16.9	27
Total	80

11. See Table A.6.

Table A.6

Class (net weight in ounces) (1)	Absolute Class Frequency (number of cans in class) (2)	Relative Class Frequency (proportion of all cans in class) (3)
15.3 to under 15.5	3	.0375
15.5 to under 15.7	5	.0625
15.7 to under 15.9	16	.2000
15.9 to under 16.1	29	.3625
16.1 to under 16.3	12	.1500
16.3 to under 16.5	8	.1000
16.5 to under 16.7	5	.0625
16.7 to under 16.9	2	.0250
Totals	80	1.0000

13. Sturgess's rule suggests seven classes $(1 + 3.3 \log 51 = 6.64)$. Dividing the data range of $39,751 − $15,971 = $23,780 by 7, suggests class widths of $3,500 (rounded as specified). The result is Table A.7.

Table A.7

Class (salary) (1)	Absolute Class Frequency (number of states in class) (2)	Relative Class Frequency (proportion of states in class) (3)
15,500 to under 19,000	5	.0980
19,000 to under 22,500	21	.4118
22,500 to under 26,000	18	.3529
26,000 to under 29,500	6	.1176
29,500 to under 33,000	0	0
33,000 to under 36,500	0	0
36,500 to under 40,000	1	.0196
Total	51	1.0000

15. See Table A.8.

Table A.8

Class (net weight in ounces) (1)	Cumulative Absolute Class Frequency (number of cans in class or lower ones) (2)	Cumulative Relative Class Frequency (proportion of cans in class or lower ones) (3)
15.3 to under 15.5	3	.0375
15.5 to under 15.7	8	.1000
15.7 to under 15.9	24	.3000
15.9 to under 16.1	53	.6625
16.1 to under 16.3	65	.8125
16.3 to under 16.5	73	.9125
16.5 to under 16.7	78	.9750
16.7 to under 16.9	80	1.0000

17. See Table A.9.

Table A.9

Classes (salary) (1)	Cumulative Absolute Class Frequency (number of states in class or higher ones) (2)	Cumulative Relative Class Frequency (proportion of states in class or higher ones) (3)
15,500 to under 19,000	51	1.0000
19,000 to under 22,500	46	.9020
22,500 to under 26,000	25	.4902
26,000 to under 29,500	7	.1373
29,500 to under 33,000	1	.0196
33,000 to under 36,500	1	.0196
36,500 to under 40,000	1	.0196

19. The missing values are 469.5, 843.1, 260.5, 140.5, totaling 1,713.6.
21. See Figures A.1 and A.2, respectively, depicting the cases of too many and too few classes.
23. The answers can vary, but some likely ones are as follows: skewed to the right—a, c, e; skewed to the left—d, f, h; rectangular—b, g, i.
25. Quite possibly the two machines are not producing parts of the same average thickness, as the double peak in the histogram suggests.
27. See Figures A.3 through A.5. Compare the answer to the answer you got for problem 21.
29. See Figure A.6.

Figure A.1

Figure A.2

Figure A.3

Figure A.4

Figure A.5

Figure A.6

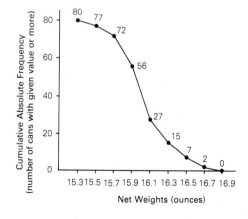

31. Table 2.G, "U.S. Net Import Reliance, 1981," is graphed in Figure A.7. Note: It is customary to order the bars by size in ascending or descending order. Any category named other is, however, always shown last, if it exists. (In this example such a category does not exist.)

Figure A.7

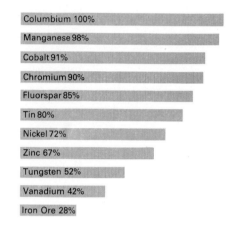

33. See Figure A.8.

Figure A.8 State and Local Government Taxes—Percent Distribution by Type: 1965 to 1982

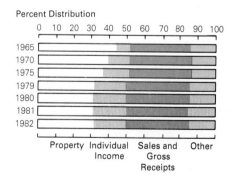

Source: U. S. Bureau of the Census.

35. See Figure A.9.

Figure A.9   Retail Store Sales by Kind of Business: 1982 and 1983

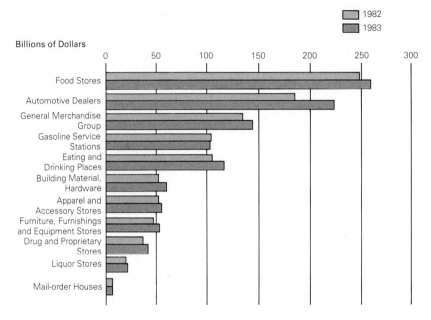

Source: U. S. Bureau of the Census.

37. See Figure A.10.

Figure A.10

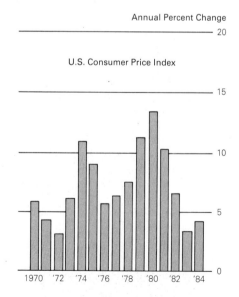

Source: U. S. Bureau of the Census.

39. Table 2.O, "U.S. Business Inventories," is graphed in Figure A.11.

Figure A.11

41. Table 2.P, "Personnel Problems at City Motors, 1986," is graphed in Figure A.12.

Figure A.12

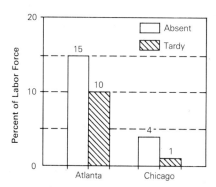

43. Table 2.R, "Car-Rental Market Shares, March 1982," is graphed in Figure A.13. Note: The 360 degrees of the circle are split in accordance with the percentages given in the table: 39.7 percent of 360° =

Figure A.13

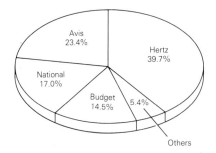

= 143° going to Hertz, 23.4 percent of 360° = 84° to Avis, and so on.

45. Table 2.S, "Fewer Farmers Feeding More People," is graphed in Figure A.14.

Chapter 3

1. a. List 1: the sample mean is $\overline{X}$ = $912.33 million; the sample median is m = $631 million; there is no mode (mo). List 2: the sample mean is $\overline{X}$ = 15.53%; the sample median is m = 15.2%; there is no mode (mo). List 3: the sample mean is $\overline{X}$ = $10,208.93 million; the sample median is m = $9,039; there is no mode (mo).

b. No, except that the population mean, median, and mode, respectively, would be symbolized by μ, **M**, and Mo.

3. The sample mean is $\overline{X}$ = 16.03 ounces; the sample median is m = 15.98 ounces; there are 6 modes: 15.9, 15.91, 15.92, 15.98, 16, and 16.01 ounces.

5. a. The population mean is μ = $22,832.34; the population median is **M** = $22,141; there is no mode.

b. Presumably, the U.S. average given in the table is a weighted average that weights each state average by that state's school-teacher population.

7. The population mean is μ = $22,808.83; the population median is **M** = $22,416.67; the population mode is Mo = $21,947.37.

9. The sample mean is $\overline{X}$ = $65.08; the sample median is m = $60.60; the sample mode is mo = $61.19.

11. We can make use of the relative frequency distribution found in the Chapter 2, problem 13 answer (Table A.7) and draw the graph, labelled A.15, that follows.

Figure A.14

Farm Population / Nonfarm Population / Total Population

1940: 31 million / 101 million / 132 million

1970: 10 million / 193 million / 203 million

Figure A.15

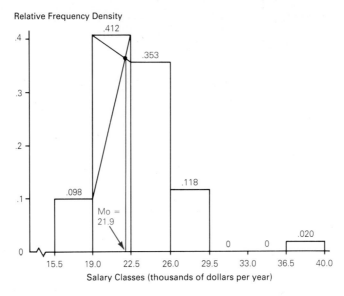

13. Answers can vary, but here is a possible one: $278,000; 17,000; 15,000; 12,000; 11,000; $9,000; 2,000; 2,000; 2,000; 2,000. Note how the *mean* equals $35,000; the *median* is $10,000; the *mode* is $2,000. Presumably, the real-estate agent trying to sell houses will use the first of these "averages"; the mayor looking for a federal grant will probably use the last one.

15. Mean: *iii, v, viii;* median: *ii* (the distributions are skewed), *iv, vii;* mode: *i, vi, vii.*

17. 8.62 (21.5) = 185.33
    57.51 (7.9) = 454.33
    42.91 (7.4) = 317.53
    ———————————————
    ΣwX = 957.19

$$\frac{\Sigma wX}{\Sigma w} = \frac{957.19}{109.04} = 8.8 \text{ percent}$$

This weighted average compares with an unweighted average of 12.3 percent.

19. The weighted mean is μ_w = 3.54 percent. The unweighted mean is μ = 3.51 percent.

21. a. Range: 16.88 − 15.33 = 1.55 ounces
    b. 1st quartile: 15.835 ounces; 3rd quartile: 16.205 ounces.
    c. 2nd decile: 15.78 ounces; 7th decile: 16.125 ounces.
    d. See Table A.10 and Figure A.16. Note: Can you see why less-than and more-than ogives, when plotted in the same graph, would intersect above the median?

Table A.10

Class (net weight in ounces)	Less-than-or-Equal Cumulative Relative Class Frequency (proportion of cans in class or lower ones)
15.3 to under 15.5	.04
15.5 to under 15.7	.10
15.7 to under 15.9	.30
15.9 to under 16.1	.66
16.1 to under 16.3	.81
16.3 to under 16.5	.91
16.5 to under 16.7	.97
16.7 to under 16.9	1.00

Figure A.16

23. a. Range
List 1: $2,250 million − $347 million =
$1,903 million
List 2:   22.1% − 11.8% = 10.3%
List 3: $15,060 − $8,210 = $6,850
b. Interquartile Range
List 1: $1,612 million − $503 million =
$1,109 million
List 2:   17.2% − 13.1% = 4.1%
List 3: $11,130 − $8,357 = $2,773
25. a. Range
Table 3.F data: 17.5 percent − 4.7 percent =
12.8 percent
Table 3.G data: 3.95 percent − 2.32 percent =
1.63 percent
Table 3.H data: 12.3 percent − 2.4 percent =
9.9 percent
b. Interquartile Range
Table 3.F data: 17 percent − 9.3 percent = 7.7
percent
Table 3.G data: 3.79 percent − 3.19 percent =
0.60 percent
Table 3.H data: 10.35 percent − 3.95 percent
= 6.4 percent
27. a. MAD = (31.6/15) = 2.1 percent. See
Table A.11.
b. See Table A.12.
Using the *long method*:

$$s^2 = \frac{105.58}{14}$$

= 7.54 percent squared.

Using the *short method*:

$$s^2 = \frac{3,721.73 - 15(15.52667)^2}{14}$$

$$= \frac{3,721.73 - 3,616.16}{14}$$

$$= \frac{105.57}{14}$$

= 7.54 percent squared.

c. $s = \sqrt{7.54 \text{ percent squared}} = 2.746$ percent.
29. Table 2.A data: MAD = .244 ounces; s^2 = .1016
squared ounces; s = .3188 ounces. Table 2.B data:
MAD = 13,200 miles; s^2 = 239,015,400 squared
miles; s = 15,460 miles. Table 2.C data: MAD =
$2,850.85; σ^2 = 14,696,250 squared dollars; σ =
$3,833.57. (Data treated as population data.)

Table A.11

$\|X - \overline{X}\|$
6.6
3.4
2.3
1.7
1.5
.5
.2
.3
.7
1.1
1.5
2.4
2.8
2.9
3.7
$\Sigma\|X - \overline{X}\| = 31.6$

Table A.12

$(X - \overline{X})^2$	X^2
43.56	488.41
11.56	357.21
5.29	316.84
2.89	295.84
2.25	289.00
.25	256.00
.04	234.09
.09	231.04
.49	219.04
1.21	207.36
2.25	196.00
5.76	171.61
7.84	161.29
8.41	158.76
13.69	139.24
$\Sigma(X - \overline{X})^2 = 105.58$	$\Sigma X^2 = 3,721.73$

31. See Table A.13 on the following page.
33. MAD = $559.677; s^2 = 1,398,516 squared dollars;
s = $1,182.589. If these were population data, one
would estimate σ^2 = 1,398,286 squared dollars
and σ = $1,182.491.
35. MAD = 8.845 months; σ^2 = 442.948 squared
months; σ = 21.046 months. If these were sample
data, one would estimate s^2 = 443.135 squared
months and s = 21.051 months.
37. The lower control limit is $\mu - 2\sigma$ = 15.394004
ounces; the upper control limit is $\mu + 2\sigma$ =
16.661256 ounces. The sample contains one reading below the lower limit and two above the upper
limit. Thus, there is reason for concern.

Table A.13

f	X	fX	X^2	fX^2
3	15.4	46.2	237.16	711.48
5	15.6	78.0	243.36	1,216.80
16	15.8	252.8	249.64	3,994.24
29	16.0	464.0	256.00	7,424.00
12	16.2	194.4	262.44	3,149.28
8	16.4	131.2	268.96	2,151.68
5	16.6	83.0	275.56	1,377.80
2	16.8	33.6	282.24	564.48
$n = 80$		$\Sigma fX = 1{,}283.2$		$\Sigma fX^2 = 20{,}589.76$

$$\overline{X} = \frac{1{,}283.2}{80} = 16.04 \text{ ounces.}$$

$$s^2 = \frac{20{,}589.76 - 80(16.04)^2}{79}$$

$$= \frac{20{,}589.76 - 20{,}582.53}{79}$$

$$= \frac{7.23}{79}$$

$$= .09 \text{ ounces squared.}$$

$$s = \sqrt{.09 \text{ ounces squared}} = .3 \text{ ounces.}$$

39. A normal frequency distribution was assumed because it would place 95.4 percent of all observations within 2 standard deviations of the mean. If the assumption did not hold, a more conservative estimate might be made using Chebyshev's theorem: At least 75 percent of the lightbulbs would have lifetimes within the range indicated by the manager.

41. Relative dispersion can be measured by the coefficient of variation, $v = (s/\overline{X})$. The three sample means were calculated in problem 1a as $912.33 million, 15.53 percent, and $10,208.93 million. The sample standard deviation for List 2 was calculated in problem 27c as 2.746 percent; the remaining two have to be calculated as shown in Table A.14. Thus, the three coefficients of variation are the following:

$$v_1 = \frac{\$598.1733 \text{ million}}{\$912.33 \text{ million}} = .656$$

$$v_2 = \frac{2.746\%}{15.53\%} = .177$$

$$v_3 = \frac{\$2{,}580.3949}{\$10{,}208.93} = .253$$

Dispersion is least by the criterion of R&D spending as a percent of sales; it is highest by the criterion of total dollar spending on R&D.

43. Aggressive growth: $4/23 = .17$
Small company growth: $6/23 = .26$

45. The range of $\overline{X} \pm 2.5s$ extends from $16.03 - (2.5 \times .3188) = 15.233$ ounces to $16.03 + (2.5 \times .3188) = 16.827$ ounces. It covers 79 of the 80 data, or 98.75 percent of them. Chebyshev's theorem gives a minimum proportion of

$$1 - \frac{1}{(2.5)^2} = 1 - \frac{1}{6.25}$$

$$= 1 - .16 = .84;$$

that is, of 84 percent. The result is consistent with Chebyshev's theorem.

Chapter 4
1. a. There are 38 univariate basic outcomes.
   b. See Figure A.17. There are 8 multivariate basic outcomes.

Table A.14

List 1	List 3
X^2	X^2
5,062,500	226,803,600
2,951,524	220,552,201
2,842,596	217,857,600
2,598,544	123,876,900
712,336	113,998,329
662,596	94,517,284
541,696	88,228,449
398,161	81,703,521
396,900	72,795,024
378,225	71,605,444
276,676	70,862,724
253,009	69,839,449
163,216	68,840,209
136,161	67,667,076
120,409	67,404,100

$$\Sigma X^2 = 17{,}494{,}549 \qquad \Sigma X^2 = 1{,}656{,}551{,}910$$

$$s^2 = \frac{17{,}494{,}549 - 15(912.33)^2}{14} \qquad s^2 = \frac{1{,}656{,}551{,}910 - 15(10{,}208.93)^2}{14}$$

$$= \frac{5{,}009{,}359}{14} = \$357{,}811.35 \text{ million squared} \qquad = \frac{93{,}218{,}134}{14} = \$6{,}658{,}438.1 \text{ squared}$$

$$s = \$598.1733 \text{ million} \qquad s = \$2{,}580.3949$$

Figure A.17

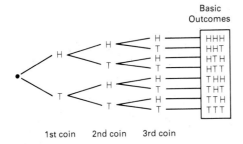

Basic
Outcomes

HHH
HHT
HTH
HTT
THH
THT
TTH
TTT

1st coin   2nd coin   3rd coin

3. See the four parts of Figure A.18 on the next page.

5. Mutually exclusive: *i, ii*; compatible: *iii, iv, v*; collectively exhaustive: *ii, iv*; not collectively exhaustive: *i, iii, v*.

7. (i) Not finding defective items in the group. (ii) Inflation not above 10 percent per year. (iii) Real GNP growth below 3 percent per year. (iv) Not drawing an ace from a deck of cards. (v) Finding at least one defective item in the group.

9. Unions: A or C; B or $\overline{C}$
Intersections: A and B; $\overline{B}$ and C
Complements: $\overline{A}$ (a nonbrick house)

11. Objective, classical: *iii, viii, x*; objective, empirical: *vi*; subjective: *i, ii, iv, v, vii, ix*.

13. Four outcomes are possible, as shown in panel (a) of Figure 4.2 in the text, if "1st toss" and "2nd toss," respectively, are replaced by "1st coin" and "2nd coin." Each person, therefore, has an *equal* chance to win. In the long run, the person proposing the bet will get $2 for every $1 paid out.

15. a. p(Avis customer) = .234, assuming that the distribution of customers equals that of their expenditures on which Figure A.13 is presumably based.

b. p(net weight 15.5 to under 15.7 oz) = .06. (The answer is *not* .10 because the table shows a *cumulative* distribution.)

17. $p(<\$100) = (900/6{,}066) = .148$; $p(\$100 \text{ to under } \$500) = (4{,}264/6{,}066) = .703$; $p(>\$500) = (902/6{,}066) = .149$.

19. $P_1^2 = 2$
$P_4^6 = 360$
$P_2^{15} = 210$
$P_7^{25} = 2{,}422{,}728{,}000$
P_{32}^{25} is an impossibility, because Formula 4.D requires $x \le n$.

21. $P_{5,5,5}^{15} = 756{,}756$; $P_{2,2,3,3}^{10} = 25{,}200$.

Figure A.18

Case i   Text Figure B

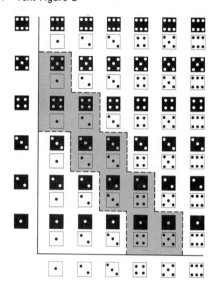

Case ii   Text Figure C

Case iii   Text Figure D

	Clubs	Diamonds	Hearts	Spades
King	♣ K	♦ K	♥ K	♠ K
Queen	♣ Q	♦ Q	♥ Q	♠ Q
Jack	♣ J	♦ J	♥ J	♠ J
10	♣ 10	♦ 10	♥ 10	♠ 10
9	♣ 9	♦ 9	♥ 9	♠ 9
8	♣ 8	♦ 8	♥ 8	♠ 8
7	♣ 7	♦ 7	♥ 7	♠ 7
6	♣ 6	♦ 6	♥ 6	♠ 6
5	♣ 5	♦ 5	♥ 5	♠ 5
4	♣ 4	♦ 4	♥ 4	♠ 4
3	♣ 3	♦ 3	♥ 3	♠ 3
Deuce	♣ 2	♦ 2	♥ 2	♠ 2
Ace	♣ A	♦ A	♥ A	♠ A

Case iv   Text Figure E

	Clubs	Diamonds	Hearts	Spades
King	♣ K	♦ K	♥ K	♠ K
Queen	♣ Q	♦ Q	♥ Q	♠ Q
Jack	♣ J	♦ J	♥ J	♠ J
10	♣ 10	♦ 10	♥ 10	♠ 10
9	♣ 9	♦ 9	♥ 9	♠ 9
8	♣ 8	♦ 8	♥ 8	♠ 8
7	♣ 7	♦ 7	♥ 7	♠ 7
6	♣ 6	♦ 6	♥ 6	♠ 6
5	♣ 5	♦ 5	♥ 5	♠ 5
4	♣ 4	♦ 4	♥ 4	♠ 4
3	♣ 3	♦ 3	♥ 3	♠ 3
Deuce	♣ 2	♦ 2	♥ 2	♠ 2
Ace	♣ A	♦ A	♥ A	♠ A

23. See Table A.15.

25.
$$\text{Entrees: } C_2^5 = \frac{5!}{2!(5-2)!} = 10.$$

$$\text{Vegetables: } C_2^4 = \frac{4!}{2!(4-2)!} = 6.$$

$$\text{Desserts: } C_2^7 = \frac{7!}{2!(7-2)!} = 21.$$

$$\text{Dinners: } C_2^3 \cdot C_2^5 \cdot C_2^4 \cdot C_2^6 \cdot C_2^7 = 3 \cdot 10 \cdot 6 \cdot 15 \cdot 21$$
$$= 56,700.$$

27. $C_3^{10} \cdot C_2^{10}$

$$= \left(\frac{10!}{3!(10-3)!}\right)\left(\frac{10!}{2!(10-2)!}\right)$$

$$= \left(\frac{10 \cdot 9 \cdot 8}{3 \cdot 2 \cdot 1}\right)\left(\frac{10 \cdot 9}{2 \cdot 1}\right)$$

$$= 120(45) = 5,400.$$

29. *Favorable outcomes* $= C_{11}^{50}$ because in every hand with the desired cards, there are 11 other cards that can be selected in C_{11}^{50} ways from the remain-

Table A.15

						Number of Combinations (x at a time out of n distinct items; no repetitions allowed.)					
x \ n	0	1	2	3	4	5	6	7	8	9	10
0	1										
1	1	1									
2	1	2	1								
3	1	3	3	1							
4	1	4	6	4	1						
5	1	5	10	10	5	1					
6	1	6	15	20	15	6	1				
7	1	7	21	35	35	21	7	1			
8	1	8	28	56	70	56	28	8	1		
9	1	9	36	84	126	126	84	36	9	1	
10	1	10	45	120	210	252	210	120	45	10	1
11	1	11	55	165	330	462	462	330	165	55	11
12	1	12	66	220	495	792	924	792	495	220	66
13	1	13	78	286	715	1,287	1,716	1,716	1,287	715	286
14	1	14	91	364	1,001	2,002	3,003	3,432	3,003	2,002	1,001
15	1	15	105	455	1,365	3,003	5,005	6,435	6,435	5,005	3,003
16	1	16	120	560	1,820	4,368	8,008	11,440	12,870	11,440	8,008
17	1	17	136	680	2,380	6,188	12,376	19,448	24,310	24,310	19,448
18	1	18	153	816	3,060	8,568	18,564	31,824	43,758	48,620	43,758
19	1	19	171	969	3,876	11,628	27,132	50,388	75,582	92,378	92,378
20	1	20	190	1,140	4,845	15,504	38,760	77,520	125,970	167,960	184,756
21	1	21	210	1,330	5,985	20,349	54,264	116,280	203,490	293,930	352,716
22	1	22	231	1,540	7,315	26,334	74,613	170,544	319,770	497,420	646,646
23	1	23	253	1,771	8,855	33,649	100,947	245,157	490,314	817,190	1,144,066
24	1	24	276	2,024	10,626	42,504	134,596	346,104	735,471	1,307,504	1,961,256
25	1	25	300	2,300	12,650	53,130	177,100	480,700	1,081,575	2,042,975	3,268,760
26	1	26	325	2,600	14,950	65,780	230,230	657,800	1,562,275	3,124,550	5,311,735
27	1	27	351	2,925	17,550	80,730	296,010	888,030	2,220,075	4,686,825	8,436,285
28	1	28	378	3,276	20,475	98,280	376,740	1,184,040	3,108,105	6,906,900	13,123,110
29	1	29	406	3,654	23,751	118,755	475,020	1,560,780	4,292,145	10,015,005	20,030,010
30	1	30	435	4,060	27,405	142,506	593,775	2,035,800	5,852,925	14,307,150	30,045,015
31	1	31	465	4,495	31,465	169,911	736,281	2,629,575	7,888,725	20,160,075	44,352,165
32	1	32	496	4,960	35,960	201,376	906,192	3,365,856	10,518,300	28,048,800	64,512,240
33	1	33	528	5,456	40,920	237,336	1,107,568	4,272,048	13,884,156	38,567,100	92,561,040
34	1	34	561	5,984	46,376	278,256	1,344,904	5,379,616	18,156,204	52,451,256	131,128,140
35	1	35	595	6,545	52,360	324,632	1,623,160	6,724,520	23,535,820	70,607,460	183,579,396
36	1	36	630	7,140	58,905	376,992	1,947,792	8,347,680	30,260,340	94,143,280	254,186,856
37	1	37	666	7,770	66,045	435,897	2,324,784	10,295,472	38,608,020	124,403,620	348,330,136
38	1	38	703	8,436	73,815	501,942	2,760,681	12,620,256	48,903,492	163,011,640	472,733,756
39	1	39	741	9,139	82,251	575,757	3,262,623	15,380,937	61,523,748	211,915,132	635,745,396
40	1	40	780	9,880	91,390	658,008	3,838,380	18,643,560	76,904,685	273,438,880	847,660,528
41	1	41	820	10,660	101,270	749,398	4,496,388	22,481,940	95,548,245	350,343,565	1,121,099,408
42	1	42	861	11,480	111,930	850,668	5,245,786	26,978,328	118,030,185	445,891,810	1,471,442,973
43	1	43	903	12,341	123,410	962,598	6,096,454	32,224,114	145,008,513	563,921,995	1,917,334,783
44	1	44	946	13,244	135,751	1,086,008	7,059,052	38,320,568	177,232,627	708,930,508	2,481,256,778
45	1	45	990	14,190	148,995	1,221,759	8,145,060	45,379,620	215,553,195	886,163,135	3,190,187,286
46	1	46	1,035	15,180	163,185	1,370,754	9,366,819	53,524,680	260,932,815	1,101,716,330	4,076,350,421
47	1	47	1,081	16,215	178,365	1,533,939	10,737,573	62,891,499	314,457,495	1,362,649,145	5,178,066,751
48	1	48	1,128	17,296	194,580	1,712,304	12,271,512	73,629,072	377,348,994	1,677,106,640	6,540,715,896
49	1	49	1,176	18,424	211,876	1,906,884	13,983,816	85,900,584	450,978,066	2,054,455,634	8,217,822,536
50	1	50	1,225	19,600	230,300	2,118,760	15,890,700	99,884,400	536,878,650	2,505,433,700	10,272,278,170

ing 50 cards. *Possible outcomes* = C_{13}^{52} because there are 52 cards in a deck and 13 cards are to be selected. Thus, the probability is

$$\frac{C_{11}^{50}}{C_{13}^{52}} = \frac{\dfrac{50!}{11!39!}}{\dfrac{52!}{13!39!}} = \frac{50!}{11!39!} \times \frac{13!39!}{52!}$$

$$= \frac{13 \cdot 12}{52 \cdot 51} = \frac{156}{2,652} \cong .059$$

31. a. The events are mutually exclusive; thus, the special law applies:

$$p(5 \text{ or } 6) = \frac{4}{36} + \frac{5}{36} = \frac{9}{36} = \frac{1}{4}$$

This result was noted in problems 3i and 12d as well.
b. The events are not mutually exclusive; thus, the general law must be used. $p(\text{black or ace}) = p(\text{black}) + p(\text{ace}) - p(\text{black and ace}) =$

$$\frac{26}{52} + \frac{4}{52} - \frac{2}{52} = \frac{28}{52} = \frac{7}{13}.$$

This result was noted in problems 3iv and 12d as well.
c. The events are not mutually exclusive; thus,

$$p(\text{inflation or recession}) = .8 + .2 - .1 = .9.$$

d. The events are not mutually exclusive; thus, the general law must be used, but the information is insufficient for determining an answer.
33. $p(A \text{ or } B \text{ or } C) = p(A) + p(B) + p(C) - p(A \text{ and } B) - p(A \text{ and } C) - p(B \text{ and } C) + p(A \text{ and } B \text{ and } C) = .2 + .1 + .05 - .08 - .02 - .07 + .05 = .23.$
35.

$$p(G|H) = \frac{240}{300} = .8$$

Alternative:

$$p(G|H) = \frac{p(H \text{ and } G)}{p(H)} = \frac{.60}{.75} = .8$$

$$p(B|H) = \frac{60}{300} = .2$$

Alternative:

$$p(B|H) = \frac{p(H \text{ and } B)}{p(H)} = \frac{.15}{.75} = .2$$

$\Bigg\} 1.0$

$$p(G|L) = \frac{10}{100} = .1$$

Alternative:

$$p(G|L) = \frac{p(L \text{ and } G)}{p(L)} = \frac{.025}{.25} = .1$$

$$p(B|L) = \frac{90}{100} = .9$$

Alternative:

$$p(B|L) = \frac{p(L \text{ and } B)}{p(L)} = \frac{.225}{.25} = .9$$

$\Bigg\} 1.0$

$$p(H|G) = \frac{240}{250} = .96$$

Alternative:

$$p(H|G) = \frac{p(H \text{ and } G)}{p(G)} = \frac{.60}{.625} = .96$$

$$p(L|G) = \frac{10}{250} = .04$$

Alternative:

$$p(L|G) = \frac{p(L \text{ and } G)}{p(G)} = \frac{.025}{.625} = .04$$

$\Bigg\} 1.0$

$$p(H|B) = \frac{60}{150} = .4$$

Alternative:

$$p(H|B) = \frac{p(H \text{ and } B)}{p(B)} = \frac{.15}{.375} = .4$$

$$p(L|B) = \frac{90}{150} = .6$$

Alternative:

$$p(L|B) = \frac{p(L \text{ and } B)}{p(B)} = \frac{.225}{.375} = .6$$

$\Bigg\} 1.0$

37. (i) $p(U|S) = (30/630) \cong .048$; conditional probability. (ii) $p(A|F) = (90/100) = .9$; conditional probability. (iii) $p(N \text{ and } H) = (100/900) \cong .11$; joint probability. (iv) $p(N|H) = (100/400) = .25$; conditional probability. (v) $p(F|S) = (200/500) = .4$; conditional probability. (vi) $p(U \text{ and } E) = (1,000/5,100) \cong .2$; joint probability. (vii) $p(A) = (3,400/5,100) \cong .67$; marginal probability.
39. Table 4.B: $p(H \text{ and } B) = p(H) \cdot p(B|H) = .75(.2) = .15$
　　　　Proof: $(60/400) = .15$
　　Table 4.C: $p(S \text{ and } U) = p(S) \cdot p(U|S) = .9(\approx.048) \cong .043$
　　　　Proof: $(30/700) \cong .043$

Table 4.D: $p(\text{M } and \text{ B}) = p(\text{M}) \cdot p(\text{B}|\text{M}) = (\approx.67)(.9) = .6$
Proof: $(180/300) = .6$

Table 4.E: $p(\text{S } and \text{ F}) = p(\text{S}) \cdot p(\text{F}|\text{S}) = (\approx.56)(.4) \cong .22$
Proof: $(200/900) \cong .22$

Table 4.F: $p(\text{U } and \text{ E}) = p(\text{U}) \cdot p(\text{E}|\text{U}) = (\approx.33)(\approx.59) = .196$
Proof: $(1{,}000/5{,}100) = .196$

41. $p(\text{S } or \text{ A}) = p(\text{S}) + p(\text{A}) - p(\text{S } and \text{ A}) = .9 + (\approx.91) - (\approx.86) = .95$

$p(\text{M } or \text{ A}) = p(\text{M}) + p(\text{A}) - p(\text{M } and \text{ A}) = (\approx.67) + (\approx.37) - (\approx.07) = (\approx.97)$

$p(\text{N } or \text{ H}) = p(\text{N}) + p(\text{H}) - p(\text{N } and \text{ H}) = (\approx.44) + (\approx.44) - (\approx.11) = (\approx.77)$

$p(\text{U } or \text{ L}) = p(\text{U}) + p(\text{L}) - p(\text{U } and \text{ L}) = (\approx.33) + (\approx.41) - (\approx.14) = (\approx.61)$

43. Table 4.B: H and B are dependent. Proof: $p(\text{H}) = .75 \neq p(\text{H}|\text{B}) = .40$.

Table 4.C: S and U are dependent. Proof: $p(\text{S}) = .9 \neq p(\text{S}|\text{U}) = .46$.

Table 4.D.: M and B are dependent. Proof: $p(\text{M}) = .67 \neq p(\text{M}|\text{B}) = .95$.

Table 4.E: S and F are dependent. Proof: $p(\text{S}) = .56 \neq p(\text{S}|\text{F}) = .40$.

Table 4.F: U and E are independent. Proof: $p(\text{U}) = .33 = p(\text{U}|\text{E}) = .33$.

45. When E is the event in question, $\bar{\text{E}}$ its complement, and R the empirical result, we have

$$p(\text{E}|\text{R}) = \frac{p(\text{E}) \cdot p(\text{R}|\text{E})}{p(\text{E}) \cdot p(\text{R}|\text{E}) + p(\bar{\text{E}}) \cdot p(\text{R}|\bar{\text{E}})}.$$

Thus,

$$p(\text{E}|\text{R}) = \frac{.3(.75)}{.3(.75) + (.7)(.1)}$$

$$= \frac{.225}{.225 + .07} = .76$$

if praised and

$$p(\text{E}|\text{R}) = \frac{.3(.25)}{.3(.25) + (.7)(.9)}$$

$$= \frac{.075}{.075 + .63} = .11$$

if critized.

Chapter 5

1. See Table A.17. For determining the probability distribution, you may wish to take another look at panel (c) of Figure 4.1.

Table A.17

Probability Distribution	
Point Value, x	Probability $p(R = x)$
5	36/108
10	48/108
20	16/108
50	4/108
100	4/108
	$\Sigma p = 108/108$ $= 1.00$

3. Case *a* could be a probability distribution; see Table A.18; case *b* could not be (it implies *negative* probability).

Table A.18

Probability Distribution Case a	
x	$p(x) = (x/10)$
1	1/10
2	2/10
3	3/10
4	4/10
	$\Sigma p = 10/10$ $= 1.00$

$\mu_R = 3; \sigma_R^2 = 1; \sigma_R = 1$

5. Case *b* could be a probability distribution; see Table A.19; case *a* could not be (it implies a sum of probabilities of less than unity).

Table A.19

Probability Distribution Case (b)	
x	$p(x) = (10-x)/40$
0	10/40
1	9/40
2	8/40
3	7/40
4	6/40
	$\Sigma p = 40/40$ $= 1.00$

$\mu_R = 1.75; \sigma_R^2 = 1.9375; \sigma = 1.3919$

7. The Box 5.A formulas apply. See Table A.20.

Table A.20

Probability Distribution			Summary Measures			
Number of Points, x	Probability, $p(R = x)$	$p \cdot x$	$x - \mu_R$	$(x - \mu_R)^2$	$p(x - \mu_R)^2$	
1	1/6	1/6	−2.5	6.25	1.042	
2	1/6	2/6	−1.5	2.25	.375	
3	1/6	3/6	−.5	.25	.042	
4	1/6	4/6	.5	.25	.042	
5	1/6	5/6	1.5	2.25	.375	
6	1/6	6/6	2.5	6.25	1.042	
	$\Sigma p = (6/6)$ $= 1.00$	$\mu_R = (21/6)$ $= 3.50$			$\sigma_R^2 = 2.918$ $\sigma_R = 1.708$	

9. The summary measures are given in Table A.21. $\mu_R = 3$ lawn mowers will be ordered.

Table A.21

$p \cdot x$	$x - \mu_R$	$(x - \mu_R)^2$	$p(x - \mu_R)^2$
0	−3	9	.45
.10	−2	4	.40
.40	−1	1	.20
.90	0	0	0
.80	1	1	.20
.50	2	4	.40
.30	3	9	.45
$\mu_R = 3.00$			$\sigma_R^2 = 2.1$ $\sigma_R = 1.45$

11. The implied probability distribution and the expected value of the random variable (cars coming in) is given in Table A.22. Hence the expected revenue is $310.

Table A.22

Probability Distribution	
x	$p(x)$
20	$(79/313) = .2524$
25	$(121/313) = .3866$
32	$(19/313) = .0607$
40	$(25/313) = .0799$
47	$(39/313) = .1246$
51	$(30/313) = .0958$
	$\Sigma p = 1.0000$

$\mu_R = \Sigma p(x) \cdot x = 30.59$

13. Using the class midpoints, one can develop Table A.23.

Table A.23

Probability Distribution	
x	$p(x)$
5,500	.05
8,500	.07
11,500	.11
14,500	.54
17,500	.11
20,500	.07
23,500	.05
	$\Sigma p = 1.00$

$\mu_R = \Sigma p(x) \cdot x = 14,500$
$\sigma_R = 3,888.44$

15. The answer, from Appendix Table C, is given in Table A.24.

Table A.24

x	p
0	.3585
1	.3774
2	.1887
3	.0596
4	.0133
5	.0022
6	.0003
7	.0000

17. Since 5 percent of a sample of 20 cars equals 1 car, the answer equals the probability of finding more than 1 defective car, or $1 - .3585 - .3774 = .2641$.

19. According to Appendix Table C, for $n = 15$, $x = 15$, and $\pi = .5$, $p = .0000$. According to the Box 5.B formula, $p = .00003052$.

21. See Figure A.19. Q = quits, S = stays.

Figure A.19

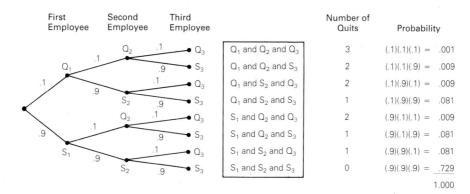

				Number of Quits	Probability
Q₁ and Q₂ and Q₃				3	(.1)(.1)(.1) = .001
Q₁ and Q₂ and S₃				2	(.1)(.1)(.9) = .009
Q₁ and S₂ and Q₃				2	(.1)(.9)(.1) = .009
Q₁ and S₂ and S₃				1	(.1)(.9)(.9) = .081
S₁ and Q₂ and Q₃				2	(.9)(.1)(.1) = .009
S₁ and Q₂ and S₃				1	(.9)(.1)(.9) = .081
S₁ and S₂ and Q₃				1	(.9)(.9)(.1) = .081
S₁ and S₂ and S₃				0	(.9)(.9)(.9) = .729
					1.000

Thus the probability distribution is given in Table A.25. This is confirmed in Appendix Table C for $n = 3$, $\pi = .1$.

Table A.25

x	$p(x)$
0	.729
1	.243
2	.027
3	.001
	$\Sigma p = 1.000$

23. See Table A.26 for answers (from Appendix Table C, $n = 15$, $\pi = .25$).
  a. For medium-sized cars, read the table directly.
  b. For compacts, the table must be read in reverse: $x = 0$ for medium-sized cars corresponds to $x = 15$ for compact cars; $x = 1$ for medium-sized cars corresponds to $x = 14$ for compacts, and so on.

Table A.26

x	p	x	p
0	.0134	7	.0393
1	.0668	8	.0131
2	.1559	9	.0034
3	.2252	10	.0007
4	.2252	11	.0001
5	.1651	12	.0000
6	.0917		

25. The answers, from Appendix Table D for $n = 100$ and $\pi = .40$, are (i) .0000, (ii) .0248 − .0148 = .01, (iii), .4621, (iv) 1 − .8689 = .1311, (v) 1 − .9729 = .0271, (vi) = .5433 − .0148 = .5285.

27. Answers (from Appendix Table C), for $n = 5$ and $\pi = .20$, are: (i) .0003, (ii) .3277, (iii) 1 − .3277 = .6723, (iv) .0512 + .0064 + .0003 = .0579, (v) .2048, (vi) .2048 + .0512 + .0064 = .2624.

29. Referring to Appendix Table D (for $n = 50$ and $\pi = .30$), which gives cumulative probabilities for tails coming up, (i) 44 or more heads is equivalent to 6 or fewer tails, $p = .0025$; (ii) exactly 40 heads is equivalent to exactly 10 tails, $p = .0789 − .0402 = .0387$; (iii) more than 37 heads is equivalent to 12 or fewer tails, $p = .2229$; (iv) 31 or fewer heads is equivalent to 19 or more tails, $p = 1 − .8594 = .1406$; (v) fewer than 28 heads is equivalent to more than 22 tails, $p = 1 − .9877 = .0123$; (vi) between 30 and 40 heads, inclusive, is equivalent to between 20 and 10 tails, inclusive,

$$p = .9522 − .0402 = .9120.$$

31. According to the Box 5.C formulas,

$$\mu_R = n \cdot \pi = 5(.20) = 1;$$

$$\sigma_R^2 = n \cdot \pi(1 − \pi)$$
$$= 5(.20)(.80) = .80;$$

$$\sigma_R = \sqrt{.80} = .89.$$

33. According to Appendix Table D, for $n = 100$ and $\pi = .30$, a. $p(R \leq 20) = .0165$; b. $p(R = 31) = p(R \leq 31) − p(R \leq 30) = .6331 − .5491 = .084$; c. $p(R \geq 47) = 1 − p(R \leq 46) = 1 − .9997 = .0003$; d. $\mu_R = 30$, $\sigma_R^2 = 21$, $\sigma_R = 4.58$.

35. According to Appendix Tables C and D, for $n = 7$ and $\pi = .05$, these answers apply: (i) $p(\text{all executed}) = p(\text{none late}) = .6983$; (ii) $p(\text{none executed}) = p(7 \text{ late}) = .0000$; (iii) $p(5 \text{ or more executed}) = p(2 \text{ or fewer late}) = .9962$; (iv) $p(6 \text{ or fewer executed}) = p(1 \text{ or more late}) = 1 - p(0 \text{ late}) = 1 - .6983 = .3017$.

37. The binomial formula suggests probabilities of .3660 and .3697, respectively.

39. a. False. According to Box 5.C, $\mu_R = n \cdot \pi$, and $\sigma_R^2 = n \cdot \pi(1 - \pi)$. Hence, $(\sigma_R^2/\mu_R) = 1 - \pi$. In this case, $1 - \pi = (.8/4) = .2$. Thus, $\pi = .8$.

b. True. According to Box 5.C, $\mu_R = n \cdot \pi$. Hence, $(\mu_R/\pi) = n$. In this case, $n = (4/.8) = 5$.

c. True. See Appendix Table C for $n = 5$ and read the $\pi = .20$ column backwards. Because .20 is the probability of *failure*, 0 failures implies 5 successes, and so on.

41. True. See Appendix Table C, section $n = 6$, $\pi = .5$, for $x = 4$.

43. True. See Appendix Table D, section $n = 20$, $\pi = .2$, for $x \leq 3$.

45. False. See Appendix Table C, section $n = 10$, $\pi = .1$, for $x = 1$. The answer is .3874 (picking 1 unsatisfactory item).

Chapter 6

1. Given the assumption of a normally distributed random variable, Appendix Table E can be used. *The 10th percentile* is that time in which 10 percent or fewer of all checks are cashed. Thus, 10 percent of the normal curve area must lie to the left of this value, and 90 percent must lie to the right of it. This means also that 40 percent of the area must lie between the 10th percentile and the (higher) mean (which is the 50th percentile). In Appendix Table E, an entry of .4000 lies between .3997 and .4015, corresponding to z values of 1.28 and 1.29, respectively. Interpolating, we find the z value corresponding to .4000 as 1.2817. Given that $z = (x - \mu)/\sigma$, it follows that $x = (\sigma \cdot z) + \mu$. In this case, with $\mu = 30$, $\sigma = 10$, and z negative (it lies *below* the mean), we have $x = 10(-1.2817) + 30 = 17.183$. Thus, 10 percent or fewer of all checks are cashed in 17.2 seconds. *The 75th percentile* is that time in which 75 percent or fewer of all checks are cashed. Thus, 75 percent of the normal-curve area must lie to the left of this value, and 25 percent of the area lies between the mean and the 75th percentile. In Appendix Table E, an entry of .2500 lies between .2486 and .2518, corresponding to z values of .67 and .68, respectively. Interpolating, we find that the z value for .2500 is .6744. Hence, $x = 10(.6744) + 30 = 36.74$. Thus, 75 percent or fewer of all checks are cashed in 36.7 seconds.

3. A raw score of 140 corresponds to a z value of $[(140 - 100)/16] = 2.5$. Appendix Table E lists a corresponding probability of .4938 for values between the mean and 2.5z. Thus, $p(z \leq 2.5) = .5000 + .4938 = .9938$; and the answer equals $1 - .9938 = .0062$.

5. The 75th percentile has to be found, as in problem 1. The z value is $+.6744$; hence, $x = 100(.6744) + 500 = 567.44$. This is also the lowest acceptable score.

The 95th percentile has to be found, as in problem 2. The z value is $+1.645$; hence, $x = 12(1.645) + 60 = 79.74$. The cutoff time should be 80 minutes.

7. The 99th percentile has to be found. It corresponds to a *positive* z value that places 49 percent of the normal curve area between itself and the (lower) mean. As in answer 6, the z value is 2.3267. Hence, $x = 2.45(2.3267) + 25.9 = 31.6$. The rating must be 31.6 miles per gallon or better.

9. a. z values are 0 to 1.5; $p = .4332$.

b. z values are -2 to 0; $p = .4772$.

c. z values are -1 to $+2$; $p = .3413 + .4772 = .8185$.

d. z values are $+3$ to ∞; $p = .5000 - .4986 = .0014$.

e. z values are $-\infty$ to -3.9; $p = .5000 - .4999 = .0001$.

f. z values are $+1$ to $+2$; $p = .4772 - .3413 = .1359$.

g. z values are -3 to -2; $p = .4986 - .4772 = .0214$.

h. z values are $-\infty$ to $+3.7$; $p = .5000 + .4999 = .9999$.

i. z values are -1.7 to ∞; $p = .4554 + .5000 = .9554$.

11. The relevant z value is $[(140,000 - 122,000)/10,000] = 1.8$; $p = .5000 - .4641 = .0359$. Only in about 4 percent of all weeks will the workload exceed 140,000 checks.

13. a. The relevant z value is $[(680,000 - 700,000)/50,000] = -.4$; the relevant range for $p(\text{tolls} \geq 680,000)$ is $-.4 \leq z \leq \infty$; thus $p = .1554 + .5000 = .6554$.

b. The multiplication law for independent events (Box 4.J) can be used; p(toll $\geq$ 680,000 in year 1 *and* toll $\geq$ 680,000 in year 2) = .6554(.6554) = .4295.

c. The 75th percentile has to be found. According to the procedure discussed in Answer 1, x = 50,000(.6744) + 700,000 = 733,720. In the best 25 percent of years, revenue will exceed $733,720.

15. a. The relevant z value is $z = [(400 - 375)/40] = .625$; thus, $p(>400 \text{ students}) = p(.625 \leq z \leq \infty) = .5000 - .2341 = .2659$. (The value of .2341 is found in Appendix Table E by interpolating between .2324 and .2357.)

b. The relevant z values are $z = [(350 - 375)/40] = -.625$ and, as before, $+.625$. Thus, $p(350 \leq \text{students} \leq 400) = p(-.625 \leq z \leq .625) = .2341 + .2341 = .4682$.

c. From (a) and (b) we know that $p(350 \geq \text{students}) = p(-\infty \leq z \leq -.625) = .5000 - .2341 = .2659$. It is no accident that the three answers add to unity; the three events are collectively exhaustive.

17. The probability for acceptable parts is $p(2.87 \leq \text{length} \leq 3.10) = p\left(\dfrac{2.87 - 3}{.15} \leq z \leq \dfrac{3.10 - 3}{.15}\right) = p(-.87 \leq z \leq .67) = .3078 + .2486 = .5564$. Thus the probability sought is $1 - .5564 = .4436$.

19. p(breakage at 73 or less) = $p\left(z \leq \dfrac{73 - 80}{7}\right) = p(z \leq -1) = .5000 - .3413 = .1587$; p(breakage at 100 or more) = $p\left(z \geq \dfrac{100 - 80}{7}\right) = p(z \geq 2.86) = .5000 - .4979 = .0021$. Thus, $50:.1587 = x:.0021$, and $x = .66$. About a single pane could take such a force.

21. p(income $\geq$ 127,000) = $p\left(z \geq \dfrac{127,000 - 89,000}{19,000}\right) = p(z \geq 2) = .5000 - .4772 = .0228$. Thus, $3:.0228 = x:1$, and $x = 131.6$. There are 132 psychiatrists.

23. Assume the normal distribution applies. Some 39.25 percent of workers produce between 17.3

and 18.1 tons; the z value corresponding to 18.1 tons is found in Appendix Table E as 1.24. But $\sigma = [(x - \mu)/z] = [(18.1 - 17.3)/1.24] = .645$ tons. Thus, the z value for 16 tons is $z = [(x - \mu)/\sigma] = [(16 - 17.3)/.645] = -2.02$. The associated probability is $p(\geq 16 \text{ tons}) = p(-2.02 \leq z \leq \infty) = .4783 + .5000 = .9783$. The answer is 97.83 percent.

25. a. $p(\leq 1 \text{ lb}) = p\left(-\infty \leq z \leq \dfrac{1 - 1.5}{.3}\right) = p(-\infty \leq z \leq -1.67) = .5000 - .4525 = .0474$.

b. $p(\leq 2 \text{ lb}) = p\left(-\infty \leq z \leq \dfrac{2 - 1.5}{.3}\right) = p(-\infty \leq z \leq 1.67) = .5000 + .4525 = .9525$.

c. $p(\geq 1.4 \text{ lb}) = p\left(\dfrac{1.4 - 1.5}{.3} \leq z \leq \infty\right) = p(-.33 \leq z \leq \infty) = .1293 + .5000 = .6293$.

d. $p(1.6 \leq \text{weight gain} \leq 1.7) = p\left(\dfrac{1.6 - 1.5}{.3} \leq z \dfrac{1.7 - 1.5}{.3}\right) = p(.33 \leq z \leq .66) = .2454 - .1293 = .1161$.

27. $p(\geq 63) = p\left(\dfrac{63 - 52}{10} \leq z \leq \infty\right) = p(1.1 \leq z \leq \infty) = .5000 - .3643 = .1357$. Also $p(\geq 93) = p\left(\dfrac{93 - 52}{10} \leq z \leq \infty\right) = p(4.1 \leq z \leq \infty) = .5000 - .5000 = 0$. At least 13.57 percent of employees can be expected to take 63 or more hours, like A. The case of B is, indeed, highly unusual.

29. a. p(time $\leq$ 11) = $p\left(-\infty \leq z \leq \dfrac{11 - 14}{1.3}\right) = p(-\infty \leq z \leq -2.31) = .5000 - .4896 = .0104$. Not impossible, but only 1 percent of electricians can do it.

b. p(time $\geq$ 17) = $p\left(\dfrac{17 - 14}{1.3} \leq z \leq \infty\right) = p(2.31 \leq z \leq \infty) = .5000 - .4896 = .0104$. Only 1 percent of all electricians need this much time or more.

31. The *95th percentile* places 5 percent of possible outflow values beyond the sought value of x. It also places 45 percent of all possible values between itself and the (lower) mean; thus Appendix Table E suggests a z value of 1.645. Since $x = (\sigma \cdot z) + \mu$, $x = [(400)(1.645)] + 1,500 = 2,158$. The answer: $2,158.

33. $p(\text{insider loans} \geq 1.9) = p\left(\dfrac{1.9 - .437}{.129} \leq z \leq \right.$

$\left. \infty\right) = p(11.34 \leq z \leq \infty) = .5000 - .5000 = 0.$

Yes, there is practically no chance of another bank like this.

35. $p(W \geq 500) = p\left(\dfrac{500 - 400}{30} \leq z \leq \infty\right) =$

$p(3.33 \leq z \leq \infty) = .5000 - .4996 = .0004.$

$p(W \leq 350) = p\left(-\infty \leq z \leq \dfrac{350 - 400}{30}\right) =$

$p(-\infty \leq z \leq -1.67) = .5000 - .4525 = .0475.$

$p(410 \leq W \leq 430) = p\left(\dfrac{410 - 400}{30} \leq z \leq\right.$

$\left.\dfrac{430 - 400}{30}\right) = p \ (.33 \leq z \leq 1) = .3413 - $

$.1293 = .212.$

37. The *10th percentile* has to be found; it places 40 percent of all P/E ratios between itself and the

(higher) mean. Appendix Table E suggests a z value of -1.2817. Since $x = (\sigma \cdot z) + \mu$, $x = [(2.1)(-1.2817)] + 9.6 = 6.9.$

39. a. The approximation of the binomial by the normal distribution can be used. In this example, $n = 20$, $\pi = .25$, and $1 - \pi = .75$. The rule of thumb [that $n \cdot \pi$ as well as $n(1 - \pi)$ equal or exceed 5] is met; the approximation, therefore, should be close. The necessary calculations are organized in Table A.27. As noted in the text, z scores are calculated as

$$z = \frac{x - (n \cdot \pi)}{\sqrt{n \cdot \pi(1 - \pi)}}$$

which in this case equals

$$z = \frac{x - [20(.25)]}{\sqrt{20(.25).75}} = \frac{x - 5}{1.9365}.$$

The corresponding probabilities come from Appendix Table E. The normal approximation to binomial probabilities is calculated from the

Table A.27

Discrete x Value	Corresponding Interval		Probabilities		Normal Approximation	Precise Binomial Probabilities
	x Scores	z Scores	Lower z Score	Upper z Score		
0	−.5 to .5	−2.84 to −2.32	.4977	.4898	.0079	.0032
1	.5 to 1.5	−2.32 to −1.81	.4898	.4649	.0249	.0211
2	1.5 to 2.5	−1.81 to −1.29	.4649	.4015	.0634	.0669
3	2.5 to 3.5	−1.29 to −.77	.4015	.2794	.1221	.1339
4	3.5 to 4.5	−.77 to −.26	.2794	.1026	.1768	.1897
5	4.5 to 5.5	−.26 to .26	.1026	.1026	.2052	.2023
6	5.5 to 6.5	.26 to .77	.1026	.2794	.1768	.1686
7	6.5 to 7.5	.77 to 1.29	.2794	.4015	.1221	.1124
8	7.5 to 8.5	1.29 to 1.81	.4015	.4649	.0634	.0609
9	8.5 to 9.5	1.81 to 2.32	.4649	.4898	.0249	.0271
10	9.5 to 10.5	2.32 to 2.84	.4898	.4977	.0079	.0099
11	10.5 to 11.5	2.84 to 3.36	.4977	.4996	.0019	.0030
12	11.5 to 12.5	3.36 to 3.87	.4996	.4999	.0003	.0008
13	12.5 to 13.5	3.87 to 4.39	.4999	.4999		.0002
14	13.5 to 14.5	4.39 to 4.91	.4999	.4999		
15	14.5 to 15.5	4.91 to 5.42	.4999	.4999		
16	15.5 to 16.5	5.42 to 5.94	.4999	.4999	.0000	
17	16.5 to 17.5	5.94 to 6.45	.4999	.4999		.0000
18	17.5 to 18.5	6.45 to 6.97	.4999	.4999		
19	18.5 to 19.5	6.97 to 7.49	.4999	.4999		
20	19.5 to 20.5	7.49 to 8.00	.4999	.4999		
					.9976	1.0000

probabilities associated with the two z scores as follows: in rows 0–4 of Table A.27 by deducting the probability of the upper z score from that of the lower z score, because the relevant area under the normal curve lies to the left of the mean; in row 5, by adding the two probabilities, because the relevant area under the normal curve straddles the mean; in the remaining rows, by deducting the probability of the lower z score from that of the upper z score, because the relevant area under the normal curve now lies to the right of the mean. The probabilities for x values above 12 are still positive, but tiny; note that the probabilities given for $x \leq 12$ add to only .9976. An interesting comparison can also be made by consulting Appendix Table C and finding the precise binomial probabilities directly in the section for $n = 20$ and $\pi = .25$. The values so found are given in the last column of Table A.27. As a comparison of the last two columns of Table A.27 shows, the normal approximation of the binomial probablities is extremely close.

b. Using the normal approximation, the answer is .0079 + .0249 + .0634 + .1221 = .2183. The precise value, found by adding the corresponding binomial probabilities, is .2251.

41. Use the normal approximation to the binomial distribution, noting that $n \cdot \pi = 500(.9) = 450$ as well as $n(1 - \pi) = 500(.1) = 50$ exceed 5. Here $\mu = n\pi = 450$; $\sigma = \sqrt{n \cdot \pi(1 - \pi)} = \sqrt{500(.1)(.9)} = 6.7082$.

a. $p(\text{attend} > 400) = p\left(\dfrac{400.5 - 450}{6.7082} \leq z \leq \infty\right)$

$= p(-7.38 \leq z \leq \infty) = .5000 + .5000 = 1.$

b. $p(\text{attend} < 450) = p\left(-\infty \leq z \leq \dfrac{449.5 - 450}{6.7082}\right) = p(-\infty \leq z \leq -.07) =$

$.5000 - .0279 = .4721.$ (Remember this is an approximation.)

c. $p(413 \leq \text{attendees} \leq 463) = p\left(\dfrac{412.5 - 450}{6.7082} \leq z \leq \dfrac{463.5 - 450}{6.7082}\right) =$

$p(-5.59 \leq z \leq 2.01) = .5000 + .4778 = .9778.$

43. Use the normal approximation to the binomial distribution, noting that $n \cdot \pi$ as well as $n \cdot \pi(1 - \pi)$ exceed 5. Here $\mu = n \cdot \pi = 100(.1) = 10$, and $\sigma = \sqrt{n \cdot \pi(1 - \pi)} = \sqrt{100(.1)(.9)} = 3.$

$p(\text{errors} < 5) = p\left(-\infty \leq z \leq \dfrac{4.5 - 10}{3}\right) =$

$p(-\infty \leq z \leq -1.83) = .5000 - .4664 = .0336.$

45. The problem amounts to finding an area under the normal curve to the left of a value that itself lies above the mean. Thus, the answer equals .5 (for the area left of the mean) plus whatever probability is associated with values between the mean and the given value of 2,075 hours. In this case, as a look at text Figure A confirms, the answer equals .5000 + .2734, or .7734.

Chapter 7

1. From Box 7.A, we can determine $\sigma_{\bar{x}} = (\sigma/\sqrt{n}) = (50/5) = 10$. Given $\sigma_{\bar{x}} = 10$ and a normal sampling distribution (because the population is normal), we can employ Appendix Table E to find z values and our answers: (a) Between $z = [(140 - 150)/10] = -1$ and $z = [(160 - 150)/10] = +1$, the probability is .3413 + .3413 = .6826. (b) Above $z = [(140 - 150)/10] = -1$, the probability is .3413 + .5000 = .8413. (c) Below $z = [(130 - 150)/10] = -2$, the probability is .5000 - .4772 = .0228. (d) Between $z = [(160 - 150)/10] = +1$ and $z = [(170 - 150)/10] = +2$, the probability is .4772 - .3413 = .1359. (e) Above $z = [(155 - 150)/10] = +.5$, the probability is .5000 - .1915 = .3085.

3. The formulas of Box 7.A apply.

a. $\mu_{\bar{x}} = 30$; $\sigma_{\bar{x}} = (3/\sqrt{10}) = .9487.$
b. $\mu_{\bar{x}} = 50$; $\sigma_{\bar{x}} = (2/\sqrt{15}) = .5164.$
c. $\mu_{\bar{x}} = 100$; $\sigma_{\bar{x}} = (10/\sqrt{30}) = 1.8257.$
d. $\mu_{\bar{x}} = 400$; $\sigma_{\bar{x}} = (8/\sqrt{100}) = .8.$

5. $\mu_{\bar{x}} = 1,200$ and $\sigma_{\bar{x}} = (250/\sqrt{36}) = 41.\overline{66}.$

$p(1,150 \leq \bar{X} \leq 1,250) = p\left(\dfrac{1,150 - 1,200}{41.\overline{66}} \leq z \leq \right.$

$\left.\dfrac{1,250 - 1,200}{41.\overline{66}}\right) = p(-1.2 \leq z \leq 1.2) = .3849$

$+ .3849 = .7698.$

7. See Table A.28, which proceeds from the fact that the possible number of equally likely samples is

$$C_2^5 = \dfrac{5!}{2!(5 - 2)!} = 10.$$

Using the Box 5.A formulas, $\mu_{\bar{x}} = 280/10 = 28$; $\sigma_{\bar{x}}^2 = \Sigma p(\bar{X} - \mu_{\bar{x}})^2 = 6.45$; $\sigma_{\bar{x}} = 2.54.$

Table A.28

Sample Number	Sample Ages	Possible Sample Means, $\overline{X}$	Probability $p(R = \overline{X})$	The Sampling Distribution of the Sample Mean	
				$\overline{X}$	$p(R = \overline{X})$
1	34, 29	31.5	.1	23.5	.1
2	34, 22	28.0	.1	25.5	.1
3	34, 30	32.0	.1	26.0	.1
4	34, 25	29.5	.1	27.0	.1
5	29, 22	25.5	.1	27.5	.1
6	29, 30	29.5	.1	28.0	.1
7	29, 25	27.0	.1	29.5	.2
8	22, 30	26.0	.1	31.5	.1
9	22, 25	23.5	.1	32.0	.1
10	30, 25	27.5	.1		
		$\Sigma\overline{X} = 280$			1.0

9. See Figure A.20. The area in question lies above z = $\dfrac{43 - 40}{4.655}$ = .64. Thus, the probability equals .5000 − .2389 = .2611 (the shaded area in Figure A.20).

11. **a.** π = .61.
    b. π = .6.
    c. μ = \$243.

13. This problem corresponds to Practice Problem 4 in the text: $n \geq 30$, $n < .05N$, σ is known. Hence,

$$\sigma_{\overline{X}} = \frac{\sigma}{\sqrt{n}} = \frac{4}{\sqrt{50}} = .566.$$

The required z value equals 1.96. Thus, the 95-percent confidence interval goes from 7.99 to 10.21 minutes, because $\mu = \overline{X} \pm z\sigma_{\overline{X}} = 9.1 \pm 1.96(.566) = 9.1 \pm 1.11$.

15. In this problem, $n \geq 30$, $n \geq .05N$, the population is normal, σ is known. Hence,

$$\sigma_{\overline{X}} = \frac{\sigma}{\sqrt{n}}\sqrt{\frac{N - n}{N - 1}} = \frac{99}{\sqrt{36}}\sqrt{\frac{75 - 36}{74}} = 11.98.$$

The required z value equals 3.08. Thus, the 99.8-percent confidence interval goes from \$699.10 to

Figure A.20

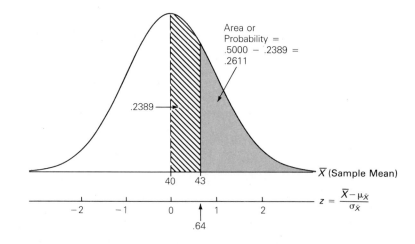

$772.90, because $\mu = \overline{X} \pm z\sigma_{\overline{X}} = 736 \pm 3.08(11.98) = 736 \pm 36.90$.

17. This problem corresponds to Practice Problem 5 in the text: $n \geq 30$, $n < .05N$, σ unknown. Hence,

$$\sigma_{\overline{X}} \cong \frac{s}{\sqrt{n}} = \frac{3,600}{\sqrt{100}} = 360.$$

The required z value equals 1.96. Thus the 95-percent confidence interval goes from $14,044.40 to $15,455.60, because $\mu = \overline{X} \pm z\sigma_{\overline{X}} = 14,750 \pm 1.96(360) = 14,750 \pm 705.60$.

19. In this problem, $n \geq 30$, $n \geq .05N$, σ unknown. Hence,

$$\sigma_{\overline{X}} \cong \frac{s}{\sqrt{n}}\sqrt{\frac{N-n}{N-1}} = \frac{.231}{\sqrt{50}}\sqrt{\frac{200-50}{200-1}} = .0284.$$

The required z value equals 1.28. Thus the 80-percent confidence interval goes from $1.303 to $1.375, because $\mu = \overline{X} \pm z\sigma_{\overline{X}} = 1.339 \pm 1.28(.0284) = 1.339 \pm .036$.

21. This problem corresponds to Practice Problem 5 in the text: $n \geq 30$, $n < .05N$, σ unknown. Hence,

$$\sigma_{\overline{X}} \cong \frac{s}{\sqrt{n}} = \frac{5.2}{\sqrt{81}} = .5777.$$

The required z value equals 3.08. Thus the 99.8-percent confidence interval goes from 12.12 to 15.68 seats, because $\mu = \overline{X} \pm z\sigma_{\overline{X}} = 13.9 \pm 3.08(.5777) = 13.9 \pm 1.7796$.

23. This problem corresponds to Practice Problem 5 in the text: $n \geq 30$, $n < .05N$, σ unknown. Hence,

$$\sigma_{\overline{X}} \cong \frac{s}{\sqrt{n}} = \frac{2.01}{\sqrt{100}} = .201.$$

The required z value (from Appendix Table E) is 2.24. Thus the 97.5-percent confidence interval goes from 5.30 lb. to 6.20 lb., because $\mu = \overline{X} \pm z\sigma_{\overline{X}} = 5.75 \pm 2.24(.201) = 5.75 \pm .45$.

25. $\pi = P + (z\sigma_P) \cong P \pm \left(1.96\sqrt{\frac{P(1-P)}{n}}\right)$

1960: $.51 \pm \left(1.96\sqrt{\frac{.51(.49)}{1,500}}\right) = .51 \pm .025$;

range is .485 to .535.

1964: $.64 \pm \left(1.96\sqrt{\frac{.64(.36)}{1,500}}\right) = .64 \pm .024$;

range is .616 to .664.

1968: $.50 \pm \left(1.96\sqrt{\frac{.5(.5)}{1,500}}\right) = .50 \pm .025$;

range is .475 to .525.

1972: $.38 \pm \left(1.96\sqrt{\frac{.38(.62)}{1,500}}\right) = .38 \pm .025$;

range is .355 to .405.

27. $\pi = .39 \pm \left(1.96\sqrt{\frac{.39(.61)}{100}}\right) = .39 \pm .10$;

range is .29 to .49.

29. $\pi = P \pm (z\sigma_P)$

$\cong P \pm \left(z\sqrt{\frac{P(1-P)}{n}}\sqrt{\frac{N-n}{N-1}}\right)$

$= .7 \pm \left(2.57\sqrt{\frac{.7(.3)}{30}}\sqrt{\frac{100-30}{99}}\right)$

$= .7 \pm .18$; range is .52 to .88.

31. $\pi = .38 \pm \left(1.22\sqrt{\frac{.38(.62)}{50}}\right) = .38 \pm .084$;

range is .30 to .46.

33. $\pi = .26 \pm \left(2.51\sqrt{\frac{.26(.74)}{50}}\sqrt{\frac{150-50}{150-1}}\right) = .26 \pm .1276$; the range is .13 to .39.

35. $\pi = .12 \pm \left(2.33\sqrt{\frac{.12(.88)}{100}}\right) = .12 \pm .0757$; the range is .04 to .20. The auditor believes that between 4 and 20 percent of all bank statements mailed to customers are in error; the procedure employed can be relied upon to give the correct answer 98 times out of 100.

37. This problem corresponds to Practice Problem 6 in the text: $n < 30$, $n \geq .05N$, the population is normal, σ is known. Hence,

$$\sigma_{\overline{X}} = \frac{\sigma}{\sqrt{n}}\sqrt{\frac{N-n}{N-1}} = \frac{3.7}{\sqrt{10}}\sqrt{\frac{35-10}{34}} = 1.00.$$

The required z value equals 2.054 (from Appendix Table E, looking for an inside value of $.96/2 = .4800$). Thus, the 96-percent confidence interval goes from 17.946 to 22.054 gallons, because $\mu = \overline{X} \pm z\sigma_{\overline{X}} = 20 \pm 2.054(1)$.

39. Part (i): This problem corresponds to Practice Problem 8 in the text: $n < 30$, $n < .05N$, the population is normal, σ is unknown. The required t value (for 19 degrees of freedom and an upper-tail area, $\alpha = .025$) is 2.093. The 95-percent confidence interval goes from 6.83 to 9.17 days, because

$$\mu = \overline{X} \pm t\frac{s}{\sqrt{n}} = 8 \pm \left[2.093\left(\frac{2.5}{\sqrt{20}}\right)\right]$$

$$= 8 \pm 1.17.$$

Part (ii): This problem corresponds to Practice Problem 7 in the text: $n < 30$, $n < .05N$, the population is *not* normal, σ is known. As in the text, $k = 4.472$. The 95-percent confidence interval now goes from 5 to 11 days, because

$$\mu = \overline{X} \pm \left(k\frac{\sigma}{\sqrt{n}}\right)$$

$$= 8 \pm \left[4.472\left(\frac{3}{\sqrt{20}}\right)\right] = 8 \pm 3.$$

41. The same procedure as in part (i) of problem 39 applies. The required t value (for 4 degrees of freedom and an upper-tail area, $\alpha = .005$) is 4.604. The 99-percent confidence interval goes from 47.3 to 56.7 knots, because

$$\mu = \overline{X} \pm \left(t\frac{s}{\sqrt{n}}\right) = 52 \pm \left(4.604\frac{2.3}{\sqrt{5}}\right)$$

$$= 52 \pm 4.7356.$$

43. This case involves two large, independent samples; we can apply the Box 7.E formula to the following data: $n_A = 290$; $n_B = 333$; $\overline{X}_A = 59$, $\overline{X}_B = 71$; $s_A = 20$; $s_B = 30$; $z = 2.57$. Hence,

$$\mu_A - \mu_B \cong -12 \pm \left(2.57\sqrt{\frac{400}{290} + \frac{900}{333}}\right)$$

$$= -12 \pm [2.57(2.02)]$$

$$= -12 \pm 5.19.$$

The difference between the mean intensities lies between -17.19 and -6.81 miles per hour. Storms are that much less intense in region A.

45. The case involves large and independent samples; hence the formula of Box 7.G applies.

$$\pi_A - \pi_B \cong (.68 - .22)$$

$$\pm \left(2.33\sqrt{\frac{(.68)(.32)}{50} + \frac{(.22)(.78)}{50}}\right)$$

$$= .46 \pm .206.$$

The difference in proportions ranges from .254 to .666.

Chapter 8

1. For $\alpha = .001$, this being a two-tailed test, the critical values in Practice Problem 1 would have been $\pm z_{\alpha/2} = \pm 3.27$ (see Appendix Table F). Hence, the decision rule would have been "Accept H_0 if $-3.27 \leq z \leq +3.27$." Given the computed value of the test statistic ($z = -1$), the null hypothesis would have been *accepted*. For $\alpha = .25$, we would have found $\pm z_{\alpha/2} = \pm 1.15$ (see Appendix Table E), with the same result.

3. *Step 1:* H_0: $\mu = 33$. H_A: $\mu \neq 33$
Step 2:

$$z = \frac{\overline{X} - \mu_0}{\sigma_{\overline{X}}}$$

Step 3: The critical value is: $\pm z_{\alpha/2} = \pm 2.24$. The decision rule is "Accept H_0 if $-2.24 \leq z \leq +2.24$."
Step 4:

$$z = \frac{34 - 33}{2/\sqrt{100}} = 5$$

H_0 should be *rejected:* the sleeves are longer.

5. *Step 1:* H_0: $\mu = 12,357$. H_A: $\mu \neq 12,357$
Step 2:

$$z = \frac{\overline{X} - \mu_0}{\sigma_{\overline{X}}}$$

Step 3: The critical value is: $\pm z_{\alpha/2} = \pm 3.27$. The decision rule is "Accept H_0 if $-3.27 \leq z \leq +3.27$."
Step 4:

$$z = \frac{13,950 - 12,357}{3,972/\sqrt{36}} = 2.41$$

H_0 should be *accepted*.

7. *Step 1:* H_0: $\mu = 500$. H_A: $\mu \neq 500$
Step 2:

$$z = \frac{\overline{X} - \mu_0}{\sigma_{\overline{X}}}$$

Step 3: The critical value can be found in Appendix Table E (by noting that $\alpha/2 = .01$, that $.5 - .01 = .49$, and that .49 corresponds to $z = 2.327$) as: $\pm z_{\alpha/2} = \pm 2.327$. The decision rule is "Accept H_0 if $-2.327 \leq z \leq +2.327$."
Step 4:

$$z = \frac{592 - 500}{101/\sqrt{32}} = 5.15$$

H_0 should be *rejected*; the cost is higher.

9. *Step 1:* H_0: $\mu \geq 60$. H_A: $\mu < 60$
Step 2:

$$z = \frac{\overline{X} - \mu_0}{\sigma_{\overline{X}}}$$

Step 3: The critical value is: $-z_\alpha = -1.64$. The decision rule is "Accept H_0 if $z \geq -1.64$."
Step 4:

$$z = \frac{101 - 60}{21/\sqrt{50}} = 13.81$$

H_0 should be *accepted*.

11. *Step 1:* H_0: $\mu \leq 50$. H_A: $\mu > 50$
Step 2:

$$z = \frac{\overline{X} - \mu_0}{\sigma_{\overline{X}}}$$

Step 3: The critical value is: $z_\alpha = 3.08$. The decision rule is "Accept H_0 if $z \leq 3.08$."
Step 4:

$$z = \frac{71 - 50}{30/\sqrt{100}} = 7$$

H_0 should be *rejected*; the mean was higher.

13. In each case, the interval limits are $\mu = \overline{X} \pm z_\alpha \sigma_{\overline{X}}$.

Problem 7: 98-percent confidence inverval, $z = 2.33$.

$$\mu = 592 \pm \left[2.33\left(\frac{101}{\sqrt{32}}\right)\right] =$$

592 ± 41.60.
The range is \$550.40 to \$633.60.

Problem 8: 99-percent confidence interval, $z = 2.57$.

$$\mu = 399 \pm$$
$$\left[2.57\left(\frac{25}{\sqrt{50}}\sqrt{\frac{450 - 50}{450 - 1}}\right)\right] = 399 \pm$$

8.576.
The range is \$390.42 to \$407.58.

Problem 9: 95-percent confidence interval, $z = 1.96$.

$$\mu = 101 \pm \left[1.96\left(\frac{21}{\sqrt{50}}\right)\right] = 101 \pm$$

5.82.
The range is 95.18 minutes to 106.82 minutes.

Problem 10: 97.5-percent confidence interval, $z = 2.24$.

$$\mu = 4.2 \pm \left[2.24\left(\frac{1}{\sqrt{64}}\right)\right] = 4.2 \pm$$

$.28$.
The range is 3.92 ppm to 4.48 ppm.

15. *Step 1:* H_0: $\pi = .25$. H_A: $\pi \neq .25$
Step 2:

$$z = \frac{P - \pi_0}{\sigma_P}$$

Step 3: The critical value is: $\pm z_{\alpha/2} = \pm 2.57$. The decision rule is "Accept H_0 if $-2.57 \leq z \leq +2.57$."
Step 4:

$$\sigma_P \cong \sqrt{\frac{P(1 - P)}{n}} = \sqrt{\frac{.19(.81)}{200}} = .0277$$
$$z = \frac{.19 - .25}{.0277} = -2.16$$

H_0 should be *accepted*.

17. a. *Step 1:* H_0: $\pi \geq .50$. H_A: $\pi < .50$
Step 2:

$$z = \frac{P - \pi_0}{\sigma_P}$$

Step 3: The critical value is: $-z_\alpha = -1.64$. The decision rule is "Accept H_0 if $z \geq -1.64$."

Step 4:

$$\sigma_P \cong \sqrt{\frac{P(1 - P)}{n}} = \sqrt{\frac{.4(.6)}{25}} = .098$$

$$z = \frac{.4 - .5}{.098} = -1.02$$

H_0 should be *accepted*.

b. Now $\sigma_P = .049$ and $z = -2.04$.

H_0 should be *rejected* in this case.

19. *Step 1:* H_0: $\pi \leq .12$. H_A: $\pi > .12$.

Step 2:

$$z = \frac{P - \pi_0}{\sigma_P}$$

Step 3: The critical value is: $z_\alpha = 1.64$. The decision rule is "Accept H_0 if $z \leq 1.64$."

Step 4:

$$\sigma_P \cong \sqrt{\frac{P(1 - P)}{n}} = \sqrt{\frac{.108(.892)}{250}} = .01963$$

$$z = \frac{.108 - .12}{.01963} = -.61$$

H_0 should be *accepted*.

21. *Step 1:* H_0: $\pi = .53$. H_A: $\pi \neq .53$

Step 2:

$$z = \frac{P - \pi_0}{\sigma_P}$$

Step 3: The critical value is: $\pm z_{\alpha/2} = \pm 1.96$. The decision rule is "Accept H_0 if $-1.96 \leq z \leq +1.96$."

Step 4:

$$\sigma_P \cong \sqrt{\frac{.51(.49)}{1,500}} = .01291$$

$$z = \frac{.51 - .53}{.01291} = -1.55$$

H_0 should be *accepted*. Note how, in the Chapter 7, problem 25 answer, a proportion of .53 also lay within the 95-percent confidence interval.

23. *Step 1:* H_0: $\mu_A - \mu_B = 0$; H_A: $\mu_A - \mu_B \neq 0$.

Step 2:

$$z = \frac{\overline{X}_A - \overline{X}_B}{\sigma_d}.$$

Step 3: The critical value is: $\pm z_{\alpha/2} = \pm 2.24$. The decision rule is "Accept H_0 if $-2.24 \leq z \leq +2.24$."

Step 4:

$$\sigma_d \cong \sqrt{\frac{s_A^2}{n_A} + \frac{s_B^2}{n_B}}$$

$$= \sqrt{\frac{(200)^2}{100} + \frac{(100)^2}{80}} = 22.9129.$$

$$z = \frac{500 - 410}{22.9129} = 3.93.$$

H_0 should be *rejected*. New York wages are higher.

25. *Step 1:* H_0: $\mu_A - \mu_B = 0$. H_A: $\mu_A - \mu_B \neq 0$

Step 2:

$$z = \frac{\overline{X}_A - \overline{X}_B}{\sigma_d}$$

Step 3: The critical value is: $\pm z_{\alpha/2} = \pm 2.81$. The decision rule is "Accept H_0 if $-2.81 \leq z \leq +2.81$."

Step 4:

$$\sigma_d \cong \sqrt{\frac{s_A^2}{n_A} + \frac{s_B^2}{n_B}} = \sqrt{\frac{64}{36} + \frac{9}{49}} = 1.4005$$

$$z = \frac{93 - 87}{1.4005} = 4.28$$

H_0 should be *rejected*. Center A scores are higher.

27. *Step 1:* H_0: $\mu_A - \mu_B \geq 0$. H_A: $\mu_A - \mu_B < 0$.

Step 2:

$$z = \frac{\overline{X}_A - \overline{X}_B}{\sigma_d}$$

Step 3: The critical value is: $-z_\alpha = -1.64$. The decision rule is "Accept H_0 if $z \geq -1.64$."

Step 4:

$$\sigma_d \cong \sqrt{\frac{s_A^2}{n_A} + \frac{s_B^2}{n_B}} = \sqrt{\frac{(18.2)^2}{100} + \frac{(3.2)^2}{100}} = 1.8479$$

$$z = \frac{48.25 - 39.95}{1.8479} = 4.49$$

H_0 should be *accepted*.

29. *Step 1*: H_0: $\pi_A - \pi_B = 0$. H_A: $\pi_A - \pi_B \neq 0$
Step 2:

$$z = \frac{P_A - P_B}{\sigma_d}$$

Step 3: The critical value is: $\pm z_{\alpha/2} = \pm 1.64$. The decision rule is "Accept H_0 if $-1.64 \leq z \leq +1.64$."
Step 4:

$$P_P = \frac{25(.72) + 171(.269)}{196} = .3265$$

$$\sigma_d \cong \sqrt{.3265(.6735)\left(\frac{1}{25} + \frac{1}{171}\right)} = .1004$$

$$z = \frac{.72 - .269}{.1004} = 4.49$$

H_0 should be *rejected*. Having a committee member's district involved made a big difference.

31. *Step 1*: H_0: $\pi_A - \pi_B \geq 0$. H_A: $\pi_A - \pi_B < 0$
Step 2:

$$z = \frac{P_A - P_B}{\sigma_d}$$

Step 3: The critical value is: $-z_\alpha = -2.33$. The decision rule is "Accept H_0 if $z \geq -2.33$."
Step 4:

$$P_P = \frac{200(.18) + 200(.19)}{400} = .185$$

$$\sigma_d \cong \sqrt{.185(.815)\left(\frac{1}{200} + \frac{1}{200}\right)} = .03883$$

$$z = \frac{.18 - .19}{.03883} = -.26.$$

H_0 should be *accepted*.

33. *Step 1*: H_0: $\pi_A \leq \pi_B$. H_A: $\pi_A > \pi_B$.
Step 2:

$$z = \frac{P_A - P_B}{\sigma_d}.$$

Step 3: The critical value is: $z_\alpha = 2.57$. The decision rule is "Accept H_0 if $z \leq 2.57$."
Step 4:

$$P_P = \frac{100(.9) + 150(.7)}{250} = .78$$

$$\sigma_d \cong \sqrt{.78(.22)\left(\frac{1}{100} + \frac{1}{150}\right)} = .0535$$

$$z = \frac{.9 - .7}{.0535} = 3.74.$$

H_0 should be *rejected*: The quality of the foreign components is better.

35. *Step 1*: H_0: $\mu = 5$. H_A: $\mu \neq 5$. Assume normal distribution of lifetimes.
Step 2:

$$t = \frac{\overline{X} - \mu_0}{s/\sqrt{n}}$$

Step 3: The critical value is: $\pm t_{.0005,15} = \pm 4.073$ (Appendix Table G, 15 degrees of freedom). The decision rule is "Accept H_0 if $-4.073 \leq t \leq +4.073$."
Step 4:

$$t = \frac{4.1 - 5}{1.6/\sqrt{16}} = -2.25$$

H_0 should be *accepted*: The seemingly low sample result can be attributed to sampling error; rats with artificial blood do live normal life spans.

37. *Step 1*: H_0: $\mu = 190$. H_A: $\mu \neq 190$. Assume normal distribution of temperatures.
Step 2:

$$t = \frac{\overline{X} - \mu_0}{s/\sqrt{n}}$$

Step 3: The critical value is: $\pm t_{.025,24} = \pm 2.064$ (Appendix Table G, 24 degrees of freedom). The decision rule is "Accept H_0 if $-2.064 \leq t \leq +2.064$."

Step 4:

$$t = \frac{193 - 190}{3/\sqrt{25}} = 5$$

H_0 should be *rejected*. The engine *is* running hot.

39. Step 1: H_0: $\mu \leq 105$. H_A: $\mu > 105$. Assume normal distribution of crankcase weights.

Step 2:

$$t = \frac{\overline{X} - \mu_0}{s/\sqrt{n}}$$

Step 3: The critical value is $t_\alpha = t_{.01,19} = 2.539$ (Appendix Table G, 19 degrees of freedom). The decision rule is "Accept H_0 if $t \leq 2.539$."

Step 4:

$$t = \frac{107 - 105}{.1/\sqrt{20}} = 89.44$$

H_0 should be *rejected*. The crankcases are unsuitable.

41. Step 1: H_0: $\mu_A - \mu_B = 0$. H_A: $\mu_A - \mu_B \neq 0$

Step 2:

$$t = \frac{\overline{X}_A - \overline{X}_B}{\sqrt{s_P^2 \left(\frac{1}{n_A} + \frac{1}{n_B}\right)}}$$

Step 3: The critical value is: $\pm t_{\alpha/2} = \pm t_{.05,40} = \pm 1.684$ (Appendix Table G with $n_A + n_B - 2$ degrees of freedom). The decision rule is "Accept H_0 if $-1.684 \leq t \leq +1.684$."

Step 4:

$$s_P^2 = \frac{(n_A - 1)s_A^2 + (n_B - 1)s_B^2}{n_A + n_B - 2}$$
$$= \frac{15(.2)^2 + 25(2.3)^2}{40} = 3.3213$$

$$t = \frac{23 - 24}{\sqrt{3.3213\left(\frac{1}{16} + \frac{1}{26}\right)}} = -1.73$$

H_0 should be *rejected*. Yarn B is stronger.

43. a. $\alpha = .05, .3954, .8808$, respectively. (In each case, find the Appendix Table E entry corresponding to $z_{\alpha/2}$, subtract it from .5000, and double the result.)

b. $\alpha = .025, .1056, .4404$, respectively. (In each case, find the Appendix Table E entry corresponding to z_α and subtract it from .5000.)

45. a. See Figure A.21. In this example, $\alpha = .05$ and is split between the two tails. Similarly, the p value is split. Thus, if the observed test statistic is $z = -.92$, the crosshatched plus dotted area on the left represents only half the p value; combined with a similar area on the right, the p value equals .3576.

Figure A.21

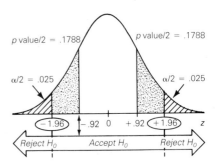

b. The p values are given in Table A.29.

Table A.29

Practice Problems	p value	End-of-Chapter Problem	p value	End-of-Chapter Problem	p value
1	.3174	3	.0000	22	.0000
4	.0124	4	.0000	23	.0000
5	.0000	5	.0160	25	.0000
6	.0000	6	.0000	26	.0000
7	.1388	7	.0000	28	.3078
AE 8.1	.6170	14	.0750	29	.0000
	.2542	15	.0308		
	.0012	21	.1212		
	.0004				
	.0910				

Note: Depending on the size of the observed test statistic, the p value can exceed α (as in Figure A.21), and H_0 is accepted, or it can fall short of α (being smaller than the crosshatched area in Figure A.21), and H_0 is rejected.

Chapter 9

1. *Step 1:*

> H_0: The two variables are independent.
> H_A: The two variables are dependent.

Step 2:

$$\chi^2 = \sum \frac{(f_o - f_e)^2}{f_e}$$

Step 3: For $\alpha = .02$ and 4 degrees of freedom, the critical value is $\chi^2_{.02,4} = 11.668$. The decision rule is "Accept H_0 if $\chi^2 \leq 11.668$."

Step 4: The χ^2 value is computed in Table A.30. H_0 should be *accepted*. Severity and location are independent variables in the case of automobile accidents.

3. *Steps 1* and *2* are formulated as in problem 1. *Step 3:* For $\alpha = .01$ and 10 degrees of freedom, the critical value is $\chi^2_{.01,10} = 23.209$. The decision rule is: "Accept H_0 if $\chi^2 \leq 23.209$." *Step 4:* According to the χ^2 value computed in Table A.31, H_0 should be *rejected*. Source of supply and type of defect are dependent variables.

5. a. $d.f. = 9$, critical $\chi^2_{.001} = 27.877$. *Reject H_0.*
b. $d.f. = 1$, critical $\chi^2_{.01} = 6.635$. *Accept H_0.*
c. $d.f. = 27$, critical $\chi^2_{.02} = 44.140$. *Reject H_0.*
d. $d.f. = 22$, critical $\chi^2_{.05} = 33.924$. *Reject H_0.*

7. *Steps 1* and *2* are formulated as in problem 1. *Step 3:* For $\alpha = .02$ and 4 degrees of freedom, the critical value is $\chi^2_{.02,4} = 11.668$. The decision rule is "Accept H_0 if $\chi^2 \leq 11.668$." *Step 4:* Actual $\chi^2 = 95.66$ *Reject H_0.* Stock price and dividend yield are dependent variables.

9. *Steps 1* and *2* are formulated as in problem 1. *Step 3:* For $\alpha = .05$ and 4 degrees of freedom, the critical value is $\chi^2_{.05,4} = 9.488$. The decision rule is "Accept H_0 if $\chi^2 \leq 9.488$." *Step 4:* Actual $\chi^2 =$

Table A.30

Row, Column	f_o	f_e	$f_o - f_e$	$(f_o - f_e)^2$	$\frac{(f_o - f_e)^2}{f_e}$
1,1	10	15	−5	25	1.667
1,2	20	20	0	0	0
1,3	20	15	5	25	1.667
2,1	10	7.5	2.5	6.25	.833
2,2	10	10	0	0	0
2,3	5	7.5	−2.5	6.25	.833
3,1	10	7.5	2.5	6.25	.833
3,2	10	10	0	0	0
3,3	5	7.5	−2.5	6.25	.833
Total	100	100	0		$\chi^2 = 6.666$

Table A.31

Row, Column	f_o	f_e	$f_e - f_e$	$(f_o - f_e)^2$	$\frac{(f_o - f_e)^2}{f_e}$
1,A	5	8.33	−3.33	11.09	1.33
1,B	0	11.67	−11.67	136.19	11.67
1,C	3	13.33	−10.33	106.71	8.01
1,D	12	25.00	−13.00	169.00	6.76
1,E	50	11.67	38.33	1,469.19	125.89
1,F	30	30.00	0	0	0
2,A	18	16.67	1.33	1.77	.11
2,B	32	23.33	8.67	75.17	3.22
2,C	42	26.67	15.33	235.01	8.81
2,D	68	50.00	18.00	324.00	6.48
2,E	0	23.33	−23.33	544.29	23.33
2,F	40	60.00	−20.00	400.00	6.67
3,A	27	25.00	2.00	4.00	.16
3,B	38	35.00	3.00	9.00	.26
3,C	35	40.00	−5.00	25.00	.63
3,D	70	75.00	−5.00	25.00	.33
3,E	20	35.00	−15.00	225.00	6.43
3,F	110	90.00	20.00	400.00	4.44
Total	600	600	0		$\chi^2 = 214.53$

96.03. *Reject H_0.* Awareness of ad and magazine type are dependent variables.

11. *Steps 1* and *2* are formulated as in problem 1. *Step 3:* For $\alpha = .10$ and 2 degrees of freedom, the critical $\chi_{.10,2}^2 = 4.605$. The decision rule is "Accept H_0 if $\chi^2 \leq 4.605$." *Step 4:* Actual $\chi^2 = 138.898$. *Reject H_0.* The type of chicken consumed and watching "Falcon Crest" are dependent variables.

13. The test must be based on the data in Table A.32:

Table A.32 Drivers in Accidents (Massachusetts sample, 1986)

Accident Type	Seat Belt Worn		Total
	Yes	No	
Serious	49	114	163
Non-Serious	222	165	387
Total	271	279	550

Steps 1 and *2* are formulated as in problem 1. *Step 3:* For $\alpha = .01$ and 1 degree of freedom, the critical $\chi_{.01,1}^2 = 6.635$. The decision rule is "Accept H_0 if $\chi^2 \leq 6.635$." *Step 4:* Actual $\chi^2 = 34.206$. *Reject H_0.* Severity of accidents and wearing seat belts are dependent variables.

15. *Steps 1* and *2* are formulated as in problem 1. *Step 3:* For $\alpha = .01$ and 6 degrees of freedom, the critical $\chi_{.01,6}^2 = 16.812$. The decision rule is "Accept H_0 if $\chi^2 \leq 16.812$." *Step 4:* Actual $\chi^2 = 136.959$. *Reject H_0.* Type of policy sold and geographic region are dependent variables.

17. See Table A.33.

19. *Step 1:* H_0: $\pi_A = \pi_B = \pi_C = \pi_D = .35$

Step 2: $\chi^2 = \Sigma \dfrac{(f_o - f_e)^2}{f_e}$

Step 3: Given $\alpha = .10$ and 3 degrees of freedom, the critical value is $\chi_{.10,3}^2 = 6.251$ (Appendix Table

Table A.33

Problem	E_{χ^2}	$\sigma_{\chi^2}^2$	σ_{χ^2}
12	3	6	2.449
13	1	2	1.414
14	3	6	2.449
15	6	12	3.464

H), making the decision rule: "Accept H_0 if $\chi^2 \leq 6.251$." *Step 4:* According to the χ^2 value computed in Table A.34, H_0 should be *rejected.* The proportions differ.

21. *Step 1:* H_0: $\pi_P = \pi_N = \pi_D = .5$. *Step 2:* χ^2 is formulated as in problem 19. *Step 3:* Given $\alpha = .05$ and 2 degrees of freedom (a contingency table would have 2 rows—heads, tails—and 3 columns—penny, nickel, dime), the critical value is $\chi_{.05,2}^2 = 5.991$. The decision rule is "Accept H_0 if $\chi^2 \leq 5.991$." *Step 4:* Actual $\chi^2 = 4$. *Accept H_0.* The coins are fair.

23. *Step 1:* H_0: $\pi_A = \pi_B = .066$. *Step 2:* χ^2 is formulated as in problem 19. *Step 3:* Given $\alpha = .01$ and 1 degree of freedom (2 rows—defective, acceptable—and 2 columns—A and B), the critical value is $\chi_{.01,1}^2 = 6.635$. The decision rule is "Accept H_0 if $\chi^2 \leq 6.635$." *Step 4:* Actual $\chi^2 = .389$. *Accept H_0.* The proportions are the same.

25. *Step 1:* H_0: $\pi_A = \pi_B = \pi_C = \pi_D = .551$. *Step 2:* χ^2 is formulated as in problem 19. *Step 3:* Given $\alpha = .001$ and 3 degrees of freedom (2 rows—last week, not last week—and 4 columns—companies A–D), the critical value is $\chi_{.001,3}^2 = 16.266$. The decision rule is "Accept H_0 if $\chi^2 \leq 16.266$." *Step 4:* Actual $\chi^2 = 22.276$. *Reject H_0.* The proportion differs among the companies.

27. H_0: $\pi_A = .2$, $\pi_B = .5$, $\pi_C = .1$, $\pi_D = .2$. Given $\alpha = .10$ and 3 degrees of freedom, the critical $\chi_{.10,3}^2 = 6.251$. The decision rule is "Accept H_0 if $\chi^2 \leq 6.251$." The *computation of χ^2* is given in

Table A.34

Row, Column	f_o	f_e	$f_o - f_e$	$(f_o - f_e)^2$	$\dfrac{(f_o - f_e)^2}{f_e}$
1,A	10	17.5	−7.5	56.25	3.21
1,B	0	17.5	−17.5	306.25	17.50
1,C	20	17.5	2.5	6.25	.36
1,D	40	17.5	22.5	506.25	28.93
2,A	40	32.5	7.5	56.25	1.73
2,B	50	32.5	17.5	306.25	9.42
2,C	30	32.5	−2.5	6.25	.19
2,D	10	32.5	−22.5	506.25	15.58
Total	200	200	0		$\chi^2 = 76.92$

Table A.35. H_0 should be *rejected*. (At $\alpha = .01$, the critical $\chi^2_{.01,3}$ would have been 11.345 and H_0 would have been *accepted*. The educational level of jurors would then have been deemed to reflect the educational level of the county population.)

Table A.35

f_o	f_e	$f_o - f_e$	$(f_o - f_e)^2$	$\dfrac{(f_o - f_e)^2}{f_e}$
12	20	-8	64	3.20
64	50	14	196	3.92
12	10	2	4	.40
12	20	-8	64	3.20
100	100	0		$\chi^2 = 10.72$

29. Dividing the year into quarters beginning with the birthday, there were 215 deaths in the first quarter, 145 in the second, 95 in the third, 45 in the fourth (preceding the *next* birthday). Without any linkage between time of death and birthday, one would expect an equal number of deaths in each quarter, here $(500/4) = 125$. Thus:
Step 1: H_0: $\pi_I = \pi_{II} = \pi_{III} = \pi_{IV} = .25$.
Step 2:

$$\chi^2 = \sum \frac{(f_o - f_e)^2}{f_e}$$

Step 3: Given $\alpha = .01$ and 3 degrees of freedom, the critical $\chi^2_{.01,3} = 11.345$. The decision rule is "Accept H_0 if $\chi^2 \le 11.345$." *Step 4:* Actual $\chi^2 = 126.4$. *Reject H_0.* There *is* a strange linkage between time of death and birthday.

31. Given 29 degrees of freedom and $\alpha = .04$, Appendix Table H yields $\chi^2_U = \chi^2_{.02,29} = 46.693$ and $\chi^2_L = \chi^2_{.98,29} = 15.574$. According to Box 9.F, the interval is

$$\frac{127.69(29)}{46.693} \le \sigma^2 \le \frac{127.69(29)}{15.574}$$

or

$$79.31 \le \sigma^2 \le 237.77.$$

33. Given 9 degrees of freedom and $\alpha = .02$, Appendix Table H yields $\chi^2_U = \chi^2_{.01,9} = 21.666$ and $\chi^2_L = \chi^2_{.99,9} = 2.088$. According to Box 9.F, the interval is

$$\frac{26(9)}{21.666} \le \sigma^2 \le \frac{26(9)}{2.088}$$

or

$$10.800 \le \sigma^2 \le 112.069.$$

The altimeters do not meet specifications.
35. a. Lower-tailed test. Critical $\chi^2_{.99,25} = 11.524$.

$$\text{Computed } \chi^2 = \frac{10(25)}{64} = 3.91.$$

H_0 should be *rejected*.
b. Lower-tailed test. Critical $\chi^2_{.98,14} = 5.368$.

$$\text{Computed } \chi^2 = \frac{.1(14)}{.64} = 2.19.$$

H_0 should be *rejected*.
c. Upper-tailed test. Critical $\chi^2_{.001,4} = 18.467$.

$$\text{Computed } \chi^2 = \frac{110(4)}{100} = 4.4.$$

H_0 should be *accepted*.
37. *Step 1:* H_0: $\sigma^2 = 100$. H_A: $\sigma^2 \ne 100$
Step 2: $\chi^2 = \dfrac{s^2(n-1)}{\sigma_0^2}$
Step 3: For $\alpha = .04$ and 13 degrees of freedom, critical χ^2 values are $\chi^2_{.98,13} = 4.765$ and $\chi^2_{.02,13} = 25.472$. The decision rule is "Accept H_0 if $4.765 \le \chi^2 \le 25.472$."
Step 4: To compute χ^2, we need the Table A.36 computations.

Table A.36

X	X^2
-17	289
22	484
5	25
0	0
8	64
-3	9
13	169
20	400
25	625
-41	1,681
0	0
-4	16
8	64
13	121
$\Sigma X = 49$	$\Sigma X^2 = 3,947$
$\bar{X} = 3.5$	

Thus,

$$s^2 = \frac{\Sigma X^2 - n\overline{X}^2}{n-1} = \frac{3{,}947 - 14(3.5)^2}{13} = 290.42$$

and

$$\chi^2 = \frac{s^2(n-1)}{\sigma_0^2} = \frac{290.42(13)}{100} = 37.75$$

H_0 should be *rejected*. The variance is now larger.

39. *Step 1*: H_0: $\sigma^2 \geq 80$. H_A: $\sigma^2 < 80$.

Step 2: $\chi^2 = \dfrac{s^2(n-1)}{\sigma_0^2}$

Step 3: Given $\alpha = .05$ and 24 degrees of freedom, Appendix Table H suggests a critical value of $\chi^2_{.95,24} = 13.848$. The decision rule is "Accept H_0 if $\chi^2 > 13.848$."

Step 4: $\chi^2 = \dfrac{64(24)}{80} = 19.2$

H_0 should be *accepted*. The claim is justified.

41. See Table A.37. Degrees of freedom $= k - 1 - 0 = 2$. For $\alpha = .05$, the critical value is $\chi^2_{.05,2} = 5.991$. H_0 should be *rejected*. The number of acceptances is either not binomially distributed or, if it is, the value of $\pi \neq .25$.

43. One would first have to calculate $\overline{X}$ and s and use these statistics as estimates of the missing parameters. This calculation could be accomplished with the help of (Chapter 9's) Table H data (repeated in the first and third columns of Table A.38 and a more or less arbitrary estimate of the midpoints of the two open-ended classes (given in the second column of Table A.38).

$$\overline{X} = \frac{\Sigma f_o X}{n} = \frac{3{,}380}{90} = 37.56.$$

$$s = \sqrt{\frac{\Sigma f_o X^2 - n\overline{X}^2}{n-1}}$$

$$= \sqrt{\frac{153{,}850 - 90(37.56)^2}{89}} = 17.379.$$

Table A.37

Number of Acceptances	f_o	Binomial Probability, p, for $n = 3$ and $\pi = .25$	$f_e = 100p$	$f_o - f_e$	$(f_o - f_e)^2$	$\dfrac{(f_o - f_e)^2}{f_e}$
0	24	.4219	42.19	−18.19	330.88	7.84
1	36	.4219	42.19	−6.19	38.32	.91
2	26	.1406	14.06	} 24.38	594.38	38.05
3	14	.0156	1.56			
Total	100	1.0000	100	0		$\chi^2 = 46.80$

Table A.38

Number of Futures Contracts Traded per Day (millions)	Class Midpoint, X	Observed Frequency, f_o	$f_o X$	X^2	$f_o X^2$
under 10	5	5	25	25	125
10 − under 20	15	9	135	225	2,025
20 − under 30	25	15	375	625	9,375
30 − under 40	35	23	805	1,225	28,175
40 − under 50	45	20	900	2,025	40,500
50 − under 60	55	8	440	3,025	24,200
60 − under 70	65	6	390	4,225	25,350
70 − under 80	75	3	225	5,625	16,875
80 and above	85	1	85	7,225	7,225
		90	3,380		153,850
		$\Sigma f_o = n$	$\Sigma f_o X$		$\Sigma f_o X^2$

These two values could be inserted in the null hypothesis, making Chapter 9's Table J look like Table A.39. After combining classes as needed, we have $k = 7$ classes. Since two parameters were estimated, we have $7 - 1 - 2 = 4$ degrees of freedom. With $\alpha = .02$, this *d.f.* gives us a critical value of $\chi^2_{.02,4} = 11.668$. Thus, the decision rule must be "Accept H_0 if $\chi^2 \leq 11.668$." The actual value of χ^2 is calculated in Table A.40. Accordingly, the null hypothesis, as amended, should be *accepted.*

45. H_0: Public opinion 3 years after inauguration is the same as 1 year after inauguration. There being $4 - 1 - 0 = 3$ degrees of freedom, the critical value is $\chi^2_{.02,3} = 9.837$. According to the χ^2 value calculated in Table A.41, H_0 should be *rejected.*

Table A.39

Number of Futures Contracts Traded per Day (millions): Upper Class Limit, x	Normal Deviate, $z = \dfrac{x - \mu}{\sigma}$ $= \dfrac{x - 37.56}{17.379}$	Area Under Standard Normal Curve Left of x	Area of Class Interval, p	Expected Frequency (if H_0 is true), $f_e = 90p$
10	−1.59	.0559	.0559	5.031
20	−1.01	.1562	.1003	9.027
30	− .44	.3300	.1738	15.642
40	.14	.5557	.2257	20.313
50	.72	.7642	.2085	18.765
60	1.29	.9015	.1373	12.357
70	1.87	.9693	.0678	6.102 ⎫
80	2.44	.9927	.0234	2.106 ⎬ 8.865
∞	∞	1.0000	.0073	.657 ⎭
			1.0000	90

Table A.40

Number of Futures Contracts Traded per Day (millions)	f_o	f_e	$f_o - f_e$	$\dfrac{(f_o - f_e)^2}{f_e}$
under 10	5	5.031	− .031	.0002
10 − under 20	9	9.027	− .027	.0001
20 − under 30	15	15.642	− .642	.0263
30 − under 40	23	20.313	2.687	.3554
40 − under 50	20	18.765	1.235	.0813
50 − under 60	8	12.357	−4.357	1.5363
60 and above	10	8.865	1.135	.1453
Total	90	90	0	$\chi^2 = 2.1449$

Table A.41

Opinion Expressed About the President's Performance	f_o (after 3 years)	f_e (1.5 times f_o after 1 year)	$f_o - f_e$	$\dfrac{(f_o - f_e)^2}{f_e}$
Does superb job	68	75	− 7	.653
Does good job	33	37.5	−4.5	.540
Does fair job	22	22.5	−.5	.011
Does lousy job	27	15	12.0	9.600
Total	150	150	0	$\chi^2 = 10.804$

Chapter 10

1. The sample means are: $\overline{X}_1 = 3.2$, $\overline{X}_2 = 3$, $\overline{X}_3 = 2.4$, $\overline{X}_4 = 2.8$, $\overline{X}_5 = 3.6$; the grand mean is $\overline{\overline{X}} = 3$. $TSS = r\Sigma(\overline{X}_j - \overline{\overline{X}})^2 = 5[(3.2 - 3)^2 + (3 - 3)^2 + (2.4 - 3)^2 + (2.8 - 3)^2 + (3.6 - 3)^2] = 4$. ESS can be calculated from Table A.42 as $ESS = 4.8 + 10 + 5.2 + 24.8 + 5.2 = 50$. The ANOVA table is given in Table A.43. The critical $F_{.025}$ for $d.f. = (4, 20)$ equals 3.51. H_0 should be *accepted*. The mean delivery times are the same.

3. H_0: The mean housing prices are the same regardless of which of 3 air pollution levels prevails:

$\mu_L = \mu_M = \mu_H$; H_A: At least one of the population means is different from the others. The sample means are $\overline{X}_L = 81.2$, $\overline{X}_M = 77$, $\overline{X}_H = 55.4$; the grand mean is $\overline{\overline{X}} = 71.2$. $TSS = r\Sigma(\overline{X}_j - \overline{\overline{X}})^2 = 5[(81.2 - 71.2)^2 + (77 - 71.2)^2 + (55.4 - 71.2)^2] = 1,916.4$. ESS can be calculated from Table A.44 as: $ESS = 3,570.8 + 1,682 + 729.2 = 5,982$. The ANOVA table is given in Table A.45. The critical $F_{.025}$ for $d.f. = (2, 12) = 5.10$ (Appendix Table I). H_0 should be *accepted*.

5. H_0: $\mu_A = \mu_B = \mu_C$. The ANOVA table is Table A.46. Given $\alpha = .025$ and $d.f. = (2, 15)$, Appendix Table

Table A.42

Observation	1 $(X_{j1} - \overline{X}_1)^2$	2 $(X_{j2} - \overline{X}_2)^2$	3 $(X_{j3} - \overline{X}_3)^2$	4 $(X_{j4} - \overline{X}_4)^2$	5 $(X_{j5} - \overline{X}_5)^2$
1	$(3 - 3.2)^2 = .04$	$(4 - 3)^2 = 1$	$(2 - 2.4)^2 = .16$	$(7 - 2.8)^2 = 17.64$	$(2 - 3.6)^2 = 2.56$
2	$(3 - 3.2)^2 = .04$	$(1 - 3)^2 = 4$	$(4 - 2.4)^2 = 2.56$	$(1 - 2.8)^2 = 3.24$	$(3 - 3.6)^2 = .36$
3	$(2 - 3.2)^2 = 1.44$	$(3 - 3)^2 = 0$	$(3 - 2.4)^2 = .36$	$(1 - 2.8)^2 = 3.24$	$(4 - 3.6)^2 = .16$
4	$(5 - 3.2)^2 = 3.24$	$(5 - 3)^2 = 4$	$(1 - 2.4)^2 = 1.96$	$(2 - 2.8)^2 = .64$	$(5 - 3.6)^2 = 1.96$
5	$(3 - 3.2)^2 = .04$	$(2 - 3)^2 = 1$	$(2 - 2.4)^2 = .16$	$(3 - 2.8)^2 = .04$	$(4 - 3.6)^2 = .16$
	4.8	10	5.2	24.8	5.2

Table A.43

Source of Variation	Sum of Squares (1)	Degrees of Freedom (2)	Mean Square (3) = (1) ÷ (2)	Test Statistic (4)
Treatments	$TSS = 4$	$c - 1 = 4$	$TMS = \dfrac{4}{4} = 1$	$F = \dfrac{TMS}{EMS} = \dfrac{1}{2.5} = .4$
Error	$ESS = 50$	$(r - 1)c = 20$	$EMS = \dfrac{50}{20} = 2.5$	
Total	$Total\ SS = 54$	$rc - 1 = 24$		

Table A.44

Observation	Low $(X_{jL} - \overline{X}_L)^2$	Moderate $(X_{jM} - \overline{X}_M)^2$	High $(X_{jH} - \overline{X}_H)^2$
1	$(120 - 81.2)^2 = 1,505.44$	$(61 - 77)^2 = 256$	$(40 - 55.4)^2 = 237.16$
2	$(68 - 81.2)^2 = 174.24$	$(59 - 77)^2 = 324$	$(55 - 55.4)^2 = .16$
3	$(40 - 81.2)^2 = 1,697.44$	$(110 - 77)^2 = 1,089$	$(73 - 55.4)^2 = 309.76$
4	$(95 - 81.2)^2 = 190.44$	$(75 - 77)^2 = 4$	$(45 - 55.4)^2 = 108.16$
5	$(83 - 81.2)^2 = 3.24$	$(80 - 77)^2 = 9$	$(64 - 55.4)^2 = 73.96$
	3,570.8	1,682	729.2

Table A.45

Source of Variation	Sum of Squares (1)	Degrees of Freedom (2)	Mean Square (3) = (1) ÷ (2)	Test Statistic (4)
Treatments	$TSS = 1{,}916.4$	$c - 1 = 2$	$TMS = 958.2$	
Error	$ESS = 5{,}982.0$	$(r - 1)c = 12$	$EMS = 498.5$	$F = \dfrac{TMS}{EMS} = \dfrac{958.2}{498.5} = 1.92$
Total	$Total\ SS = 7{,}898.4$	$(rc) - 1 = 14$		

Table A.46

Source of Variation	Sum of Squares	Degrees of Freedom	Mean Square	Test Statistic
Treatments	3,078.11	2	1,539.06	$F = 38.58$
Error	598.33	15	39.89	
Total	3,676.44	17		

Table A.47

Source of Variation	Sum of Squares	Degrees of Freedom	Mean Square	Test Statistic
Treatments	.67	2	.33	$F = .0018$
Error	3,322.29	18	184.57	
Total	3,322.95	20		

Table A.48

Source of Variation	Sum of Squares	Degrees of Freedom	Mean Square	Test Statistic
Treatments	107.11	2	53.56	$F = .36$
Error	2,208.67	15	147.24	
Total	2,315.78	17		

I suggests "Accept H_0 if $F \leq 4.77$." H_0 should be *rejected*. Worker productivity differs. The results are valid if the three output-per-worker populations are normal and have equal variances.

7. H_0: $\mu_A = \mu_B = \mu_C$. The ANOVA table is Table A.47. Given $\alpha = .05$ and $d.f. = (2,18)$, Appendix Table I suggests "Accept H_0 if $F \leq 3.55$." H_0 should be *accepted*. The results are valid if the three reading-score populations are normal and have equal variances.

9. H_0: $\mu_A = \mu_B = \mu_C$. The ANOVA table is Table A.48. Given $\alpha = .01$ and $d.f. = (2,15)$, Appendix Table I suggests "Accept H_0 if $F \leq 6.36$." H_0 should be *accepted*. The results are valid if the three consumption populations are normal and have equal variances.

11. H_0: $\mu_A = \mu_B = \mu_C = \mu_D$. The ANOVA table is Table A.49.

Given $\alpha = .025$ and $d.f. = (3,36)$, Appendix Table I suggests "Accept H_0 if $F \leq 3.51$ (interpolated)." H_0 should be *rejected*. The incidence of crime differs among the districts. The results are valid if the four police-call populations are normal and have equal variances.

13. H_0: $\mu_A = \mu_B = \mu_C = \mu_D$. The ANOVA table is Table A.50.

Given $\alpha = .05$ and $d.f. = (3,20)$, Appendix Table I suggests "Accept H_0 if $F \leq 3.10$." H_0 should be *rejected*. The average debt-to-equity ratios differ among the industries. The results are valid if the four debt-to-equity ratio populations are normal and have equal variances.

15. H_0: $\mu_A = \mu_B = \mu_C$. The ANOVA table is Table A.51. Given $\alpha = .025$ and $d.f. = (2, 12)$. Appendix Table I suggests: "Accept H_0 if $F \leq 5.10$." H_0 should be *accepted*. The mean yields on the 3 types of soil

are the same. The results are valid if the three populations are normal and have equal variances.

17. H_0: $\mu_A = \mu_B = \mu_C = \mu_D = \mu_E$. The ANOVA table is Table A.52.

Given $\alpha = .025$ and $d.f. = (4, 20)$. Appendix Table I suggests: "Accept H_0 if $F \leq 3.51$." H_0 should be *accepted*. The average quality is the same on the 5 surfaces. The results are valid if the five populations are normal and have equal variances.

19. H_0: $\mu_A = \mu_B = \mu_C$. The ANOVA table is Table A.53.

Given $\alpha = .05$ and $d.f. = (2,9)$, Appendix Table I suggests: "Accept H_0 if $F \leq 4.26$." H_0 should be *accepted*. The average tensile strength is the same for the 3 production methods. The results are valid if the three populations are normal and have equal variances.

21. H_0: $\mu_A = \mu_B = \mu_C$. The ANOVA table is Table A.54.

Given $\alpha = .01$ and $d.f. (2,9)$. Appendix Table I suggests: "Accept H_0 if $F \leq 8.02$." H_0 should be *accepted*. The average price is the same for the 3 sta-

Table A.49

Source of Variation	Sum of Squares	Degrees of Freedom	Mean Square	Test Statistic
Treatments	1,065.88	3	355.29	$F = 10.56$
Error	1,211.10	36	33.64	
Total	2,276.98	39		

Table A.50

Source of Variation	Sum of Squares	Degrees of Freedom	Mean Square	Test Statistic
Treatments	1.70	3	.568	$F = 8.80$
Error	1.29	20	.065	
Total	3.00	23		

Table A.51

Source of Variation	Sum of Squares	Degrees of Freedom	Mean Square	Test Statistic
Treatments	7.60	2	3.80	$F = .36$
Error	128.40	12	10.70	
Total	136.00	14		

Table A.52

Source of Variation	Sum of Squares	Degrees of Freedom	Mean Square	Test Statistic
Treatments	1,333.6	4	333.40	$F = 3.22$
Error	2,068.4	20	103.42	
Total	3,402.0	24		

Table A.53

Source of Variation	Sum of Squares	Degrees of Freedom	Mean Square	Test Statistic
Treatments	20,000	2	10,000.00	$F = .38$
Error	240,000	9	26,666.67	
Total	260,000	11		

Table A.54

Source of Variation	Sum of Squares	Degrees of Freedom	Mean Square	Test Statistic
Treatments	389,616.7	2	194,808.40	$F = 7.87$
Error	222,850.0	9	24,761.11	
Total	612,466.7	11		

Table A.55

Source of Variation	Sum of Squares	Degrees of Freedom	Mean Square	Test Statistic
Treatments	172.29	3	57.43	$F = 2.11$
Error	217.91	8	27.24	
Total	390.20	11		

Table A.56

Source of Variation	Sum of Squares	Degrees of Freedom	Mean Square	Test Statistic
Treatments	2,457.33	2	1,228.67	$F = 91.62$
Error	201.17	15	13.41	
Total	2,658.50	17		

tion types. The results are valid if the three populations are normal and have equal variances.

23. H_0: $\mu_A = \mu_B = \mu_C = \mu_D$. The ANOVA table is Table A.55.

Given $\alpha = .025$ and $d.f. = (3,8)$. Appendix Table I suggests: "Accept H_0 if $F \leq 5.42$." H_0 should be

accepted. Average findings are the same among the 4 laboratories. The results are valid if the four populations are normal and have equal variances.

25. H_0: $\mu_A = \mu_B = \mu_C$. The ANOVA table is Table A.56.

Given $\alpha = .05$ and $d.f. = (2,15)$. Appendix Table I suggests: "Accept H_0 if $F \leq 3.68$." H_0 should be rejected. The average tensile strength differs among the 3 processes. The results are valid if the three populations are normal and have equal variances.

27. H_0: $\mu_A = \mu_B = \mu_C$. The ANOVA table is Table A.57. Given α = .01 and $d.f.$ = (2, 9). Appendix Table I suggests: "Accept H_0 if $F \leq$ 8.02." H_0 should be accepted. The average weight is the same in the 3 areas. The results are valid if the three populations are normal and have equal variances.

29. H_0: $\mu_A = \mu_B = \mu_C$. The ANOVA table is Table A.58. Given α = .025 and $d.f.$ = (2, 15). Appendix Table I suggests: "Accept H_0 if $F \leq$ 4.77." H_0 should be accepted. The average score is the same among the 3 types of work. The results are valid if the three populations are normal and have equal variances.

31. H_0: $\mu_A = \mu_B = \mu_C$. The ANOVA table is Table A.59. Given α = .05 and $d.f.$ = (2, 15). Appendix Table I suggests: "Accept H_0 if $F \leq$ 3.68." H_0 should be accepted. The average score is the same for the 3 types of work. The results are valid if the three populations are normal and have equal variances.

33. H_0: $\mu_A = \mu_B = \mu_C = \mu_D$. The ANOVA table is Table A.60.

Given α = .05 and $d.f.$ = (3, 20). Appendix Table I suggests: "Accept H_0 if $F \leq$ 3.10." H_0 should be rejected. The average number of strikes differs among the 4 industries. The results are valid if the four populations are normal and have equal variances.

35. H_0: $\mu_A = \mu_B = \mu_C$. The ANOVA table is Table A. 61. Given α = .025 and $d.f.$ = (2, 15). Appendix Table I suggests: "Accept H_0 if $F \leq$ 4.77." H_0 should be accepted. The average yield is the same among the 3 tree types. The results are valid if the three populations are normal and have equal variances.

37. H_0: $\mu_A = \mu_B = \mu_C$. The ANOVA table is Table A.62. Given α = .01 and $d.f.$ = (2,6). Appendix Table I suggests: "Accept H_0 if $F \leq$ 10.92." H_0 should be rejected. The average price differs among the 3 brands. The results are valid if the three populations are normal and have equal variances.

39. H_0: $\mu_A = \mu_B = \mu_C$. The ANOVA table is Table A.63. Given α = .05 and $d.f.$ = (2, 15). Appendix Table I suggests: "Accept H_0 if $F \leq$ 3.68." H_0 should be rejected. The average price differs among the 3 states. The results are valid if the three populations are normal and have equal variances.

Table A.57

Source of Variation	Sum of Squares	Degrees of Freedom	Mean Square	Test Statistic
Treatments	54.65	2	27.33	$F = 6.72$
Error	36.58	9	4.06	
Total	91.23	11		

Table A.58

Source of Variation	Sum of Squares	Degrees of Freedom	Mean Square	Test Statistic
Treatments	218.11	2	109.06	$F = .47$
Error	3,509.50	15	233.97	
Total	3,727.61	17		

Table A.59

Source of Variation	Sum of Squares	Degrees of Freedom	Mean Square	Test Statistic
Treatments	560.78	2	280.39	$F = 2.04$
Error	2,063.50	15	137.57	
Total	2,624.28	17		

Table A.60

Source of Variation	Sum of Squares	Degrees of Freedom	Mean Square	Test Statistic
Treatments	64.46	3	21.49	$F = 5.15$
Error	83.50	20	4.18	
Total	147.96	23		

Table A.61

Source of Variation	Sum of Squares	Degrees of Freedom	Mean Square	Test Statistic
Treatments	5,340.11	2	2,670.05	$F = 1.00$
Error	40,046.17	15	2,669.75	
Total	45,386.28	17		

Table A.62

Source of Variation	Sum of Squares	Degrees of Freedom	Mean Square	Test Statistic
Treatments	44,960.23	2	22,480.11	$F = 384.64$
Error	350.67	6	58.44	
Total	45,310.89	8		

Table A.63

Source of Variation	Sum of Squares	Degrees of Freedom	Mean Square	Test Statistic
Treatments	2,170,767	2	1,085,383.5	$F = 7.55$
Error	2,157,308	15	143,820.6	
Total	4,328,075	17		

41. The Box 10.B formula applies, the appropriate t value being $t_{.025,18}$. Thus,

$$\mu_A = 77.43 \pm \left(2.101\sqrt{\frac{184.57}{7}}\right) = 77.43 \pm 10.79.$$
$$\mu_B = 77.57 \pm 10.79.$$
$$\mu_C = 77.86 \pm 10.79.$$

Hence,
$$66.64 \leq \mu_A \leq 88.22$$
$$66.78 \leq \mu_B \leq 88.36$$
$$67.07 \leq \mu_C \leq 88.65$$

43. The Box 10.B formula applies, the appropriate t value being $t_{.005,15}$. Thus,

$$\mu_A = 18 \pm \left(2.947\sqrt{\frac{147.24}{6}}\right) = 18 \pm 14.60.$$
$$\mu_B = 23.33 \pm 14.60.$$
$$\mu_C = 18.33 \pm 14.60.$$

Hence,
$$3.40 \leq \mu_A \leq 32.60$$
$$8.73 \leq \mu_B \leq 37.93$$
$$3.73 \leq \mu_C \leq 32.93$$

45. The Box 10.B formula applies, the appropriate t value being $t_{.025,20}$. Thus,

$$\mu_A = .150 \pm \left(2.086\sqrt{\frac{.065}{6}}\right) = .150 \pm .217.$$

$$\mu_B = .300 \pm .217.$$

$$\mu_C = .117 \pm .217.$$

$$\mu_D = .783 \pm .217.$$

Hence,

$$-.067 \le \mu_A \le .367$$
$$.083 \le \mu_B \le .517$$
$$-.100 \le \mu_C \le .334$$
$$.566 \le \mu_D \le 1.000$$

Chapter 11

1. a. See Figure A.22 for the scatter diagram.
   b. $\hat{Y}_X = 52.89 - 1.12\,X.$
   c. The relationship does make sense. According to the estimated regression line, the incumbent party can expect to capture 52.89 percent of the vote if there is no change in the election-year unemployment rate (point B) but to gain (or lose) about one percentage point in the vote for every one percentage point difference between the election-year unemployment rate and that of the previous year. If unemployment is down 3 percentage points, the predicted vote is 56.3

percent (point A); if unemployment is up 3 percentage points, the predicted vote is 49.5 percent (point C).

3. $\hat{Y}_X = 452.6558 + 96.69197\,X.$
5. $\hat{Y}_X = 57.08909 + 17.37734\,X.$
7. $\hat{Y}_X = 17.46059 - .07219\,X.$
9. $\hat{Y}_X = 10.47237 - .33201\,X.$
11. $\hat{Y}_X = .35432 + .02364\,X.$
13. $\hat{Y}_X = 228.8908 + 19.70607\,X.$
15. $\hat{Y}_X = 1.23145 + .35778\,X.$
17. $\hat{Y}_X = 36.94287 + 4.21079\,X.$
19. $\hat{Y}_X = -72.75798 + .68403\,X.$
21. The *point estimate* is $\hat{Y}_5 = 52.89 - 1.12(5) = 47.29$. For the *98-percent prediction intervals,* the Box 11.B and 11.D formulas apply (with $t_{.01,22} = 2.508$ and $s_{Y\cdot X} = 6.021974$):

$$\mu_{Y\cdot 5} = 47.29$$

$$\pm \left[2.508(6.021974)\sqrt{\frac{1}{24} + \frac{(5-.13)^2}{159.86} \quad 24(.0169)}\right]$$

$$= 47.29 \pm 6.59$$

Hence, $40.7 \le \mu_{Y\cdot 5} \le 53.88$.

$$I_{Y\cdot 5} = 47.29 \pm 16.48$$

Hence, $30.81 \le I_{Y\cdot 5} \le 63.77$.
23. The *point estimate* is $\hat{Y}_5 = 452.6558 + 96.69197(5) = 936.12$. For the *90-percent predic-*

Figure A.22

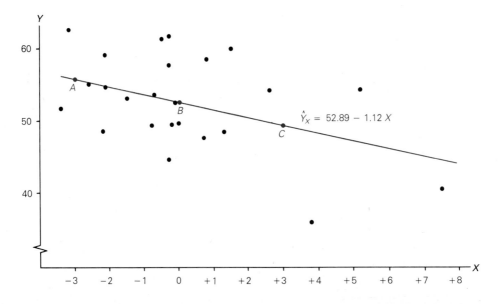

tion intervals, the Box 11.B and 11.D formulas apply (with $t_{.05,8} = 1.86$ and $s_{Y \cdot X} = 263.3122$):

$\mu_{Y \cdot 5} = 936.12$

$$\pm \left[1.86(263.3122)\sqrt{\frac{1}{10} + \frac{(5 - 5.7)^2}{433 - 10(5.7)^2}} \right]$$

$= 936.12 \pm 158.35.$

Hence, $777.77 \leq \mu_{Y \cdot 5} \leq 1{,}094.47.$

$I_{Y \cdot 5} = 936.12 \pm 514.72$

Hence, $421.40 \leq I_{Y \cdot 5} \leq 1{,}450.84.$

25. The *point estimate* is $\hat{Y}_{11} = 57.08909 + 17.37734(11) = 248.24.$ For the *98-percent prediction intervals,* the Box 11.B and 11.D formulas apply (with $t_{.01,12} = 2.681$ and $s_{Y \cdot X} = 56.83893$):

$\mu_{Y \cdot 11} = 248.24$

$$\pm \left[2.681(56.83893)\sqrt{\frac{1}{14} + \frac{(11 - 7.71)^2}{1{,}016 - 14(7.71)^2}} \right]$$

$= 248.24 \pm 55.01.$

Hence, $193.23 \leq \mu_{Y \cdot 11} \leq 303.25.$

$I_{Y \cdot 11} = 248.24 \pm 162.01$

Hence, $86.23 \leq I_{Y \cdot 11} \leq 410.25.$

27. The *point estimate* is $\hat{Y}_{140} = 17.46059 - .07219(140) = 7.35.$ For the *99.5-prediction intervals,* the Box 11.B and 11.D formulas apply (with $t_{.0025,14} = 3.326$ and $s_{Y \cdot X} = 1.41038$):

$\mu_{Y \cdot 140} = 7.35$

$$\pm \left[3.326(1.41038)\sqrt{\frac{1}{16} + \frac{(140 - 74.69)^2}{114{,}377 - 16(74.69)^2}} \right]$$

$= 7.35 \pm 2.26$

Hence, $5.09 \leq \mu_{Y \cdot 140} \leq 9.61.$

$I_{Y \cdot 140} = 7.35 \pm 5.21$

Hence, $2.14 \leq I_{Y \cdot 140} \leq 12.56.$

29. The *point estimte* is $\hat{Y}_{14} = 10.47237 - .33201(14) = 5.82.$ For the *99.9-percent prediction intervals,* the Box 11.B and 11.D formulas apply (with $t_{.0005,7} = 5.408$ and $s_{Y \cdot X} = .8641118$):

$\mu_{Y \cdot 14} = 5.82$

$$\pm \left[5.408(.8641118)\sqrt{\frac{1}{9} + \frac{(14 - 9.69)^2}{896.54 - 9(9.69)^2}} \right]$$

$= 5.82 \pm 3.21$

Hence, $2.61 \leq \mu_{Y \cdot 14} \leq 9.03.$

$I_{Y \cdot 14} = 5.82 \pm 5.67$

Hence, $.15 \leq I_{Y \cdot 14} \leq 11.49.$
No, the relationship doesn't make sense. One would expect, all else being equal, that higher interest rates lower investment demand, hence aggregate demand, hence output and employment. Thus higher interest rates should be associated with higher, not lower unemployment. The value of slope b should be positive. Presumably, other things were not equal as this reasoning assumes.

31. The *point estimate* is $\hat{Y}_{25} = .35432 + .02364(25) = .945.$ For the *95-percent prediction intervals,* the Box 11.B and 11.D formulas apply (with $t_{.025,8} = 2.306$ and $s_{Y \cdot X} = .07485795$):

$\mu_{Y \cdot 25} = .945$

$$\pm \left[2.306(.07485795)\sqrt{\frac{1}{10} + \frac{(25 - 22.49)^2}{6{,}923.354 - 10(22.49)^2}} \right]$$

$= .945 \pm .056$

Hence, $.889 \leq \mu_{Y \cdot 25} \leq 1.001.$

$I_{Y \cdot 25} = .945 \pm .181$

Hence, $.764 \leq 1.126.$

33. The Box 11.F and 11.G formulas apply. Given $t_{.025,22} = 2.074,$

$$\beta = -1.118 \pm \left(2.074 \frac{6.021974}{\sqrt{159.86 - 24(.13)^2}} \right)$$

$= -1.118 \pm .989$

Hence, $-2.107 \leq \beta \leq -.129.$

$$\alpha = 52.891 \pm \left(.989 \sqrt{\frac{159.86}{24}} \right)$$

$= 52.891 \pm 2.552$

Hence, $50.339 \leq \alpha \leq 55.443.$

35. See Table A.64.

Table A.64

Problem	95-percent Confidence Interval For	
	β	α
3	$38.29 \leq \beta \leq 155.09$	$68.36 \leq \alpha \leq 836.95$
5	$8.22 \leq \beta \leq 26.54$	$-20.94 \leq \alpha \leq 135.11$
7	$-.09 \leq \beta \leq -.05$	$15.85 \leq \alpha \leq 19.07$

37. See Table A.65.

Table A.65

Problem	95-percent Confidence Interval For	
	β	α
9	$-.62 \leq \beta \leq -.05$	$7.63 \leq \alpha \leq 13.31$
11	$.02 \leq \beta \leq .03$	$.25 \leq \alpha \leq .46$
13	$8.64 \leq \beta \leq 30.77$	$11.30 \leq \alpha \leq 446.48$

39. H_0: $\beta = 0$ and H_A: $\beta \neq 0$. Given $n - 2 = 22$ degrees of freedom, Appendix Table G implies a critical value of $\pm t_{.025,22} = \pm 2.074$. Hence the decision rule is "Accept H_0 if $-2.074 \leq t \leq +2.074$." The actual value of t is

$$t = \frac{b}{\frac{s_{Y \cdot X}}{\sqrt{\Sigma X^2 - n\overline{X}^2}}} = \frac{-1.11836}{\frac{6.021974}{\sqrt{159.86 - 24(.13)^2}}}$$
$$= -2.345.$$

H_0 should be *rejected*.

41. See Table A.66

Table A.66

Problem	$r^2 = \dfrac{RSS}{Total\ SS}$	$r = \sqrt{r^2}$
1	.20	$-.45$
3	.65	.80
5	.59	.77
7	.82	$-.91$
9	.52	$-.72$

Problem 1: Only 20 percent of the variation in Y (the presidential voting percentage captured by the incumbent party) can

be explained by the association of Y with X (the change in the election-year unemployment rate for which the incumbent party might be blamed).

43. a. For $F_{.05}$ and $d.f. = (1,22)$, Appendix Table I suggests "Accept H_0 if $F \leq 4.30$." The ANOVA table is Table A.67.

Table A.67

Source of Variation	Sum of Squares	Degrees of Freedom	Mean Square	Test Statistic
Regression	199.41	1	199.41	$F = 5.50$
Error	797.81	22	36.26	
Total	997.22	23		

Reject H_0 for problem 1, as in problem 39.
b. For $F_{.05}$ and $d.f. = (1,8)$, Appendix Table I suggests "Accept H_0 if $F \leq 5.32$." The ANOVA table is Table A.68.

Table A.68

Source of Variation	Sum of Squares	Degrees of Freedom	Mean Square	Test Statistic
Regression	1,010,663.0	1	1,010,663.00	$F = 14.58$
Error	554,666.6	8	69,333.33	
Total	1,565,329.6	9		

Reject H_0 for problem 3.
45. $\hat{Y}_X = -26.79262 + 15.14805\ X$.
   $s_{Y \cdot X} = 21.38662$
   $r^2 = .8072$
   For $F_{.05}$ and $d.f. = (1,16)$, Appendix Table I suggests "Accept H_0 if $F \leq 4.49$." The ANOVA table is Table A.69.

Table A.69

Source of Variation	Sum of Squares	Degrees of Freedom	Mean Square	Test Statistic
Regression	30,648.28	1	30,648.28	$F = 67.01$
Error	7,318.20	16	457.39	
Total	37,966.48	17		

H_0 should be *rejected*.

Chapter 12

1. $\hat{Y} = 27.3595 + .0337 X_1 - 1.2383 X_2$
$(4.4477)\ (3.5919)\ (-3.7682)$
$R^2 = .8027 \quad \bar{R}^2 = .7869$
The ANOVA table is Table A.70.

Table A.70

Source of Variation	Sum of Squares	Degrees of Freedom	Mean Square	Test Statistic
Regression	370.01	2	185.00	$F = 50.85$
Error	90.95	25	3.64	
Total	460.96	27		

Critical $F_{.05}$ for $d.f. = (2,25) = 3.39$. *Reject H_0:* $\beta_1 = \beta_2 = 0$.

3. $\hat{Y} = 4.0231 + .0848 X_1 - 3.8750 D_1$
$(1.2556)\ (3.9120)\ (-2.1685)$
$R^2 = .8012 \quad \bar{R}^2 = .7651.$
The ANOVA table is Table A.71.

Table A.71

Source of Variation	Sum of Squares	Degrees of Freedom	Mean Square	Test Statistic
Regression	332.37	2	166.19	$F = 22.17$
Error	82.46	11	7.50	
Total	414.83	13		

Critical $F_{.05}$ for $d.f. = (2,11) = 3.98$. *Reject H_0:* $\beta_1 = \beta_2 = 0$.

5. $\hat{Y} = 313.5316 + 1.5120 X_1 - 4.9085 X_2$
$(2.6985)\ (1.9284)\ (-2.0893)$
$R^2 = .7014 \quad \bar{R}^2 = .6161$
The ANOVA table is Table A.72.

Table A.72

Source of Variation	Sum of Squares	Degrees of Freedom	Mean Square	Test Statistic
Regression	67,121.21	2	33,560.60	$F = 8.22$
Error	28,568.80	7	4,081.26	
Total	95,690.01	9		

Critical $F_{.05}$ for $d.f. = (2,7) = 4.74$. *Reject H_0:* $\beta_1 = \beta_2 = 0$.

7. $\hat{Y} = -29.7684 + 6.6804 X_1 + 9.9322 X_2$
$(-.3222)\ (.9656)\ (21.5428)$
$R^2 = .9752 \quad \bar{R}^2 = .9710$
The ANOVA table is Table A.73.

Table A.73

Source of Variation	Sum of Squares	Degrees of Freedom	Mean Square	Test Statistic
Regression	14,071,500	2	7,035,750	$F = 235.60$
Error	358,358.8	12	29,863.23	
Total	14,429,858.8	14		

Critical $F_{.05}$ for $d.f. = (2,12) = 3.89$. *Reject H_0:* $\beta_1 = \beta_2 = 0$.

9. $\hat{Y} = -111.4061 + .9286 X_1 + .6844 X_2$
$(-2.2822)\ (.8427)\ (1.8663)$
$R^2 = .8824 \quad \bar{R}^2 = .8354.$
The ANOVA table is Table A.74.

Table A.74

Source of Variation	Sum of Squares	Degrees of Freedom	Mean Square	Test Statistic
Regression	1,275.99	2	638.00	$F = 18.76$
Error	170.01	5	34.00	
Total	1,446.00	7		

Critical $F_{.05}$ for $d.f. = (2,5) = 5.79$. *Reject H_0:* $\beta_1 = \beta_2 = 0$.

11. $\hat{Y} = 96.2144 + .0891 X_1 + 43.2533 D_1$
$(1.8197)\ (.2389)\ (1.6825)$
$R^2 = .3695 \quad \bar{R}^2 = .2293.$
The ANOVA table is Table A.75.

Table A.75

Source of Variation	Sum of Squares	Degrees of Freedom	Mean Square	Test Statistic
Regression	4,399.75	2	2,199.88	$F = 2.64$
Error	7,508.92	9	834.32	
Total	11,908.67	11		

Critical $F_{.05}$ for $d.f. = (2,9) = 4.26$. *Accept H_0:* $\beta_1 = \beta_2 = 0$.

13. $\hat{Y} = 142.4233 - 1.8289X_1 - 127.3630D_1$
$(3.3380)\ (-.1294)\qquad (-2.4030)$
$R^2 = .4963\qquad \bar{R}^2 = .3523$
The ANOVA table is Table A.76.

Table A.76

Source of Variation	Sum of Squares	Degrees of Freedom	Mean Square	Test Statistic
Regression	40,544.16	2	20,272.08	$F = 3.45$
Error	41,156.00	7	5,879.43	
Total	81,700.16	9		

Critical $F_{.05}$ for *d.f.* $= (2,7) = 4.74$. *Accept* H_0: $\beta_1 = \beta_2 = 0$.

15. $\hat{Y} = 11.4382 - .0328\,X_1 + 1.0701\,D_1$
$(2.5444)(-.5995)\qquad (.7003)$
$R^2 = .1092\qquad \bar{R}^2 = -.3362$
The ANOVA table is Table A.77.

Table A.77

Source of Variation	Sum of Squares	Degrees of Freedom	Mean Square	Test Statistic
Regression	.51	2	.26	$F = .25$
Error	4.20	4	1.05	
Total	4.71	6		

Critical $F_{.05}$ for *d.f.* $= (2,4) = 6.94$. *Accept* H_0: $\beta_1 = \beta_2 = 0$.

17. $\hat{Y} = 500 - 1\,X_1 - .00027\,D_1$
$(\infty)\qquad (\infty)\,(-2.66185)$
$R^2 = 1\qquad \bar{R}^2 = 1$
The ANOVA table is Table A.78.

Table A.78

Source of Variation	Sum of Squares	Degrees of Freedom	Mean Square	Test Statistic
Regression	170,930.8	2	85,465.4	$F = 5.7$ trillion
Error	.0	9	.0	
Total	170,930.8	11		

Critical $F_{.05}$ for *d.f.* $= (2,9) = 4.26$. *Reject* H_0: $\beta_1 = \beta_2 = 0$.

19. $\hat{Y} = 230.6815 - 4.1441X_1 + 144.7237D_1$
$(1.4056)\quad (-.3970)\qquad (.8267)$
$R^2 = .1171\qquad \bar{R}^2 = -.1352$
The ANOVA table is Table A.79.

Table A.79

Source of Variation	Sum of Squares	Degrees of Freedom	Mean Square	Test Statistic
Regression	67,376.22	2	33,688.11	$F = .46$
Error	508,152.20	7	72,593.18	
Total	575,528.42	9		

Critical $F_{.025}$ for *d.f.* $= (2,7) = 6.54$. *Accept* H_0: $\beta_1 = \beta_2 = 0$.

21. $\hat{Y} = 23.1063 - .0138X_1 + .0143X_2$
$(2.4007)\ (-.7788)\quad (.0611)$

$\qquad + \ .0114X_3 - 1.6115\,X_4 + 20.7406\ D_1$
$\qquad (.6883)\ (-1.6771)\qquad (2.5624)$

$R^2 = .9121\qquad \bar{R}^2 = .8022$
The ANOVA table is Table A.80.

Table A.80

Source of Variation	Sum of Squares	Degrees of Freedom	Mean Square	Test Statistic
Regression	2,307.62	5	461.52	$F = 8.30$
Error	222.38	4	55.59	
Total	2,530	9		

Critical $F_{.025}$ for *d.f.* $= (5,4) = 9.36$. *Accept* H_0: $\beta_1 = \beta_2 = \beta_3 = \beta_4 = \beta_5 = 0$.

23. $\hat{Y} = 101.7044 - 10.1697X_1 - .6381X_2$
$(4.5223)\quad (-2.5110)\ (-.0966)$
$R^2 = .4800\qquad \bar{R}^2 = .3315$
The ANOVA table is Table A.81.

Table A.81

Source of Variation	Sum of Squares	Degrees of Freedom	Mean Square	Test Statistic
Regression	3,285.95	2	1,642.98	$F = 3.23$
Error	3,559.10	7	508.44	
Total	6,845.05	9		

Critical $F_{.01}$ for *d.f.* = (2,7) = 9.55. *Accept* H_0: $\beta_1 = \beta_2 = 0$.

25. See Table A.82.

Table A.82

Problem	Sample Standard Error of the Estimate of Y
1	1.9074
2	281.3748
3	2.7380
4	46.9302
5	63.8847
6	1.2469
7	172.8098
8	.8179

27. See Table A.83.

Table A.83

Problem	Sample Standard Error of the Estimate of Y
17	.000122
18	16.2215
19	269.4312
20	20.6976
21	7.4562
22	.1293
23	22.5487
24	28.9192

29. a. Box 12.B applies. The DAYSAL-2 computer program shows a point estimate of 60.76 and an interval of $41.81 \leq \mu_{Y \cdot 12} \leq 79.71$.
   b. Box 12.C applies. The DAYSAL-2 computer program shows a point estimate of 60.76 and an interval of $4.18 \leq I_{Y \cdot 12} \leq 117.35$.
31. Using the same procedure as in problem 29, the point estimate is 42.83, and the intervals are $23.39 \leq \mu_{Y \cdot 12345} \leq 62.26$ and $14.44 \leq I_{Y \cdot 12345} \leq 71.22$.
33. Using the same procedure as in problem 29, the point estimate is 180.95, and the intervals are $-79.32 \leq \mu_{Y \cdot 12} \leq 441.22$ and $-507.36 \leq I_{Y \cdot 12} \leq 869.26$.
35. Box 12.D applies. The answer to problem 17 implies sample standard errors of $s_{b_1} = 0$ and $s_{b_2} =$

.0001 (coefficient divided by its t value); a critical $t_{\alpha/2}$ for 9 degrees of freedom is 2.262 (Appendix Table G). Hence
$\beta_1 = 1 \pm [2.262(0)]$ and $1 \leq \beta_1 \leq 1$.
$\beta_2 = -.00027 \pm [2.262(.0001)]$ and $-.0005 \leq \beta_2 \leq -.00004$.
37. Using the same procedure as in problem 35,
$\beta_1 = -.0328 \pm [2.776(.0547)]$ and $-.1846 \leq \beta_1 \leq +.1190$.
$\beta_2 = 1.0701 \pm [2.776(1.5281)]$ and $-3.1719 \leq \beta_2 \leq +5.3121$.
39. Given the Box 12.I formula,

$$r^2_{Y1 \cdot 2} = \frac{.845 - .072}{1 - .072} = .833.$$

$$r^2_{Y2 \cdot 1} = \frac{.845 - .715}{1 - .715} = .456.$$

41. a. *Caution:* The parenthetical values are standard errors, thus, t values must be calculated by dividing the coefficients by these standard errors. In the order given, these ratios equal 1.59, -1.07, .39, .06, $-.28$, .67. They are hardly significant. Indeed, the computer shows corresponding p values of .073, .156, .353, .476, .394, .260. (The probabilities of getting the results by pure chance are 7.3 percent, 15.6 percent, and so on.)
   b. We can calculate the F statistic and compare it with a critical value of, say $F^{5,9}_{.05} = 3.48$ (Appendix Table I). (There are $m = 5$ degrees of freedom associated with the numerator, or *RSS*, and $n - m - 1 = 15 - 5 - 1 = 9$ degrees of freedom associated with the denominator, or *ESS*.) Thus,

$$RSS = 9,773 - 8,337 = 1,436$$

$$RMS = (1,436/5) = 287.2$$

$$EMS = (8,337/9) = 926.\overline{33}$$

$$F = (287.2/926.\overline{33}) = .31$$

This result does not show statistical significance. Indeed, the associated p value is .897.
43. a. $\hat{C} = -11.46 + .9589\ DPI$; t values in parentheses. $(-.95)\ (11.7010)$
   b. It is the marginal propensity to consume. For each $1 change in disposable personal income, personal consumption expenditures change by about 96 cents.

c. See Figure A.23.

Figure A.23

C = −11.46 + .9589 Y

d. The equation for *war years only* is

$\hat{C}_w = -12.62 + .8565\ DPI$; t values in parentheses.
$(-.53)\ (4.8781)$

The equation for *peace years only is*

$\hat{C}_p = -1.39 + .9413\ DPI$; t values in parentheses.
$(-.37)\ (38.7450)$

The aggregate consumption function's intercept as well as slope declined during the war. (See again Figure A.23).

e. $s_{C \cdot Y}$ = 12.23 (all years), 4.473 (war years only), 3.561 (peace years only).

f. The ANOVA tables are given in Parts A–C of Table A.84.
All three regressions are highly significant; $r^2 = (RSS/\text{Total } SS) = .93$ (all years), .92 (war years), and .996 (peace years).

45. The *ratios* of b/t are 8.563, .491, .498, 1.053, .517, .181, .0001117, .01309, .01399, .03963. (These are computer printouts; your answer may differ slightly because of rounding.)

Chapter 13

1. The missing values are: $I_t = .94$, $Y_t = \$26,647.50$, $S_t = .40$, $C_t = .89$. Because, in this model, $Y_t = T_t \cdot C_t \cdot S_t \cdot I_t$, the missing I_t (row 1) is found as

Table A.84

Source of Variation	Sum of Squares	Degrees of Freedom	Mean Square	Test Statistic
A. All Years				
Regression	20,489.693	1	20,489.693	F = 136.91
Error	1,496.547	10	149.6547	
Total	21,986.24	11		
Given a critical $F_{.05}$ for *d.f.* = (1,10) = 4.96, *reject* H_0: $\beta_1 = 0$.				
B. War Years				
Regression	476.17637	1	476.17637	F = 23.80
Error	40.02113	2	20.010565	
Total	516.1975	3		
Given a critical $F_{.05}$ for *d.f.* = (1,2) = 18.51, *reject* H_0: $\beta_1 = 0$.				
C. Peace Years				
Regression	19,039.721	1	19,039.721	F = 1,501.17
Error	76.09928	6	12.683213	
Total	19,115.82	7		
Given a critical $F_{.05}$ for *d.f.* = (1,6) = 5.99, *reject* H_0: $\beta_1 = 0$				

$Y_t/(T_t \cdot C_t \cdot S_t)$. The Y_t equation applies directly in row 2, while the missing S_t (row 3) is found as $Y_t/(T_t \cdot C_t \cdot I_t)$ and the missing C_t (row 4) as $Y_t/(T_t \cdot S_t \cdot I_t)$.

3. a. The 10-year moving-averages series (going from 1944–1981 in successive *columns*) is

172.47	278.89	402.535
193	283.34	428.7
213.2	287.88	462.21
232.085	292.62	501.13
247.25	297.895	544.995
256.755	304.175	594.915
260.805	312.125	651.22
261.61	320.595	716.425
262.465	329.145	797.585
264.615	339.56	897.7851
267.76	352.09	1017.16
271.22	366.67	1154.125
274.8	383.02	

b. Between 1980 and 1981, the debt rose at a rate of 136.965 per year; hence a 1989 forecast is $1{,}154.125 + [8(136.965)] = 2{,}249.845$ (billions of dollars).

5. a. The 6-year moving-averages series (going from 1962–1981 in successive *columns*) is

1480.333	1618.083	1669.833
1468.333	1778	1641.667
1456.417	1823.333'	1632.417
1457.167	1779.917	1586.833
1450.25	1759.083	1523.083
1431.333	1758.75	1475.25
1471.583	1723.25	

b. Between 1980 and 1981, housing starts *fell* at a rate of 47.833; hence a 1989 forecast is $1{,}475.25 - [8(47.833)] = 1{,}092.586$ (thousand units).

7. a. The 14-year moving-averages series (going from 1954–1976 in successive *columns*) is

67.6536	92.6357	127.525
69.9214	96.6429	132.7286
72.3929	100.575	138.275
75.08923	104.6678	143.9893
77.9964	109.1071	149.8
81.2	113.6929	155.6571
84.7607	118.2036	161.5214
88.6250	122.7143	

b. Between 1975 and 1976, the service-industry contribution to GNP rose at a rate of 5.8643; hence a 1989 forecast is $161.5214 + [13(5.8643)] = 237.7573$ (billion 1972 dollars).

9. a. The 4-year moving-averages series (going from 1949–1981 in successive *columns*) is

8.825	12.55	18.225
9.3375	12.7	20.75
9.7375	12.9	25.25
10.25	13.1	30.925
10.725	13.375	37.1375
11.4125	13.7375	42.825
12.1875	14.1625	48.800
12.7375	14.725	58.925
12.9125	15.475	76.1625
12.8	16.25	95.9125
12.6	17.0625	110.75

b. Between 1980 and 1981, the mining industry contribution to GNP rose at a rate of 14.8375 per year; hence a 1989 forecast is $110.75 + [8(14.8375)] = 229.45$ (billion 1972 dollars).

11. a. The 4-year moving averages series (going from 1962–1984 in successive *columns*) is

24.8	57.7	184.325
26.25	66.6875	207.8125
27.6875	77.35	236.925
29.8625	91.1625	270.2125
33.1125	108.8375	301.0375
37.5625	128.475	325.2125
43	148.0125	345.2625
49.7125	165.925	

b. Between 1983 and 1984, the transfer payments rose at a rate of 20.05; hence a 1989 forecast is $345.2625 + [5(20.05)] = 445.5125$ (billion dollars).

13. The 1987 forecast is 2,015.02 billion dollars. The use of $\alpha = .8$ is preferable: The error sum of squares is 2,774,340 when $\alpha = .2$ is used; it is only 427,407.3 when $\alpha = .8$ is used.

15. The 1985 forecast is 1,740.84 thousand units. The use of $\alpha = .9$ is preferable: The error sum of squares is 2,932,269 when $\alpha = .1$ is used; it is only 2,620,507 when $\alpha = .9$ is used.

17. The 1984 forecast is 204.28 billions of 1972 dollars. The use of $\alpha = .7$ is preferable: The error sum of squares is 7,467.688 when $\alpha = .3$ is used; it is only 1,668.609 when $\alpha = .7$ is used.

19. The 1984 forecast is 113.58 billions of 1972 dollars. The use of $\alpha = .5$ is preferable: The error sum of squares is 7,887.674 when $\alpha = .4$ is used; it is only 6,105.464 when $\alpha = .5$ is used.

21. The 1987 forecast is 373.38 billion dollars. The use of $\alpha = .8$ is preferable: The error sum of squares is 174,257.9 when $\alpha = .15$ is used; it is only 12,445.12 when $\alpha = .8$ is used.

23. The 1985 forecast is 1,948.88 thousand units. Note the last row of Table A.85. (If you have done it, see problem 14 for comparison).

25. The regression (with t values in parentheses) is

$$\hat{Y}_X = 113.7242 + 25.6180 X \quad \text{and } r^2 = .64.$$
$$(-1.4677) \quad (9.0193)$$

The 1989 ($X = 50$) forecast is 1,167.1758.

27. a. $s_{Y \cdot X_1} = 272.6152$

   b. $-265.74 \le \alpha \le 38.29$
   $20.05 \le \beta \le 31.19$

   Note: This is a *large-sample* test, use $z_{\alpha/2} = 1.96$.

   c. The critical $z_{\alpha/2} = \pm 1.96$. The coefficients, standard errors, and t values are
   for a: -113.72, 77.48, -1.47. H_0 is accepted.
   for b: 25.62, 2.84, 9.02. H_0 is rejected.

29. We require X values measured in months from a zero base, each of them centered in the middle of a month. If $X = 0$ were placed in the middle of the first month of 1968, or at mid-January, this would shift the base backward in time by 5.5 months (the original base was mid-1968, at the end of the 6th month). The original equation shows sales rising annually by $b = 47$ units; hence, they are rising by $(b/12) = (47/12)$ units per month, which is the slope of the new regression line for monthly data. Hence, the new intercept equals 200 (or mid-1968 trend sales) minus 5.5 $(47/12)$, or 178.46 units. The transformed regression line (measuring sales monthly at annual rates) is $\hat{Y}_X = 178.46 + 3.92X$ (where X is measured in months and $X = 0$ is located at mid-January, 1968). Note how this equation accurately predicts the 200-unit annual trend sales of 1968 that constitute the intercept of the annual-data regression line in Figure 15.10: $\hat{Y}_{5.5} = 178.46 + 3.92(5.5) = 200$.

Table A.85

Year, t	Actual Value, Y_t	Forecast Value, F_t	Error, $Y_t - F_t$
1959	1554.00		
1960	1296.00		
1961	1365.00	1296.00	69.00
1962	1493.00	1047.66	445.34
1963	1635.00	854.77	780.23
1964	1561.00	726.57	834.43
1965	1510.00	637.18	872.82
1966	1196.00	586.53	609.47
1967	1322.00	533.93	788.07
1968	1545.00	530.71	1014.29
1969	1500.00	590.69	909.31
1970	1469.00	676.54	792.46
1971	2085.00	782.40	1302.60
1972	2379.00	991.38	1387.62
1973	2058.00	1264.37	793.63
1974	1353.00	1509.70	-156.70
1975	1171.00	1653.74	-482.74
1976	1548.00	1745.86	-197.86
1977	2002.00	1858.55	143.45
1978	2036.00	2011.11	24.89
1979	1760.00	2152.81	-392.81
1980	1313.00	2237.03	-924.03
1981	1100.00	2231.17	-1131.17
1982	1072.00	2159.34	-1087.34
1983	1713.00	2048.41	-335.41
1984	1751.00	1999.25	-248.25
1985		1948.88	

Chapter 14

1. See Table A.86. The maximin action is A_1; the maximax action is A_3.

3. The maximin action is A_4; the maximax action is A_4 also.

5. The maximin action is A_2; the maximax action is A_3.

7. See Table A.87. (The cell entries show revenue minus cost equaling profit, in thousands of dollars.) The maximin action is A_1; the maximax action is A_5.

9. See Table A.88. The minimax action is A_3; the minimin action is A_1.

11. The minimax action is A_1; the minimin action is A_5.

Table A.86

Actions	Events E_1 = low demand	E_2 = moderate demand	E_3 = high demand	Row Minimum	Row Maximum
A_1 = small project	4	5	6	④ maximin	6
A_2 = medium project	1	6	· 10	1	10
A_3 = large project	−5	0	30	−5	㉚ maximax

Table A.87

Actions / Events	E_0 Demand = 0	E_1 Demand = 1	E_2 Demand = 2	E_3 Demand = 3	E_4 Demand = 4	E_5 Demand = 5
A_1 = build 1	48 − 60 = −12	90 − 60 = 30	90 − 60 = 30	90 − 60 = 30	90 − 60 = 30	90 − 60 = 30
A_2 = build 2	96 − 120 = −24	138 − 120 = 18	180 − 120 = 60	180 − 120 = 60	180 − 120 = 60	180 − 120 = 60
A_3 = build 3	144 − 180 = −36	186 − 180 = 6	228 − 180 = 48	270 − 180 = 90	270 − 180 = 90	270 − 180 = 90
A_4 = build 4	192 − 240 = −48	234 − 240 = −6	276 − 240 = 36	318. − 240 = 78	360 − 240 = 120	360 − 240 = 120
A_5 = build 5	240 − 300 = −60	282 − 300 = −18	324 − 300 = 24	366 − 300 = 66	408 − 300 = 108	450 − 300 = 150

Table A.88

Actions	Events E_1 = Recession	E_2 = Inflation	E_3 = Stagflation	Row Maximum	Row Minimum
A_1 = direct-mail campaign	75	20	40	75	⑳ minimin
A_2 = newspaper campaign	50	60	30	60	30
A_3 = TV campaign	55	40	45	㊶ minimax	40

13. The minimax action is A_2; the minimin action is A_4.

15. See Table A.89. The minimax-regret action is A_3.

17. The maximum-likelihood action is A_1.

19. See Table A.90. The expected-monetary-value action is A_2.

21. The optimal action is to sell the rights for $125 million, as Figure A.24 shows.

23. See Table A.91. The expected-utility action is A_1.

25. $EVPI$ = $55 million − $42.5 million = $12.5 million, where $55 million is computed as $.5(\$100$ million) + $.5(\$10$ million).

27. Given prior probabilities of $p(E_1) = .6$ and $p(E_2) = .4$, we can compute the following. *If the advice is R_1:*

$$p(E_1|R_1) = \frac{.6(.8)}{.6(.8) + .4(.3)} = .8$$

$$p(E_2|R_1) = \frac{.4(.3)}{.4(.3) + .6(.8)} = .2$$

$$\begin{aligned} p(R_1) &= p(E_1 \text{ and } R_1) \\ &+ p(E_2 \text{ and } R_1) \\ &= .6(.8) + .4(.3) = .6 \end{aligned}$$

Figure A.24

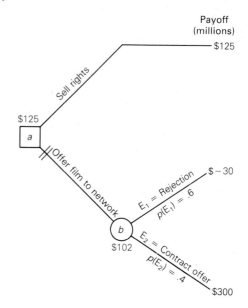

Table A.89

Actions	Events			Row Maximum
	E_1 = low demand	E_2 = moderate demand	E_3 = high demand	
A_1 = small run	15 − 15 = 0	20 − 20 = 0	69 − 25 = 44	44
A_2 = medium run	15 − 6 = 9	20 − 20 = 0	69 − 35 = 34	34
A_3 = large run	15 − (−10) = 25	20 − 0 = 20	69 − 69 = 0	ⓐ25

Table A.90

Actions	Events		EMV
	E_1 = Success at New Location; $p(E_1) = .5$	E_2 = Failure at New Location; $p(E_2) = .5$	
A_1 = move East	50	10	30
A_2 = move West	100	0	㊿50 Optimum

Table A.91

Actions	Events		EU
	E_1 = Success at New Location; $p(E_1) = .5$	E_2 = Failure at New Location; $p(E_2) = .5$	
A_1 = move East	$\sqrt{50} = 7.07$	$\sqrt{10} = 3.16$	⑤.12 Optimum
A_2 = move West	$\sqrt{100} = 10.00$	$\sqrt{0} = 0$	5.00

Figure A.25

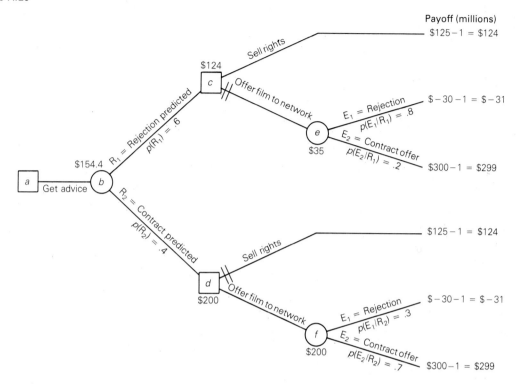

If the advice is R_2:

$$p(E_1|R_2) = \frac{.6(.2)}{.6(.2) + .4(.7)} = .3$$

$$p(E_2|R_2) = \frac{.4(.7)}{.4(.7) + .6(.2)} = .7$$

$$p(R_2) = p(E_1 \text{ and } R_2) + p(E_2 \text{ and } R_2)$$
$$= .6(.2) + .4(.7) = .4$$

On the basis of the preceding calculations, we can construct Figure A.25.

The optimal strategy is: if R_1 is received, sell the rights (and take the $125 million minus the $1 million fee); if R_2 is received, offer the film to the network (and earn an expected $200 million).

29. See Table A.92 (in millions of dollars).

Table A.92

First-Stage Event Probability	Second-Stage Event Probability	Joint Probability	Payoff	Joint Probability × Payoff
$p(R_1) = .69$	$p(E_1|R_1) = .91$	$.69(.91) = .63$	$-\$2.1$	$.63(-2.1) = -\$1.32$
$p(R_1) = .69$	$p(E_2|R_1) = .09$	$.69(.09) = .06$	11.9	$.06(11.9) = .71$
$p(R_2) = .31$	$p(E_1|R_2) = .23$	$.31(.23) = .07$	-2.1	$.07(-2.1) = -.15$
$p(R_2) = .31$	$p(E_2|R_2) = .77$	$.31(.77) = .24$	11.9	$.24(11.9) = 2.86$
		1.00		Expected payoff (S_4): $\$2.10$

α risk see *type I error*
absolute class frequency 41, 70
absolute frequency distribution 39–42, 44–45, 47–52, 61, 69, 71, 73–74, 81, 87, 106–7, 111
acceptance number 267, 294
acceptance region 267, 289
acceptance sampling plans 186–87, 215
action branches 482–83, 492, 504, 509
action points 482–83, 492, 504, 509
actions 480–81, 483–87, 489, 491, 493–94, 500
addition law of probability 134–37, 140, 144, 147–48, 150, 154, 161
additive time-series model 457, 475
adjusted sample coefficient of multiple determination 430, 437, 441–45
alternative hypothesis 262, 295
analysis of covariance 293
analysis of residuals 447
analysis of variance 293, 301, 342–63, 396–97, 413, 427–28, 437
analytical statistics 2–3, 113–526
ANOVA see *analysis of variance*
ANOVA table 349–50, 357–62, 397, 413, 428, 441–44, 446–47, 477
approximation
    of binomial probability distribution by normal probability distribution 208–10
arithmetic chart 56, 58
arithmetic mean 77–85, 94, 102–8, 241, 247, 249–50, 254, 260, 262–63, 269, 274–78, 280–82, 284–88, 291–92, 301, 342–63, 458–62, 473, 475–76, 478
    of random variable 169–70, 172, 183–84, 188–92, 197–99
arithmetic summary measures see *measures of central tendency, measures of dispersion,* or *measures of shape*
array 38–39, 68
assumptions underlying regression analysis 376–78, 422
authorship, disputed 64–65, 73
average see *arithmetic mean*
average deviation measures of dispersion 92–101, 108–10

β analysis 399
β risk see *type II error*

backward-elimination method 447
backward induction 491, 504, 509, 515
bar chart 56–57, 71, 72, 74
basic outcome 115–19, 158
Bayesian estimation 260
Bayes's postulate 490
Bayes's strategy 519
Bayes's theorem 151–53, 163–64, 489, 500–5, 513–15, 518, 525–26
Bernoulli process 172–92
bias 19–25, 27–28, 47, 221, 235, 289, 292
bimodal frequency distribution 88–89
binomial coefficient 176–77
binomial formula 175–77, 208
binomial probability distribution 154, 172–92, 208–10, 322–24, 337–38
binomial probability tables 181–83, 190–92, Appendix Tables C and D
bivariate data set 4–5
bivariate simple events 118
blocking 345

capitalization 438
census 10–11, 25, 28, 32, 34
central limit theorem 226–29, 260
chance error see *random error*
chances see *probability, defined*
chance variable see *random variable*
charts 56–60, 71–72, 74, 102–3, 109–10
Chebyshev's theorem 104, 110, 244, 257, 260
chi-square distributions 308–10, 351, Appendix Table H
chi-square statistic 305–7, 315
chi-square technique 302–39
classes 39–42, 69, 73–74
class frequency see *absolute class frequency, cumulative class frequency,* and *relative class frequency*
classical approach
    to defining probability 124–28
    to hypothesis testing 261–301
classical time-series model 451–57, 475
class mark 52
class width 40, 74
clustered random samples 18–19, 34
cobweb cycles 490
codes, secret 35, 43, 60–62, 73, 110
coding data 7
coefficient of correlation 394–96, 401, 413

coefficient of determination 391, 393–94, 401–2, 413, 428–30, 437, 441–46
coefficient of multiple correlation 430
coefficient of multiple determination 428–30, 437, 445–46
coefficient of partial correlation 433, 445
coefficient of partial determination 432, 445
coefficient of variation 100–1
coin-toss experiments 35
collection of data 4–35
collectively exhaustive events 120, 122–23, 136–37, 159–60
column chart 56–57, 72, 74
combinations 129, 132–34, 161, 164, 176, 187
compatible events 120, 159
complementary events 120, 122–23, 137, 160
completely randomized design 345
composite event 118–21, 158
computer programs 35, 73–74, 111, 164, 192, 217, 260, 301, 339, 363, 413, 423–24, 433–37, 447, 478, 526
conditional mean of Y 377–79, 382–84, 412–13, 423–24, 436, 445
conditional probability 138–43, 149–51, 162, 502–5, 513–14
conditional probability distribution of Y 377, 422
conditional standard deviation of Y 377, 422
confidence bands 383, 386
confidence intervals 237–60, 293–94, 297, 316–17, 336, 353–55, 362–63, 379, 382–84, 387–89, 412–13, 425–26, 436, 445, 447, 477
confidence levels 238–41, 271
confidence limits 238
confounding 293, 301
consistent estimators 232–33, 235, 293, 378
consumer price index *(CPI)* 12–13
contingency-table test 302–3, 331–34, 339
continuous quantitative variables 6–7, 30–32, 40, 47–52, 193–217
control charts 102–3, 109–10
convenience sample 12
correlation analysis 293, 301, 329, 365, 391–96, 413–14, 423, 428–33, 437, 447
correlation matrix 413, 447

costs of sampling 10–11, 26–27
counting techniques 13, 21, 129–34, 161, 164, 176, 187, 200
criterion of insufficient reason 490
critical values 267, 294, Appendix Table F
cross-section data 450
cryptanalysis 60–62, 73
cumulative class frequency 45
cumulative frequency distribution 44–45, 55, 69, 71, 73–74, 93, 108
cumulative probability distribution 167–68, 182–83
curvilinear regression 413
cut-off points 267, 294
cyclical component (of time series) 451–57, 475

data coding 7
data collection 4–35
data presentation 38–111
data set 4–5
data types 7–9, 32, 450
datum 4–5
deciles 92–108
decision branches 482–83, 492, 504, 509
decision forks 482–83, 492, 504, 509
decision making under uncertainty 479–526
decision nodes 482–83, 492, 504, 509
decision points 482–83, 492, 504, 509
decision rules 264
decision theory 479–526
decision trees 481–83, 491–92, 504, 508–9, 516–17, 524, 526
deductive reasoning 2
degrees of freedom 245, 308, 323, 348–52
dependent events 142–43, 150, 163, 230–31, 302–12
dependent variables 365–69
descriptive statistics 1–2, 38–111
deterministic relationship 366–67
dichotomous qualitative variables 6, 30–32
direct relationship 366, 368
discrete quantitative variables 6–7, 30–32, 39–40, 165–92, 208–10
disjoint events see *mutually exclusive events*
distance measures of dispersion 91–92, 108
dominant action 481
draft lottery 27
dummy variables 447
Durbin-Watson test 447

econometric models 477
efficiency of sample information 507–8
efficient estimators 232–34, 293, 378
elementary events 118
elementary units 4–5, 30–31
empirical approach to defining probability 126–28, 164
equal-likelihood criterion 490
error
    in experiments 348
    in forecasting 463, 466
    in hypothesis tests 270–74, 292, 295, 301, 343, 512
    in surveys 19–25, 34, 47, 210–11, 221, 235, 237, 260, 348
error mean square 349–51

error of acceptance 271–73, 292, 295, 301, 343, 512
error of rejection 270–73, 292, 295, 301, 512
error sum of squares 348–49, 393, 396–97, 430
estimated partial-regression coefficients 416, 418–21, 447
estimated regression coefficients 373, 378, 387–91, 399–400, 413, 447
estimated regression line 372–73, 376–78, 381, 386, 398–400, 404, 406–12, 434
estimation 220–60, 293, 298, 301, Appendix Table F
estimators 220, 232–35, 293
event branches 482–83, 492, 504, 509
event points 482–83, 492, 504, 509
events 480–81, 483–87, 489, 491, 493–94, 500
exact hypothesis 262
expected-monetary-value (criterion) 489–92, 494–95, 498, 503, 509, 511, 515–17, 524, 526
expected opportunity loss (criterion) 489, 492–93
expected regret value 489, 492–93
expected utility (criterion) 489, 494–500, 503, 516
expected value of random variable 169–70, 172, 183–84, 188–92, 197–99
expected value of perfect information 506–8, 517–18, 524
expected value of sample information 507
experimental designs 293, 301, 345, 362, 519
experimental error 348
experimental units 345
experiments 9–10, 34–35, 293, 301, 328–29, 397–99, 415, 472
explained deviation of *Y* 392
explained variable 365–69
explained variance 348–51
explained variation 346, 392–93, 396–97, 429–30
explanatory variable 365–69
exponential probability distribution 51–52, 70, 196, 217
exponential smoothing 462–68, 476–78
extensive-form analysis 508–11, 524
extrapolation 462, 474
extrasensory perception 154

F distributions 350–52, Appendix Table I
F statistic 350–51
factorial designs 293–301
factorial products see *factorials*
factorials 13, 21, 129–30, 164, 200
fair gamble 494
fiducial intervals 293–94
first quartile 92–93
flow variables 450
forecasting 221, 450–51, 457–71, 475–78
fractiles 92–93
frame 4–5
frequency curves 52–54, 71, 76, 88–89, 91, 194–217
frequency density 49
frequency distribution see *absolute frequency distribution, cumulative*

frequency distribution, and *relative frequency distribution*
frequency histograms 47–54, 63–65, 70, 74, 87, 107, 194
frequency polygons 52–53, 71

gambles 494–500
game theory 525–26
general addition law of probability 134–37, 161
general multiplication law of probability 134, 137–42, 144, 148–49, 163
geometric mean 88
Gompertz growth curve 468
goodness-of-fit tests 302, 321–29, 337–39
grand mean 346
graphs 38–74

harmonic mean 88
histograms see *frequency histograms*
homoscedasticity 343, 422
hypergeometric probability distribution 172, 192
hypothesis 261
hypothesis testing 261–339, 352–63, 389–91, 396–97, 413, 427–28, 437, 447, 477, 512–15, 519, Appendix Table F

inadmissible action 481
incompatible events 118–20, 159–60
independent events 142–44, 163, 230–31, 302–12, 331–34, 339
independent variables 365–69
index numbers 477
inductive reasoning 2
inexact hypothesis 262
inferential statistics, see *analytical statistics*
information 505–8, 517–18, 524
interaction 363
interfractile ranges 91–92
interquartile range 92–93, 108
intersection of events 121–23, 160, 163
interval data 8, 32
interval estimates 237–60, 293–94, 297, 316–17, 336, 353–55, 362–63, 379, 382–84, 387–89, 412–13, 425–26, 436, 447, 477
inverse relationship 366, 368
irregular component (of time series) 451, 454, 456–57, 475

joint probability 139–40, 149–50, 162, 304, 502–5, 510
judgment sample 12, 24, 28

Kruskal-Wallis test 339
kurtosis 110

Laplace criterion 490
Latin-square design 293, 301
law of large numbers 128
laws of probability 134–50

LE-type frequency distribution 44–45, 55, 69, 71, 73–74
least-squares technique 213, 418–47
line graphs 56, 58, 65–66, 73–74
Literary Digest case 27–28
logarithmic scale 56, 58
logistic (Pearl-Reed) growth curve 468
lower-tailed hypothesis tests 267–68, 275–78, 280–82, 285–86, 301, 317–18, Appendix Table F

maps 56, 59
marginal probability 138, 140, 143, 149–51, 162
marginal utility of money 496–99
matched-pairs samples 249–52, 254, 260, 281–82, 287–88
matrix algebra 423–25, 447
maximax 482, 485, 488, 522–23
maximin 482, 484, 488, 522–23
maximum-likelihood criterion 489–90, 524
ME-type frequency distribution 44–45, 55, 71, 73–74
mean see *arithmetic mean, geometric mean* and *harmonic mean*
mean absolute deviation 92–94, 108–9
mean of squared deviations see *variance*
mean squared error 235
measurement 7
measures of central tendency or location 75–90, 92, 94, 102–8, 111
measures of dispersion or variability 75–77, 90–111
measures of shape 76–77, 111
median 77, 83–85, 92, 105–7
median class 84
method of least squares 372–447
minimax 482, 484, 523–24
minimax regret 482, 486–88, 524
minimin 482, 485–86, 523–24
mixed time-series model 457, 475
modal class 86
mode 77, 86–88, 105–7
more-than-fair gamble 494
moving averages (series) 458–62, 473, 475–76, 478
multicollinearity 447
multimodal frequency distribution 88–89
multinomial qualitative variable 6, 30–32
multiple correlation analysis 414, 428–33, 437, 447
multiple regression analysis 365, 414–28, 433–47
multiplication law of probability 134, 137–42, 144, 148–49, 154, 163
multiplicative time-series model 457, 475
multivariate data set 4–5
multivariate simple events 118
mutually exclusive events 118–20, 159–60

nominal data 7, 32
nonparametric statistics 526
nonparametric tests 339
nonresponse bias 23–25, 27–28, 289
nonsampling error see *bias*
nonsense correlation 396, 402
normal curve see *normal probability distribution*

normal equations 418
normal-form analysis 508, 510–11, 524
normal probability distribution 51, 70, 99–102, 154–55, 181, 196–217, 226–27, 241, 257, 260, 275–84, 309, 324–26, 338, 351, 377, 422, Appendix Tables E and F
not statistically significant result 266, 289
null hypothesis 262, 295
number-of-runs test 339

objective probability 122–28, 160–61, 164
observational studies see *surveys*
observed significance level see *p values*
ogives 52–53, 55, 71, 92–93, 108
one-factor ANOVA 345–63
one-sided hypothesis 263
one-tailed hypothesis tests 267–68, 275–78, 280–82, 285–87, 292, 300–1, 317–20, 390, Appendix Table F
one-way ANOVA 345–63
open-ended classes 42, 80, 91
operating-characteristics curve 301
opportunity loss 486, 489, 492–93
optimal sample size 260–301
optimal strategy 503–12
ordered array 38–39, 68
ordinal data 8, 32
outcome space 115–18
outliers 25

p values 293–94, 300–1, 427
parallel-path strategy 185
parameter 77–78, 220, 305
partial correlation 431–33, 445
partition (of sample space) 122
Pascal's triangle 155–57, 176, 190
payoffs 480–81, 483–87, 489, 491, 493–95, 500, 510–11
payoff table 481, 484–87, 489, 491, 493–94, 500
Pearson's sample coefficient of correlation 394–96, 401, 413
percentile 92, 108, 206, 214
permutations 129–32, 161, 164, 176, 187
pictograms 59–60, 65, 67, 73
pie charts 56, 59, 72, 74
point estimates 221, 236, 257, 378–79, 412
Poisson probability distribution 172, 192, 339
polygons 52–53, 71
pooled variance 287
population 5–6
population coefficient of correlation 394
population coefficient of determination 394
population regression line 376–78
population standard error of the estimate 378
posterior analysis 500–509
posterior expected value of perfect information 506–7
posterior probability 151–53
power (of a hypothesis test) 272
predicted variable 365–69
prediction intervals 379, 382–87, 412–13, 423–25, 436, 445, 447, 477
predictor variables 365–69
preposterior analysis 508–13, 524
present value 524

presentation of data 38–111
prior analysis 489–500, 506–9, 524
prior expected value of perfect information 506–8, 524
prior probability 151–53, 489–500, 506–9, 524
probability (defined) 124–28, 164
probability density functions 194–217, Appendix Tables E and F
probability distributions see *binomial distribution, chi-square distribution, exponential distribution, F distribution, hypergeometric distribution, normal distribution, Poisson distribution, t distribution, and uniform distribution*
probability intervals 315–16
probability samples see *random samples*
probability space 115–18
probability theory 114–64
proportion 77, 101, 110, 241, 253–54, 260, 301–2, 312–15, 334–36, 339
pruning (decision trees) 492

qualitative variables 5–7, 30–32, 39, 165, 302–12, 331–34, 339
quality control 63
quantitative variables 5–7, 30–32, 39–40, 47–52, 165–217
quartiles 92, 108
quota sampling 24, 28

random error 19–21, 210–11, 221, 237, 260, 348
random events 114, 118–23, 136–37, 142–44, 150, 158–63
random experiments 114–34, 158–60, 166–217
randomization 345
randomized block design 293, 301, 345
randomized group design 345
random numbers 14, 16–17, 27, 33, 35, 73, Appendix Table A
random samples 13–29, 33–34
random variables 166–217
random-walk hypothesis 35
range 91–92, 108
ratio data 8–9, 32
rectangular probability distribution 51–52, 70
regressand 365–69
regression analysis 293, 329, 364–91, 396–428, 433–47, 469–71, 477
regression coefficients 373, 378, 387–91, 399–400, 413, 447
regression effect 404, 473
regression fallacy 405
regression line 366–78
regression plane 416–17
regression sum of squares 392–93, 396–97, 429–30
regressors 365–69
regret 486, 489, 492–93
regret table 486
rejection region 267
relative class frequency 42, 70
relative efficiency (of a statistic) 234
relative frequency distribution 42–45, 50–51, 61–62, 69, 71, 73–74

residual variation 348
response bias 22–25
risk 480; see also α *risk* and β *risk*
risk aversion 497–98, 515
risk neutrality 497–99
risk seeking 497–99
rounding 46–47
runs test 339

St. Petersburg paradox 495–97
sample 5–6
sample coefficient of correlation 394–96, 401, 413
sample coefficient of determination 391, 393–94, 401–2, 413
sample coefficient of multiple correlation 430
sample coefficient of multiple determination 428–30, 437, 445–46
sample coefficient of partial correlation 433, 445
sample coefficient of partial determination 432, 445
sample points 118
sample regression line 372–73, 376–78, 381, 386, 398–400, 404, 406–12, 434
sample space 114–18, 123, 158–59
sample standard error of the estimate of *Y* 379–81, 421–22, 436, 444–46, 477
sample surveys see *sampling*
sampling 10–29, 32–34
sampling distributions 220–60, 265, 307–10
sampling error see *random error*
scatter diagrams 370–71, 375, 398, 401, 406, 413, 416–17, 446–47, 468–69
seasonal component (of time series) 451, 454–57, 475
seasonal indices 477
second quartile 92
selection bias 21–22, 24, 27
semilogarithmic scale 56, 58
sequential analysis 512–13, 519, 526
sequential sampling 512–13, 519, 526
signed-rank test 339
significance level (of hypothesis test) 266–67, 297–301, 515
sign test 339
simple correlation analysis 365, 391–96, 413, 423
simple event 118–21, 158
simple random sample 11–17, 33
simple regression analysis 364–91, 396–413
single-parameter exponential smoothing 462–65, 476–78
size of hypothesis test 266–67, 297–301
skewness 51, 70, 110, 179–81, 308–9
Slutsky-Yule effect 461, 473
smoothing constant 462, 476–78
smoothing technique 458, 460, 462–68, 476–78
special addition law of probability 134, 136–37, 140, 144, 147–48, 150, 161
special multiplication law of probability 143–44, 147, 149
spurious correlation 396, 402

standard deviation 92, 96–104, 108–10, 172, 184–85, 188–91, 197–99, 329, see also *standard error*
standard error
    of regression coefficients 413, 446–47
    of the estimate of *Y* 378–81, 421–22, 436, 444–46
    of the sample mean 225
    of the sample proportion 225
standard normal curve 198–217, Appendix Table E
standard normal deviates 198–217, Appendix Table F
standard normal tables 198–217, Appendix Tables E and F
standard scores 101–2, 109
state-of-nature branches 482–83, 492, 504, 509
state-of-nature forks 482–83, 492, 504, 509
state-of-nature nodes 482–83, 492, 504, 509
state-of-nature points 482–83, 492, 504, 509
states of nature 480–81, 483–87, 489, 491, 493–94, 500
statistic 77–78, 220
statistical control 415
statistically dependent events see *dependent events*
statistically independent events see *independent events*
statistically significant result 266, 278, 289
statistical map 56, 59
statistics
    defined 1–4
    analytical 2–3, 113–526
    descriptive 1–2, 38–111
stepwise multiple regression 447
stochastic relationship 366–67
stock variable 450
stratified random samples 17–18
Student's *t* distribution 245–55, 260, 284–93, 351, 353–55, 362–63, Appendix Table G
Sturgess's rule 40, 69
subjective probability 128–29, 150–53, 160–61, 164, 513–14
summary measures 75–111
    for probability distributions 169–72, 183–85, 188–91, 197–99, 223–26, 230–31, 257, 310, 334
summation notation 79, 167
surveys 9–29, 34, see also *census, sampling*
systematic error see *bias*
systematic random samples 17, 33–34

t distribution 245–55, 260, 284–93, 351, 353–55, 362–63, Appendix Table G
t distribution table, see Appendix Table G
t ratios 413, 426–27, 434, 441–47
t values see *t ratios*
table construction 38–74
test statistic 264
third quartile 92–93
three-parameter exponential smoothing 468
three-way ANOVA 362–63
time-series analysis 450–78
time-series data 450

time-series line graphs 56, 58, 65–66, 73–74, 478
total deviation of *Y* 392
total sum of squares 349, 392–93, 396–97, 429–30
total utility of money 494–500
total variation of *Y* 392–93, 396–97, 429–30
treatments 345
treatments mean square 348–51
treatments sum of squares 347
treatments variation 346
tree diagrams 130, 148–50, 158, see also *decision trees*
trend component (of time series) 451–62, 465–72, 474–77
trend-smoothing constant 466
true regression coefficients 376, 378, 387–91, 399–400, 412–13, 425–26, 436–37, 445, 447
true regression line 376–78
Tukey's *HSD* test 355–56, Appendix Table J
two-factor ANOVA 345, 362–63
two-parameter exponential smoothing 465–67, 477–78
two-sided hypothesis 263
two-tailed hypothesis tests 267–68, 274–75, 278–92, 300–1, 320–21, 389–91, 413, 427–28, 477, Appendix Table F
two-way ANOVA 345, 362–63
type I error 270–73, 292, 295, 301, 343, 512
type II error 271–73, 292, 295, 301, 512

unbiased estimators 232–34, 293, 378
uncertainty 3–4, see also *decision making under uncertainty*
unconditional probability 138, 140, 143, 149–51, 162, 304, 502–5, 513–14
unexplained deviation of *Y* 393
unexplained variance 349–51
unexplained variation 348, 393, 396–97, 430
unfair gamble 494
uniform probability distribution 51–52, 70, 196, 217, 339
union of events 121–23, 160
univariate data set 4–5
univariate simple events 118, 158
universe see *population*
upper-tailed hypothesis tests 268–69, 277–78, 284, 286–87, 292, 301, 319–20, 390, Appendix Table F
utility-of-money function 496
utility theory 525

value of information 506–8, 517–18, 524
variables 4–6 see also *qualitative variables, quantitative variables,* and *random variables*
variance 92, 94–96, 98–99, 108–10, 170–72, 184–85, 188–91, 197–99, 232–33, 302, 315–21, 336–37, 339, see also *analysis of variance*
Venn diagrams 122–23

weighted mean 88–90, 107–8, 169–70
Wilcoxon rank-sum test 339
Wilcoxon signed-rank test 339

NAME INDEX

Adams, William J. 260
Aitchison, John 525
Allen, D. M. 447
Andersen, Leonall C. 401
Anderson, V. L. 362
Antonovitz, Frances 518
Archimedes 213
Armstrong, J. Scott 477

Bacon, Francis 64, 73
Bayes, Joshua 518
Bayes, Thomas 150–51, 490, 518–19, 505, 525–26
Bell, D. E. 525
Berkson, Joseph 525
Bernoulli, Daniel 495–97, 499–500
Bernoulli, James 155, 187
Bodily, Samuel E. 516
Box, George E. P. 525
Brainerd, B. 73
Brown, B. W. Jr. 403
Brown, George F. Jr. 438
Brown, Rex V. 516, 525
Brown, Robert G. 477
Bryan, Michael F. 400
Bryson, Maurice C. 28

Cady, F. B. 447
Campbell, Donald T. 472
Campbell, Stephen K. 110
Carroll, Mavis B. 399
Chatterjee, S. 447
Chebyshev, Pafnuty 103–4
Cochran, William G. 34, 362, 440
Cohen, Daniel 217
Collins, Janet 146–47
Collins, Malcolm 146–47
Cox, G. M. 362
Crossley, A. M. 28

Daniel, C. 447
Darwin, Charles 330, 403
David, F. N. 155, 157, 164, 187
Davies, Owen L. 254, 362
Deming, W. Edwards 63
Dewey, Thomas 28

Disraeli, Benjamin 65
Doll, R. 403
Draper, N. 447
Dunnett, Charles W. 291

Efron, B. 525

Fairley, W. 146
Faraday, Michael 255
Feller, William 164, 192
Fermat, Pierre de 144, 148, 155, 187
Fienberg, Stephen E. 27
Fisher, Ronald A. 255, 293–95, 301, 328–29, 350, 362, 403, 418, 440, 513, 518–19, 525, Appendix Table H
Folks, J. Leroy 154, 525
Fourier, Jean Baptiste Joseph 28
Freedman, David A. 210

Gallup, George H. 28
Galton, Francis 210, 217, 329–30, 403–5, 473
Gauss, Carl F. 196, 210, 212–13, 404
Geis, Irving 65, 73
Gombauld, Antoine 147
Gorman, J. W. 447
Gossett, William S. 245, 255, 293
Greenlees, John S. 292
Guinness, Arthur 255

Hafer, R. W. 400
Hall, Ray O. 73
Hamilton, Alexander 64–65
Hartley, H. O. Appendix Tables G, I, J
Helmert, Friedrich 329
Henry II, King of England 274
Heyne-Hafer, Gail 400
Horim, Moshe Ben 186
Howard, R. A. 212, 517
Huff, Darrell 65, 73, 146–47, 154, 157, 164, 474

Jay, John 64
Jones, Grover C. 190

Kahr, Andrew S. 525
Keeney, R. L. 525
Kempthorne, Oscar 440, 525
Kjetsaa, Geir 73
Knight, Frank 480
Kohler, Heinz 110, 192, 217, 260, 301, 339, 362, 413, 438, 447, 477, 490, 524–25
Kramp, Chrétien 200
Kryukov, Fedor 73
Kutner, M. H. 413

La Brecque, Mort 211
Landon, Alfred M. 27
Lapin, Lawrence L. Appendix Table D
Laplace, Pierre Simon de 28, 124, 213
Leibniz, Gottfried Wilhelm von 187
Levy, Haim 186
Lieberman, Gerald J. 144
Lincoln, Abraham 474
Lindley, Dennis V. 525
Luce, Robert D. 525

Madison, James 64–65
Marlowe, Christopher 64
Marshall, John M. 292
Matheson, J. E. 212, 517
McCarthy, Philip J. 28
McClave, James T. 447
McLean, R. A. 362
Mendel, Gregor 328–29
Mendenhall, William 447
Menges, G. 525
Méré, Chevalier de 144, 147–48, 155
Mises, Richard von 164
Moivre, Abraham de 134, 154–55, 196, 518
Moore, D. S. 34.
Morgenstern, Oskar 34, 477, 500, 519, 525
Morris, W. T. 525
Mosteller, Frederick 28, 34, 65, 73, 146

Nelson, Richard R. 186
Neter, John 27, 413
Neumann, John von 500, 525
Newton, Isaac 154, 213
Neyman, Jerzy 294–95, 513, 519, 525
Noon, John 518
North, D. W. 212, 517

Olson, Charles L. 400

Parten, Mildred B. 28
Pascal, Blaise 144, 148, 155–57, 187
Pearson, E. S. 294–95, Appendix Tables G, I, J
Pearson, Karl 213, 255, 294–95, 329–30, 339, 404, 513, 519, 525
Peterson, C. 525
Picconi, Mario J. 400
Pisani, Robert 210
Poisson, Siméon D. 28
Pratt, Fletcher 73
Price, B. 447
Price, Richard 578
Purves, Roger 210

Quetelet, Adolphe 28–29, 196

Raiffa, Howard 525–26
Roe, Terry 518
Roosevelt, Franklin D. 27
Roper, Elmo B. 28
Rosen, Kenneth T. 437

St. Paul 64
Savage, Leonard J. 164, 526
Schlaifer, Robert 164, 526
Shakespeare, William 64, 73
Shimi, I. N. 526
Sholokhov, Mikhail A. 73
Sielaff, Theodore J. 27, 34
Silverman, Lester P. 438
Slutsky, Eugen E. 473
Smith, H. 447
Smith, Laurence Dwight 73
Snedecor, George W. 350, 418, 440
Sommers, A. 34
Stephan, Frederick F. 28
Stigler, Stephen M. 292, 526
Street, Elisabeth 399
Student *see* Gossett, William S.
Sturgess 40

Tanur, Judith M. 27, 34, 63, 65, 144, 291, 399, 401, 403, 472
Tiao, George C. 525
Trujillo, Joseph 146–47
Truman, Harry S. 28
Tsokos, Chris P. 526
Twain, Mark 474

Ulvila, Jacob W. 516

Venn, John, 122–23

Wald, Abraham 512–13, 519, 525–26
Walker, Helen M. 213, 217, 413
Wallace, David L. 73
Wallace, Henry A. 440
Wasserman, William 413
Weldon, W. F. R. 330
Winkler, Robert L. 526
Wonnacott, Ronald J. 413
Wonnacott, Thomas H. 413
Wood, F. S. 447
Wyckoff, Paul Gary 439

Yates, Frank Appendix Table H
Yett, Donald E. 292
Yule, G. Udny 473

Zellner, Arnold 526

a the vertical intercept of the sample regression line (estimator of α)

$A \cup B$ the union of events A, B (A *or* B (or both) occur)

$A \cap B$ the intersection of events A, B (A *and* B occur simultaneously)

b the slope of the sample regression line (estimator of β)

b_1 and b_2 estimated partial regression coefficients

C_x^n combinations for x at a time out of n distinct items

c_x^n binomial coefficient; number of permutations of x successes and $n - x$ failures in n Bernoulli trials

C_t the cyclical component in time-series analysis, measured against time, t

D the difference between 2 sample·observations in a matched-pairs sample

$\overline{D}$ the average of differences between 2 sample observations in a matched-pairs sample

d the difference between 2 sample means (or proportions) from independent samples

d_1 the difference between the frequency density of one class and the preceding class

d_2 the difference between the frequency density of one class and the following class

$d.f.$ degrees of freedom

E the statistic that is being used as an estimator

e the constant 2.71828; a sampling error

EMS error mean square (unexplained variance)

EMV expected monetary value

EOL expected opportunity loss

ESI efficiency of sample information *(EVSI/EVPI)*

ESS error sum of squares (unexplained variation)

EU expected utility

$EVPI$ expected value of perfect information

$EVSI$ expected value of sample information

E_R expected value of random variable, R

F the sum of frequencies up to, but not including those of the median class; the test statistic that is the ratio of explained variance to unexplained variance (e.g., TMS/EMS or RMS/EMS)

f class frequency

$f(x)$ any function of x

$f(z)$ the standard normal probability density function

f_e expected frequency

f_o observed frequency

F_t the current period's forecast value (of a time series)

F_{t+1} next period's forecast value (of a time series)

HSD Tukey's honestly significant difference

H_0 the null hypothesis

H_A the alternative hypothesis

I_t the irregular component in time-series analysis, measured against time, t

$I_{Y \cdot X}$ expected individual value of Y, given X

$I_{Y \cdot 12}$ expected individual value of Y, given X_1 and X_2

k desirable number of classes in Sturgess's rule; number of standard deviations in Chebyshev's theorem; number of means in Tukey's HSD test

L lower limit of median (or modal) class

m sample median; number of independent variables in a regression

MAD mean absolute deviation

Mo population mode

mo sample mode

N the total number of observations in a population

n the total number of observations in a sample; number of trials in a random experiment

$n!$ n factorial

OL opportunity loss

p a probability value that measures the likelihood that a test-statistic value could occur by chance, given that the null hypothesis is true

$p(A)$ probability of event A

$p(A|B)$ probability of event A, given B

$p(R = x)$ probability of random variable (R) being equal to a given possible value, x

P sample proportion

P_p pooled estimator of the population proportion

P_x^n permutations for x at a time out of n distinct items

$P_{x_1, x_2, \ldots x_k}^n$ permutations for n at a time out of n nondistinct items, there being k kinds of items and x_1 units of kind 1, x_2 units of kind 2, and so forth

q_α key number in Tukey's HSD test

RSS regression sum of squares (variation explained by regression)

RMS regression mean square (variance explained by regression)

R name of some random variable; sample coefficient of multiple correlation

r sample coefficient of correlation

R^2 sample coefficient of multiple determination (*RSS/Total SS* in multiple regression)

r^2 sample coefficient of determination (*RSS/Total SS* in simple regression)

$\overline{R}^2$ adjusted sample coefficient of multiple determination

$\overline{r}^2$ adjusted sample coefficient of simple determination

$r_{Y2 \cdot 1}$ sample coefficient of partial correlation

$r_{Y2 \cdot 1}^2$ sample coefficient of partial determination